# Particle and Astroparticle Physics

## Series in High Energy Physics, Cosmology, and Gravitation



This series of books covers all aspects of theoretical and experimental high energy physics, cosmology and gravitation and the interface between them. In recent years the fields of particle physics and astrophysics have become increasingly interdependent and the aim of this series is to provide a library of books to meet the needs of students and researchers in these fields.

*Other recent books in the series:*

**Joint Evolution of Black Holes and Galaxies**
M Colpi, V Gorini, F Haardt, and U Moschella *(Eds.)*

**Gravitation: From the Hubble Length to the Planck Length**
I Ciufolini, E Coccia, V Gorini, R Peron, and N Vittorio *(Eds.)*

**Neutrino Physics**
K Zuber

**The Galactic Black Hole: Lectures on General Relativity and Astrophysics**
H Falcke, and F Hehl *(Eds.)*

**The Mathematical Theory of Cosmic Strings: Cosmic Strings in the Wire Approximation**
M R Anderson

**Geometry and Physics of Branes**
U Bruzzo, V Gorini and, U Moschella *(Eds.)*

**Modern Cosmology**
S Bonometto, V Gorini and, U Moschella *(Eds.)*

**Gravitation and Gauge Symmetries**
M Blagojevic

**Gravitational Waves**
I Ciufolini, V Gorini, U Moschella, and P Fré *(Eds.)*

**Classical and Quantum Black Holes**
P Fré, V Gorini, G Magli, and U Moschella *(Eds.)*

**Pulsars as Astrophysical Laboratories for Nuclear and Particle Physics**
F Weber

**The World in Eleven Dimensions: Supergravity, Supermembranes and M-Theory**
M J Duff *(Ed.)*

**Particle Astrophysics**
*Revised paperback edition*
H V Klapdor-Kleingrothaus, and K Zuber

**Electron-Positron Physics at the Z**
M G Green, S L Lloyd, P N Ratoff, and D R Ward

Series in High Energy Physics, Cosmology, and Gravitation

# Particle and Astroparticle Physics

Utpal Sarkar
*Physical Research Laboratory*
*Ahmedabad, India*

CRC Press
Taylor & Francis Group
Boca Raton London New York

CRC Press is an imprint of the
Taylor & Francis Group, an **informa** business
A TAYLOR & FRANCIS BOOK

CRC Press
Taylor & Francis Group
6000 Broken Sound Parkway NW, Suite 300
Boca Raton, FL 33487-2742

First issued in paperback 2019



ISBN-13: 978-1-58488-931-1 (hbk)
ISBN-13: 978-0-367-38810-2 (pbk)



**Library of Congress Cataloging-in-Publication Data**

Sarkar, U. (Utpal)
Particle and astroparticle physics / Utpal Sarkar.
p. cm. -- (Series in high energy physics, cosmology and gravitation)
Includes bibliographical references and index.
ISBN-13: 978-1-58488-931-1
ISBN-10: 1-58488-931-4
1. Particles (Nuclear physics) 2. Nuclear astrophysics. I. Title.

QC793.S27 2008
539.7'2--dc22 2007046150

**Visit the Taylor & Francis Web site at**
**http://www.taylorandfrancis.com**

**and the CRC Press Web site at**
**http://www.crcpress.com**

# Preface

Our knowledge of elementary particles and their interactions has reached an exciting juncture. We have learned that there are four fundamental interactions: strong, weak, electromagnetic and gravitational. The weak and the electromagnetic interactions are low energy manifestations of the electroweak interaction. The standard model of particle physics is the theory of the strong and the electroweak interactions, while general theory of relativity is the theory of the gravitational interaction. We now believe that at higher energies these forces may unify, yielding yet another unified theory. In this quest, many new possible theories beyond the standard model have been proposed in recent times and attempts are being made to unify these theories with gravity.

To name a few, the grand unified theories would unify three of the four forces, supersymmetry would unify fermions with bosons, Kaluza–Klein theories would unify gravity with internal gauge symmetries with extra dimensions, superstring gives us hope of unifying all of these concepts into an ultimate theory of everything. There have also been very fascinating developments with extra dimensions, e.g., gravity could become strong at very low energy, the geometry is warped leading to new phenomenology, or the occurrence of electroweak symmetry breaking from boundary conditions in higher dimensions. In parallel, newer activities are coming up in the areas of astroparticle physics, which studies the cosmological consequences of particle physics and relates them with astrophysical observations. Some topics in cosmology, such as dark matter, dark energy or cosmological constant, matter–antimatter asymmetry and models of inflation are also concerns of particle physics models and may expose themselves through some low energy phenomenology.

While delivering talks or discussing with colleagues at different places in the world, I realized the need for a book that provides an introduction to such wide varieties of topics in one place. The main motivation of this book is to present the recent developments in these diverse areas of particle physics and astroparticle physics in a coherent manner, providing the required background materials. Since many of these newer results may change with future findings, a major part of the book covers the essential ingredients in a systematic and concise manner, which can help the readers to follow any other related developments in the field.

The book is divided into five parts. The first part provides a working knowledge of group theory and field theory. The second part summarizes the standard model of particle physics including some extensions such as the neutrino physics and CP violation. The next part contains an introduction to grand unification and supersymmetry. In the fourth part of the book, an introduction to the general theory of relativity, higher dimensional theories of gravity and superstring theory are discussed. Then the various newer ideas and models with extra dimensions including low scale gravity are introduced. The last part of the book deals with astroparticle physics, which studies the interplay between particle physics and cosmology. After an introduction to cosmology, some specialized topics such as baryogenesis, dark matter and dark energy and models of extra dimensions are introduced.

Efforts of many people went into the book directly or indirectly. I would first remember my collaboration with the late Prof. Abdus Salam, whose teachings changed the course of my scientific career. Some other senior collaborators, E. Ma, E.A. Paschos, J.C. Pati and A. Raychaudhuri, influenced me very much. I earnestly thank my collaborators from all over the world, from whom I have learned many things that have enriched the book to a large extent. I owe sincere thanks to Prof. H.V. Klapdor-Kleingrothaus, MPI, Heidelberg, who inspired me to join him in writing a book, which however did not materialize. I profusely thank the Alexander von Humboldt Foundation for their support to visit places in Germany at different times and acknowledge hospitality at the Institut für physik at Universität Dortmund, DESY at Hamburg, Max-Planck-Institut für Kernphysik at Heidelberg and the University of California at Riverside, where parts of the book were written.

I have included references, which are directly related and contain details. Some original articles have also been referenced. The list is by no means complete. I tried to be as careful as possible with the contents of the book, but some errors may still remain. I would appreciate if the readers would bring them to my notice, so I can make them available at http://www.prl.res.in/~utpal/books. At this site I shall also keep a glossary, writeups of any new developments in the field and additional references that I think may help the readers. While writing the book I realized that I could continue to add more materials and improve the book indefinitely. So, I had to conclude at one point with the hope of providing further inputs in this link.

Ahmedabad
December 2007

**Utpal Sarkar**

# Contents

# Introduction

At present our knowledge of elementary particles and their interactions is passing through an interesting phase. We believe that there are four fundamental interactions: the strong, the weak, the electromagnetic and the gravitational. The standard model of particle physics explains the three gauge interactions: the strong, weak and the electromagnetic interactions, which originate from some internal symmetries. The gravitational interaction originates from space–time symmetry and is described by the general theory of relativity.

The standard model has two parts: the strong interaction and the electroweak interaction. The strong interaction, which is experienced by the quarks, is a gauge theory and the carriers of the interaction are eight massless gluons. There are eighteen quarks that come in six flavors: up, down, charm, strange, top and bottom, and each of these flavored quarks comes in three colors: red, green and blue. A gauge theory originates from gauge invariance, i.e., any symmetry of the state vectors in some internal space. The state vectors represent the particles that undergo the particular interaction. In this case the state vectors correspond to the quarks of different colors and the interaction is the strong interaction. The carriers (the gluons) of the interaction ensure that the gauge transformations at two different points in space do not destroy the invariance of the Lagrangian. These carriers are spin-1 bosons, called the gauge bosons. Since the exchange of these gauge bosons gives rise to the particular interaction, one may say that the interaction is mediated by the gauge bosons. The range of the force is restricted by the mass of the gauge bosons; for massless gauge bosons, the force has infinite range.

The electroweak interaction is a spontaneously broken gauge theory, acting on both quark flavors and leptons. Every state has two particles: two quarks with different flavors or two leptons. There are four gauge bosons, three of which $W^{\pm}, Z$ become massive after the spontaneous symmetry breaking and the fourth one $\gamma$ (photon) remains massless. The electromagnetic interaction is mediated by the photon and any charged particle undergoes electromagnetic interaction. The spontaneous symmetry breaking means that the Lagrangian is invariant under the symmetry transformation of the state vectors, but the vacuum or the minimum energy state does not respect this symmetry. For any gauge theory, spontaneous symmetry breaking makes the mediating gauge bosons massive.

While talking about the gauge interactions, we mentioned quarks and leptons. These are the elementary or fundamental particles, which are the building blocks of all matter. They are spin-1/2 fermions. There are three quarks with charge $+2/3$ $(u,c,t)$ and three quarks with charge $-1/3$ $(d,s,b)$. Each of them carries a color quantum number which can have three values. These eighteen quarks interact with each other through the strong interaction. They also experience the weak, electromagnetic and gravitational forces, but these forces are negligible compared with the strong force. It is not possible to see a free quark because they are confined. They may exist as baryons, which are composed of three quarks so the quarks carry a baryon number $1/3$, or as mesons, which are quark–antiquark pairs. The most common baryons are the nucleons (protons and neutrons) that reside in the nucleus. A proton is made of *uud* quarks and a neutron is made of *udd* quarks. The strong nuclear force between the nucleons can be derived from the strong interaction that acts on the quarks.

The three negatively charged particles, $e^-, \mu^-, \tau^-$, and the corresponding neutrinos, $\nu_e, \nu_\mu, \nu_\tau$, together are called leptons. While the charged leptons undergo weak, electromagnetic and gravitational interactions, the neutral leptons or the neutrinos undergo only weak and gravitational interactions. The gravitational interaction is too weak compared with other interactions at our present energy. The leptons do not take part in the strong interaction.

We now come to the fourth interaction, the gravitational interaction. The gravitational interaction is described by the general theory of relativity, which is based on the equivalence principle. The basic idea of the theory is to draw an equivalence between the gravitational force due to a massive object and some curvature in space. In other words, the gravitational interaction can be viewed as a theory originating from the space–time symmetry. The gravitational interaction is mediated by spin-2 bosons, the gravitons. Although this is the weakest of all the interactions, it governs the motion of the stars, planets and also the falling bodies on Earth.

This completes our brief discussion about the standard model and the general theory of relativity as the theory of the strong, weak, electromagnetic and gravitational interactions. Although we have very little experimental evidence for any new physics, there are many theoretical considerations that make us believe that there should be new horizons that are waiting for us to explore. We shall now mention some of the exciting ideas that will be discussed later in this book.

The success of the electroweak theory makes us think that at very high energies all the three interactions could become part of a single theory, called the grand unified theory. The coupling constants of the strong, weak and the electromagnetic interactions evolve with energy and seem to meet at a point at very high energy. At higher energies all three interactions will be described by only one theory, the grand unified theory, with only one interaction strength. Although the idea of grand unification is highly interesting, it has not yet been established experimentally. Our accelerators can only reach the electroweak unification scale and verify the standard model beyond doubt. But our quest for theories beyond the standard model continues since it may not be possible to search for any new physics unless we know what we are looking for.

Several theoretically fascinating ideas have evolved over the past couple of decades for physics beyond the standard model. Since the grand unification scale is about 14 orders of magnitude higher than the electroweak symmetry breaking scale, this gives rise to a new theoretical inconsistency called the gauge hierarchy problem. As a solution to this problem supersymmetry has been proposed. Supersymmetry is a symmetry between a fermion and a boson. As a result, all the known particles will have their superpartners in the supersymmetric standard model of strong and electroweak interactions. Since a solution of the gauge hierarchy problem requires all these particles to be as light as 100 GeV to few TeV, these particles should be observed in the next generation accelerators.

One of the most important theoretical challenges is to unify gravity with the standard model. This leads to higher dimensional theories near the Planck scale, where gravity needs to be quantized. The criterion for any higher dimensional theories is to reproduce the standard model and the general theory of relativity at energies below the Planck scale. The theory should also have the known fermions at low energies. Finally the theory should be consistent and free from any unwanted infinities such as the anomalies that can make the theory unstable. Taking all these points into consideration the ten dimensional superstring theory emerged as the most promising theory for unifying the strong, electroweak and the gravitational interactions. Superstring theory also appears to be the most consistent theory of quantum gravity. Although it may not be possible to establish the superstring theory from experiments, an enormous amount of theoretical effort went into the development of the subject.

Another interesting idea predicts new extra space-dimensions at around the electroweak symmetry breaking scale, which are seen only by gravity and not the other interactions. This became possible as a consequence of the duality conjecture and the brane solutions. This class of models with extra dimensions allows a low Planck scale and also predicts new physics in the next generation accelerators. Grand unification and all new physics should also become accessible to the next generation accelerators in this class of theories. All these theories tend to solve the cosmological problems from a different approach and hence some of these results could also be verified by astrophysical observations.

There are also results from cosmology that require new physics beyond the standard model. It has been established that a large fraction of the matter in the universe is in the form of dark energy and dark matter. Only a small fraction of the matter is the visible matter. The required dark matter candidate has to come from some new physics. It could be the lightest supersymmetric particle or some new particle. The explanation of the dark energy of the universe is also a challenging question for particle physics. Models with extra dimensions provide some hope to this problem, so any indication of dimensions beyond the usual four space–time dimensions is most welcome. The baryon asymmetry of the universe also requires new physics beyond the standard model. The lepton number violation required for the neutrino masses seems to provide a natural explanation to this problem.

One thus studies the possible theoretical extensions of the standard model and then considers its cosmological consequences and looks for consistency. Although we have been enriching our theoretical concepts very rapidly, we may not be able to verify them in the laboratory. Then cosmology remains to be the only testing ground for these theories. According to the big-bang theory, the universe was extremely hot at early times. So, the new physics at very high energy may have some signatures imprinted in our present day cosmological observations. Thus, the interplay of particle physics and cosmology has opened up a new era of astroparticle physics.

In this book an attempt is made to introduce these new concepts in a coherent approach. First, we shall provide some background on field theory and group theory, which form the backbone for modern particle physics. In the second part, we discuss the standard model of particle physics. We discuss the neutrino physics and *CP* violation in some detail, since there are some new discoveries in these topics during the past few years and these topics play a crucial role in astroparticle physics. In the next part, we discuss grand unification and supersymmetry. Then we introduce general theory of relativity and extend it to higher dimensions and to supergravity. All these concepts are then used in developing superstring theory. The new ideas with extra dimensions are then introduced. Since each of these ideas has many virtues and limitations, an attempt is made to explain the concepts, which will be useful to readers even if these models are modified by future findings. In the last part of the book, we provide an introduction to cosmology including the recent results. Some of the topics in astroparticle physics are then discussed in the subsequent chapters. Most of the discussions in earlier chapters may have application to these topics. Finally we conclude with a brief epilogue.

# Part I

# Formalism

# 1

# *Particles and Fields*

To understand the nature of interactions of any particles at very high energies, one needs to treat the particles as quantum fields and study them in Quantum field theory (QFT), which started as an elegant theory of relativistic quantum electrodynamics (QED) [1, 2, 3]. Although the beauty of the theory was plagued with all kinds of apparent infinities, the QFT was accepted because of its predictability. The infinities are accepted as our inability to deal with interactions at very high energies, not as a problem with the theory. Proper mathematical prescriptions, called renormalization, to take care of the infinities are then introduced to maintain the predictive power of the theory even at high energies. So far no discrepancy between the predictions of QFT and experiments has been noticed, which makes it one of the building blocks for our present knowledge of particle physics.

In the first two chapters we present a brief introduction to quantum field theory and renormalization. We shall try to present the basic idea and some key points, which may help in understanding some of the concepts we shall be discussing in later chapters. For details, the readers may consult any textbooks on quantum field theory. The books we consulted, while writing this chapter, are listed in the bibliography.

We follow the convention of natural units, that the Plancks constant $\hbar = 1$ and the velocity of light is $c = 1$. The Boltzmann constant is also chosen to be $k_B = 1$. In general one should be careful in applying these units. To go to the classical limit from the domain of quantum mechanics one takes the limit $\hbar \to 0$, which becomes difficult with natural units. Similarly for going from relativistic to nonrelativistic limits one has to be careful in working with the natural units. By fixing the Boltzmann constant, one has to be careful while applying microscopic theory of statistical mechanics to macroscopic quantities. Fortunately, in particle physics we are mostly concerned about interactions at very high energies and at very short distances, where we can safely apply the relativistic and quantum mechanical formalism, which is the quantum field theory. The Boltzmann constant appears in cosmology, where we restrict ourselves only to macroscopic quantities such as temperature. This justifies the choice of natural units in particle physics. The main consequence of the natural units is to have a unifying unit for length, time, mass, temperature and energy:

$$[length] = [time] = [mass]^{-1} = [temperature]^{-1} = [energy]^{-1}.$$

Masses of the particles are thus given by their rest energy $(mc^2)$ in units of $eV$ or inverse Compton wavelength $(mc/\hbar)$ in units of $\text{cm}^{-1}$.

## 1.1 Action Principle

To study the time development of any particle one may start from the action principle, which states that in classical mechanics, the motion of any particle is determined uniquely by minimizing the action

$$\delta S = \delta \int_{t_1}^{t_2} L(q_i, \dot{q}_i)\, dt = 0, \tag{1.1}$$

where $L(q_i, \dot{q}_i)$ is the Lagrangian. In quantum mechanics one needs to sum over all possible paths that are allowed by the uncertainty principle.

The classical equations of motion can be obtained by taking the variation of both the position $q_i$ and the velocity $\dot{q}_i$ and applying the boundary conditions $\delta q_i(t_1) = 0$ and $\delta q_i(t_2) = 0$. The minimization of the action then results in the Euler–Lagrange equations of motion

$$\frac{\delta L}{\delta q_i} - \frac{d}{dt}\frac{\delta L}{\delta \dot{q}_i} = 0. \tag{1.2}$$

For a particle moving under the influence of a potential energy $V(q_i)$, the Lagrangian can be written as

$$L(q_i, \dot{q}_i) = \frac{1}{2} m\, \dot{q}_i^2 - V(q_i), \tag{1.3}$$

so the classical equations of motion become

$$m \frac{d^2 q_i}{dt^2} = -\frac{\partial V(q_i)}{\partial q_i}. \tag{1.4}$$

In this Lagrangian formalism the positions $q_i$ and the velocities $\dot{q}_i$ are considered as independent variables that identify the particle.

In the Hamiltonian formalism, one defines the momentum $p_i = \delta L / \delta \dot{q}_i$ and then the Hamiltonian as

$$H(q_i, p_i) = p_i \dot{q}_i - L(q_i, \dot{q}_i). \tag{1.5}$$

The Hamiltonian corresponding to the Lagrangian of equation 1.3 then becomes

$$H(q_i, p_i) = \frac{p_i^2}{2m} + V(q_i). \tag{1.6}$$

The first term gives the kinetic energy, the second term is the potential energy and the total energy is given by the Hamiltonian for the system. The equations of motion in terms of the Hamiltonian are given by

$$\frac{dq_i}{dt} = \frac{\delta H}{\delta p_i}, \quad \text{and} \quad -\frac{dp_i}{dt} = \frac{\delta H}{\delta q_i}. \tag{1.7}$$

The time variation of any field $F$ is given by

$$\frac{dF}{dt} = \frac{\partial F}{\partial t} + \{H, F\} \tag{1.8}$$

where the Poisson bracket for the two fields $A$ and $B$ is defined as

$$\{A,B\} = \frac{\partial A}{\partial p_i}\frac{\partial B}{\partial q_i} - \frac{\partial A}{\partial q_i}\frac{\partial B}{\partial p_i}.$$

The transition to quantum mechanics is possible in many ways, which are all equivalent.

The main conceptual difference between classical and quantum mechanics is the uncertainty principle:

$$\Delta q_i \Delta p_i \sim \hbar. \tag{1.9}$$

Mathematical equivalence of the uncertainty principle is the replacement of fields with their corresponding operators, so the Poisson bracket may be replaced by the commutator of two operators

$$\{A,B\} \to \frac{i}{\hbar}[A,B].$$

Thus, the commutation relation between $p_i$ and $q_i$ is given by

$$[p_i, q_j] = -i\hbar\delta_{ij}. \tag{1.10}$$

This can be done either by considering matrix representations of these operators or by defining the operators:

$$p_i \to -i\hbar\frac{\partial}{\partial q_i} \quad \text{and} \quad E \to i\hbar\frac{\partial}{\partial t},$$

which acts on the wave function or the state vector of the particle, $\psi$, to give the eigenvalue of the operator. The wave function of the particle can be obtained by solving the Schrödinger equation

$$H\psi = E\psi \quad \Longrightarrow \quad \left(-\frac{\hbar^2}{2m}\frac{\partial^2}{\partial q_i^2} + V(q_i)\right)\psi = \left(i\hbar\frac{\partial}{\partial t}\right)\psi. \tag{1.11}$$

Then $\rho = |\psi|^2$ is interpreted as the probability of finding the particle at any given point $q_i$ with momentum $p_i$, where the experimental error in measuring $q_i$ and $p_i$ satisfies the uncertainty principle.

For a relativistic free particle represented by $\phi$, satisfying

$$E^2 = \mathbf{p}^2 + m^2,$$

(in natural units, $\hbar = c = 1$), the corresponding equation of motion becomes

$$\frac{\partial^2\phi}{\partial t^2} - \nabla^2\phi + m^2\phi = 0 \quad \text{or} \quad \partial_\mu\partial^\mu\phi + m^2\phi = 0. \tag{1.12}$$

This is known as the Klein–Gordon equation [4]. The derivatives are defined as

$$\partial_\mu = \left(\frac{\partial}{\partial t}, \frac{\partial}{\partial x^i}\right) \quad \text{and} \quad \partial^\mu = g^{\mu\nu}\partial_\nu = \left(\frac{\partial}{\partial t}, -\frac{\partial}{\partial x^i}\right) \tag{1.13}$$

where the metric is given by

$$g_{\mu\nu} = g^{\mu\nu} = \begin{pmatrix} 1 & 0 & 0 & 0 \\ 0 & -1 & 0 & 0 \\ 0 & 0 & -1 & 0 \\ 0 & 0 & 0 & -1 \end{pmatrix}. \tag{1.14}$$

For a free particle with wave function $\phi = Ne^{-ip_\mu x^\mu}$, the probability density and the energy eigenvalues are then given by

$$\rho = |N|^2 E \quad \text{and} \quad E = \pm\left(\mathbf{p}^2 + m^2\right)^{1/2}. \tag{1.15}$$

This appears to be inconsistent. The negative energy with negative probability density cannot represent any physical system, and hence, this extension to relativistic quantum mechanics was abandoned.

At this point a relativistic equation, linear in $x_i$ and $t$, was proposed by Dirac [5]:

$$(i\gamma_\mu \partial^\mu - m)\psi = 0, \tag{1.16}$$

where $\gamma_\mu$ are $4 \times 4$ matrices given by

$$\gamma^0 = \gamma_0 = \begin{pmatrix} \mathbf{1} & \mathbf{0} \\ \mathbf{0} & -\mathbf{1} \end{pmatrix} \quad \gamma^i = -\gamma_i = \begin{pmatrix} \mathbf{0} & \sigma^i \\ -\sigma^i & \mathbf{0} \end{pmatrix}, \tag{1.17}$$

where each of the elements are $2 \times 2$ matrices: $\mathbf{1} = \begin{pmatrix} 1 & 0 \\ 0 & 1 \end{pmatrix}$, $\mathbf{0} = \begin{pmatrix} 0 & 0 \\ 0 & 0 \end{pmatrix}$ and $\sigma^i$ are the Pauli Matrices, given by

$$\sigma^1 = \begin{pmatrix} 0 & 1 \\ 1 & 0 \end{pmatrix} \quad \sigma^2 = \begin{pmatrix} 0 & -i \\ i & 0 \end{pmatrix} \quad \sigma^3 = \begin{pmatrix} 1 & 0 \\ 0 & -1 \end{pmatrix}. \tag{1.18}$$

The $\gamma$-matrices satisfy the anticommutation relations

$$\{\gamma_\mu, \gamma_\nu\} = \gamma_\mu\gamma_\nu + \gamma_\nu\gamma_\mu = 2g_{\mu\nu} \tag{1.19}$$

and the Pauli matrices satisfy the commutation relation

$$[\sigma^i, \sigma^j] = i\varepsilon^{ijk}\sigma^k, \tag{1.20}$$

where $\varepsilon^{ijk}$ is the totally antisymmetric tensor with $\varepsilon^{123} = 1$. For completeness we also define

$$\gamma^5 = \gamma_5 = i\gamma^0\gamma^1\gamma^2\gamma^3 = -\frac{i}{4!}\varepsilon_{\mu\nu\rho\sigma}\gamma^\mu\gamma^\nu\gamma^\rho\gamma^\sigma = \begin{pmatrix} \mathbf{0} & \mathbf{1} \\ \mathbf{1} & \mathbf{0} \end{pmatrix}, \tag{1.21}$$

where $\varepsilon_{\mu\nu\rho\sigma} = -\varepsilon^{\mu\nu\rho\sigma}$ is the totally antisymmetric tensor with $\varepsilon^{0123} = 1$. The $\gamma^5$ matrix satisfies $(\gamma^5)^\dagger = \gamma^5$, $(\gamma^5)^2 = 1$ and $\{\gamma^5, \gamma^\mu\} = 0$.

The Dirac equation could explain particles with spin-1/2 and there is no negative probability density. However, the energy of any particle could be positive or negative. The negative energy solution was then interpreted by the hole theory, which assumes an infinite sea of negative energy states, which are all filled up. Thus, any positive energy particles cannot usually go into these states due to the Pauli exclusion principle. However, when any negative energy particle with mass $m$ absorbs a photon of energy more than $2m$, the negative energy particle can come out of the negative energy sea and propagate as a positive energy particle creating a hole in the negative energy sea. The propagation of the hole will then appear as an antiparticle with mass $m$ and positive energy, whose quantum numbers are opposite to those of the particle. The process of a photon of energy greater than $2m$ creating a particle and an antiparticle is called the pair creation. Similarly, any particle can fall into the hole in the negative energy sea releasing energy. In this process of particle–antiparticle annihilation, the particle and the antiparticle corresponding to the hole will disappear, releasing a photon with energy $2m$ or more depending on the kinetic energy of the particle.

Later the Klein–Gordon equation was also revived with the suggestion to multiply the probability density by electric charge and define it as charge density, which could be negative. The second problem of negative energy of the Klein–Gordon equation could not be solved by the concept of holes because it is not possible to saturate any negative energy states by bosons because there is no exclusion principle for bosons. Nevertheless, it would be possible to interpret the negative energy states as positive energy antiparticles even in this case.

Thus, we have the Dirac equation to explain the motion of any relativistic spin-1/2 fermions and the Klein–Gordon equation to explain the motion of any relativistic integer spin bosons. At the level of quantum mechanics both of these equations are consistent and describe the motion of any single relativistic particle. However, these equations suffer from other problems and lead to inconsistency, which could be solved in quantum field theory.

The main problem with the relativistic quantum mechanics to describe a relativistic particle is that the creation of particle–antiparticle pairs from vacuum cannot be taken care of. Even when the energy of the particles is less than the energy required for pair creation, it is possible to have virtual pair creation and annihilation which will influence the motion of the particle. The virtual particle–antiparticle pairs can exist for a short period of time, allowed by the uncertainty principle. Thus, any single particle equation of motion is inadequate to explain any quantum relativistic theory.

Another problem encountered by any relativistic quantum mechanics is the violation of causality. In both nonrelativistic and relativistic quantum mechanics, it is possible for a state to propagate between two points separated by space-like intervals, violating causality. In quantum field theory the propagation of any particle in a space-like interval would appear as propagation of an antiparticle in the opposite direction. Thus, the amplitudes of a particle and an antiparticle propagating between two points, separated by space-like interval, will cancel each other making the theory consistent.

## Classical Field Theory

We start our discussion with classical field theory, which describes a system with infinite degrees of freedom. In field theories any interaction can be treated as collision while the initial and final states are treated as fields of free particles with infinite degrees of freedom at each space–time point. The ensemble of free particles are described by fields $\phi(x)$ with infinite degrees of freedom. The Lagrangian of the system depends on both of the fields as well as their space–time derivatives $\partial_\mu \phi(x)$ and can be written as spatial integral of a Lagrangian density, so the action is given by

$$S = \int L \, dt = \int \mathcal{L}(\phi, \partial_\mu \phi) d^4x. \tag{1.22}$$

The principle of least action would then give us the Euler–Lagrange equations of motion. Taking the variation of the action, if we apply the boundary condition that the variations of the fields vanish at the end-points, we get the equations of motion

$$\partial_\mu \frac{\delta \mathcal{L}}{\delta \partial_\mu \phi} - \frac{\delta \mathcal{L}}{\delta \phi} = 0. \tag{1.23}$$

In the Hamiltonian formalism the conjugate momentum is defined as

$$p(\mathbf{x}) = \frac{\partial L}{\partial \dot{\phi}(\mathbf{x})} = \frac{\partial}{\partial \dot{\phi}(\mathbf{x})} \int \mathcal{L} d^3x = \int \pi(\mathbf{x}) d^3x, \tag{1.24}$$

where $\pi(\mathbf{x}) = \partial \mathcal{L} / \partial \dot{\phi}(\mathbf{x})$ is called the momentum density. The Hamiltonian is then defined as

$$H = \int \left( \pi(\mathbf{x}) \dot{\phi}(\mathbf{x}) - \mathcal{L} \right) d^3x = \int \mathcal{H} d^3x. \tag{1.25}$$

Explicit Lorentz invariance makes the Lagrangian formalism more convenient compared with the Hamiltonian formalism in many problems of quantum field theory.

## Noether's Theorem

The symmetry of the action determines the conserved quantities in the theory. The Noether's theorem states that corresponding to any symmetry of the action, there exists a conserved current and, hence, a conserved charge [6]. We are familiar with the invariance of the action under space–time translation that ensures conservation of momentum and energy, while the conservation of angular momentum is an outcome of rotational symmetry of the action. There could be other internal symmetries of the action, which would give us some new conservation laws.

Consider the infinitesimal continuous transformation of the field

$$\phi(x) \to \phi'(x) = \phi(x) + \alpha \delta \phi(x),$$

where $\alpha$ is an infinitesimal parameter and $\delta \phi$ is some variation of the field configuration. The variation of the action due to this transformation of the field is then given by

$$\delta S = \int d^4x \left( \frac{\delta \mathcal{L}}{\delta \phi} \alpha \delta \phi + \frac{\delta \mathcal{L}}{\delta(\partial_\mu \phi)} \alpha \delta(\partial_\mu \phi) \right)$$
$$= \int d^4x \alpha \partial_\mu \left( \frac{\delta \mathcal{L}}{\delta \partial_\mu \phi} \delta \phi \right). \tag{1.26}$$

We used the Euler–Lagrange equations of motion to arrive at the last expression. The invariance of the action with respect to the variation of the field configuration implies invariance of the Lagrangian up to a 4-divergence

$$\mathcal{L}(x) \to \mathcal{L}(x) + \alpha \partial_\mu t^\mu(x),$$

where $t^\mu$ is some 4-vector. Combining the two we get the conservation of current as a result of invariance of the action

$$\partial_\mu j^\mu = 0, \quad \text{where} \quad j^\mu = \frac{\delta \mathcal{L}}{\delta \partial_\mu \phi} \delta \phi - t^\mu. \tag{1.27}$$

The conservation law also implies that the corresponding charge

$$Q = \int d^3x j^0 \tag{1.28}$$

is a constant in time $dQ/dt = 0$. Thus, the symmetry of the action implies a conservation principle.

To obtain the conservation principle corresponding to a space–time translation, consider the infinitesimal translation

$$x^\mu \to x^\mu + a^\mu$$

and, hence,

$$\delta \phi(x) = \phi(x+a) - \phi(x) = a^\mu \partial_\mu \phi(x).$$

The Lagrangian, being a scalar, transforms the same way

$$\mathcal{L} \to \mathcal{L} + a^\nu \partial_\mu (\delta^\mu_\nu \mathcal{L}),$$

which gives an additional term. Variation of the action then gives the conserved current

$$T^\mu{}_\nu = \frac{\delta \mathcal{L}}{\delta \partial_\mu \phi} \partial_\nu \phi - \mathcal{L} \delta^\mu{}_\nu, \qquad \text{with } \partial_\mu T^\mu{}_\nu = 0, \tag{1.29}$$

which is the energy-momentum tensor. The energy–momentum 4-vector $P^\mu \equiv (E, P^i)$ can then be defined as

$$P^\mu = \int d^3x \, T^\mu{}_0, \tag{1.30}$$

which is conserved as $dP^\mu/dt = 0$, if the action is invariant under space–time translation. $P^i$ is the physical momentum carried by the field and $E$ is the energy.

Let us now consider a generalized Lorentz transformation:

$$\delta x^\mu \to \varepsilon^\mu{}_\nu x^\nu .$$

The field and the Lagrangian will transform as

$$\begin{aligned} \delta\phi(x) &= \varepsilon^\mu{}_\nu x^\nu \partial_\mu \phi(x) \\ \delta\mathscr{L} &= \varepsilon^\mu{}_\nu x^\nu \partial_\mu \mathscr{L}, \end{aligned} \tag{1.31}$$

and the conserved current becomes

$$\mathscr{M}^{\rho,\mu\nu} = T^{\rho\nu} x^\mu - T^{\rho\mu} x^\nu, \qquad \text{with } \partial_\rho \mathscr{M}^{\rho,\mu\nu} = 0. \tag{1.32}$$

The corresponding conserved charge

$$M^{\mu\nu} = \int d^3x\, \mathscr{M}^{0,\mu\nu},$$

with $dM^{\mu\nu}/dt = 0$, generates the Lorentz group. The 3-dimensional rotational invariance of the action thus give us the conservation of angular momentum.

## 1.2 Scalar, Spinor and Gauge Fields

The purpose of field theory is to understand the behaviour of particles and their interactions. So, we start with the descriptions of scalar, spinor and gauge fields, then discuss how they are quantized, and finally write down the free-field propagators.

### Klein–Gordon Fields

Although we have not seen any fundamental scalar particles with spin-0, because of simplicity we start our discussions about quantum field theory with scalar field theory. The Lagrangian for a real scalar field $\phi$ is given by

$$\mathscr{L} = \frac{1}{2}(\partial_\mu \phi)^2 - \frac{1}{2} m^2 \phi^2, \tag{1.33}$$

where $m$ is the mass of the particle. The Euler–Lagrange equation then gives us the Klein–Gordon equation (see equation (1.12)). The corresponding Hamiltonian is given by

$$H = \int d^3x \mathscr{H} = \int d^3x [\frac{1}{2}\pi^2 + \frac{1}{2}(\nabla\phi)^2 + \frac{1}{2} m^2 \phi^2], \tag{1.34}$$

where

$$\pi(x) = \frac{\delta \mathscr{L}}{\delta \dot{\phi}(x)} = \dot{\phi}(x)$$

is the canonical momentum density conjugate to $\phi$. The theory is quantized by demanding the commutation relation among the conjugate fields

$$[\phi(\mathbf{x},t),\pi(\mathbf{y},t)] = i\delta^3(\mathbf{x}-\mathbf{y}), \tag{1.35}$$

while $\phi(x)$ and $\pi(x)$ commute with themselves.

A free-particle solution to the Klein–Gordon equation may be written as

$$\begin{aligned} \phi(x) &= \int \frac{d^3k}{(2\pi)^{3/2}} \frac{1}{\sqrt{2\omega_k}} \left[a(k)e^{-ik\cdot x} + a^\dagger(k)e^{ik\cdot x}\right] \\ \pi(x) &= \int \frac{d^3k}{(2\pi)^{3/2}} (-i)\sqrt{\frac{\omega_k}{2}} \left[a(k)e^{-ik\cdot x} - a^\dagger(k)e^{ik\cdot x}\right], \end{aligned} \tag{1.36}$$

where $k\cdot x = k_\mu x^\mu = (Et - \mathbf{k}\cdot\mathbf{x})$, $\omega_k = \sqrt{\mathbf{k}^2+m^2}$, and $a_k$ and $a_k^\dagger$ are the annihilation and creation operators satisfying the commutation relation

$$\left[a(k), a^\dagger(k')\right] = \delta^3(\mathbf{k}-\mathbf{k}'), \tag{1.37}$$

so $[\phi(\mathbf{x},t),\pi(\mathbf{x}',t)] = i\delta^3(\mathbf{x}-\mathbf{x}')$. In terms of the ladder operators, the Hamiltonian $H$ and the momentum $\mathbf{p}$ are given by

$$\begin{aligned} H &= \int d^3k\omega_k \left[a^\dagger(k)\, a(k) + \frac{1}{2}\right] \\ \mathbf{p} &= \int d^3k\mathbf{k} \left[a^\dagger(k)\, a(k) + \frac{1}{2}\right]. \end{aligned} \tag{1.38}$$

The factor $1/2$ in both energy and momentum is divergent. This divergent part corresponding to the zero-point energy is taken out for consistency. In the $x$ space this corresponds to moving the creation operator to the left of annihilation operator, which is done by normal-ordering

$$:\phi_1\phi_2: \;\equiv\; a^\dagger(k_1)a^\dagger(k_2) + a^\dagger(k_1)a(k_2) + a^\dagger(k_2)a(k_1) + a(k_1)a(k_2).$$

This normal-ordering or dropping out the factor of $1/2$ in the momentum space will remove the infinities.

The vacuum can be defined as

$$a(k)|0\rangle = 0 \tag{1.39}$$

so all other states can be obtained from this state. A single-particle state can be written as

$$a^\dagger(k)|0\rangle = |k\rangle. \tag{1.40}$$

Similarly we can construct multiparticle states

$$a^\dagger(k_1)a^\dagger(k_2)\cdots a^\dagger(k_N)|0\rangle = |k_1,k_2,\cdots k_N\rangle. \tag{1.41}$$

In general, a multiparticle state can have $n(k_i)$ number of particles with momentum $k_i$. The number operator, defined as

$$N = \int d^3k_i \, a^\dagger(k_i) \, a(k_i), \tag{1.42}$$

gives the number $n(k_i)$, when it acts on a multiparticle state

$$\begin{aligned} N|n(k_1)n(k_2)\cdots n(k_m)\rangle &= N \left[ \prod_{i=1}^{m} \frac{\left(a^\dagger(k_i)\right)^{n(k_i)}}{\sqrt{n(k_i)!}} \right] |0\rangle \\ &= \left( \sum_{i=1}^{m} n(k_i) \right) |n(k_1)n(k_2)\cdots n(k_m)\rangle. \end{aligned} \tag{1.43}$$

If we define $\langle k| = \langle 0|a(k)$ with $\langle 0|0\rangle = 1$, then the norm should be positive $\langle k|k'\rangle = \delta^3(\mathbf{k}-\mathbf{k}')$ for any physical state. There could be ghost states with negative norms, but these should cancel out to maintain the unitarity of the theory.

In case of a charged scalar, we combine two real scalars to form a complex field

$$\phi = \frac{1}{\sqrt{2}}(\phi_1 + i\phi_2), \tag{1.44}$$

which satisfies the action given by the Lagrangian density

$$\mathscr{L} = \partial_\mu \phi^\dagger \partial^\mu \phi - m^2 \phi^\dagger \phi. \tag{1.45}$$

For a free field we can expand in terms of its Fourier components as

$$\begin{aligned} \phi(x) &= \int \frac{d^3k}{(2\pi)^{3/2}} \frac{1}{\sqrt{2\omega_k}} (a(k)e^{-ik\cdot x} + b^\dagger(k)e^{ik\cdot x}) \\ \phi^\dagger(x) &= \int \frac{d^3k}{(2\pi)^{3/2}} \frac{1}{\sqrt{2\omega_k}} (a^\dagger(k)e^{ik\cdot x} + b(k)e^{-ik\cdot x}). \end{aligned} \tag{1.46}$$

These operators satisfy the Bose commutation relation

$$[a(k), a^\dagger(k')] = [b(k), b^\dagger(k')] = \delta^3(\mathbf{k}-\mathbf{k}'). \tag{1.47}$$

All other pairs of operators commute.

This theory has a symmetry

$$\phi \to e^{i\theta}\phi, \qquad \text{and} \qquad \phi^\dagger \to e^{-i\theta}\phi^\dagger,$$

and the corresponding Noether current and charge are respectively given by

$$\begin{aligned} J_\mu &= i\phi^\dagger \partial_\mu \phi - i\partial_\mu \phi^\dagger \phi \\ Q &= \int d^3k [a^\dagger(k)a(k) - b^\dagger(k)b(k)] = N_a - N_b, \end{aligned} \tag{1.48}$$

where $N(a)$ and $N(b)$ are the number operators for $a$ and $b$ type oscillators. This theory leads to negative probability if $j^0$ is the probability making it apparently inconsistent. A proper interpretation [7] is that $a(k)$ and $b(k)$ are annihilation operators for particles and antiparticles (with opposite charge), respectively, while $a^\dagger(k)$ and $b^\dagger(k)$ are the corresponding creation operators.

After describing the states, we now proceed to describe how these particles propagate in space–time. We first introduce a source term $j(x)$ in the Klein–Gordon equation (1.12)

$$\partial_\mu \partial^\mu \phi + m^2 \phi = j(x) \tag{1.49}$$

which corresponds to the Lagrangian

$$\mathscr{L} = \frac{1}{2}(\partial_\mu \phi)^2 - \frac{1}{2}m^2\phi^2 + j(x)\phi(x). \tag{1.50}$$

In the absence of the source term, the field can be described by equation (1.36). In the presence of the source term, the solution of the Klein–Gordon equation can be constructed using retarded Green's function

$$\phi(x) = \phi_0(x) - \int d^4x \Delta_F(x-y) j(y), \tag{1.51}$$

where $\phi_0(x)$ is the free-particle wave function, which satisfies Klein–Gordon equation in the absence of the source term and the propagator $\Delta_F(x-y)$ satisfies

$$(\partial_\mu \partial^\mu + m^2)\Delta_F(x-y) = -\delta^4(x-y). \tag{1.52}$$

To solve for the propagator, we take the Fourier transformation

$$\Delta_F(x-y) = \int \frac{d^4k}{(2\pi)^4} e^{-ik(x-y)} \Delta_F(k) \tag{1.53}$$

and solve for $\Delta_F(k)$. The solution does not allow us to perform the integral over $k$ at the points $k_\mu^2 = m^2$. One prescription to resolve this ambiguity is to shift the poles by $i\varepsilon$, which corresponds to proper boundary conditions. Then we can write

$$\Delta_F(k) = \frac{1}{k^2 - m^2 + i\varepsilon}. \tag{1.54}$$

We would like to relate this propagator with the correlation functions. First we define the $\theta$-function which will allow us to define the time-ordered product and interpret the particles and antiparticles properly. We define the $\theta$-function as

$$\begin{aligned} \theta(x^0 - y^0) &= -\lim_{\varepsilon \to 0} \frac{1}{2\pi i} \int_{-\infty}^{\infty} \frac{e^{-i\omega(x^0-y^0)} d\omega}{\omega + i\varepsilon} \\ &= \begin{cases} 1 & \text{if } x^0 > y^0 \\ 0 & \text{otherwise} \end{cases}. \end{aligned} \tag{1.55}$$

We can then write the expression for the propagator as

$$
\begin{aligned}
i\,\Delta_F(x-y) &= \theta(x^0-y^0)\int\frac{d^3k\,e^{-ik(x-y)}}{(2\pi)^3 2\omega_k}+\theta(y^0-x^0)\int\frac{d^3k\,e^{ik(x-y)}}{(2\pi)^3 2\omega_k}\\
&= \theta(x^0-y^0)\langle 0|\phi(x)\phi(y)|0\rangle+\theta(y^0-x^0)\langle 0|\phi(y)\phi(x)|0\rangle\\
&\equiv \langle 0|\,T\phi(x)\phi(y)\,|0\rangle,
\end{aligned}
\tag{1.56}
$$

where the time-ordered operator $T$ is defined as

$$
T\phi(x)\phi(y)=\begin{cases}\phi(x)\phi(y) & \text{if } x^0>y^0\\ \phi(y)\phi(x) & \text{if } y^0>x^0\end{cases}
\tag{1.57}
$$

so it ensures that the operators follow in order, the one with the latest time component appearing to the left. The boundary conditions now imply that the positive energy solution is moving forward in time, while the negative energy solution is moving backward in time. Since the quantum numbers of any antiparticle is opposite to the particle, the negative energy solution moving backward can be interpreted as an antiparticle moving forward in time. The boundary conditions considered also ensure that the propagator vanishes for any space-like separations, satisfying the microscopic causality.

**Dirac Fields**

A Dirac field [5] represents a spin-1/2 particle and satisfies the Dirac equation (see equation (1.16)). The corresponding Lorentz invariant Lagrangian is given by

$$
\mathscr{L}=\bar{\psi}(i\gamma^\mu\partial_\mu-m)\psi,
\tag{1.58}
$$

where $\bar{\psi}=\psi^\dagger\gamma^0$, so variation with respect to $\bar{\psi}$ gives equation 1.16. The field $\psi$ is a 4-dimensional spinor and can be expressed in terms of the basis spinors

$$
u_1(0)=\begin{pmatrix}1\\0\\0\\0\end{pmatrix};\quad u_2(0)=\begin{pmatrix}0\\1\\0\\0\end{pmatrix};\quad v_1(0)=\begin{pmatrix}0\\0\\1\\0\end{pmatrix};\quad v_2(0)=\begin{pmatrix}0\\0\\0\\1\end{pmatrix}.
\tag{1.59}
$$

These basis states may be acted upon by the Lorentz boost matrix

$$
S(\Lambda)=\sqrt{E+m}\begin{pmatrix}\mathbf{1} & \frac{\sigma\cdot\mathbf{p}}{E+m}\\ \frac{\sigma\cdot\mathbf{p}}{E+m} & \mathbf{1}\end{pmatrix}
\tag{1.60}
$$

to get the momentum dependent spinors

$$
\begin{aligned}
u_\alpha(p)&=S(\Lambda)u_\alpha(0)\\
v_\alpha(p)&=S(\Lambda)v_\alpha(0).
\end{aligned}
\tag{1.61}
$$

These spinors satisfy the equations of motion

$$\begin{aligned} (\not{p}-m)u(p) &= 0; \qquad (\not{p}+m)v(p) = 0; \\ \bar{u}(p)(\not{p}-m) &= 0; \qquad \bar{v}(p)(\not{p}+m) = 0. \end{aligned} \tag{1.62}$$

where $\not{p} = \gamma \cdot p = \gamma^{\mu} p_{\mu}$. The $u$ spinors represent particles with positive energy and moving forward in time, while the $v$ spinors represent particles with negative energy and moving backward in time, which is an antiparticle moving forward in time.

A spinor with a spin $s$ may be represented by $u^s(p)$ and $v^s(p)$, where the spin-vector $s_{\mu} \equiv (0, \mathbf{s})$ satisfies the Lorentz invariant conditions

$$s_{\mu}^2 = -1 \qquad \text{and} \qquad p_{\mu} s^{\mu} = 0. \tag{1.63}$$

The spin projection operator may then be defined as

$$P^s = \frac{1+\gamma_5 \not{s}}{2}, \tag{1.64}$$

where $\not{s} = s_{\mu}\gamma^{\mu}$, so the spinors satisfy

$$\begin{aligned} P^s u^s(p) = u^s(p), \qquad & P^s v^s(p) = v^s(p) \\ P^{-s} u^s(p) = P^{-s} & v^s(p) = 0. \end{aligned} \tag{1.65}$$

We normalized these spinors as

$$\begin{aligned} \bar{u}^r(p) u^s(p) &= 2m\delta^{rs} \\ \bar{v}^r(p) v^s(p) &= -2m\delta^{rs}. \end{aligned} \tag{1.66}$$

Then they satisfy certain completeness conditions

$$\begin{aligned} \sum_s u_{\alpha}^s(p)\bar{u}_{\beta}^s(p) - v_{\alpha}^s(p)\bar{v}_{\beta}^s(p) &= 2m\delta_{\alpha\beta} \\ \sum_s u_{\alpha}^s(p)\bar{u}_{\beta}^s(p) &= (\not{p}+m)_{\alpha\beta} = [\Lambda_+(p)]_{\alpha\beta} \\ \sum_s v_{\alpha}^s(p)\bar{v}_{\beta}^s(p) &= (\not{p}-m)_{\alpha\beta} = -[\Lambda_-(p)]_{\alpha\beta}. \end{aligned} \tag{1.67}$$

We defined two operators $[\Lambda_{\pm}(p)]_{\alpha\beta}$, which project out the positive and negative solutions and satisfy $\Lambda_{\pm}^2 = 2m\Lambda_{\pm}$, $\Lambda_+\Lambda_- = 0$ and $\Lambda_+ + \Lambda_- = 2m$.

To quantize the theory, we define the conjugate momentum

$$\pi(x) = \delta\mathcal{L}/\delta\psi(x) = i\psi^{\dagger},$$

and to satisfy the spin-statistics postulate, we define equal time anticommutation relations

$$\begin{aligned} \{\psi_a(x), \psi_b^{\dagger}(y)\} &= \delta^3(\mathbf{x}-\mathbf{y})\delta_{ab} \\ \{\psi_a(x), \psi_b(y)\} &= \{\psi_a^{\dagger}(x), \psi_b^{\dagger}(y)\} = 0. \end{aligned} \tag{1.68}$$

We can now expand the Dirac fields in terms of ladder operators

$$\psi(x) = \int \frac{1}{\sqrt{2E_p}} \frac{d^3p}{(2\pi)^{3/2}} \sum_s \left( b^s(p)u^s(p)e^{-ipx} + d^{s\dagger}(p)v^s(p)e^{ipx} \right)$$
$$\bar{\psi}(x) = \int \frac{1}{\sqrt{2E_p}} \frac{d^3p}{(2\pi)^{3/2}} \sum_s \left( b^{s\dagger}(p)\bar{u}^s(p)e^{ipx} + d^s(p)\bar{v}^s(p)e^{-ipx} \right). \quad (1.69)$$

The four terms represent annihilation of particles $[b^s u^s]$, creation of antiparticles $[d^{s\dagger} v^s]$, creation of particles $[b^{s\dagger}\bar{u}^s]$ and annihilation of antiparticles $[d^s\bar{v}^s]$. These creation and annihilation operators satisfy the anticommutation relations

$$\{a^r(p), a^{s\dagger}(p')\} = \{b^r(p), b^{s\dagger}(p')\} = \delta^{rs}\delta^3(\mathbf{p}-\mathbf{p}'), \quad (1.70)$$

and all other anticommutators vanish. The Hamiltonian and the momentum are now given by

$$H = \int d^3p\, p_0 \sum_s [b^{s\dagger}(p)b^s(p) + d^{s\dagger}(p)d^s(p)]$$
$$\mathbf{p} = \int d^3p\, \mathbf{p} \sum_s [b^{s\dagger}(p)b^s(p) + d^{s\dagger}(p)d^s(p)]. \quad (1.71)$$

We have normal-ordered the operators to remove the zero-point energy, which also removed the negative energy eigenvalues of the Hamiltonian since the ladder operators anticommute.

The Dirac Lagrangian has a symmetry $\psi \to e^{i\theta}\psi$ and $\bar{\psi} \to \bar{\psi}e^{-i\theta}$, and the corresponding current $j_\mu = \bar{\psi}\gamma_\mu\psi$ is conserved. The associated normal-ordered charge becomes

$$Q = \int d^3x\, :\psi^\dagger\psi: \; = \int d^3p \sum_s [b^{s\dagger}(p)b^s(p) - d^{s\dagger}(p)d^s(p)]. \quad (1.72)$$

Since antiparticles have opposite charges compared with particles, the negative sign gives proper explanation. The anticommutation of the ladder operator also ensures Pauli exclusion principle, since $d^{s\dagger}(k)d^{s\dagger}(k)|0\rangle = 0$. Any multiparticle state, created by any operator satisfying anticommutation relation, will be of the form

$$\prod_{i=1}^{N} d^{r_i\dagger}(p_i) \prod_{j=1}^{M} b^{r_j\dagger}(p_j)\, |0\rangle. \quad (1.73)$$

Thus, two particles with same quantum numbers cannot be in any single quantum state [8].

To define a Dirac propagator, we introduce a source term $j(x)$ in the Dirac equation

$$(i\partial\!\!\!/ - m)\psi(x) = j(x). \quad (1.74)$$

The Dirac propagator then satisfies

$$(i\partial\!\!\!/ - m)S_F(x-y) = \delta^4(x-y). \quad (1.75)$$

Using Fourier transformation we can solve for the propagator

$$S_F(x-y) = \int \frac{d^4p}{(2\pi)^4} e^{-ip(x-y)} \frac{\not{p}+m}{p^2-m^2+i\varepsilon}. \tag{1.76}$$

Proceeding in the same way as for the Klein–Gordon fields, we can write

$$iS_F(x-y) = \langle 0|T\psi(x)\bar{\psi}(y)|0\rangle. \tag{1.77}$$

While taking time-ordered products of fermions, interchanging the fields would introduce an additional negative sign.

**Quantum Electrodynamics**

We started with quantization of spin-0 scalar particles and then discussed the spin-1/2 fermions. We complete this discussion with spin-1 electromagnetic fields. We start with the Maxwell's equation in the relativistic form

$$\partial_\mu F^{\mu\nu} = j^\nu, \tag{1.78}$$

where the four-vector current $j^\nu \equiv (\rho, \mathbf{j})$ satisfies the current conservation equation

$$\partial_\mu j^\mu = 0. \tag{1.79}$$

The electromagnetic field tensor $F^{\mu\nu}$ is defined in terms of the electric ($\mathbf{E}$) and the magnetic ($\mathbf{B}$) fields or the the scalar ($\phi$) and the vector ($\mathbf{A}$) potentials $A_\mu \equiv (\phi, \mathbf{A})$ as

$$F^{\mu\nu} = \begin{pmatrix} 0 & -E^1 & -E^2 & -E^3 \\ E^1 & 0 & -B^3 & B^2 \\ E^2 & B^3 & 0 & -B^1 \\ E^3 & -B^2 & B^1 & 0 \end{pmatrix} = \partial_\mu A_\nu - \partial_\nu A_\mu. \tag{1.80}$$

The electric and the magnetic fields can then be defined in terms of the potentials as

$$F^{0i} = -E^i, \qquad \text{and} \qquad F^{ij} = -\varepsilon^{ijk} B^k, \tag{1.81}$$

where $\varepsilon^{ijk}$ is the totally antisymmetric tensor. The Lagrangian that gives the Maxwell's equation as the Euler–Lagrange equation of motion is given by

$$\mathscr{L} = -\frac{1}{4} F_{\mu\nu} F^{\mu\nu} + A_\nu j^\nu. \tag{1.82}$$

We shall again discuss this Lagrangian and its origin as a $U(1)$ gauge theory in section 4.4.

Quantization of the Maxwell's equation has to be done by properly taking care of gauge invariance. Let us consider the interaction of a fermion $\psi$ with the electromagnetic field $A_\mu$. The four-vector current with the fermions can be written as $j^\mu = e\bar{\psi}\gamma^\mu\psi$, which is conserved because of the symmetry $\psi \to e^{i\theta}\psi$. If we promote $\theta(x)$ to a local variable, then invariance of the Lagrangian requires

$$\partial_\mu \to D_\mu = \partial_\mu + ieA_\mu,$$

where $A_\mu \to A_\mu - \partial_\mu \theta(x)$. The freedom to add $\partial_\mu \theta(x)$ to $A_\mu$ does not allow us to construct any propagator. We, thus, work in a particular gauge to avoid this problem of infinite redundancy. This is conveniently done either by constraining the field $A_\mu$ or by adding a term $-\frac{1}{2\alpha}(\partial_\mu A^\mu)^2$ for arbitrary $\alpha$ in the Lagrangian.

While working in a particular gauge, existence of $\theta$ should be verified. For example, in the Landau gauge $\partial_\mu A^\mu = 0$, there should be $\theta$ which allows this gauge condition, which is

$$\theta = -\frac{1}{\partial^2}\partial_\mu A^\mu.$$

Similarly for the Coulomb gauge $\Delta \cdot \mathbf{A} = 0$, there should be a $\theta$ that satisfies $\Delta \cdot \mathbf{A}' = \Delta \cdot (\mathbf{A} + \Delta\theta) = 0$, which is given by

$$\theta = -\int \frac{d^3x'}{4\pi|\mathbf{x}-\mathbf{x}'|}\Delta' \cdot \mathbf{A}(x').$$

In the Coulomb gauge only the physical states are allowed to propagate by extracting the longitudinal modes of the field from the beginning. So we shall work in the Coulomb gauge. The $A_0$ is eliminated by the equations of motion, and the longitudinal mode is gauged away leaving the two transverse modes of the photons.

We can then write down the modified canonical commutation relations, which takes care of the fact that $A_i$ is divergence free. This is given by

$$[A_i(\mathbf{x},t), \pi_j(\mathbf{y},t)] = -i\tilde{\delta}_{ij}(\mathbf{x}-\mathbf{y}), \tag{1.83}$$

where we modified the delta function to make it transverse

$$\tilde{\delta}_{ij}(\mathbf{x}-\mathbf{y}) = \int \frac{d^3k}{(2\pi)^3} e^{i\mathbf{k}\cdot(\mathbf{x}-\mathbf{y})}\left(\delta_{ij} - \frac{k_i k_j}{\mathbf{k}^2}\right). \tag{1.84}$$

We can then write down the decomposition of the fields in terms of their Fourier modes as

$$\mathbf{A}(x) = \int \frac{d^3k}{(2\pi)^{3/2}\sqrt{2k_0}} \sum_{\lambda=1}^{2} \varepsilon^\lambda(k)\left[a^\lambda(k)e^{-ik\cdot x} + a^{\lambda\dagger}(k)e^{ik\cdot x}\right], \tag{1.85}$$

where the polarization vector $\varepsilon^\lambda(k)$ satisfies

$$\varepsilon^\lambda \cdot \mathbf{k} = 0$$
$$\varepsilon^\lambda(k) \cdot \varepsilon^{\lambda'}(k) = \delta^{\lambda\lambda'}. \tag{1.86}$$

The polarization vector thus ensures that the fields have only the transverse modes. The Fourier moments then satisfy

$$[a^\lambda(k), a^{\lambda'\dagger}(k')] = \delta^{\lambda\lambda'}\delta^3(\mathbf{k}-\mathbf{k}'). \tag{1.87}$$

Thus, there are only positive norm states in the Coulomb gauge. The Hamiltonian and the momentum are given by

$$H = \frac{1}{2}\int d^3x(\mathbf{E}^2+\mathbf{B}^2) = \int d^3k\ \omega \sum_{\lambda=1}^{2}[a^{\lambda\dagger}(k)a^{\lambda}(k)]$$
$$\mathbf{P} = \int d^3x(:\mathbf{E}\times\mathbf{B}:) = \int d^3k\,\mathbf{k} \sum_{\lambda=1}^{2}[a^{\lambda\dagger}(k)a^{\lambda}(k)]. \tag{1.88}$$

We have positive definite energy after normal-ordering. The propagator is now given by

$$iD_F^{tr}(x-y)_{\mu\nu} = \langle 0|TA_\mu(x)A_\nu(y)|0\rangle = i\int \frac{d^4k}{(2\pi)^4}\frac{e^{-ik\cdot(x-y)}}{k^2+i\varepsilon}\sum_\lambda \varepsilon_\mu^\lambda(k)\varepsilon_\nu^\lambda(k). \tag{1.89}$$

The main difference between the propagators for scalars or fermions and the gauge bosons is the presence of the polarization tensor. This propagator is not Lorentz invariant, although by working in the Coulomb gauge we had the advantage that all states are physical. However, this is not a problem since the Lorentz invariance violating part will vanish from the full S-matrix. The sum over the polarization vectors can be expressed as

$$\sum_\lambda \varepsilon_\mu^\lambda(k)\varepsilon_\nu^\lambda(k) = -g_{\mu\nu} + (\text{Lorentz noninvariant part}). \tag{1.90}$$

The Lorentz noninvariant part will vanish from any scattering amplitude as a consequence of gauge invariance. Thus, we can write the propagator as

$$iD_F^{tr}(x-y)_{\mu\nu} = \langle 0|TA_\mu(x)A_\nu(y)|0\rangle = \int \frac{d^4k}{(2\pi)^4}\, e^{-ik\cdot(x-y)}\frac{-ig_{\mu\nu}}{k^2+i\varepsilon}. \tag{1.91}$$

This form of the propagator will be used for all calculations.

We shall now discuss the Gupta–Bleuler covariant quantization method [9]. In this case the theory remains Lorentz invariant explicitly although gauge invariance is broken by the Lagrangian

$$\mathscr{L} = -\frac{1}{4}F_{\mu\nu}F^{\mu\nu} - \frac{1}{2\alpha}(\partial_\mu A^\mu)^2 \tag{1.92}$$

for arbitrary $\alpha$. Then it is possible to get the propagator

$$D_{\mu\nu} = -[g_{\mu\nu} - (1-\alpha)\partial_\mu\partial_\nu/\partial^2]/\partial^2, \tag{1.93}$$

which has the propagating ghost states violating unitarity. In the Feynman gauge, $\alpha = 1$, the equation of motion reads $\partial^2 A_\mu = 0$ and the conjugate field now becomes a four-vector $\pi^\mu(x) = \dot{A}^\mu(x)$, so the commutation relations becomes

$$[A_\mu(x), \pi^\nu(y)] = i\delta_\mu^\nu\delta^3(\mathbf{x}-\mathbf{y}). \tag{1.94}$$

The Fourier decomposition of the field can be given in terms of the polarization four-vector $\varepsilon_\mu^\lambda$ as

$$A_\mu(x) = \int \frac{d^3k}{(2\pi)^{3/2}\sqrt{2\omega_k}} \left[a^\lambda(k)\varepsilon_\mu^\lambda(k)e^{-ik\cdot x} + a^{\lambda^\dagger}(k)\varepsilon_\mu^\lambda(k)e^{ik\cdot x}\right]. \tag{1.95}$$

The ladder operators now satisfy the commutation relation

$$[a^\lambda(k), a^{\lambda'^\dagger}(k)] = -g^{\lambda\lambda'}\delta^3(\mathbf{k}-\mathbf{k}') \tag{1.96}$$

and the propagator becomes

$$\langle 0|TA_\mu(x)A_\nu(y)|0\rangle = -ig_{\mu\nu}\Delta_F(x-y), \tag{1.97}$$

where $\Delta_F$ is the propagator for a massless scalar field. This is the same as that given in equation (1.91). In this formalism one needs to remove the ghosts or the negative norm states that appear because the metric has both signs. This can be achieved by acting on the physical states by the destruction part of the constraint,

$$(\partial_\mu A^\mu)^+|\Psi\rangle = 0.$$

This is equivalent to the condition $k^\mu a_\mu(k)|\psi\rangle = 0$. This ensures that the negative norm ghost states are explicitly removed from the system.

## 1.3 Feynman Diagrams

After describing how the scalars, fermions and electromagnetic fields propagate in free space, we would like to know how they interact with each other and give graphical representations of different interactions in terms of Feynman diagrams. We mentioned about the interactions of fermions with electromagnetic fields through the term $e\bar{\psi}\gamma_\mu\psi$. Similarly, there could be interactions of fermions with scalars ($g\bar{\psi}\psi\phi$) or of fermions with pseudo-scalars ($g\bar{\psi}\gamma^5\psi\phi$) or of scalars with electromagnetic fields ($eA^\mu\phi^*\partial\phi$ or $e^2|\phi|^2A^\mu A_\mu$) or self interactions of the fields ($\phi^4 or A^4$). To preserve causality we consider interactions of fields at the same space–time point.

### Interaction picture

We represent the interaction part of the Hamiltonian by

$$H_{Int} = \int d^3x \mathscr{H}_{Int}[\phi(x)] = -\int d^3x \mathscr{L}_{Int}[\phi(x)]. \tag{1.98}$$

For a $\phi^4$ theory the interaction Lagrangian would be $\mathscr{L} = (\lambda/4!)\phi^4(x)$. The complete Hamiltonian will have the Klein–Gordon Hamiltonian and the interaction part

$$H = H_0 + H_{Int} = \int d^3x[\frac{1}{2}\dot{\phi}^2 + \frac{1}{2}(\nabla\phi)^2 + \frac{1}{2}m^2\phi^2 + \frac{\lambda}{4!}\phi^4(x)].$$

Starting from the propagator of the free theory

$$\langle 0|T\phi(x)\phi(y)|0\rangle_{free} = i\Delta_F(x-y) = \int \frac{d^4k}{(2\pi)^4}\frac{i\,e^{-ik(x-y)}}{k^2-m^2+i\varepsilon},$$

we would like to find the two-point correlation function or the two-point Green's function

$$\langle\Omega|T\ \phi(x)\phi(y)|\Omega\rangle,$$

where $|\Omega\rangle$ is the vacuum state in the presence of the interaction, which is different from $|0\rangle$. This analysis can then be generalized to correlation functions with several field operators.

It is not easy to solve the problem in a most general way. We thus assume that the interactions are not too strong, so one can treat them as small perturbations. This means that instead of taking the interactions at all the points simultaneously, to the leading order we shall treat them as instantaneous interactions and consider the particles as free particles propagating in space–time for the rest of the time. So, the initial and the final states are always free-particle states. This will give us the leading order term. The next term in the perturbation series will treat the interactions at two different points, again instantaneously. Summing over all such terms would then give us the complete two-point function, which can then be generalized to the formalism of Feynman diagrams.

The mathematical description of this notion of treating the interaction as perturbation series was provided by Dyson's generalization of the Heisenberg's formalism. Consider a matrix element in the Heisenberg picture

$$\langle\psi_2|O(t)|\psi_1\rangle = \langle\psi_2|e^{iHt}O(0)e^{-iHt}|\psi_1\rangle,$$

where the state vectors are constant in time and the field operators obey the Heisenberg's equation of motion

$$\dot{O}(t) = -i[O(t),H].$$

Dyson considered the time evolution of the operators $O$ as

$$\dot{O}_I(t) = -i[O_I(t),H_0], \quad \text{so that} \quad O_I(t) = e^{iH_0(t-t_0)}O(t_0)e^{-iH_0(t-t_0)}. \tag{1.99}$$

This will allow us to treat the time evolution of the field operators as free-fields so we can construct the field for any instant of time in this interaction picture as

$$\begin{aligned}\phi_I(\mathbf{x},t) &= e^{iH_0(t-t_0)}\phi(\mathbf{x},t_0)e^{-iH_0(t-t_0)} \\ &= \int \frac{d^3k}{(2\pi)^{3/2}}\frac{1}{\sqrt{2\omega_k}}\left[a(k)e^{-ik\cdot x}+a^\dagger(k)e^{ik\cdot x}\right]_{x^0=t-t_0}.\end{aligned} \tag{1.100}$$

To preserve the matrix elements, the state vectors will now become time dependent

$$\langle\psi_2|O(t)|\psi_1\rangle = {}_I\langle\psi_2|O_I(t)|\psi_1\rangle_I. \tag{1.101}$$

The state vectors in the Heisenberg picture and the interaction picture are related by the unitary time-evolution operator $U(t,t_0)$:

$$|\psi(t)\rangle_I = U(t,t_0)|\psi(t_0)\rangle = e^{iH_0(t-t_0)}e^{-iH(t-t_0)}|\psi\rangle\,, \tag{1.102}$$

which obeys the initial condition $U(t_0,t_0) = 1$ and its time evolution is given by

$$i\frac{\partial}{\partial t}U(t,t_0) = e^{iH_0(t-t_0)}(H - H_0)e^{-iH(t-t_0)} = H_I(t)U(t,t_0)\,, \tag{1.103}$$

where

$$H_I(t) = e^{iH_0(t-t_0)}H_{Int}e^{-iH_0(t-t_0)} = \int d^3x\frac{\lambda}{4!}\phi_I^4$$

is the interaction Hamiltonian in the interaction picture.

The time evolution of the interaction picture propagator can be written in the form of an integral equation as

$$U(t,t_0) = 1 + (-i)\int_{t_0}^{t} dt_1 H_I(t_1)U(t_1,t_0). \tag{1.104}$$

Successive iterations can then give us an expression for the time-evolution operator as a power series in the coupling constant $\lambda$

$$\begin{aligned}U(t,t_0) &= 1 + (-i)\int_{t_0}^{t} dt_1 H_I(t_1) + (-i)^2\int_{t_0}^{t} dt_1\int_{t_0}^{t_1} dt_2 H_I(t_1)H_I(t_2)\\ &\quad + (-i)^3\int_{t_0}^{t} dt_1\int_{t_0}^{t_1} dt_2\int_{t_0}^{t_2} dt_3 H_I(t_1)H_I(t_2)H_I(t_3) + \cdots.\end{aligned} \tag{1.105}$$

This expression can be simplified using the time-ordering symbol since $H_I(t)$ enters in time-ordered sequence, which allows us to write

$$\begin{aligned}&\int_{t_0}^{t} dt_1\int_{t_0}^{t_1} dt_2\cdots\int_{t_0}^{t_{n-1}} dt_n H_I(t_1)H_I(t_2)\cdots H_I(t_n)\\ &\qquad = \frac{1}{n!}\int_{t_0}^{t} dt_1 dt_2\cdots dt_n T\{H_I(t_1)H_I(t_2)\cdots H_I(t_n)\}\end{aligned} \tag{1.106}$$

so we can express the interaction picture propagator in a compact form:

$$\begin{aligned}U(t,t_0) &= 1 + (-i)\int_{t_0}^{t} dt_1 H_I(t_1) + \frac{(-i)^2}{2!}\int_{t_0}^{t} dt_1 dt_2 T\{H_I(t_1)H_I(t_2)\}\\ &\equiv T\left\{\exp\left[-i\int_{t_0}^{t} dt' H_I(t')\right]\right\},\end{aligned} \tag{1.107}$$

which satisfies $U(t_1,t_2)U(t_2,t_3) = U(t_1,t_3)$ and $U(t_1,t_3)U^\dagger(t_2,t_3) = U(t_1,t_2)$.

We shall now discuss the ground state $|\Omega\rangle$ of the Hamiltonian $H$. If $E_n$ are the eigenvalues of $H$, we can write

$$e^{-iHT}|0\rangle = e^{-iE_0T}|\Omega\rangle\langle\Omega|0\rangle + \sum_{n\neq 0} e^{-iE_nT}|n\rangle\langle n|0\rangle, \tag{1.108}$$

where $E_0 = \langle\Omega|H|\Omega\rangle$ and $\langle 0|H_0|0\rangle = 0$. Since $E_n > E_0$, we can eliminate all terms with $n \neq 0$ on the right by taking the limit $T \to \infty(1-i\varepsilon)$. This will then give

$$|\Omega\rangle = \lim_{T\to\infty(1-i\varepsilon)} \left[e^{-iE_0(t_0-(-T))}\langle\Omega|0\rangle\right]^{-1} U(t_0,-T)|0\rangle. \tag{1.109}$$

Since $T$ is very large we shift by a constant amount $t_0$ to get this expression. The two-point correlation function can then be given by

$$\langle\Omega|\phi(x)\phi(y)|\Omega\rangle = \lim_{T\to\infty(1-i\varepsilon)} \frac{\langle 0|U(T,x^0)\phi_I(x)U(x^0,y^0)\phi_I(y)U(y^0,-T)|0\rangle}{\langle 0|U(T,-T)|0\rangle}. \tag{1.110}$$

Taking the time-ordered product we can express the correlation function as

$$\langle\Omega|T[\phi(x)\phi(y)]|\Omega\rangle = \lim_{T\to\infty(1-i\varepsilon)} \frac{\langle 0|T\left[\phi_I(x)\phi_I(y)e^{-i\int_{-T}^{T}dtH_I(t)}\right]|0\rangle}{\langle 0|T\left[e^{-i\int_{-T}^{T}dtH_I(t)}\right]|0\rangle}. \tag{1.111}$$

This formula can be generalized to any higher correlation function with arbitrary numbers of field operators. In practice we may expand the exponential function in power series and consider only the first few terms as long as the perturbation theory holds because higher order terms are suppressed by powers of the coupling constant.

**Wick's Theorem**

We have defined two types of ordering, normal-ordered products that takes care of the infinities originating from the zero-point energy and the time-ordered product that enters in the propagators of the free-fields. The Wick's theorem relates these two types of ordering with the Feynman propagators [10]. We demonstrate this with scalar fields, although the theorem is applicable for fermions also. In case of fermion field operators, the ordering that requires interchange of two fermions will introduce a negative sign.

Let us decompose a scalar field in terms of its positive and negative frequency components:

$$\begin{aligned}
\phi(x) &= \phi^+(x) + \phi^-(x) \\
\phi^+(x) &= \int \frac{d^3k}{(2\pi)^{3/2}} \frac{1}{\sqrt{2\omega_k}} a(k) e^{-ik\cdot x} \\
\phi^-(x) &= \int \frac{d^3k}{(2\pi)^{3/2}} \frac{1}{\sqrt{2\omega_k}} a^\dagger(k) e^{ik\cdot x}.
\end{aligned} \tag{1.112}$$

These fields satisfy $\phi^+(x)|0\rangle = 0$ and $\langle 0|\phi^-(x) = 0$. The normal-ordering means all negative frequency components appear to the left,

$$:\phi(x)\phi(y) := \phi^+(x)\phi^+(y) + \phi^-(x)\phi^+(y) + \phi^-(y)\phi^+(x) + \phi^-(x)\phi^-(y)],$$

so the vacuum expectation value (*vev*) of a normal-ordered product vanishes

$$\langle 0| : \phi(x)\phi(y) : |0\rangle = 0.$$

Let us define the contraction of two fields as $C[\phi(x)\phi(y)]$, which is a c-number. For two particles, the Wick's theorem states

$$T[\phi(x)\phi(y)] =: \phi(x)\phi(y) : + C[\phi(x)\phi(y)]. \tag{1.113}$$

Taking the *vev* on both sides, the c-number can be identified with the propagator

$$C[\phi(x)\phi(y)] = \langle 0|T[\phi(x)\phi(y)]|0\rangle = i\Delta_F(x-y)$$

so we can write

$$T[\phi(x)\phi(y)] =: \phi(x)\phi(y) : +i\Delta_F(x-y). \tag{1.114}$$

For the time-ordered product of $n$ fields, the Wick's theorem reads

$$T[\phi(x_1)\phi(x_2)\cdots\phi(x_n)] \\ =: \phi(x_1)\phi(x_2)\cdots\phi(x_n) : + \ : \text{all possible contractions} : \tag{1.115}$$

Let us explain with one example. We shall simplify notation as $\phi(x_i) \equiv \phi_i$ and $i\Delta_F(x_i - x_j) = \Delta_{ij}$. For $n = 4$ we then have

$$\begin{aligned} T[\phi_1\phi_2\phi_3\phi_4] = &: \phi_1\phi_2\phi_3\phi_4 : +\Delta_{12} : \phi_3\phi_4 : +\Delta_{13} : \phi_2\phi_4 : +\Delta_{14} : \phi_2\phi_3 : \\ &+\Delta_{23} : \phi_1\phi_4 : +\Delta_{24} : \phi_1\phi_3 : +\Delta_{34} : \phi_1\phi_2 : \\ &+\Delta_{12}\Delta_{34} + \Delta_{13}\Delta_{24} + \Delta_{14}\Delta_{23}. \end{aligned} \tag{1.116}$$

This theorem can be proved by method of induction. Since the *vev* of any normal-ordered product vanishes, we can denote the expression

$$\langle 0|T[\phi_1\phi_2\phi_3\phi_4]|0\rangle = \Delta_{12}\Delta_{34} + \Delta_{13}\Delta_{24} + \Delta_{14}\Delta_{23}$$

by a sum of products of Feynman propagators.

**Feynman's Diagrams**

The last expression can be represented by Feynman diagrams, as

$\langle 0|T\{\phi_1\phi_2\phi_3\phi_4\}|0\rangle =$

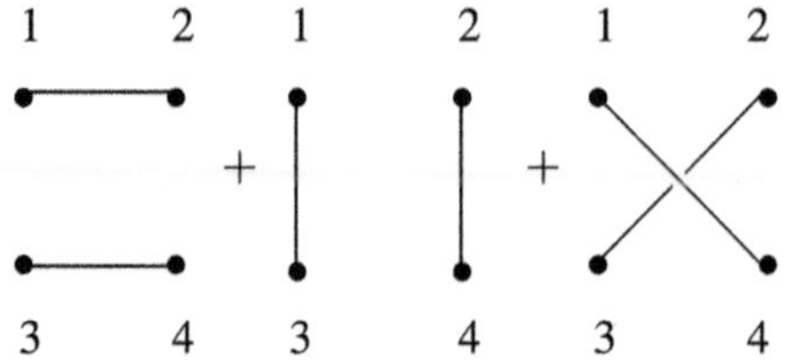

These diagrams show particles created at a point and annihilated at another point, which can happen in three different ways shown by separate diagrams. Total amplitude for the process will be summed over all the diagrams. In this case since all the diagrams give same amplitudes, there will be a factor of 3, which is the symmetry factor.

Let us now discuss a slightly more involved case when there are interactions given by the fields at the same points. Consider a theory with the interaction Hamiltonian $H_I(t) = \int d^3x(\lambda/4!)\phi^4(x)$. We are interested in evaluating the two-point correlation function $\langle\Omega|T\{\phi(x)\phi(y)\}|\Omega\rangle$. As we shall see later, the numerator can be factored out into two parts: one with propagators connected to external points $x$ and $y$, called the connected diagrams, and the other part that is not connected to external legs, called the disconnected diagrams. The denominator comprises only the disconnected part of the diagrams, so the two-point correlation function may be evaluated by summing over only the connected diagrams. For the present we shall discuss only the numerator, given by

$$\langle 0|T\left\{\phi(x)\phi(y) + \phi(x)\phi(y)\left[-i\int dtH_I(t)\right] + \cdots\right\}|0\rangle. \tag{1.117}$$

The first term in the series is the free-particle propagator $\langle 0|T\phi(x)\phi(y)|0\rangle = i\Delta_F(x-y)$. The second term contains the interactions at only one point and is the leading term in the perturbation series representing the interactions. We can express this term in terms of propagators using the Wick's theorem as

$$\begin{aligned}
&\langle 0|T\left\{\phi(x)\phi(y)\left(\frac{-i\lambda}{4!}\right)\int d^4z\phi(z)\phi(z)\phi(z)\phi(z)\right\}|0\rangle \\
&\quad = 3\cdot\left(\frac{-i\lambda}{4!}\right)\Delta_{xy}\int d^4z\Delta_{zz}\Delta_{zz} + 12\cdot\left(\frac{-i\lambda}{4!}\right)\int d^4z\Delta_{xz}\Delta_{yz}\Delta_{zz}
\end{aligned} \tag{1.118}$$

Previously we introduced the notation, $i\Delta_F(x-y) = \Delta_{xy}$. The factors 3 and 12 are the symmetry factors, which are the different possible contractions that can give the same diagrams. As discussed above, for the four $\phi_i$ three possible contractions are possible. The propagators $\Delta_{zz}$ represent loops, since the starting and end points are same. Thus, this process can be represented by the two diagrams

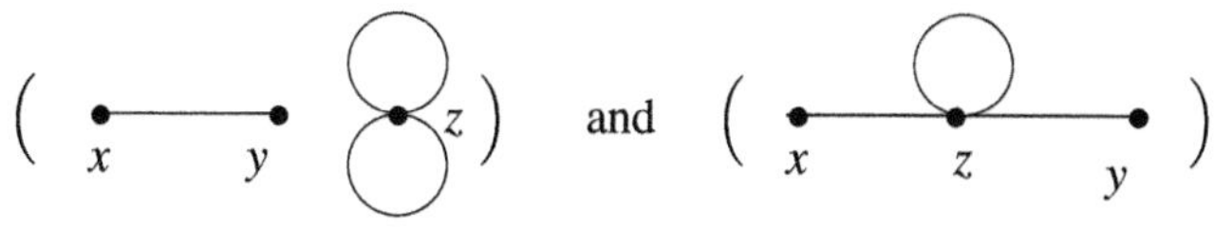


The first diagram is the disconnected diagram, while the second one is a connected diagram. When we consider both the numerator and the denominator, the disconnected diagrams will get eliminated from the final expression of the correlation function.

The next term in the perturbative expansion is

$$\langle 0|T\left\{\phi(x)\phi(y)\frac{1}{2!}\left(\frac{-i\lambda}{4}\right)^2\int d^4z\int d^4w\phi^4(z)\phi^4(w)\right\}|0\rangle.$$

This will contain terms such as $\Delta_{xy}\Delta_{zz}^2\Delta_{ww}^2$ and $\Delta_{xz}\Delta_{yz}\Delta_{zz}\Delta_{ww}^2$. Continuing the same way we get a series, which will factor out in two parts, containing terms of the form

$$[\Delta_{xy};\ \Delta_{xz}\Delta_{yz}\Delta_{zz}\ \cdots] \quad \text{and} \quad \left[\Delta_{zz}^2;\ \Delta_{zz}^2\Delta_{ww}^2;\ \Delta_{zw}^2\Delta_{zz}\Delta_{ww};\ \cdots\right].$$

The first term contains the connected diagrams and the second term contains only disconnected diagrams. This part cancels the denominator of the correlation function exactly. Thus, the two-point correlation function contains only the connected diagrams. Explicit dependence on $T$ also disappears in the large time $T$ limit.

Considering all the factors, we can now write the rules for calculating the two-point correlation function:

$$\langle\Omega|T\{\phi(x)\phi(y)\}|\Omega\rangle = \begin{bmatrix}\text{Sum over all possible}\\ \text{connected diagrams}\end{bmatrix}, \tag{1.119}$$

where connected diagrams mean topologically distinct diagrams including loops connected to the external points. Feynman rules give the analytic expressions for each of the diagrams. We can now write the Feynman rules for any particular interaction Hamiltonian and then calculate the different diagrams. For the $\phi^4$ theory we just discussed, the rules are

1. For each propagator: 

$= i\Delta_F(x-y)$;

2. For each vertex: 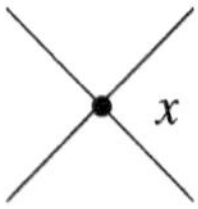

$= (-i\lambda)\int d^4x$;

3. For external points: 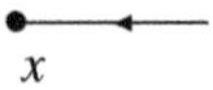

$= 1$;

4. Divide by the symmetry factor.

Instead of these position-space Feynman rules, it is more convenient to work with momentum-space Feynman rules, given by

1. For each propagator: 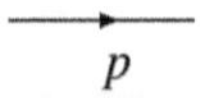

$= \dfrac{i}{p^2-m^2+i\varepsilon}$;

2. For each vertex: 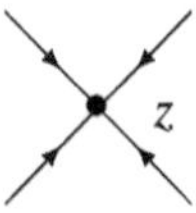

$= (-i\lambda)$;

3. For external points: 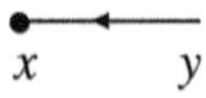

$= e^{-ip\cdot x}$;

4. Impose momentum conservation at each vertex;

5. For each internal momentum corresponding to an internal loop associate: $\int \frac{d^4p}{(2\pi)^4}$;

6. Divide by the symmetry factor: number of ways one can permute the internal lines and vertices, leaving the external lines fixed.

**Feynman rules for Fermions**

To find out Feynman rules for fermions, we shall have to generalize the Wick's theorem for fermions. The time-ordered and normal products have been defined in section 1.2, where we stated that a negative sign should be introduced every time any two fields are interchanged, but otherwise these definitions are similar to scalars. The Wick's theorem then gives

$$T\left[\psi(x)\bar{\psi}(y)\right] =: \psi(x)\bar{\psi}(y) : +C\left[\psi(x)\bar{\psi}(y)\right], \tag{1.120}$$

where the contraction of the fields gives

$$C\left[\psi(x)\bar{\psi}(y)\right] = iS_F(x-y); \qquad C\left[\psi(x)\psi(y)\right] = C\left[\bar{\psi}(x)\bar{\psi}(y)\right] = 0.$$

The Wick's theorem for any number of fields then reads

$$T\left[\psi_1\bar{\psi}_2\psi_3\cdots\right] = \ :\left[\psi_1\bar{\psi}_2\psi_3\cdots\right]: + :\left[\text{all possible contractions}\right]: . \tag{1.121}$$

To find out the Feynman rules for the fermions let us now consider the Yukawa interactions giving the interactions between a scalar $\phi$ and a fermion $\psi$, where the interaction Hamiltonian is given by

$$\mathscr{H}_I = f\ \bar{\psi}\psi\phi. \tag{1.122}$$

Let us consider a specific two-particle scattering process,

$$\psi(p) + \psi(k) = \psi(p') + \psi(k').$$

The leading order contribution comes from $H_I^2$ term in the perturbation expansion (see equation (1.105)). Using Wick's theorem the fields can be contracted and acting on the initial and final states by the uncontracted fields, the matrix element (which shall be discussed in the next section) for this process is given by

$$(-if)^2 \int \frac{d^4q}{(2\pi)^4} \frac{i}{q^2 - m_\phi^2} (2\pi)^8 \delta(p' - p - q)\delta(k' - k + q)\bar{u}(p')u(p)\bar{u}(k')u(k),$$

where $q$ is the momentum carried by the scalar propagator. We can get this expression starting with the Feynman diagram for the scattering process and applying Feynman rules. The Feynman diagram for this process is

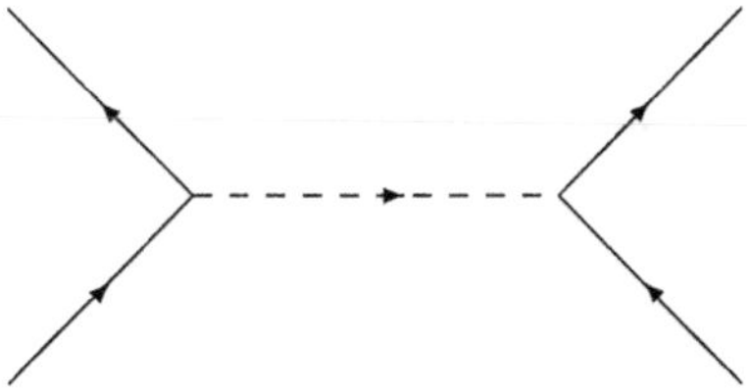

From now on we shall follow the standard convention of dashed line for scalars, solid line for fermions and wavy lines for gauge bosons. We can get the matrix element by applying the Feynman rule for this interaction:

1. Associate fermions propagators for internal fermions:

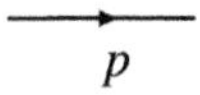


$$= \frac{i}{\not p - m_\psi + i\varepsilon} = \frac{i(\not p + m_\psi)}{p^2 - m_\psi^2 + i\varepsilon};$$

2. Associate scalar propagators for internal scalar lines:

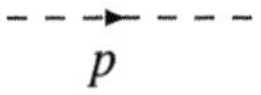


$$= \frac{i}{p^2 - m_\phi^2 + i\varepsilon};$$

3. For each vertex:

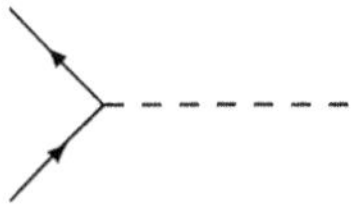

$$= (-if);$$

4. Insert a factor $(-1)$ for each closed fermion loop;
5. Insert a relative factor $-1$ for interchange of identical external fermion lines.
6. For each internal momentum corresponding to an internal loop associate: $\int \frac{d^4p}{(2\pi)^4};$
7. Topologically equivalent diagrams with internal fermion lines with arrows in clockwise or anticlockwise directions should not be counted twice.
8. Insert a factor $u(p,s)$ $[\bar{v}(p,s)]$ for external electron [positron] line entering a graph or a factor $\bar{u}(p,s)$ $[v(p,s)]$ for external electron [positron] line leaving a graph, respectively. The direction of positron lines is opposite to electron lines, so for incoming positron, momentum will be leaving the diagram.

Similarly for the interaction Hamiltonian

$$\mathscr{H}_I = -ie\bar{\psi}\gamma^\mu \psi A_\mu ,$$

we can use Wick's theorem and express the correlation function in terms of fermion and gauge boson propagators. The corresponding Feynman rules are similar to the previous case:

1. Associate fermions propagators for internal fermions:

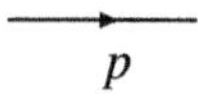


$$= \frac{i}{\not{p} - m + i\varepsilon} = \frac{i(\not{p} + m)}{p^2 - m^2 + i\varepsilon};$$

2. Associate photon propagators for internal photons:

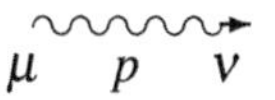


$$= -\frac{ig_{\mu\nu}}{p^2 + i\varepsilon};$$

3. For each vertex:

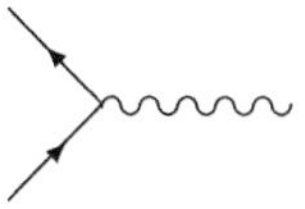

$$= (-ie\gamma_\mu);$$

4. Insert a factor $(-1)$ for each closed fermion loop;
5. Insert a relative factor $-1$ for interchange of identical external fermion lines.
6. For each internal momentum corresponding to an internal loop associate: $\int \frac{d^4p}{(2\pi)^4}$;
7. Topologically equivalent diagrams with internal fermion lines with arrows in clockwise or anticlockwise directions should not be counted twice.
8. Insert a factor $u(p,s)$ $[\bar{v}(p,s)]$ for external electron [positron] line entering a graph or a factor $\bar{u}(p,s)$ $[v(p,s)]$ for external electron [positron] line leaving a graph, respectively. The direction of positron lines is opposite to electron lines, so for incoming positron, momentum will be leaving the diagram.

We can now generalize the result of the two-point correlation function to any higher correlation function:

$$\langle\Omega|T\{\phi(x_1)\phi(x_2)\cdots\phi(x_n)\}|\Omega\rangle = \begin{bmatrix}\text{Sum over all connected diagrams} \\ \text{with } n \text{ external lines}\end{bmatrix}. \quad (1.123)$$

Once we formulated how to calculate the correlation functions through Feynman diagram technique, the next task would be to make use of this technique in calculating any experimentally measurable quantity such as scattering cross-sections or decay rates.

## 1.4 Scattering Processes and Cross-Section

We have developed the propagators for the scalar, fermion and vector fields. In the absence of any interactions, we can describe these particles by their respective propagators. When any interaction is turned on, we treat the interaction as small perturbations to the free propagator assuming that the interactions are not strong. To the first order this amounts to treating the interaction to be instantaneous, so for the rest of the time the free propagators can describe these particles. At higher orders, several instantaneous interactions are taken into consideration. As long as the strength of the interaction is small, higher order contributions will be smaller and we shall be able to describe the nature of interaction taking the first few terms in the perturbation expansions.

### Cross-Section

Different experiments measure the cross-section of various processes, which can be compared with measurements in different environments with differing beam sizes and intensity. Simply speaking the cross-section measures an effective size of the target particle. If $\rho_{B,T}$ are the number densities of the incident beam of particles of type $B$ and of the target particles of type $T$, which are at rest, $\ell_{B,T}$ are the lengths of the bunches of these particles; A is the cross-sectional area, then the cross-section is defined as

$$\sigma \equiv \frac{\text{Number of scattering events}}{\rho_B\, \ell_B\, \rho_T\, \ell_T\, A}. \tag{1.124}$$

The total number of events ($N$) will then be given by

$$N = \sigma\, \ell_B\, \ell_T \int d^2x\, \rho_B(x)\, \rho_T(x).$$

Since the beam density is not constant in practice, one needs to integrate over the beam area. When we are interested in the cross-section with definite momenta, we have to work with differential cross-section $d\sigma/d^3p_1 \cdots d^3p_n$. Integrating over the small regions of momentum–space gives us the final-state cross-section for scattering into that region of final-state momentum–space.

For the decays of any unstable particles ($X$), we define the decay rate as

$$\Gamma \equiv \frac{\text{Number of decays per unit time}}{\text{Number of decaying particles } (X) \text{ present}}. \tag{1.125}$$

The lifetime $\tau_X$ (half-life $\tau \cdot \ln 2$) of the particle is then

$$\tau_X \equiv \frac{1}{\sum \Gamma(X \to \text{all possible final states})}.$$

In a scattering experiment, an unstable particle shows up as a resonance. Near the resonance, the scattering amplitude is given by the Breit–Wigner formula

$$f(E) \propto \frac{1}{E - E_0 + i\Gamma/2}. \tag{1.126}$$

The cross-section has a resonance peak $\sigma \propto 1/((E-E_0)^2 + \Gamma^2/4)$, whose width gives the decay rate of the unstable state. In relativistic quantum mechanics, the Breit–Wigner formula can be generalized to $1/(p^2 - m^2 + im\Gamma)$.

To calculate the cross-section, we introduce the $S$-matrix

$$S_{fi} = \langle f|S|i\rangle = \delta_{fi} - i(2\pi)^4\delta^4(P_f - P_i)\mathscr{T}_{fi}, \tag{1.127}$$

where $\delta_{fi}$ corresponds to no interactions and $\mathscr{T}_{fi}$ is the transition matrix between the initial state $|i\rangle$ and the final state $|f\rangle$ in the presence of interactions. The initial and final states are collections of free asymptotic states at $t \to -\infty$ and $t \to \infty$, respectively. The $S$-matrix is unitary.

The differential cross-section is then given by

$$\begin{aligned} d\sigma &= \frac{1}{J}\frac{|S_{fi}|^2 dN_f}{VT} \\ &= \frac{1}{J}(2\pi)^4\delta^4(P_f - P_i)|\mathscr{T}_{fi}|^2 dN_f, \end{aligned} \tag{1.128}$$

where $VT = (2\pi)^4\delta^4(0)$, $J$ is the flux of incident particles, and

$$dN_f = \prod_{i=1}^{N_f} V\frac{d^3p}{(2\pi)^3}$$

is the number of states with momentum between $\mathbf{p}$ and $\mathbf{p} + \delta\mathbf{p}$.

For a scattering process $1 + 2 \to 3 + 4 + \cdots$, the incident flux $J$ can be given in the center of mass frame ($\mathbf{p}_1 = -\mathbf{p}_2$) as $J = [(p_1 \cdot p_2)^2 - m_1^2 m_2^2]^{1/2}/(VE_1E_2)$, so the differential cross-section is given by

$$\begin{aligned} d\sigma &= \frac{(2\pi)^4|\mathscr{M}_{fi}|^2\delta^4(P_f - P_i)}{(2E_1)(2E_2)|v_1 - v_2|}\prod_f \frac{d^3p_f}{(2\pi)^3 2E_{p_f}} \\ &= \frac{(2\pi)^4|\mathscr{M}_{fi}|^2\delta^4(P_f - P_i)}{4[(p_1 . p_2)^2 - m_1^2 m_2^2]^{1/2}}\prod_f \frac{d^3p_f}{(2\pi)^3 2E_{p_f}} \end{aligned} \tag{1.129}$$

in a collinear frame where $\mathscr{T}_{fi} = \prod_{j=1}^{N}(2E_{p_j}V)^{-1/2}\mathscr{M}_{fi}$. For two final-state particles, we can simplify this formula to

$$\left(\frac{d\sigma}{d\Omega}\right)_{CM} = \frac{1}{(2E_1)(2E_2)|v_1 - v_2|}\frac{|\mathbf{p}_1|}{(2\pi)^4 4E_{CM}}|\mathscr{M}(p_1, p_2 \to p_3, p_4)|^2. \tag{1.130}$$

The matrix elements may be read from the Feynman diagram.

The probability of the decay of a single particle can then be evaluated in the same way:

$$P_{tot} = \sum_f \int dN_f |S_{fi}|^2 = \sum_f \int dN_f |\mathscr{T}_{fi}|^2[(2\pi)^4\delta^4(P_f - P_i)]^2, \tag{1.131}$$

so after proper normalization the final differential decay rate becomes

$$d\Gamma(i \to \{f\}) = \frac{(2\pi)^4}{2m_i}\prod_f \frac{d^3p_f}{(2\pi)^3 2E_f}|\mathscr{M}_{fi}|^2\delta^4(P_f - P_i). \tag{1.132}$$

It is assumed that the decaying particle is at rest, so we could replace $(2E_i)^{-1}$ by $(2m_i)^{-1}$. The lifetime of the particle will be $\tau = 1/\Gamma$. In this formalism the main task is then to evaluate the matrix elements $\mathscr{M}_{fi}$ for any given Feynman diagram.

### LSZ Reduction

A direct connection between the scaterring matrix and the correlation function was given by Lehmann, Symanzik and Zimmermann [11], and is known as the LSZ reduction formula. This will allow us to calculate any physical scattering process by calculating the correlation functions using Feynman rules.

Consider a scalar free-field whose two-point function in momentum space is given by the propagator

$$\int d^4 x e^{ipx} \langle \Omega | T[\phi(x)\phi(0)] | \Omega \rangle = \frac{i}{p^2 - m^2 + i\varepsilon}.$$

This also represents the asymptotic state of the scalar field at time $t \to \pm\infty$ and has a simple pole at the mass of the one-particle state.

If we are interested in calculating the S-matrix element for a $n_i$-body $\longrightarrow n_f$-body scattering process, we start with a correlation function with $n_i + n_f$ Heisenberg fields. If we Fourier transform the correlation function with respect to any one of the $n_i + n_f$ fields, it will be possible to factor out a term with a pole, which represents a free-field and may be associated with the asymptotic propagator of the corresponding field. Continuing in the same way it will be possible to write the $(n_i + n_f)$-point correlation function as a product of asymptotic states corresponding to all the $(n_i + n_f)$ fields. The S-matrix element of our interest will then be given by the coefficient of these asymptotic states. This can be expressed as

$$\prod_{i=1}^{n_i} \int d^3 x_i e^{ip_i \cdot x_i} \prod_{j=1}^{n_f} \int d^3 y_j e^{ik_j \cdot y_j} \langle \Omega | T[\phi(x_1) \cdots \phi(x_{n_i}) \phi(y_1) \cdots \phi(y_{n_f})] | \Omega \rangle$$
$$= \prod_{i=1}^{n_i} \frac{i}{p_i^2 - m^2 + i\varepsilon} \prod_{j=1}^{n_f} \frac{i}{k_j^2 - m^2 + i\varepsilon} S_{fi}. \tag{1.133}$$

The momenta of the incoming particles are denoted by $p_i$ and those of the outgoing particles as $k_j$ and $S_{fi}$ is the S-matrix for the scattering process. Effects of renormalization have not been included in this discussion but will be considered in the next chapter.

### Optical Theorem

We shall now discuss one more important aspect of field theory, how to deal with unstable particles. Since the asymptotic states cannot be defined for any unstable states, LSZ reduction is not possible for any decaying particles. This is solved by using the optical theorem in conjunction with the Cutkowsky rules [12]. In this approach the decays of any unstable particle are related to the imaginary part of the two-point correlation function of the decaying particles with all possible intermediate states, so Feynman rules can be used to calculate the decay rate.

Consider a process, $a_1 + a_2 \to [f_i] \to b_1 + b_2$, which can go through several intermediate channels, where $[f_i]$ denotes different multiparticle intermediate states. The optical theorem states that the imaginary part of the amplitude for the forward scattering $a_1 + a_2 \to b_1 + b_2$ can be obtained as a product of the amplitudes for the transition to the different intermediate states $a_1 + a_2 \to [f_i]$ and $b_1 + b_2 \to [f_i]$ and by summing over all possible intermediate states $\sum_f [f_i]$:

$$2\,\mathrm{Im}\,\{a_1(p_1) + a_2(p_2) \to b_1(k_1) + b_2(k_2)\}$$

$$= \sum_{[f]} \int d\Pi_f \left\{ \sum_i a_i(p_i) \to [f_i(q_i)] \right\} \cdot \left\{ [f_i(q_i)] \to \sum_i b_i(k_i) \right\}.$$

This can be expressed as

$$-i[\mathcal{M}_{ba} - \mathcal{M}^*_{ab}] = \sum_f \left( \prod_i \int \frac{d^3 q_i}{(2\pi)^3} \frac{1}{2E_i} \right) \mathcal{M}^*_{fb} \mathcal{M}_{fa}\,(2\pi)^4 \delta^4(P_a - P_f), \quad (1.134)$$

with an overall delta function $(2\pi)^4 \delta^4(P_a - P_b)$, and $q_i$ are the momenta of the intermediate states $[f_i]$. This result can be generalized to any number of initial and final states $a$ and $b$. The most interesting case is when $a$ and $b$ correspond to a single unstable particle.

In a simpler case $p_i = k_i$, this formula reduces to

$$\mathrm{Im}[\mathcal{M}_{aa}] = 2\,E_{CM}\,p_{CM}\,\sigma_{tot}(a \to f), \quad (1.135)$$

which has the interpretation that the loss of particles of type $a$, given by the imaginary part of the amplitude, is proportional to the scattering cross-section for $a \to f$.

In the Feynman diagram language, this theorem implies that the intermediate states are the states with real momenta, so the diagram can be cut in two parts. Cutkowsky [12] provided, in a most general way, the algorithm to cut the Feynman diagrams, which led to a general proof of the optical theorem to all orders in perturbation theory. The algorithm recommends:

> Cut a diagram in all possible ways which allow the cut propagators to be on mass shell and perform the loop integrals after replacing $i/(p^2 - m^2 + i\varepsilon)$ by $2\pi\delta(p^2 - m^2)$ and sum over all possible cuts.

In addition to applying the result for other processes, we can also apply this result to calculate the decay rate of any unstable particle. Consider a two-point correlation function including all higher loops. The imaginary part of the amplitude can be given by

$$\mathrm{Im}\mathcal{M}_{aa} = \frac{1}{2} \sum_f \int d\Pi_f\, |\mathcal{M}_{fa}|^2. \quad (1.136)$$

This states that the imaginary part of the two-point correlation function for the propagator of the field $a$ with mass $m$ gives the depletion of the state as it propagates, which depends on all possible final states $f$, to which it can decay. The decay rate can then be written as

$$\Gamma = \frac{1}{2m} \sum_f \int d\Pi_f\, |\mathcal{M}_{fa}|^2. \quad (1.137)$$

This is the result given in equation (1.132). Thus, although the asymptotic states for any unstable particle cannot be defined, it is possible to calculate the decay rate of the particle by calculating the two-point function using Feynman diagrams.

**Møller Scattering**

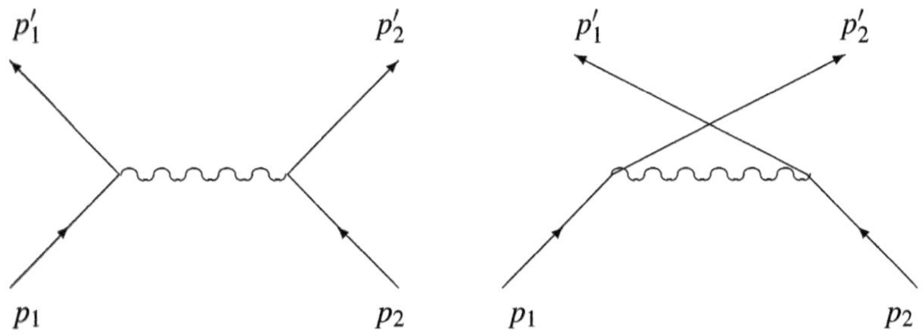


**FIGURE 1.1**
Møller scattering of two electrons

We shall work out the cross-section for the electron–electron scattering. To the lowest order there are two Feynman graphs contributing to the process, which are shown in figure 1.4. Using the Feynman rules we can write the matrix element for these diagrams as

$$i\mathcal{M}_{fi} = (-ie)^2 \bar{u}(p_1')\gamma^\mu u(p_1) \frac{-ig_{\mu\nu}}{(p_1' - p_1)^2} \bar{u}(p_2')\gamma^\nu u(p_2)$$

and

$$i\mathcal{M}_{fi} = \frac{ie^2 g_{\mu\nu}}{(p_1 - p_2')^2} \bar{u}(p_2')\gamma^\mu u(p_1)\bar{u}(p_1')\gamma^\nu u(p_2).$$

To write the matrix element squared, we shall use the relation

$$\begin{aligned} |\bar{u}(p')\Gamma u(p)|^2 &= [\bar{u}(p')\Gamma u(p)][u^\dagger(p)\Gamma^\dagger \bar{u}^\dagger(p')] \\ &= [\bar{u}(p')\Gamma u(p)][\bar{u}(p)\gamma_0\Gamma^\dagger\gamma_0 u(p')] \\ &= \text{Tr}\left[(\not{p}' + m)\,\Gamma\,(\not{p} + m)\gamma_0\Gamma^\dagger\gamma_0\right], \end{aligned} \tag{1.138}$$

where $u_\alpha(p)\bar{u}_\beta(p) = (\not{p} + m)_{\alpha\beta}$. In the center of mass frame ($\mathbf{p}_1 = -\mathbf{p}_2$), the differential cross-section will then be given by

$$d\sigma = \frac{(2\pi)^4\delta^4(p_1' + p_2' - p_1 - p_2)}{4[(p_1 \cdot p_2)^2 - m^4]^{1/2}} \int \frac{d^3p_1'}{(2\pi)^3 2E_1'} \frac{d^3p_2'}{(2\pi)^3 2E_2'} |\mathcal{M}_{fi}|^2 \tag{1.139}$$

with the matrix element squared given by

$$|\mathcal{M}_{fi}|^2 = \frac{e^4}{4}\left[\frac{\mathrm{Tr}[\gamma_\mu(\not{p}_1+m)\gamma_\nu(\not{p}_1{}'+m)]\,\mathrm{Tr}[\gamma^\mu(\not{p}_2+m)\gamma^\nu(\not{p}_2{}'+m)]}{(p_1'-p_1)^2(p_1'-p_1)^2}\right.$$
$$-\frac{\mathrm{Tr}[\gamma_\mu(\not{p}_1+m)\gamma_\nu(\not{p}_2{}'+m)]\,\mathrm{Tr}[\gamma^\mu(\not{p}_2+m)\gamma^\nu(\not{p}_1{}'+m)]}{(p_1'-p_1)^2(p_2'-p_2)^2}$$
$$\left.+(p_1'\leftrightarrow p_2')\quad\right]. \tag{1.140}$$

In this expression contributions from both the diagrams have been included. To take the trace of the gamma matrices, we can make use of the following identities:

$$\begin{aligned}\gamma^\mu\gamma_\mu &= 4\\ \gamma^\sigma\gamma^\mu\gamma_\sigma &= -2\gamma^\mu\\ \gamma^\sigma\gamma^\mu\gamma^\nu\gamma_\sigma &= 4g^{\mu\nu}\\ \gamma^\sigma\gamma^\mu\gamma^\nu\gamma^\rho\gamma_\sigma &= -2\gamma^\rho\gamma^\nu\gamma^\mu.\end{aligned} \tag{1.141}$$

We shall also make use of the trace operations:

$$\begin{aligned}\mathrm{Tr}\,[\gamma^5\gamma^\mu] = \mathrm{Tr}\,[\sigma^{\mu\nu}] &= \mathrm{Tr}\,[\gamma^\mu\gamma^\nu\gamma^5] = 0\\ \mathrm{Tr}\,[\gamma^\mu\gamma^\nu] &= 4g^{\mu\nu}\\ \mathrm{Tr}\,[\gamma^\mu\gamma^\nu\gamma^\rho\gamma^\sigma] &= 4[g^{\mu\nu}g^{\rho\sigma}-g^{\mu\rho}g^{\nu\sigma}+g^{\mu\sigma}g^{\nu\rho}]\\ \mathrm{Tr}\,[\gamma^5\gamma^\mu\gamma^\nu\gamma^\rho\gamma^\sigma] &= 4i\varepsilon^{\mu\nu\rho\sigma}.\end{aligned} \tag{1.142}$$

Trace of any odd numbers of $\gamma$-matrices vanishes. These relations can be applied to products of $\not{a} = a_\mu\gamma^\mu$, for example,

$$\begin{aligned}\mathrm{Tr}\,[\not{a}\not{b}] &= 4(a\cdot b)\\ \mathrm{Tr}\,[\not{a}\not{b}\not{c}\not{d}] &= 4[(a\cdot b)(c\cdot d)-(a\cdot c)(b\cdot d)+(a\cdot d)(b\cdot c)].\end{aligned}$$

After taking the trace we work in the center of mass frame and choose $p_1$ and $p_2$ to be along the $z$-axis. If the scattered particles make an angle $\theta$ with the incident particles, then we can express products of all the momenta in terms of the center of mass energy $E$, electron momentum $\mathbf{p}$ and $\theta$ as

$$\begin{aligned}p_1\cdot p_2 &= 2E^2-m^2\\ p_1\cdot p_1' &= E^2(1-\cos\theta)+m^2\cos\theta\\ p_1\cdot p_2' &= E^2(1+\cos\theta)-m^2\cos\theta\end{aligned} \tag{1.143}$$

The differential cross-section for the electron–electron scattering per unit of solid angle is then given by [13]

$$\frac{d\sigma}{d\Omega} = \frac{\alpha^2}{4E^2}\left[\frac{(2E^2-m^2)^2}{(E^2-m^2)^2}\left(\frac{4}{\sin^4\theta}-\frac{3}{\sin^2\theta}\right)+\left(1+\frac{4}{\sin^4\theta}\right)\right], \tag{1.144}$$

where we decomposed the momentum volume element $d^3p = d\Omega p^2 dp$, and is written in terms of the fine structure constant $\alpha = e^2/4\pi = 1/137$. Any interaction can thus be written as a Feynman diagram and the amplitude can be read using the Feynman rules.

**Bhabha Scattering**

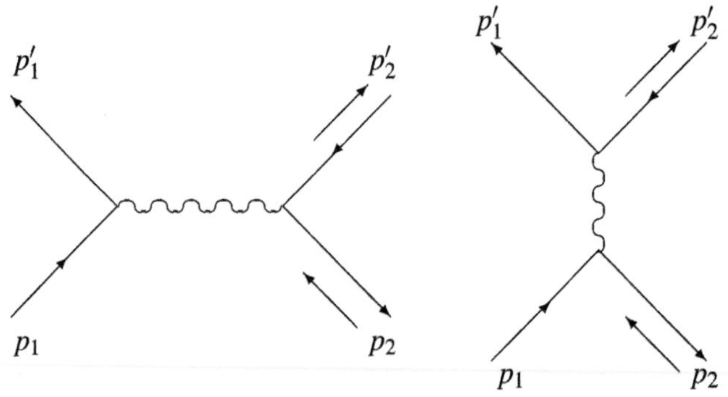


**FIGURE 1.2**
Bhabha scattering of an electron from a positron

Let us consider one more application of the electron–positron scattering

$$e^-(p_1) + e^+(-p_2) \to e^-(p'_1) + e^+(-p'_2)$$

as shown in figure 1.4. Using the Feynman rules we can write the matrix elements for the two diagrams as

$$i\mathcal{M}_{fi} = \frac{ie^2}{(p'_1 - p_1)^2}\bar{u}(p'_1)\gamma^\mu u(p_1)\bar{v}(p_2)\gamma_\mu v(p'_2)$$

and

$$i\mathcal{M}_{fi} = \frac{ie^2}{(p_1 + p_2)^2}\bar{v}(p_2)\gamma^\mu u(p_1)\bar{u}(p'_1)\gamma_\mu v(p'_2).$$

Proceeding in the same way we can calculate the differential cross-section as [14]

$$\frac{d\sigma}{d\Omega} = \frac{\alpha}{2E^2}\left\{\frac{5}{4} - \frac{8E^4 - m^4}{E^2(E^2 - m^2)(1 - \cos\theta)} + \frac{(2E^2 - m^2)^2}{2(E^2 - m^2)^2(1 - \cos\theta)^2}\right.$$
$$\left. + \frac{1}{8}(-1 + 2\cos\theta + \cos^2\theta) + \frac{m^2}{4E^2}(1 - \cos\theta)(2 + \cos\theta) + \frac{m^4}{8E^4}\cos^2\theta\right\}. \quad (1.145)$$

Although the final results for these two cases are different because of the kinematics, the matrix elements are very much similar.

### $e^+e^- \to \mu^-\mu^+$ and $e^-\mu^- \to e^-\mu^-$ Scattering

The similarity in the matrix element squared of the electron–electron and electron–positron scattering has some significance. Consider the processes $e^+ + e^- \to \mu^- + \mu^+$ and $e^- + \mu^- \to e^- + \mu^-$. The corresponding diagrams and the matrix elements are given by

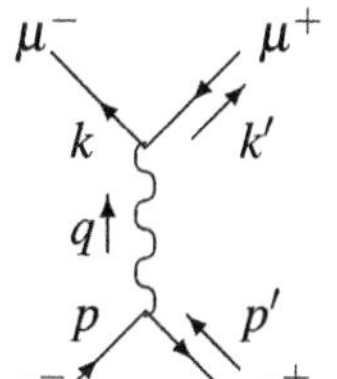


$$i\mathcal{M}_1 = \frac{ie^2}{q^2}\bar{v}(p')\gamma^\mu u(p)\bar{u}(k)\gamma_\mu v(k')$$

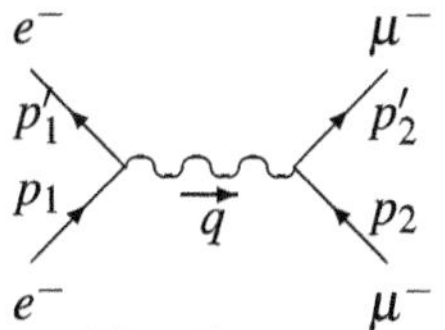


$$i\mathcal{M}_2 = \frac{ie^2}{q^2}\bar{u}(p_1')\gamma^\mu u(p_1)\bar{u}(p_2')\gamma_\mu u(p_2)$$

After taking the average and sum over spins, the matrix elements squared of the two processes are given by

$$\frac{1}{4}\sum_{spins}|\mathcal{M}_1|^2 = \frac{8e^4}{q^4}\left[(p\cdot k)(p'\cdot k') + (p\cdot k')(p'\cdot k) + m_\mu^2(p\cdot p')\right] \quad \text{and}$$
$$\frac{1}{4}\sum_{spins}|\mathcal{M}_2|^2 = \frac{8e^4}{q^4}\left[(p_1\cdot p_2')(p_1'\cdot p_2) + (p_1\cdot p_2)(p_1'\cdot p_2') - m_\mu^2(p_1\cdot p_1')\right]. \tag{1.146}$$

At this stage it would be interesting to compare the matrix elements of the two processes because there is a symmetry in the problem. Both expressions become identical if we replace $p \to p_1,\ p' \to -p_1',\ k \to p_2',\ k' \to -p_2$. This is an example of a more general result known as *crossing symmetry*. For completeness we shall first find out the scattering cross-section for these processes and then come back to this point again.

We work in the center of mass frame for purpose of demonstration. Taking the incident direction to be the $z$-direction, we can then write the four-momenta for the $e^+e^- \to \mu^-\mu^+$ scattering as

$$p = (E, E\hat{z}), \qquad p' = (E, -E\hat{z}),$$
$$k = (E, \mathbf{k})), \qquad k' = (E, -\mathbf{k}),$$
$$|\mathbf{k}| = \sqrt{E^2 - m_\mu^2}, \qquad \mathbf{k}\cdot\hat{z} = |\mathbf{k}|\cos\theta,$$

so the relevant quantities for the calculation of the squared matrix elements are

$$q^2 = (p+p')^2 = 4E^2, \qquad p\cdot p' = 2E^2,$$
$$p\cdot k = p'\cdot k' = E^2 - E|\mathbf{k}|\cos\theta, \qquad p\cdot k' = p'\cdot k = E^2 + E|\mathbf{k}|\cos\theta.$$

The final squared matrix element is then given by

$$\frac{1}{4}\sum_{spins}|\mathcal{M}|^2 = e^4\left[\left(1+\frac{m_\mu^2}{E^2}\right)+\left(1-\frac{m_\mu^2}{E^2}\right)\cos^2\theta\right]$$

and the differential scattering cross-section for the unpolarized $e^+e^- \to \mu^-\mu^+$ scattering in the center of mass frame is given by

$$\frac{d\sigma}{d\Omega} = \frac{\alpha^2}{4E_{CM}^2}\left(1-\frac{m_\mu^2}{E^2}\right)^{1/2}\left[\left(1+\frac{m_\mu^2}{E^2}\right)+\left(1-\frac{m_\mu^2}{E^2}\right)\cos^2\theta\right], \tag{1.147}$$

where the center of mass energy $E_{CM} = 2E$.

In the case of $e^-\mu^- \to e^-\mu^-$ scattering, we can choose the four-momenta for the initial and final states as

$$p_1 = (k, k\hat{z}), \qquad p_2 = (E, -k\hat{z}),$$
$$p_1' = (k, \mathbf{k})), \qquad p_2' = (E, -\mathbf{k}),$$
$$E^2 = k^2 + m_\mu^2, \qquad \mathbf{k}\cdot\hat{z} = k\cos\theta,$$

with the center of mass energy $E_{CM} = E + k$. The final differential cross-section for unpolarized electron–muon scattering is given by

$$\frac{d\sigma}{d\Omega} = \frac{\alpha^2}{2k^2E_{CM}^2(1-\cos\theta)^2}[E_{CM}^2 + (E+k\cos\theta)^2 - m_\mu^2(1-\cos\theta)]. \tag{1.148}$$

The cross-sections for the two processes are, thus, much different.

### Crossing Symmetry and Mandelstam Variables

We notice the similarity between the matrix elements for the processes $e^+e^- \to \mu^-\mu^+$ and $e^-\mu^- \to e^-\mu^-$ in equation (1.146) under the exchange $p \to p_1,\ p' \to -p_1',\ k \to p_2',\ k' \to -p_2$. This is a more general result known as the *crossing symmetry*, which states that the $S$-matrices for any two processes become equal (to be precise, can be analytically continued to each other), if these processes are identical except for the exchange of a particle with momentum $p$ in the initial state with an antiparticle in the final state with momentum $k = -p$. This can be expressed by

$$\mathcal{M}[\phi(p) + \cdots \to \cdots] = \mathcal{M}[\cdots \to \bar{\phi}(k) + \cdots],$$

where $\bar{\phi}(k = -p)$ is the antiparticle of $\phi(p)$. However, the kinematics for the two cases are different. This is a consequence of

$$\sum u(p)\bar{u}(p) = \not{p} + m = -(\not{k} - m) = -\sum v(k)\bar{v}(k).$$

The minus sign may be taken care of by introducing a minus sign for every pair of crossed fermions.

We shall now introduce the Mandelstam variables that can simplify application of crossing relations [15]. For a $2 \to 2$ scattering process

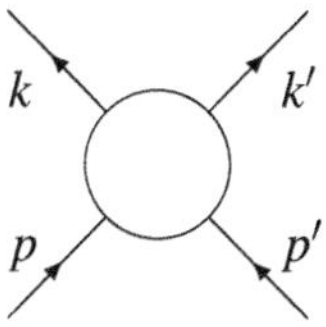


we define the Mandelstam variables as

$$\begin{aligned} s &= (p+p')^2 = (k+k')^2 \\ t &= (k-p)^2 = (k'-p')^2 \\ u &= (k'-p)^2 = (k-p')^2. \end{aligned} \tag{1.149}$$

If all the momenta were coming in, then all signs in these definitions would be positive.

When a $2 \to 2$ scattering is mediated by a single virtual particle $\phi$ with mass $m_\phi$, the angular dependence of cross-section in the different channel becomes

$s$-channel

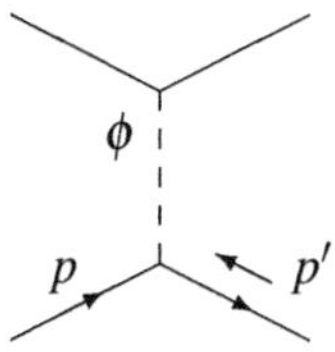


$$\mathcal{M} \propto \frac{1}{s - m_\phi^2}$$

$t$-channel

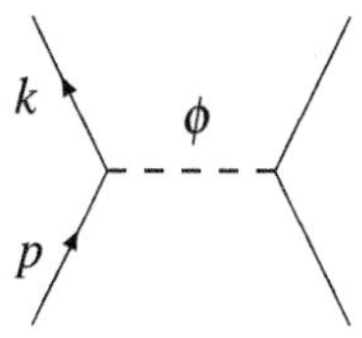


$$\mathcal{M} \propto \frac{1}{t - m_\phi^2}$$

$u$-channel

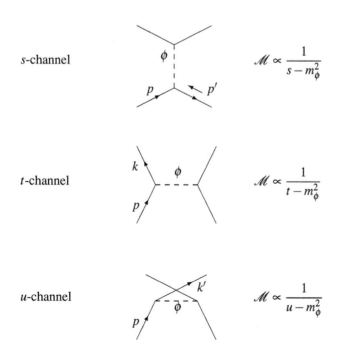

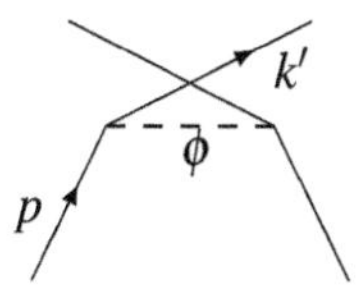


$$\mathcal{M} \propto \frac{1}{u - m_\phi^2}$$

In the massless limit of $e^+e^- \to \mu^+\mu^-$ scattering, the Mandelstam variables are $s = (p+p')^2 = q^2$, $t = -2p\cdot k = -2p'\cdot k'$ and $u = -2p\cdot k' = -2p'\cdot k$, which give

$$\frac{1}{4}\sum_{spins} |\mathscr{M}|^2 = \frac{2e^4}{s^2}(t^2+u^2).$$

Let us now consider the massless limit of the $e^-\mu^- \to e^-\mu^-$ scattering. The crossing relations $p \to p_1,\;\; p' \to -p_1',\;\; k \to p_2',\;\; k' \to -p_2$ would imply that the $s$ and $t$ variables of the $e^+e^- \to \mu^+\mu^-$ process have become $t$ and $s$ variables, respectively, with $u$ remaining the same. So, for the process $e^-\mu^- \to e^-\mu^-$ we now have

$$\frac{1}{4}\sum_{spins} |\mathscr{M}|^2 = \frac{2e^4}{t^2}(s^2+u^2).$$

Some processes may receive contributions from more than one type channel. We noticed, Møller scattering received contributions from $t$- and $u$-channel diagrams, while Bhabha scattering received contributions from $s$- and $t$-channel diagrams.

In the center of mass frame for particles of mass $m$, we can write

$$\begin{aligned} p &= (E, p\hat{z}), & p' &= (E, -p\hat{z}), \\ k &= (E, \mathbf{p})), & k' &= (E, -\mathbf{p}), \end{aligned}$$

so the Mandelstam variables becomes

$$s = (2E)^2, \quad t = -2p^2(1-\cos\theta), \quad u = -2p^2(1+\cos\theta).$$

Thus, $t \to 0$ as $\theta \to 0$, while $u \to 0$ as $\theta \to \pi$. Another useful identity with the Mandelstam variables is

$$s+t+u = \sum_{i=1}^{4} m_i^2. \tag{1.150}$$

The sum extends over all 4-external particles.

So far we restricted ourselves to only tree-level or the lowest order Feynman diagrams for the different processes. In the next chapter we shall consider higher-order radiative corrections and discuss how to take care of such divergences to make the theory consistent.

# 2

# *Renormalization*

In the last chapter we restricted ourselves to only tree-level processes, which are the lowest order Feynman diagrams for the different processes. We shall now consider higher-order radiative corrections, which are essentially quantum corrections. Since the uncertainty principle would allow creation and annihilation of virtual particles, there will be higher order loops contributing to the different processes. The main problem with such higher-order processes are the infinities. QED was accepted as a consistent theory only after these infinities could be properly handled and the predictions of the theory could be tested with precision measurements [2, 3].

In any consistent theory all the infinities that appear in higher-order processes are absorbed in the redefinition of bare parameters, which are not measurable quantities. Proper treatment to remove these unwanted infinities is called *renormalization.* In other words, the theory starts with the bare parameters that are assumed to be infinite, then higher order corrections cancel these infinities making the resultant theory finite and experimentally testable.

The renormalization program is rather technical and involved. In this short chapter we shall only introduce the concept of renormalization and state how to figure out if any theory is renormalizable. We shall also demonstrate, to one-loop order, how to take care of these infinities, so the theory becomes consistent. We shall first describe some radiative processes and prescribe how to deal with the problems and then demonstrate with examples of the $\phi^4$ theory and QED how the renormalization program is carried out.

## 2.1 Divergences

Consider the scattering of an electron with any heavier charged particle. For example, the electron muon scattering we studied will get higher order corrections due to the virtual photon exchange or virtual pair production and annihilation by the photon propagator. These processes for $eX$ scattering, where $X$ is any heavier particle, are given by the diagrams in figure 2.1.

The interference of these diagrams with the tree-level diagrams gives the first order corrections, which are suppressed by a factor $\alpha$. There will be other diagrams contributing to order-$\alpha$, but they can be neglected as long as $X$ is heavier than the electrons. These diagrams enter in many processes and sometimes give new effects.

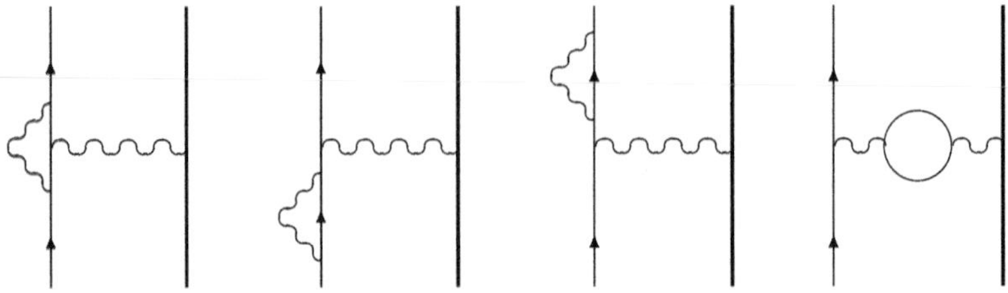

**FIGURE 2.1**
One-loop radiative corrections to *eX* scattering

The first diagram is known as the vertex diagram, the second and the third as self-energy diagrams, and the last one as vacuum polarization. These diagrams are all divergent, but when the divergent part is properly removed, they yield finite corrections and provide us with a consistent theory.

The first three diagrams contain infrared divergences. We encountered infrared divergence first in classical theory of bremsstrahlung, where an electron emits radiation as it passes by a nucleus. The amplitude for the process diverges in the limit where the photon energy vanishes. The tree-level diagrams for bremsstrahlung are shown in figure 2.2. Although there is no classical solution to this problem, it was found that the infrared divergence coming from the vertex diagram cancels this infrared divergence of bremsstrahlung. It may appear strange that the divergence of the classical treatment is cancelling the divergence from the one-loop radiative corrections which are essentially quantum corrections. But the reason behind this cancellation is that the origin of infrared divergence in bremsstrahlung is quantum mechanical.

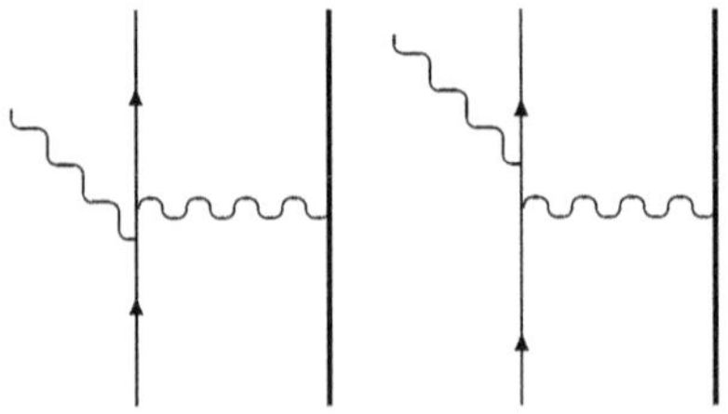

**FIGURE 2.2**
Bremsstrahlung diagram for *eX* scattering

Let us try to understand the infrared divergence of the bremsstrahlung process more carefully. There will be uncertainty, which is of quantum mechanical origin, in the measurement of energy of the soft photons, whose energy tends to zero. This will allow all the photons with energy lower than the uncertainty in energy measurement to emit without being detected. Since photons are massless, there could be an infinite number of photons below any given momentum. This infinite number of photons gives us the divergence even in a classical treatment. Since these soft photons cannot be detected, these processes have to be included in the cross-sections for scattering without radiation, which includes the vertex diagram. In other words, the uncertainty principle will not allow us to determine if these soft photons are real or virtual. So, by calculating the vertex correction and the bremsstrahlung together, we can eliminate this divergence.

**Bremsstrahlung**

The bremsstrahlung process has two parts: an elastic scattering of an electron from a nucleus and the emission of photons from the electron during the interaction. The matrix element for the bremsstrahlung process (shown in figure 2.2) can, thus, be written using the Feynman rules as

$$\begin{aligned} i\mathcal{M} = & -ie\bar{u}(p')\left(\mathcal{M}_0(p',p-k)\frac{i}{\not{p}-\not{k}-m+i\varepsilon}\gamma^\mu\varepsilon^*_\mu(k)\right. \\ & \left. + \gamma^\mu\varepsilon^*_\mu(k)\frac{i}{\not{p}'+\not{k}-m+i\varepsilon}\mathcal{M}_0(p'+k,p)\right)u(p). \end{aligned}$$

In this expression $\mathcal{M}_0$ includes the interaction of the electrons with the external field, which is essentially the contribution from the elastic scattering. In our discussion we shall not specify this part of the interaction, which is finite.

In our notation $p$ is the momentum of the incoming electron, $k$ is the momentum carried by the photon, and $p'$ is the momentum of the outgoing electron. For soft photons $|\mathbf{k}| \ll |\mathbf{p}' - \mathbf{p}|$, we can approximate

$$\mathcal{M}_0(p',p-k) = \mathcal{M}_0(p'+k,p) = \mathcal{M}_0(p',p)$$

and simplify

$$\begin{aligned} \frac{1}{\not{p}-\not{k}-m+i\varepsilon}\,\gamma^\mu\varepsilon^*_\mu(k)u(p) &= \frac{-1}{p\cdot k}\,p^\mu\varepsilon^*_\mu(k)u(p) \\ \bar{u}(p)\gamma^\mu\varepsilon^*_\mu(k)\,\frac{1}{\not{p}'+\not{k}-m+i\varepsilon} &= \bar{u}(p)\,\frac{1}{p'\cdot k}\,p'^\mu\varepsilon^*_\mu(k), \end{aligned}$$

so the amplitude in the soft-photon approximation becomes

$$i\mathcal{M} = e\,\bar{u}(p')\mathcal{M}_0(p',p)u(p)\left(\frac{p'\cdot\varepsilon^*}{p'\cdot k} - \frac{p\cdot\varepsilon^*}{p\cdot k}\right). \tag{2.1}$$

The cross-section also factors out as the cross-section for the elastic scattering (without bremsstrahlung) times a factor coming from soft bremsstrahlung:

$$d\sigma(p\to p'+\gamma) = d\sigma(p\to p')\int\frac{d^3k}{(2\pi)^3}\frac{e^2}{2k}\sum_\lambda\left|\frac{p'\cdot\varepsilon(\lambda)}{p'\cdot k} - \frac{p\cdot\varepsilon(\lambda)}{p\cdot k}\right|^2. \tag{2.2}$$

We sum over the two photon polarization states $\lambda = 1,2$ and perform the integral up to the energy $|\mathbf{q}| = |\mathbf{p} - \mathbf{p}'|$, at which time the soft-photon approximation breaks down. The theory is consistent for the upper limit, but the integral diverges at the lower limit, which is the infrared divergence we discussed. To see the nature of divergence and to compare with other diagrams, we introduce a small photon mass $\mu$ and write down the final form of the cross-section as [16]

$$d\sigma(p \to p' + \gamma) \approx d\sigma(p \to p')\frac{\alpha}{\pi}\log\left(\frac{-q^2}{\mu^2}\right).\log\left(\frac{-q^2}{m^2}\right), \tag{2.3}$$

where $q^2 = (p' - p)^2$. The $q^2$ dependence, known as the Sudakov double logarithm, is physical. The problem with $\mu \to 0$ is the infrared divergence of the bremsstrahlung process, which will be solved when we include vertex corrections to the elastic scattering.

This infrared divergence problem is generic to any process in which massless particles are emitted from an initial or final leg that is on mass shell. If a soft photon of momentum $k$ is emitted from an on-shell electron with momentum $p$, the momentum integration diverges. Since $k$ is small and $p^2 = m^2$, the propagator

$$\frac{1}{(p-k)^2 - m^2 + i\varepsilon} \sim \frac{1}{-2p\cdot k}$$

gives the infrared divergence. However, this infrared divergence is not really a problem, since it cancels with divergences in the radiative diagrams.

### Degree of Divergence and Renormalizability

The main problem with higher order processes is the ultraviolet divergences. One may look at the problem from the point of view that at higher momentum the theory may not remain the same and, hence, our lack of knowledge of very high energy interactions may be manifesting as the unknown ultraviolet divergences, which we eliminate by extracting the infinite contribution from the finite contribution by means of some regularization and then by redefining some bare parameters to remove the infinite contributions. In some theories it is not possible to absorb all the infinities in the bare parameters. Renormalizable theories are ones that allow us to remove the infinities. We shall now give some broad descriptions of the types of divergences one may encounter and when the theory is renormalizable.

We try to understand the type of divergences by counting the powers of momenta $p$ in any Feynman graph and calculating the degree of divergence $D$. We count the degree of divergence $D$ by counting (i) boson propagators, each contributing $p^{-2}$; (ii) fermion propagators, each contributing $p^{-1}$; (iii) loop integrals, contributing loop integration with $p^4$; and (iv) every vertex with $n$ derivatives contributing at most $n$ powers of $p$. If $D$ is 0 or positive, then the graph diverges. $D = 0$ gives to logarithmically divergent, $D = 1$ gives linearly divergent, and $D = 2$ gives quadratically divergent integrals.

We now state a criterion for the renormalizability of any theory.

> If any theory allows graphs, whose degree of divergence $D$ depends only on the external legs (internal loops do not contribute), we can collect all $N$-point loop graphs into one term. In such theories, if there are finite numbers of classes of divergent $N$-point graphs which can be absorbed in the bare parameters of the theory, the theory is renormalizable.

In a super-renormalizable theory, adding internal loops decreases $D$. The degree of divergence of any theory is determined in terms of the mass dimensions of the various fields and coupling constants. We start with a dimensionless action. The volume element $d^4x$ has dimension $[M]^{-4}$ (or equivalently $[L]^4$ in the natural unit system $\hbar = 1$ and $c = 1$). Thus, the Lagrangian should have a dimension $[M]^4$ to keep the action dimensionless. The kinetic energy term of a scalar field, $(\partial_\mu \phi)^2$ then means $\phi$ has a dimension $[M]$ since the derivative $\partial_\mu$ has a dimension $[M]$. Similarly the kinetic energy term of a gauge field $F_{\mu\nu}F^{\mu\nu}$ would give us the dimension of a vector field or a gauge boson to be $[M]$. The kinetic energy term of a fermion $\bar{\psi} i \partial\!\!\!/ \psi$ would give us the dimension of any fermion $[M]^{3/2}$.

> *In any renormalizable theory coupling constants do not have any mass-dimensions.*

In a $d$-dimensional theory (space–time dimension $d$), the fields will have different mass dimensions. Since the action should be dimensionless and the integration is over a $d$-dimensional space $d^dx$, it would mean that

$$[\phi] = [M]^{(d-2)/2}; \quad [\psi] = [M]^{(d-1)/2} \quad \text{and} \quad [A_\mu] = [M]^{(d-2)/2}. \tag{2.4}$$

This also includes the mass-dimensions in four space–time dimensions.

Let us consider a non-renormalizable theory, which allows an interaction $\mathscr{L}_I$ with coupling constant $[g] = [M]^{-n}$, $n > 0$. For $n = 2$, the interaction could be a four-fermion interaction $g(\bar{\psi}_a \psi_b)(\bar{\psi}_c \psi_d)$ or a six-scalar interaction of the form $g\phi^6$. Let us consider an $N$-point function in this theory. If we insert the interaction $\mathscr{L}_I$, the N-point function will have one extra coupling constant $g$, which will reduce the mass-dimension of the $N$-point function. To maintain the mass-dimension of the graph, every coupling constant $g$ should be accompanied by $k^n$, which makes the graph more divergent by a factor of $k^n$. Thus, by going to higher orders in perturbation theory, we make the graph arbitrarily divergent.

Gravity, supergravity, four-fermion interactions and massive vector theory with non-Abelian gauge group are thus, examples of non-renormalizable theories. A $\phi^4$ model, Yukawa interactions, gauge theories, massive vector Abelian theory and spontaneously broken gauge theories are examples of renormalizable theories. In addition, there are super-renormalizable theories, in which the degree of divergence goes down so the theory has only a few divergent graphs. $\phi^3$ theory is an example of super-renormalizable theory. Finally, there are finite theories such as the super Yang–Mills theory and the superstring theory, which are free of any divergences to all orders.

## 2.2 Radiative Corrections

We have mentioned three types of radiative corrections: *vertex*, *self-energy* and *vacuum polarizations* (see figure 2.1). These three types of diagrams are distinct and have special characteristics. We shall now study these diagrams and then proceed to discuss some general features of renormalization.

The loop in the *vacuum polarization* graph (figures 2.1 and 2.3) contains two fermion propagators that give two powers in $p^{-1}$, while the loop integral gives $d^4p$. So the graph has a degree of divergence two and, hence, diverges quadratically in the ultraviolet region. For a *vertex correction* (figures 2.1 and 2.4), the loop has two fermions giving two powers in $p^{-1}$, and a photon propagator giving a factor $p^{-2}$, so the loop integral $d^4p$ makes it logarithmically divergent. The *self-energy diagram* (figures 2.1 and 2.5) is also a logarithmically divergent graph in the ultraviolet region, although it has one fermion propagator with $p^{-1}$ and a photon propagator giving a factor $p^{-2}$ making the degree of divergence three.

The one-loop diagrams are relevant to check the consistency of the theory. If the theory is renormalizable, then it should be possible to absorb the infinities in the redefinition of some bare parameters of the theory. In addition, the finite part of the diagram may contribute to some physical processes. For example, if there is any *CP* violation in the theory, it should appear in the interference of the tree-level and one-loop diagrams. In electrodynamics, the vertex corrections contribute to the anomalous magnetic moment of the electron while all the three corrections contribute to the Lamb shift. We shall now discuss these graphs in more detail.

### Vacuum Polarization

We mentioned about the vacuum polarization graph, also known as the *photon self-energy* graph, in figure 2.1 of the *eX* scattering. The loop diagrams are much more difficult to deal with than the tree-level Feynman diagrams. Our main interest now is to understand how to deal with the divergent parts of these integrals. This is rather involved, so we shall not present all the details; however, we shall try to explain the general prescriptions that are used in dealing with these diagrams and describe the methods.

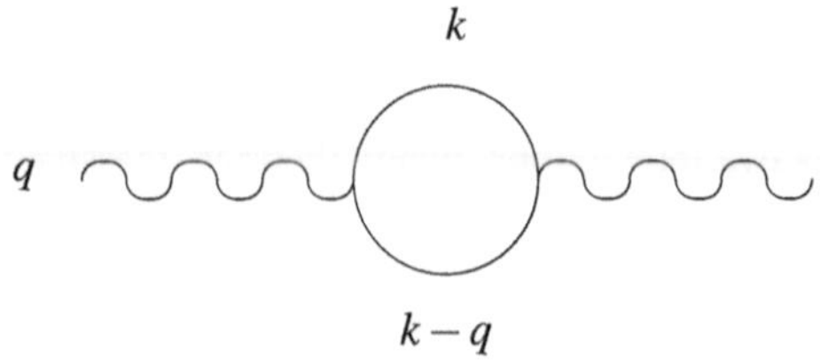


**FIGURE 2.3**
One-loop vacuum polarization correction to photon propagator.

The relevant part of the diagram for understanding the divergent character is shown in figure 2.3. This diagram will add contributions to the photon propagator. We shall now discuss only the divergent part of the process and, therefore, not include the external photon lines. The loop contribution of the process is given by

$$\begin{aligned} i\Pi_{\mu\nu,m}(q) &= e^2 \int \frac{d^4k}{(2\pi)^4} \operatorname{tr}\left[\gamma_\mu \frac{i}{\not{k}-m+i\varepsilon}\gamma_\nu \frac{i}{\not{k}-\not{q}-m+i\varepsilon}\right] \\ &= 4e^2 \int \frac{d^4k}{(2\pi)^4} \frac{i}{k^2-m^2+i\varepsilon} \frac{i}{(k-q)^2-m^2+i\varepsilon} \\ &\quad \times \left[k_\mu (k-q)_\nu + k_\nu (k-q)_\mu - g_{\mu\nu}(k^2 - k\cdot q - m^2)\right]. \end{aligned} \tag{2.5}$$

This integral is quadratically divergent.

One can regularize the divergent integrals by different methods, which essentially means that we add new contributions to the integral, which allows us to evaluate the integral, and then take the limit in which the new contributions vanish. This will allow us to take out the finite part from the infinite part of the integral. In this chapter we shall present two popular approaches of regularization, the Pauli–Villars method and the dimensional regularization method, to evaluate the different divergent integrals. In this present case we shall use the Pauli–Villars method [17], which introduces new fictitious ghost fermions in the theory with mass $M$. This will allow us to treat the integral in a covariant manner, since we are treating the fictitious particles in the same way as the ordinary particles. Introduction of the new particles will provide us a cut-off for the divergences and allow us to perform the integration in terms of the new mass parameters $M$. At the end we take infinite $M$ limit to remove these new states.

Before introducing the Pauli–Vilars regulator fields, we simplify the integration by introducing auxiliary variables

$$\frac{i}{k^2-m^2+i\varepsilon} = \int_0^\infty d\alpha e^{i\alpha(k^2-m^2+i\varepsilon)} . \tag{2.6}$$

To simplify $\Pi_{\mu\nu,m}$, we thus replace the electron propagators by two additional integrals with the new auxiliary fields $\alpha_1$ and $\alpha_2$. A more convenient approach to deal with these kinds of integrals is by using Feynman variables, which we shall discuss while evaluating other divergent integrals. The integration over the momenta can then be easily performed using

$$\begin{aligned} \int \frac{d^4p}{(2\pi)^4} e^{ip^2\beta} &= \frac{-i}{16\pi^2\beta^2} \\ \int \frac{d^4p}{(2\pi)^4} p_\mu p_\nu e^{ip^2\beta} &= \frac{g_{\mu\nu}}{32\pi^2\beta^3}, \end{aligned} \tag{2.7}$$

where $\beta = \alpha_1 + \alpha_2$ and $p = k - q\alpha_2/\beta$. The amplitude can then be written as

$$i\Pi_{\mu\nu,m} = (q^2 g_{\mu\nu} - q_\mu q_\nu)\, i\Pi_1 + g_{\mu\nu}\, i\Pi_2, \tag{2.8}$$

where

$$\Pi_1 = -\frac{e^2}{2\pi^2}\int_0^\infty\int_0^\infty \frac{\alpha_1\alpha_2\, d\alpha_1 d\alpha_2}{\beta^4} e^{f(\alpha_1,\alpha_2)}$$

$$\Pi_2 = -\frac{e^2}{4\pi^2}\int_0^\infty\int_0^\infty i\frac{d\alpha_1 d\alpha_2}{\beta^3}[f(\alpha_1,\alpha_2)-1]e^{f(\alpha_1,\alpha_2)} \tag{2.9}$$

with $f(\alpha_1,\alpha_2) = -i(m^2-i\varepsilon)\beta + iq^2\alpha_1\alpha_2/\beta$. We can now add a similar expression for the Pauli–Villars regulator $\Pi_{\mu\nu,M}$, so in the modified amplitude

$$\Pi_{\mu\nu} = \Pi_{\mu\nu,m} - \Pi_{\mu\nu,M}$$

the quadratically divergent parts coming from $\Pi_{\mu\nu,m}$ and $\Pi_{\mu\nu,M}$ cancel for fixed $M$ and one can show that $\Pi_2$ in $\Pi_{\mu\nu,m}$ cancels the $\Pi_2$ in $\Pi_{\mu\nu,M}$ and $\Pi_2 = 0$ in $\Pi_{\mu\nu}$.

To perform the $\Pi_1$ integral, we use

$$\int_0^\infty \frac{d\rho}{\rho}\delta\left(1-\frac{\beta}{\rho}\right) = 1 \tag{2.10}$$

and after some manipulation obtain

$$\Pi_1 = -\frac{e^2}{2\pi^2}\int_0^\infty\int_0^\infty d\alpha_1 d\alpha_2 \alpha_1\alpha_2\delta(1-\beta)\int_0^\infty \frac{d\rho}{\rho}e^{i\rho(-a(m)+i\varepsilon)}\,, \tag{2.11}$$

where $a(m) = m^2 - q^2\alpha_1\alpha_2$. The $a(m)$ independent terms, including the uncontrollable divergent part, cancels out with the contributions from the Pauli–Villars contributions. Finally, we arrive at the expression for the divergent part in the limit $q^2 \to 0$:

$$\Pi_1 = -\frac{\alpha}{3\pi}\log\frac{M^2}{m^2} \tag{2.12}$$

where $\alpha = e^2/4\pi$. This is equivalent to saying that the tree-level photon propagator has been corrected by an infinite factor

$$-i\frac{g_{\mu\nu}}{q^2} \longrightarrow -i\frac{g_{\mu\nu}}{q^2}\left(1-\frac{\alpha}{3\pi}\log\frac{M^2}{m^2}\right) = -i(Z_3)\frac{g_{\mu\nu}}{q^2}, \tag{2.13}$$

when one-loop vacuum polarization corrections are included. This infinite factor can be absorbed in the redefinition of the charge of the electron. Since the photon couples with two external electrons with coupling strength $e_0$ (we denote the bare unrenormalized coupling with a subscript 0), the factor $Z_3$ can be absorbed in the redefinition of the bare coupling constant. Thus, to the first order, the renormalized charge of electron is given by

$$e = e_0\sqrt{Z_3}. \tag{2.14}$$

The infinities in the bare coupling $e_0$ cancel the infinities coming from the loop-diagram, giving us a finite renormalized charge $e$. The energy dependence of the fine structure constant is given by

$$\alpha_{eff}^{-1}(M) = \alpha^{-1} - \frac{1}{3\pi}\log\frac{M^2}{m^2}. \tag{2.15}$$

At very small distances the virtual electron positron pairs surround the electrons, which screen the electric charge and the effective charge becomes large.

**Vertex Correction**

The next divergent diagram we shall consider is the vertex diagram, which contains both infrared singularities and ultraviolet singularities. The divergence in the infrared region cancels that of the bremsstrahlung diagram, while the divergence in the ultraviolet region renormalizes the vertex function. In addition, the finite part of the diagram contributes to the anomalous magnetic moment of the electron.

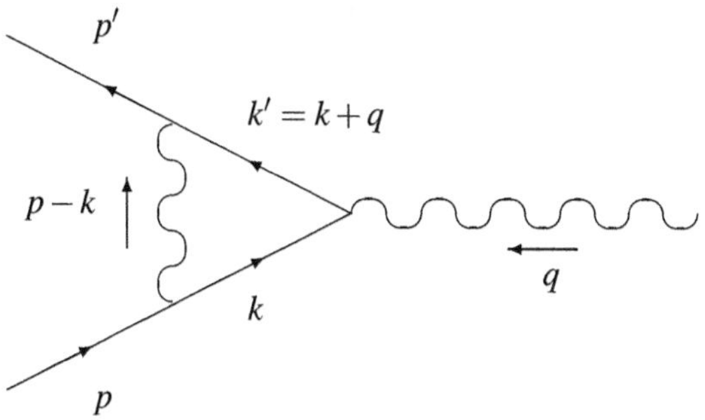


**FIGURE 2.4**
Correction to electron vertex function.

The relevant part of the diagram is shown in figure 2.4. This diagram modifies the electromagnetic vertex with two external electron lines and the photon propagator

$$-ie\bar{u}(p')\gamma^{\mu}u(p)A_{\mu}(p'-p) \longrightarrow -ie\bar{u}(p')\Gamma^{\mu}(p',p)u(p)A_{\mu}(p'-p), \qquad (2.16)$$

where the most general form of $\Gamma^{\mu}(p',p)$ can be written as

$$\Gamma^{\mu}(p',p) = \gamma^{\mu}\, F_1(q^2) + \frac{i\sigma^{\mu\nu}q_{\nu}}{2m}\, F_2(q^2), \qquad (2.17)$$

where $-i\sigma^{\mu\nu} = (1/2)[\gamma^{\mu},\gamma^{\nu}]$, and we introduce two unknown functions of $q^2$, called form-factors, $F_1$ and $F_2$. The main task of calculating the vertex correction is to find out these two form-factors.

The tree-level contributions give $F_1 = 1$ and $F_2 = 0$. So we write $\Gamma^{\mu} = \gamma^{\mu} + \delta\Gamma^{\mu}$ to distinguish the one-loop vertex correction $\delta\Gamma^{\mu}$. From figure 2.4, we can then write

$$\begin{aligned} V^{\mu} &= \bar{u}(p')\delta\Gamma^{\mu}(p',p)u(p) \\ &= (-ie)^2 \int \frac{d^4k}{(2\pi)^4} \frac{-ig_{\alpha\beta}}{(k-p)^2+i\varepsilon} \bar{u}(p')\gamma^{\alpha} \frac{i}{\not{k}'-m+i\varepsilon}\gamma^{\mu}\frac{i}{\not{k}-m+i\varepsilon}\gamma^{\beta}u(p) \end{aligned}$$

$$= 2ie^2 \int \frac{d^4k}{(2\pi)^4} \frac{\bar{u}(p')[\not{k}\gamma^\mu \not{k}' + m^2\gamma^\mu - 2m(k+k')^\mu]u(p)}{((k-p)^2+i\varepsilon)(k'^2-m^2+i\varepsilon)(k^2-m^2+i\varepsilon)}. \quad (2.18)$$

This logarithmically divergent integral is difficult to calculate. We shall present some crucial steps and then the final result.

The first step is to put the three factors in the denominator into one. This is conveniently done with the Feynman parameters, which are some auxiliary parameters that are to be integrated out. For a factor of two terms, we can use the identity

$$\frac{1}{ab} = \int_0^1 dx \frac{1}{[ax+b(1-x)]^2} = \int_0^1 dxdy\delta(x+y-1)\frac{1}{(ax+by)^2}. \quad (2.19)$$

This is a simpler case of a more general identity

$$\frac{1}{a_1 a_2 \cdots a_n} = \int_0^1 dx_1 \cdots dn_n\, \delta\left(\sum x_i - 1\right) \frac{(n-1)!}{[\sum a_i x_i]^n}. \quad (2.20)$$

This will allow us to write the denominator of $V^\mu$ as

$$\frac{1}{((k-p)^2+i\varepsilon)(k'^2-m^2+i\varepsilon)(k^2-m^2+i\varepsilon)} = \int_0^1 dxdydz\delta(x+y+z-1)\frac{2}{D^3},$$

where

$$D = \ell^2 - \Delta + i\varepsilon; \quad \ell \equiv k + yq - zp; \quad \text{and} \quad \Delta \equiv -xyq^2 + (1-z)^2 m^2.$$

To simplify the numerator we use the identities

$$\int \frac{d^4\ell}{(2\pi)^4} \frac{\ell^\mu}{D^3} = 0$$

$$\int \frac{d^4\ell}{(2\pi)^4} \frac{\ell^\mu \ell^\nu}{D^3} = \int \frac{d^4\ell}{(2\pi)^4} \frac{\ell^2\, g^{\mu\nu}}{4\, D^3}, \quad (2.21)$$

which states that the integral depends only on the magnitude of $\ell$. So we can write the numerator only in terms of $\ell^2$. We further use the Ward identity [18], which is an extension of the charge conservation and reads $q_\mu \Gamma^\mu = 0$. The vertex contribution then simplifies to

$$V^\mu = 2ie^2 \int \frac{d^4\ell}{(2\pi)^4} \int_0^1 dx\, dy\, dz\, \delta(x+y+z-1)$$
$$\frac{2}{D^3}\bar{u}(p')\left[\gamma^\mu \cdot \left(\frac{-\ell^2}{2} + A\right) + \frac{i\sigma^{\mu\nu}q_\nu}{2m} \cdot B\right] u(p), \quad (2.22)$$

with $A = (1-x)(1-y)q^2 + (1-4z+z^2)m^2$ and $B = 2m^2 z(1-z)$. We shall then perform the momentum integral.

To evaluate the integral, we make use of Wick's rotation trick and work with Euclidian four-momentum variable $\ell_E$, where

$$\ell^0 \equiv i\ell_E^0; \quad \text{and} \quad \ell = \ell_E, \quad (2.23)$$

so $\ell_E^0$ varies between $-\infty$ to $\infty$. The integral then becomes

$$\int \frac{d^4\ell}{(2\pi)^4} \frac{1}{(\ell^2-\Delta)^m} = \frac{i}{(-1)^m} \int \frac{d^4\ell_E}{(2\pi)^4} \frac{1}{(\ell_E^2+\Delta)^m}. \tag{2.24}$$

One can then work in four-dimensional spherical coordinate to evaluate the integral. However, for divergent integrals we have to adopt some regularization technique to deal with the integral.

The integral is divergent, and hence, we introduce the Pauli–Villars regulators. We shall replace the photon propagator with

$$\frac{1}{(k-p)^2+i\varepsilon} \longrightarrow \frac{1}{(k-p)^2+i\varepsilon} - \frac{1}{(k-p)^2+\Lambda^2+i\varepsilon}.$$

Here $\Lambda$ is the large mass of the fictitious heavy photons. The denominator will then be replaced by

$$\Delta \longrightarrow \Delta_\Lambda = -xyq^2 + (1-z)^2 m^2 + z\Lambda^2.$$

This will make the momentum integral finite for any fixed value of $\Lambda$. We can then perform the integral after making Wick's rotation

$$\int \frac{d^4\ell}{(2\pi)^4} \left( \frac{\ell^2}{(\ell^2-\Delta)^3} - \frac{\ell^2}{(\ell^2-\Delta_\Lambda)^3} \right) = \frac{i}{(4\pi)^2} \log\left(\frac{\Delta_\Lambda}{\Delta}\right). \tag{2.25}$$

Any $\Lambda$ dependent contribution to the finite part of the integral is ignored. Finally we arrive at

$$\begin{aligned} V^\mu &= \frac{\alpha}{2\pi} \int_0^1 dx\, dy\, dz\, \delta(x+y+z-1) \\ &\quad \bar{u}(p') \left[ \frac{\gamma^\mu}{\Delta} \left( \log(z\Lambda^2) + A \right) + \frac{i\sigma^{\mu\nu} q_\nu}{2m} \frac{B}{\Delta} \right] u(p). \end{aligned} \tag{2.26}$$

This gives the one-loop correction to the form-factors. The first form-factor $F_1$ contains both the infrared and ultraviolet divergences. The infrared divergence comes out to be exactly the same as the infrared divergence in the bremsstrahlung process and they cancel each other. The ultraviolet divergence appears in $\delta F_1(q^2=0)$, where $\delta F_1$ is the one-loop vertex correction to $F_1$. $F_1(q^2=0) + \delta F_1(q^2=0)$ gives the charge of the electron in units of $e$ and, hence, should be 1. We know that the tree-level contribution is $F_1(q^2=0)=1$, so we normalize the vertex function by

$$\delta F_1(q^2) \to \delta F_1(q^2) - \delta F_1(q^2=0), \tag{2.27}$$

which eliminates the ultraviolet divergent contribution. Since the electric charge is the only parameter entering in the vertex function, this infinite contribution cancels the infinity of the bare charge. We shall come back to this point again when we discuss renormalization of QED.

The second form-factor $F_2(q^2)$ is finite, and for zero momentum it is

$$F_2(q^2=0) = \frac{\alpha}{2\pi}. \tag{2.28}$$

The magnetic moment of the electron is given by

$$\mu = g\left(\frac{e}{2m}\right)\mathbf{S}, \tag{2.29}$$

where the Landè $g$-factor is given by

$$g = 2[F_1(q^2=0) + F_2(q^2=0)]. \tag{2.30}$$

At the tree-level $F_1 = 1$ and $F_2 = 0$; hence, $g = 2$. At the one-loop level, the normalization gives $F_1(q^2=0) = 1$. The correction to the $g$-factor thus comes only from the correction $F_2(q^2=0)$, which gives the anomalous magnetic moment of the electron [19]

$$a_e \equiv \frac{g-2}{2} = \frac{\alpha}{2\pi} = 0.0011614. \tag{2.31}$$

This anomalous magnetic moment has been measured experimentally $a_e(expt) = 0.0011597$.

The first form-factor $F_1(q^2)$ also contains the infrared divergence. In the limit of $q^2 \to 0$, we have

$$\int_0^1 dx\, dy\, dz \delta(x+y+z-1)\frac{1-4z+z^2}{\Delta(q^2=0)} = \int_0^1 dz \frac{-2}{m^2(1-z)} + \cdots.$$

While discussing the bremsstrahlung process, we have seen that the infrared divergence appears because photons are massless. If we assign a small mass $\mu$ to the photon, this problem may be tackled, since it adds a term $z\mu^2$ to $\Delta$. After a tedious calculation, one can write the contribution from $F_1$ of the vertex diagram to the scattering cross-section to be

$$\frac{d\sigma}{d\Omega} \approx \left(\frac{d\sigma}{d\Omega}\right)_0 \left[1 - \frac{\alpha}{\pi}\log\left(\frac{-q^2}{\mu^2}\right).\log\left(\frac{-q^2}{m^2}\right)\right]. \tag{2.32}$$

Comparing equation (2.32) with the scattering cross-section for the bremsstrahlung process given in equation (2.3), it is clear that the infrared divergent parts of the two diagrams cancel out when the cross-sections for the two processes are added together. Although the processes appear to be different, since it is not possible to distinguish the virtual and real soft photons experimentally due to quantum uncertainty principle, we have to add these processes, and then the final result becomes free of any infrared divergences.

### Self-Energy Correction

We shall now discuss the self-energy diagram. This diagram in figure 2.5 gives correction to the free-fermion propagator $i/(\not{p} - m_0 + i\varepsilon)$. Since the bare parameter in the fermion propagator is the fermion mass, it renormalizes the fermion mass. We shall, therefore, denote the bare electron mass as $m_0$ to distinguish it from the renormalized physical mass $m$.

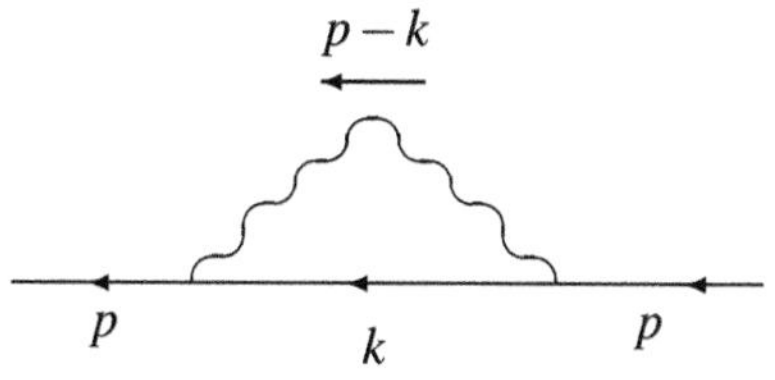


**FIGURE 2.5**
Electron self-energy diagram.

Using Feynman rules we can write the contribution of the diagram of figure 2.5

$$\frac{i}{\not p - m_0} \cdot [-i\Sigma_2(p)] \cdot \frac{i}{\not p - m_0}, \tag{2.33}$$

where the factor $\Sigma_2(p)$ comes from the loop integral and is given by

$$-i\Sigma_2(p) = -e^2 \int \frac{d^4k}{(2\pi)^4} \gamma^\mu \frac{i}{\not k - m_0 + i\varepsilon} \gamma_\mu \frac{i}{(p-k)^2 - \mu^2 + i\varepsilon}. \tag{2.34}$$

This integral is logarithmically divergent. The infrared divergent part has been taken care of by adding a finite mass $\mu$ to the photon. For the ultraviolet divergence we shall use the method of dimensional regularization to regulate the infinite integral.

We first use the Feynman parameters (see equation (2.20)) to simplify the denominator and write

$$-i\Sigma_2(p) = -e^2 \int_0^1 dx \int \frac{d^4\ell}{(2\pi)^4} \frac{-2x\not p + 4m_0}{[\ell^2 - \Delta + i\varepsilon]^2}, \tag{2.35}$$

where $\ell = k - xp$ and $\Delta = -x(1-x)p^2 + x\mu^2 + (1-x)m_0^2$. We then make Wick's rotation (see equation (2.23)) and work with Euclidian 4-momentum variable $\ell_E$:

$$-i\Sigma_2(p) = -ie^2 \int_0^1 dx \int \frac{d^4\ell_E}{(2\pi)^4} \frac{-2x\not p + 4m_0}{[\ell_E^2 + \Delta]^2}. \tag{2.36}$$

We shall evaluate this divergent integral using the dimensional regularization method.

*Dimensional Regularization*: In the dimensional regularization method we compute the Feynman integral at a sufficiently small dimension $d$, in which the integral converges, and then take the limit $d \to 4$. Consider the Wick's rotated integral of equation (2.35). We shall first evaluate the integral in $d$ dimensions:

$$\int \frac{d^d\ell_E}{(2\pi)^d} \frac{1}{[\ell_E^2 + \Delta]^2} = \int \frac{d\Omega_d}{(2\pi)^d} \int_0^\infty d\ell_E \frac{\ell_E^{d-1}}{[\ell_E^2 + \Delta]^2}, \tag{2.37}$$

where the area of a unit sphere in $d$ dimensions is given by

$$\int d\Omega_d = \frac{2\pi^{d/2}}{\Gamma(d/2)}, \tag{2.38}$$

where $\Gamma(1/2) = \sqrt{\pi}$ and $\Gamma(n+1) = n\Gamma(n)$.

The second factor can be integrated out exactly in terms of beta function by substituting $z = \Delta/(\ell^2 + \Delta)$ and is given by

$$\int_0^\infty d\ell_E \frac{\ell_E^{d-1}}{[\ell_E^2 + \Delta]^2} = \frac{1}{2}\Delta^{(d-4)/2}\Gamma\left(2 - \frac{d}{2}\right)\Gamma\left(\frac{d}{2}\right). \tag{2.39}$$

The behavior near $d = 4$ can be expressed in terms of a variable $\varepsilon = 4 - d$ in the limit $\varepsilon \to 0$. For the $\Gamma$-function one can approximate it as

$$\lim_{d\to 4}\Gamma\left(2 - \frac{d}{2}\right) = \lim_{d\to 4}\frac{1}{(2 - \frac{d}{2})}\Gamma\left(3 - \frac{d}{2}\right) = \lim_{\varepsilon\to 0}\frac{2}{\varepsilon}.$$

However, a more appropriate procedure will be to consider a power series expansion of the $\Gamma$-function and then take the limit. This form of the approximation gives

$$\lim_{\varepsilon\to 0}\Gamma(\varepsilon/2) = \frac{2}{\varepsilon} - \gamma + O(\varepsilon), \tag{2.40}$$

where $\gamma \approx 0.5772$ is the Euler–Mascheroni constant. Then the leading terms come from

$$\lim_{\varepsilon\to 0}\Gamma(\frac{\varepsilon}{2})\Delta^{-\varepsilon/2} = \lim_{\varepsilon\to 0}\Gamma(\frac{\varepsilon}{2})e^{-(\varepsilon/2)\log\Delta} = \frac{2}{\varepsilon} - \log\Delta + \cdots.$$

Only the first term is relevant for understanding the divergent nature of the integral. Finally we can write the integral as

$$\lim_{d\to 4}\int \frac{d^d\ell_E}{(2\pi)^d}\frac{1}{[\ell_E^2 + \Delta]^2} = \frac{1}{(4\pi)^2}\left[\frac{2}{\varepsilon} - \log\Delta - \gamma + \log(4\pi) + O(\varepsilon)\right]. \tag{2.41}$$

If we compare this result with the Pauli–Villars method, the pole $1/\varepsilon$ in dimensional regularization corresponds to a logarithmic divergence $\log\Lambda^2$.

For using dimensional regularization one can use some more general formulae:

$$\int \frac{d^d\ell_E}{(2\pi)^d}\frac{1}{[\ell_E^2 + \Delta]^n} = \frac{1}{(4\pi)^{d/2}}\Delta^{(d-2n)/2}\frac{\Gamma\left(n - \frac{d}{2}\right)}{\Gamma(n)} \tag{2.42}$$

$$\int \frac{d^d\ell_E}{(2\pi)^d}\frac{\ell_E^2}{[\ell_E^2 + \Delta]^n} = \frac{1}{(4\pi)^{d/2}}\frac{d}{2}\Delta^{(d+2-2n)/2}\frac{\Gamma\left(n - \frac{d}{2} - 1\right)}{\Gamma(n)}. \tag{2.43}$$

There are $d$ gamma matrices $\gamma^\mu$ satisfying

$$\{\gamma^\mu, \gamma^\nu\} = 2g^{\mu\nu}, \quad \mathrm{tr}\mathrm{I} = 4, \quad g_{\mu\nu}g^{\mu\nu} = d.$$

We replace $\ell^\mu\ell^\nu \to \ell^2 g^{\mu\nu}/d$ for any numerator of a symmetric integrand. The contraction identities in $d = 4 - \varepsilon$ will become

$$\begin{aligned}
\gamma^\mu\gamma^\nu\gamma_\mu &= -(2-\varepsilon)\gamma^\nu \\
\gamma^\mu\gamma^\nu\gamma^\rho\gamma_\mu &= 4g^{\nu\rho} - \varepsilon\gamma^\nu\gamma^\rho \\
\gamma^\mu\gamma^\nu\gamma^\rho\gamma^\sigma\gamma_\mu &= -2\gamma^\sigma\gamma^\rho\gamma^\nu + \varepsilon\gamma^\nu\gamma^\rho\gamma^\sigma.
\end{aligned}$$

We list these results for completeness although they will not be needed for the self-energy diagram.

The required result for the self-energy contribution can, therefore, be expressed as

$$\Sigma_2(p) = \frac{\alpha}{4\pi}\int_0^1 dx(2m_0 - x\not{p})\left[\frac{2}{\varepsilon} - \log\Delta - \gamma + \log(4\pi) + O(\varepsilon)\right]. \tag{2.44}$$

Unlike the vertex or the vacuum polarization contributions, we cannot renormalize any bare coupling constant with this one-loop contribution of the self-energy diagram. We have to consider all other related diagrams to see how this infinite contribution can be absorbed by any bare coupling constant. In this case, the mass of the electron is renormalized.

We first define a one-particle irreducible (1PI) diagram, which is a diagram with more than one part, each connected by only a single fermion line. In other words, by removing any single fermion line, if the diagram cannot be split in two parts, we call this a 1PI diagram. Let the contribution of sum over all 1PI diagrams be $\Sigma$. A self-energy diagram is a 1PI diagram to leading order, and hence, $\Sigma = \Sigma_2$ to order $\alpha$.

Consider the Fourier transform of a two-point function, it can be written as a series

(2.45)

We shall now compute the series to order $\alpha$. The total contribution of the series will be given by

$$\begin{aligned}\frac{i}{\not{p}-m} &= \frac{i}{\not{p}-m_0} + \frac{i}{\not{p}-m_0}\cdot[-i\Sigma_2(p)]\cdot\frac{i}{\not{p}-m_0} \\ &\quad + \frac{i}{\not{p}-m_0}\cdot[-i\Sigma_2(p)]\cdot\frac{i}{\not{p}-m_0}\cdot[-i\Sigma_2(p)]\cdot\frac{i}{\not{p}-m_0} + \cdots \\ &= \frac{i}{\not{p}-m_0-\Sigma_2(\not{p})},\end{aligned} \tag{2.46}$$

where $m$ is the renormalized mass after taking care of all the infinities coming from the series 2.45. Thus, although the infinity in the self-energy diagram cannot be absorbed in any bare parameter, the sum of the series 2.45 relates the infinite contribution of the self-energy diagram with the bare mass of the electron. The shift in mass is then given by

$$\delta m = m - m_0 = \Sigma_2(\not{p}=m) \approx \Sigma_2(\not{p}=m_0), \tag{2.47}$$

which can be obtained from equation (2.44) as

$$\delta m = \lim_{\varepsilon\to 0}\frac{\alpha m_0}{4\pi}\int_0^1 dx(2-x)\left[\frac{2}{\varepsilon} - \log(\frac{\Delta}{4\pi}) - \gamma + O(\varepsilon)\right] = \frac{3\alpha}{4\pi\varepsilon}m_0. \tag{2.48}$$

This is the correction to the bare mass coming from the self-energy diagrams to order $\alpha$. In general the 1PI blobs in the series 2.45 should include all the 1PI diagrams to all orders. In that case, we can replace $\Sigma_2(p)$ by $\Sigma(p)$, which is the sum of all the 1PI diagrams with two external fermions. In that case the mass renormalization will include contributions from diagrams to all orders. Comparing with the results from Pauli–Villars regularization, the divergence is found to be logarithmic and not linear. This is because the diagram is proportional to the mass of the electron. Since the diagram vanishes if electrons are massless, the symmetry ensures that one of the momentum in the numerator does not contribute to the divergence.

## 2.3 Renormalization of QED

We demonstrated how to deal with the infinities in quantum field theory and how some of the infinities can be cancelled with the infinities of the bare parameters of any theory, which cannot be measured. The next task will be to construct a renormalizable theory, in which all the divergences have been taken care of properly through a prescribed renormalization program. The steps one follows for such construction are

- Find out all the divergent diagrams order by order in perturbation theory. Superficial power counting method may be adopted for the purpose.
- Use some regularization technique to evaluate the infinite integrals in terms of some large parameters $\Lambda$. After completion of the renormalization, we take the infinite limit of the parameter $\Lambda$ and verify that the theory is consistent.
- Take care of the infinities by redefining the bare parameters in the theory [2, 3], as we did with the mass and charge of electron after evaluating the vertex and self-energy diagrams.
- Another equivalent, more appropriate, approach introduces counterterms (developed by Bogoliubov, Parasiuk, Hepp and Zimmerman [20]) also known as the BPHZ approach. In this approach we add counterterms to the Lagrangian, whose coefficients are chosen so they cancel the divergences to that order. These counterterms are proportional to the original action. Any renormalizable theory has finite numbers of counterterms. In this approach one can proceed with renormalized perturbation calculations with the new set of Feynman rules.
- At the end of these exercises one can verify by method of induction that the theory is renormalizable to all higher orders. This can be done using Weinberg's theorem [21] that any graph converges if the degree of divergence of the graph and all its subgraphs is negative.

**Ultraviolet Divergences and Power Counting**

The main problem in field theory is the ultraviolet divergence, which needs to be regularized and then absorbed in some bare coupling through renormarlization. So, the first task is to find the nature of divergences in any graph, which is done by power counting or by counting the superficial degree of divergence $D$. Since every fermion line enters with $k^{-1}$, the boson line contributes $k^{-2}$ and a loop integral $d^4k$; by counting the overall power of $k$, we can tell if any diagram diverges for large $k$. Let us explain this for QED.

Although we shall be discussing QED with only electrons and photons, we can as well generalize it to any other charged fermions interacting with photons. If we denote the numbers of external fermions [photons] by $E_f[E_\gamma]$, numbers of fermion photon] propagators by $P_f[P_\gamma]$, number of vertices by $V$, and number of loops by $L$, then the superficial degree of divergence is given by

$$D = 4L - P_f - 2P_\gamma. \tag{2.49}$$

This simply means that in a given loop there are $L$ loop integrals, each giving a factor $d^4k$ and, hence, $4L$ numbers of $k$ in the numerator; $P_f$ internal fermion propagators $i/(\not{k} - m + i\varepsilon)$, each contributing one $k$ in the denominator; and $P_\gamma$ numbers of photon propagators $-ig^{\mu\nu}/(k^2 + i\varepsilon)$, each contributing two powers of $k$ in the denominator.

*If $D$ is 0 or positive, the graph diverges.*

In general, the graph with positive $D$ diverges as $\Lambda^D$, where $\Lambda$ is some large parameter, although there are exceptions. For $D = 0$ the graph diverges logarithmically. For the vacuum polorization diagram, $D = 2$, and hence, it is quadratically divergent; for vertex function diagram, $D = 0$ and it is logarithmically divergent; while for the self-energy diagram, $D = 1$ and it is linearly divergent. However, actual calculations show that all three graphs are logarithmically divergent. For the vacuum polarization graph, the Ward–Takahashi identity reduces the degree of divergence. For the self-energy graph the Lorentz invariance and chiral symmetry ensures that the diagram is proportional to electron mass, so there is one extra momentum in the denominator making it logarithmically divergent. On the other hand, if any diagram has some subdiagrams, then the actual divergence could be worse than predicted by $D$, and for a tree-level diagram with $D = 0$, there are no divergences.

Any vertex can have two fermions and a photon, so the number of vertices can be given in terms of numbers of internal propagators and external lines:

$$V = 2P_\gamma + E_\gamma = \frac{1}{2}(2P_f + E_f),$$

since only one end of an external line connects to a vertex. Any loop will have an independent momentum, which is the same as the number of internal propagators minus the delta function appearing at the vertices modulo the overall momentum conservation. Thus, we can write

$$L = P_f + P_\gamma - V + 1.$$

With these relations we can express the superficial degree of divergence as

$$D = 4 - \frac{3}{2}E_f - E_\gamma. \tag{2.50}$$

Thus, for QED we can characterize all the divergent diagrams from the number of external lines:

*No external lines* ($D = 4$): These diagrams do not contribute to any scattering processes.

*One external photon* ($D = 3$): Vanishes due to symmetries.

*Two external photons* ($D = 2$): Vacuum polarization diagrams that were discussed earlier. Ward–Takahashi identity softens the divergence and makes it logarithmically divergent.

*Three external photons* ($D = 1$): These are called anomaly diagrams, which can come with three fermion lines forming a triangle with the photons attached to the vertices. Such diagrams are divergent for chiral theories only. Since QED has no axial couplings, this diagram vanishes. For QED there will be loops with fermion momenta flowing in one direction, which cancels a similar diagram with momenta flowing in the opposite direction.

*Four external photons* ($D = 0$): This diagram also vanishes by gauge symmetry.

*Two external fermions* ($D = 1$): This is the self-energy diagram. The diagram is proportional to fermion masses, and is logarithmically divergent.

*Two external fermions and an external photon* ($D = 0$): This is the logarithmically divergent vertex diagram.

After characterizing the divergent diagrams the next task is to take care of them, which is done through renormalization.

**Renormalization**

After identifying the divergent graphs, first we have to regularize the integrals and express the divergent part in terms of some large scale. We demonstrated this with the Pauli–Villars method and the dimensional regularization method for the vacuum polarization, the vertex correction and the self-energy diagrams. One has to ensure that the different regularization schemes give the same result.

Once the regularization is done the large scale dependent divergent contributions have to be absorbed into the Lagrangian by rescaling the fields. Then we can write the Lagrangian in two parts, the renormalized Lagrangian and the counterterms, with all the divergences included in the counterterms. The next step will be to write the physical renormalized masses and coupling constant and write the Feynman rules in terms of new renormalized quantities and the new parameters. Any amplitude at this order will then be finite. At higher orders there will be additional divergences, which can again be removed by adjusting the counterterms because renormalizability ensures that there are only a finite number of counterterms.

We start with the QED Lagrangian with bare parameters

$$\mathscr{L} = -\frac{1}{4}(F^{\mu\nu})^2 + \bar{\psi}(i\partial\!\!\!/ - m_0)\psi - e_0\, \bar{\psi}\gamma^\mu \psi A_\mu. \tag{2.51}$$

Let us study the contributions of the divergent diagrams. Consider the self-energy diagram that we calculated and the final result given by equation (2.44). This is the leading order contribution to any 1PI diagrams with two external fermions. The sum of contributions from all such 1PI diagrams to leading order is given in equation (2.46), which can be generalized to higher orders, and the complete propagator can be written as

$$iS'_F = iS_F(p\!\!\!/) + iS_F(p\!\!\!/)[-i\Sigma(p\!\!\!/)]iS_F(p\!\!\!/) + \cdots = \frac{i}{p\!\!\!/ - m_0 - \Sigma(p\!\!\!/) + i\varepsilon}. \tag{2.52}$$

$\Sigma(p\!\!\!/)$ is the sum over 1PI graphs with two external electrons, which can be Taylor expanded around the renormalized mass $m$ as

$$\Sigma(p\!\!\!/) = \Sigma(m) + (p\!\!\!/ - m)\Sigma' + \tilde{\Sigma}(p\!\!\!/), \tag{2.53}$$

where the renormalized mass $m$ is defined in the same way as equations (2.47) and (2.48):

$$m = m_0 + \Sigma(m).$$

Then the renormalized propagator is given by

$$iS'_F = \frac{iZ_2}{p\!\!\!/ - m - Z_2\tilde{\Sigma}(p\!\!\!/) + i\varepsilon} = \frac{iZ_2}{p\!\!\!/ - m + i\varepsilon} + \cdots, \tag{2.54}$$

where $Z_2 = 1/[1 - \Sigma'(m)]$ contains the divergence in the propagator of a electron.

The vacuum polarization graph contributes to the propagator of the photon, which is expressed as in equation (2.13):

$$-i\frac{g_{\mu\nu}}{q^2} \longrightarrow -i(Z_3)\frac{g_{\mu\nu}}{q^2}.$$

We can then renormalize the electron and photon wave functions as

$$\psi \to \sqrt{Z_2}\,\psi_r \quad \text{and} \quad A^\mu \to \sqrt{Z_3}\,A^\mu_r, \tag{2.55}$$

so the S-matrix does not contain any of these infinities. The bare Lagrangian of equation (2.51) will now be modified to

$$\mathscr{L} = -\frac{1}{4}Z_3(F^{\mu\nu}_r)^2 + Z_2\bar{\psi}_r(i\partial\!\!\!/ - m_0)\psi_r - e_0 Z_2 Z_3^{1/2}\, \bar{\psi}_r\gamma^\mu \psi_r A_{r\mu}. \tag{2.56}$$

Let us now consider the vertex corrections. We normalized the form factor $F_1(q^2)$ in equation (2.27) to eliminate the infinite contribution. This is equivalent to redefining the vertex function, defined in equation (2.17), as

$$\Gamma_\mu(p,p') = \gamma^\mu\, F_1(q^2) + \frac{i\sigma^{\mu\nu}q_\nu}{2m}\, F_2(q^2) \longrightarrow \frac{1}{Z_1}\tilde{\Gamma}_\mu(p,p'). \tag{2.57}$$

This will then give the renormalized electric charge

$$e = Z_2 Z_3^{1/2} Z_1^{-1}\, e_0. \tag{2.58}$$

With these renormalized parameters, we can now write the Lagrangian in two parts:

$$\mathscr{L} = \mathscr{L}_r + \mathscr{L}_{ct}\,,$$

where the renormalized Lagrangian $\mathscr{L}_r$ and the Lagrangian with the counterterms $\mathscr{L}_{ct}$ are given by

$$\begin{aligned}\mathscr{L}_r &= -\frac{1}{4}(F_r^{\mu\nu})^2 + \bar{\psi}_r(i\not{\partial} - m)\psi_r - e\,\bar{\psi}_r\gamma^\mu\psi_r A_{r\mu}\\ \mathscr{L}_{ct} &= -\frac{1}{4}\delta_3(F_r^{\mu\nu})^2 + \bar{\psi}_r(i\delta_2\not{\partial} - \delta_m)\psi_r - e\delta_1\,\bar{\psi}_r\gamma^\mu\psi_r A_{r\mu}\,,\end{aligned} \tag{2.59}$$

where $\delta_3 = Z_3 - 1$, $\delta_2 = Z_2 - 1$, $\delta_m = Z_2 m_0 - m$, and $\delta_1 = Z_1 - 1$. If we write the Lagrangian to order $\alpha$, then the renormalized Lagrangian will not contain any infinities to this order. The new Feynman rules for this Lagrangian will be given in terms of the renormalized mass and electric charge of the electron. If we calculate correlation functions with the renormalized Feynman rules, then to one-loop order there will not be any divergences. However, at higher orders there will be new corrections which can be taken care of by adjusting the parameters of the counterterms again.

The most important feature of this procedure is that the same counterterms can cancel divergences to all orders and no new counterterms have to be introduced. In any nonrenormalizable theories, we have to introduce new counterterms at all orders of perturbation, which makes the theory unpredictable. Moreover, the original symmetry of the theory will not be preserved by the new counterterms. For any renormalizable theory all counterterms respect the symmetry of the original Lagrangian, and hence, the symmetry is preserved to all orders of perturbation theory.

## 2.4 Renormalization in $\phi^4$ Theory

We shall now discuss the renormalization in $\phi^4$ theory. The simplicity of the model clarifies certain points, which are not transparent in QED. The bare Lagrangian is given by

$$\mathscr{L} = \frac{1}{2}(\partial_\mu\phi)^2 - \frac{1}{2}m_0^2\phi^2 - \frac{\lambda_0}{4!}\phi^4. \tag{2.60}$$

There are two bare parameters in the model, $m_0$ and $\lambda_0$.

We shall first find the divergent diagrams in this case. If there are $E$ number of external lines, $P$ number of propagators, $V$ vertices, and $L$ loops in a diagram, then we get the relations

$$L = P - V + 1 \qquad \text{and} \qquad 4V = E + 2P\,.$$

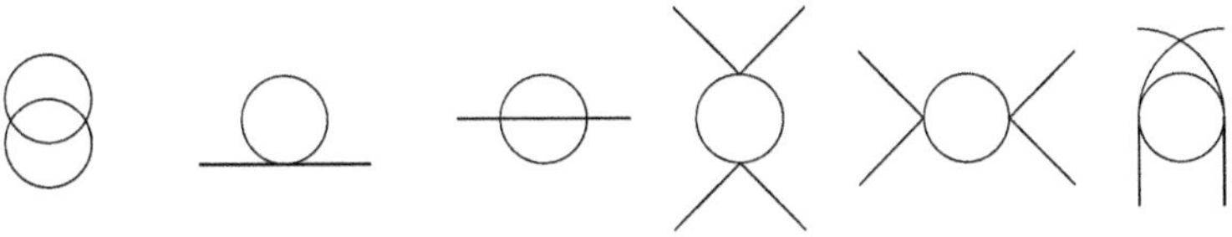

**FIGURE 2.6**
Loop diagrams in $\phi^4$ theory.

Each propagator will carry a different momentum, from which delta functions in each of the vertices reduce $V$ modulo an overall momentum giving us the number of independent momenta and, hence, the number of loops. Since only one end of the external legs is attached to any vertices and in each vertex there are four lines, we get the second relation. Since each propagator contributes $p^{-2}$ and loop integral gives $p^4$, the superficial degree of divergence is given by

$$D = 4L - 2P = 4 - E. \tag{2.61}$$

Again the superficial degree of freedom depends only on the number of external legs.

Let us now consider loop diagrams allowed in this theory. A few simple loop diagrams are given in figure 2.6. Although scalar particles are normally represented with dashed lines, for convenience of drawing we represented them using solid lines in this diagram. The first graph without external lines does not contribute to any scattering processes. The second graph with two external lines is a one-loop diagram with $L = 1, V = 1, P = 1, E = 2$ and $D = 2$, while the third diagram is a two-loop diagram with $L = 2, V = 2, P = 3, E = 2$, and, as expected, has the same superficial degree of divergence $D = 2$ as the one-loop diagram. All graphs with two external lines have $D = 2$ and, hence, are quadratically divergent. The last three diagrams are one-loop diagrams, which are logarithmically divergent since they have $L = 1, V = 2, P = 2, E = 4$, and $D = 0$.

We restrict our discussion to one-loop only. Let us consider the second diagram in figure 2.6, which is the one-loop contribution with two external legs. This is given by

$$-i\Sigma_0(p^2) = \frac{-i\lambda_0}{2} \int \frac{d^4k}{(2\pi)^4} \frac{i}{k^2 - m_0^2 + i\varepsilon}. \tag{2.62}$$

The subscript 0 in $-i\Sigma_0(p^2)$ indicates that we have written this contribution with bare parameters. Let us now consider the series

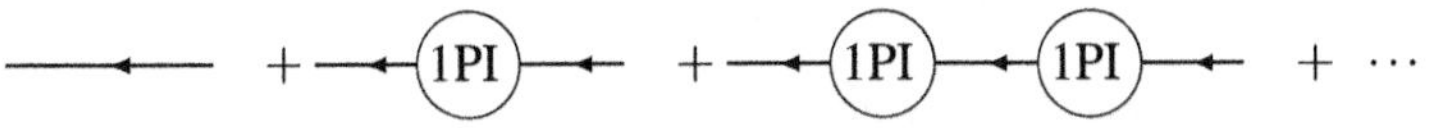


(2.63)

with each blob containing a one-loop diagram contributing $-i\Sigma_0$ to the propagator

$$\begin{aligned} i\Delta'(p) &= i\Delta_F(p) + i\Delta_F(p)[-i\Sigma_0(p^2)]i\Delta_F(p) + \cdots \\ &= \frac{i}{p^2 - m_0^2 - \Sigma_0(p^2)}. \end{aligned} \tag{2.64}$$

As in the case of QED, we can now Taylor expand $\Sigma(p^2)$ around $\Sigma(m^2)$ as

$$\Sigma(p^2) = \Sigma(m^2) + (p^2 - m^2)\Sigma'(m^2) + \cdots, \tag{2.65}$$

so the propagator can be written as

$$i\Delta'(p) = \frac{iZ_\phi}{p^2 - m^2} + [\text{terms regular in } (p^2 - m^2)]. \tag{2.66}$$

The physical mass $m$ and the renormalization factor $Z_\phi$ are given by

$$m^2 = m_0^2 + \Sigma(m^2) \quad \text{and} \quad Z_\phi = \frac{1}{1 - \Sigma'(m^2)}. \tag{2.67}$$

We shall now renormalize the wave function $\phi$ to remove $Z_\phi$ from the two-point function

$$\phi \longrightarrow \sqrt{Z_\phi}\, \phi_r. \tag{2.68}$$

The Lagrangian then becomes

$$\mathscr{L} = \frac{1}{2} Z_\phi (\partial_\mu \phi_r)^2 - \frac{1}{2} m_0^2 Z_\phi \phi_r^2 - \frac{\lambda_0}{4!} Z_\phi^2 \phi_r^4. \tag{2.69}$$

Before we define the renormalized coupling constant, we shall consider the other one-loop logarithmically divergent diagrams of figure 2.6, which contribute to the vertex function.

Let us consider the last three diagrams of figure 2.6. If the momenta of the incoming particles are $p_1$ and $p_2$, and that of the outgoing particles are $p_3$ and $p_4$, then the amplitudes of these processes can be written in terms of the Mandelstam variables $(p_1 + p_2)^2 = p^2 = s$, $(p_1 + p_3)^2 = t$ and $(p_1 + p_4)^2 = u$. The contribution of the fourth diagram is given by

$$-i\lambda_0^2 V_0(p^2) = -i\lambda_0^2 V_0(s) = \frac{i\lambda_0^2}{2} \int \frac{d^4k}{(2\pi)^4} \frac{i}{k^2 - m_0^2 + i\varepsilon} \frac{i}{(k+p)^2 - m_0^2 + i\varepsilon}, \tag{2.70}$$

where $k + p$ and $-k$ are the momenta of the scalars in the loop. The fifth and sixth diagrams of figure 2.6 will then give contributions $i\lambda^2 V_0(t)$ and $i\lambda^2 V_0(u)$. The integral $V_0$ is logarithmically divergent. The total contribution then becomes

$$i\Gamma_0 = \lambda_0 - i\lambda_0^2 [V_0(s) + V_0(t) + V_0(u)]. \tag{2.71}$$

We shall then renormalize it with another divergent factor $Z_\lambda$

$$\Gamma_0 \longrightarrow Z_\lambda^{-1} \Gamma_0.$$

The renormalized coupling constant can then be defined as

$$\lambda_0 \longrightarrow \lambda = Z_\phi^2 Z_\lambda^{-1}\, \lambda_0, \tag{2.72}$$

so this takes care of the infinities in the vertex function as well as from the Lagrangian after wave function renormalization.

We shall now eliminate the bare mass and coupling constant and write the Lagrangian with the counterterms

$$\begin{aligned}
\mathscr{L} &= \mathscr{L}_r + \mathscr{L}_{ct} \\
\mathscr{L}_r &= \frac{1}{2}(\partial_\mu \phi_r)^2 - \frac{1}{2}m^2\phi_r^2 - \frac{\lambda}{4!}\phi_r^4 \\
\mathscr{L}_{ct} &= \frac{1}{2}\delta_Z(\partial_\mu \phi_r)^2 - \frac{1}{2}\delta_m\phi_r^2 - \delta_\lambda\frac{\lambda}{4!}\phi_r^4, \qquad (2.73)
\end{aligned}$$

where $\delta_Z = Z_\phi - 1$, $\delta_m = m_0^2 Z_\phi - m^2$, and $\delta_\lambda = Z_\lambda - 1$. Thus, as per the BPHZ formalism, we constructed the counterterms to one-loop order. We shall now evaluate the one-loop diagrams with the renormalized parameters and new Feynman rules and confirm that the theory has no divergences to first order.

Let us first consider the second diagram of figure 2.6. The physical mass $m$ is now identified as the pole in the propagator, and the amplitude for the process will be

$$-i\Sigma(p^2) = \frac{-i\lambda}{2}\int \frac{d^4k}{(2\pi)^4}\frac{i}{k^2 - m^2 + i\varepsilon} + i(p^2\delta_Z - \delta_m). \qquad (2.74)$$

The last term comes from the counterterm that contributes to the amplitude. To evaluate this integral, we go to Euclidean system and then use dimensional reduction method and get

$$\begin{aligned}
-i\Sigma(p^2) &= -\frac{\lambda}{2}\int \frac{d^4k_E}{(2\pi)^4}\frac{i}{k_E^2 + m^2} + i(p^2\delta_Z - \delta_m) \\
&= -\lim_{d\to 4}\frac{i\lambda}{(4\pi)^{d/2}}\int dk_E \frac{k_E^{d-1}}{k_E^2 + m^2} + i(p^2\delta_Z - \delta_m) \\
&= -\frac{i\lambda m^2}{2(4\pi)^2}\lim_{d\to 4}\Gamma\left(1 - \frac{d}{2}\right) + i(p^2\delta_Z - \delta_m). \qquad (2.75)
\end{aligned}$$

The divergence of the integral can be seen after taking the approximation

$$\lim_{d\to 4}\Gamma\left(1 - \frac{d}{2}\right) = \lim_{d\to 4}\frac{\Gamma(3 - \frac{d}{2})}{(2 - \frac{d}{2})(1 - \frac{d}{2})} = -\lim_{\varepsilon\to 0}\frac{2}{\varepsilon},$$

where $\varepsilon = 4 - d$. Since the integral does not depend on $p^2$, we set

$$\delta_Z = 0 \quad \text{and} \quad \delta_m = -\frac{\lambda m^2}{2(4\pi)^2}\lim_{d\to 4}\Gamma\left(1 - \frac{d}{2}\right), \qquad (2.76)$$

which makes the the one-loop contribution vanish.

The one-loop integrals with four external lines should also vanish. If we decide to set $\lambda$ to be the magnitude of scattering at zero momentum, then the vertex should be

$-i\lambda$ at $s = 4m^2$, $t = u = 0$. This will then dictate us to choose, following equation (2.71),

$$\delta_\lambda = i\lambda[V(4m^2) + 2V(0)].$$

We shall now ensure that

$$i\Gamma = \lambda - i\lambda^2[V(s) + V(t) + V(u)] + \lambda\delta_\lambda \tag{2.77}$$

remains finite. We shall use dimensional regularization for evaluating the divergent integral $V$.

To compute $V$, we shall use Feynman parameters $x$, shift the variables to $\ell = k + xp$, Wick rotate to Euclidian space, and then perform dimensional regularization:

$$\begin{aligned} V(p^2) &= \frac{1}{2}\int \frac{d^4k}{(2\pi)^4} \frac{1}{k^2 - m^2 + i\varepsilon} \frac{1}{(k+p)^2 - m^2 + i\varepsilon} \\ &= \frac{i}{2}\int_0^1 dx \lim_{d\to 4} \int \frac{d^d\ell_E}{(2\pi)^d} \frac{1}{[\ell_E^2 - x(1-x)p^2 + m^2]^2} \\ &= \frac{i}{2} \lim_{d\to 4} \frac{\Gamma\left(2 - \frac{d}{2}\right)}{(4\pi)^{d/2}} \int_0^1 dx \frac{1}{[m^2 - x(1-x)p^2]^{2-d/2}}. \end{aligned} \tag{2.78}$$

Thus, the counterterm becomes

$$\delta_\lambda = -\frac{\lambda}{2} \lim_{d\to 4} \frac{\Gamma\left(2 - \frac{d}{2}\right)}{(4\pi)^{d/2}} \frac{1}{m^{4-d}} \int_0^1 dx \left(2 + \frac{1}{[1 - 4x(1-x)]^{2-d/2}}\right). \tag{2.79}$$

If we now substitute these two expressions in equation 2.77, the divergent contributions that appear in the limit $d \longrightarrow 4$ cancel out and the finite part is given by

$$i\Gamma = \lambda\left\{1 + \frac{\lambda}{32\pi^2}\int_0^1 dx \log\left(\frac{f(s)\cdot f(t)\cdot f(u)}{[1 - 4x(1-x)]}\right)\right\}, \tag{2.80}$$

where $f(q) = [m^2 - x(1-x)q]/m^2$, with $q = s, t, u$.

We have demonstrated the general principle of dealing with the infinities in field theory with explicit examples of one-loop diagrams for QED and $\phi^4$ theory. At higher loops the complications will be much more. But at least we understand the theory now, and the predictions are verified with an extremely high degree of precision.

# 3

# *Group Theory*

Symmetry plays the most important role in our understanding of nature. Many diverse phenomena may be explained by a single theory if there is some symmetry behind these phenomena. If two particles behave the same way in an environment, one can postulate that some symmetry is relating these similar behaviors. Group theory is the mathematical tool to study a set of objects, which have some underlying symmetry.

In particle physics it is now impossible to understand the modern developments without some basic understanding of group theory. The strong, the weak and the electromagnetic interactions are explained by gauge theories based on different symmetry groups and extensions of the standard model of electroweak interaction are often based on some extended symmetry groups. Although discrete groups are also used, the continuous groups are more frequently used in particle physics. I shall, therefore, restrict our discussions to continuous groups keeping in mind the applications we encounter in studying models of particle physics. In this short chapter, only a working knowledge on some aspects of group theory that are required for the rest of the book will be provided.

## 3.1 Matrix Groups

In this section we present the basics of group theory and different types of matrix groups. We shall concentrate on continuous groups and our emphasis will be on aspects of group theory that have applications in particle physics.

### Definition of a Group

A *group* $(G)$ is a collection of elements $\{A, B, C, ...\}$, on which a product operation $(\bullet)$ is defined that combines different elements satisfying the four group postulates:

1. *Closure:* The product of two elements of the group corresponds to a unique element of the group:

$$A \bullet B \subset G. \tag{3.1}$$

2. *Identity:* One of the elements ($I$) of the group, known as the identity element, satisfies

$$A \bullet I = I \bullet A = A \tag{3.2}$$

for any $A$ belonging to $G$.

3. *Inverses:* For any element $A$ in $G$, there exists an inverse element $A^{-1}$ in $G$, such that

$$A \bullet A^{-1} = I. \tag{3.3}$$

4. *Associativity:* The product of three or more elements of $G$ should not depend on the order of product

$$A \bullet (B \bullet C) = (A \bullet B) \bullet C = A \bullet B \bullet C. \tag{3.4}$$

**Discrete and Continuous Groups**

When there are finite numbers of elements in the group, the group is called a *finite group*. If there are denumerable infinite numbers of elements, an *infinite discrete group* is formed. When the elements form a continuum, the group is said to be a *continuous group*.

1. The elements $\pm 1, \pm i$ form a *finite group* $Z_4$ when the product operation is multiplication.

2. The *additive group of the integers* is formed by the set of integers when the product operation is addition. The zero corresponds to the identity element of the group.

3. The infinite set of matrices $\begin{pmatrix} \cos\theta & \sin\theta \\ -\sin\theta & \cos\theta \end{pmatrix}$ produced by the continuous variations of $\theta$ from 0 to $2\pi$ forms a *continuous group*.

As an example, a theory invariant under the $Z_4$ symmetry means that the Lagrangian is invariant under the operation of this group. If any complex scalar field $\phi$ transforms under the operation of this symmetry group as $\phi \to -i\phi$, then the allowed interaction terms in the Lagrangian will be given by

$$\mathcal{L} = m^2 \phi^\dagger \phi + \lambda (\phi^\dagger \phi)^2 + \lambda' (\phi^4 + \phi^{\dagger 4}). \tag{3.5}$$

The quadratic ($\phi^2$) and cubic ($\phi^3$) couplings are not allowed by the $Z_4$ symmetry. If there are two fields $\phi_1$ and $\phi_2$ transforming under $Z_4$ as $\phi_1 \to i\phi_1$ and $\phi_2 \to -i\phi_2$, then the quadratic terms allowed by the symmetry will be $\phi_1\phi_2 + \phi_1^\dagger\phi_1 + \phi_2^\dagger\phi_2$. There will also be quartic interactions, but all trilinear couplings will be forbidden.

**Matrix Groups**

Any set of $n \times n$ square matrices can form a *regular matrix group*; if the matrices are nonsingular, the matrix multiplication satisfies associative law and the set contains all the elements required for closure including the identity matrix

$$I = \text{diag}[1,1,1,...] = \begin{pmatrix} 1 & 0 & 0 & \cdots \\ 0 & 1 & 0 & \cdots \\ 0 & 0 & 1 & \cdots \\ \vdots & \vdots & \vdots & \ddots \end{pmatrix}. \tag{3.6}$$

*Regular matrix groups* could be *finite* or *infinite*, *discrete* or *continuous*. A matrix $A(n)$ of degree $n$ can have real $(R)$ or complex $(C)$ elements, and it can act on an *n-dimensional real vector space* $R^n$ spanned by the vector $\mathbf{x} \equiv (x_1, x_2, ..., x_n)$ or on an *n-dimensional complex vector space* $C^n$ spanned by the vector $\mathbf{z} \equiv (z_1, z_2, ..., z_n)$. The action of the matrix $A(n)$ on the real or complex vectors produces the transformations $\mathbf{x} \to \mathbf{x}'$ or $\mathbf{z} \to \mathbf{z}'$, respectively.

Let us consider a *continuous matrix group* of degree 2. The elements of the group

$$A = \begin{pmatrix} a_{11} & a_{12} \\ a_{21} & a_{22} \end{pmatrix}$$

will be restricted by the nonsingular condition $\det A = 0$, and writing the matrix elements as $a_{ij} = \delta_{ij} + \alpha_{ij}$, the identity matrix would correspond to $\alpha_{ij} = 0$. Continuous variations of the parameter $\alpha_{ij}$, which are otherwise unbounded and limited only by the nonsingular condition, will then generate the *continuous matrix group* with elements parameterized by $\alpha_{ij}$. For different types of continuous matrix groups, the number of parameters can be different. In general, the elements of a continuous matrix group can be represented by matrices of degree $n$:

$$A_n \equiv A_n(\alpha_1, ..., \alpha_r), \tag{3.7}$$

which depends on $r$ parameters $\alpha_i, i = 1, ..., r$. The identity element would correspond to $A_n(0, ..., 0) = I$.

The *matrix groups* can be of different types, depending on whether the group elements are

$$\begin{array}{ll} \textit{Symmetric} & A = A^T \\ \textit{Skew–Symmetric} & A + A^T = 0 \\ \textit{Orthogonal} & A^T = A^{-1} \\ \textit{Unitary} & A^\dagger = A^{-1} \\ \textit{Real} & A = A^* \\ \textit{Imaginary} & A = -A^* \\ \textit{Hermitian} & A = A^\dagger \\ \textit{Skew–Hermitian} & A = -A^\dagger. \end{array} \tag{3.8}$$

The matrix elements could be real or imaginary. Some of the important continuous matrix groups are:

*General Linear Group* $GL(n,C)$: The elements of the *complex general linear group* are $n \times n$ invertible matrices with complex elements. Any matrix is characterized by $n^2$ complex elements and, hence, $2n^2$ real parameters. If the matrix elements are real, we get the subgroup $GL(n,R)$ of $GL(n,C)$.

*Special Linear Group* $SL(n,C)$: Restricting the determinants of the group elements of $GL(n,C)$ to $+1$, we get the *complex special linear group* $SL(n,C)$. When the elements are real, the group becomes $SL(n,R)$ so $SL(n,R) \subset SL(n,C) \subset GL(n,C)$ and $SL(n,R) \subset GL(n,R)$.

*Unitary Groups* $U(n)$: The *Hermitian form* $z_1z_1^* + z_2z_2^* + \cdots + z_nz_n^*$ remains invariant under the action of the elements of an $n^2$-parameter *unitary group* $U(n)$ formed by the unitary matrices $A$ of degree $n$. The elements $a_{ij}$ of the unitary matrices $A$ are bounded $|a_{ij}|^2 \leq 1$ because of the unitarity condition $A^\dagger A = I$ or $a_{ik}a_{kj}^* = \delta_{ij}$.

The *unitary matrices* that leave the *Hermitian form* $-z_1z_1^* - \cdots - z_pz_p^* + z_{p+1}z_{p+1}^* + \cdots + z_{p+q}z_{p+q}^*$ invariant are denoted as $U(p,q)$, so that $U(n,0) = U(0,n) = U(n)$.

*Special Unitary Groups* $SU(n)$: When the determinant of a *unitary group* is restricted to $+1$, the resulting $(n^2-1)$-parameter group is called *special unitary group* $SU(n)$. Similarly, $SU(p,q)$ groups are obtained by restricting the elements of $U(p,q)$ groups to have determinant $+1$.

*Orthogonal Groups* $O(n)$: The group $O(n,C)$ is formed by the *complex orthogonal matrices* $A$ of degree $n$, which has $n(n-1)$ complex parameters. The orthogonality condition $A^TA = I$ implies $|A| = \pm 1$, and hence, there are two disconnected pieces of the group. *Special complex orthogonal groups* $SO(n,C)$ are formed by the orthogonal matrices of determinant $+1$, which leaves the *quadratic form* $z_1^2 + \cdots + z_n^2$ invariant and $SO(n,C) \subset O(n,C)$.

*Special Orthogonal Groups* $SO(n)$: The $n(n-1)$-parameter group formed by real orthogonal matrices of determinant $+1$ is called the *special orthogonal group* $SO(n,R)$ or simply $SO(n)$, which leaves the *real quadratic form* $x_1^2 + \cdots + x_n^2$ invariant and $SO(n) \subset O(n,R)$. The $SO(p,q)$ group ensures invariance of the *quadratic form* $-x_1^2 - \cdots - x_p^2 + x_{p+1}^2 + \cdots + x_{p+q}^2$.

*Symplectic Groups* $Sp(2n)$: Given two vectors $\mathbf{x} = (x_1, ..., x_n, x_1', ..., x_n')$ and $\mathbf{y} = (y_1, ..., y_n, y_1', ..., y_n')$ (in general, $x_i$ and $y_i$ are complex numbers) the *nondegenerate skew–symmetric bilinear form* $\sum_{1=1}^n (x_iy_i' - x_i'y_i)$ is left invariant by the *symplectic group* $Sp(2n,C)$ formed by the $2n(2n+1)$-parameter complex matrices. The $n(2n+1)$-parameter $Sp(2n,R)$ groups are formed with real matrices. The unitary subset of the group $Sp(2n,C)$ is called the *unitary symplectic group* $Sp(2n) = U(2n) \cap Sp(2n,C)$, which is also the $n(2n+1)$-parameter group.

To study group properties with matrices, we need the exponential function of a matrix, which is defined as

$$e^A = I + A + \frac{A^2}{2!} + \frac{A^3}{3!} + \cdots = \sum_{p=0}^{\infty} \frac{A^p}{p!}, \tag{3.9}$$

where $A^0 = I$. This series is convergent when the elements $(a_{ij})$ of the matrix $A$ are bounded from above $|a_{ij}| \leq \mu$. For the exponential matrices, it can be shown that

- $e^{A+B} = e^A e^B$ when $A$ and $B$ commutes,
- $Be^A B^{-1} = e^{BAB^{-1}}$,
- $e^{A^*} = (e^A)^*$,
- $e^{A^T} = (e^A)^T$,
- $e^{A^\dagger} = (e^A)^\dagger$,
- $e^{-A} = (e^A)^{-1}$, and
- $\det e^A = e^{tr\, A}$.
- If $\lambda_1, \ldots, \lambda_n$ are *characteristic roots* of $A$, then the *characteristic roots* of $e^A$ are $e^{\lambda_1}, \ldots, e^{\lambda_n}$.
- $e^A$ is *orthogonal* when $A$ is *skew–symmetric*, and
- $e^A$ is *unitary* when $A$ is *skew–Hermitian.*

## 3.2 Lie Groups and Lie Algebras

Lie groups and Lie algebras are widely used in particle physics. In this section we give a brief introduction to the subject and proceed to discuss unitary groups that will be used afterwards.

### Lie Groups

In equation (3.7) the element $A$ of a continuous group $G$ was expressed in terms of $r$ continuous parameters

$$A(\alpha) \equiv A_n(\alpha_1, \ldots, \alpha_r), \tag{3.10}$$

which includes the *identity element* $A(\mathbf{0})$. The *closure* of the group requires

$$A(\gamma) = A(\alpha)A(\beta) \equiv A(f(\alpha, \beta)) \implies \gamma = f(\alpha, \beta),$$

where $\gamma$ should be a continuously differentiable function of $\alpha$ and $\beta$, and satisfy $\gamma = f(\gamma, \mathbf{0}) = f(\mathbf{0}, \gamma)$. The *inverse element* $A(\alpha)^{-1} = A(\alpha')$ exists if $\alpha'$ is a continuously differentiable function of $\alpha$. The continuous group under consideration is referred to as a *Lie group*, if the associative condition

$$A(\alpha)(B(\beta)C(\gamma)) \equiv (A(\alpha)B(\beta))C(\gamma) \quad \Longrightarrow \quad f[\alpha, f(\beta,\gamma)] = f[f(\alpha,\beta),\gamma] \tag{3.11}$$

is also satisfied.

Let us now consider an infinitesimal transformation in the neighbourhood of the identity element. We can write any element of the group $A(\alpha)$ lying near the identity element by Taylor expansion

$$\begin{aligned} A(\alpha) &= A(\mathbf{0}) + \alpha_i \left(\frac{\partial A}{\partial \alpha_i}\right)_{\alpha_i=0} + \frac{1}{2}\alpha_i\alpha_j \left(\frac{\partial A}{\partial \alpha_i}\right)_{\alpha_i=0} \left(\frac{\partial A}{\partial \alpha_j}\right)_{\alpha_j=0} + O(\alpha^3) \\ &= A(\mathbf{0}) + \alpha_i X_i + \frac{1}{2}\alpha_i\alpha_j X_i X_j + O(\alpha^3), \end{aligned} \tag{3.12}$$

where sum over repeated indices are implied and $X_i = (\partial A/\partial \alpha_i)_{\alpha_i=0}$ are the *infinitesimal group generators* of the group $G$. For a group parameterized by $r$ continuous parameters $\alpha_i, i = 1,...,r$, there are $r$ generators $X_i$. The group is then called an *r-parameter Lie group*. The inverse element can be expanded as

$$A(\alpha)^{-1} = A(\mathbf{0}) - \alpha_i X_i + \frac{1}{2}\alpha_i\alpha_j X_i X_j + O(\alpha^3), \tag{3.13}$$

so $A(\alpha)^{-1}A(\alpha) = A(\mathbf{0}) + O(\alpha^2)$.

We now define the commutator $A(\beta,\gamma)$ of two group elements $A(\beta)$ and $A(\gamma)$ near the identity. The combined group action $A(\beta)^{-1}$ and then $A(\beta)$ brings the state to its identity element. Similarly, the action $A(\gamma)^{-1}$ and then $A(\gamma)$ is equivalent to the action $A(\mathbf{0})$. However, if the group actions $A(\beta)$ and $A(\gamma)$ do not commute near the identity, then the commutator can be defined as

$$A(\beta,\gamma) = A(\beta)^{-1}A(\gamma)^{-1}A(\beta)A(\gamma) = A(\mathbf{0}) + \beta_i\gamma_j[X_i, X_j], \tag{3.14}$$

where $[X_i, X_j] = X_iX_j - X_jX_i$ is the commutator of the group generators. Since the commutator takes the state to another neighbourhood of the identity element, we can write

$$A(\beta,\gamma) = A(\alpha) = A(\mathbf{0}) + \alpha_k X_k + \ldots, \tag{3.15}$$

and then comparing terms, we get

$$[X_i, X_j] = C_{ij}^k X_k, \tag{3.16}$$

where $\alpha_k = C_{ij}^k \beta_i \gamma_j$ and $C_{ij}^k$ are the *structure constant of the infinitesimal Lie group*. The *structure constants* satisfy

$$C_{ij}^k = -C_{ji}^k, \tag{3.17}$$

and the Jacobi identity

$$[[X_i, X_j], X_m] + [[X_j, X_m], X_i] + [[X_m, X_i], X_j] = 0 \tag{3.18}$$

implies

$$C_{ij}^k C_{mn}^l + C_{jm}^k C_{in}^l + C_{mi}^k C_{jn}^l = 0. \tag{3.19}$$

For finite group transformations, the group generators and the structure constants also remain the same.

We now consider generation of finite group elements, which can be done by repeated application of infinitesimal group transformations. Any group elements that are connected to the identity element through continuous transformations can be expressed as

$$A(\alpha) = e^{\alpha^i X_i}. \tag{3.20}$$

The sum over repeated index $i = 1, ..., r$ is implied for an $r$-parameter group generated by the generators $X_i$. This generalization to finite group transformation can be verified by expanding the exponential function and comparing it with the leading order terms of the infinitesimal transformation, or can be verified with some example.

**Lie Algebras**

An $r$-parameter *Lie group* can be generated by $r$ generators $X_i$, which span an $r$-dimensional vector space characterized by $\alpha_i X_i$, where $\alpha_i$ are real numbers. The algebra of the $r$-parameter vector space is governed by the properties of the generators given in equations (3.16), (3.17) and (3.19). Thus, $X_i$ form the *Lie algebra* of the corresponding *Lie group*. If $\alpha_i$ are finite and bounded, the group is *compact*. All complex Lie algebras are *noncompact*.

A *Lie algebra* may be defined for an $r$-dimensional vector space $A$ over a field $K$ by the condition that for a pair of vectors $X$ and $Y$, there corresponds a vector $Z = [X, Y]$ such that

$$[\alpha X + \beta Y, Z] = \alpha [X, Z] + \beta [Y, Z] \tag{3.21}$$

$$[X, Y] + [Y, X] = 0 \tag{3.22}$$

$$[X, [Y, Z]] + [Y, [Z, X]] + [Z, [X, Y]] = 0 \tag{3.23}$$

for all $\alpha, \beta, ... \in K$ and all $X, Y, Z, ..., \in A$. Under a transformation of basis $X'_k = a_k^l X_l$, where $a_k^l$ is a nonsingular matrix, the structure function transforms as

$$C'^{\,m}_{kl} = a_k^p a_l^q C_{pq}^n (a_m^n)^{-1}. \tag{3.24}$$

The algebraic property of the generators is more frequently used to understand the symmetry properties in particle physics.

A mapping $p$ of $A$ into $A'$ is *homomorphism* of $A$ into $A'$ if

$$p(\alpha X + \beta Y) = \alpha p X + \beta p Y \quad \text{for any } (X, Y \in A, \ \ \alpha, \beta \in K)$$

and

$$p[X,Y] = [p(X), p(Y)] \quad \text{for any} \quad (X, Y \in A).$$

If the mapping is one to one, then $p$ is an *isomorphism* of $A$ into $A'$. The *Lie algebras* of two different *Lie groups* could be *locally isomorphic* and have the same structure constant. In this case the groups are called *locally isomorphic* in the neighbourhood of the identity element.

We now summarize a few properties of the *Lie groups*:

When all the elements of a Lie group commute with each other, it is called an *Abelian group*. Generators of an Abelian group also commute with each other.

A *subgroup H* of $G$ contains elements of $G$ which satisfy the group postulates by themselves. A subalgebra is Abelian if all generators of the subalgebra commute with each other.

An *ideal* or *invariant subalgebra* of $A$ is formed by a subset $P$ if the commutators of the generators $X_a \in P$ with any other generators $X_i \in A$ are also generators of $P$, i.e., $[X_a, X_i] \in P$ or $[X_a, X_i] = C^b_{ai} X_b$, $(a, b \in P, i \in A)$. For a *proper ideal*, $A$ must contain elements that do not belong to $P$.

A *Lie algebra A* can be split into a *direct sum of Lie subalgebras* $A = A_1 \oplus A_2 \oplus \cdots \oplus A_n$, if any pair of subalgebras is orthogonal to each other $A_i \cap A_j = 0$. In this case, the group $G$ split into a *direct product group* $G = G_1 \otimes G_2 \otimes \cdots \otimes G_n$, where the subalgebras $A_i$ correspond to the *simple groups* $G_i$.

A *simple Lie algebra* contains no *proper ideals*, while a *semi-simple Lie algebra* contains no *Abelian ideals* except the null element $\{\mathbf{0}\}$. A *semi-simple Lie algebra A* can be written as a direct sum $A = A_1 \oplus A_2 \oplus \cdots \oplus A_n$ of its ideals $A_i$, which also form *simple Lie algebras*.

A *metric tensor* or *Killing form* is a symmetric tensor defined for any Lie group as

$$g_{mn} = g_{nm} = C^k_{ml} C^l_{nk}. \tag{3.25}$$

The condition det $|g_{mn}| \neq 0$ ensures that the Lie algebra is *semi-simple*.

The *Casimir operator* of a *Lie algebra* is defined as

$$C = g^{mn} X_m X_n, \tag{3.26}$$

which commutes with all the elements of the Lie algebra. The *Casimir operator* is defined for any *semi-simple Lie algebra*.

**Representations of Groups**

In a physical system any transformation is associated with an operator, which, acting on the state vector, transforms into another state vector. In quantum mechanics these operators are unitary operators or are represented by matrices. The symmetry group of the state vectors is, thus, mapped into a set of operators that gives a realization of the group. The mapping of the elements of a group ($X \in G$) to a set of operators ($D(X)$) would imply that the mapping preserves the multiplication law

$$D(X)D(Y) = D(X \cdot Y) \quad \forall (X,Y \in G).$$

$D(X)$ is then a representation of the element $X$ of the group $G$.

An $r$-parameter *Lie group* is determined by $r$ matrices $D_p$, which satisfies

$$[D_k, D_l] = C_{kl}^{m} D_m, \tag{3.27}$$

where $C_{kl}^{m}$ is the *structure function* of the group.

The operators $D(X)$ now represent linear transformations on a vector space $V$ spanned by the state vectors. The basis vectors of this vector space $V$ form the fundamental representation of the group, which is also the lowest dimensional nontrivial representation of the group. If the fundamental representation is $N$-dimensional, then the operators are $N \times N$ matrices. In other words, if $V$ is a $N$-dimensional vector space, then the operators $D(X)$ are $N \times N$ matrices. The dimension of the representation is given by the number of rows or columns of the matrix $D(X)$. Two representations $D(X)$ and $E(X)$ are equivalent if $MD(X)M^{-1} = E(X)$ for any constant matrix $M$.

An important representation is the *adjoint representation* of the Lie algebra $ad(X)$. The *adjoint representation* defines a linear transformation of the *Lie algebra A* onto itself:

$$ad(X): \quad Z \to [X,Z] \ \ \forall \ (Z \in A).$$

This means that for any $K \in A$,

$$[ad(Y), ad(Z)]K = ad([Y,Z])K.$$

For any $SU(n)$ group, the $n \times n$ matrices representing the generators of the group belong to the *adjoint representation* of the group, which acts on the *fundamental representations* of $n$-dimensions.

An *irreducible representation* does not have any invariant subspaces of $V$ apart from the identity. Any *reducible representation* can be expressed as the direct sum of *irreducible subrepresentations*. If any representation $D(X)$ is reducible, we can express $D(X)$ as

$$MD(X)M^{-1} = \begin{pmatrix} D_1(X) & 0 & \cdots \\ 0 & D_2(X) & \cdots \\ \vdots & \vdots & \ddots \end{pmatrix} \qquad \forall X \in G,$$

where $M$ is a nonsingular matrix.

The complex conjugate $D^*(X)$ of $D(X)$ is also a representation of $G$, and it is irreducible if $D(X)$ is irreducible. If $D(X)$ and $D^*(X)$ are not equivalent, then $D(X)$ is complex. If they are equivalent

$$D(X) = CD^*(X)C^{-1},$$

then $D(X)$ is *real positive* or *real* if the constant matrix is symmetric ($C = C^T$), or else $D(X)$ is *real negative* or *pseudo-real* if the constant matrix is antisymmetric ($C = -C^T$). For real $D(X)$ there exists a transformation $R(X) = UD(X)U^{-1}$, such that $R(X) = R^*(X)$. This is not possible for a pseudo-real $D(X)$. However, there exists a transformation $R(X) = UD(X)U^{-1}$, such that $ZR(X) = R^*(X)Z$ where $Z = \begin{pmatrix} A & 0 & \cdots \\ 0 & A & \cdots \\ \vdots & \vdots & \ddots \end{pmatrix}$ with $A = \begin{pmatrix} 0 & -1 \\ 1 & 0 \end{pmatrix}$ as the diagonal entries and all other entries are equal to 0.

Since we are familiar with unitary operators in Hilbert space, we shall now discuss some of the group properties when the operators are unitary. It is convenient to write equation (3.20) as

$$A(\theta) = e^{i\theta^k Z_k} \tag{3.28}$$

with real parameters $\theta^k$, so the *infinitesimal group generators* are related by $Z_k = -iX_k$ and $Z_k = -i(\partial A/\partial\theta_k)_{\theta_k=0}$ are Hermitian. The generators now satisfy the Lie algebra

$$[Z_k, Z_l] = iC^p_{kl} Z_p\,, \tag{3.29}$$

where $C^p_{kl}$ is the *structure constant.* We can now define a *matrix representation* of the group, where the structure constants generate the *adjoint representation* $T_k$ of the algebra

$$C^p_{kl} = i\,(T_k)^p_l\,, \tag{3.30}$$

which satisfies

$$[T_k, T_l] = iC^p_{kl} T_p\,. \tag{3.31}$$

For a semi-simple group, we normalize the generators as

$$\mathrm{tr}(T_k T_l) = \lambda\,\delta_{kl}\,. \tag{3.32}$$

Once the generators are normalized for the fundamental representation, the value of $\lambda$ for all other irreducible representations is determined uniquely. For any representations $R$, $\lambda(R)$ is called the *index* for the representation. In this basis the structure constant becomes

$$C^p_{kl} = -i\lambda^{-1}\mathrm{tr}\,(T^p[T_k, T_l])\,, \tag{3.33}$$

implying it is completely antisymmetric in all three indices.

## 3.3 $SU(2)$, $SU(3)$ and $SU(n)$ Groups

*Special unitary groups* are the most common groups used in particle physics. The standard model requires knowledge of $SU(2)$ and $SU(3)$, while grand unified theories requires knowledge of $SU(5)$. There are other extensions of the standard model that require knowledge of special unitary groups.

$SU(n)$ is a continuous matrix group. The group is generated by $n \times n$ matrices, which are in the adjoint representation of the group and acts on the $n$-dimensional *fundamental representation* of the group. There are $(n-1)$ diagonal generators, which determine the rank of the group to be $(n-1)$. It is an $(n^2-1)$-parameter group with elements

$$U = e^{i\alpha_k T_k}, \tag{3.34}$$

where sum over repeated index $k = 1, \ldots, n^2 - 1$ is implied. The $n^2 - 1$ generators $T_k$ of the group are $n \times n$ traceless Hermitian matrices.

### $SU(2)$ Group

We start our discussions with the simplest non-Abelian group $SU(2)$. It is the group of $2 \times 2$ unitary matrices with determinant $+1$. Thus, the generators are $2 \times 2$ traceless Hermitian matrices. Out of the 8 elements of a $2 \times 2$ complex matrix, Hermiticity eliminates 4 elements and traceless condition leaves 3 independent elements. Thus, $SU(2)$ is a 3-parameter group. Only one of the generators is diagonal, so the rank of the group is 1.

A convenient choice for the generators of the group is

$$J_a = \frac{1}{2}\,\tau_a \qquad a = 1,2,3, \tag{3.35}$$

where $\tau_a$ are the Pauli matrices

$$\tau_1 = \begin{pmatrix} 0 & 1 \\ 1 & 0 \end{pmatrix}, \quad \tau_2 = \begin{pmatrix} 0 & -i \\ i & 0 \end{pmatrix}, \quad \tau_3 = \begin{pmatrix} 1 & 0 \\ 0 & -1 \end{pmatrix}. \tag{3.36}$$

The generators satisfy the commutation relation

$$[J_a, J_b] = i f_{abc} J_c\,, \tag{3.37}$$

where $f_{abc} = \varepsilon_{abc}$ is the totally antisymmetric Levi–Civita symbol with $\varepsilon_{123} = 1$. The complex conjugates $J_a^*$ of the generators $J_a$ are related to each other by the transformations

$$M J_a^* M^{-1} = -J_a, \quad \text{where } M = i\tau_2 = \varepsilon_{ij} = \begin{pmatrix} 0 & 1 \\ -1 & 0 \end{pmatrix}.$$

Thus, all representations of $SU(2)$ are *pseudo-real.*

This Lie algebra is the same as that of the angular momentum, and hence, we shall assume some results from our experience with angular momentum. The *quadratic Casimir invariant* of the group corresponds to the total angular momentum operator

$$J^2 = J_1^2 + J_2^2 + J_3^2, \tag{3.38}$$

which commutes with all other generators of the group.

The fundamental representation of the group is a 2-dimensional representation, *doublet*, on which the generators $J_a$ act. The two basis states are

$$u_+ = \begin{pmatrix} 1 \\ 0 \end{pmatrix} \quad \text{and} \quad u_- = \begin{pmatrix} 0 \\ 1 \end{pmatrix}, \tag{3.39}$$

so the diagonal generator, $J_3$, acting on these states, gives eigenvalues $+1/2$ and $-1/2$, respectively. Any state $\psi$ that transforms as a *doublet* under the group $SU(2)$ can be written as

$$\psi = \psi_1 u_+ + \psi_2 u_- = \begin{pmatrix} \psi_1 \\ \psi_2 \end{pmatrix}. \tag{3.40}$$

In case of angular momentum, the states $u_\pm$ are written in terms of their total angular momentum and their eigenvalues, $u_+ = |1/2,\ +1/2\rangle$ and $u_- = |1/2,\ -1/2\rangle$.

The *index* $\lambda$ of the *fundamental representation* is $\lambda = \mathrm{tr}(J_a J_a) = 1/2$ for all $a$ (not sum over $a$). This gives the normalization of the operators. In fact, we shall follow the convention that for any $SU(n)$ the generator of the *fundamental representation* is normalized to $1/2$. This will imply that the index for the *adjoint representation* of $SU(n)$ is $n$.

The *raising and lowering operators* can be defined as

$$J_+ = (J_1 + i\, J_2) = \begin{pmatrix} 0 & 1 \\ 0 & 0 \end{pmatrix} \tag{3.41}$$

and

$$J_- = (J_1 - i\, J_2) = \begin{pmatrix} 0 & 0 \\ 1 & 0 \end{pmatrix}, \tag{3.42}$$

respectively, so

$$J_+ u_+ = 0, \quad J_- u_+ = u_-,$$

$$J_+ u_- = u_+, \quad J_- u_- = 0.$$

In terms of the raising and lowering operators, the commutation relations become

$$[J_+, J_-] = 2J_3 \qquad [J_\pm, J_3] = \mp J_\pm. \tag{3.43}$$

The raising and lowering operators commute with the *quadratic Casimir invariant* or the total angular momentum operator.

It is possible to construct any state vector of the higher-dimensional representations starting from the fundamental representation by taking direct products of the states. Consider two doublet states

$$\psi = \psi_1 u_+ + \psi_2 u_-$$

$$\phi = \phi_1 u_+ + \phi_2 u_-. \tag{3.44}$$

If we take a direct product of these states, different combinations of these states will belong to different irreducible representations. We would like to find out which combinations of the component states belong to which irreducible representations of the group $SU(2)$. There are several ways to find these irreducible states. One may write down the combinations, $\psi_i\phi_j$, and act on them by the operators of the group and see which combinations are irreducible and which are the $J^2$ and $J_3$ quantum numbers of the states. For example,

$$J_3(\psi_1\phi_1) = (\psi_1\phi_1)$$

implies that the state $(\psi_1\phi_1)$ corresponds to $J = 1, J_3 = 1$. We can act on this state by lowering operators to get other states with $J = 1, J_3 = +1, 0, -1$, and $J = 0, J_3 = 0$. The procedure is exactly the same as addition of angular momentum in quantum mechanics. Finally, we get

$$2 \otimes 2 = 1 \oplus 3,$$

i.e., the two doublets combines to give one singlet representation given by

$$\Phi_1 \equiv \frac{1}{\sqrt{2}}(\psi_1\phi_2 - \psi_2\phi_1)$$

and a triplet representation with three components

$$\Phi_3 \equiv \begin{pmatrix} \psi_1\phi_1 \\ \frac{1}{\sqrt{2}}(\psi_1\phi_2 + \psi_2\phi_1) \\ \psi_2\phi_2 \end{pmatrix}.$$

The *singlet* is an antisymmetric combination of the two states and can be written as $\Phi_1 = \frac{1}{\sqrt{2}}\varepsilon_{ij}\psi_i\phi_j$, where $\varepsilon_{ij}$ is the totally antisymmetric tensor with $\varepsilon_{12} = 1$

$$\varepsilon = \begin{pmatrix} 0 & 1 \\ -1 & 0 \end{pmatrix},$$

and the *triplet* correspond to the three symmetric states.

It is possible to express the *triplet* state in two equivalent representations. Since both the representations are widely used depending on convenience, we shall discuss them both. The *basis vector* for the *triplet representation* could be chosen to be

$$v_+ = \begin{pmatrix} 1 \\ 0 \\ 0 \end{pmatrix}, \quad v_0 = \begin{pmatrix} 0 \\ 1 \\ 0 \end{pmatrix}, \quad v_- = \begin{pmatrix} 0 \\ 0 \\ 1 \end{pmatrix}, \tag{3.45}$$

so we can write

$$\Phi_3 = \psi_1\phi_1 v_+ + \frac{1}{\sqrt{2}}(\psi_1\phi_2 + \psi_2\phi_1)v_0 + \psi_2\phi_2 v_-.$$

In this basis, the generators of the group will be $3 \times 3$ matrices, given by

$$J_1 = \frac{1}{\sqrt{2}}\begin{pmatrix} 0 & 1 & 0 \\ 1 & 0 & 1 \\ 0 & 1 & 0 \end{pmatrix}, \quad J_2 = \frac{1}{\sqrt{2}}\begin{pmatrix} 0 & -i & 0 \\ i & 0 & -i \\ 0 & i & 0 \end{pmatrix}, \quad J_3 = \begin{pmatrix} 1 & 0 & 0 \\ 0 & 0 & 0 \\ 0 & 0 & -1 \end{pmatrix}, \tag{3.46}$$

which satisfies the Lie algebra of $SU(2)$.

It is also possible to consider the basis vector for the *triplet state* in a $2 \times 2$ representation as

$$\begin{aligned} T_1 &= -J_+\varepsilon = -(J_1 + i\,J_2)\varepsilon = \begin{pmatrix} 1 & 0 \\ 0 & 0 \end{pmatrix}, \\ T_2 &= \frac{1}{\sqrt{2}} J_3 \varepsilon = \frac{1}{\sqrt{2}} \begin{pmatrix} 0 & 1 \\ 1 & 0 \end{pmatrix}, \\ T_3 &= J_-\varepsilon = (J_1 - i\,J_2)\varepsilon = \begin{pmatrix} 0 & 0 \\ 0 & 1 \end{pmatrix}, \end{aligned}$$

where $\varepsilon$ is the totally antisymmetric matrix. In this case we can write the combination of the two states as a triplet in a compact form

$$\Phi_3 = \psi_i^T (T_a)_{ij} \phi_j.$$

Sometimes the normalization factors are absorbed in the definition of the coupling constant and $T_a$ is replaced by the symmetric matrix $\tau_a \varepsilon$, where $J_a$ and $\tau_a$ were defined in equations (3.35) and (3.36).

Any irreducible representation can be written as $|j,m\rangle$, where

$$J^2 |j,m\rangle = j(j+1)|j,m\rangle \quad \text{and} \quad J_3 |j,m\rangle = m|j,m\rangle.$$

The representation has dimension $2j+1$. The lowering and raising operators act on these states as

$$J_\pm \, |j,m\rangle = [(j \mp m)(j \pm m + 1)]^{1/2} \, |j, m \pm 1\rangle,$$

where $|m| \le j$. It is possible to graphically represent the different irreducible representations for convenience. For $SU(2)$ it will be one-dimensional with equally spaced points representing the different values of $m$. The lowering and raising operators $J_+$ and $J_-$ take from one of the points to its adjacent points:

$$\begin{array}{llcccc} j = \frac{1}{2}, & m = & -\frac{1}{2} & \frac{1}{2} & & \\ & & \otimes & \otimes & & \\ j = 1, & m = & -1 & 0 & 1 & \\ & & \otimes & \otimes & \otimes & \\ j = \frac{3}{2}, & m = & -\frac{3}{2} & -\frac{1}{2} & \frac{1}{2} & \frac{3}{2} \\ & & \otimes & \otimes & \otimes & \otimes \end{array}$$

$$\begin{array}{llccccccc} & & & & \leftarrow J_- & & J_+ \rightarrow & & \\ j = j, & m = & -j & \cdots & m-1 & m & m+1 & \cdots & j \\ & & \otimes & \cdots & \otimes & \otimes & \otimes & \cdots & \otimes \end{array}$$

## $SU(3)$ Group

The $SU(3)$ group is an 8-parameter rank-2 group of $3 \times 3$ unitary matrices. The fundamental representation is 3-dimensional and the basis states can be chosen to be

$$v_1 = \begin{pmatrix} 1 \\ 0 \\ 0 \end{pmatrix}, \quad v_2 = \begin{pmatrix} 0 \\ 1 \\ 0 \end{pmatrix}, \quad v_3 = \begin{pmatrix} 0 \\ 0 \\ 1 \end{pmatrix}, \tag{3.47}$$

so any vector can be written as $\psi = \psi_k v_k$. The special unitary matrices that act on these states can be written as

$$U(\alpha_a) = e^{i\alpha_a T_a}, \tag{3.48}$$

where $\alpha_a, a = 1, ..., 8$ are real numbers and $T_a$ are the 8 Hermitian generators of the group. A conventional choice for the generators is given in terms of the Gell–Mann $\lambda$-matrices,

$$T_a = \frac{1}{2}\lambda_a \tag{3.49}$$

with

$$\lambda_1 = \begin{pmatrix} 0 & 1 & 0 \\ 1 & 0 & 0 \\ 0 & 0 & 0 \end{pmatrix} \quad \lambda_2 = \begin{pmatrix} 0 & -i & 0 \\ i & 0 & 0 \\ 0 & 0 & 0 \end{pmatrix} \quad \lambda_3 = \begin{pmatrix} 1 & 0 & 0 \\ 0 & -1 & 0 \\ 0 & 0 & 0 \end{pmatrix}$$

$$\lambda_4 = \begin{pmatrix} 0 & 0 & 1 \\ 0 & 0 & 0 \\ 1 & 0 & 0 \end{pmatrix} \quad \lambda_5 = \begin{pmatrix} 0 & 0 & -i \\ 0 & 0 & 0 \\ i & 0 & 0 \end{pmatrix}$$

$$\lambda_6 = \begin{pmatrix} 0 & 0 & 0 \\ 0 & 0 & 1 \\ 0 & 1 & 0 \end{pmatrix} \quad \lambda_7 = \begin{pmatrix} 0 & 0 & 0 \\ 0 & 0 & -i \\ 0 & i & 0 \end{pmatrix} \quad \lambda_8 = \frac{1}{\sqrt{3}}\begin{pmatrix} 1 & 0 & 0 \\ 0 & 1 & 0 \\ 0 & 0 & -2 \end{pmatrix}.$$

The generators are normalized to $\mathrm{tr}(T_a T_b) = 1/2\ \delta_{ab}$, and they satisfy the commutation relation

$$[T_a, T_b] = i\, f_{abc}\, T_c, \tag{3.50}$$

where $f_{abc}$ is totally antisymmetric, and is given by

$$f_{123} = 1, \quad f_{458} = f_{678} = \sqrt{3/2}$$

$$f_{147} = f_{165} = f_{246} = f_{257} = f_{345} = f_{376} = 1/2.$$

The different states can be identified by the eigenvalues of the two diagonal generators $T_3$ and $T_8$. Thus, we can represent the basis vectors with the eigenvalues of the two diagonal generators as

$$v_1 \equiv \left| \frac{1}{2}, \frac{1}{2\sqrt{3}} \right\rangle, \quad v_2 \equiv \left| -\frac{1}{2}, \frac{1}{2\sqrt{3}} \right\rangle, \quad v_3 \equiv \left| 0, -\frac{1}{\sqrt{3}} \right\rangle.$$

For a complex conjugate state the corresponding basis vectors would be

$$v_1^* \equiv \left| -\frac{1}{2}, -\frac{1}{2\sqrt{3}} \right\rangle, \quad v_2^* \equiv \left| \frac{1}{2}, -\frac{1}{2\sqrt{3}} \right\rangle, \quad v_3^* \equiv \left| 0, \frac{1}{\sqrt{3}} \right\rangle.$$

Any higher-dimensional representations can be constructed from products of these basis vectors.

The generators

$$T_1 \equiv I_1, \; T_2 \equiv I_2, \; T_3 \equiv I_3$$

form an $SU(2)$ subalgebra. Two more $SU(2)$ embeddings are also possible in $SU(3)$ with the generators

$$T_4 \equiv V_1, \;\; T_5 \equiv V_2, \;\; (\sqrt{3}T_8 + T_3) \equiv V_3$$

and

$$T_6 \equiv U_1, \;\; T_7 \equiv U_2, \;\; (\sqrt{3}T_8 - T_3) \equiv U_3,$$

but they are not orthogonal to each other. For each of these $SU(2)$ embeddings one can construct the raising and lowering operators and with these generators construct the different irreducible states. We, thus, define

$$I_\pm = T_1 \pm iT_2 = I_1 \pm iI_2, \quad V_\pm = T_4 \pm T_5 = V_1 \pm V_2,$$
$$U_\pm = T_6 \pm T_7 = U_1 \pm U_2.$$

To compare between different notations, we further define

$$H_1 = T_3 \quad \text{and} \quad H_2 = \frac{2}{\sqrt{3}} T_8.$$

The generators $I_\pm$ interchange the states $v_1$ and $v_2$ and change the $I_3$ quantum number by $\pm 1$. Similarly, $V_\pm$ interchange the states $v_1$ and $v_3$, $U_\pm$ interchange between $v_2$ and $v_3$.

Since any state can be identified by the two quantum numbers given by the eigenvalues $x$ and $y$ of the generators $H_1$ and $H_2$, it is possible to plot every state in an irreducible representation in the $H_1 - H_2$ plane. In terms of the $(x,y)$ quantum numbers, the three basis vectors correspond to the points

$$v_1(1/2, 1/3), \quad v_2(-1/2, 1/3), \quad v_3(0, -2/3),$$

and the conjugate states correspond to the points

$$v_1^*(-1/2, -1/3), \quad v_2^*(1/2, -1/3), \quad v_3^*(0, 2/3).$$

Since the $I, V, U$ raising and lowering operators relate the different basis states, the action of these operators can be given by

$$I_+ : (x \to x + 1, \;\; y \to y);$$
$$V_+ : (x \to x - 1/2, \; y \to y + 1);$$
$$U_+ : (x \to x + 1/2, \; y \to y + 1).$$

Thus, any irreducible representation of $SU(3)$ can be given by two numbers $(p,q)$. If we represent a state in the fundamental representation, it will be a triangle $(1,0)$, while a conjugate state will be an inverted triangle $(0,1)$. Any higher-dimensional representations will be a hexagon with three sides of length $p$ and the other three sides of length $q$. Any representations $(p,0)$ or $(0,q)$ will be of triangular shape. The length of the sides will be given by the numbers $p$ and $q$, which will, thus, determine the multiplicity of the irreducible representation. Although this is a simple way to determine the irreducible representations, it is not easy to use this technique for higher groups, and this technique is not adequate for many applications.

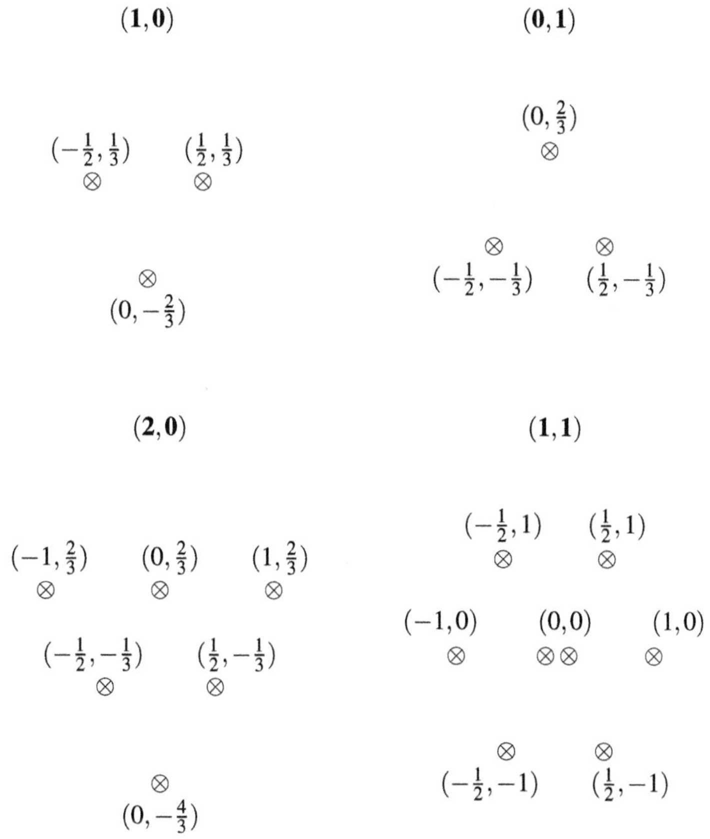


## $SU(n)$ Group

We shall now discuss some general methods to study an $SU(n)$ group. One can generalize the concept of the raising and the lowering operators for any $SU(n)$ group, which allows us to work in the Cartan–Weyl representation. Consider an $r$-element

Lie algebra of rank-$l$. There will be $l$ generators that commute with each other, which we denote as $H_s, s = 1, ..., l$. $H_s$ will then span an $l$-dimensional subspace $C$ of the $r$-dimensional vector space. The set $H_s$ is called the Cartan subalgebra $C$. If the generators $H_s$ of the Cartan subalgebra is diagonalized in the basis $|\mu, D\rangle$ for the irreducible representation $D$, then we can write

$$H_s|\mu, D\rangle = \mu_s|\mu, D\rangle\,, \tag{3.51}$$

where $\mu$ is the weight vector with components $\mu_s$.

For an $SU(n)$ group there could be $l = n-1$ commuting generators that form the subspace $C$. For $SU(3)$, there are two: $H_1$ and $H_2$. The remaining $r - l = n(n-1)$ even number of generators that span the $n(n-1)$-dimensional vector subspace can be written as raising and lowering operators. In the adjoint representation the states $|X_a\rangle$ correspond to the generators $(X_a)$ and the action of the generators on these states is

$$X_a|X_b\rangle = |[X_a, X_b]\rangle.$$

The states belonging to the Cartan subalgebra commute with each other, and hence, these are the states with weight zero $H_s|H_l\rangle = 0$. Diagonalizing the remaining $n(n-1)$-dimensional space gives the $n(n-1)/2$ states $|T_a\rangle$ that satisfies

$$H_s|T_a\rangle = a_s|T_a\rangle, \tag{3.52}$$

and the corresponding generators would satisfy the commutation relation

$$[H_s, T_a] = a_s T_a\,. \tag{3.53}$$

These weights $a_s$ are called roots. These generators are not Hermitian $T_a^\dagger = T_{-a}$. They are the lowering and raising operators.

For every nonvanishing raising operator $T_a$, there is a corresponding nonvanishing lowering operator $T_{-a}$, which can be rephrased as: if $T_a$ is a raising operator and $T_b$ is the corresponding lowering operator, then $a+b=0$. In this basis, the commutator relations of the generators are given by

$$\begin{aligned}
[H_s, H_l] &= 0 \quad \text{for } s, l \in C \\
[H_s, T_a] &= f_{saa} T_a = a_s T_a \quad \text{for } s \in C \\
[T_a, T_b] &= \sum_{s \in C} f_{abs} H_s = \sum_{s \in C} a_s H_s \quad \text{for } (a+b) = 0 \\
[T_a, T_b] &= \begin{cases} f_{aba+b} T_{a+b} & \text{if } (a+b) \neq 0 \text{ is a root}. \\ 0 & \text{otherwise} \end{cases}
\end{aligned} \tag{3.54}$$

Application of $T_a$ on the state $|\mu, D\rangle$ with weight vector $\mu$ and irreducible dimensions $D$ would raise one of the weights $\mu_i$. Repeated application of $T_a$ or $T_{-a}$ will, thus, eventually make it vanish. If $p$ repeated action of $T_a$ or $q$ repeated action of $T_{-a}$ annihilates the state, we can write

$$T_a|\mu + pa, D\rangle = 0, \quad T_{-a}|\mu - qa, D\rangle = 0$$

for positive integers $p$ and $q$. Operating on these states with raising and lowering operators, we can obtain the entire spectrum of the irreducible representations.

We shall next study the tensor method in $SU(n)$. Let us consider a vector in the fundamental representation of the group $\psi_k = \{\psi_1, ..., \psi_n\}$, which can be mapped into another vector in the same vector space by the unitary transformations generated by the generators of the group given in equation (3.34. Writing the transformation matrix as $U_{kl}$, the transformation can be written as

$$\psi_k \to \psi'_k = U_{kl}\psi_l. \tag{3.55}$$

The fundamental representation of dimension $n$ is also referred to as simply **n**. Similarly for a conjugate representation $\psi_k^*$, which is referred to as $\mathbf{n}^*$, the corresponding transformation becomes

$$\psi_k^* \to \psi'^*_k = U_{kl}^*\psi_l^* = \psi_l^* U_{lk}^\dagger. \tag{3.56}$$

For a representation with both regular as well as conjugate indices, it becomes difficult to distinguish the different indices. It is, thus, convenient to use a convention

$$\psi^k \equiv \psi_k^*, \quad U_k{}^l \equiv U_{kl} \quad \text{and} \quad U^k{}_l = U_{kl}^*.$$

In this notation

$$\psi_k \to \psi'_k = U_k{}^l \psi_l \qquad \psi^k \to \psi'^k = U^k{}_l \psi^l. \tag{3.57}$$

Higher rank tensors can be formed with the upper and lower indices, which would correspond to higher-dimensional representations formed from the fundamental representation and its conjugate.

We now define the Kronecker delta function as

$$\delta_k{}^l = \delta_{kl} = \begin{cases} 1 & \text{if } k = l \\ 0 & \text{otherwise.} \end{cases} \tag{3.58}$$

The unitarity condition then translates into

$$U^k{}_l\, U^l{}_m = \delta^k{}_m \quad \text{and} \quad \delta^k{}_l = U^k{}_m\, U^n{}_l\, \delta^m{}_n.$$

Thus, the Kronecker delta function is an invariant tensor. The product of a vector $\psi_k$ and its conjugate $\psi^k$ will, in general, be a $3 \times 3$ matrix, from which the trace will form a separate 1-dimensional representation, which is invariant under any $SU(n)$ transformations. The singlet state can be projected out with the Kronecker delta function, which defines the $SU(n)$ invariant scalar product

$$(\psi, \phi) = \psi^k\, \phi_l\, \delta^l{}_k = \psi^k \phi_k. \tag{3.59}$$

In general, the Kronecker delta function can contract indices reducing the rank of a tensor

$$\psi^{abc..}_{pqr..}\, \delta^p_a = \psi^{bc..}_{qr..}.$$

When all the indices of any tensor are contracted, the tensor becomes invariant under any group transformations. It is then said to be a singlet of the group $SU(n)$.

We shall now define the Levi–Civita symbol for the group $SU(n)$ as

$$\varepsilon^{k_1..k_n} = \varepsilon_{k_1..k_n} = \begin{cases} 1 & \text{for } (\mathrm{k_1..k_n}) \equiv (1..\mathrm{n}) \text{ or even permutations} \\ -1 & \text{for } (\mathrm{k_1..k_n}) \text{ to be odd permutations of } (1..\mathrm{n}) \\ 0 & \text{otherwise.} \end{cases} \tag{3.60}$$

It is an invariant tensor and can be used to relate the regular states with conjugate states

$$\psi^{k_r...k_n} = \varepsilon^{k_1 k_2...k_n} \phi_{k_1 k_2...k_{r-1}}.$$

The Levi–Civita symbol allows contraction of indices, and when it contracts all the indices of a tensor, it results in an invariant singlet.

## 3.4 Group Representations

In this section we shall try to explain how to find the dimensions of any irreducible representations and the decomposition of products of any two representations in terms of irreducible representations. We shall then discuss how these representations may be related to the representations of any semi-simple subgroup.

### Irreducible Representations

Any group representations are formed by combining the regular states and conjugate states as direct products. If we write a composite state with several indices, in general, the state will be reducible. First we identify the properties of the composite states that make them reducible.

Consider first the case of $SU(3)$, and then we shall generalize the result to other groups. There are two invariant tensors in $SU(3)$, the Kronecker delta $\delta_l^k$ and the Levi–Civita symbol $\varepsilon^{ijk} = \varepsilon_{ijk}$. Thus, if any two upper or two lower indices of a tensor are antisymmetric, then we can contract them with the Levi–Civita symbol. For example, the antisymmetric tensor $\psi_{ij}$ can be written as $\phi^k = \varepsilon^{ijk}\psi_{ij}$, which is a conjugate vector. Similarly since all states are traceless, the trace of any tensor can be written with the delta function, which is equivalent to contracting the indices with the delta function.

With this introduction we can now work out the product decompositions for the group $SU(3)$ and find the irreducible representations. Let us start with two vectors $\psi_k$ and $\phi_k$ belonging to the fundamental representations **3**. We can write the product as

$$\psi_i\phi_j = \frac{1}{2}(\psi_i\phi_j + \psi_j\phi_i) + \frac{1}{2}(\psi_i\phi_j - \psi_j\phi_i) = \Psi_{\{ij\}} + \Phi^k.$$

Thus, the product can be written as a rank-2 symmetric tensor with two symmetric lower indices $\Psi_{\{ij\}}$ and another conjugate vector state $\Phi^k = \frac{1}{2}\varepsilon^{ijk}\psi_i\phi_j$. This product decomposition can be written as

$$\mathbf{3} \times \mathbf{3} = \mathbf{6} + \mathbf{3}^*.$$

Let us next consider the product of a vector $\psi_k$ belonging to the fundamental representation **3** and its conjugate $\phi^k$ belonging to a $\mathbf{3}^*$. One possible combination is the scalar product, which is the trace and corresponds to a invariant singlet state, which is obtained by contracting the indices with a Kronecker delta function. Thus, we can write the product as

$$\psi_k \phi^l = [\psi_k \phi^l - \frac{1}{3} \delta_k^l \psi_m \phi^m] + \frac{1}{3} \delta_k^l \psi_m \phi^m.$$

The product decomposition, thus, becomes

$$\mathbf{3} \times \mathbf{3}^* = \mathbf{1} + \mathbf{8}.$$

Essentially we take out the traceless symmetric tensor and contract the remaining components with the delta function or the Levi–Civita symbol and write the irreducible representations.

For an $SU(n)$ group we generalize the concept of symmetric and antisymmetric states along with the traceless condition and find the irreducible representations. Let us first start with a tensor $\psi_{kl}$ and consider its transformation under the action of the group generators

$$\psi_{kl} \to \psi'_{kl} = U_k^p U_l^q \psi_{pq}.$$

If the tensor $\psi_{kl}$ is symmetric, then it remains symmetric even after the transformation. If the tensor is antisymmetric $\psi_{kl} = -\psi_{lk}$, then the transformed tensor is also antisymmetric $\psi'_{pq} = -\psi'_{qp}$. In other words, the permutation operator commutes with the generators of $SU(n)$.

Thus, we can generalize the product decomposition rule of two fundamental representations $\psi_k$ and $\phi_k$ as

$$\psi_i \phi_j = \frac{1}{2}(\psi_i \phi_j + \psi_j \phi_i) + \frac{1}{2}(\psi_i \phi_j - \psi_j \phi_i).$$

The two states belonging to the fundamental representations **n** combine to a rank-2 symmetric state with components $n(n+1)/2$ and an antisymmetric tensor of dimension $n(n-1)/2$, so the product decomposition for these states can be written as

$$\mathbf{n} \times \mathbf{n} = \frac{\mathbf{n(n+1)}}{\mathbf{2}} + \frac{\mathbf{n(n-1)}}{\mathbf{2}}.$$

Similarly the product of a vector in the fundamental representation and a conjugate vector decompose into a singlet and an adjoint representation

$$\psi_k \phi^l = [\psi_k \phi^l - \frac{1}{n} \delta_k^l \psi_m \phi^m] + \frac{1}{n} \delta_k^l \psi_m \phi^m,$$

so the product decomposition reads

$$\mathbf{n} \times \mathbf{n}^* = (\mathbf{n}^2 - \mathbf{1}) + \mathbf{1}.$$

For any group $SU(n)$, the adjoint representation has a dimension $n^2 - 1$. In general if the symmetries of the indices can be kept track of, it may be possible to determine the irreducible representations from the product of two tensor states. This is conventionally done in a pictorial representation, known as the Young Tableaux.

The main rules for the Young tableaux are the following:

A tensor with $r$ lower indices is denoted by $r$ boxes.

If the tensor is symmetric in all the indices, the boxes are all placed in a row.

If all the indices of a tensor are antisymmetric, the boxes are all in a column.

Numbers of rows do not increase while going from top to bottom. Any lower rows should have lesser numbers of boxes.

Indices appearing in the same row are subject to symmetrization.

Indices appearing in the same column are subject to antisymmetrization.

Indices for conjugate states are lowered using the Levi–Civita symbol. So, a conjugate state in the fundamental representation will be represented by $n-1$ boxes in a column.

If the length of the rows of a Young tableaux is denoted by $(f_1, f_2, ..., f_{n-1})$, then the Young tableaux can be characterized by $(\lambda_1, ..., \lambda_{n-1})$, and its conjugate state is represented by, $(\lambda_{n-1}, ..., \lambda_1)$, where $\lambda_1 = f_1 - f_2, \lambda_2 = f_2 - f_3, ...$, and $\lambda_{n-1} = f_{n-1}$.

Antisymmetrization implies that there can be at most $n-1$ rows in a column. If there are $n$ rows in a column, the Levi–Civita symbol can contract all the $n$ antisymmetric indices leaving a trivial singlet representation of the group.

When two Young tableaux combine, the resultant tableaux should preserve the symmetry property of the boxes in the original tableaux.

Once the tableaux are formed, associate two numbers with each box:

1. Associate a number $D_i$ to all the boxes, which gives the distance to the box in the first column, first row, counting +1 for each step toward right and $-1$ for each step downward.

2. Associate a number $h_i$ with every box, which counts the numbers of boxes on the right in the same row + numbers of boxes below the box in the same column + 1 (for the box itself).

$$D_i: \begin{array}{|c|c|c|c|}\hline 0 & 1 & 2 & 3 \\ \hline -1 & 0 & 1 \\ \cline{1-3} -2 & -1 & 0 \\ \cline{1-3} -3 & -2 \\ \cline{1-2}\end{array} \qquad\qquad h_i: \begin{array}{|c|c|c|c|}\hline 7 & 6 & 4 & 1 \\ \hline 5 & 4 & 2 \\ \cline{1-3} 4 & 3 & 1 \\ \cline{1-3} 2 & 1 \\ \cline{1-2}\end{array}$$

Then, the dimension of the irreducible representation is given by

$$d = \prod_i (n + D_i)/h_i. \tag{3.61}$$

We shall now demonstrate these rules with examples.

Consider a rank-2 tensor $\psi_{ij}$. If it is symmetric, it is represented by $\begin{array}{|c|c|}\hline i & j \\ \hline\end{array}$, and we can assign for $i:\ D_i = 0, h_i = 2$; and for $j:\ D_i = 1, h_i = 1$, so $d = n(n+1)/2$. Similarly, if the tensor $\psi_{ij}$ is antisymmetric, it is represented by $\begin{array}{|c|}\hline i \\ \hline j \\ \hline\end{array}$, and we can assign for $i:\ D_i = 0, h_i = 2$; and for $j:\ D_i = -1, h_i = 1$, so $d = n(n-1)/2$. The conjugate of the fundamental representation is denoted by $n-1$ boxes in one column. The dimension of the representation will be $n$ as expected.

Let us consider a few more examples. Another simple example is the product of the fundamental representation with its conjugate state given by

$$\begin{array}{|c|}\hline \mathrm{n} \\ \hline\end{array} \times \begin{array}{|c|}\hline 1 \\ \hline 2 \\ \hline \vdots \\ \hline n-1 \\ \hline\end{array} = \begin{array}{|c|c|}\hline 1 & \mathrm{n} \\ \hline 2 \\ \cline{1-1} \vdots \\ \cline{1-1} n-1 \\ \cline{1-1}\end{array} + \begin{array}{|c|}\hline 1 \\ \hline 2 \\ \hline \vdots \\ \hline n-1 \\ \hline \mathrm{n} \\ \hline\end{array}$$

This corresponds to a product decomposition

$$\mathbf{n} \times \mathbf{n}^* = (\mathbf{n^2} - \mathbf{1}) + \mathbf{1},$$

where an adjoint representation and a singlet irreducible representation are formed.

Let us now consider the product of two rank-3 tensors with mixed symmetries, represented by the Young tableaux $\begin{array}{|c|c|}\hline a & b \\ \hline c \\ \cline{1-1}\end{array}$. The product decomposition then reads

$$
\begin{array}{|c|c|}\hline 1a & 1b \\ \hline 1c \\ \cline{1-1}\end{array}
\times
\begin{array}{|c|c|}\hline 2a & 2b \\ \hline 2c \\ \cline{1-1}\end{array}
=
\begin{array}{|c|c|c|c|}\hline 1a & 1b & 2a & 2b \\ \hline 1c & 2c \\ \cline{1-2}\end{array}
+
\begin{array}{|c|c|c|c|}\hline 1a & 1b & 2a & 2b \\ \hline 1c \\ \cline{1-1} 2c \\ \cline{1-1}\end{array}
$$
$$
+
\begin{array}{|c|c|c|}\hline 1a & 1b & 2a \\ \hline 1c & 2b & 2c \\ \hline\end{array}
+
\begin{array}{|c|c|c|}\hline 1a & 1b & 2a \\ \hline 1c & 2b \\ \cline{1-2} 2c \\ \cline{1-1}\end{array}
$$
$$
+
\begin{array}{|c|c|c|}\hline 1a & 1b & 2a \\ \hline 1c & 2c \\ \cline{1-2} 2b \\ \cline{1-1}\end{array}
+
\begin{array}{|c|c|}\hline 1a & 1b \\ \hline 1c & 2a \\ \hline 2b & 2c \\ \hline\end{array}
\tag{3.62}
$$

In case of $SU(3)$, this product decomposition amounts to

$$\mathbf{8} \times \mathbf{8} = \mathbf{27} + \mathbf{10} + \mathbf{10}^* + \mathbf{8} + \mathbf{8} + \mathbf{1}.$$

Some combinations are not considered, where the boxes $\boxed{2a}$ and $\boxed{2c}$ were appearing in a row, implying a symmetric combination. Since they were in antisymmetric combination in the original representations, these states were not included. For example, $\begin{array}{|c|c|c|c|c|}\hline 1a & 1b & 2a & 2b & 2c \\ \hline 1c \\ \cline{1-1}\end{array}$ is not allowed because the boxes $\boxed{2a}$ and $\boxed{2c}$ were in antisymmetric combination in the original state.

**Representations of Subalgebras**

It is often useful to write the irreducible representations of any group in terms of the irreducible representations of it subgroups. As an example consider the $SU(2)$ subgroup of $SU(3)$ generated by the $I$-spin

$$SU(3) \supset SU(2) \times U(1),$$

where the Abelian $U(1)$ group is generated by the remaining diagonal generator of $SU(3)$. The vectors $v_1$ and $v_2$ (defined in equation (3.47)) will form the fundamental doublet representation of the group $SU(2)$, and $v_3$ will be a singlet. The $U(1)$ quantum numbers will ensure the traceless condition, and hence, we can write

$$\mathbf{3} = (\mathbf{2}, \mathbf{N}) + (\mathbf{1}, -\mathbf{2N}),$$

where $N$ is the normalization factor of the $U(1)$ quantum numbers. If we want the normalization of $U(1)$ to be the same as the normalization of $SU(2)$ or $SU(3)$, which is $\mathrm{tr} T_f^2 = 1/2$, then we get $N = 1/2\sqrt{3}$. However, for simplicity we shall choose $N = 1$.

Let us now consider the product decomposition

$$
\begin{aligned}
\mathbf{3} \times \mathbf{3} = \mathbf{3}^* + \mathbf{6} &= (\mathbf{2} \times \mathbf{2}, \mathbf{2}) + (\mathbf{1} \times \mathbf{1}, -\mathbf{4}) + (\mathbf{2} \times \mathbf{1}, -\mathbf{1}) + (\mathbf{1} \times \mathbf{2}, -\mathbf{1}) \\
&= (\mathbf{1} + \mathbf{3}, \mathbf{2}) + (\mathbf{1}, -\mathbf{4}) + (\mathbf{2}, -\mathbf{1}) + (\mathbf{2}, -\mathbf{1}).
\end{aligned}
$$

Since the sextet is symmetric and the antitriplet is antisymmetric, the sextet should contain the $SU(2)$ triplet, a symmetric combination of the two states $(2,-1)$ and $(1,-4)$, while the antitriplet will contain the $SU(2)$ antisymmetric singlet and an antisymmetric combination of the two $SU(2)$ doublets

$$\mathbf{6} = (\mathbf{3},\mathbf{2}) + (\mathbf{2},-\mathbf{1}) + (\mathbf{1},-\mathbf{4}) \qquad\qquad \mathbf{3}^* = (\mathbf{2},-\mathbf{1}) + (\mathbf{1},\mathbf{2}).$$

Similarly taking the product of **3** and $\mathbf{3}^*$, we can get the decomposition of the octet

$$\mathbf{8} = (\mathbf{3},\mathbf{0}) + (\mathbf{1},\mathbf{0}) + (\mathbf{2},-\mathbf{3}) + (\mathbf{2},\mathbf{3}).$$

This tells us how the generators of $SU(3)$ decompose under the $SU(2)$ subgroup. The $(3,0)$ corresponds to the three generators $I_\pm, I_3$ which generate the group $SU(2)$ and $(1,0)$ corresponds to the diagonal generator $T_8$ that generates the Abelian group $U(1)$. The raising and lowering operators $U_\pm$ and $V_\pm$ are doublets under $SU(2)$ and have definite $T_8$ eigenvalues $\pm 3$, so they correspond to the states $(2,\pm 3)$.

These results can be generalized for any group $SU(n)$, and the decompositions of the irreducible representations in terms of its $SU(n-1)\times U(1)$ subgroup can be written

$$\begin{aligned}
\mathbf{n} &= (\mathbf{n}-\mathbf{1},\mathbf{1}) + (\mathbf{1},-(\mathbf{n}-\mathbf{1})),\\
\mathbf{n}^* &= ((\mathbf{n}-\mathbf{1})^*,-\mathbf{1}) + (\mathbf{1},\mathbf{n}-\mathbf{1})\\
\mathbf{n}(\mathbf{n}+\mathbf{1})/\mathbf{2} &= (\mathbf{n}(\mathbf{n}-\mathbf{1})/\mathbf{2},\mathbf{2}) + (\mathbf{n}-\mathbf{1},-\mathbf{n}+\mathbf{2}) + (\mathbf{1},-\mathbf{2n}+\mathbf{2}),\\
\mathbf{n}(\mathbf{n}-\mathbf{1})/\mathbf{2} &= ((\mathbf{n}-\mathbf{1})(\mathbf{n}-\mathbf{2})/\mathbf{2},\mathbf{2}) + (\mathbf{n}-\mathbf{1},-\mathbf{n}+\mathbf{2})\\
\mathbf{n}^2-\mathbf{1} &= (\mathbf{n}(\mathbf{n}-\mathbf{2}),\mathbf{0}) + (\mathbf{1},\mathbf{0}) + (\mathbf{n}-\mathbf{1},\mathbf{n}) + ((\mathbf{n}-\mathbf{1})^*,-\mathbf{n}).
\end{aligned} \tag{3.63}$$

We shall also present the decompositions of the irreducible representations of $SU(n)$ under the subgroup $SU(n-m)\times SU(m)\times U(1)$:

$$\begin{aligned}
\mathbf{n} &= (\mathbf{n}-\mathbf{m},\mathbf{1},\mathbf{m}) + (\mathbf{1},\mathbf{m},-(\mathbf{n}-\mathbf{m}))\\
\mathbf{n}^* &= ((\mathbf{n}-\mathbf{m})^*,\mathbf{1},-\mathbf{m}) + (\mathbf{1},\mathbf{m}^*,(\mathbf{n}-\mathbf{m}))\\
\frac{\mathbf{n}(\mathbf{n}+\mathbf{1})}{\mathbf{2}} &= \left(\frac{(\mathbf{n}-\mathbf{m})(\mathbf{n}-\mathbf{m}+\mathbf{1})}{\mathbf{2}},\mathbf{1},\mathbf{2m}\right)\\
&+ \left(\mathbf{1},\frac{\mathbf{m}(\mathbf{m}+\mathbf{1})}{\mathbf{2}},-\mathbf{2}(\mathbf{n}-\mathbf{m})\right) + (\mathbf{n}-\mathbf{m},\mathbf{m},-\mathbf{n}+\mathbf{2m})\\
\frac{\mathbf{n}(\mathbf{n}-\mathbf{1})}{\mathbf{2}} &= \left(\frac{(\mathbf{n}-\mathbf{m})(\mathbf{n}-\mathbf{m}-\mathbf{1})}{\mathbf{2}},\mathbf{1},\mathbf{2m}\right)\\
&+ \left(\mathbf{1},\frac{\mathbf{m}(\mathbf{m}-\mathbf{1})}{\mathbf{2}},-\mathbf{2}(\mathbf{n}-\mathbf{m})\right) + (\mathbf{n}-\mathbf{m},\mathbf{m},-\mathbf{n}+\mathbf{2m})\\
\mathbf{n}^2-\mathbf{1} &= ((\mathbf{n}-\mathbf{m})^2-\mathbf{1},\mathbf{1},\mathbf{0}) + (\mathbf{1},\mathbf{m}^2-\mathbf{1},\mathbf{0}) + (\mathbf{1},\mathbf{0})\\
&+ (\mathbf{n}-\mathbf{m},\mathbf{m}^*,\mathbf{n}) + ((\mathbf{n}-\mathbf{m})^*,\mathbf{m},-\mathbf{n}).
\end{aligned} \tag{3.64}$$

We have not considered the normalization of the group $U(1)$, which is a multiplicative factor for all the $U(1)$ quantum numbers. The decomposition of the fundamental representation under the subgroup $SU(n-m) \times SU(m) \times U(1)$ means that a vector $\psi_k, k = 1,...,n$ can be written as a direct sum $\psi_a \oplus \psi_p, a \in SU(n-m), p \in SU(m)$. The relative $U(1)$ quantum numbers ensure traceless condition. The generators in the adjoint representation $(\mathbf{n^2-1})$ of $SU(n)$ now include generators of the subgroups $SU(n-m)$, $SU(m)$ and $U(1)$ and the lowering and raising operators that take a state $\psi_a \in SU(n-m)$ to a state $\psi_p \in SU(m)$, which transforms nontrivially under both the groups.

The decompositions of the irreducible representations in terms of its subgroup can also be useful in determining the invariants for the different representations. Let us first determine the *index* for the irreducible representations of the group $SU(2)$. For any irreducible representation denoted by the quantum numbers $|j,m\rangle$ and hence of dimension $n = 2j+1$, the index is given by

$$\lambda(n) = j^2 + (j-1)^2 + \cdots + (-j+1)^2 + (-j)^2.$$

So, the index for some of the representations is

$$\lambda(2) = \frac{1}{2}, \quad \lambda(3) = 2, \quad \lambda(4) = 5, \quad \lambda(5) = 10.$$

We can now write the index for any irreducible representations of $SU(3)$ with our knowledge of the index for representations of $SU(2)$. We follow the prescription that the index for any irreducible representations $R_n$ of $SU(n)$ can be obtained in terms of the irreducible representations $R^m_{n-1}$ of $SU(n-1)$ as $\lambda(R_n) = \sum_m \lambda(R^m_{n-1})$ if $R_n = \sum_m R^m_{n-1}$. Let us consider the example of $SU(3)$. Ignoring the $U(1)$ quantum numbers, we can write the decomposition of $SU(3)$ representations in terms of the $SU(2)$ representations as

$$\mathbf{3} = \mathbf{2} + \mathbf{1}, \qquad \mathbf{6} = \mathbf{3} + \mathbf{2} + \mathbf{1}, \qquad \mathbf{8} = \mathbf{3} + \mathbf{2} + \mathbf{2} + \mathbf{1}.$$

Then the index for the $SU(3)$ representations will be $\lambda(3) = \lambda(2) + \lambda(1) = 1/2 + 0 = 1/2$, as expected, since for all $SU(n)$ groups, fundamental representations are normalized to $1/2$. The $SU(3)$ index for **6** and **8** are thus

$$\begin{aligned} \lambda(6) &= \lambda(3) + \lambda(2) + \lambda(1) = 2 + 1/2 + 0 = 5/2 \\ \text{and} \quad \lambda(8) &= \lambda(3) + \lambda(2) + \lambda(2) + \lambda(1) = 3. \end{aligned} \tag{3.65}$$

As expected, for the adjoint representations of $SU(n)$, the index is $\lambda(n^2-1) = n$. Similarly the group factor for the triangle anomaly for any representations of any group can be obtained in terms of the anomaly or index of the representations of its subgroup.

## 3.5 Orthogonal Groups

Any groups generated by orthogonal matrices

$$O^T O = 1 \quad \text{or} \quad O_{ij} O_{kj} = \delta_{ik}$$

are called the *orthogonal groups*. For orthogonal groups there are two disconnected vector spaces, which correspond to $\det O = \pm 1$. We restrict our discussions to only $\det O = +1$, which is called the *special orthogonal group*.

We are familiar with orthogonal groups in many ways. Consider a 2-dimensional rotation group $O(2)$ generated by

$$O = e^{i\sigma_2\theta} = \begin{pmatrix} \cos\theta & \sin\theta \\ -\sin\theta & \cos\theta \end{pmatrix},$$

where $\sigma_2 = \begin{pmatrix} 0 & -i \\ i & 0 \end{pmatrix}$ is the Pauli matrix. This is a 1-parameter group; the generator commutes with itself, and hence, it is equivalent to the *Abelian group* $U(1)$.

The group $SO(3)$ is the rotation group in 3-dimensional space with determinant $+1$. The group transformations on the 3-vectors $x_i, i = 1,2,3$, given by

$$x_i \to x_i' = O_{ij}x_j$$

leave the real quadratic form $x_1^2 + x_2^2 + x_3^2$ invariant and preserve the distance. The group elements are generated by the three angular momentum operators

$$O(\theta_a) = e^{J_a\theta_a},$$

where the generators are real antisymmetric matrices and can be given by

$$J_1 = \begin{pmatrix} 0 & 1 & 0 \\ -1 & 0 & 0 \\ 0 & 0 & 0 \end{pmatrix}, \quad J_2 = \begin{pmatrix} 0 & 0 & -1 \\ 0 & 0 & 0 \\ 1 & 0 & 0 \end{pmatrix}, \quad J_3 = \begin{pmatrix} 0 & 0 & 0 \\ 0 & 0 & 1 \\ 0 & -1 & 0 \end{pmatrix}.$$

These generators satisfy the commutation relation

$$[J_a, J_b] = f_{abc}J_c,$$

where the structure function $f_{abc}$ is a completely antisymmetric tensor. Thus, the Lie algebra satisfied by the generators of $SO(3)$ is the same as that of the generators of the group $SU(2)$, so these two groups are *locally isomorphic*.

Another convention is to replace $J_a$ by $iT_a$ and write $T_a$ as imaginary antisymmetric matrices; however, we shall restrict ourselves to the present convention with $J_a$. Let us now consider the group $SO(4)$, which is locally isomorphic to the Lorentz group $O(3,1)$. The Lorentz group $SO(3,1)$ preserves the distance $x^2 + y^2 + z^2 - t^2$ and, hence, allows disconnected regions in space–time because of the relative sign in the metric, which is not present in the $SO(4)$ group which preserves $x^2 + y^2 + z^2 + t^2$. But locally they satisfy the same Lie algebra. The group $SO(4)$ is isomorphic to the semi-simple group $SU(2) \times SU(2)$, which we shall discuss later. Another isomorphism between orthogonal and unitary groups is that the groups $SO(6)$ and $SU(4)$ are also isomorphic to each other.

Let us write the operator, generating the $SO(n)$ group, as

$$O(\theta_a) = e^{\theta_{ab}M_{ab}},$$

where $\theta_{ab}$ is antisymmetric and the generators are real antisymmetric matrices $M_{ab} = -M_{ba}$ given by

$$[M_{ab}]_{ij} = \delta_{ai}\delta_{bj} - \delta_{bi}\delta_{aj}\,, \tag{3.66}$$

and they satisfy the commutation relation

$$[M_{ab}, M_{cd}] = \delta_{bc}M_{ad} - \delta_{ac}M_{bd} - \delta_{bd}M_{ac} + \delta_{ad}M_{bc}. \tag{3.67}$$

The commuting generators could be chosen to be $M_{2j-1,2j}, j = 1,2,...,N$ for both the groups $SO(2N+1)$ and $SO(2N)$.

The Lorentz group is locally isomorphic to the $SO(4)$ group. The generators of the Lorentz group are

$$M_{\mu\nu} \equiv L_{\mu\nu} = x_\mu p_\nu - x_\nu p_\mu\,,$$

which satisfy the same algebra as the $SO(n)$ group generators. There are six generators of the group $SO(4)$ that are given by

$$M_a = \varepsilon_{abc}M_{bc} \qquad N_a = M_{0a}\,,$$

where $\varepsilon_{abc}$ is the totally antisymmetric invariant tensor of the group. These generators satisfy the algebra

$$[M_a, M_b] = \varepsilon_{abc}M_c \quad [N_a, N_b] = \varepsilon_{abc}M_c$$

$$[M_a, N_a] = 0 \quad [M_a, N_b] = \varepsilon_{abc}N_c.$$

In another basis these generators can be written as direct sum of two $SU(2)$ algebras, which are

$$J_a = \frac{1}{2}(M_a + N_a), \quad \text{and} \quad K_a = \frac{1}{2}(M_a - N_a). \tag{3.68}$$

These generators now satisfy the commutation relation

$$[J_a, J_b] = \varepsilon_{abc}J_c, \quad [K_a, K_b] = \varepsilon_{abc}K_c \quad \text{and} \quad [K_a, J_b] = 0. \tag{3.69}$$

$J_a$ and $K_a$ then generates two distinct $SU(2)$ groups, which are direct product groups, and hence, $SO(4) \supset SU(2) \times SU(2)$.

In terms of the $SU(2) \times SU(2)$ subgroup of $SO(4)$, the different representations are given by

$$S \equiv (0,0) \quad V \equiv (1/2, 1/2) \quad \Gamma \equiv (1/2, 0) + (0, 1/2).$$

A doublet under the group $SU(2)$ is represented by its quantum number $1/2$. The scalars ($S$) transform trivially under the group. There is a vector representation ($V$) of dimension 4, which is the fundamental representation of the group. The four-momentum belongs to the vector representation. There is also a 4-dimensional spinor representation of the group $\Gamma$, which has two distinct parts. The usual $\gamma_5$ matrix projects out these two states, and they correspond to the left-handed spinor and the right-handed spinor.

All representations of $SU(n)$ can be constructed as products of the fundamental representations. However, for the orthogonal groups $SO(n)$, there are two distinct representations, the usual vector representations of dimension $n$ and a spinor representation of dimension $2^{N-1}$ for both $n = 2N$ and $n = 2N - 1$. To discuss the spinor representations let us introduce the Clifford algebra, which entails the anticommutation relations ($\{X,Y\} = XY + YX$)

$$\{\Gamma_i, \Gamma_j\} = 2\delta_{ij}, \quad i, j = 1, 2, ..., n \tag{3.70}$$

satisfied by the set of $N$ matrices $\Gamma_i$. These matrices can give us a representation of the group $SO(n)$

$$M_{ab} = \frac{1}{4}[\Gamma_a, \Gamma_b], \tag{3.71}$$

such that $M_{ab}$ satisfy the $SO(n)$ algebra and

$$[M_{ab}, \Gamma_c] = \delta_{bc}\Gamma_a - \delta_{ac}\Gamma_b \tag{3.72}$$

The $\Gamma$-matrices form a set of $n$ tensors that transform according to vector representation of $SO(n)$.

Let us first consider the case $n = 2N$. It is possible to find a complex spinorial representation of dimension $2^N$. But it is possible to define a generator

$$\Gamma_{n+1} = \Gamma_1\Gamma_2...\Gamma_n, \tag{3.73}$$

which commutes with all the generators, that projects out two components of the $2^N$ spinor. We can define the projection operators as

$$P_- = \frac{1}{2}(1 - \Gamma_{n+1}) \quad P_+ = \frac{1}{2}(1 + \Gamma_{n+1}),$$

which obey $P_\pm^2 = P_\pm$ and $P_- + P_+ = 1$. That leaves two independent $2^{N-1}$ component spinors in an $SO(n = 2N)$ group. The main difference between the groups $SO(2N)$ and $SO(2N+1)$ is that the two spinors of different helicity of the group $SO(2N)$ transform like one spinor representation of the group $SO(2N+1)$, and hence, the spinor of $SO(2N+1)$ is of dimension $2^N$. This is because the generator $\Gamma_{n+1}$ satisfies a similar anticommutation relation with other generators, and it belongs to the group $SO(2N+1)$. It is the $2N+1$-th element of the $SO(2N+1)$ vector. Thus, all representations of the group $SO(2N+1)$ are vector-like, and it is not possible to distinguish the left-handed fermions from the right-handed fermions. For example, the $SO(10)$ group has a 10-dimensional vector representation and a 16-dimensional spinor representation. The left-handed fermions and antifermions belong to the 16-dimensional representation and the right-handed fermions and antifermions belong to the $16^*$-dimensional representation. However, in $SO(11)$ this is not possible, because both the 16 and $16^*$ representations of $SO(10)$ are part of the same 32-dimensional representation of $SO(11)$.

Construction of the irreducible representations of any $SO(n)$ group is not as simple as that of the $SU(n)$ groups. We shall, thus, discuss only a few simple representations. Consider the rotation group $SO(3)$ in 3-dimensions, with which we are most familiar. There is a 3-dimensional vector representation $\psi_k, k = 1,2,3$. One can combine two vectors to construct higher dimensional representations. For this purpose we shall make use of the invariants of the group, $\varepsilon_{ijk}$ and $\delta_{ij}$. The product of two vectors can then be decomposed as

$$\begin{aligned}\psi_i\phi_j &= [\frac{1}{2}(\psi_i\phi_j + \psi_j\phi_i) - \frac{1}{3}\delta_{ij}\psi_i\phi_j] + \frac{1}{2}\varepsilon_{ijk}\psi_i\phi_j + \frac{1}{3}\delta_{ij}\psi_i\phi_j \\ &= T_{ij} + V_k + S,\end{aligned}$$

where $T_{ij}$ is the symmetric traceless rank-2 tensor; $V_k$ is a vector, which comes from the cross product of two vectors; and $S$ is a scalar, which comes from the scalar product of two vectors. The adjoint representation belongs to the antisymmetric representation of dimension-3.

Let us now consider the case of $SO(10)$. The vector belongs to a 10-dimensional representation. When two vectors combine, it gives a scalar, an antisymmetric tensor in the adjoint representation and a symmetric rank-2 tensor. The product decomposition then becomes

$$\mathbf{10} \times \mathbf{10} = \mathbf{1} + \mathbf{45} + \mathbf{54}.$$

The product decomposition of the spinors does not have any analogy with lower-dimensional groups. It is given by

$$\mathbf{16} \times \mathbf{16} = \mathbf{10} + \mathbf{120} + \mathbf{126} \qquad \mathbf{16} \times \mathbf{16}^* = \mathbf{1} + \mathbf{45} + \mathbf{210}.$$

The adjoint representation is a 45-dimensional representation.

Construction of the irreducible representations of any $SO(n)$ group is not as simple as that of the $SU(n)$ groups. Because of this it is convenient to study the orthogonal groups in terms of their unitary subgroups. We shall discuss the $SO(10)$ group in some more detail in terms of its subgroup while discussing the $SO(10)$ grand unified theories.

# Part II

# Standard Model and Beyond

# 4

# *Symmetries in Nature*

If we do not try to find any correlation between different phenomena, then for every event we have to find a different law. But if we can correlate these events, then we find that only a very few laws can explain most of these phenomena. These correlations lead us to the concept of symmetry. Consider a spherical ball. One can visualize it by taking its projection from all orientations, which will give us an infinite set of data. However, if we consider its spherical symmetry, we can visualize it from only one direction and get all the infinite projections from all orientations by making spherical transformations. Children develop such a notion of symmetry in their mind just by observing. But they will not try to understand that the spherical symmetry of the object means that this object has the minimum surface area and any loop on the surface of the sphere can be contracted to a point. Neither do they try to find out the meaning of spherical transformations. So, the concept of symmetry can be understood at different levels and can have different applications as well. The mathematical tool to study symmetries is the group theory which we discussed in chapter 3.

We encounter several types of symmetries in nature. These symmetries could be continuous or discrete and global or local. All fundamental interactions can originate from local continuous symmetries and can be explained by gauge theories. On the other hand, it is very common to have global symmetries in nature. Interesting possibilities emerge, when any symmetry of the Lagrangian is not respected by the vacuum, leading to spontaneous symmetry breaking. All these notions form the basis for the standard model of particle physics.

## 4.1 Discrete Symmetries

When a system remains invariant under a finite or denumerable infinite number of transformations, the system is said to possess discrete symmetry. A term $x^2$ is symmetric under a $Z_2$ symmetry generated by $x \rightarrow -x$; $x^3$ is symmetric under $Z_3$ symmetry $x \rightarrow \omega x$, where $\omega$ is the cube root of unity $\omega^3 = 1$ and $1 + \omega + \omega^2 = 0$; and $x^4$ is invariant under $x \rightarrow ix$, which is a $Z_4$ symmetry. In particle physics such simple discrete symmetries are very common, although in some cases more complex discrete symmetries including non-Abelian discrete symmetries, whose generators do not commute, are also used.

Let us consider a simple example of $Z_2$ discrete symmetry. When a scalar field $\phi(x)$ transforms under a $Z_2$ symmetry as $\phi \to -\phi$, the Lagrangian $\mathcal{L}(\phi)$ will remain invariant under this $Z_2$ transformation if the system is invariant under this $Z_2$ discrete symmetry. Thus, the most general scalar interactions in the Lagrangian can be written as

$$\mathcal{L}_s = \frac{1}{2} m_\phi^2 \phi^2 + \frac{1}{4!} \lambda \phi^4. \tag{4.1}$$

The cubic term will not be allowed by the $Z_2$ symmetry. Similarly the discrete symmetry can be extended to several fields, each of which will transform under the discrete symmetry in a specified manner and any term in the Lagrangian should be invariant under the given discrete symmetry. As an example consider a $Z_3$ symmetry, under which the scalar fields $\phi_1$ and $\phi_2$ transform as $\phi_1 \to \omega \phi_1$ and $\phi_2 \to \omega^2 \phi_2$. The $Z_3$ invariant scalar interactions in the Lagrangian will be given by

$$\mathcal{L}_s = m^2 \phi_1 \phi_2 + \mu_1 \phi_1^3 + \mu_2 \phi_2^3 + \lambda \phi_1^2 \phi_2^2. \tag{4.2}$$

Several interactions that are allowed otherwise are now forbidden by the $Z_3$ symmetry.

We now discuss some important discrete symmetries:

| | | |
|---|---|---|
| parity | $\mathcal{P}$ : | $(\mathbf{x},t) \to (-\mathbf{x},t)$ |
| charge conjugation | $\mathcal{C}$ : | $e \to -e$ |
| time – reversal | $\mathcal{T}$ : | $(\mathbf{x},t) \to (\mathbf{x},-t)$ |

and products of these symmetries. In weak interactions, $C$ and $P$ are maximally broken, but $CP$ and $T$ are weakly broken. However, the product $CPT$ is always conserved for consistency of the quantum field theory.

**Parity $P$**

We shall study the transformations of scalars, spinors and vectors under the parity operation. We start with a complex scalar field $\phi(x)$, which can be expressed as

$$\phi(x) = \int \frac{d^3k}{(2\pi)^{3/2}} \frac{1}{\sqrt{2\omega_k}} (a(k)e^{-ik\cdot x} + b^\dagger(k)e^{ik\cdot x}), \tag{4.3}$$

in terms of the operators $a(k)$ and $b(k)$, which satisfy

$$[a(k), a^\dagger(k')] = [b(k), b^\dagger(k')] = \delta^3(\mathbf{k} - \mathbf{k}'), \tag{4.4}$$

and all other pairs of operators commute. Under parity operation, $a(k) \to \pm a(-k)$ and $b(k) \to \pm b(-k)$ and the scalar field transforms as

$$\mathcal{P} : \phi(r,t) \to \pm \phi(-r,t), \tag{4.5}$$

for which we need to change the summation variable $k \to -k$.

The transformation of a fermion $\psi(\mathbf{x},t)$ under the operation of parity is given by

$$\mathscr{P}\psi(\mathbf{x},t)\mathscr{P}^\dagger = \eta_p\gamma^0\psi(-\mathbf{x},t), \tag{4.6}$$

where $|\eta_p| = 1$ is a phase. The left-handed field $\psi_L = \frac{1}{2}(1-\gamma_5)\psi$ and the right-handed field $\psi_R = \frac{1}{2}(1+\gamma_5)\psi$ transform into each other under parity

$$\begin{aligned} \frac{1}{2}(1-\gamma_5)\psi = \psi_L \xrightarrow{\mathscr{P}} \psi_R = \frac{1}{2}(1+\gamma_5)\psi \\ \frac{1}{2}(1-\gamma_5)\psi^c = (\psi^c)_L = (\psi_R)^c \xrightarrow{\mathscr{P}} (\psi_L)^c = (\psi^c)_R = \frac{1}{2}(1+\gamma_5)\psi^c. \end{aligned} \tag{4.7}$$

The charge conjugations of the fields $\psi$ are defined as $\psi \xrightarrow{\mathscr{C}} \psi^c = C\bar{\psi}^T = C\gamma_0\psi^*$, where $C = -i\gamma_2\gamma_0$. A left-handed field transforms under parity into a right-handed field, so left-handed and right-handed fields are treated the same way if any interaction conserves parity.

A vector field $A_\mu$ and the momentum vector $P_\mu$ transform under parity operation as

$$\{A^\mu(x,t), P^\mu\} \xrightarrow{\mathscr{P}} \{A_\mu(-x,t), P_\mu\}. \tag{4.8}$$

This can be seen from the transformation of the source term $j^\mu$. The charge distribution $\rho$ remains unchanged but $\mathbf{j} \to -\mathbf{j}$ under parity since the current flows in the opposite direction. Thus,

$$\{j^\mu, \bar{\psi}(x,t)\gamma^\mu\psi(x,t)\} \xrightarrow{\mathscr{P}} \{j_\mu, \bar{\psi}(-x,t)\gamma_\mu\psi(-x,t)\}. \tag{4.9}$$

An axial vector field will transform with a negative sign. Strong and electromagnetic interactions are invariant under parity, but weak interaction violates parity strongly.

### Charge Conjugation *C*

Under charge conjugation, *C*, a particle transforms into an antiparticle. For a scalar particle, charge conjugation implies interchange of the creation and annihilation operators for particles and antiparticles

$$a(k) \leftrightarrow b(k) \qquad a^\dagger(k) \leftrightarrow b^\dagger(k), \tag{4.10}$$

so the scalar field transforms under charge conjugation as

$$\mathscr{C} : \phi(r,t) \to \pm\phi^\dagger(r,t). \tag{4.11}$$

The transformation of a fermion can be derived from the invariance of the Dirac equation. Since the electric charge changes sign under charge conjugation, the invariance of the Dirac equation dictates that the charge conjugation of a fermion field should be

$$\psi \xrightarrow{\mathscr{C}} \psi^c = C\bar{\psi}^T = C\gamma_0\psi^*, \qquad \text{where } C = -i\gamma_2\gamma_0 \tag{4.12}$$

and

$$\mathscr{C}\psi(x,t)\mathscr{C}^\dagger = \eta_c C\bar{\psi}^T(x,t), \tag{4.13}$$

where $|\eta_c| = 1$ is another phase. The transformations of the left-handed and right-handed fermions are given by

$$\psi_L \xrightarrow{\mathscr{C}} \psi^c{}_L = (\psi_R)^c \quad \text{and}$$
$$\psi_R \xrightarrow{\mathscr{C}} \psi^c{}_R = (\psi_L)^c \,. \tag{4.14}$$

Finally the transformations of the four-momenta $P_\mu$, vector field $A_\mu(x,t)$, source current $j^\mu(x)$, and the current $\bar{\psi}(x,t)\gamma^\mu\psi$ are given by

$$\{P_\mu, A_\mu(x,t)\} \xrightarrow{\mathscr{C}} \{P_\mu, -A_\mu(x,t)\}$$
$$\{j_\mu, \bar{\psi}(x,t)\gamma_\mu\psi(x,t)\} \xrightarrow{\mathscr{C}} \{-j_\mu, -\bar{\psi}(x,t)\gamma_\mu\psi(x,t). \tag{4.15}$$

Under weak interaction, both $P$ and $C$ are violated strongly, although $CP$ remains almost conserved. On the other hand, the strong, the electromagnetic, and the gravitational interactions conserve both $P$ and $C$ independently.

Under $CP$ operation, the various fields transform as

$$\psi_L \xrightarrow{\mathscr{CP}} \psi^c{}_R = (\psi_L)^c$$
$$\psi_R \xrightarrow{\mathscr{CP}} \psi^c{}_L = (\psi_R)^c$$
$$\{P^\mu, A^\mu(x,t)\} \xrightarrow{\mathscr{CP}} \{P_\mu, -A_\mu(-x,t)\}$$
$$\bar{\psi}(x,t)\gamma^\mu\psi(x,t) \xrightarrow{\mathscr{CP}} -\bar{\psi}(-x,t)\gamma_\mu\psi(-x,t). \tag{4.16}$$

Thus, $CP$ operation relates all the interaction terms to their Hermitian conjugates, except for the coupling constants. As a result, $CP$ violation can come from the imaginary part of the couplings. Consider the Yukawa coupling of a fermion field $\psi$ with a scalar $\phi$ and the mass term

$$\mathscr{L} = m\bar{\psi}_L\psi_R + m^*\bar{\psi}_R\psi_L + f\bar{\psi}_L\psi_R\phi + f^*\bar{\psi}_R\psi_L\phi^\dagger \,. \tag{4.17}$$

We have written the Hermitian conjugate terms explicitly, the second and the fourth terms. Under the operation of $CP$, $\bar{\psi}_L\psi_R \to \bar{\psi}_R\psi_L$, and $\phi \to \phi^\dagger$, $CP$ invariance implies $m = m^*$ and $f = f^*$. Any complex phase in the Yukawa couplings or in the mass or the scalar potential then implies $CP$ violation.

$CP$ violation was first observed in $K^\circ - \overline{K^\circ}$ oscillations [22], where strangeness changes by 2 units. Since Kaons are produced through the strong interactions, they carry definite strangeness quantum numbers, which allows us to distinguish between a $K^\circ (\equiv d\bar{s}$ with $S = 1)$ and its antiparticle $\overline{K^\circ} (\equiv s\bar{d}$ with $S = -1)$. But the decays of the kaons are through the weak interaction, violating strangeness quantum number. This allows $K^\circ$ and $\overline{K^\circ}$ to decay into two or three pions, which are even or odd under $CP$, respectively. Virtual pion exchange will then convert the $K^\circ$ into a $\overline{K^\circ}$

$$[K^\circ, \overline{K^\circ}] \leftrightarrow [2\pi \text{ or } 3\pi] \leftrightarrow [\overline{K^\circ}, K^\circ]. \tag{4.18}$$

Strangeness number is changed by two units in $K^\circ - \overline{K^\circ}$ oscillations, although $K$-decays violate strangeness by one unit.

The $CP$ transformation takes $K^\circ \to \overline{K^\circ}$, so we can define the states with definite $CP$ eigenvalues

$$|K_1\rangle = \frac{1}{\sqrt{2}}[|K^\circ\rangle + |\overline{K^\circ}\rangle] \quad CP = +1 \tag{4.19}$$

$$|K_2\rangle = \frac{1}{\sqrt{2}}[|K^\circ\rangle - |\overline{K^\circ}\rangle] \quad CP = -1. \tag{4.20}$$

Since pions are odd under $CP$, the $CP$ even state $|K_1\rangle$ can decay only to $2\pi$ while the $CP$ odd state $|K_2\rangle$ can decay only to $3\pi$.

If there is $CP$ violation in the weak interaction Hamiltonian appearing as a complex phase in the Yukawa couplings, a small fraction of $CP$-even state $|K_1\rangle$ can evolve into $|K_2\rangle$. The physical states that evolve with time are then given by

$$|K_S\rangle = \frac{1}{\sqrt{1+|\varepsilon|^2}}[|K_1\rangle + \varepsilon|K_2\rangle] \tag{4.21}$$

$$|K_L\rangle = \frac{1}{\sqrt{1+|\varepsilon|^2}}[|K_2\rangle + \varepsilon|K_1\rangle]. \tag{4.22}$$

The amount of $CP$ violation is characterized by the parameter $\varepsilon$. These states have definite lifetimes of $0.9 \times 10^{-10}$ s and $0.5 \times 10^{-7}$ s, respectively.

**Time-Reversal $T$**

The time-reversal symmetry implies that any reaction rate in the forward direction (in time) should be the same as in the backward direction. Thus, the time-reversal operator is defined as

$$\mathscr{T}\psi(x,t)\mathscr{T}^\dagger = \eta_T T\psi^*(x,-t) \tag{4.23}$$

with the phase $\eta_T = \pm 1$ and $T = i\gamma^1\gamma^3 = -i\gamma^5 C$. Since time-reversal implies the exchange of positive and negative energy solutions, particles have to be interchanged with antiparticles. In the spinor space, this is done with $T$, which satisfies $T\gamma^\mu T^{-1} = \gamma_\mu^T = \gamma^{\mu *}$ and $T = T^\dagger = T^{-1} = -T^*$. Strong and electromagnetic interactions are invariant under the $T$ transformation, but the time-reversal symmetry is broken in the weak interaction very weakly.

The time-reversal operator $\mathscr{T}$ has the unique property that it is antiunitary. Consider the operation of $\mathscr{T}$ on the time-evolution operator

$$\mathscr{T} e^{iH(t_1-t_2)} \mathscr{T}^{-1} = e^{-iH(t_1-t_2)}. \tag{4.24}$$

This implies that the Hamiltonian does not commute with the operation $\mathscr{T}$ and it should be antiunitary, i.e., for any two states $\phi$ and $\psi$, the time-reversal operator satisfies

$$<\mathscr{T}\phi|\mathscr{T}\psi> = <\psi|\phi>.$$

The time-reversal transformation of a scalar field is defined as

$$\mathscr{T} : \phi(r,t) \rightarrow \pm\phi^\dagger(r,-t). \tag{4.25}$$

Under time-reversal, the annihilation operators transform into creation operators as $a(k) \rightarrow \pm a^\dagger(-k)$ and $b(k) \rightarrow \pm b^\dagger(-k)$. The vector field transforms under $T$ as $A_\mu(\mathbf{x},t) \xrightarrow{\mathscr{T}} A_\mu(\mathbf{x},-t)$.

The product of all the discrete symmetries, $CPT$, is a foundation stone in quantum field theory and it is always conserved. Under the $CPT$ transformation, a fermion transforms as

$$(\mathscr{C}\mathscr{P}\mathscr{T})\ \psi(x,t)\ (\mathscr{C}\mathscr{P}\mathscr{T})^\dagger = i\gamma_5\psi(-x,-t), \tag{4.26}$$

implying that a positive frequency particle could be replaced by a negative frequency antiparticle with their momenta reversed and multiplied by $i\gamma_5$. For a scalar field the $CPT$ transformation reads

$$\phi(r,t) = +\phi(-r,-t). \tag{4.27}$$

Since this relation can be obtained by redefining the summation variables $k \rightarrow -k$ and $\varepsilon \rightarrow -\varepsilon$, this implies $a(k) \rightarrow b^\dagger(k)$ and $b(k) \rightarrow a^\dagger(k)$, i.e., interchange of particle and antiparticle.

The Lagrangian and the Hamiltonian of any system transform under $\mathscr{C}\mathscr{P}\mathscr{T}$ as

$$\begin{aligned}(\mathscr{C}\mathscr{P}\mathscr{T})\ \mathscr{L}(x)\ (\mathscr{C}\mathscr{P}\mathscr{T})^{-1} &= \mathscr{L}(-x)\\ (\mathscr{C}\mathscr{P}\mathscr{T})\ \mathscr{H}(x)\ (\mathscr{C}\mathscr{P}\mathscr{T})^{-1} &= \mathscr{H}(-x).\end{aligned} \tag{4.28}$$

This general invariance is stated by the $CPT$ theorem [23] that under the combined transformation of parity, charge conjugation and time-reversal, any system remains invariant. $CPT$ and local Lorentz invariance may, thus, be considered as the most sacred symmetries in nature, and they are the two foundation stones of quantum field theory.

## 4.2 Continuous Symmetries

Continuous symmetries are very common in particle physics. When a new result is observed, one tries to extend the existing model with some new particles and imposing additional symmetries. If the extended model explains the result, one tries to find predictions of the new model and then test them. When some of these predictions are verified and few of these predictions cannot be explained by other similar models, the model is accepted. However, our limitations with high energy accelerators does not allow us to test all these models, and we have to study several alternative models simultaneously to look for their testable signatures in the upcoming accelerators or cosmology.

A continuous symmetry consists of an infinite number of states that are related by some continuous transformations of a parameter in a finite range. For example, a state vector $|\psi(\theta)\rangle$ could be symmetric under rotation, so by varying the angle of rotation $\theta$ continuously in the range 0 to $2\pi$, we can relate all the infinite set of state vectors parameterized by $\theta$.

Any continuous symmetry of any system implies some conserved quantity. On the other hand, if there is any conserved quantity, it is also possible to associate with it some symmetry of nature. Consider a system described by the Hamiltonian $H$. If there is a conserved quantity, the corresponding operator $O$ will commute with the Hamiltonian $H$, $HO = OH$, and it will be possible to obtain a complete set of eigenfunctions $\psi_i$, which are simultaneous eigenstates of both $O$ and $H$. The operator $O$, acting on a state $\psi_i$, will give the corresponding conserved quantity $O\psi_i = q_i\psi_i$. Since the time evolution of the state $\psi_i$ is governed by the Hamiltonian, in the absence of any new interaction, this state cannot evolve into another state $\psi_j$ with a different quantum number $O\psi_j = q_j\psi_j$.

In field theory, the symmetry principle and its connection with conserved quantities is stated by the Noether theorem [6]. It states that if a Lagrangian density ($\mathscr{L}$) is invariant under any symmetry of the fields $\phi_i \to \phi_i + \delta\phi_i$, there is always a conserved current $J_\mu$, which is called the Noether current. The corresponding charge is given by

$$Q = \int d^3x J_0(x), \tag{4.29}$$

which satisfies

$$\frac{dQ}{dt} = 0. \tag{4.30}$$

This gives the corresponding conservation laws.

We are familiar with the local Lorentz invariance, which is a manifestation of the space–time symmetry and the corresponding conserved quantities are energy, momentum and angular momentum. There are also internal symmetries in nature, which are the symmetries of the system under the transformations of some internal degrees of freedom. Any space–time independent continuous symmetries of the state vectors are called global symmetries. When such continuous symmetries depend on the space–time coordinate of the state vector, it is called a local symmetry and the corresponding theory is called the gauge theory. In this section we shall restrict ourselves to only global symmetries.

Let us consider a particle, represented by a wave function $\psi(x,t)$, at a position $x$ and time $t$. This function $\psi(x)$ satisfies the Schrödinger equation

$$\left[-\frac{1}{2m}\frac{\partial^2}{\partial x^2} + V(x)\right]\psi(x,t) = i\frac{\partial}{\partial t}\psi(x,t). \tag{4.31}$$

Consider a symmetry transformation on the wave function

$$\psi(x,t) \to e^{-iq\alpha}\psi(x,t), \tag{4.32}$$

where $\alpha$ is the parameter defining the symmetry. If $\alpha$ is independent of $x$ and the potential $V(x)$ is invariant under this transformation, then this theory possesses a global symmetry generated by the parameter $\alpha$. In general, there could be several states $\psi_i$, $i = 1,\ldots,n$ and the transformations of these states,

$$\psi(x,t) \to e^{-iq(\alpha\cdot T)}\psi(x,t), \tag{4.33}$$

could be generated by more than one parameters $\alpha_i$, and the generators of the transformation $T_{ij}$ could be noncommuting. When the generators of any transformation commute, it corresponds to an Abelian symmetry. When the symmetry is generated by a set of noncommuting generators, the system possesses a non-Abelian global continuous symmetry.

We shall first consider an Abelian $U(1)$ symmetry. If there exists a complex scalar field $\phi(x)$, which transforms under the $U(1)$ group as

$$\phi \to \phi' = e^{i\alpha}\phi, \tag{4.34}$$

we can write the $U(1)$ invariant Lagrangian as

$$\mathscr{L} = (\partial_\mu\phi^*)(\partial^\mu\phi) - \mu^2(\phi^*\phi) - \lambda(\phi^*\phi)^2. \tag{4.35}$$

Both $\phi$ and $\phi^*$ or the real and the imaginary components of $\phi$,

$$\phi \equiv \frac{1}{\sqrt{2}}[\phi_r + i\phi_i],$$

will have equal masses as long as the $U(1)$ symmetry is exact. In terms of the components, the Lagrangian becomes

$$\mathscr{L} = \frac{1}{2}\left[(\partial_\mu\phi_r)^2 + (\partial_\mu\phi_i)^2\right] - \frac{1}{2}\mu^2(\phi_r^2 + \phi_i^2) - \frac{\lambda}{4}(\phi_r^2 + \phi_i^2)^2. \tag{4.36}$$

The $O(2)$ symmetry corresponding to the transformation of the components, which is equivalent to this $U(1)$ symmetry, is given by

$$\begin{aligned}\phi_r \to \phi_r' &= \phi_r\cos\alpha - \phi_i\sin\alpha\\ \phi_i \to \phi_i' &= \phi_r\sin\alpha + \phi_i\cos\alpha.\end{aligned} \tag{4.37}$$

The $U(1)$ or the $O(2)$ symmetry of the Lagrangian would imply the same conserved currents

$$\begin{aligned}J_\mu &= i\left[(\partial_\mu\phi^*)\phi - (\partial_\mu\phi)\phi^*\right]\\ &= -(\partial_\mu\phi_r)\phi_i + (\partial_\mu\phi_i)\phi_r.\end{aligned} \tag{4.38}$$

We now extend this model with more particles.

Consider two complex scalar fields $\phi_i$, $i = 1,2$ and a couple of fermions $\psi_p$, $p = 1,2$. The transformations of these fields under the $U(1)$ symmetry is given by

$$\begin{aligned}\phi_i \to \phi_i' &= e^{iq_{si}\alpha}\phi_i\\ \psi_p \to \psi_p' &= e^{iq_{fp}\alpha}\psi_p.\end{aligned} \tag{4.39}$$

The $U(1)$ charges or the $U(1)$ quantum numbers $q_{si}$ and $q_{fp}$ could be positive or negative integers. For simplicity, we assume that both the left-handed and the right-handed fermions carry the same $U(1)$ charges although, in general, they can be different. The $U(1)$ invariant Lagrangian will contain the kinetic energy terms for all the fields

$$\mathscr{L} = \sum_{i=1,2} \left[(\partial_\mu \phi_{si}^*)(\partial^\mu \phi_{si})\right] + \sum_{p=1,2} \left[\bar{\psi}_{fp}\, i\gamma^\mu \partial_\mu\, \psi_{fp}\right], \tag{4.40}$$

and the interaction terms will depend on the $U(1)$ charges of the fields. Let us explain with an example. The charge assignment of the fields $q_{s1} = 1, q_{s2} = 2$, $q_{f1} = 1$ and $q_{f2} = 2$ would give us the $U(1)$ invariant interaction Lagrangian

$$\begin{aligned}\mathscr{L}_{int} = &-\mu_1^2(\phi_1^*\phi_1) - \mu_2^2(\phi_2^*\phi_2) - M\phi_1\phi_1\phi_2^* - \lambda_1(\phi_1^*\phi_1)^2 - \lambda_2(\phi_2^*\phi_2)^2 \\ &+ m_1\bar{\psi}_1\psi_1 + m_2\bar{\psi}_2\psi_2 + h\bar{\psi}_1\psi_2\phi_1^* + H.c.\end{aligned} \tag{4.41}$$

Invariance under $U(1)$ implies that the sum of the $U(1)$ charges must vanish for any allowed terms. The sum of the charges for the term $\bar{\psi}_1\psi_2\phi_1^*$ is $-q_{f1} + q_{f2} - q_{s1} = 0$, so under the $U(1)$ symmetry transformation

$$\bar{\psi}_1\psi_2\phi_1^* \to e^{-i\alpha}\cdot e^{2i\alpha}\cdot e^{-i\alpha}\bar{\psi}_1\psi_2\phi_1^* = \bar{\psi}_1\psi_2\phi_1^*,$$

this term remains invariant.

We now consider an example of a non-Abelian symmetry $SU(2)$. Particles can belong to any of the representations of the group. A scalar field $\phi_s(x)$ transforming as singlet under $SU(2)$ will have one component and will transform to itself $\phi_s \to \phi_s$. A scalar field $\phi_d(x)$ belonging to a doublet representation of $SU(2)$ will have two components

$$\phi_d = \begin{pmatrix}\phi_+ \\ \phi_-\end{pmatrix},$$

and it will transform under the operation of $SU(2)$ group as

$$\phi_d \to \phi_d' = e^{i\alpha^a \tau^a_{ij}}\phi_d.$$

$\tau^a$, $a = 1,2,3$ are the Pauli matrices. Any representation of the group is characterized by the number of its components, so a singlet is **1** and a doublet is **2**, etc. The components of $\phi_d$ are characterized by their $\tau^2$ and $\tau^3$ eigenvalues. Thus, $\phi_+ \equiv [2,+1/2]$ and $\phi_- \equiv [2,-1/2]$.

Let us consider a model with one complex doublet scalar field $\phi$ and a real triplet scalar field $T$:

$$\text{doublet} \quad \phi \equiv \begin{pmatrix}\phi_+ \\ \phi_-\end{pmatrix} \quad \text{triplet} \quad T \equiv \begin{pmatrix}T_+ \\ T_0 \\ T_-\end{pmatrix}.$$

In addition to the kinetic energy terms for these fields, the interaction part of the Lagrangian or the scalar potential is given by

$$\begin{aligned}\mathscr{V} = &\,\mu_\phi^2\, \phi^\dagger\phi + \mu_T^2\, T^2 + M\,(\phi\phi + \phi^\dagger\phi^\dagger)\,T + \lambda_\phi\,(\phi^\dagger\phi)^2 - \lambda_T\, T^4 \\ &+ \lambda_{\phi T}\,(\phi^\dagger\phi)T^2 + \tilde{\lambda}_{\phi T}\,(\phi^\dagger T)(\phi T).\end{aligned} \tag{4.42}$$

A singlet combination of $\phi\phi$ vanishes because this term is symmetric under the exchange of the two fields while an antisymmetric combination of the two doublets makes a singlet. Similarly a triplet combination of $T \cdot T$ also vanishes. All components of a field belonging to a particular representation of the $SU(2)$ group always have the same mass. $SU(2)$ invariance also implies that the sum of the $\tau_3$ quantum numbers of the component fields in any allowed interaction term must vanish. For example without going into the details, one can directly infer that $\phi_+\phi_+T_-$ is allowed ($j_3 = 1/2 + 1/2 - 1$), but $\phi_+\phi_-T_-$ is forbidden.

## 4.3 Symmetry Breaking

The symmetries we observe in nature could be exact or broken. Symmetries arising from the Lorentz invariance are exact, which implies conservation of energy and momentum. There are also some approximate global symmetries in nature corresponding to baryon number, lepton number, strangeness numbers, etc. Strangeness numbers are respected by the strong interaction but not by the weak interaction, while baryon and lepton numbers are broken by some new interactions at very high energies.

If any symmetry of the Lagrangian is exact, the corresponding charge will be exactly conserved. When any symmetry is broken, there will be violation of the corresponding charge. When one part of the Lagrangian is invariant under a given symmetry operation, but some other parts of the Lagrangian do not respect the symmetry, the symmetry is said to be explicitly broken. The constant coefficients or the coupling constants of these symmetry breaking terms dictate the amount of symmetry breaking. In other words, some of the interactions are invariant under the given symmetry, while some interactions violate the symmetry. In addition to the explicit symmetry breaking, it is also possible that all the terms in the Lagrangian are invariant under a given symmetry, but the minimum energy states do not respect the symmetry. In other words, the zero-point energy or the vacuum state is not invariant under the symmetry of the Lagrangian. As a result, the physical states will not experience this symmetry of the Lagrangian. This is called spontaneous symmetry breaking. The phenomenology of the explicit and the spontaneous symmetry breaking are different, and hence, they are applicable to different scenarios.

Consider a Lagrangian with the symmetry $\phi \to -\phi$. This symmetry will not allow any trilinear scalar interaction term $\mu\phi^3$, where $\mu$ is some parameter of mass dimension 1. If we now write the Lagrangian including this trilinear term, the symmetry of the Lagrangian will be explicitly broken. In this case the theory ceases to be renormalizable, which means that all interaction terms that are not protected by this symmetry will be radiatively generated. These infinite contributions may not be absorbed by redefining any bare parameters because the original theory did not have such bare parameters. Thus, large symmetry breaking interactions cannot be eliminated naturally in these theories, except by fine tuning of parameters. We shall next discuss some of the features of the spontaneous symmetry breaking.

**Spontaneous Symmetry Breaking**

In the case of spontaneous symmetry breaking, the symmetry of the Lagrangian is not the symmetry of the vacuum state or the minimum energy state. If the vacuum state corresponds to a nonvanishing value $v$ of any field $\phi$ ($\langle\phi\rangle = v$), then any physical field can be written as $\phi^{phys} = \phi - v$. $v$ is called the vacuum expectation value (*vev*) of the field $\phi$. Although the physical field $\phi^{phys}$ will no longer have the symmetry of the field $\phi$ and, hence, the original symmetry of the Lagrangian will not be reflected in the interactions of the physical field, the theory remains renormalizable. Any new interaction terms that are not allowed by the symmetry of the original Lagrangian may not be generated radiatively, and hence, all the infinite contributions coming after including the radiative corrections can be absorbed by redefining some bare parameter of the theory. If the *vev* of any field is not invariant under any symmetry, that symmetry will be spontaneously broken. Since we do not want to break the Lorentz group, the field should not transform nontrivially under the Lorentz group. So only a scalar field can have nonvanishing *vev*.

We now discuss an example of a spontaneous symmetry breaking. Consider a theory with a scalar field $\phi$, which is invariant under a discrete symmetry $Z_2$ under which $\phi \to -\phi$. The $Z_2$ invariant scalar potential for this scalar field is given by

$$V(\phi) = \frac{\mu^2}{2}\phi^2 + \frac{\lambda}{4}\phi^4 \tag{4.43}$$

with positive definite $\lambda$. Depending on the sign of the $\mu^2$ term, the minimum of this potential would correspond to

$$\begin{aligned} &<\phi^2> = 0 && \text{when } \mu^2 > 0 \\ &<\phi^2> = v^2 = \frac{-\mu^2}{\lambda} > 0 && \text{when } \mu^2 < 0\,. \end{aligned} \tag{4.44}$$

The second case with $\mu^2 < 0$ represents the spontaneous symmetry breaking, since the vacuum now corresponds to a nonvanishing value of $\phi$. If the vacuum corresponds to any one of the minima, $+v$ or $-v$, then any field residing on this vacuum will not find the theory invariant under the $Z_2$ symmetry, although the Lagrangian is still symmetric. When the vacuum picks up a value $<\phi>= v$, this is termed as the vacuum expectation value (*vev*) of $\phi$. Once the field $\phi$ acquires a *vev*, the physical field $\phi_{phys}$ can be expanded around its *vev* as

$$\phi_{phys} = \phi - v. \tag{4.45}$$

The potential $V(\phi)$ will now have cubic term when written in terms of the physical field $\phi_{phys}$, and hence, the $Z_2$ symmetry $\phi \to -\phi$ will be spontaneously broken. So, although the Lagrangian is symmetric under the $Z_2$ symmetry $\phi \to -\phi$, the vacuum

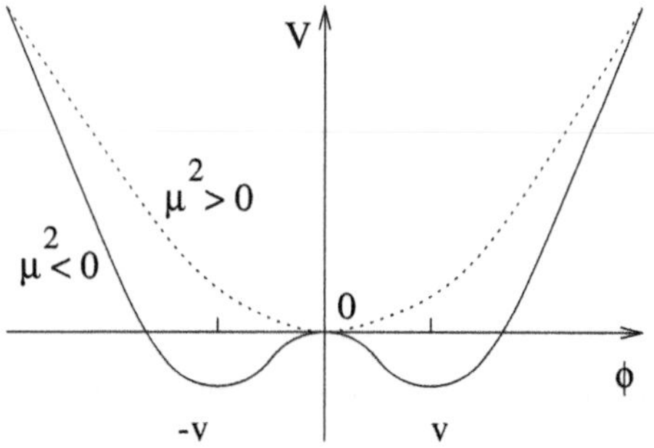


**FIGURE 4.1**
Shape of the potential for $\mu^2 > 0$ and $\mu^2 < 0$. For $\mu^2 < 0$, the minima correspond to $\phi = \pm v$. When the field settles at one of these minima, the $\phi \to -\phi$ symmetry of the Lagrangian is lost.

or the minimum energy state will not be invariant. In other words, when the Lagrangian is expressed in terms of the physical fields, it will no longer be invariant under the $Z_2$ symmetry $\phi^{phys} \to -\phi^{phys}$.

We now study a Lagrangian that is invariant under a *continuous symmetry*, say a $U(1)$ symmetry. For spontaneous symmetry breaking, we introduce a complex scalar field $\phi$, which is invariant under the $U(1)$ symmetry $\phi \to \exp^{i\alpha}\phi$. The scalar potential of this complex field may now be taken to be of the form

$$V(\phi) = \frac{\mu^2}{2}\phi^\dagger\phi + \frac{\lambda}{4}(\phi^\dagger\phi)^2 \tag{4.46}$$

with positive definite $\lambda$. Similar to the previous example, the minimum corresponds to the origin $\phi = 0$ for $\mu^2 > 0$. But for $\mu^2 < 0$, the minima of the potential would correspond to all the points that satisfy

$$< \phi^\dagger\phi >= v^2 = \frac{-\mu^2}{\lambda} > 0. \tag{4.47}$$

The vacuum will now settle to only one point, say $< \phi >= v$, which is termed the *vev* of $\phi$. This *vev* of $\phi$ will break the $U(1)$ symmetry spontaneously.

It is now convenient to express $\phi$ in terms of the real ($\eta$) and imaginary ($\sigma$) components of the physical field around the minimum $\phi = v$ as

$$\phi = (\eta + v)\exp^{i\sigma/v}, \tag{4.48}$$

where the *vev* $v$ is a real quantity. Around the minimum $\phi = v$, we can write this as $\eta + i\sigma = \phi - v$. An expansion of the Lagrangian around the minimum would allow us to write the scalar potential as

$$V(\eta) = \frac{\mu^2}{2}v^2 + \frac{\lambda}{4}v^4 - \mu^2\,\eta^2 + \lambda v\,(\eta^2+\sigma^2)\,\eta + \frac{\lambda}{4}\,(\eta^2+\sigma^2)^2. \tag{4.49}$$

The $U(1)$ symmetry $\eta \to \exp^{i\alpha}\eta$ is then spontaneously broken by the physical field. One important feature of the spontaneous symmetry breaking is that the field $\sigma$ remains massless and is called the *Nambu–Goldstone boson* [24] or the *Goldstone boson*. In case of $\mu^2 > 0$, the minimum is at $\phi = 0$ and the mass term reads

$$\frac{\mu^2}{2}\phi^\dagger\phi = \frac{\mu^2}{2}[(\text{Re}\ \phi)^2 + (\text{Im}\ \phi)^2],$$

and hence, both Re $\phi$ and Im $\phi$ have the same mass. But for $\mu^2 < 0$, the mass term $\phi^\dagger\phi$ can never include the field $\sigma$, and hence, it remains massless. It is also clear from these discussions that the Nambu–Goldstone boson can have only derivative couplings.

When the original theory is non-Abelian, the residual symmetry after spontaneous symmetry breaking will be determined by the representation to which the scalar field belongs and which component of the scalar field acquires a *vev*. Consider an $SU(3)$ gauge symmetry, in which the scalar field that acquires a *vev* belongs to the fundamental triplet representation. If the third component of the triplet Higgs ($T$) acquires a *vev*

$$v = < \begin{pmatrix} 0 \\ 0 \\ T_3 \end{pmatrix} >, \tag{4.50}$$

the vacuum will not be able to distinguish between the states 1 and 2 even after the symmetry breaking, although it distinguishes the third state from others. As a result, the $SU(2)$ subgroup (whose fundamental representation contains the states 1 and 2) still remains unbroken, although the $SU(3)$ symmetry of the Lagrangian is now broken. Out of the 8 generators of the $SU(3)$ group $\lambda^i$, $i = 1,\ldots,8$, the first three generators $\lambda^i = \begin{pmatrix} \sigma_i & 0 \\ 0 & 0 \end{pmatrix}$, $i = 1,2,3$ correspond to the residual symmetry group $SU(2)$. Corresponding to the remaining five generators $\lambda^{4,5,6,7,8}$, there will be 5 Nambu–Goldstone bosons. The triplet complex scalar field $T$ has six real components, out of which one will remain as the physical field $\eta$

$$T = \begin{pmatrix} 0 \\ 0 \\ \eta + v \end{pmatrix},$$

while all other 5 real components will become Nambu–Goldstone bosons.

This is a generic feature of any global symmetry breaking that gives rise to massless Nambu–Goldstone bosons. Consider any theory with a global symmetry $\mathcal{G}_1$, which is generated by $N_1$ generators. Let this theory be spontaneously broken, and after symmetry breaking the theory remains invariant under a global symmetry $\mathcal{G}_2$, which is generated by $N_2$ generators. Then there will be $N_1 - N_2$ Nambu–Goldstone bosons.

Massless Nambu–Goldstone bosons could be a problem for some theories from phenomenological considerations, particularly since it would imply a long range force in classical physics. However, their spin dependence would make detection very difficult [25], and they can solve some other problems. A very special and interesting case corresponds to models with pseudo-Nambu–Goldstone bosons (pNGB), when the Nambu–Goldstone bosons receive a tiny but calculable mass.

Let us consider a model with global symmetry $\mathscr{G}$, (i.e., the Lagrangian is invariant under $\mathscr{G}$). If we now introduce a new interaction term in the Lagrangian, which breaks this symmetry $\mathscr{G}$ explicitly, then that term will make the Nambu–Goldstone boson massive. To see this, consider an example of a model with a scalar field $\phi$ and a $U(1)$ global symmetry $\phi \to \exp^{i\alpha}\phi$. The scalar potential is given by equation (4.46). If $\mu^2 < 0$, then the $U(1)$ symmetry will be spontaneously broken and there will be a massless Nambu–Goldstone boson, which is the imaginary part of the physical field.

If we now introduce an explicit symmetry breaking term

$$m^2(\phi\phi + \phi^*\phi^*),$$

then the Nambu–Goldstone boson will no longer remain massless even for $\mu^2 < 0$. We can expand the field $\phi$ around its *vev* $\langle\phi\rangle = v$ as

$$\phi = (\eta + v)\exp^{i\sigma/v} \approx \eta + v + i\sigma - \frac{\sigma^2}{2v},$$

and find the mass of $\sigma$ that comes out to be $m^2$. The field $\sigma$ is now called a pseudo-Nambu–Goldstone boson (pNGB). If the explicit symmetry breaking term has a dimension less than 4, then radiative corrections cannot destabilize any result that is derived from the symmetry argument. For example the mass of the pNGB remains finite. These are called *soft symmetry breaking* terms. On the other hand, if the explicit symmetry breaking terms have dimension 4, then they introduce infinite radiative corrections to the mass of the pNGB and destabilize the theory.

Soft symmetry breaking terms appear in many models of particle physics. They may have origin in some higher symmetric theory. For example, any spontaneous gauge symmetry breaking can give rise to some soft terms. The advantage with the soft terms is that they do not introduce new divergences in the theory, and hence, the renormalizability of the original theory is not jeopardized. Any term in the Lagrangian with coefficients having mass dimension is a soft term. For example, $m^2\phi_1^\dagger\phi_2$ and $m\bar{\psi}\psi$ are soft terms, but $\bar{\psi}\psi\phi$ and $\phi^4$ are not soft terms, since the mass-dimension of the scalar fields is 1 while the mass-dimension of the fermion fields is 3/2.

In case of local symmetry, the spontaneous symmetry breaking plays a different role. When any local symmetry is broken spontaneously by the *vev* of a scalar field, the Nambu–Goldstone boson no longer remain massless. One of the massless gauge bosons of the symmetric theory makes the Nambu–Goldstone boson its longitudinal mode and becomes massive [26]. Thus, the massless gauge bosons of the symmetric theory can become massive during any spontaneous symmetry breaking. Since the starting Lagrangian is invariant under the local symmetry, the theory remains renormalizable even after spontaneous symmetry breaking. From the name of one of the inventors, the scalar field that breaks the symmetry is called a *Higgs scalar* [27].

## 4.4 Local Symmetries

When a symmetry transformation depends on the space–time position, it is called a *local symmetry* and the theory is called a *gauge theory*. Since a particle at one point transforms in a different way in comparison with another particle located at a different point, a new gauge particle is introduced to make the theory invariant under the local symmetry transformation. Lorentz invariance requires this particle to have spin-1 and this vector particle is called a *gauge boson*.

If the local symmetry is generated by $n$ generators, there will be $n$ gauge bosons. If the different generators commute with each other, it is called an Abelian gauge theory. When the generators of the local symmetry do not commute with each other, it is called a non-Abelian gauge theory. All gauge theories are renormalizable and the gauge bosons are massless. As a result the range of these interactions is extended to infinity. When a gauge theory is spontaneously broken by the vacuum expectation value of a Higgs scalar, whose minimum energy state does not respect the symmetry although the Lagrangian is still symmetric, the theory remains renormalizable although the gauge bosons become massive. The renormalizability of the gauge theory or the spontaneously broken gauge theory makes these theories predictable, since the parameters of these theories are finite and calculable and, hence, can be compared with experiments. Although these theories are consistent classically, consistency of these theories at the quantum level requires that these theories should be free from anomalies [28]. Anomalies are one-loop diagrams with fermions in the loop and gauge bosons attached at the vertices. In four dimensions, an anomaly is a triangle diagram and it depends on the fermion contents of the theory. As a result, appropriate choice of the fermion representations allows us to make a theory anomaly free. The strong, weak and electromagnetic interactions can be explained by renormalizable spontaneously broken gauge theories.

### Abelian Gauge Theory

An Abelian gauge theory is invariant under a local symmetry transformation generated by Abelian or commuting generators. The interactions among the fermions and scalars are mediated by massless gauge bosons in this theory. There are as many gauge bosons as the number of generators of the local symmetry group. Consider a local symmetry transformation,

$$\psi_i'(x) = \exp^{-ig\mathcal{O}\alpha(x)}\psi_i(x), \tag{4.51}$$

where $\alpha(x)$ depends on the space–time coordinate $x$, $g$ sets the unit of the charge, and the operator $\mathcal{O}$ gives the quantum number in units of $g$ (i.e., $\mathcal{O}\psi_i = q_i\psi_i$). Since the fields $\psi(x_1)$ and $\psi(x_2)$ will have different transformations, the invariance of the theory would require gauge fields that can communicate between these two points $x_1$ and $x_2$.

As an example, let us start with the Lagrangian density for the fermion field $\psi_i$:

$$\mathscr{L}(\psi,\partial\psi) = i\sum_i \overline{\psi_i}\gamma^\mu \partial_\mu \psi_i + \sum_{i,j} m_{ij}\overline{\psi_i}\psi_j. \tag{4.52}$$

We assume the condition $m_{ij} = 0$ for $i \neq j$, which will guarantee that the system is invariant under the transformation generated by $\mathscr{O}$. The Lagrangian transforms as

$$\mathscr{L} \xrightarrow{\mathscr{O}} \mathscr{L}' = i\sum_i \overline{\psi_i}\gamma^\mu \partial_\mu \psi_i + g\mathscr{O}\partial_\mu \alpha(x)\sum_i \overline{\psi_i}\gamma^\mu \partial_\mu \psi_i + \sum_i m_{ii}\overline{\psi_i}\psi_i. \tag{4.53}$$

This Lagrangian is not invariant under the transformation generated by $\mathscr{O}$, since $\partial_\mu \alpha(x)$ does not vanish. To compensate for this additional contribution, we introduce a new vector field $A_\mu$, which transforms as

$$A_\mu(x) \xrightarrow{\mathscr{O}} A'_\mu(x) = A_\mu(x) + \partial_\mu \alpha(x). \tag{4.54}$$

We can then define an invariant derivative,

$$D_\mu \psi_i(x) = \left[\partial_\mu + ig\mathscr{O}A_\mu(x)\right] \psi_i(x), \tag{4.55}$$

so $i\sum_i \overline{\psi_i}\gamma^\mu D_\mu \psi_i$ transforms to itself. If we now replace the ordinary derivatives from all the terms with the invariant derivative, the Lagrangian becomes invariant. Since the derivative of the wave function ($\partial \psi_i$) corresponds to the change in the wave function at two different space–time points, the introduction of the gauge field ($A_\mu$) implies that this gauge field carries this information about the gauge transformation from one point to the other. In other words, when two fields are transformed at two different points, the gauge fields carry this information from one particle to the other and compensate the difference, making the theory invariant under this transformation. Since the gauge fields convey this information to all points including the points at infinity, this field cannot have any mass. From the transformations of the gauge fields it is also possible to verify that the mass term $m^2 A_\mu A_\mu$ for this field is not invariant under the gauge symmetry. Thus, in all unbroken gauge theories, gauge bosons are massless.

The gauge invariance will allow those terms in the Lagrangian that are invariant under the gauge transformation. The final gauge invariant Lagrangian includes the kinetic energy term for the gauge fields, making the gauge field a dynamical variable and is given by

$$\mathscr{L} = i\sum_i \overline{\psi_i}\gamma^\mu D_\mu \psi_i + \sum_i m_{ii}\overline{\psi_i}\psi_i - \frac{1}{4}F_{\mu\nu}F_{\mu\nu}. \tag{4.56}$$

The field–strength tensor is defined by

$$F_{\mu\nu} = \partial_\mu A_\nu - \partial_\nu A_\mu, \tag{4.57}$$

which satisfies the equation of motion

$$D_\mu F_{\mu\nu} = j_\nu \tag{4.58}$$

and the identity

$$D_\mu F_{\nu\lambda} + D_\nu F_{\lambda\mu} + D_\lambda F_{\mu\nu} = 0, \tag{4.59}$$

with the external current $j_\nu$ satisfying $D_\mu j_\mu = 0$.

The simplest and most common gauge theory is the quantum electrodynamics (QED). It is an Abelian $U(1)$ gauge theory with only one generator $\mathscr{O}$. In this case $|g| = e$ is the charge of the electron, and $\mathscr{O}$ acting on the wave function of any particle gives the charge of the particle. $A_\mu$ is the usual four-vector potential and $F_{\mu\nu}$ is the field–strength tensor. The field equations and the identity give the four Maxwell equations. There is only one gauge boson corresponding to the generator of the theory, which is the massless photon.

### Non-Abelian Gauge Theory

In a *non-Abelian gauge theory* the symmetry is generated by more than one generator, which do not commute with each other [30]. The operators generating the symmetry are now associated with the generators of a non-Abelian group. Consider a non-Abelian group $SU(2)$, generated by the three *Pauli matrices* $\tau_i, i = 1,2,3$, satisfying

$$[\tau_i, \tau_j] = i\varepsilon_{ijk}\tau_k, \tag{4.60}$$

where $\varepsilon_{ijk}$ is the *Levi–Civita symbol* ($+1$ for even permutation of 123; $-1$ for odd permutation; 0 if any two indices are the same). Let the particles $u$ and $d$ belong to the two-dimensional fundamental representation of the group

$$\Psi = \begin{pmatrix} u \\ d \end{pmatrix},$$

and the state vector transform as

$$\Psi \equiv \begin{pmatrix} u \\ d \end{pmatrix} \to \exp^{-ig\tau_i\alpha_i(x)} \begin{pmatrix} u \\ d \end{pmatrix}. \tag{4.61}$$

It is now convenient to write the generators of the group in the Cartan–Weyl basis $[\tau_+, \tau_-, \tau_3]$, where $\tau_\pm = \tau_1 \pm i\tau_2$ are the two raising and lowering generators, so the covariant derivative may be defined as

$$D_\mu = \partial_\mu + ig\sum_i \tau_i A^i_\mu = \partial_\mu + ig(\tau_+ A^+_\mu + \tau_- A^-_\mu + \tau_3 A^0_\mu), \tag{4.62}$$

where we define $A^\pm_\mu = A^1_\mu \mp iA^2_\mu$ and $A^0_\mu = A^3_\mu$. As with the Abelian gauge theory, we now replace all derivatives with the covariant derivatives and can write the Lagrangian including the kinetic energy terms of the gauge fields as

$$\mathscr{L}\left[\Psi, \partial_\mu\Psi\right] \longrightarrow \mathscr{L}\left[\Psi, D_\mu\Psi\right] - \frac{1}{4}F_{\mu\nu i}F^i_{\mu\nu}, \tag{4.63}$$

with the field–strength tensor

$$F_{\mu\nu i} = D_\mu A_{\nu i} - D_\nu A_{\mu i} + i\varepsilon_{ijk}A^j_\mu A^k_\nu \tag{4.64}$$

satisfying the field equation and the identity

$$D_\mu F^i_{\mu\nu} = J^i_\nu$$
$$\varepsilon_{\mu\nu\lambda} D_\mu F_{\nu\lambda} = 0, \tag{4.65}$$

where the current $J^i_\mu$ satisfies the conservation equation

$$D_\mu J^i_\mu = 0. \tag{4.66}$$

Equations 4.65 are the generalization of the Maxwell's equations for non-Abelian gauge theory and are called the *Yang–Mills equations* [30].

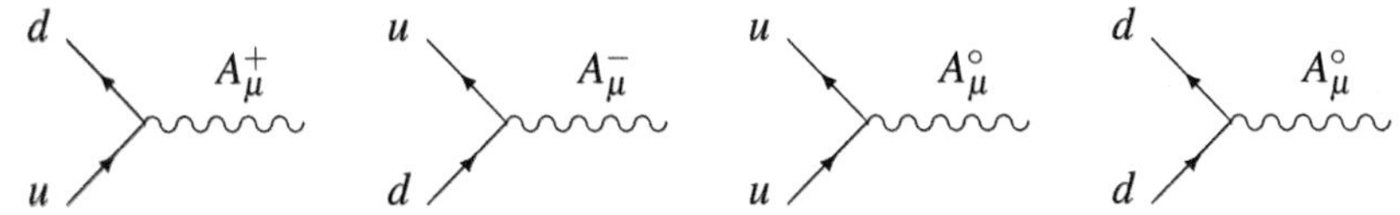


**FIGURE 4.2**
Interactions of non-Abelian gauge bosons with the fermions.

In a non-Abelian gauge theory, there is a gauge boson corresponding to each of the generators of the group. The interactions of the gauge bosons with the fermions are given by

$$\mathscr{L} = -ig\overline{\Psi}^i \gamma^\mu A^a_\mu \tau^a_{ij} \Psi^j. \tag{4.67}$$

These interactions are represented by the Feynman diagrams in figure 4.2. The gauge bosons $A^\pm_\mu$ mediate the interaction between the two states ($\bar{u}\gamma_\mu d$ or $\bar{d}\gamma_\mu u$) with different $SU(2)$ quantum numbers $\pm 1/2$, while the gauge boson $A^3_\mu$ mediates the interaction between two states ($\bar{u}\gamma_\mu u$ or $\bar{d}\gamma_\mu d$) with the same $SU(2)$ quantum number. Let us generalize this to the $SU(3)$ group. There are 8 generators of the group and hence, there will be 8 gauge bosons in an $SU(3)$ gauge theory. The three states belonging to the fundamental representation of the group are $u, v, w$. Two of the gauge bosons, which are associated with the two diagonal generators of the group, will mediate interactions among any one of the states ($\bar{u}\gamma_\mu u$ or $\bar{v}\gamma_\mu v$ or $\bar{w}\gamma_\mu w$), while the remaining six gauge bosons will mediate interactions between the different states ($\bar{u}\gamma_\mu v$, $\bar{v}\gamma_\mu w$, $\bar{w}\gamma_\mu u$, $\bar{u}\gamma_\mu w$, $\bar{w}\gamma_\mu v$ or $\bar{v}\gamma_\mu u$). Since the mass terms are not gauge invariant, all the gauge bosons are massless, and hence, these interactions have infinite range.

### Higgs Mechanism

If gauge boson mass terms are included in a gauge theory, the theory does not remain renormalizable. The spontaneous symmetry breaking of a gauge theory, which is known as the *Higgs mechanism*, can make the gauge bosons massive without destroying the renormalizability [27]. In the *Higgs mechanism*, the *vev* of a scalar field breaks the gauge symmetry spontaneously. Although the interactions of the Higgs scalar are invariant under the gauge symmetry, the minimum of the scalar potential does not respect the gauge symmetry. Since the Lagrangian is gauge invariant, the theory remains renormalizable, but the vacuum breaks the gauge symmetry allowing a mass term for the gauge boson.

We shall consider a $U(1)$ gauge theory, in which the fermion $\psi$ transforms locally as

$$\psi(x) \to \psi'(x) = \exp^{-ig\mathscr{O}q(x)}\psi(x).$$

The corresponding Lagrangian will be given by

$$\mathscr{L} = i\bar{\psi}\gamma^{\mu}D_{\mu}\psi + m\bar{\psi}\psi - \frac{1}{4}F^{\mu\nu}F_{\mu\nu}, \tag{4.68}$$

where

$$D_{\mu} = \partial_{\mu} + ig\mathscr{O}A_{\mu}(x)$$

is the covariant derivative and

$$F_{\mu\nu} = \partial_{\mu}A_{\nu} - \partial_{\nu}A_{\mu}$$

is the field–strength tensor that satisfies the equation of motion

$$\partial_{\mu}F_{\mu\nu} = j_{\nu}.$$

The mass term for the gauge boson $A_{\mu}$ is not allowed by the gauge symmetry; hence, it remains massless and hence the range of this interaction is infinity.

We now introduce a scalar field in the theory $\phi(x)$ that transforms as

$$\phi(x) \to \phi'(x) = \exp^{-ig\mathscr{O}q(x)}\phi(x).$$

The gauge invariant Lagrangian is given by

$$\mathscr{L}(\phi) = D^{\mu}\phi^{\dagger}(x)D_{\mu}\phi(x) - \frac{1}{2}\mu^2\phi^{\dagger}\phi - \frac{1}{4}\lambda(\phi^{\dagger}\phi)^2. \tag{4.69}$$

When $\mu^2 < 0$, the scalar potential will have a continuum of minima satisfying

$$\langle\phi^{\dagger}\phi\rangle = v^2 = -\mu^2/\lambda > 0.$$

The vacuum will pick up any one of these minima and the field will acquire a *vev* $\langle\phi\rangle = v$. If we now expand the physical field $\eta$ around the *vev* of the scalar field $v$

$$\phi = (\eta + v)\exp^{i\sigma/v},$$

the kinetic energy term for the scalar field will contain a mass term for the gauge boson,

$$g^2v^2A^{\mu}A_{\mu} \subset D^{\mu}\phi^{\dagger}D_{\mu}\phi.$$

Unlike the case of spontaneous global symmetry breaking, in this case there is no physical Nambu–Goldstone boson. The imaginary part of the scalar field ($\sigma$) has now become the longitudinal mode of the gauge boson $A_\mu$.

In case of non-Abelian gauge symmetry, it is possible to introduce the proper Higgs scalar field with appropriate interactions, leaving a residual gauge symmetry that we require after the spontaneous symmetry breaking. The gauge bosons corresponding to the generators of the coset space of the original symmetry group and the residual symmetry group will now become massive. In other words, if the original theory has a symmetry $\mathscr{G}_1$ (with $n_1$ generators), which breaks down to a lower subgroup $\mathscr{G}_2 \subset \mathscr{G}_1$ (with $n_2$ generators) by the *vev* of a scalar field $\phi$, there will be $n_1 - n_2$ massive gauge bosons, while $n_2$ number of gauge bosons will still remain massless.

Let us consider some examples of symmetry breaking. If we want to break an $SU(n)$ group to its subgroup $SU(n-1)$, we should consider a scalar field belonging to $\mathbf{n}$ representation of the group $SU(n)$. The gauge bosons corresponding to the coset space $SU(n)/SU(n-1)$ will now become massive after the symmetry breaking. When the *vev* of a scalar field belonging to the adjoint representation of the gauge group induces the spontaneous symmetry breaking, the rank of the residual gauge group is not changed. If we start with an $SU(n)$ gauge group, the residual group will be $SU(n-1) \times U(1)$ or $SU(n-m) \times SU(m) \times U(1)$. For example, a scalar field $\phi$ belonging to the adjoint representation of the group $SU(5)$ can break the $SU(5)$ symmetry into its subgroup

$$SU(5) \to SU(4) \times U(1) \quad \text{or} \quad SU(5) \to SU(3) \times SU(2) \times U(1)$$

if the *vev* of the scalar field is of the form

$$\langle\phi\rangle = \begin{pmatrix} 1 & 0 & 0 & 0 & 0 \\ 0 & 1 & 0 & 0 & 0 \\ 0 & 0 & 1 & 0 & 0 \\ 0 & 0 & 0 & 1 & 0 \\ 0 & 0 & 0 & 0 & -4 \end{pmatrix} \quad \text{or} \quad \begin{pmatrix} 1 & 0 & 0 & 0 & 0 \\ 0 & 1 & 0 & 0 & 0 \\ 0 & 0 & 1 & 0 & 0 \\ 0 & 0 & 0 & -\frac{3}{2} & 0 \\ 0 & 0 & 0 & 0 & -\frac{3}{2} \end{pmatrix},$$

respectively. Depending on the components of the scalar fields, the phenomenology of the spontaneous symmetry breaking may also change. In the case of electroweak symmetry breaking, if the symmetry breaking is done only by a triplet Higgs scalar, the $\rho$ parameter will not be consistent. It is also possible to put constraint on the mass of the triplet Higgs scalar from such consideration.

### Anomalies

An *anomaly* is a quantum effect that breaks the classical symmetry of any theory. In the language of field theory, anomalies are loop diagrams with infinite contributions so the renormalizability of the theory is lost. Consider a chiral gauge theory, in which the bare mass terms of any fermions are prevented and, hence, axial currents are conserved. In any chiral gauge theories, there are divergent one-loop triangle diagrams (see figure 4.3), which are divergent and are known as Adler–Bardeen–Jackiw anomalies [28]. The anomaly diagrams involve fermions in the triangle loop.

Consistency of any gauge theory requires that the sum over all the fermions should cancel these anomaly diagrams and that the theory be anomaly free. The contribution of the left-handed fermions and the right-handed fermions come with opposite signs, and hence, any vectorial theory is always free of anomaly. When the left-handed and right-handed particles transform in a different manner under any gauge theory, one takes care of anomaly cancellation by ensuring that the sum over all the fermions cancels the anomaly. If there are nonvanishing anomalies in a chiral gauge theory, there will be infinite one-loop diagrams that will destroy the conservation of the axial current. Since there are no bare mass terms of the fermions, these infinities can not be removed by introducing any counterterms.

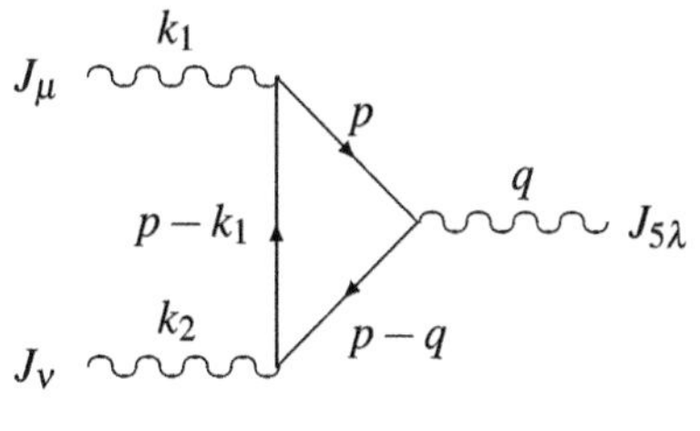


**FIGURE 4.3**
One-loop triangle anomaly diagram.

Let us consider a gauge theory with chiral fermions and both vector and axial vector gauge bosons, $A^\mu$ and $A_5^\mu$, respectively. Their interactions are given by

$$\begin{aligned} J_\mu A^\mu &= \bar{\psi}\gamma_\mu\psi A^\mu = (\overline{\psi_L}\gamma_\mu\psi_L + \overline{\psi_R}\gamma_\mu\psi_R)A^\mu \\ J_{5\mu}A_5^\mu &= \bar{\psi}\gamma_\mu\gamma_5\psi A_5^\mu = (-\overline{\psi_L}\gamma_\mu\psi_L + \overline{\psi_R}\gamma_\mu\psi_R)A_5^\mu . \end{aligned} \tag{4.70}$$

The gauge invariance will ensure that the vector current is conserved

$$\partial^\mu J_\mu = 0.$$

However, the Dirac equations for the fermions would give

$$\partial^\mu J_{5\mu} = 2im\ \bar{\psi}\gamma_5\psi,$$

implying that the axial current vanishes only when the fermions are massless $m = 0$. In other words, the chiral symmetry that keeps the fermions massless implies that this theory is invariant at the tree level under the axial symmetry

$$\psi = \exp^{i\varepsilon(x)\gamma_5}\psi.$$

We shall now demonstrate that this conservation of the axial current is valid only at the tree level. Once we introduce the infinite contributions from the one-loop anomalies, the axial currents are no longer conserved, even in the limit of $m = 0$.

The matrix element for the one-loop triangle anomaly diagram of figure 4.3 is given by

$$T_{\mu\nu\lambda}(k_1,k_2,q) = -\int \frac{d^4p}{(2\pi)^4}\,\mathrm{Tr}\left[\frac{i}{\not p - m}\gamma_\lambda\gamma_5\frac{i}{\not p - \not q - m}\gamma_\nu\frac{i}{\not p - \not k_1 - m}\gamma_\mu\right] + [k_1 \leftrightarrow k_2; \mu \leftrightarrow \nu]\,. \tag{4.71}$$

Simple power counting would tell us that the integral is linearly divergent. This will complicate certain manipulation of the integral. For example, usually any shift of variables $x+a \to x$ does not modify an integral:

$$I(a) = \int_{-\infty}^{+\infty} dx\,[f(x+a) - f(x)] = 0\,. \tag{4.72}$$

However, when the integral diverges linearly, this difference does not vanish and instead it is given by

$$I(a) = \int_{-\infty}^{+\infty} dx\left[af'(x) + \frac{a^2}{2}f''(x) + \cdots\right] = a\,[f(\infty) - f(-\infty)] + \cdots\,. \tag{4.73}$$

We shall now calculate certain identities without including the $I(a)$ contributions due to the shift in the variables of integration, and then see how the results are modified by including these infinities.

Starting from the matrix element for the anomaly, it is possible to derive the Ward identities

$$\begin{aligned} k_1^\mu T_{\mu\nu\lambda} &= 0 \\ k_2^\nu T_{\mu\nu\lambda} &= 0 \\ q^\lambda T_{\mu\nu\lambda} &= 2\,m\,T_{\mu\nu}\,, \end{aligned} \tag{4.74}$$

where

$$T_{\mu\nu}(k_1,k_2,q) = -\int \frac{d^4p}{(2\pi)^4}\,\mathrm{Tr}\left[\frac{i}{\not p - m}\gamma_5\frac{i}{\not p - \not q - m}\gamma_\nu\frac{i}{\not p - \not k_1 - m}\gamma_\mu\right] + [k_1 \leftrightarrow k_2; \mu \leftrightarrow \nu]\,. \tag{4.75}$$

We make use of the identities

$$\begin{aligned} \not k_i &= (\not p - m) - (\not p - \not k_i - m) = (\not p + \not k_i - m) - (\not p - m) \quad (i = 1,2) \\ \not q\gamma_5 &= (\not p - m)\gamma_5 - (\not p - \not q - m)\gamma_5 = (\not p - m)\gamma_5 - \gamma_5(\not p - \not q - m) - 2m\gamma_5 \end{aligned} \tag{4.76}$$

and shift the variables of integration from $(\not p - \not k_1) \to \not p$ or $(\not p - \not q)$ to arrive at the Ward identities.

It is possible to take care of the infinities in any of the standard ways. In the Pauli–Villars method, the infinite integrals $I(a)$ are regularized by subtracting similar integrals with very large fermion mass $M$ and then taking the limit $M \to \infty$. In the Ward identity, this prescription modifies to

$$2\, m\, T_{\mu\nu}(m) \to 2\, M\, T_{\mu\nu}(M).$$

$T_{\mu\nu}(M)$ can now be evaluated by ignoring $k_1$ and $k_2$ in the integrand, and finally we obtain the anomaly in the axial vector current

$$q^{\lambda} T_{\mu\nu\lambda} = 2\, m\, T_{\mu\nu} - \frac{1}{2\pi^2} \varepsilon_{\mu\nu\rho\sigma} k_1^{\rho} k_2^{\sigma}. \tag{4.77}$$

In the coordinate space the axial vector anomaly is expressed as

$$\partial^{\mu} J_{5\mu} = 2i\, m\, \bar{\psi}\gamma_5 \psi + \frac{g^2}{8\pi^2} \widetilde{F}_{\mu\nu} F^{\mu\nu}, \tag{4.78}$$

where $\widetilde{F}_{\mu\nu} = \frac{1}{2}\varepsilon_{\mu\nu\rho\sigma}F^{\rho\sigma}$ and $g$ is the gauge coupling constant. There is no contribution from anomaly to the vector currents.

The anomalies make any theory nonrenormalizable and, hence, unpredictable. Thus, any consistent gauge theory must be free of all anomalies. If fermions couple to the gauge fields through the current

$$J_{\mu}^{a} = \bar{\psi}_R \gamma_{\mu} T_R^a \psi_R + \bar{\psi}_L \gamma_{\mu} T_L^a \psi_L,$$

where $T^a$ defines the matrix representations for the fermions, then every fermion will be associated with a group factor

$$\mathscr{A} = \mathrm{Tr}\left[ (T_L^a T_L^b + T_L^b T_L^a) T_L^c - (T_R^a T_R^b + T_R^b T_R^a) T_R^c \right]. \tag{4.79}$$

The trace is over all the fermions in each representations. The condition for cancellation of anomalies is then given by

$$\sum \mathscr{A} = 0,$$

where the sum is extended over all fermion representations, since anomalies do not depend on the masses of the fermions.

We now present a simple algorithm to calculate anomalies in 4 and higher dimensions [29]. If the fermions belong to a representation $\mathscr{R}_r$ of $\mathscr{G}$ in 4 dimensions, the group factor contributing to the triangle anomaly will be

$${}_3\mathscr{A} = \mathrm{tr}\, [T^a(\mathscr{R}_r) T^b(\mathscr{R}_r) T^c(\mathscr{R}_r)], \tag{4.80}$$

where $T^a(\mathscr{R}_1)$ are the generators of the representation. In $2d$ dimensions, the anomaly will have $n$ fermions in the loop, where $n = d+1$, so we need to calculate an anomaly

$${}_n\mathscr{A} = \mathrm{tr}\, [T^{a1}(\mathscr{R}_r) T^{a2}(\mathscr{R}_r) \cdots T^{an}(\mathscr{R}_r)].$$

In odd dimensions, there is no anomaly, since there are no chiral fermions. For some representations in higher dimensions, there are additional factorized contributions of the form $\left(\text{tr}\left[T^{a1}(\mathscr{R}_r)T^{a2}(\mathscr{R}_r)\cdots\right]\right)^2$, but we shall not discuss these reducible group factors separately.

Consider the subgroup $\mathscr{G}_1 \times \mathscr{G}_2 \subset \mathscr{G}$, where $\mathscr{G}_2 = U(1)$ is an Abelian subgroup of $\mathscr{G}$. We can then decompose any representations of $\mathscr{G}$ under $\mathscr{G}_1 \times \mathscr{G}_2$ subgroup as

$$\mathscr{R} = \sum_i (r_i, f_i), \tag{4.81}$$

where $f_i$ represents the $\mathscr{G}_2 \equiv U(1)$ quantum numbers. $f_i$ is normalized so the anomaly for the fundamental representation of any group is 1. The decomposition of the fundamental $m$-dimensional representation of $SU(m)$ under $SU(m-1) \times U(1)$ will then become $m = (m-1, 1) + (1, -m+1)$. The product decomposition rules then give the decompositions of the other representations.

We can then use the formulas

$$
\begin{aligned}
{}_3\mathscr{A}(\mathscr{R}) = \sum_i \lambda(r_i)\cdot f_i \qquad\qquad & {}_4\mathscr{A}(\mathscr{R}) = \sum_i {}_3\mathscr{A}(r_i)\cdot f_i \\
{}_n\mathscr{A}(\mathscr{R}) = & \sum_i {}_{(n-1)}\mathscr{A}(r_i)\cdot f_i
\end{aligned}
\tag{4.82}
$$

to calculate the anomalies for the representation $\mathscr{R}$ of the group $\mathscr{G}$ in terms of the invariants of the subgroup $\mathscr{G}_1 \times \mathscr{G}_2$. Here $\lambda(r_i)$ is the index for the representation $r_i$ of $\mathscr{G}_1$ (discussed in section 3.4). For verification of the calculated anomaly, one can use the relation

$$ {}_n\mathscr{A}(\mathscr{R}) = \sum_i {}_n\mathscr{A}(r_i). \tag{4.83}$$

For completeness we also present a couple of useful relations

$$
\begin{aligned}
{}_n\mathscr{A}(\mathscr{R}_1 + \mathscr{R}_2) &= {}_n\mathscr{A}(\mathscr{R}_1) + {}_n\mathscr{A}(\mathscr{R}_2) \\
{}_n\mathscr{A}(\mathscr{R}_1 \times \mathscr{R}_2) &= {}_n\mathscr{A}(\mathscr{R}_1)D(\mathscr{R}_2) + {}_n\mathscr{A}(\mathscr{R}_2)D(\mathscr{R}_1).
\end{aligned}
\tag{4.84}
$$

Thus, by writing the decomposition of any representation under its subgroup containing a properly normalized $U(1)$ factor, it will be possible to calculate the irreducible group factor entering into the expression for anomalies.

As an example, let us calculate the anomaly of 6-dimensional representation of the group $SU(3)$, which decomposes under $SU(2) \times U(1)$ as $6 = (3,2) + (1,-4) + (2,-1)$. Then the required anomaly will be ${}_3\mathscr{A}(6) = \lambda(3)\cdot 2 + \lambda(2)\cdot(-1) = 7$. For the cancellation of anomalies in any theory, we require only this factor, and hence, these results will be extremely useful while building models in many extensions of the standard model.

# 5

# *The Standard Model*

The standard model of particle physics is the theory of the strong, the electromagnetic and the weak interactions. The strong interaction is the interaction among the quarks of different colors (there are three colors) and flavors (six flavors $u, c, t, d, s, b$), and they are mediated by eight gluons. It is an $SU(3)$ gauge theory, called quantum chromodynamics (QCD), and the three colored states of every flavor belong to the triplet representation of the $SU(3)_c$. The gluons are the eight gauge bosons corresponding to the generators of the group $SU(3)_c$. Colored states are confined, and hence, only color singlet states can exist in nature as free particles. Known color singlet states are baryons (made of three quarks) or mesons (made of quark–antiquark pairs). The strong nuclear force is the force between the protons and neutrons, which is a manifestation of the underlying $SU(3)_c$ interactions among the quarks, although with our present knowledge of the many-body problem it is almost impossible to derive the nuclear potential starting from QCD without any assumption.

Interactions of charged particles are governed by electromagnetic interaction. It is described by quantum electrodynamics (QED), which is a $U(1)_Q$ gauge theory, where $Q$ corresponds to the electric charge. Electromagnetic interaction is mediated by the photon, which is the gauge boson corresponding to the generator of the $U(1)_Q$ group.

The weak interaction describes nuclear beta decay, and at low energy it is given by an effective four fermion interaction. Since the effective operator has dimension 6, the coupling constant has inverse mass-squared dimension. This effective theory fails at higher energies but is found to emerge from another renormalizable theory, the electroweak interaction, which unifies both the weak and the electromagnetic interactions.

The electroweak theory is described by an $SU(2)_L \times U(1)_Y$ interaction, which is spontaneously broken down to $U(1)_Q$ at around 100 GeV. The spontaneous symmetry breaking makes three of the gauge bosons $W^{\pm}, Z$ heavy, leaving only one gauge boson massless, which is the photon. Since the $W^{\pm}, Z$ bosons are massive, at low energies they appear as internal propagators in the effective four fermion interactions. Under the $SU(2)_L$ group, the left-handed quarks and leptons are doublets, while the right-handed fields are singlets. Thus, the mass terms for the fermions are forbidden before the electroweak symmetry breaking. However, after the electroweak phase transition, the *vevs* of the Higgs scalar that breaks the electroweak symmetry give masses to the quarks and leptons. The problem of neutrino mass is somewhat different, which we shall discuss in the next chapter. With this introduction we shall now describe the standard model.

## 5.1 Quantum Chromodynamics

The forces between the nucleons (protons and neutrons) have to be much stronger than the electromagnetic forces to keep the positively charged protons in the nucleus. This force is called the strong nuclear force, which is mediated by the pions or the $\pi$-meson. This is an effective theory at low energy. As we probe higher and higher energies, it becomes clear that the nucleons and pions are part of a larger family, called the baryons and mesons, respectively. The large number of baryons and mesons are again made of only a few quarks and antiquarks [31]. The interaction between the quarks and antiquarks, mediated by gluons, is called the strong interaction. At low energy the strong interaction keeps the quarks and antiquarks together inside the baryons and the mesons.

All baryons are made of three quarks, and mesons are made of quark–antiquark pairs. There are three quarks $u,c,t$ with charge $+2/3$ and three quarks $d,s,b$ with charge $-1/3$. These six are called flavors of quarks. Each of these six flavors of quarks can have one of the three colors $r,g,b$, so there are 18 quarks in total. The proton is made of three quarks, each carrying different colors $p \equiv u^r u^g d^b$; the neutron is similarly made of $n \equiv u^r d^g d^b$; and the $\pi$-mesons or the pions ($\pi^+$, $\pi^0$, $\pi^-$) are made of quark–antiquark pairs, each of them carrying the same color ($\pi^+ \equiv u^i \bar{d}^i$, $\pi^0 \equiv u^i \bar{u}^i$ or $d^i \bar{d}^i$, $\pi^- \equiv d^i \bar{u}^i$, with $i = r,g,b$). There are 8 gluons that mediate the interactions between the quarks and they carry color quantum numbers. The strong interaction between the quarks and antiquarks keeps them together to form nucleons and mesons, and the strong nuclear force is an effective theory, which should in principle be derivable from the interactions between the quarks.

The strong interaction that binds the quarks and antiquarks in the baryons and mesons is described by an $SU(3)_c$ gauge theory, called quantum chromodynamics (QCD). The three colored quarks belong to the 3-dimensional fundamental representation of the group,

$$\Psi^i_\alpha \equiv \begin{pmatrix} q^i_1 \\ q^i_2 \\ q^i_3 \end{pmatrix}, \tag{5.1}$$

where $\alpha = 1,2,3$ is the color index corresponding to $r,g,b$ and $i = u,d,c,s,t,b$ is the flavor index. Only the color singlet states could exist in nature, which are the baryons $\varepsilon^{\alpha\beta\rho} q^i_\alpha q^j_\beta q^k_\rho$ and the mesons $\delta^\alpha_\beta \bar{q}^{\beta i} q^j_\alpha$.

The eight gauge bosons $G^a_\mu, a = 1...8$, called the gluons, belong to the adjoint representation of the $SU(3)_c$ group. These gauge bosons are associated with the eight generators of the group $T^a$, which can be interpreted as the raising and lowering operators: $T^{1\pm} = T^1 \pm iT^2$, connecting the states $q^i_1 \leftrightarrow q^i_2$; $T^{2\pm} = T^4 \pm iT^5$, connecting the states $q^i_2 \leftrightarrow q^i_3$; $T^{3\pm} = T^6 \pm iT^7$, connecting the states $q^i_1 \leftrightarrow q^i_3$; and the diagonal generators defining the quantum numbers of any state: $T^3 = \text{diag}[+\frac{1}{2},\ -\frac{1}{2},\ 0]$ and $T^8 = \frac{1}{2\sqrt{3}}\,\text{diag}[+1,\ +1,\ -2]$. The generators satisfy the commutation relation $[T^a, T^b] = i f_{abc} T^c$.

The interactions of the quarks and the gluons are governed by the $SU(3)$ gauge invariant Lagrangian, given by

$$\mathscr{L}[\psi, \partial\psi] = \overline{\psi}_\alpha \, i\gamma^\mu D_\mu \psi_\alpha + m_{\alpha\beta}\overline{\psi}_\alpha \psi_\beta - \frac{1}{4} G^a_{\mu\nu} G^{\mu\nu a}, \tag{5.2}$$

with the covariant derivative and the field-strength tensors defined as

$$D_\mu = \partial_\mu + i g_s T^a G^a_\mu \qquad \text{or} \qquad (D_\mu)_{bc} = \partial_\mu \delta_{bc} + i g_s (T^a)_{bc} G^a_\mu$$

$$G^a_{\mu\nu} = \partial_\mu G^a_\nu - \partial_\nu G^a_\mu + i f^{abc} G^b_\mu G^c_\nu .$$

Here, $g_s$ is the gauge coupling constant, which is related to the strong fine structure constant $\alpha_s = g_s^2/4\pi$.

The term $g_s \overline{\psi}^i_\alpha \gamma^\mu G^a_\mu T^a_{\alpha\beta} \psi^i_\beta$ gives the gauge interactions of the quarks and antiquarks. It tells us how the gluons interact with the quarks and transforms one colored quark to another keeping the flavor unchanged. In an Abelian gauge theory such as the electromagnetic interaction, there is only one commuting generator, and hence, there are no structure functions $f^{abc}$. As a result there are no self-interactions of the photon. The main difference between the electromagnetic interaction and the strong interaction arises from the term $i f^{abc} G^b_\mu G^c_\nu$, which allows the three- and four-gluon interactions.

For a consistent perturbation theory, we have to include another term in the Lagrangian

$$\mathscr{L}_g = \mathscr{L}_{gauge-fixing} + \mathscr{L}_{ghost}. \tag{5.3}$$

The gauge-fixing term ensures validity of the perturbative expansion and consistency of the definition of the gluon propagator. A class of covariant gauges could be fixed by

$$\mathscr{L}_{gauge-fixing} = -\frac{1}{2\lambda} (\partial^\mu A^a_\mu)^2.$$

This has to be supplemented by the ghost Lagrangian

$$\mathscr{L}_{ghost} = \partial_\mu \eta^{a\dagger} \left( D^\mu_{ab} \eta^b \right),$$

for consistency in any non-Abelian gauge theory. The ghost $\eta^a$ is a complex scalar field that obeys Fermi statistics.

In gauge theories, renormalization makes the gauge coupling constants energy dependent [32]. Although at the tree-level the gauge coupling is a constant quantity, the quantum corrections or the higher-order loop integrals make it dependent on energy. Thus, the value of the coupling constant will depend on the energy at which it has been measured. The divergent one-loop diagrams, which affect the gauge coupling constant, are given in figure 5.1: the quark self-energy diagram, the gluon self-energy diagrams and the quark–gluon vertex diagrams. Most of these diagrams are present in QED, except for the ones involving the three- and four-gluon vertices, which makes QCD so different from QED. All these diagrams are divergent, and

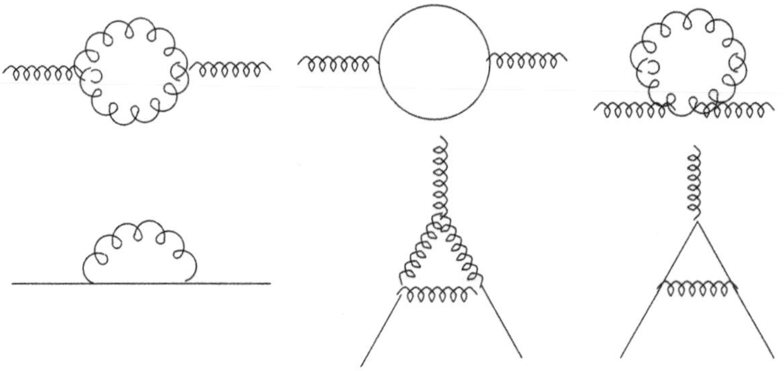

**FIGURE 5.1**
One-loop quark and gluon self-energy diagrams and vertex diagrams contributing to the gauge coupling constant renormalization.

hence, we need to introduce counterterms to cancel these infinities. Since the theory is renormalizable, all these infinities could be removed by redefining the bare coupling constants. After removing all the infinities with a proper renormalization procedure, the renormalized gauge coupling constant now satisfies the renormalization group equation [32]

$$\mu \frac{dg_s}{d\mu} = \beta(g_s) = \frac{g_s^3}{16\pi^2}\left(-11 + \frac{2n_F}{3}\right), \tag{5.4}$$

where $n_F$ is the number of quark flavors and $\mu$ is some arbitrary mass parameter used to evaluate the divergent integrals, which is the energy scale at which the coupling constant has to be evaluated.

For $n_F = 6$ the coefficient $\beta(g_s) < 0$, which implies that the strong coupling constant $g_s$ decreases with increasing mass (momentum) scale $\mu$, so the theory is asymptotically free. This asymptotic freedom is one of the most important features of any non-Abelian gauge theory, in particular for the QCD. So, although quarks are required to be confined [33]; this asymptotic freedom allows us to probe them with high-energy projectiles inside a hadron and treat them as free point particles [34]. Since the confinement does not allow free quarks, any low energy probes can find only the hadrons and mesons. Only very high energy projectiles can see the quarks in any hadrons and allow us to study QCD. This is done by probing a proton or a nucleon with very high energy electrons or neutrinos with the exchange of very high momentum. Consider an electron with incident energy $E$, which is scattered from a proton. To understand the experiment, the parton model was proposed by Feynman [34]. Feynman proposed an intuitive parton picture, which assumes that the proton is made of free constituent point-like particles called *partons*. This model was then developed by Bjorken and Paschos [34]. In this picture, when an electron is scattered by a parton, the fraction $x$ of the transverse momentum of the proton is carried by the parton. A measurement of the outgoing electron energy and its scattering angle determines this fraction $x$. For large momentum transfers ($q^2 \to \infty$), the structure functions of the proton become a scale-invariant function depending only on the fraction $x$, which is given by the ratio $x = (q^2/2mE)$. This is known as Feynman scaling. Further studies of this scaling phenomenon identified these partons as quarks.

At higher energies, the probing electrons scatter from the three valence quarks of the proton, and also the sea quarks and the gluons. Quantum fluctuations will create pairs of quarks and antiquarks and also gluons, which forms the sea of quarks and antiquarks and the gluons. As the energy increases, more sea quark–antiquark pairs are formed. These sea partons, coming from quantum corrections or higher order effects, appear as a screening and lead to small scaling violation, which has also been tested experimentally to a fairly good accuracy. In all experiments a large uncertainty is introduced because of our lack of complete understanding of QCD. The formation of hadrons, the confinement mechanism, and the scaling violation are among the few less understood problems in QCD. There could also be some nonperturbative effects introducing more complications. Thus, in many measurements the largest uncertainty comes from hadronic corrections. Moreover, the theoretical calculations of higher-order corrections are also very difficult.

## 5.2 $V - A$ Theory

The weak interaction has brought many surprises. The first one is the existence of this interaction. If this interaction were not present at all, maybe nothing would have changed at least at the macroscopic level. We can explain all matter around us and their interactions in terms of the electromagnetic and the gravitational interactions, and strong interaction explains why protons could stay together inside the nucleus in spite of the electromagnetic repulsive force. The first evidence of weak interaction appeared in the beta decay, in which a neutron converts into a proton and an electron is emitted – a simple, but fairly puzzling, process. A neutrino associated with this decay was postulated, but the neutrino was discovered much later. The mass of the neutrinos was discovered after more than sixty years. We still do not know if the neutrinos are Dirac or Majorana particles.

In the beginning the weak interaction could be explained by an effective four-point interaction at low energies [35]. It was not a renormalizable theory. Consider for example, the neutron decay $n \to p + e^- + \bar{\nu}_e$. Fermi generalized the concept of electrodynamics to explain this interaction. In the coupling of a vector current to the radiation field, $\bar{\Psi}\gamma_\mu \Psi A^\mu$, the 4-vector potential of the radiation field $A^\mu$ is replaced by another vector current. This leads to Fermi's $\beta$ decay Lagrangian for an effective four-fermion interaction

$$\mathscr{L}_\beta = \frac{G_F}{\sqrt{2}} \, [\bar{\Psi}_p \gamma_\mu \Psi_n] \, [\bar{\Psi}_e \gamma^\mu \Psi_\nu]. \tag{5.5}$$

Since this effective interaction has four fermions and mass dimension six, the effective coupling has an inverse mass-squared dimension and it is determined experimentally to be $G_F = 1.17 \times 10^{-5}$ GeV$^{-2}$.

The four-fermion could explain many observed characteristics of the beta decay, but some modifications were needed to explain several other weak interactions. For this purpose this interaction was then generalized to include all possible forms of the interactions,

$$\mathscr{L}_\beta = \frac{G_F}{\sqrt{2}} \sum_{i=S,P,V,A,T} C_i[\bar{\Psi}_p \Gamma_i \Psi_n]\,[\bar{\Psi}_e \Gamma^i \Psi_\nu]. \tag{5.6}$$

The interactions could now be scalar (S, $\Gamma_i = 1$), pseudo-scalar (P, $\Gamma_i = \gamma_5$), vector (V, $\Gamma_i = \gamma_\mu$), axial-vector (A, $\Gamma_i = \gamma_\mu\gamma_5$), or tensor (T, $\Gamma_i = \sigma_{\mu\nu}$). The coupling constants have to be determined from experiments. For example, for the Fermi transitions, $[\Delta J = 0]$, the interaction could be $S$ or $T$, while for the Gamow–Teller transitions, $[\Delta J = 0,\ \pm 1$, but $J = 0 \not\rightarrow J = 0]$, the interaction could be $A$ or $T$. Although this phenomenological Lagrangian could explain the nuclear beta decays at low energies, this is far from the actual theory of the weak interaction.

The next major step in the history of weak interactions was the discovery of parity nonconservation. The two particles $\tau$ and $\theta$ with the same mass, charge, spin (0) and lifetime were found to decay into states with odd and even parity, respectively,

$$\tau \rightarrow \pi^+\pi^+\pi^- \quad \text{and} \quad \theta \rightarrow \pi^+\pi^0.$$

This was known as the $\tau - \theta$ puzzle. As an explanation to this puzzle the possibility of parity nonconservation was suggested, so both particles ($\tau$ and $\theta$) could be identified as one particle $K^+$ [36]. Possible tests of parity violation were also proposed in other weak interactions. Soon parity violation was observed [37] through a measurement of the asymmetry in the beta decay of the polarized $^{60}$Co nuclei and later in $\pi$ decays.

The parity violation would then require a pseudoscalar term in the four-fermion Lagrangian of the form $C_i'[\bar{\Psi}\Gamma_i\Psi]\,[\bar{\Psi}\Gamma_i\gamma_5\Psi]$. Including this parity violating term, the Lagrangian can be written in the form

$$\mathscr{L}_\beta = \frac{G_F}{\sqrt{2}} \sum_{i=S,P,V,A,T} [\bar{\Psi}_p \Gamma_i \Psi_n]\,[\bar{\Psi}_e \Gamma^i (C_i - C_i'\gamma_5)\Psi_\nu]. \tag{5.7}$$

The coefficients $C_i$ and $C_i'$ were determined through a series of observations and analysis of different weak interaction processes. Finally the Lagrangian for the weak interaction was given a simple form

$$\mathscr{L}_\beta = \frac{G_F}{\sqrt{2}} [\bar{\Psi}_p \gamma_\mu (1 - \lambda\gamma_5)\Psi_n]\,[\bar{\Psi}_e \gamma^\mu (1 - \gamma_5)\Psi_\nu], \tag{5.8}$$

where $\lambda = 1.25$ was determined experimentally. The deviation from 1 can be attributed to strong interaction effects, and for the weak interactions, a $V - A$ form of the interaction was established [38].

The $V - A$ theory [38] could successfully explain all weak interactions. All charged processes such as beta decay, muon decay or charged pion decay could be explained in the $V-A$ theory by an effective Lagrangian density in which a charged weak current $J_\alpha$ is coupled to its Hermitian conjugate at a single point,

$$\mathscr{L}_\beta = -\frac{G_F}{2\sqrt{2}}[J_\alpha J^{\alpha\dagger} + J^{\alpha\dagger} J_\alpha]. \tag{5.9}$$

The charged weak current is universal in nature and contains two parts,

$$J_\alpha = J_{\ell\alpha} + J_{q\alpha},$$

where the leptonic part contains interactions among the leptons

$$J_{\ell\alpha} = \bar{\Psi}_e \gamma^\alpha (1-\gamma_5)\Psi_{\nu_e} + \bar{\Psi}_\mu \gamma^\alpha (1-\gamma_5)\Psi_{\nu_\mu} + \bar{\Psi}_\tau \gamma^\alpha (1-\gamma_5)\Psi_{\nu_\tau}, \tag{5.10}$$

and the hadronic part of the current contains the charged weak interactions among the hadrons.

The weak hadronic interactions are also of the $V-A$ form, except for the hadronic corrections. The deviation from the $V-A$ form due to the strong hadronic corrections is parameterized by $\lambda$. The hadronic part of the beta decay includes conversion of a neutron ($udd$) into a proton ($uud$), which is equivalent to a down-quark converting into a up-quark, so the transition $\bar{\Psi}_p \gamma^\alpha (1-\gamma_5)\Psi_n$ can be equivalently written as $\bar{\Psi}_u \gamma^\alpha (1-\gamma_5)\Psi_d$. The charged weak hadronic current is then given by

$$J_{q\alpha} = \bar{\Psi}_d \gamma^\alpha (1-\gamma_5)\Psi_u + \bar{\Psi}_s \gamma^\alpha (1-\gamma_5)\Psi_c + \bar{\Psi}_b \gamma^\alpha (1-\gamma_5)\Psi_t. \tag{5.11}$$

The states $\Psi_d$, $\Psi_s$ and $\Psi_b$ are admixtures of different mass eigenstates of $d, s$ and $b$, which will give rise to quark mixing. We shall come back to this discussion of quark mixing later.

The $V-A$ structure of the interaction implies that only the left-handed fermions take part in the interaction, and hence, both the parity and the charge conjugation are explicitly broken. This can be seen from

$$\frac{1}{2}\bar{\Psi}_1 \gamma^\alpha (1-\gamma_5)\Psi_2 = \overline{\frac{(1-\gamma_5)}{2}\Psi_1}\gamma^\alpha \frac{(1-\gamma_5)}{2}\Psi_2 = \overline{\Psi}_{1L}\gamma^\alpha \Psi_{2L}.$$

The $CPT$ would imply that the Hermitian conjugate term will involve the $CP$ conjugate states of the left-handed fermions $(\Psi_L)^c = (\Psi^c)_R$ in the weak interactions, but the right-handed fields $\Psi_R$ and their charge conjugate states $(\Psi_R)^c = (\Psi^c)_L$ should not enter in the $V-A$ theory.

Although the $V-A$ theory was a highly successful, effective low-energy theory, at higher energies the theory becomes inconsistent. Consider the $\nu_e + e^- \to \nu_e + e^-$ scattering, the $V-A$ theory gives a differential cross-section $(d\sigma/d\Omega) = (G_F^2/\pi^2)\omega^2$, where $\omega$ is the neutrino energy in the center of mass frame. However, if we expand the scattering amplitude in partial waves, $f(\theta) = \omega^{-1}\sum_{J=0}^{\infty}(J + 1/2)M_J P_J(\cos\theta)$, then for the Fermi contact interaction of zero range, we obtain the differential cross-section to be $(d\sigma/d\Omega) = (1/2\omega)^2|M_0|^2$, where only the $J = 0$ partial wave enters. Conservation of probability then gives the unitarity bound

$(d\sigma/d\Omega) \leq (1/2\omega)^2$, so at energies $\omega \geq (\pi G_F)^{1/2}/4^{1/4}$ of around 300 GeV, the $V-A$ theory is in conflict with unitarity.

The disagreement of the $V-A$ theory with unitarity at around 300 GeV implies that the weak interaction should be explained by some more fundamental theory at around 100 GeV. The low-energy limit of this theory should reproduce the $V-A$ theory as an effective theory. The intermediate vector boson theory was then proposed, which has the charged vector bosons $W^{\pm}$ with charge $\pm 1$ and masses of around $m_W \sim 100$ GeV. These vector bosons couple to the charged currents

$$\mathscr{L}_{cc} = g[J_\alpha W^{\alpha -} + J^{\alpha \dagger} W_\alpha^{+}] \tag{5.12}$$

with the dimensionless coupling constant $g$, which can be related to the Fermi coupling constant through $G_F \sim g^2/m_W^2$.

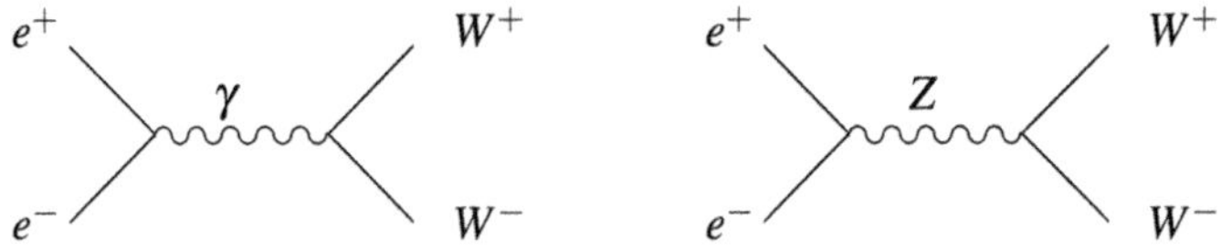


**FIGURE 5.2**
$e^+e^- \to W^+W^-$ scattering mediated by $\gamma$ and $Z$.

Although the intermediate vector boson theory solves the problem of conflict with unitarity, with only charged vector bosons $W^{\pm}$, it appeared to be incomplete. The scattering process $e^+e^- \to W^+W^-$ through photon exchange (see figure 5.2) becomes divergent in this theory. The consistency of the theory then requires the existence of a neutral vector boson $Z$, which couples to a neutral current of the form $\bar{\Psi}_e \gamma_\alpha (1-\gamma_5)\Psi_e Z^\alpha$. Similar currents should exist for all quarks and leptons. If the mass of this neutral gauge boson is similar to the mass of the charged bosons $W^{\pm}$, then the scattering $e^+e^- \to W^+W^-$ takes place through exchange of both the photon and the neutral gauge boson (as shown in figure 5.2), and the divergent contributions of these diagrams cancel, leading to a finite scattering cross-section.

In spite of its phenomenological success, the main problem with the intermediate vector boson theory is that the theory is not renormalizable and there is no explanation of the origin of the masses of the vector bosons. These problems have been solved in the standard model of electroweak interaction, which is a spontaneously broken gauge theory. After the electroweak symmetry breaking, the gauge bosons of the weak interactions as well as the fermions become massive. At energies above 100 GeV the low-energy theories of the electromagnetic interaction and the weak interaction become part of only one theory, the electroweak theory, which is a renormalizable theory.

## 5.3 Electroweak Theory

The electromagnetic and the weak interactions become unified at around 100 GeV into the electroweak theory. The electroweak theory [39] is a spontaneously broken gauge theory with the symmetry breaking pattern

$$SU(2)_L \times U(1)_Y \longrightarrow U(1)_Q,$$

where $Q$ is the electric charge and $Y$ is called the hypercharge and they are related by

$$Q = T_{3L} + Y, \tag{5.13}$$

where $T_{3L}$ is the diagonal generator of the group $SU(2)_L$. The components of any $SU(2)_L$ representation differ in their $T_{3L}$ quantum number by $\pm 1$; thus, the raising and the lowering generators carry electric charge $\pm 1$. This implies that the three gauge bosons of the group $SU(2)_L$ have electric charges $\pm 1, 0$. The gauge boson corresponding to the $U(1)_Y$ hypercharge is also electrically neutral. Before the symmetry breaking, all the four gauge bosons were massless, but after the symmetry breaking, three of these gauge bosons become massive. After the symmetry breaking, only a combination of the two neutral gauge bosons remains massless, which is the photon. The residual $U(1)_Q$ gauge theory can, thus, be identified with the electromagnetic interaction. The three gauge bosons corresponding to the broken gauge symmetry become massive, and they allow effective four-fermion weak interactions, which is the $V-A$ theory. At energies above the symmetry breaking phase transition, the electroweak theory is an $SU(2)_L \times U(1)_Y$ gauge theory. All left-handed fermions are doublets under the $SU(2)_L$ group while all right-handed fermions are singlets. Only after the symmetry breaking could they combine to give masses to the fermions.

### Particle Content

In the standard model, the quarks and the leptons are the only fermions. In addition to these fermions, there are the gauge bosons corresponding to the symmetry group $SU(2)_L \times U(1)_Y$ of the electroweak interactions. In the standard model, one writes the interactions including the strong interaction, so the symmetry group becomes $SU(3)_c \times SU(2)_L \times U(1)_Y$.

Only the left-handed fermions transform under the group $SU(2)_L$, while the right-handed fermions are all singlets. The electric charge of the different particles determines their hypercharge quantum numbers, once we assign the $SU(2)_L$ transformation. The left-handed up ($u_{\alpha L}$) and down ($d_{\alpha L}$) quarks form a doublet $Q_{i\alpha L}$. These fields carry an $SU(3)_c$ quantum number $\alpha = 1, 2, 3$ since the quarks take part in strong interactions. The weak interactions do not distinguish between the three generations, and hence, under the weak interactions, all three generations of particles behave in the same way. Thus, the left-handed charm quark ($c_{\alpha L}$) and the strange quark ($s_{\alpha L}$) form a doublet and so do the top quark ($t_{\alpha L}$) and the bottom quark ($b_{\alpha L}$). The left-handed neutrinos $\nu_{eL}$ and the electron $e_L^-$ also form a doublet under $SU(2)_L$. They do not carry any $SU(3)_c$ quantum numbers, since they do not take part in the strong interactions. There is no right-handed neutrino in the standard model. $\nu_{\mu L}$ and $\mu_L^-$ and also $\nu_{\tau L}$ and $\tau_L^-$ form doublets under $SU(2)_L$. The $SU(3)_c \times SU(2)_L \times U(1)_Y$ quantum numbers are given in table 5.1.

**TABLE 5.1**
Quantum numbers of the fermions under the gauge group $\mathcal{G}[3_c - 2_L - 1_Y] \equiv SU(3)_c \times SU(2)_L \times U(1)_Y$.

| | 1st generation | 2nd generation | 3rd generation | $\mathcal{G}[3_c - 2_L - 1_Y]$ quantum numbers |
|---|---|---|---|---|
| $Q_{\alpha iL}$ | $\begin{pmatrix} u_{\alpha L} \\ d_{\alpha L} \end{pmatrix}$ | $\begin{pmatrix} c_{\alpha L} \\ s_{\alpha L} \end{pmatrix}$ | $\begin{pmatrix} t_{\alpha L} \\ b_{\alpha L} \end{pmatrix}$ | $(3,2,1/6)$ |
| $U_{\alpha R}$ | $u_{\alpha R}$ | $c_{\alpha R}$ | $t_{\alpha R}$ | $(3,1,2/3)$ |
| $D_{\alpha R}$ | $d_{\alpha R}$ | $s_{\alpha R}$ | $b_{\alpha R}$ | $(3,1,-1/3)$ |
| $\Psi_{iL}$ | $\begin{pmatrix} \nu_{eL} \\ e_L^- \end{pmatrix}$ | $\begin{pmatrix} \nu_{\mu L} \\ \mu_L^- \end{pmatrix}$ | $\begin{pmatrix} \nu_{\tau L} \\ \tau_L^- \end{pmatrix}$ | $(1,2,-1/2)$ |
| $E_R^-$ | $e_R^-$ | $\mu_R^-$ | $\tau_R^-$ | $(1,1,-1)$ |

In addition to the fermions and the gauge bosons, there are also Higgs scalars in the theory, which breaks the $SU(2)_L \times U(1)_Y$ symmetry. The Higgs scalar is a doublet under $SU(2)_L$ with two components and its transformation under the group $SU(3)_c \times SU(2)_L \times U(1)_Y$ can be written as

$$\Phi \equiv \begin{pmatrix} \phi^+ \\ \phi^0 \end{pmatrix} \equiv [1,2,+1/2]. \tag{5.14}$$

After the symmetry breaking by $\Phi$, the $U(1)_Q$ group remains unbroken which gives us the electromagnetic interaction, although both $SU(2)_L$ and $U(1)_Y$ groups are broken.

### Interactions

We shall now write the gauge-invariant and renormalizable Lagrangian for the electroweak interaction. The kinetic energy terms, the interactions of the gauge fields, and the interactions of the gauge fields and matter fields (see figure 5.3) are given by

$$
\begin{aligned}
\mathscr{L} = & -\frac{1}{4}G_{\mu\nu p}G^{p}_{\mu\nu} - \frac{1}{4}W_{\mu\nu m}W^{m}_{\mu\nu} - \frac{1}{4}B_{\mu\nu}B_{\mu\nu} \\
& +\overline{Q}_{\alpha iL}\, i\gamma_\mu \left[\delta_{\alpha\beta}\delta_{ij}\partial_\mu + ig_s\delta_{ij}(T_{\alpha\beta})^p G^p_\mu \right. \\
& \left. +ig\delta_{\alpha\beta}(\sigma_{ij})^m W^m_\mu + i\frac{g'}{6}\delta_{\alpha\beta}\delta_{ij}B_\mu\right] Q_{\beta jL} \\
& +\overline{U}_{\alpha R}\, i\gamma_\mu \left[\delta_{\alpha\beta}\partial_\mu + ig_s(T_{\alpha\beta})^p G^p_\mu + 2i\frac{g'}{3}\delta_{\alpha\beta}B_\mu\right] U_{\beta R} \\
& +\overline{D}_{\alpha R}\, i\gamma_\mu \left[\delta_{\alpha\beta}\partial_\mu + ig_s(T_{\alpha\beta})^p G^p_\mu - i\frac{g'}{3}\delta_{\alpha\beta}B_\mu\right] D_{\beta R} \\
& +\overline{\Psi}_{iL}\, i\gamma_\mu \left[\delta_{ij}\partial_\mu + ig(\sigma_{ij})^m W^m_\mu - i\frac{g'}{2}\delta_{ij}B_\mu\right] \Psi_{jL} \\
& +\overline{E}_R\, i\gamma_\mu \left[\partial_\mu - ig'B_\mu\right] E_R\,.
\end{aligned}
\tag{5.15}
$$

Here $i,j = 1,2$ and $\alpha,\beta = 1,2,3$ are $SU(2)_L$ and $SU(3)_c$ indices, respectively, in the fundamental representations; $m,n = 1,2,3$ and $p,q = 1,\ldots,8$ are the $SU(2)_L$ and $SU(3)_c$ indices, respectively, in the adjoint representations; $G_{\mu\nu p}$ and $W_{\mu\nu m}$ are the field strength tensors of the groups $SU(3)_c$ and $SU(2)_L$, respectively; $T^p$ and $\sigma^m$ are the generators of the groups $SU(3)_c$ and $SU(2)_L$, respectively; and $B_{\mu\nu p}$ and $B_\mu$ are the field strength tensor and the generator for the $U(1)_Y$ group, respectively. The gauge coupling constant for the groups $SU(3)_c, SU(2)_L$ and $U(1)_Y$ are $g_s$, $g$ and $g'$, respectively.

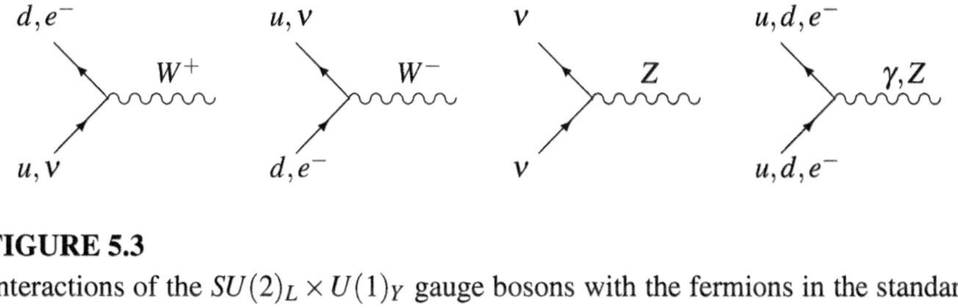


**FIGURE 5.3**
Interactions of the $SU(2)_L \times U(1)_Y$ gauge bosons with the fermions in the standard model.

In the standard model there are 12 massless gauge bosons, 8 gluons $G^p_\mu$, 3 gauge bosons $W^{\pm,0}_\mu$ corresponding to the group $SU(2)_L$, and the gauge boson $B_\mu$ for $U(1)_Y$. After the symmetry breaking, three of these gauge bosons will become massive, leaving the gluons and the photon massless. Since the left-handed fields are doublets under $SU(2)_L$ while the right-handed fields are singlets, the fermions are also massless before the symmetry breaking.

**Symmetry Breaking**

An explicit symmetry breaking can give masses to the gauge bosons and the fermions, but that will make the theory nonrenormalizable. So, the symmetry is broken spontaneously by giving a vacuum expectation value to the Higgs scalar $\phi$. We shall first write the scalar potential so the vacuum is no longer invariant under the symmetry groups. This will then break the symmetry spontaneously. The spontaneous symmetry breaking maintains the renormalizability of the theory and will give masses to the gauge bosons as well as to the fermions.

The Lagrangian involving the Higgs field can be written as

$$\mathscr{L}(\phi) = \mathscr{L}_{\phi KE} + \mathscr{L}_{Yuk} - V(\phi), \tag{5.16}$$

where the first term contains the kinetic energy and gauge interaction of the Higgs scalar field $\phi$,

$$\mathscr{L}_{\phi KE} = \left| \partial_\mu \phi + i g \sigma^m W_\mu^m \phi + i \frac{g'}{2} B_\mu \phi \right|^2 ; \tag{5.17}$$

the second term contains the Yukawa couplings

$$\begin{aligned} \mathscr{L}_{Yuk} = & \; h^u_{ab} \, \overline{Q}_{La} U_{Rb} \, \tilde{\phi} + h^d_{ab} \, \overline{Q}_{La} D_{Rb} \, \phi \\ & + h^e_{ab} \, \overline{\Psi}_{La} E_{Rb} \, \phi + H.c. \, , \end{aligned} \tag{5.18}$$

where $a,b = 1,2,3$ is the generation index; and the last term is the scalar potential

$$V(\phi) = -\frac{\mu^2}{2} \phi^\dagger \phi + \frac{\lambda}{4} (\phi^\dagger \phi)^2. \tag{5.19}$$

In the above we defined

$$\tilde{\phi} = i\tau_2 \phi^* = \begin{pmatrix} \phi_0^* \\ -\phi^- \end{pmatrix} \equiv [1, 2, -\frac{1}{2}]. \tag{5.20}$$

Since all $SU(2)$ representations are pseudo-real ($2 = \bar{2}$), both $\phi$ and $\tilde{\phi}$ transform under $SU(2)_L$ as doublets. However, they carry opposite $U(1)_Y$ quantum numbers.

The mass parameter of the Higgs scalar $\mu^2$ is taken to be positive-definite, so there is a vacuum expectation value (*vev*) of the Higgs scalar at the stationary point of the Lagrangian

$$\langle \phi^\dagger \phi \rangle = v^2 = \frac{\mu^2}{\lambda}. \tag{5.21}$$

We now perform an $SU(2)_L \times U(1)_Y$ gauge transformation to a unitary gauge in which $\phi^+ = 0$ and $\phi^\circ$ is real, with a positive *vev*

$$\langle \phi^+ \rangle = 0 \quad \text{and} \quad \langle \phi^\circ \rangle = v > 0. \tag{5.22}$$

This *vev* will then break the groups $SU(2)_L$ and $U(1)_Y$. However, an Abelian subgroup, which is generated by a linear combination of $T_{3L}$ and $Y$, will remain invariant. Since $\phi^\circ$ is electrically neutral, it leaves the $U(1)_Q$ group unbroken, which generates quantum electrodynamics.

**Gauge Bosons and Their Couplings**

The interactions of the scalar $\phi$ with the gauge bosons $W_\mu^{\pm,0}$ and $B_\mu$ make the gauge bosons corresponding to the broken groups massive. The two charged components of the $SU(2)_L$ gauge bosons

$$W_\mu^\pm = \frac{1}{\sqrt{2}}(W_\mu^1 \mp W_\mu^2) \tag{5.23}$$

get masses

$$m_W = \frac{vg}{2}, \tag{5.24}$$

and one combination of the neutral gauge bosons, given by

$$Z_\mu = \cos\theta_w W_\mu^3 - \sin\theta_w B_\mu\,, \tag{5.25}$$

gets mass

$$m_Z = \frac{v\sqrt{g^2+g'^2}}{2}. \tag{5.26}$$

These particles $W^\pm$ and $Z$ have been discovered in the proton–antiproton collider [40]. The remaining physical gauge boson, the photon,

$$A_\mu = \sin\theta_w W_\mu^3 + \cos\theta_w B_\mu, \tag{5.27}$$

remains massless along with the gluons and becomes the gauge boson corresponding to the unbroken $U(1)_Q$ and $SU(3)_c$ gauge symmetries.

The weak mixing angle $\theta_w$ is defined to relate the coupling constant of the $SU(2)_L$ and $U(1)_Y$ gauge theories with that of $U(1)_Q$ by

$$\sin\theta_w = \frac{e}{g} = \frac{g'}{(g^2+g'^2)^{1/2}}, \tag{5.28}$$

where $|e|$ is the charge of an electron. Comparison of the four-fermion interaction of the muon decay in this theory with the $(V-A)$ theory gives

$$g^2/m_W^2 = 4\sqrt{2}G_F. \tag{5.29}$$

Using the value of the Fermi constant $G_F = 1.16639(1)\times 10^{-5}\ \text{GeV}^{-2}$, we can now get the value of the *vev* of $\phi$ to be

$$v = \frac{2m_W}{g} = 2^{-1/4}G_F^{-1/2} = 246\ \text{GeV}. \tag{5.30}$$

In addition to the symmetry breaking, this *vev* gives masses to the fermions.

The interactions of the fermions with the physical gauge bosons can now be written as

$$\mathscr{L}_{gauge} = -\frac{g}{\sqrt{2}}(J_{CC}^\mu W_\mu^+ + J_{CC}^{\mu\,\dagger} W_\mu^-) - \frac{gg'}{e}J_{NC}^\mu Z_\mu - eJ_Q^\mu A_\mu, \tag{5.31}$$

where the charged current, the neutral current and the electromagnetic current are given by

$$J^{\mu}_{CC} = \bar{\psi}_L \gamma^{\mu} \frac{1}{2}(\sigma_1 + i\sigma_2)\psi_L$$
$$J^{\mu}_{NC} = \bar{\psi}\gamma^{\mu}\frac{1}{2}(c_V - c_A\gamma^5)\psi$$
$$J^{\mu}_{Q} = \bar{\psi}\gamma^{\mu} Q\psi, \tag{5.32}$$

respectively, where $c_V = T_{3L} - 2Q\sin^2\theta_w$ and $c_A = T_{3L}$. The right-handed charged fermions also take part in the neutral current interactions, since the neutral current contains couplings of the hypercharge gauge boson $B_{\mu}$.

We now introduce another important parameter, the $\rho$ parameter,

$$\rho = \frac{m_W^2}{m_Z^2 \cos^2\theta_w}. \tag{5.33}$$

For a doublet Higgs scalar, $\rho = 1$, which is the case in the standard model. If the symmetry group is broken by a triplet, the $\rho$ parameter would have been different from unity. The precision measurements gave $\rho = 1.0109 \pm 0.0006$ for $m_t = 175 \pm 5$ GeV and the Higgs mass around $m_Z$, which constrains the possible *vev* of a triplet Higgs scalar to be less than $\sim O(1)$ GeV.

## 5.4 Fermion Masses and Mixing

The standard model does not allow any fermion mass terms since the left-handed fermions are doublets while the right-handed fields are singlets, so the mass terms are not invariant under $SU(2)_L$. However, after the symmetry breaking, the Yukawa couplings (equation (5.18)) can give fermion masses when the Higgs scalar field $\phi$ acquires a *vev*

$$\begin{aligned}\mathcal{L}_{Yuk} &= h^u_{ab}\overline{Q}_{La}U_{Rb}\begin{pmatrix} <\phi_0^*>=v \\ <\phi^->=0\end{pmatrix} \\ &\quad + \left(h^d_{ab}\overline{Q}_{La}D_{Rb} + h^e_{ab}\overline{\Psi}_{La}E_{Rb}\right)\begin{pmatrix} <\phi^+>=0 \\ <\phi^0>=v\end{pmatrix} + H.c. \\ &= m^u_{ab}\overline{U}_{La}U_{Rb} + m^d_{ab}\overline{D}_{La}D_{Rb} + m^e_{ab}\overline{\ell}_{La}E_{Rb} + H.c.\,,\end{aligned} \tag{5.34}$$

where

$$Q_{La} = \begin{pmatrix} U_{La} \\ D_{La}\end{pmatrix} = \begin{pmatrix} u_L, c_L, t_L \\ d_L, s_L, b_L\end{pmatrix}$$
$$\Psi_{La} = \begin{pmatrix} \nu_{La} \\ \ell_{La}\end{pmatrix} = \begin{pmatrix} \nu_e, \nu_{\mu}, \nu_{\tau} \\ e_L, \mu_L, \tau_L\end{pmatrix}$$

$$\begin{aligned} U_{Ra} &= (u_R, c_R, t_R) \\ D_{Ra} &= (d_R, s_R, b_R) \\ E_{Ra} &= (e_R, \mu_R, \tau_R) \ ; \end{aligned} \tag{5.35}$$

and the up, down and the charged lepton masses are

$$m^u_{ab} = h^u_{ab} v; \quad m^d_{ab} = h^d_{ab} v; \quad m^e_{ab} = h^e_{ab} v. \tag{5.36}$$

Since there are no right-handed neutrinos in the theory and there are no Higgs triplet that can induce a Majorana mass term $m_M \overline{\psi_L{}^c} \psi_L$, neutrinos are massless in the standard model. We shall discuss the neutrino masses at length in the next chapter.

In general, the quark and lepton masses are not diagonal. In the case of leptons we can make unitary transformations on both the neutrino fields and the charged lepton fields and make the charged lepton mass matrix diagonal. Since neutrinos are massless in this model, this transformation will not affect the charged current interactions. So, it is convenient to work in the basis where the charged lepton masses are diagonal.

## Quark Mixing Matrix

In general, the up and the down quark mass matrices $m^u_{ab}$ and $m^d_{ab}$ are not diagonal. If we now diagonalize the mass matrices, in terms of the quark states with definite masses, which are the physical eigenstates, the charged current interaction will not be diagonal. If we define the physical states or the mass eigenstates $U'_{[L,R]}$ and $D'_{[L,R]}$, these states will be distinct from the weak eigenstates $U_{[L,R]}$ and $D_{[L,R]}$, which enter in the charged current interaction.

In the matrix form, the quark masses can be written as

$$\mathscr{L}_M = \overline{U}_L M_u U_R + \overline{D}_L M_d D_R, \tag{5.37}$$

where $U_{[L,R]}$ and $D_{[L,R]}$ are the matrix forms of $U_{a[L,R]}$ and $D_{a[L,R]}$ and the $3 \times 3$ mass matrix $M_{[u,d]}$ has the elements $m^{[u,d]}_{ab}$. As mentioned earlier, the mass matrices $M_u$ and $M_d$ are not diagonal in this weak basis, in which the charged current interactions of the quarks are given by

$$\mathscr{L}^q_{cc} = \frac{g}{\sqrt{2}} \left[ \overline{U}_{mL} \gamma_\mu D_{mL} W^+_\mu + \overline{D}_{mL} \gamma_\mu U_{mL} W^-_\mu \right]. \tag{5.38}$$

We shall now diagonalize the mass matrices $M_u$ and $M_d$ to get the physical eigenstates.

The masses $M_u$ and $M_d$ are not Hermitian and, hence, cannot be diagonalized by unitary transformations. However, they can always be diagonalized by bi-unitary transformations,

$$\begin{aligned} \mathscr{U}^\dagger_{uL} M_u \mathscr{U}_{uR} &= \hat{M}_u \\ \mathscr{U}^\dagger_{dL} M_d \mathscr{U}_{dR} &= \hat{M}_d, \end{aligned} \tag{5.39}$$

where $\hat{M}_u$ and $\hat{M}_d$ are diagonal matrices and the left-transformation matrices $\mathscr{U}_L$ and the right-transformation matrices $\mathscr{U}_R$ can be obtained by solving

$$\begin{aligned}\mathscr{U}_{AL}^\dagger M_A M_A^\dagger \mathscr{U}_{AL} &= \hat{M}_A \hat{M}_A^\dagger \\ \mathscr{U}_{AR}^\dagger M_A^\dagger M_A \mathscr{U}_{AR} &= \hat{M}_A^\dagger \hat{M}_A,\end{aligned} \tag{5.40}$$

where $A$ means either up ($u$) or down ($d$) quarks for the entire equations. The mass eigenstates are now given in terms of the weak eigenstates by

$$\begin{aligned}U'_{[L,R]} &= \mathscr{U}_{u[L,R]} U_{[L,R]} \\ D'_{[L,R]} &= \mathscr{U}_{d[L,R]} D_{[L,R]}.\end{aligned} \tag{5.41}$$

The mass eigenstates are the physical states with definite mass eigenvalues

$$\mathscr{L}_M = \overline{U'}_L \hat{M}_u U'_R + \overline{D'}_L \hat{M}_d D'_R. \tag{5.42}$$

In any weak interaction, particles are created in the weak basis, but they will evolve with time as physical eigenstates. For example, a down quark could be produced in any weak interaction, but if the physical states are mixtures of down and strange quarks, then during its propagation the original down quark may become a strange quark.

The charged-current interactions can now be written in the mass basis as

$$\mathscr{L}_{cc}^q = \frac{g}{\sqrt{2}} \left[\overline{U'}_{aL} \gamma_\mu V_{ai} D'_{iL} W_\mu^+ + \overline{D'}_{iL} \gamma_\mu V_{ai}^* U'_{aL} W_\mu^-\right], \tag{5.43}$$

where the unitary mixing matrix

$$V_{ai} = [\mathscr{U}_{uL}]_{am} [\mathscr{U}_{dL}^\dagger]_{im} \tag{5.44}$$

is called the Cabibbo–Kobayashi–Maskawa (CKM) mixing matrix [41]. The CKM matrix $V_{ai}$ is a product of the two unitary matrices, one entering in the diagonalization of the up quark and the other in the down quark mass matrix. The right transformation matrices do not enter in the charged current interactions. Thus, the CKM matrix $V_{ai}$ will appear in interaction vertices where a down-quark is annihilated and an up-quark is created and the opposite for $V_{ai}^*$. The neutral-current interaction is diagonal in both the weak basis and the physical basis, since the mixing matrices of the same field appear in the interaction. Similarly the neutral Higgs couplings are also diagonal. So, there are no flavor-changing neutral currents in the standard model [42].

**Parameterization of the CKM Matrix**

The mixing matrix appearing in the charged current interactions can be interpreted in several ways. One convention is to consider the mass eigenstates and the physical eigenstates of the up quarks to be the same, let us denote them as $u, c, t$. Let us now denote the physical down quark eigenstates as $d, s, b$, which have definite masses. The CKM mixing matrix will imply that in any charged current weak interaction of an up quark, say $u$, a combination of $d,s,b$ will appear. If the CKM matrix were diagonal, then only $d$ would have appeared, but due to the quark mixing, both $s$ and $b$ will also mix with the $d$ quark in the interactions of $u$. The weak eigenstate will be an admixture of all three mass eigenstates $d,s,b$.

There are four independent parameters in the three generation CKM mixing matrix, three mixing angles and a complex phase, which gives rise to $CP$ violation. In general all the elements of the CKM matrix could be complex, but phase transformations of the quark fields could remove most of these phases, except one for a three generation scenario. In general, for an $N$ generation scenario, $V$ can be expressed by $N(N-1)/2$ rotation angles and $(N-1)(N-2)/2$ independent phases, after removing $(2N-1)$ phases by rephasing the quark fields.

The CKM matrix could be parameterized in several ways for convenience of studying certain physical observables. One such parameterization, suggested by Kobayashi and Maskawa, can be written as

$$V = \begin{pmatrix} c_1 & -s_1c_3 & -s_1s_3 \\ s_1c_2 & c_1c_2c_3 - s_2s_3\exp^{i\delta} & c_1c_2s_3 + s_2c_3\exp^{i\delta} \\ s_1s_2 & c_1s_2c_3 + c_2s_3\exp^{i\delta} & c_1s_2s_3 - c_2c_3\exp^{i\delta} \end{pmatrix}, \tag{5.45}$$

where, $s_1 = \sin\theta_i, c_i = \cos\theta_i (i = 1,2,3)$ are the three mixing angles and $\delta$ is a $CP$ violating phase. Wolfenstein proposed another phenomenological parameterization [43], in which the matrix elements are expanded in terms of a small parameter, the Cabibbo mixing angle $\lambda = \sin\theta_c$ and unitarity is used to determine the remaining elements

$$V = \begin{pmatrix} 1-\frac{1}{2}\lambda^2 & \lambda & A\lambda^3(\rho - i\eta) \\ -\lambda & 1-\frac{1}{2}\lambda^2 & A\lambda^2 \\ A\lambda^3(1-\rho-i\eta) & -A\lambda^2 & 1 \end{pmatrix}. \tag{5.46}$$

This matrix is approximately unitary and there are four parameters $A, \lambda, \rho$ and $\eta$.

The most accurately determined quark mixing parameter is the Cabibbo angle $\lambda$ [44]. From the measurement of the hyperon decay and $K \to \pi e\nu$ decay, we have $|V_{us}| = \lambda = 0.2196 \pm 0.0023$. Combined analysis [44] of inclusive and exclusive $B$ decays gives $|V_{cb}| = 0.0402 \pm 0.0019$, which implies $A = 0.819 \pm 0.035$. End-point lepton energy spectra in the semi-leptonic decays $B \to X_u \ell \nu_\ell$ and exclusive semileptonic decays $B \to \pi\ell\nu_\ell$ or $\rho\ell\nu_\ell$ give $|V_{ub}/V_{cb}| = 0.090 \pm 0.025$ yielding $\sqrt{\rho^2+\eta^2} = 0.423 \pm 0.064$ [45].

Some combination of parameters could be constrained by the unitarity conditions of the CKM matrix. This is very useful for an understanding of the $CP$ violation. Let us consider one of the unitarity conditions of the CKM matrix

$$V_{ud}V_{ub}^* + V_{cd}V_{cb}^* + V_{td}V_{tb}^* = 0. \tag{5.47}$$

The form of equation (5.46) then allows us to write this equation as

$$\frac{V_{ub}^*}{\lambda V_{cb}} + \frac{V_{td}}{\lambda V_{cb}} = 1. \tag{5.48}$$

In the $\rho - \eta$ space this will give a triangle, which is one of the six unitary triangles. The shape of the triangle gives the allowed values of $\rho$ and $\eta$ [46].

From the various weak decays of the relevant quarks or the deep inelastic neutrino scattering, it is possible to determine the individual matrix elements. Assuming three generations and using unitarity, the experimentally allowed range of the matrix elements are given by [44],

$$|V| = \begin{pmatrix} 0.9742 - 0.9757 & 0.219 - 0.226 & 0.002 - 0.005 \\ 0.219 - 0.225 & 0.9734 - 0.9749 & 0.037 - 0.043 \\ 0.004 - 0.014 & 0.035 - 0.043 & 0.9990 - 0.9993 \end{pmatrix}. \quad (5.49)$$

The errors in the individual elements are not independent, since different elements are related by unitarity.

### CP Violation

Any complex phase in the Yukawa couplings or the mass terms give rise to *CP* violation. In the CKM matrix there is only one phase for three generations, which gives the *CP* violation. Since any observable should not depend on the phase transformations of the quarks, it is convenient to define a quantity that will be independent of any phase transformations and can be a measure of *CP* violation [47]. Any phase transformations of the up-quark and the down-quark

$$U_{La} \to e^{i\alpha_a} U_{La} \quad \text{and} \quad D_{Li} \to e^{i\alpha_i} D_{Li} \quad (5.50)$$

would transform the quark mixing matrix $V_{ai}$ as

$$V_{ai} \longrightarrow e^{i(\alpha_a - \alpha_i)} V_{ai}. \quad (5.51)$$

Thus, the rephasing invariant quantity that can be defined as a measure of the *CP* violation is given by

$$J_{abij} = \text{Im}\left[V_{ai} V_{bj} V^*_{aj} V^*_{bi}\right] , \quad (5.52)$$

not summed over repeated indices. This is called the Jarlskog invariant. In the case of three generations, using unitarity it can be shown that all the nine different combinations of $J$ are equal, i.e., for any choice of $abij$ we get the same expression for $J$. This rephasing invariant measure determines the amount of *CP* violation in the standard electroweak model uniquely.

There is thus one *CP* violating phase in the quark mixing matrix for three generations. *CP* violation has been observed in the $K^\circ - \overline{K^\circ}$ oscillation, in $K$-decays, as well as in the $B$-decays, but all of them could be explained by this single *CP* phase in the CKM matrix. But in models of two Higgs doublets or other extensions, there could be newer sources of *CP* violation. There is also a *CP* phase coming from the strong interaction. In the neutrino sector, there is no *CP* violation in the standard model because neutrinos are massless. However, when neutrino masses are introduced, there will be *CP* violation and that may even be related to the matter–antimatter asymmetry of the universe. Further studies of *CP* violation could provide information about the physics beyond the standard model.

# 6

# *Neutrino Masses and Mixing*

In a nuclear beta decay, a neutron decays into a proton, an electron and a neutrino. In the beginning neutrinos were not observed in the beta decay; they were postulated to satisfy conservation laws. Since the electrons emitted in a beta decay are not mono-energetic, energy and momentum conservation ruled out the possibility of a two-body decay of neutrons into a proton and an electron. Pauli postulated [48] the new particle, the neutrino, to explain the energy distribution of electrons. Conservation of angular momentum implied that the neutrinos are spin-1/2 fermions.

Neutrinos are neutral particles with very tiny mass. They interact with matter only through weak and gravitational interactions and, hence, very weakly. Since the strength of gravitational interaction is too small compared with the weak interaction, neutrinos interact with matter mostly through weak interactions and, hence, play a very special role in particle physics. Although detection of neutrinos is extremely difficult, experiments with neutrinos brought many surprises. The tiny mass of the neutrinos could be established almost sixty years after the discovery of the neutrinos. On the other hand, the neutrinos could travel a very long distance in the universe without interacting with anything else and, hence, play an important role in astrophysics and cosmology. If we understand all the properties of neutrinos, it may be possible to say how they contributed to nucleosynthesis. Neutrinos could have been instrumental in generating a baryon asymmetry of the universe, explaining the problem of the cosmological constant or contributing to the dark matter of the universe.

During the past few years, many new developments have taken place with neutrinos. The atmospheric and the solar neutrino deficits established that the neutrinos have very small mass, and neutrinos of one flavor oscillate into another kind during their propagation in space. There are three types of active neutrinos and two of the mixing angles are large while the third one is small. These results have been confirmed by laboratory experiments. However, we have yet to answer many more questions about the neutrinos. For example, we do not know if neutrinos are Dirac or Majorana particles, what their absolute mass is, what the value of the third mixing angle is, if there is *CP* violation in the leptonic sector, and what makes the neutrinos so light compared with charged fermions. These questions become more interesting in many extensions of the standard model, such as the supersymmetric theories or the models with extra dimensions.

## 6.1 Dirac and Majorana Neutrinos

One of the most important question for the neutrinos is whether they are Dirac or Majorana particles. All charged particles are Dirac particles, but a neutral particle like a neutrino could be either a Dirac or a Majorana particle. If neutrinos are Majorana particles [49], this could naturally explain why they are much lighter than any charged fermions. Lepton number will be violated if neutrinos are Majorana particles, which may have several interesting consequences in cosmology.

The main difference between Dirac and Majorana neutrinos lies in their lepton number violation. If neutrinos are Dirac particles, lepton number will be conserved, while for a Majorana neutrino, lepton number is violated by two units since they are their own antiparticles. Any gauge interactions in the standard model conserve lepton number. Only the mass term in the Lagrangian can distinguish between a Dirac and a Majorana particle.

A Dirac particle is a spin-1/2 particle that satisfies the Dirac equation (equation (1.16))

$$i\gamma_\mu \partial_\mu \psi^D - m_D \psi^D = 0. \tag{6.1}$$

Here $\psi^D$ is the *Dirac spinor* and has four complex components. This 4-dimensional spinor representation of the Lorentz group $O(3,1)$, which is locally isomorphic to $O(4)$, can be decomposed under its subgroup $SU(2)_L \times SU(2)_R$ as

$$4 = (2,1) + (1,2).$$

The chiral projection operator $(1 \pm \gamma_5)/2$ can project the left-handed and right-handed components of the spinor representations, which are the $(2,1)$ and $(1,2)$ representations, respectively:

$$\psi_L = \frac{1}{2}(1-\gamma_5)\psi^D \subset (2,1) \quad \text{and} \quad \psi_R = \frac{1}{2}(1+\gamma_5)\psi^D \subset (1,2).$$

These states satisfy $\gamma_5 \psi_L = -\psi_L$ and $\gamma_5 \psi_R = \psi_R$ and the identitites $\psi_L \psi_R = {\psi^c}_L \psi_R = \psi_L {\psi_L}^c = 0$. Under charge conjugation ($\mathscr{C}$) and $CP$ conjugation ($\mathscr{C}\mathscr{P}$), these components transform as

$$\begin{aligned} &\psi_L \xrightarrow{\mathscr{C}} {\psi^c}_L \qquad && \psi_L \xrightarrow{\mathscr{C}\mathscr{P}} {\psi^c}_R = (\psi_L)^c \\ &\psi_R \xrightarrow{\mathscr{C}} {\psi^c}_R \qquad && \psi_R \xrightarrow{\mathscr{C}\mathscr{P}} {\psi^c}_L = (\psi_R)^c. \end{aligned} \tag{6.2}$$

Out of the four states ($\psi_L, \psi_R, {\psi^c}_L$ and ${\psi^c}_R$), only two of the states are independent. $CPT$ invariance implies that any theory will contain both the particles and their $CPT$ mirror images, which are the $CP$ conjugates. So, if any theory contains a $\psi_L$, then $CPT$ invariance would imply the existence of ${\psi^c}_R = {\psi_L}^c$ as well.

In terms of the chiral components of the Dirac field $\psi_D$, the Dirac equation can be reduced to two coupled equations

$$\begin{aligned} i\gamma_\mu \partial_\mu \psi_L - m_D \psi_R &= 0 \\ i\gamma_\mu \partial_\mu \psi_R - m_D \psi_L &= 0. \end{aligned} \tag{6.3}$$

These follow from the Lagrangian

$$\mathscr{L} = \mathscr{L}_{KE} + \mathscr{L}_D, \tag{6.4}$$

where the kinetic energy term is

$$\mathscr{L}_{KE} = i\bar{\psi}_D \gamma_\mu \partial_\mu \psi_D = i\bar{\psi}_L \gamma_\mu \partial_\mu \psi_L + i\bar{\psi}_R \gamma_\mu \partial_\mu \psi_R. \tag{6.5}$$

The kinetic energy term, thus, treats the left-handed and the right-handed fields independently, while the Dirac mass term connects the left-handed fields to the right-handed fields. When $m_D = 0$, the left-handed and the right-handed components of $\psi_D$, i.e., the Weyl spinors [50] $\psi_L$ and $\psi_R$, can be treated as independent.

A Lagrangian with a nonzero $m_D$, which is the Dirac mass term, represents a Dirac particle. Since massive particles propagate more slowly than light, a left-handed Dirac particle will become a right-handed particle in another frame moving faster than the particle. Thus, all four states ($\psi_L$, $\psi_R$, and their $\mathscr{CP}$ conjugate states) should be present in any theory if any particle is a Dirac particle.

There are four independent solutions of the Dirac equation (two positive energy $u_{p,\pm s}$ and two negative energy solutions $v_{p,\pm s}$), so we may write (see section 1.2 for details)

$$\begin{aligned} \psi_D(x) &= \sum_{p,s} \frac{1}{\sqrt{2\varepsilon}} \left( a_{ps} u_{ps} \exp^{-ipx} + b^\dagger_{ps} v_{ps} \exp^{ipx} \right) \\ \bar{\psi}_D(x) = \psi^\dagger_D \gamma^0 &= \sum_{p,s} \frac{1}{\sqrt{2\varepsilon}} \left( a^\dagger_{ps} \bar{u}_{ps} \exp^{ipx} + b_{ps} \bar{v}_{ps} \exp^{-ipx} \right), \end{aligned} \tag{6.6}$$

where $s^2 = -1$ is the covariant spin vector, $\bar{u} = u^\dagger \gamma^0$, $\bar{v} = v^\dagger \gamma^0$ and the bispinors $u_{ps}$ and $v_{ps}$ are suitably normalized. The coefficients $a_{ps}$ and $b_{ps}$ are the annihilation operators, and the coefficients $a^\dagger_{ps}$ and $b^\dagger_{ps}$ are the creation operators for the particles and antiparticles, respectively.

The Dirac mass term comes from the Lagrangian

$$\mathscr{L}_D = -m_D \bar{\psi}_D \psi_D = -m_D \bar{\psi}_R \psi_L - m_D \bar{\psi}_L \psi_R, \tag{6.7}$$

where the second term is the Hermitian conjugate ($H.c.$) of the first term in the last expression. In the standard model it is convenient to work with the states $\psi_L$ and $\psi_R$ as independent states; $CPT$ invariance will then include the states ${\psi^c}_R$ and ${\psi^c}_L$. However, in grand unified theories it is convenient to consider the states $\psi_L$ and ${\psi^c}_L$ as independent states, so they can be put into a single representation of the grand unification group

$$\Psi_L \equiv \begin{pmatrix} \psi \\ \psi^c \end{pmatrix}_L .$$

$CPT$ invariance will then include the $CP$ conjugate states in the theory, which are the right-handed particles

$$\Psi_R \equiv \begin{pmatrix} \psi^c \\ \psi \end{pmatrix}_R .$$

The states $\Psi_L$ and $\Psi_R$ can then belong to the representations $\mathscr{R}(\mathscr{G})$ and $\overline{\mathscr{R}}(\mathscr{G})$, respectively, of any unifying group $\mathscr{G}$ that commutes with the Lorentz group. One can then write the Dirac masses as

$$\mathscr{L}_D = -m_D \bar{\psi}_R \psi_L + H.c. = m_D \psi_L^T C^{-1} {\psi^c}_L + H.c. \tag{6.8}$$

For *CPT* conservation, the Hermitian conjugate of every mass term (containing the right-handed fields) will always be present. From now on we shall not mention them explicitly, except when we discuss *CP* violation. A more general form of the Dirac mass term is

$$\begin{aligned}\mathscr{L}_D &= \frac{1}{2}(\psi_L \quad {\psi^c}_L)^T C^{-1} \begin{pmatrix} 0 & m_D \\ m_D & 0 \end{pmatrix} \begin{pmatrix} \psi_L \\ {\psi^c}_L \end{pmatrix} \\ &= \frac{1}{2}\Psi_L^T C^{-1} M_D \Psi_L,\end{aligned} \tag{6.9}$$

where $\Psi_L^T = (\psi \quad \psi^c)_L^T$. Thus, the Dirac mass term combines the two Weyl spinors, $\psi_L$ and ${\psi^c}_L$ (or $\psi_R$) to form a Dirac particle.

It is also possible to write the mass terms for each of the Weyl spinors keeping the mass matrix diagonal, as

$$\mathscr{L}_D = \frac{1}{2}(\psi_L \quad {\psi^c}_L)^T C^{-1} \begin{pmatrix} m_L & 0 \\ 0 & m_R \end{pmatrix} \begin{pmatrix} \psi_L \\ {\psi^c}_L \end{pmatrix}, \tag{6.10}$$

which gives two massive states, $\psi_L$ with mass $m_L$ and ${\psi^c}_L$ (or $\psi_R$) with mass $m_R$:

$$\begin{aligned}\mathscr{L}_L &= -\frac{1}{2} m_L \overline{\psi_L{}^c} \psi_L = \frac{1}{2} m_L \psi_L^T C^{-1} \psi_L = -\frac{1}{2} m_L \overline{\psi_R} {\psi_R}^c \\ \mathscr{L}_R &= -\frac{1}{2} m_R \overline{\psi_R{}^c} \psi_R = \frac{1}{2} m_R \psi_R^T C^{-1} \psi_R = -\frac{1}{2} m_R \overline{\psi_L} {\psi_L}^c.\end{aligned} \tag{6.11}$$

The two masses $m_L$ and $m_R$ are now independent. These states will not be part of one Dirac particle; instead they will become two Majorana particles [49], satisfying the Majorana conditions

$${\psi_L}^c = {\psi^c}_R = \eta_{LC}\psi_L \quad \text{and} \quad {\psi_R}^c = {\psi^c}_L = \eta_{RC}\psi_R, \tag{6.12}$$

where $\eta_{LC}$ and $\eta_{RC}$ are two new phases, called the creation phases or the Majorana phases. The overall phase can be chosen to be real, but the relative phase gives rise to a new source of *CP* violation. For any number of Majorana particles, the overall phase can always be chosen to be real. Since the Majorana condition relates the charge conjugate state of the particle with itself, it corresponds to a real spinor and, hence, has four real components. Thus, a Dirac particle can be formed from two Majorana particles, which are combined by the Dirac mass term.

For a Majorana particle, the creation and the annihilation operators for the particles and antiparticles are related by the Majorana condition, $a_{ps} = b_{ps} = f_{ps}$ and $a_{ps}^\dagger = b_{ps}^\dagger = f_{ps}^\dagger$. A Majorana field can then be written as

$$\psi_M(x) = \sum_{p,s} \frac{1}{\sqrt{2\varepsilon}} \left( f_{ps} u_{ps} \exp^{-ipx} + \eta_C^* f_{ps}^\dagger v_{ps} \exp^{ipx} \right). \tag{6.13}$$

It can then be checked that $\overline{\psi^c_M}\gamma_\mu \psi^c_M = -\overline{\psi_M}\gamma_\mu \psi_M$, and hence, there is no vector current for any Majorana particle. The number operator also vanishes, $\int d^3x \overline{\psi_M}\gamma_0 \psi_M = 0$ for the Majorana particles, so there is no conservation of the number of particles.

The Majorana mass terms combine a left-handed particle $\psi_L$ to its $CP$ conjugate state ${\psi^c}_R$ and a right-handed particle $\psi_R$ to its $CP$ conjugate state ${\psi^c}_L$. A particle with Majorana mass will also move with a velocity less than the speed of light. So, an observer moving with a higher velocity than a Majorana particle will see a left-handed particle $\psi_L$ to be a right-handed antiparticle ${\psi^c}_R$.

A real four-component Majorana particle can be written in the Weyl basis as

$$\psi_M = \psi_L + \eta_C^* {\psi^c}_R, \tag{6.14}$$

where $|\eta_C|^2 = 1$ is the Majorana phase, discussed earlier. A Majorana mass term can be written as

$$\mathscr{L}_M = -\frac{1}{2} m_M \bar{\psi}^c_M \psi_M = \frac{1}{2} m_M \psi^T_M C^{-1} \psi_M. \tag{6.15}$$

If the fields carry any $U(1)$ quantum number, the Majorana mass term for these fields breaks the $U(1)$ symmetry by two units. The neutrinos carry lepton number, and hence, any Majorana mass terms for the neutrinos break lepton number by two units.

In general, the mass matrix for a system comprised of both the left-handed and the right-handed particles can be written as

$$\begin{aligned}\mathscr{L}_{mass} &= \frac{1}{2}(\psi \quad \psi^c)^T_L C^{-1} \begin{pmatrix} m_L & m_D \\ m_D & m_R \end{pmatrix} \begin{pmatrix} \psi \\ \psi^c \end{pmatrix}_L \\ &= \frac{1}{2} \Psi^T_L C^{-1} M \Psi_L, \end{aligned} \tag{6.16}$$

where $\Psi^T_L = (\psi \quad \psi^c)^T_L$ and $M$ is a $2 \times 2$ symmetric mass matrix:

$$\Psi^T_L C^{-1} M \Psi_L = \Psi^T_L M^T (C^T)^{-1} \Psi_L = \Psi^T_L C^{-1} M^T \Psi_L. \tag{6.17}$$

If the particle carries any $U(1)$ charge which is not broken, such as electric charge, this matrix $M$ can only have nonvanishing off-diagonal terms. This will then correspond to a Dirac particle. Thus, any charged particles are Dirac particles.

For neutrinos the diagonal terms in the mass matrix can be nonvanishing [51], which breaks lepton number since neutrinos and antineutrinos carry opposite lepton numbers. Thus, this general mass matrix will correspond to two independent Majorana particles, which are two independent combinations of $\psi_L$ and ${\psi^c}_L$ (or $\psi_R$). Only when the diagonal elements vanish will this matrix represent a single Dirac particle. Thus, depending on the structure of the mass matrix, one can say whether these two particles combine to form a Dirac particle or represent two Majorana particles. This discussion can be generalized to any number of particles. If there are $n$ number of Majorana particles $\Psi_{iL}$, $(i = 1, \cdots, n)$ (some of which could be antiparticles $\psi_{mL} = {\chi^c}_{mL}$, i.e., $\psi_m = \chi^c_m$), one can write the mass matrix for these particles as

$$\mathscr{L}_{mass} = \frac{1}{2} \Psi^T_{iL} C^{-1} M_{ij} \Psi_{jL}, \tag{6.18}$$

where $\Psi_L^T = (\,\psi_1 \quad \cdots \quad \psi_n\,)_L^T$ and $M_{ij}$ is a $n \times n$ symmetric mass matrix. In this case, some of the states could pair up to form Dirac particles, while the remaining ones remain as physical Majorana particles.

Let us elaborate this point with an example of three Majorana neutrinos $\Psi_{iL} \equiv (\,\nu_e \quad \nu_\mu \quad \nu_\tau\,)_L$ with a mass matrix:

$$\mathscr{L}_{mass} = \frac{1}{2}\Psi_{iL}^T C^{-1} \begin{pmatrix} 0 & m & 0 \\ m & 0 & 0 \\ 0 & 0 & M \end{pmatrix} \Psi_{jL}. \tag{6.19}$$

The states $\nu_{eL}$ and $\nu_{\mu L}$ will combine to form a Dirac particle, while the state $\nu_{\tau L}$ will correspond to a Majorana particle. The physical left-handed and right-handed components of the Dirac particle correspond to

$$\chi_L = \nu_{eL} \quad \text{and} \quad {\chi^c}_L = \nu_{\mu L} \text{ or } \chi_R = (\nu_{\mu L})^c\,, \tag{6.20}$$

so an observer from a faster moving frame will see the left-handed electron neutrino ($\nu_{eL}$) to be a right-handed muon antineutrino (${\nu^c}_{\mu R}$). The $\tau$-neutrino will be a Majorana particle.

Let us now come back to the case of $n$ Majorana neutrinos. The kinetic energy terms or any other interactions will treat these states as $n$ independent Majorana states. Only the mass terms can tell us if the states correspond to different physical Majorana neutrinos, some of the states combine to give a Dirac neutrino, or if they correspond to pseudo-Dirac neutrinos or some combinations of these states [52, 53]. We shall again elaborate this point with the example of $\psi_L$ and ${\psi^c}_L$. The real symmetric $2 \times 2$ mass matrix $M^T = M$ can be diagonalized by the orthogonal transformations $O$:

$$M_d = O^T M O = \begin{pmatrix} m_1 & 0 \\ 0 & m_2 \end{pmatrix}, \tag{6.21}$$

where

$$m_{1,2} = \frac{1}{2}\left[ m_L + m_R \mp \sqrt{(m_R - m_L)^2 + 4m_D^2} \right] \tag{6.22}$$

and $M_d$ is diagonal. In this diagonal basis $\Psi_{dL}^T = \Psi_L^T O = (\,\psi_{1L} \quad \psi_{2L}\,)$, the Lagrangian can be written as

$$\mathscr{L}_{mass} = \frac{1}{2}\Psi_L^T\, C^{-1} M \Psi_L = \frac{1}{2}\Psi_{dL}^T\, C^{-1} M_d \Psi_{dL}. \tag{6.23}$$

The transformation relates the weak interaction basis $\Psi_L$ to the diagonal physical mass basis $\Psi_{dL}$ through

$$\Psi_{dL} = \begin{pmatrix} \psi_1 \\ \psi_2 \end{pmatrix}_L = O^T \Psi_L = \begin{pmatrix} \cos\theta & -\sin\theta \\ \sin\theta & \cos\theta \end{pmatrix} \begin{pmatrix} \psi \\ \psi^c \end{pmatrix}_L, \tag{6.24}$$

with the mixing angle $\tan 2\theta = 2m_D/(m_R - m_L)$.

The values of $m_L$, $m_R$ and $m_D$ determine if the physical particles are Majorana, Dirac or pseudo-Dirac particles. For $m_1 \neq m_2$, the mass term becomes

$$\mathscr{L}_{mass} = \frac{1}{2} m_1 \psi_{1L}^T C^{-1} \psi_{1L} + \frac{1}{2} m_2 \psi_{2L}^T C^{-1} \psi_{2L}. \tag{6.25}$$

This corresponds to two physical Majorana particles $\psi_{(1,2)L}$ with masses $m_{1,2}$. A special case of this is $m_D = 0$, when the physical states are

$$\psi_{1L} = \psi_L \quad \text{and} \quad \psi_{2L} = \psi^c{}_L. \tag{6.26}$$

In this case the two Majorana particles are the left-handed and the right-handed neutrinos. When $m_D \neq 0$, the two physical Majorana particles $\psi_{(1,2)L}$ become combinations of the left-handed particle and the antiparticle.

The mass matrix with $m_L = m_R = 0$ represents one physical Dirac particle

$$\mathscr{L}_{mass} = \frac{1}{2} m_D \psi_{1L}^T C^{-1} \psi_{1L} - \frac{1}{2} m_D \psi_{2L}^T C^{-1} \psi_{2L} = m_D \psi_L^T C^{-1} \psi^c{}_L, \tag{6.27}$$

with the mass eigenvalues $m_1 = -m_2 = m_D$. The states are now related by

$$\psi_L = \frac{1}{\sqrt{2}}[\psi_{1L} + \psi_{2L}] \quad \text{and} \quad \psi^c{}_L = \frac{1}{\sqrt{2}}[\psi_{1L} - \psi_{2L}]. \tag{6.28}$$

Since the diagonal terms of the mass matrix vanish, all the $U(1)$ charges carried by these particles are conserved. In a more general case, if any $U(1)$ symmetry of the weak eigenstates is not broken by the mass matrix and if two of the mass eigenvalues are equal with opposite sign, then they form a Dirac particle.

When the mass eigenvalues are equal but any symmetry of the weak eigenstates is violated by the mass matrix, the two Majorana states combine to form a Dirac particle at the tree-level; but one-loop corrections would break the degeneracy, and these states will behave as two Majorana particles. This is the case corresponding to $m_L = m_R = m$ and $m_D = 0$. The mass eigenstates are equal, $m_1 = m_2 = m$ and we can write the mass term as

$$\mathscr{L}_{mass} = \frac{1}{2} m \psi_{1L}^T C^{-1} \psi_{1L} + \frac{1}{2} m \psi_{2L}^T C^{-1} \psi_{2L} = m \chi_L^T C^{-1} \chi^c{}_L, \tag{6.29}$$

where

$$\chi_L = \frac{1}{\sqrt{2}}[\psi_{1L} - i\psi_{2L}] \quad \text{and} \quad \chi^c{}_L = \frac{1}{\sqrt{2}}[\psi_{1L} + i\psi_{2L}]. \tag{6.30}$$

Then the particle $\chi_L$ appears as a Dirac particle and if it carries any U(1) charge, it is conserved. However, if the fields $\psi_{1L}$ and $\psi_{2L}$ carry any U(1) charges, they are not conserved. So, these particles are different from the usual Dirac particles and are called pseudo-Dirac particles. Consider the case of neutrinos. The neutrinos carry lepton number. But if they are pseudo-Dirac particles, the physical state will be an admixture of a lepton and an antilepton and, hence, will not carry any definite lepton number. So, lepton number will be violated. In this case there will be one-loop self-energy type diagrams with the charged leptons in the loop. The corrections to the masses of the two states $\psi_{1L}$ and $\psi_{2L}$ will be proportional to the corresponding mass squared of the charged leptons $e_1$ and $e_2$, and hence, the mass degeneracy will be broken leading to two Majorana particles. However for a Dirac particle, the $U(1)$ symmetry of the weak eigenstates guarantees the equality of the masses of the two states even when higher order corrections are included.

A pseudo-Dirac particle is, thus, a Dirac particle at the tree-level, but after including radiative corrections it becomes two almost degenerate Majorana particles. Since Dirac, Majorana, or pseudo-Dirac neutrinos have almost similar gauge interactions, modulo some small lepton-number violating components for the Majorana and pseudo-Dirac neutrinos, it is not possible to distinguish between a pseudo-Dirac and two almost degenerate Majorana particles. The neutrinoless double beta decay experiment will be able to tell us if the neutrinos are Majorana or pseudo-Dirac particles and if there is lepton number violation in nature.

## 6.2 Models of Neutrino Masses

The atmospheric and the solar neutrinos have now established that neutrinos have tiny masses, orders of magnitude smaller than the masses of the charged fermions. Combined with some laboratory experiments we have a fairly good understanding of two of the mass squared differences and two large mixing angles. For the absolute mass scale of the neutrino mass matrix and the third mixing angles, only upper limit exists.

In the standard model, neutrinos are massless. The left-handed neutrinos $\nu_{iL}$, $i = e, \mu, \tau$ transform under the standard model gauge group $SU(3)_c \times SU(2)_L \times U(1)_Y$ as $(1,2,-1)$. Thus, it is not possible to write a Majorana mass term for the neutrinos. On the other hand, there are no right-handed neutrinos in the standard model, which would allow a Dirac neutrino. If we introduce right-handed neutrinos $\nu_{iR}$ transforming as $(1,1,0)$, then it will be possible to write the Yukawa couplings that can give a Dirac mass to the neutrinos:

$$\mathscr{L}_{mass} = \frac{1}{2} f_{ij} \overline{\psi_{iL}} \psi_{jR} \phi. \tag{6.31}$$

When the scalar doublet field $\phi$ acquires a *vev*, $\langle\phi\rangle = v$, neutrinos get a Dirac mass $m_{Dij} = f_{ij} v$. Since all the charged fermions receive mass in the same way, it is natural that this mass term is also of the same order of magnitude. Then what makes them so light?

A natural explanation for the tiny neutrinos mass is seeded into the standard model. Consider the most general dimension-5 effective lepton-number violating operator in the standard model that can contribute to the Majorana masses of the neutrinos [54]

$$\mathscr{L}_{Maj} = \Lambda^{-1}(\nu\phi^\circ - e\phi^+)^2. \tag{6.32}$$

Here $\Lambda$ is some lepton-number violating heavy scale in the theory coming from some extensions of the standard model, and $\phi$ is the usual Higgs doublet scalar. At this stage we do not specify the origin of this term. We expect some extensions of the standard model will generate this effective low-energy term. The electroweak symmetry breaking then induces a Majorana mass to the neutrinos:

$$\mathscr{L}_{Maj} = m_\nu \nu_{iL}^T C^{-1} \nu_{jL}, \tag{6.33}$$

with $m_\nu = v^2/\Lambda$. Thus, a large lepton-number violating scale $\Lambda$ would explain naturally why $m_\nu$ is much smaller than the charged fermion masses. This also suggests that a Majorana mass of the neutrinos is more natural than a Dirac mass. The lepton number violation at very high scale may have interesting cosmological consequences, which we shall discuss later. We shall now briefly discuss some of the generic mechanisms of realizing the effective dimension-5 lepton-number violating operators that can give us tiny Majorana neutrino mass naturally in some extensions of the standard model.

**See-Saw Mechanism**

The see-saw mechanism of neutrino masses is the first and most popular generic mechanism of obtaining a tiny neutrino mass in an extension of the standard model with right-handed neutrinos [55]. Three right-handed neutrinos $N_{iR}, i = 1,2,3$ are included, which are singlets under the standard model gauge group and electrically neutral. Inclusion of the right-handed neutrinos will allow the Yukawa couplings

$$\mathscr{L}_{Yuk} = h_{i\alpha} \bar{N}_{Ri} \phi \ell_{L\alpha}, \tag{6.34}$$

where $\ell_{L\alpha}, \alpha = 1,2,3$ are the left-handed leptons, $\phi$ is the standard model Higgs doublet and $h_{\alpha i}$ are the Yukawa couplings. In general, the Yukawa couplings could be complex giving rise to *CP* violation. This *CP* violation can contribute to the generation of a lepton asymmetry of the universe and can be observed in the neutrino oscillation experiments. For the present we shall assume $h_{\alpha i}$ to be real. After the electroweak symmetry breaking by the *vev* of the Higgs doublet, this Yukawa coupling will induce a Dirac mass term like the charged fermions

$$m_{D\alpha i} = h_{\alpha i} \langle \phi \rangle. \tag{6.35}$$

This Dirac mass will be of the same order of magnitude as that of the charged fermions, which is unacceptable.

If we now allow all possible interactions of the singlet right-handed neutrinos and do not impose lepton number conservation, then we can write the Majorana mass term of the right-handed neutrinos

$$\mathscr{L}_{Maj} = M_{ij}\, \overline{(N_{Ri})^c}\, N_{Rj} = M_{ij}\, N_{iR}^T C^{-1} N_{jR} = M_{ij}\, N^{cT}_{iL}\, C^{-1}\, N^c{}_{jL}. \tag{6.36}$$

Since the right-handed neutrinos are singlets under the standard model, the Majorana masses $M_{ij}$ of the right-handed neutrinos are not protected by any symmetry and it could be large. The conservation of lepton number in the Yukawa couplings of the neutrinos implies that the right-handed neutrinos carry lepton number $+1$, and hence, the Majorana mass term has a lepton number $+2$ breaking lepton number explicitly. Combining the Dirac and Majorana mass matrices for both the left-handed and right-handed neutrinos, we can write

$$\begin{aligned}\mathcal{L}_{mass} &= m_{D\alpha i}\, \nu_{L\alpha}^T\, C^{-1}\, {N^c}_{iL} + M_i\, {N^c}_{iL}^T\, C^{-1}\, {N^c}_{iL} + H.c. \\ &= \left(\nu_\alpha \quad N_i^c\right)_L^T\, C^{-1} \begin{pmatrix} 0 & m_{D\alpha i} \\ m_{Di\alpha} & M_i \end{pmatrix} \begin{pmatrix} \nu_\alpha \\ N_i^c \end{pmatrix}_L . \end{aligned} \tag{6.37}$$

Without loss of generality we assume that the Majorana mass matrix of the right-handed neutrinos is real and diagonal ($M_{ij} = M_i\delta_{ij}$). Since the right-handed neutrinos are sterile with respect to the standard model interactions, it is possible to diagonalize the right-handed neutrino mass matrix without affecting any of the left-handed interactions. There could be new sources of *CP* violation in the Majorana mass matrix, but for the present discussion we shall ignore them.

The indices $i = 1,2,3$ and $\alpha = 1,2,3$ correspond to the right-handed and left-handed neutrinos, respectively. The mass matrix is thus a $6\times 6$ matrix, and the physical masses can be obtained by diagonalizing this mass matrix. Without any knowledge of the Dirac neutrino mass matrix $m_D$, it is not possible to study this matrix in a general way. However, we are now interested in some general features, which can be extracted in a simple way. Consider a $2\times 2$ mass matrix of the similar form

$$M_\nu = \begin{pmatrix} 0 & m_D \\ m_D & M \end{pmatrix} . \tag{6.38}$$

When there is no mixing or very small mixing between the different generations, for every generation there will be a mass matrix of this form. We can now diagonalize this matrix with the assumption $m_D \ll M$, and the eigenvalues are given by

$$m_1 = -\frac{m_D^2}{M} \quad \text{and} \quad m_2 = M. \tag{6.39}$$

The sign of the neutrino masses can be absorbed in the phases of the physical fields. Although the overall sign does not matter, for the three generations the relative signs can play an important role in deciding the mass matrix.

The physical states corresponding to the two eigenvalues are now given by

$$\begin{pmatrix} \psi_1 \\ \psi_2 \end{pmatrix} = \begin{pmatrix} \cos\theta & -\sin\theta \\ \sin\theta & \cos\theta \end{pmatrix} \begin{pmatrix} \nu \\ N^c \end{pmatrix}, \tag{6.40}$$

where $\tan\theta = 2m_D/M$. For every generations there are now two physical Majorana particles: $\psi_1$ with mass $m_1$ and $\psi_2$ with mass $m_2$ and $m_1 \ll m_2$. The lighter particle is mostly the left-handed neutrino with a tiny admixture of the right-handed neutrino, while the heavy particle is mostly the right-handed neutrino. If we now assume $M \sim 10^9$ GeV and the Dirac masses $m_D$ are of the order of a few GeV, the two mass eigenvalues become of the order of $m_1 \sim 1$ eV and $m_2 \sim 10^9$ GeV. Thus, by increasing the large mass scale, we can make the left-handed neutrinos much lighter than the Dirac masses of the charged fermions. This mechanism is called the *see-saw mechanism*.

Since the neutrinos are Majorana particles in this mechanism, they will violate total lepton number and, hence, contribute to the neutrinoless double beta decay. Although both the states will contribute to the neutrinoless double beta decay, the contributions of the heavier states vanish because of a phase space factor in the nuclear matrix element. The effective mass $< m >= m_{ee} = \sum_{i=1}^{2} U_{ei}^2 m_i F(m_i, A)$ contributing to the neutrinoless double beta decay has the factor from the nuclear matrix element $F(m_i, A)$, which vanishes for large $m_i$, and hence, we get $< m >\sim m_1$. For three generations, we have to block diagonalize the neutrino mass matrix and consider only the light neutrino masses in calculating the amount of neutrinoless double beta decay.

In a three generation scenario, the see-saw mechanism gives three light neutrino state only when the determinant of the right-handed neutrinos are nonvanishing. If any of the right-handed neutrino Majorana mass eigenvalue vanishes, the determinant of the right-handed neutrino mass matrix will also vanish. In this case some of the left-handed neutrinos will combine with the right-handed neutrinos to form Dirac neutrinos. The number of Dirac neutrinos will depend on the rank of the mass matrix $M$. For $n$ generations, if the rank of $M$ is $r$, then there will be $2r$ Majorana neutrinos and $n-r$ Dirac neutrinos. The magnitude of the Dirac neutrino masses is expected to be of the order of the charged fermion masses, while there will be $r$ light Majorana neutrinos. The heavy Dirac and Majorana neutrinos with masses $m_D$ and $M$, respectively, will not contribute to the neutrinoless double beta decay.

### Triplet Higgs Model

Another realization of the effective operator (6.32) to give neutrino mass utilizes a triplet Higgs scalar field. The standard model is now extended to include an $SU(2)_L$ triplet Higgs scalar $\xi \equiv [\xi^{++},\ \xi^{+},\ \xi^{0}]$, which transforms under $SU(3)_c \times SU(2)_L \times U(1)_Y$ as $[1, 3, +1]$. We assume that there is no right-handed neutrino in the model; it contains just the minimal particle content of the standard model and one triplet Higgs scalar $\xi$.

The couplings of the triplet Higgs scalar with the left-handed doublet leptons $\ell_i \equiv [\nu_i,\ l_i]$, given by

$$\mathscr{L}_{Yuk} = f_{ij}\ \xi\ \ell_i\ \ell_j = f_{ij}\ \left[\xi^\circ \nu_i \nu_j + \xi^+(\nu_i l_j + \nu_j l_i) + \xi^{++} l_i l_j\right], \tag{6.41}$$

can give Majorana masses to the neutrinos if $\xi^\circ$ acquires a nonzero tiny *vev* $\langle\xi^\circ\rangle = u$. Since the *vev* of the $\xi^\circ$ affects the $\rho$-parameter of the precision electroweak tests, its *vev* should be less than a few GeV, but otherwise there is no constraint on $u$. $u$ does not affect any other charged fermions, so it can be as small as possible.

The minimal extension of the standard model to accommodate a triplet Higgs scalar may or may not have a global lepton number symmetry. In the original triplet Higgs model [56], lepton number was assumed to be a global symmetry as in the standard model. This would restrict some couplings of the triplet Higgs scalar, and the *vev* of the field $\xi$ would break the global lepton number symmetry spontaneously. However, this model has a couple of problems, which rules out this version of the model. We start with the most general Higgs potential with a doublet and a triplet Higgs

$$V = \frac{1}{2}m^2\phi^\dagger\phi + M^2\xi^\dagger\xi + \frac{1}{4}\lambda_1(\phi^\dagger\phi)^2 + \frac{1}{4}\lambda_2(\xi^\dagger\xi)^2 + \frac{1}{2}\lambda_3(\xi^\dagger\xi)(\phi^\dagger\phi) + \frac{1}{2}\mu\xi^\dagger\phi\phi. \tag{6.42}$$

In the original model, global lepton number symmetry was imposed to make $\mu = 0$, so the Yukawa couplings of the triplet Higgs scalar will imply that $\xi$ has a lepton number 2. Then the *vev* of the triplet Higgs $<\xi>= u$ will break lepton number spontaneously. The fields Im$\phi^0$ and Im$\xi^0$ will now remain massless; the former will become the longitudinal mode of the Z-boson while the latter will become a Nambu–Goldstone boson, the Majoron, associated with the spontaneous breaking of the global lepton number symmetry. The *vev* $u$ has to be less than a few keV to suppress the process $\gamma + e \to e+$Majoron. This implies a very severe fine tuning making the model technically unnatural. However, the most stringent constraint comes from the LEP data, which rules out the possibility of Z decaying into Majorons.

Both these problems could be solved in a variant of the model [57, 58] by choosing $\mu \neq 0$ and a heavy mass of the triplet Higgs scalar $M \sim \mu \gg <\phi>= v$. Lepton number is now violated explicitly by the couplings of the triplet Higgs $\mu\xi^\dagger\phi\phi$, and this will give a large mass to the Majoron. The mass matrix of the scalars $\sqrt{2}\mathrm{Im}\phi^0$ and $\sqrt{2}\mathrm{Im}\xi^0$ can be written as

$$\mathcal{M}^2 = \begin{pmatrix} -4\mu u & 2\mu v \\ 2\mu v & -\mu v^2/u \end{pmatrix}. \tag{6.43}$$

This leaves one combination of these fields as massless, which will become the longitudinal mode of the Z boson. The other combination now corresponds to the "would-be" Majoron and becomes massive with mass of the order of the triplet Higgs mass $M$. Thus, the Z boson cannot decay into this would-be Majoron. In this case there is also a natural explanation for the smallness of the *vev* of the Higgs scalar. The minimization of the scalar potential now gives

$$u = \frac{-\mu v^2}{M^2}, \tag{6.44}$$

so the mass of the left-handed neutrinos becomes

$$m^\nu_{ij} = f_{ij}u = -f_{ij}\frac{\mu v^2}{M^2}. \tag{6.45}$$

Another way to understand this smallness of the Majorana mass is to integrate out the heavy triplet fields to get an effective nonrenormalizable term

$$\mathcal{L}_{eff} = -\frac{f_{ij}\mu}{M^2}[\phi\phi\ell_i\ell_j], \tag{6.46}$$

which then gives the neutrino mass through the *vev* of the Higgs doublet $\phi$. The lepton-number violating scale can now be as high as $10^9$ GeV or preferably more to give an eV neutrino naturally, which is also the correct scale for leptogenesis in this model [57, 58] (see section 16.3).

There are many advantages of the triplet Higgs model of neutrino masses. The neutrino mass matrix is now directly proportional to the Yukawa couplings $f_{ij}$. This will mean that by finding the Yukawa couplings, we can find the neutrino mass matrix. In models with extra dimensions or models where one explains the cosmological constant, the triplet Higgs scalar has a mass of around a few TeV. In these models the same-sign dilepton signals with negligible standard model background in the colliders can detect the triplet Higgs scalar. Then the decays of the triplet Higgs will give us all the Yukawa couplings, and from the collider experiments it will be possible to determine the neutrino mass matrix. This may also help us understand the matter–antimatter asymmetry through leptogenesis.

Although the see-saw mechanism and the triplet Higgs mechanism are two different extensions of the standard model, if one extends the standard model to make it left–right symmetric, both of these mechanisms get embedded in the model. However, the realizations of these mechanisms are different. In the left–right symmetric model $B-L$ is a local gauge symmetry. Hence, lepton number violation requires breaking of this local $B-L$ symmetry spontaneously. Thus, all lepton-number violating effects enter through the breaking of this local symmetry.

### Left–right Symmetric models

The low-energy observed parity violation is accommodated in the standard model by treating the left-handed particles preferentially. A natural extension of the standard model should treat the left-handed and the right-handed fermions in the same way. With this motivation the left–right symmetric extension of the standard model was proposed.

The left–right symmetric extensions of the standard model gauge group is extended to a gauge group $SU(2)_L \times SU(2)_R \times U(1)_{B-L}$ [59, 60]. The $SU(2)_R \times U(1)$ group is broken at some high energy, giving our low-energy electroweak theory with only unbroken $SU(2)_L \times U(1)_Y$. The left-handed fermions are now doublets under $SU(2)_L$, while the right-handed fermions are doublets under $SU(2)_R$. The extra $U(1)$ group quantum numbers come out to be the same as the $B-L$ quantum numbers of the usual quarks and leptons. The transformations of the quarks and the leptons are given by

$$\begin{aligned} Q_L &= \begin{pmatrix} u_L \\ d_L \end{pmatrix} \equiv [3,2,1,\frac{1}{3}] \quad Q_R = \begin{pmatrix} u_R \\ d_R \end{pmatrix} \equiv [3,1,2,\frac{1}{3}] \\ \ell_L &= \begin{pmatrix} \nu_L \\ e_L \end{pmatrix} \equiv [1,2,1,-1] \quad \ell_R = \begin{pmatrix} \nu_R \\ e_R \end{pmatrix} \equiv [1,1,2,-1]. \end{aligned} \tag{6.47}$$

The electric charge and the $U(1)_Y$ hypercharge are related to the quantum numbers of the $U(1)_{(B-L)}$ and the $SU(2)_R$ groups through the relations

$$Q = T_{3L} + Y = T_{3L} + T_{3R} + \frac{B-L}{2}. \tag{6.48}$$

Thus, $B-L$ is now a local gauge symmetry, so any $B-L$ violating phenomenon such as the neutrino mass or the neutron–antineutron oscillation would be possible only after this symmetry is broken spontaneously.

The left–right symmetric models have an added attractive feature of breaking the parity symmetry spontaneously. The $SU(2)_L$ gauge bosons $W_L$ and the $SU(2)_R$ gauge bosons $W_R$ are not parity eigenstates, but under parity transformation $W_L \leftrightarrow W_R$. Since left-handed and right-handed fermions are related by the parity operation, a discrete $Z_2$ symmetry relating the group $SU(2)_L \leftrightarrow SU(2)_R$ can now be identified with the parity operator of the Lorentz group. Thus, the spontaneous breaking of the left–right symmetry will also break parity spontaneously. The invariance of the Lagrangian under the parity symmetry will ensure that the gauge coupling constants for the two $SU(2)$ groups are the same before the $SU(2)_R$ group is broken at a scale $M_R$,

$$g_L(M > M_R) = g_R(M > M_R). \tag{6.49}$$

After the left–right symmetry breaking, the gauge coupling constants for the two $SU(2)$ groups can be different. In some left–right symmetric models parity is broken explicitly, while in other models parity is broken spontaneously before the left–right symmetry breaking.

The choice of the Higgs scalars that break the left–right symmetry determines some of its phenomenology. We shall first consider the most popular choice of the Higgs scalar and later mention some other combinations of Higgs scalars and the corresponding phenomenology. The conventional choice of Higgs scalar breaks the group in two stages [61, 62]:

$$\begin{aligned} SU(3)_c \times SU(2)_L \times\ & SU(2)_R \times U(1)_{(B-L)} && [\equiv \mathcal{G}_{LR}] \\ & \stackrel{M_R}{\rightarrow} SU(3)_c \times SU(2)_L \times U(1)_Y && [\equiv \mathcal{G}_{std}] \\ & \stackrel{m_W}{\rightarrow} SU(3)_c \times U(1)_Q && [\equiv \mathcal{G}_{em}]. \end{aligned}$$

In this case the same Higgs that breaks the left–right symmetry can give masses to the neutrinos. The $SU(2)_R$ symmetry is now broken by a triplet Higgs scalar ($\Delta_R$), which transforms under $\mathcal{G}_{LR}$ as $[1,1,3,-2]$. The discrete symmetry $Z_2$ will imply the existence of another triplet scalar ($\Delta_L$), which transforms under $\mathcal{G}_{std}$ as $[1,3,1,-2]$. However, the *vev* of the left-handed triplet $\Delta_L$ is constrained by the precision measurement to be less than a few eV. Both triplet scalars have one doubly charged component, one singly charged component, and a neutral component, $\Delta \equiv [\Delta^{++}\ \Delta^{+}\ \Delta^{0}]$. Only the neutral components can acquire *vevs*,

$$<\Delta_R^0> = v_R \sim M_R \qquad <\Delta_L^0> = v_L \ll m_W. \tag{6.50}$$

Although we have already assumed $v_R \gg v_L$, this comes naturally from the minimization of the scalar potential.

The electroweak symmetry can be broken by a $SU(2)_L$ doublet $[1,2,0,1]$, but this cannot give masses to the fermions. So, a bi-doublet Higgs scalar $\Phi$, which transforms under $\mathcal{G}_{LR}$ as $[1,2,2,0]$ is considered for electroweak symmetry breaking. There are two neutral components of $\Phi$, both of which acquire nonzero *vevs* $v$ and $v'$. Without loss of generality we assume for simplicity $v \gg v'$ and neglect $v'$, which takes care of the flavor changing neutral currents.

To see the condition for minimization of the scalar potential, we now write the scalar potential with all of these fields replaced by their *vevs* and then minimize to find the condition on the *vevs* of the various fields:

$$\begin{aligned} V(v_L, v_R, v) = & -\mu^2 v^2 - m_\Delta^2 (v_L^2 + v_R^2) + \rho_1 (v_L^4 + v_R^4) \\ & + \rho_2 v_L^2 v_R^2 + \lambda v^4 + \alpha v^2 (v_L^2 + v_R^2) + \beta v^2 v_L v_R. \end{aligned} \tag{6.51}$$

The last term is the most crucial one. This term allows the breaking of discrete parity symmetry and the $SU(2)_R$ without breaking the $SU(2)_L$ group.

The minimization of this potential with respect to $v_L$ and $v_R$ gives

$$\begin{aligned} -2m_\Delta^2 v_L + 4\rho_1 v_L^3 + 2\rho_2 v_L v_R^2 + 2\alpha v^2 v_L + \beta v^2 v_R = 0 \\ -2m_\Delta^2 v_R + 4\rho_1 v_R^3 + 2\rho_2 v_R v_L^2 + 2\alpha v^2 v_R + \beta v^2 v_L = 0. \end{aligned} \tag{6.52}$$

These equations can be combined to get the condition

$$(v_L^2 - v_R^2)[(4\rho_1 - 2\rho_2) v_L v_R - \beta v^2] = 0. \tag{6.53}$$

This has two solutions: the first one is a parity conserving, $v_L = v_R$, which cannot explain the parity violation at low energies. The other solution is

$$v_L = \frac{\beta v^2}{(4\rho_1 - 2\rho_2) v_R}. \tag{6.54}$$

In this case if $SU(2)_R$ and the left–right parity are broken at a very large scale $v_R \gg v$, we get $v_L \ll v$. Thus, the parity violating solution gives a natural explanation of the hierarchy $v_R \gg v \gg v_L$.

The Yukawa couplings of the fermions will have two parts; the first part will contain terms with the bi-doublets while the second part has the couplings with the triplet Higgs scalars. The first part will give the Dirac masses while the second part gives the Majorana masses. The complete Yukawa couplings can be written as

$$\mathcal{L}_{Yuk} = \mathcal{L}_{Dir} + \mathcal{L}_{Maj}, \tag{6.55}$$

with
$$\mathcal{L}_{Dir} = h_{ia}\, \overline{Q}_{Li} Q_{Ra}\, \Phi + f_{ia}\, \bar{\ell}_{Li} \ell_{Ra}\, \Phi, \tag{6.56}$$

and
$$\mathcal{L}_{Maj} = f_{Lij} \ell_{Li}^T C^{-1} \ell_{Lj} \Delta_L + f_{Rab} \ell_{Ra}^T C^{-1} \ell_{Ra} \Delta_R, \tag{6.57}$$

where $i, j = 1,2,3$ are the indices for the three left-handed generations and $a, b = 1,2,3$ are the indices for the three right-handed generations. This indicates that the neutrino Dirac masses are of the same order of magnitude as the charged lepton masses $m_D \sim m_e \sim f_{ij}v$. The neutrino mass matrix can now be written including both the Dirac and Majorana mass contributions in the basis $(\nu_{Li} \quad \nu_{Ra}{}^c)$ as

$$M_\nu = \begin{pmatrix} m_L & m_D \\ m_D & M_N \end{pmatrix} = \begin{pmatrix} f_{Lij}\, v_L & f_{ia}\, v \\ f_{ia}\, v & f_{Rab}\, v_R \end{pmatrix}. \tag{6.58}$$

If the determinant of the heavy right-handed neutrino mass matrix is nonvanishing, this mass matrix gives three heavy Majorana fields with masses of the order of $\sim O(v_R)$, which are mostly the right-handed neutrinos. The other three neutrinos remain light with masses

$$m_{\nu Lij} = m_L - \frac{m_D^2}{M_N} = \left( \frac{\beta f_{Lij}}{4\rho_1 - 2\rho_2} - \frac{f_{ia}^2}{f_{Rab}} \right) \frac{v^2}{v_R}. \tag{6.59}$$

These light states are mostly the left-handed neutrinos with a small admixture of the right-handed neutrinos. There are now two contributions to the light neutrino masses, one coming from the *vev* of the $SU(2)_L$ triplet Higgs scalar $\Delta_L$ and the other coming from the see-saw contribution. The second term is also called type I see-saw, which is the usual see-saw contribution, while the first term coming from the *vev* of the triplet Higgs scalar is called type II see-saw. Sometimes type II see-saw includes both the contributions. The minimization of the potential ensures that both the contributions are of the same order of magnitude.

We shall consider one more combination of the Higgs scalars, which can give us the required symmetry breaking. In this case there are no triplet Higgs scalars, all Higgs scalars are doublets under the $SU(2)$ groups [63, 64]. The $SU(2)_R \times U(1)_{B-L}$ group is broken by a right-handed Higgs doublet $\chi_R \equiv [1,1,2,+1]$, which acquires a *vev* $\langle \chi_R \rangle = v_R$. The left–right parity then ensures that there is another left-handed doublet Higgs scalar field $\chi_L \equiv [1,2,1,+1]$, which acquires a small *vev* $\langle \chi_L \rangle = v_L$. We retain the bi-doublet scalar $\Phi$ to give masses to the charged fermions.

The scalars $\chi_{L,R}$ break lepton number by one unit, so they cannot contribute to the masses to the neutrinos. However, with these fields we can write effective dimension-5 terms, which can contribute to the neutrino masses,

$$\mathscr{L}_\nu = \frac{h_L}{M_h} \ell_L \ell_L \chi_L \chi_L + \frac{h_R}{M_L} \ell_R \ell_R \chi_R \chi_R. \tag{6.60}$$

$M_h$ is some new scale, which could be around the left–right symmetry breaking scale $v_R$. This effective operator may be realized by introducing a singlet fermion $S \equiv [1,1,1,0]$ with mass $M_S$. The neutrino masses can then be generated by the Yukawa couplings of the neutrinos and the singlet fermion $S$

$$\mathscr{L}_S = f_{ia} \bar{\ell}_{Li} \ell_{Ra} \Phi f_{iL} \bar{\ell}_{Li} S \chi_L + f_{aR} \bar{\ell}_{Ra} S \chi_R + M_S SS. \tag{6.61}$$

When the scalars acquire *vevs*, they contribute to the neutrino mass matrix, which can be written in the basis $(\nu_{Li} \quad \nu_{Ra}{}^c \quad S)$ as

$$M_\nu = \begin{pmatrix} 0 & f_{ia} v & f_{iL} v_L \\ f_{ia} v & 0 & f_{aR} v_R \\ f_{iL} v_L & f_{aR} v_R & M_S \end{pmatrix}. \tag{6.62}$$

The right-handed neutrinos get an effective Majorana mass of $M_N \sim f_{aR}^2 v_R^2 / M_S$, while the left-handed neutrinos remain light,

$$m_{\nu ij} = -\frac{f_{ia} f_{iL} v v_L}{f_{aR} v_R} + \frac{f_{ia} f_{jb}}{f_{aR} f_{bR}} \frac{M_S v^2}{v_R^2}. \tag{6.63}$$

The second term is the so-called double see-saw contribution, which looks like the usual see-saw contribution when the expression for $M_N$ is substituted.

The first term is distinct from the see-saw mechanisms and is independent of $M_S$. This contribution has some interesting phenomenology and is called the type III see-saw mechanism [64]. Although for simplicity we considered only one singlet $S$, in any realistic model three singlet neutrinos $S_p$ ($p = 1,2,3$) are required to have a rank-3 light neutrino mass matrix. In this model parity could be spontaneously broken by a singlet Higgs scalar or explicitly broken by giving different masses to $\chi_L$ and $\chi_R$, to get the required solution from minimization of the potential. A natural assumption for the mass scales is $v_L \lesssim v$, since $v$ breaks the electroweak symmetry and gives masses to the fermions, and both $v_R$ and $M_S$ are heavy. If $M_S < v_R$, then some of the heavy neutrinos become naturally almost degenerate, which may have some interesting consequences.

There are now three additional gauge bosons in the theory. The two charged right-handed gauge bosons become massive $M_{W_R} \sim v_R$. One combination of the neutral gauge bosons corresponding to $SU(2)_R$ and $U(1)_{B-L}$ also becomes heavy $M_{Z_R} \sim v_R$. Although low-energy phenomenology gives a bound on these gauge bosons of around 1 TeV [44], most models of neutrino masses and leptogenesis require the left–right symmetry breaking scale to be around $v_R \sim 10^9$ GeV or above. In some grand unified theories this scale cannot be lower than $10^{14}$ GeV. The left–right symmetric models have many interesting phenomenology. We shall discuss some of the consequences of the $B-L$ violation in section 8.5.

### Radiative Models

Radiative models of neutrino masses were proposed to explain the smallness of the neutrino masses without invoking a very large scale in the theory. Several radiative models can naturally explain the smallness of the neutrino mass with only TeV scale new particles. Since the scale of lepton number violation in this case is around a few TeV, all new physics is now in the next generation accelerator energy range. For example, the new Higgs scalars of these models could be observed in the near future. However, these models suffer from a serious drawback that they allow very fast lepton number violation, which can then erase any matter–antimatter asymmetry in the universe.

In a radiative model with minimal extension of the standard model, one extends the standard model with a singlet charged Higgs scalar $\eta_+$, which transforms as $[1,1,+1]$ [65]. This model is known as the Zee model, which requires two doublet Higgs scalars, $\phi_1$ and $\phi_2$, for explicit lepton number violation. The Yukawa couplings of the singlet scalar then violate lepton number:

$$\mathscr{L}_{\Delta L \neq 0} = f_{ij}\, \ell_{iL} \ell_{jL}\, \eta_+ + \mu \varepsilon_{ab}\, \phi_a \phi_b\, \eta_+^\dagger, \tag{6.64}$$

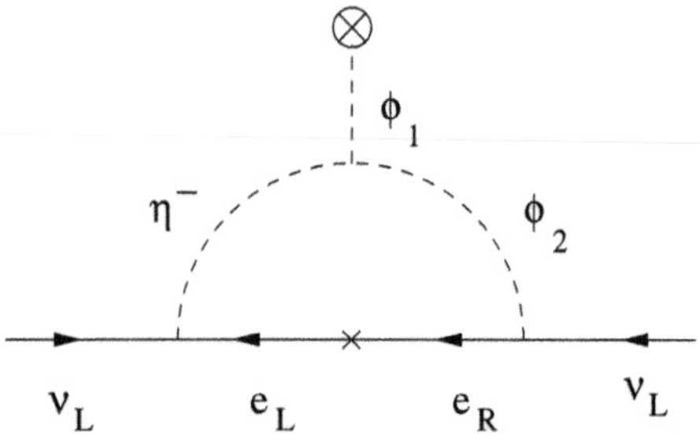


**FIGURE 6.1**
One-loop diagram generating a neutrino mass in a Zee model.

where $\varepsilon_{ab}$ is the totally antisymmetric tensor and the Yukawa coupling $f_{ij}$ is antisymmetric in the indices $i$ and $j$.

Unlike the triplet Higgs mechanism, the singlet charged scalar $\eta_+$ cannot acquire a *vev* and give masses to the neutrinos. It can give a mass to the neutrinos only through the one-loop diagram of figure 6.1. The *vev* of the standard model Higgs doublet allows mixing of the charged singlet Higgs with a charged component of the other Higgs doublet, which is crucial for this mass term to be nonvanishing.

The antisymmetric coupling of the new charged singlet Higgs $\eta_+$ with the leptons makes the diagonal terms of the mass matrix vanish. The neutrino mass matrix is now given by

$$m^{\nu}_{ij} = A f_{ij}(m_i^2 - m_j^2), \tag{6.65}$$

where $A$ is a constant and $m_i$ is the mass of the charged leptons $i = e, \mu, \tau$. The vanishing of the diagonal terms and the smallness of $m_e$ compared with $m_\mu$ and $m_\tau$ make this model quite interesting and predictive. Considering only the present bounds on the neutrino masses, this model is quite attractive, although there are some limitations of this model. For example, leptogenesis becomes difficult, which we shall discuss later.

## 6.3 Neutrino Oscillations

Two neutral states with slightly different mass and with same quantum numbers can oscillate into each other. If they differ in any quantum numbers, that symmetry will be broken in the process. Examples of matter oscillation include $K^\circ - \overline{K^\circ}$ and $B^\circ - \overline{B^\circ}$ oscillations. However, in any matter–antimatter oscillation $C$ and $CP$ are also violated, but in neutrino oscillation [66] one of the neutrino species with definite lepton flavor $L_e, L_\mu$, or $L_\tau$ is transformed into another species violating lepton flavor. This is possible when the neutrinos have mass and the mass difference is not too large. The time taken for these transitions to occur is related to the oscillation length or the distance that a neutrino species can travel without getting changed to another species. The uncertainty principle relates the oscillation time and the mass squared difference between the two states. Another crucial requirement for the neutrino oscillations is the violation of neutrino flavors through mixing of mass eigenstates [67].

The neutrinos of definite flavor are produced in any charged-current interaction. If any process involves an electron, we know that a $\nu_e$ has been produced. When this neutrino propagates, it is the physical state that propagates with time, and after some time it has a probability to be partly in a different flavor state, say, $\nu_\mu$ or $\nu_\tau$. If we detect this neutrino at a distance, we may find a neutrino with a different flavor.

If the two states have the same masses, then both of them will propagate the same way and there will not be any oscillation. If the two states have different masses but the flavor states are the same as the physical states (no mixing), then the flavor states will have definite evolution and they will not change to the other state. Thus, the main criteria for the neutrino oscillation are that there should be a mass difference between the different neutrino states and that the mass eigenstates are different from the flavor eigenstates, which is given by the neutrino mixing matrix.

### Neutrino Mixing

Let us assume that in some extensions of the standard model, neutrinos have masses, which are given by

$$\mathscr{L}_{mass} = m_i \nu_{iL}^T C^{-1} \nu_{iL} \equiv m_i \nu_{iL} \nu_{iL}, \tag{6.66}$$

where $\nu_{iL}$ are the physical states with Majorana mass $m_i$. We shall not worry about the origin of this mass in this discussion.

In general, the physical states are not the weak eigenstates that appear in the charged current and neutral current interactions of the neutrinos

$$\begin{aligned} \mathscr{L}_{CC} &= -\frac{g}{\sqrt{2}} \sum_{\alpha=e,\mu,\tau} \bar{\nu}_{\alpha L} \gamma_\rho l_{\alpha L} W^\rho \\ \mathscr{L}_{NC} &= -\frac{g}{2\cos\theta_w} \sum_{\alpha=e,\mu,\tau} \bar{\nu}_{\alpha L} \gamma_\rho \nu_{\alpha L} Z^\rho, \end{aligned} \tag{6.67}$$

where the $l_{\alpha L}$ are physical charged leptons with masses $m_\alpha$. So, the $\nu_{\alpha L}$ are the neutrinos that have definite lepton flavor $\alpha$.

The weak eigenstates are related to the physical or mass eigenstates by

$$\nu_{\alpha L} = \sum_{i=1}^{3} U_{\alpha i} \nu_{iL}. \tag{6.68}$$

$U_{\alpha i}$ is called the PMNS (Pontecorvo–Maki–Nakagawa–Sakata) mixing matrix [67]. The mass matrix in the flavor basis, i.e., the basis in which the charged lepton mass matrix is diagonal, is then given by

$$M_{\alpha\beta}^{\nu} = U_{\alpha i} M_{ij}^{\text{diag}} U_{\beta j}^T, \tag{6.69}$$

where $M_{ij}^{\rm diag} = m_i \delta_{ij}$. It is convenient to parameterize this mixing matrix $U_{\alpha i}$ in terms of 3 mixing angles

$$U = \begin{pmatrix} c_1 c_3 & -s_1 c_3 & -s_3 \\ s_1 c_2 - c_1 s_2 s_3 & c_1 c_2 + s_1 s_2 s_3 & -c_3 s_2 \\ s_1 s_2 + c_1 c_2 s_3 & c_1 s_2 - s_2 c_2 s_3 & c_2 c_3 \end{pmatrix}, \tag{6.70}$$

where $s_i = \sin\theta_i$ and $c_i = \cos\theta_i$. If there is *CP* violation, there will be complex phases in this matrix. This mixing is crucial for neutrino oscillations which violate lepton flavors $(L_e, L_\mu, L_\tau)$.

**Two Generation Vacuum Oscillation**

We shall now discuss the neutrino oscillations. First we shall assume a two neutrino scenario for simplification. Let us consider two weak eigenstates $|\nu_e>$ and $|\nu_\mu>$. But the mass matrix in this basis is not diagonal. After diagonalizing the mass matrix, we can write the two physical eigenstates as $\nu_1$ and $\nu_2$ with masses $m_1$ and $m_2$. The two weak eigenstates or the flavor eigenstates can now be expressed in terms of the mass eigenstates as

$$\begin{pmatrix} |\nu_e> \\ |\nu_\mu> \end{pmatrix} = \begin{pmatrix} \cos\theta & \sin\theta \\ -\sin\theta & \cos\theta \end{pmatrix} \begin{pmatrix} |\nu_1> \\ |\nu_2> \end{pmatrix} = \begin{pmatrix} c & s \\ -s & c \end{pmatrix} \begin{pmatrix} |\nu_1> \\ |\nu_2> \end{pmatrix}, \tag{6.71}$$

where we defined for the mixing angles $c = \cos\theta$ and $s = \sin\theta$. In this section we shall assume that there is no CP violation, and hence, the mixing matrices are real and orthogonal.

The physical states will evolve with time as

$$|\nu_{1,2}(t)>= \exp^{-iE_{1,2}t} |\nu_{1,2}(0)>, \tag{6.72}$$

where for the ultra-relativistic neutrinos with momentum $(p_i \gg m_i)$, the energy can be approximated as

$$E_i = p_i + \frac{m_i^2}{2p_i}. \tag{6.73}$$

If we start with a $\nu_e$ beam at time $t = 0$, after a time $t = t$, both $\nu_e$ and $\nu_\mu$ will be present in the beam.

After a time $t$, the weak eigenstates will evolve into

$$\begin{aligned} \begin{pmatrix} |\nu_e(t)> \\ |\nu_\mu(t)> \end{pmatrix} &= \begin{pmatrix} c & s \\ -s & c \end{pmatrix} \begin{pmatrix} \exp^{-iE_1 t} |\nu_1(0)> \\ \exp^{-iE_2 t} |\nu_2(0)> \end{pmatrix} \\ &= \begin{pmatrix} c & s \\ -s & c \end{pmatrix} \begin{pmatrix} \exp^{-iE_1 t} & 0 \\ 0 & \exp^{-iE_2 t} \end{pmatrix} \begin{pmatrix} c & -s \\ s & c \end{pmatrix} \begin{pmatrix} |\nu_e(0)> \\ |\nu_\mu(0)> \end{pmatrix}. \end{aligned} \tag{6.74}$$

If we thus start with $|\nu_e(0)>$, after a time $t$, we shall have a beam $|\nu_e(t)>$. The probability of $|\nu_\mu(0)>$ in this beam of $|\nu_e(t)>$ is the probability of finding a $\nu_\mu$ after a time $t$, starting with a $\nu_e$ beam, which is

$$P_{\nu_e \to \nu_\mu} = |<\nu_\mu(0)|\nu_e(t)>|^2 = \sin^2(2\theta)\sin^2\left(\frac{\Delta m^2 L}{4E}\right), \tag{6.75}$$

where $\Delta m^2 = |m_2^2 - m_1^2|$, $E$ is the average energy of the neutrino beam, and $L$ is the length traversed by the beam before a $\nu_\mu$ detection was taking place. Similarly the survival probabilities for a $\nu_e(\nu_\mu)$ to be detected in a $\nu_e(\nu_\mu)$ beam after a time $t$ are

$$P_{\nu_e \to \nu_e} = P_{\nu_\mu \to \nu_\mu} = 1 - \sin^2(2\theta)\sin^2\left(\frac{\Delta m^2 L}{4E}\right). \tag{6.76}$$

Thus, the two parameters $\sin^2(2\theta)$ and $\Delta m^2$ determine the oscillation probability.

The oscillation length can be defined as

$$L_0 = 2.47 \, \frac{E \text{ (GeV)}}{\Delta m^2 \text{ (eV}^2)} \text{ km}, \tag{6.77}$$

which is the distance, after which the neutrino returns to the initial state, i.e., the phase becomes $2\pi$. We can also write equation 6.75 in a more convenient form

$$\begin{aligned} P_{\nu_e \to \nu_\mu} &= \sin^2(2\theta) \, \sin^2\left(1.27 \, \frac{\Delta m^2(\text{eV}^2) \, L(\text{km})}{E(\text{GeV})}\right) \\ &= \sin^2(2\theta)\sin^2\left(\frac{\pi L}{L_0}\right). \end{aligned} \tag{6.78}$$

Most of the neutrino oscillation experiments analyze and present their results in terms of the parameters of a two neutrino oscillation in one of the above forms, assuming that the dominant mixing is between these two neutrinos.

The distance $L$ at which a significant oscillation can be seen should be close to the oscillation length $L \sim L_0$. For any shorter distance, the neutrinos will not get enough time to oscillate. On the other hand, if the source to detector distance is too large, $L \gg L_0$, many oscillations will take place within this distance and we get an average value

$$P_{\nu_e \to \nu_\mu}(L) = \frac{1}{2}\sin^2(2\theta). \tag{6.79}$$

In neutrino oscillations neutrino flavor is violated but the total lepton number is conserved. If there is helicity flip in the oscillation, total lepton numbers can be violated, but such $\Delta L = 2$ neutrino oscillation processes (originally proposed in [66]) will be highly suppressed by a factor of $m_\nu/E$. This is the so-called neutrino–antineutrino oscillation, which is impossible to see with our present detectors.

In a neutrino oscillation experiment the distance between the source and the detector $L$ and the energy of the neutrino beam $E$ determines what is the lowest mass-squared difference that can be measured in the experiment

$$\Delta m^2 \sim \frac{E}{L}. \tag{6.80}$$

The sensitivity ranges of the different experiments are presented in table 6.1.

**TABLE 6.1**
Neutrino sources and an estimate of their sensitivity for neutrino oscillations. SBL and LBL correspond to short baseline and long baseline.

| Source | Energy $E$ GeV | Distance $L$ km | $\Delta m^2$ eV$^2$ |
|---|---|---|---|
| Reactor SBL | $10^{-3}$ | 0.1 | 0.01 |
| Reactor LBL | $10^{-3}$ | 1 | $10^{-3}$ |
| Accelerators SBL | 1 | 1 | 1 |
| Accelerators LBL | 1 | $10^3$ | $10^{-3}$ |
| Atmosphere | 1 | $10^4$ | $10^{-4}$ |
| Sun | $10^{-3}$ | $10^8$ | $10^{-11}$ |

**Three Generation Scenario**

There are at least three light neutrinos. So, in any detail analysis one has to consider a three generation scenario. We shall thus generalize our previous discussion to include the third neutrino. We start with the general relation between the mass eigenstates and the flavor eigenstates of equation (6.68). The physical states $|\nu_i(t)>$, which are the states with masses $m_i$, evolve with time as

$$|\nu_i(t)>= \exp^{-iE_it}|\nu_i(0)> . \tag{6.81}$$

The states $|\nu_i>$ have momentum $p_i$ and energy $E_i = p_i + m_i^2/(2p_i)$, assuming $p_i \gg m_i$. After a time $t$, the weak eigenstates will evolve into

$$\begin{aligned} |\nu_\alpha(t)> &= \sum_i U_{\alpha i}\exp^{-iE_it}|\nu_i(0)> \\ &= \sum_i U_{\alpha i}\exp^{-iE_it}U^*_{\beta i}|\nu_\beta(0)> . \end{aligned} \tag{6.82}$$

As in the earlier case, the oscillation probability of a $\nu_\beta$ to be found in a neutrino beam of $\nu_\alpha$ after a time $t$ will then be given by

$$\begin{aligned} P_{\nu_\alpha\to\nu_\beta} &= \left|<\nu_\beta(0)|\nu_\alpha(t)>\right|^2 \\ &= \sum_{ij}\left|U_{\alpha i}U^*_{\beta i}U^*_{\alpha j}U_{\beta j}\right|\cos[(E_i - E_j)t - \phi_{\alpha\beta ij}], \end{aligned} \tag{6.83}$$

where $\phi_{\alpha\beta ij} = \arg[U_{\alpha i}U^*_{\beta i}U^*_{\alpha j}U_{\beta j}]$ is a phase, which vanishes when there is no CP violation. The oscillation length can be defined as

$$L^0_{ij} = \frac{4\pi E}{\Delta m^2_{ij}} = 2.47\frac{E(\text{GeV})}{\Delta m^2_{ij}(\text{eV}^2)}\ \text{km}, \tag{6.84}$$

where $\Delta m_{ij}^2 = \left|m_i^2 - m_j^2\right|$, so the oscillation probability can be written as

$$P_{\nu_\alpha \to \nu_\beta} = \sum_{ij} \left|U_{\alpha i} U_{\beta i}^* U_{\alpha j}^* U_{\beta j}\right| \cos\left(\frac{2\pi L}{L_{ij}^0}\right). \tag{6.85}$$

The CP phase has not been included in this expression.

The condition for neutrino oscillation is that the masses should not be degenerate and the mixing matrices should not be diagonal. In a three generation scenario, there are two independent mass-squared differences, and we need at least one of them to be nonvanishing for the neutrino oscillation to be observed. If only one mass-squared difference is nonzero, we get the two-generation result. Similarly, if only one of the mixing angle is nonvanishing, we get a two-generation scenario. A general analysis of the three-generation case is rather complicated. But for the purpose of demonstration we present a hierarchical mass structure for a three-generation model $m_1 < m_2 < m_3$. In this approximation, using the unitarity of the mixing matrix, we can write the oscillation probability as

$$P_{\nu_\alpha \to \nu_\beta} = \left|\delta_{\alpha\beta} + \sum_{i=2}^{3} U_{\beta i} U_{\alpha i}^* \left[\exp^{-i\Delta m_{i1}^2 L/2E} - 1\right]\right|^2. \tag{6.86}$$

Since $\Delta m_{31}^2 - \Delta m_{32}^2 = \Delta m_{21}^2$, only two of the combinations enter in this expression. Unitarity fixes one of the indices, which we chose to be 1. In this case only $\Delta m_{31}^2$ and $\Delta m_{21}^2$ enter into the discussion. However, one may also take the 2 or 3 as the preferred index and sum over the other two indices, but the final result would remain unchanged.

Let us consider an experiment in which the choice of $L$ and $E$ is such that only $\Delta m_{31}^2$ is relevant and $\Delta m_{31}^2 L/(2E) \ll 1$. The oscillation probability ($\alpha \neq \beta$) and the survival probability then become

$$\begin{aligned} P_{\nu_\alpha \to \nu_\beta} &= 2|U_{\beta 3}|^2 |U_{\alpha 3}|^2 \left(1 - \cos\frac{\Delta m_{31}^2 L}{2E}\right) \\ P_{\nu_\alpha \to \nu_\alpha} &= 1 - 2|U_{\alpha 3}|^2 (1 - |U_{\alpha 3}|^2) \left(1 - \cos\frac{\Delta m_{31}^2 L}{2E}\right). \end{aligned} \tag{6.87}$$

As expected, only the largest mass-squared difference appears in the expression and the dependence of $L/E$ is the same as in the two-generation case. There are four parameters entering in the expression, one of which (say, $|U_{\tau 3}|$) can be eliminated leaving three independent parameters $|U_{e3}|, |U_{\mu 3}|$ and $\Delta m_{31}^2$.

In experiments where the $\Delta m_{21}^2$ is relevant ($\Delta m_{31}^2 L/2E \geq 1$), the $\Delta m_{31}^2$ disappears from the expression after averaging over the distance between the source and the detector. In this case the relevant quantity is the survival probability, which becomes

$$P_{\nu_\alpha \to \nu_\alpha} = |U_{\alpha 3}|^4 + (1 - |U_{\alpha 3}|^2)^2 [1 - \frac{1}{2}\sin^2 2\bar{\theta}_{12}(1 - \cos^2\frac{\Delta m_{21}^2 L}{2E})], \tag{6.88}$$

where

$$\sin^2 \bar{\theta}_{12} = \frac{|U_{\alpha 2}|^2}{\sum_{i=1}^{2} |U_{\alpha i}|^2}. \tag{6.89}$$

The only common parameter between this probability and the previous one is $|U_{e3}|^2$. Since experimentally $|U_{e3}|^2$ has been constrained by CHOOZ [68] to be extremely small and the mass-squared differences required to explain the solar and atmospheric neutrino problems do not overlap, these two experiments could thus be treated independently for an order of magnitude estimate. However, for a detailed analysis one needs to study a three-generation case for the allowed range of $|U_{e3}|^2$.

**Neutrino Oscillation in Matter**

The neutrino oscillation depends on the mass-squared difference and the mixing angle. For the present let us restrict ourselves to only two generations. Then there are only one mass-squared difference and one mixing angle. It was noticed that the mass of one of the neutrinos could vary while propagating through matter, due to its interaction with the media. For example, if a beam of $\nu_e$ and $\nu_\mu$ is propagating in matter with varying density, the $\nu_e$ will interact with matter through charged-current interactions and its mass can vary while $\nu_\mu$ will not have any effect. The mass-square difference can then vary, and this can lead to an interesting phenomenon of resonant oscillation, called the Mikheyev–Smirnov–Wolfenstein (MSW) effect, which is required to solve the solar neutrino problem [69]. The matter effect was first pointed out by Wolfenstein, but some factors were corrected and the resonance effect was noticed by others [70]. Later this effect was applied to the solar neutrinos by Mikheyev and Smirnov.

While propagating in matter, the $\nu_e$ interacts with matter through charged-current interaction. This contributes an additional potential to the electron neutrinos, proportional to the electron number density $N_e$,

$$V = \sqrt{2} G_F N_e. \tag{6.90}$$

This will contribute to the energy of the electron neutrinos, generating an effective mass of $\nu_e$

$$m_e^2 \to m_e^2 + A = m_e^2 + 2\sqrt{2}\, G_F\, N_e\, E. \tag{6.91}$$

Now consider a simple situation with two neutrinos $\nu_e$ and one other neutrino $\nu_X$, where $X$ could be $\mu, \tau$ or some other light neutrinos that do not interact through the weak interaction, are singlets under $SU(2)_L$, and are called the sterile neutrinos.

We shall now study the effect of this potential on the neutrino masses. If the effective mass eigenvalues are $m_1$ and $m_2$ and the mass eigenstates are related to the weak eigenstates through the orthogonal mixing matrix

$$O = \begin{pmatrix} \cos\theta & \sin\theta \\ -\sin\theta & \cos\theta \end{pmatrix}, \tag{6.92}$$

the mass-squared matrix of $\nu_e$ and $\nu_X$ can be written as

$$M_\nu^2 = O^T (M_\nu^{\text{diag}})^2 O + \begin{pmatrix} A & 0 \\ 0 & 0 \end{pmatrix}, \tag{6.93}$$

where

$$(M_\nu^{\text{diag}})^2 = \begin{pmatrix} m_1 & 0 \\ 0 & m_2 \end{pmatrix}^2. \tag{6.94}$$

We then define $\Delta = |m_1^2 - m_2^2|$ and $m_0 = (m_1^2 + m_2^2 + A)$ and write the mass-squared matrix as

$$M_\nu^2 = \frac{m_0}{2}\begin{pmatrix} 1 & 0 \\ 0 & 1 \end{pmatrix} + \frac{1}{2}\begin{pmatrix} A - \Delta\cos 2\theta & \Delta\sin 2\theta \\ \Delta\sin 2\theta & -A + \Delta\cos 2\theta \end{pmatrix}. \tag{6.95}$$

The eigenvalues of this matrix are

$$m_{\nu 1,2} = \frac{m_0}{2} \pm \frac{1}{2}\sqrt{(\Delta\cos 2\theta - A)^2 + \Delta^2 \sin^2 2\theta}, \tag{6.96}$$

and the effective mixing angle $\tilde{\theta}$ becomes

$$\tan 2\tilde{\theta} = \frac{\Delta \sin 2\theta}{\Delta\cos 2\theta - A}. \tag{6.97}$$

Thus, the effective physical masses of the two neutrinos as well as their mixing angles depend on the electron density. This is very important inside the sun, where the matter density is too high near the core and reduces as the neutrino propagates outward. We shall discuss this later.

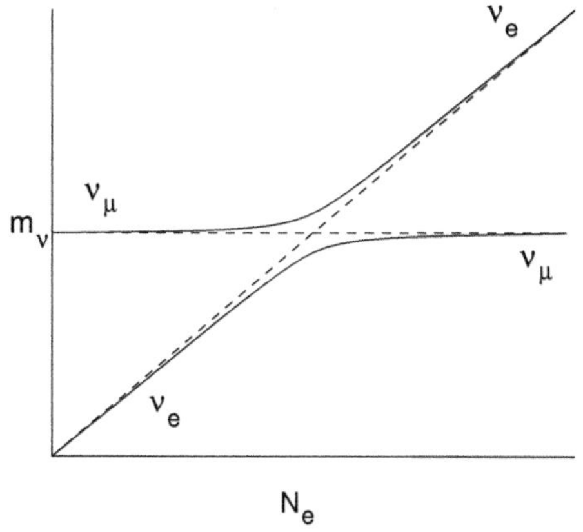


**FIGURE 6.2**
Resonant conversion of $\nu_e$ in matter. Solid lines are the physical mass eigenstates.

The $\nu_e$ mass varies with the electron number density $N_e$, but since the quantum mechanical level-crossing is not allowed adiabatically and the mass-squared difference never vanishes, all the $\nu_e$ are completely converted into $\nu_\mu$.

The effective electron neutrino mass in the matter varies with varying electron density. Let us consider a scenario in which an electron neutrino propagates from an extremely dense media, such as the core of the sun, to rarer media such as the surface of the sun. Let us also assume that near the core of the sun, the $\nu_e$ effective mass is higher than the effective mass of $\nu_\mu$. Since both $\nu_e$ and $\nu_\mu$ are combinations of the mass eigenstates $m_1$ and $m_2$, one cannot talk about the mass of $\nu_e$ or $\nu_\mu$. However, if the mixing angle is small, $\nu_e$ can be identified mostly with $\nu_1$ with a small admixture of $\nu_2$, and then one can say that the effective mass of $\nu_e$ is $m_1$. As $\nu_e$ comes out from the core of the sun to the surface, the $\nu_e$ mass varies with the electron number density $N_e$. When the $\nu_e$ mass approaches the $\nu_\mu$ mass and the mass-squared difference approaches the point $\Delta \cos 2\theta$, a resonant oscillation takes place at $A \sim \Delta \cos 2\theta$. At this point, the two masses are supposed to be the same and equal to $\Delta \sin \theta$, so they cross each other. But in quantum mechanics the phenomenon of level crossing does not allow two quantum levels to cross adiabatically. If we now minimize the mass-squared difference with respect to $A$, we find that the minimum never vanishes and the two levels never cross each other. So, a total conversion of levels takes place and all the $\nu_e$ become $\nu_\mu$. If there are any small numbers of $\nu_\mu$, they may get converted to $\nu_e$.

The minimum corresponds to a point

$$A = \Delta \cos 2\theta, \tag{6.98}$$

where a total conversion of the states take place. Since $A$ depends on both matter density and the energy of the neutrinos, any $\nu_e$ with energy higher than a critical value, will be completely converted into a $\nu_\mu$ during its propagation inside the sun and will come out as $\nu_\mu$. This is the MSW effect, demonstrated in figure 6.2. As the $\nu_e$ propagate inside the sun from the dense core to its surface with very low matter density, the effective mass of $\nu_e$ reduces until it approaches the cross-over point. Near this point all of these $\nu_e$ are completely converted into $\nu_\mu$. Since the mass of $\nu_\mu$ does not depend on the matter density, it remains constant. All of the $\nu_\mu$ will also be completely converted into $\nu_e$ near the cross-over point. We explained the MSW effect with adiabatic approximation and with only two neutrinos for purpose of demonstration. In practice one needs to consider the nonadiabatic region of the parameter space and all the neutrino species, which makes it more complicated. However, the basic idea remains the same.

## 6.4 Summary of Experiments

Experiments with neutrinos are extremely difficult since they interact with matter very weakly. Almost sixty years after its discovery we discovered that the neutrinos have small masses. First the electron neutrino flux from the sun was observed to be lower than the predictions in the chlorine experiment. Although this result had several uncertainties, it initiated further activities in the field. Subsequently a deficiency in the atmospheric muon neutrino flux was observed, which established that the neutrinos have tiny masses. Solar neutrino experiments as well as laboratory experiments then confirmed this result.

We shall now present a brief summary of the results and then provide a few details about some of the important results. The atmospheric neutrino results imply that $\nu_\mu$ oscillates into $\nu_\tau$ with a mass-squared difference and mixing angle [71]:

$$\Delta m^2_{atm} \simeq (1.5 - 3.4) \times 10^{-3}\ \text{eV}^2, \quad \sin^2 2\theta_{atm} > 0.92, \tag{6.99}$$

with a best fit of $\Delta m^2_{atm} = 2.1 \times 10^{-3}$ eV$^2$ with $\sin^2 2\theta_{atm} = 1$. A global fit to all the solar neutrino experiments, combined with the KamLAND result [72], implies that $\nu_e$ oscillates into a combination of the states $\nu_\mu$ and $\nu_\tau$ with

$$\Delta m^2 = 7.9^{+0.6}_{-0.5} \times 10^{-5}\ \text{eV}^2 \quad \text{and} \quad \tan^2\theta = 0.40^{+0.10}_{-0.07}. \tag{6.100}$$

These results also prefer a three-generation solution, although a sterile neutrino is not yet ruled out.

If we assume that there are three left-handed Majorana neutrinos and the physical neutrinos $\nu_i$, $i = 1,2,3$ with masses $m_i$ are related to the weak eigenstates $\nu_\alpha$, $\alpha = e, \mu, \tau$ by the mixing matrix $U_{\alpha i}$, then the mixing matrix can be parameterized as

$$U = \begin{pmatrix} c_1 c_3 & -s_1 c_3 & -s_3 \\ s_1 c_2 - c_1 s_2 s_3 & c_1 c_2 + s_1 s_2 s_3 & -c_3 s_2 \\ s_1 s_2 + c_1 c_2 s_3 & c_1 s_2 - s_2 c_2 s_3 & c_2 c_3 \end{pmatrix}, \tag{6.101}$$

where $s_i = \sin\theta_i$ and $c_i = \cos\theta_i$. Nonobservation of oscillations of $\nu_e$ to some other neutrinos at the CHOOZ experiment provided strong constraint on the third mixing angle $s_3$. For a mass-squared difference of $7 \times 10^{-4}$ eV$^2$ or more, the CHOOZ bound [68] on the mixing angle is $\sin^2\theta_{13} < 0.1$. The WMAP constraint [73] on the sum of all three neutrino masses is $\sum m_i < 0.69$ eV, which may be softened in some cases. The Heidelberg–Moscow neutrinoless double beta decay experiment [74] reported $m_{ee} = 0.1 - 0.9$ eV, but this result is yet to be confirmed. The first positive indication of neutrino oscillation was reported by LSND collaboration [75] that there is $\nu_\mu \to \nu_e$ oscillation with mass-squared difference $\Delta m^2 > 0.2$ eV and $\sin^2\theta \sim (0.001 - 0.04)$. This result would have required at least one sterile neutrino. But this result has been ruled out by the MiniBOONE experiment at Fermilab [76]. After this brief summary we shall now discuss some of these experiments.

**Atmospheric Neutrinos**

The neutrinos produced in our atmosphere mostly from pion and subsequent muon decays,

$$\pi \to \mu + \bar{\nu}_\mu \quad \text{and} \quad \mu \to e + \bar{\nu}_e + \nu_\mu\,,$$

were measured at Kamiokande and super-Kamiokande detectors. Although these detectors could distinguish between $\nu_\mu$ and $\nu_e$, since the absolute flux of atmospheric $\nu_\mu$ and $\nu_e$ are not well-known, these experiments measured a double ratio

$$R = \frac{(N_\mu/N_e)_{\text{obs}}}{(N_\mu/N_e)_{\text{MC}}}. \tag{6.102}$$

This compares the observed ratio of numbers of $\nu_\mu$ to numbers of $\nu_e$, with the monte carlo predictions of this ratio. After a run of about 848 days, super-Kamiokande experiment presented

$$\begin{aligned} R &= 0.680^{+0.023}_{-0.022} \pm 0.053 \quad \text{subGeV data} \\ R &= 0.678^{+0.042}_{-0.039} \pm 0.080 \quad \text{multiGeV data}. \end{aligned} \tag{6.103}$$

Departure of this double ratio from unity at 3-4$\sigma$ established that the atmospheric neutrino anomaly is due to $\nu_\mu$-$\nu_\tau$ oscillations [71].

Measurements of the zenith angle distributions for fully contained single-ring $e$-like and $\mu$-like events, multiring $\mu$-like events, partially contained events, and upward going muons provide more information. Since neutrinos coming from vertical direction travel less distance compared with the horizontal directions, the strong zenith angle dependence of the muon neutrinos confirms oscillation of muon neutrinos. The zenith angle dependence of the electron neutrinos is not so prominent. Similarly the up–down asymmetry is measured, which is defined as

$$A = \frac{U-D}{U+D}, \tag{6.104}$$

$U$ being the number of up-coming events (which travel about 13000 km through the earth) and $D$ is the down-going events (which travel about 20 km). The multiGeV muon events show an asymmetry of

$$A_\mu = 0.311 \pm 0.043 \pm 0.010, \tag{6.105}$$

although no such asymmetry was observed for $\nu_e$.

CHOOZ result [68] on nonobservation of $\bar{\nu}_e$ disappearance in the Chooz underground neutrino detector in the Ardennes, France, at a distance of 1 km away from the Chooz B nuclear power station, is then taken into account and states:

$$\sin^2 2\theta_{eX} < 0.1 \quad \text{or} \quad \Delta m^2_{eX} < 7 \times 10^{-4}\ \text{eV}^2 \quad \text{for } \sin^2 2\theta_{eX} = 1.$$

This rules out $\nu_\mu \to \nu_e$ oscillation as a solution to the atmospheric neutrino problem. $\nu_\mu \to \nu_s$ oscillation also fails to explain the atmospheric neutrino anomaly. Finally it is established that the atmospheric neutrino anomaly can be explained if $\nu_\mu$ oscillates into $\nu_\tau$ with maximal mixing. Thus, the physical eigenstates will be almost equal admixtures of $\nu_\mu$ and $\nu_\tau$. The mass-squared difference between these two physical eigenstates and the mixing angle is given by

$$\Delta m^2_{atm} \simeq (1.5 - 3.4) \times 10^{-3}\ \text{eV}^2, \quad \sin^2 2\theta_{atm} > 0.92, \tag{6.106}$$

with a best fit of $\Delta m^2_{atm} = 2.1 \times 10^{-3}$ eV$^2$ with $\sin^2 2\theta_{atm} = 1$. This result was found to be consistent with the long-baseline laboratory experiment: *K2K* [77]. 107 few-GeV $\nu_\mu$-events from the KEK accelerator were detected at a distance of $L = 250$ km at Kamioka over expected $151^{+12}_{-10}$ (syst) events.

**Solar Neutrinos**

The first indication of neutrino oscillation was obtained from solar neutrinos, when the solar model predictions of $\nu_e$ flux were found to differ from the observed $\nu_e$ flux in the chlorine experiment. It is now established that the $\nu_e$ coming from the sun oscillates into a combination of $\nu_\mu$ and $\nu_\tau$ with fairly large mixing angles. This also confirms that the source of energy in the sun and other stars is nuclear fusion.

There are two methods of hydrogen fusion into helium inside the sun, the *pp-cycle* and the *CNO-cycle*, which produce energy and release neutrinos

$$4p \to {}^4He + 2e^+ + 2\nu_e + 26.73 \text{ MeV}. \tag{6.107}$$

The dominant contribution comes from the pp-cycle:

| Reactions | % | | $E_\nu$ in MeV |
|---|---|---|---|
| $p+p \to {}^2\text{H}+e^+ + \nu_e$ | 99.6 % | *pp* | $\leq 0.42$ |
| $p+e^- +p \to {}^2\text{H}+\nu_e$ | 0.4 % | *pep* | 1.44 |
| ${}^2\text{H}+p \to {}^3\text{He}+\gamma$ | | | |
| ${}^3\text{He}+{}^3\text{He} \to {}^4\text{He}+2\,p$ | 85 % | | |
| ${}^3\text{He}+p \to {}^4\text{He}+e^+ + \nu_e$ | $2\times10^{-5}$% | hep | $\leq 18.77$ |
| ${}^3\text{He}+{}^4\text{He}+ \to {}^7\text{Be}+\gamma$ | 15 % | | |
| ${}^7\text{Be}+e^- \to {}^7\text{Li}+\nu_e$ | 99.87 % | ${}^7$Be | 0.861 |
| ${}^7\text{Li}+p \to 2\,{}^4\text{He}$ | | | |
| ${}^7\text{Be}+p \to {}^8\text{B}+\gamma$ | 0.13 % | | |
| ${}^8\text{B} \to {}^8\text{Be}^* + e^+ + \nu_e$ | | ${}^8$B | $\leq 14.06$ |
| ${}^8\text{Be}^* \to 2\,{}^4\text{He}$ | | | |

Only about 1.6% of the energy in the sun comes from the CNO-cycle:

$$\begin{aligned}
{}^{12}\text{C}+p &\to {}^{13}\text{N}+\gamma \\
{}^{13}\text{N} &\to {}^{13}\text{C}+e^+ + \nu_e \;\; (E_\nu \leq 1.2\,\text{MeV}) \\
{}^{13}\text{C}+p &\to {}^{14}\text{N}+\gamma \\
{}^{14}\text{N}+p &\to {}^{15}\text{O}+\gamma \\
{}^{15}\text{O} &\to {}^{15}\text{N}+e^+ + \nu_e \;\; (E_\nu \leq 1.73\,\text{MeV}) \\
{}^{15}\text{N}+p &\to {}^{12}\text{C}+\alpha \\
{}^{15}\text{N}+p &\to {}^{16}\text{O}+\gamma \\
{}^{16}\text{O}+p &\to {}^{17}\text{F}+\gamma \\
{}^{17}\text{F} &\to {}^{17}\text{O}+e^+ + \nu_e \;\; (E_\nu \leq 1.74\,\text{MeV}) \\
{}^{17}\text{O}+p &\to {}^{14}\text{N}+\alpha.
\end{aligned}$$

Here C, N and O act as catalysts to fuse the hydrogen into helium.

The predictions of the solar neutrino flux are based on the so-called *standard solar model*. Using the basic equations of stellar evolution (hydrodynamic equilibrium is assumed), the energy produced by the nuclear reactions given above is transported by radiation and convection. The main uncertainties in the input parameters comes from some nuclear cross-sections and the core temperature of the sun. The standard solar models predict fluxes of electron neutrinos $\nu_e$ on earth of about $10^{10}$ cm$^{-2}$ $s^{-1}$ [78]. But since the detection of the neutrinos is extremely difficult because of their very small cross-section, this flux leads to very few events in the detectors. With a cross-section of about $10^{-45}$ cm$^2$, about $10^{30}$ target atoms are required to produce one event per day. So a very large physical size of the detectors is required to detect the solar neutrinos.

It took many years to understand the solar neutrino problem. In the first experiment at Homestake gold mine [79], chlorine was used to detect the neutrinos through the reaction

$$^{37}\mathrm{Cl} + \nu_e \rightarrow {}^{37}\mathrm{Ar} + e^- . \tag{6.108}$$

By measuring its radioactivity, $^{37}$Ar is detected. This reaction cannot detect the *pp* neutrinos, since it has a threshold of 0.814 MeV. This experiment reported, from their 20 years of measurement, an average counting rate of $2.56 \pm 0.16 \pm 0.16$ SNU (solar neutrino unit), which is less than the predictions of the standard solar model of $7.7 \pm 1.2$ SNU [80]. One SNU is defined as $10^{-36}$ captures per target atom per sec. This was the first experiment to report the solar neutrino problem, indicating the possibility of neutrino oscillations. The detection of neutrinos in this experiment from the sun also established that the energy source of sun is nuclear fusion which is associated with neutrino emission.

The solar neutrino experiments, the Kamiokande and the super-Kamiokande, used water Cerenkov detectors [81], which are real-time detectors and can also measure the zenith angle dependence of the events. They measure all the flavors of neutrinos, but for the $\nu_\mu$ and $\nu_\tau$, the scattering rate is 6 times smaller than the $\nu_e - e$ scattering cross-section. The threshold of these experiments is rather high, 7.0 MeV and 5.5 MeV, respectively, but the statistics compensates for this and they expect about 20 to 30 signals every day, which are mostly the $^8B$ neutrinos. The observed flux at Kamiokande and super-Kamiokande is $(2.80 \pm 0.19 \pm 0.33) \times 10^6$ cm$^{-2}$s$^{-1}$ and $(2.44 \pm 0.05\,^{+0.09}_{-0.07}) \times 10^6$ cm$^{-2}$s$^{-1}$ as opposed to the theoretical predictions of $(5.15\,^{+1.00}_{-0.72}) \times 10^6$ cm$^{-2}$s$^{-1}$. The large statistics at the super-Kamiokande experiment coming from 11,240 events in about 825 days allow the measurement of the energy spectrum of the recoil electrons and also the day/night asymmetry.

The most sensitive and advanced solar neutrino experiment is the Sudbury neutrino observatory [82], which is an imaging water Cerenkov detector, filled with an ultra pure $D_2O$ target in an acrylic vessel and surrounded by photo-multiplier tubes. This is then shielded by ultra-pure $H_2O$ and construction material, which acts as a cosmic ray veto. By using a $D_2O$ target, the SNO detector can detect electron neutrinos ($\nu_e$) through charged current (CC) interactions, whereas through the neutral current (NC) interactions and elastic scattering (ES), it can simultaneously measure the flux of all active neutrinos ($\nu_x$, $x = e, \mu, \tau$) from $^8B$ decay in the sun:

$$\begin{aligned} \nu_e + d &\to p + p + e && \text{(charged current)}, \\ \nu_x + d &\to \nu_x + p + n && \text{(neutral current)}, \\ \nu_x + e &\to \nu_x + e && \text{(elastic scattering)} \end{aligned} \tag{6.109}$$

The neutrino fluxes deduced from the CC, NC and ES interactions are [82]

$$\Phi^{CC} = 1.59^{+0.08}_{-0.07}\,(stat)\ ^{+0.06}_{-0.08}(syst) \times 10^6 \text{ cm}^{-2}\text{ s}^{-1} \tag{6.110}$$

$$\Phi^{NC} = 5.21 \pm 0.27(stat) \pm 0.38(syst) \times 10^6 \text{ cm}^{-2}\text{ s}^{-1} \tag{6.111}$$

$$\Phi^{ES} = 2.21^{+0.31}_{-0.26}\,(stat) \pm 0.10(syst) \times 10^6 \text{ cm}^{-2}\text{ s}^{-1}. \tag{6.112}$$

The difference between the CC and ES fluxes is $(0.62 \pm 0.40) \times 10^6$ cm$^{-2}$ s$^{-1}$, which is about $1.6\sigma$. The discrepancy is more prominent in the CC data, when compared with the NC data. The ratio of the two fluxes is [82]

$$\frac{\Phi^{CC}}{\Phi^{NC}} = 0.306 \pm 0.026(stat) \pm 0.024(syst). \tag{6.113}$$

This ratio is expected to be 1 if all the neutrinos from the sun are $\nu_e$. The observed deviation thus establishes the fact that a fraction of $\nu_e$ from the sun has reached us as $\nu_x$ ($\nu_X$ could be $\nu_\mu$ or $\nu_\tau$). An estimate of the $\nu_e$ flux and the fluxes of the nonelectron-neutrino component then implies neutrino oscillation at a $5\sigma$ c.l. SNO is the only solar neutrino experiment which proves in a solar-model independent way that solar neutrinos undergo oscillations. Comparing the day–night energy spectra, day–night asymmetries of different classes of events are also estimated. No substantial distortion of the neutrino energy spectrum compared with the theoretical expectation has been found.

The simplest explanation for the deficiency in the number of detected solar electron neutrinos is neutrino oscillations. Unlike for the atmospheric neutrinos, for the solar neutrinos there could be two distinct solutions. The first possibility involves the electron neutrinos coming from the sun oscillating into a $\nu_\mu$ or a $\nu_\tau$ during their passage to Earth. This is called the vacuum oscillation. All the experiments together rule out this explanation.

The second possibility involves resonant matter effect (MSW effect, discussed in section 6.3). When the neutrinos come out from the core of the sun, they pass through varying matter density. The matter density at the core is maximum and on the surface almost zero. As the $\nu_e$ from the core pass through this varying density, their effective mass reduces due to their interaction with matter. However, as they reach the $\nu_\mu$ mass, a total conversion takes place and all the $\nu_e$ become $\nu_\mu$, since quantum mechanical level crossing is forbidden. The lighter $\nu_e$ continue without any such conversion, although they may oscillate during their passage to Earth. This is called the MSW solution of the solar neutrino problem. All present results indicate that this is the solution to the solar neutrino problem with the parameter space specified by a global fit to all the solar neutrino experiments including the most sensitive result from SNO.

The mass-squared difference and the mixing angle for the $\nu_e$ to a combination of $\nu_\mu$ and $\nu_\tau$, as obtained from the solar neutrino analysis, are further improved by the results from the laboratory neutrino oscillation experiment: Kamioka Liquid scintillator AntiNeutrino Detector (KamLAND) [72]. $\bar{\nu}_e$ emitted from several distant power reactors are detected at KamLAND using the inverse beta decay $\bar{\nu}_e + p \rightarrow e^+ + n$. The reactors are located at a typical distance of about 180 km. Detecting the $e^+$ in coincidence with the delayed 2.2 MeV $\gamma$-ray from neutrons captured on a proton reduces the background. The detector is made of 1 kton of ultra-pure liquid scintillator (LS) contained in a 13 m diameter spherical balloon made of thin nylon film and surrounded by a stainless steel containment vessel, on which 1879 photo multiplier tubes (PMTs) are mounted. The whole setup is surrounded by a 3.2 kton water Cerenkov detector.

The $\bar{\nu}_e$ flux detected at KamLAND is smaller than expected from the neighbouring reactors, which is the first evidence for neutrino oscillations from any laboratory experiments. The two-neutrino oscillation analysis of the data from KamLAND and a global fit of all solar neutrino experiments gives [72]

$$\Delta m^2 = 7.9^{+0.6}_{-0.5} \times 10^{-5}\ \text{eV}^2 \quad \text{and} \quad \tan^2\theta = 0.40^{+0.10}_{-0.07}. \tag{6.114}$$

This mass-squared difference and mixing angle is for oscillations of $\nu_e$ to a combination of $\nu_\mu$ and $\nu_\tau$. Maximal mixing is rejected at the equivalent of $5.4\sigma$. The atmospheric and solar neutrinos have thus provided us the mass-squared difference between $\nu_e$, $\nu_\mu$ and $\nu_\tau$ and two mixing angles. Purely sterile neutrino solutions are ruled out at $4.7\sigma$. This means that the $\nu_e$ oscillating to a sterile neutrino with no other oscillations cannot explain the solar neutrino problem. However, $\nu_e$ oscillating into mostly active neutrinos with a small admixture of sterile neutrinos is still not ruled out. The active-sterile admixture is constrained to $\sin^2\eta < 0.19$ [0.56] at $1\sigma$ [$3\sigma$].

**Absolute Mass of the Neutrinos**

The atmospheric and the solar neutrinos provide us with the information about two mass-squared differences and two mixing angles. The CHOOZ results constrain the third mixing angle ($s_3$ in equation (6.101)) to a very low value, consistent with zero. Thus, for a three generation scenario, the most important unknown parameter is the absolute mass of the neutrinos.

The direct measurement of the neutrino mass comes from the nuclear beta decay:

$$n \rightarrow p + e^- + \bar{\nu}. \tag{6.115}$$

The end-point of the electron energy spectrum is shifted if neutrinos have masses:

$$N(E)dE \sim p_e^2 F(Z,E)(E_0 - E)[(E_0 - E)^2 - m_\nu^2]^{1/2} dE, \tag{6.116}$$

where $E_0$ corresponds to the end-point energy and $m_\nu$ is the mass of the electron neutrino $\nu_e$. By measuring the deviation toward the end of the energy spectrum of the electrons in a tritium beta decay, the present upper limit on the mass of the $\nu_e$ is obtained: $m_{\nu_e} < 2.8$ eV [44]. For the $\nu_\mu$ and the $\nu_\tau$ the corresponding bounds from accelerator experiments are $m_{\nu_\mu} < 190$ keV and $m_{\nu_\tau} < 18.2$ MeV [44].

This bound can be improved by the neutrinoless double beta decay experiment, if neutrinos are Majorana particles. Any positive result in neutrinoless double beta decay would also imply that there is lepton number violation in nature, which is a fundamental question. Lepton number is conserved in all gauge interactions. Since the Majorana mass of the neutrinos has to be very small, it is extremely difficult to find out if there is lepton number violation in nature from any other experiment. On the other hand, if the neutrinoless double beta decay can confirm that neutrinos are Majorana particles, this will have many interesting consequences in particle physics and cosmology. Even the constraints from the neutrinoless double beta decay can provide us with strong bounds on many lepton-number violating processes.

In a neutrinoless double beta decay, the proton number of the decaying nucleus changes by two units, keeping the mass number unchanged and two electrons (or positrons) are emitted. This is an extremely rare process and the measurement is extremely difficult. During beta decay, an even-even nucleus would go to an odd-odd nucleus. When the binding energy of the odd–odd nuclei is smaller than that of the parent nuclei, ordinary beta decay is forbidden energetically. However, a $2\nu\beta\beta$ decay, in which an even-even nucleus decays to another even-even nucleus, may still be allowed:

$$(A,Z) \rightarrow (A,Z+2) + e^- + e^- + \bar{\nu} + \bar{\nu}. \tag{6.117}$$

This second order process is extremely weak and is called two-neutrino double beta decay ($2\nu\beta\beta$). There are about 35 candidates for double beta decay, but only about ten cases have so far been observed [44]. This process conserves lepton number and is allowed in the standard model. On the other hand, if the neutrinos are Majorana particles, a neutrinoless double beta decay is possible ($0\nu\beta\beta$):

$$(A,Z) \rightarrow (A,Z+2) + e^- + e^-, \tag{6.118}$$

in which two neutrons decay into two electrons without any neutrinos

$$n + n \rightarrow p + p + e^- + e^-. \tag{6.119}$$

This process violates lepton number by $\Delta L = 2$ and directly measures the amount of total lepton number violation. The total kinetic energies of the two electrons is constant and is given by the $Q$ value of the decay in the case of $0\nu\beta\beta$ decay. This measurement does not have any background from the $2\nu\beta\beta$ decay or $2M\beta\beta$ decay (where $M$ represents Majorons in models of spontaneously broken lepton number), since in the $2\nu\beta\beta$ decay the total kinetic energy of the electrons have a distribution because neutrinos or the Majorons carry away part of the energy.

In the standard model, the $0\nu\beta\beta$ decay is forbidden since neutrinos are massless. However, in extensions of the standard model, neutrinos may have masses and Majorana neutrinos are more natural than the Dirac neutrinos. If neutrinos have Majorana masses, the measurement of the $0\nu\beta\beta$ decay will be able to measure this mass.

Consider an extension of the standard model that accommodates a small neutrino mass

$$\mathscr{L}_{mass} = m_i \nu_i \nu_i, \tag{6.120}$$

where $\nu_i, i = 1,2,3$ are the physical states with Majorana mass $m_i$. The physical eigenstates could be different from the weak eigenstates $\nu_\alpha, \alpha = e, \mu, tau$, and they are related by the PMNS mixing matrix $U_{\alpha i}$ (see equation (6.68)).

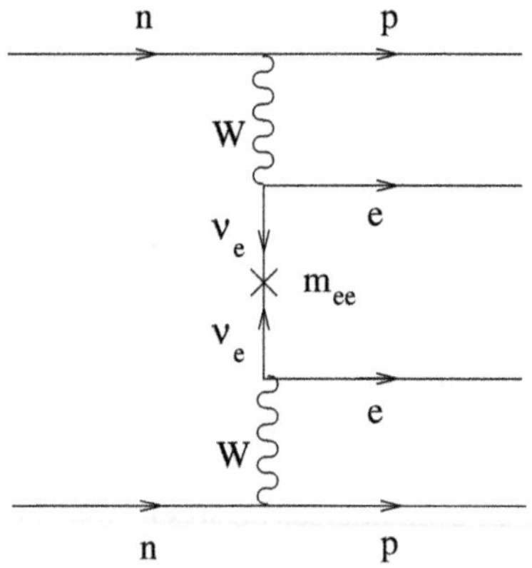


**FIGURE 6.3**
Diagram for $0\nu\beta\beta$ decay through exchange of a Majorana neutrino.

The total lepton number ($L = L_e + L_\mu + L_\tau$) violation is possible when neutrinos are Majorana particles, in which case there will be neutrinoless double beta decay through the diagram given in figure 6.3. Since the neutrinos are virtual intermediate particles, the amplitude of this process is proportional to an effective mass [53]

$$< m >= \sum_{i=1}^{3} U_{ei}^2 m_i = m_{ee} = M_{ee}^{\nu}, \tag{6.121}$$

where $m_{ee}$ is the (11) element of the neutrino mass matrix when written in the flavor basis. Thus, the neutrinoless double beta decay measures the Majorana mass of the neutrinos and its occurrence implies lepton number violation in nature.

The amplitude for the neutrinoless double beta decay is derived from the half-life measurement and is given by

$$A_{(\beta\beta)_{0\nu}} = \frac{G_F^2}{2\pi^3} \sum_{i=1}^{3} U_{ei}^2 m_i F(m_i, A) M^{0\nu}, \tag{6.122}$$

where $M^{0\nu}$ is the nuclear matrix element,

$$F(m_i, A) = \frac{< \exp^{-m_i r}/r >}{< 1/r >} \tag{6.123}$$

is the phase space factor, $r$ is the distance between the two neutrons within the nucleus, and $R \geq r/2$ is the radius of the initial nucleus. The average is with respect to the two-nucleon correlation function appropriate to a nucleus of atomic number $A$. $F(m_i, A)$ can be approximated by a step function [83],

$$\begin{aligned} F(m_i, A) &= 1 \quad \text{for} \quad \mathrm{m_i} \ll 10\,\mathrm{MeV} \\ F(m_i, A) &= 0 \quad \text{for} \quad \mathrm{m_i} \gg 10\,\mathrm{MeV}. \end{aligned} \tag{6.124}$$

Thus, any heavy physical states will not contribute to the amplitude of the neutrinoless double beta decay. In the case of the see-saw mechanism the right-handed neutrinos are heavy and cannot cancel the contribution of the light left-handed neutrinos to the neutrinoless double beta decay. As a result, it is sufficient to work with the effective Majorana mass matrices of the light left-handed neutrinos to study the neutrinoless double beta decay.

The most sensitive double beta decay experiment is the Heidelberg–Moscow double beta decay experiment in the Gran Sasso underground laboratory which was running for over a decade. The experiment operated five enriched (to 86%) high purity $^{76}Ge$ detectors, with a total mass of 11.5 kg, the active mass of 10.96 kg being equivalent to a source strength of 125.5 mol $^{76}Ge$ nuclei. It looked for the decay mode

$$^{76}Ge \rightarrow {}^{76}Se + 2\,e^- + (2\,\nu).$$

The high resolution of the $Ge$ detector $(3-4$ keV) leads to no background from the $2\nu\beta\beta$ decay, in which the total energy carried by the two electrons is less than the $Q$ value. The $Q$ value for this decay is known from earlier measurements with high precision to be $Q_{\beta\beta} = 2039.006\ (50)$ keV. This is very useful in identifying the experimental signature of the $0\nu\beta\beta$ decay mode, which is a peak at the $Q$-value of the decay.

The data from this experiment was accumulated in the five enriched detectors over the period August 1990 through May 2003. An analysis of this data identified a line at $Q_{\beta\beta}$, with 29 events in the full spectrum corresponding to a $4.2\sigma$ evidence ($> 5\sigma$ for the pulse shape selected spectrum) for the neutrinoless double beta decay [74]. The half-life for the $0\nu\beta\beta$ decay is reported to be

$$T^{0\nu}_{1/2} = 1.19^{+0.37}_{-0.23} \times 10^{25}\ \mathrm{y} \qquad (1\,\sigma)\,. \tag{6.125}$$

Including the uncertainty in the nuclear matrix elements, this corresponds to an effective Majorana neutrino mass of the neutrino

$$< m >= (0.1 - 0.9)\ \mathrm{eV} \qquad (99.73\%\ \mathrm{c.l.}). \tag{6.126}$$

This is the first reported evidence for lepton number violation in nature. However, this result relies on a very few events that could not be reproduced, and hence, needs to be confirmed by future experiments.

Since the neutrinoless double beta decay measures the amount of lepton number violation, the present upper bound on the lifetime of the $0\nu\beta\beta$ decay can constrain several lepton-number violating processes [84]. In some extensions of the standard model in which lepton number violation is allowed, there will be new diagrams contributing to the neutrinoless double beta decay. Bounds on the half-life of the $0\nu\beta\beta$ decay can then constrain the parameters entering in the new diagrams and, hence, the extensions of the standard model. In the left–right symmetric models, the $0\nu\beta\beta$ decay constraint on the right-handed charged gauge bosons $W_R$ mass could be as strong as 1.2 TeV. Right-handed heavy neutrino masses could also be constrained severely by the $0\nu\beta\beta$ decay, although the bounds depend on details of the model. Indirect bounds on the masses and couplings of the exotic scalar bilinears, which couple to two standard model fermions, are also possible. Some composite models could also be constrained by the $0\nu\beta\beta$ decay limit. The present limit on the $0\nu\beta\beta$ decay implies too small lepton number violation, so the inverse double beta decay process $e^-e^- \to W^-W^-$ may not be observed even at ILC.

There are also cosmological bounds on the sum of all the physical neutrino masses. The latest and the strongest of these bound comes from the Wilkinson Microwave Anisotropy Probe (WMAP) and the 2dF Galaxy Redshift Survey (2dFGRS) [73]. Long before the structure formation, the neutrinos decouple and hang around freely until they become nonrelativistic, smoothening out structures on the smallest scales. The power spectrum of density fluctuations within the scale when the neutrinos were still relativistic is suppressed as [85, 86] $\Delta P_m/P_m \approx -8\Omega_\nu/\Omega_m$. This provide us with a bound on the sum over all neutrino masses. The 2dFGRS experiment gave a bound of 1.8 eV [87], which is further improved with the result from WMAP giving a bound $\Omega_\nu h^2 < 0.0076$ (95% confidence range), which translates into [73]

$$\sum_{i=e,\mu,\tau} m_{\nu_i} \leq 0.69 \text{ eV}. \tag{6.127}$$

There are, however, some uncertainties in this bound and it may be relaxed to 1.0 eV. If there are sterile neutrinos, if the neutrinos can decay to some scalars before they become nonrelativistic, or if scale invariance in the primordial power spectrum is broken, then some more uncertainties may creep in this bound [88].

### Other Results

There are several other experiments and cosmological constraints that added to our knowledge about neutrinos. The measurements of the invisible Z-width give the number of light neutrinos to be $N_\nu = 3.00 \pm 0.06$, and the standard model fits to all LEP data give $N_\nu = 2.994 \pm 0.012$ [44]. This does not constrain the number of sterile neutrinos, which do not take part in weak interactions, but they are constrained by cosmological considerations. The baryon density $\Omega_b h^2$ and the baryon to photon ratio $\eta$ have been measured by WMAP independently, which implies an expected primordial $^4$He abundance of

$$Y = 0.249 \pm 0.013\ (N_{eff} - 3).$$

The present value of the helium abundance allows only $N_{eff} < 3.0$ at 95% c.l. Taking all possible uncertainties and allowing for maximum systematic error, one may stretch this number to 3.4, but a fourth neutrino with large mixing with other active neutrinos is not allowed when WMAP results are combined with other measurements including the big-bang nucleosynthesis.

There is one result from the Liquid Scintillator Neutrino Detector (LSND) [75] at the Los Alamos National Laboratory, which claimed a mass-squared difference of $\Delta m^2 > 0.2$ eV with small mixing ($\sin^2 2\theta \sim 0.001 - 0.04$) between $\nu_e$ and $\nu_\mu$. When combined with other neutrino oscillation results, this result would imply existence of a fourth neutrino. However, this result has been ruled out by the MiniBOONE experiment at Fermilab [76].

Another source of information about neutrinos is supernovae [89]. During a supernovae explosion neutrinos are released. The typical energy of escaping neutrinos is $\sim 10$ to 15 MeV, and about $10^{56}$ neutrinos are released in a few milliseconds. The neutrinos released from SN1987A have been monitored by several experiments involved in detecting proton decay [90]. Supernova SN1987A is a type II supernova, observed in the Large Magellanic Cloud at a distance of about 50 kpc (about $10^{21}$ m) with a mass of about 20 $M_\odot$. The detection of neutrinos from a type II supernova is the first proof that the cores of massive stars collapse to neutron star densities. Since the neutrinos traveled over 50 kpc in $5 \times 10^5$ y, one can derive a bound on their lifetime, which rules out an explanation of the solar neutrino deficit as neutrino decay. There is also bound on the sum of the masses, magnetic moment and charge of the neutrinos coming from supernovae. Supernovae results could be used to derive constraints on axion coupling and mass, any new force coupling to neutrinos and the characteristic scale of any new extra dimensions.

## Implications

All the results now suggest that there are three active neutrinos that undergo weak interactions. There may be sterile neutrinos that do not interact with other charged fermions, but their mixing with the active neutrinos should be small. We shall not include such sterile neutrinos in the following discussion. The atmospheric and the solar neutrinos provide us with two mass-squared differences and two mixing angles. The third mixing angle is severely constrained by the CHOOZ data. The absolute mass of the neutrinos is still not determined, although the WMAP and the neutrinoless double beta decay result give strong constraints. Although none of these results provide us with any clue about the origin of the neutrino masses, at the phenomenological level we shall be able to talk about the three generation neutrino mass matrix fairly well.

Without loss of generality, we shall work in the basis in which the charged lepton mass matrix is diagonal. The weak eigenstates of the neutrinos $\nu_\alpha$, $\alpha = e, \mu, \tau$, are then defined as the states with diagonal charged current interactions. When the neutrino mass matrix is written in this basis, the mass matrix is not diagonal. We can then diagonalize the neutrino mass matrix and define the physical neutrino states $\nu_i$, $i = 1, 2, 3$ as the states with masses $m_i$. In general, $m_i$ could be complex when there is *CP* violation, otherwise they could be positive or negative. Due to neutrino mixing, these states are not the weak eigenstates. A solution to the atmospheric neutrino anomaly requires that the physical neutrino states $\nu_2$ and $\nu_3$ are the states with maximal mixing ($\sin^2 2\theta_{23} \sim 1$) between the states $\nu_\mu$ and $\nu_\tau$. This implies that both $\nu_2$ and $\nu_3$ contain almost equal admixtures of $\nu_\mu$ and $\nu_\tau$. The mass-squared difference is then $m_2^2 - m_3^2 \sim \Delta m^2_{atm} \sim 2.1 \times 10^{-3}$ eV$^2$. Because of the maximal mixing it is not meaningful to talk about the mass-squared difference between $\nu_\mu$ and $\nu_\tau$. The mass-squared difference implies that at least one of the masses should be as heavy as $m_{atm} = \sqrt{\Delta m^2_{atm}} \sim 0.046$ eV. However, if neutrino masses are almost degenerate $m_2 \approx m_3 \approx m_{deg}$, then we may have $m_{deg} > m_{atm}$ satisfying $m_2 - m_3 = \Delta m^2_{atm}/(2m_{deg})$.

Let us consider a $2 \times 2$ mass matrix in the basis $[\nu_\mu \;\; \nu_\tau]$,

$$M_\nu = \begin{pmatrix} a & b \\ b & c \end{pmatrix}. \tag{6.128}$$

The atmospheric neutrino problem implies that this matrix should provide a mixing angle $\sin^2 2\theta_{23}$ and mass-squared difference $\Delta m^2_{atm}$. If we diagonalize this matrix, the mass eigenvalues and the mixing angle are given by,

$$\begin{aligned} m_{1,2} &= \frac{1}{2}\left[a + c \mp \sqrt{(a-c)^2 + 4b^2}\right] \\ \tan 2\theta &= \frac{2b}{c-a}. \end{aligned} \tag{6.129}$$

Two solutions to this problem allow an almost maximal mixing angle and explain the atmospheric neutrino anomaly:

(a) $a = c$ and $b \neq 0$. In this case the mixing angle is maximal, $\theta = \pi/4$, and the mass eigenstates are $m_{1,2} = a \pm b$, so the mass-squared difference is $m_1^2 - m_2^2 = 4ab$. In particular, when $b = c = a$, one of the eigenvalues vanishes, $m_1 = 0$ and $m_2 = 2b$ gives the mass-squared difference.

(b) $b \gg a, c$. Consider now $a = 0$, so the mass-squared difference is $m_1^2 - m_2^2 = 2bc$. The large observed mixing angle then implies $b/c > 1.3$, which corresponds to almost degenerate mass eigenvalues.

The low-energy mass matrix can allow one of these scenarios quite naturally in the see-saw mechanism of neutrino masses [91]. For demonstrating the basic idea, let us consider a $3 \times 3$ mass matrix, in which the first two neutrinos are light and the third one is the lightest of the right-handed neutrinos. The elements in the Dirac mass terms are assumed to be of the same order of magnitude and the left-handed neutrinos receive masses only through the see-saw mechanism. Then the mass matrix in the basis $[\nu_2 \;\; \nu_3 \;\; N_1]$ is

$$M_\nu = \begin{pmatrix} 0 & 0 & a \\ 0 & 0 & b \\ a & b & c \end{pmatrix} \tag{6.130}$$

and will automatically ensure a maximal mixing among the light neutrinos $\nu_2$ and $\nu_3$, which are the states with $\nu_\mu$ and $\nu_\tau$.

In three generation scenarios solar neutrinos determine the mixing angle $\sin\theta_{12} = s \sim 0.55$, while $\sin\theta_{23} = s_2 \sim 1/\sqrt{2}$ is fixed by the atmospheric neutrinos. $\sin\theta_{13} = s_3$ satisfies the CHOOZ constraints and the value zero is consistent with experiments. However, if we need *CP* violation in the mass matrix, $s_3$ has to be nonvanishing. In the simplified situation with $s_3 = 0$, the mixing matrix takes the form,

$$U = \begin{pmatrix} c & -s & 0 \\ s/\sqrt{2} & c/\sqrt{2} & -1/\sqrt{2} \\ s/\sqrt{2} & c/\sqrt{2} & +1/\sqrt{2} \end{pmatrix}, \tag{6.131}$$

where $c^2 = 1 - s^2$. For a generic $s$, the neutrino mass matrix now becomes

$$M_\nu = \begin{pmatrix} 2\varepsilon & \delta & \delta \\ \delta & m_3/2+\varepsilon' & -m_3/2+\varepsilon' \\ \delta & -m_3/2+\varepsilon' & m_3/2+\varepsilon' \end{pmatrix}, \tag{6.132}$$

with $\varepsilon = (m_1\, c^2 + m_2\, s^2)/2$, $\delta = (m_1 - m_2)\, c\, s/\sqrt{2}$ and $\varepsilon' = (m_1\, s^2 + m_2\, c^2)/2$.

The atmospheric neutrinos determine the mass-squared difference $m_2^2 - m_3^2 = m_{atm}^2$, while the solar neutrinos determine the second mass-squared difference $m_1^2 - m_2^2 = m_{sol}^2 \sim 7.0 \times 10^{-5}$ eV$^2$. The overall or the absolute mass is the only unknown quantity in this mass matrix. The reported value of the neutrinoless double beta decay [74] result $0.1 < m_{ee} < 0.9$ with

$$m_{ee} = \sum_i |U_{ei}|^2\, m_i = m_1\, c^2 + m_2\, s^2 + m_3\, u^2 \tag{6.133}$$

can constrain the absolute mass to a large extent. The WMAP result $|m_1| + |m_2| + |m_3| = m_s \lesssim 0.69$ eV [73] or $m_s \lesssim 1.0$ eV, [88] can add to this bound in contributing toward our understanding of the absolute mass.

The mass-squared differences required for the atmospheric and solar neutrinos allow different three generation models, which may be classified as:

**Hierarchical**: The masses satisfy a hierarchical pattern $m_1 < m_2 < m_3$ so $m_3 = m_{atm} = \sqrt{\Delta m_{atm}^2}$, $m_2 = m_{sol} = \sqrt{\Delta m_{sol}^2}$ and $m_1 < m_{sol}$.

**Degenerate**: All three neutrino masses are approximately equal $m_1 \approx m_2 \approx m_3 \approx m_0$, where $m_0$ is the overall mass, and the mass-squared differences are as required by the solar and atmospheric neutrinos.

**Inverted Hierarchical**: Two of the neutrinos are degenerate and heavier than the third one, $m_1 \approx m_2 > m_3$. Solar neutrinos require $\Delta m_{12}^2 = \Delta m_{sol}^2$ while $m_1 \sim m_{atm}$, so $\Delta m_{23}^2 = \Delta m_{atm}^2$.

**Partially Degenerate**: Two neutrinos are degenerate satisfying $\Delta m_{12}^2 = \Delta m_{sol}^2$ and the third neutrino is heavier, $m_{sol} < m_1 \approx m_2 < m_3 = m_{atm}$.

All the four scenarios mentioned above are solutions of the solar and atmospheric neutrino problems. WMAP cannot differentiate among these solutions [92], while the neutrinoless double beta decay [74] allows only the degenerate case. In other words, if the mass range reported by the Heidelberg–Moscow neutrinoless double beta decay experiment is confirmed, all the solutions except for the degenerate neutrino solution will be ruled out. However, obtaining a degenerate mass matrix with the required mixing angle is more difficult compared with hierarchical mass matrix.

In summary, the atmospheric and the solar neutrinos have provided us with the most crucial information that neutrinos are massive. With more information from these experiments and laboratory experiments, we have almost determined the neutrino mass matrix for a three generation scenario. Although a small admixture of a sterile neutrino with the active neutrinos is not completely ruled out, a sterile neutrino is not required to explain any of the existing experimental results. If we restrict to only a three generation scenario, the unknown parameters remaining to be determined would be the third mixing angle ($U_{e3}$, which should be very small to be consistent with the CHOOZ data), the *CP* phases, and the absolute mass of the neutrinos.

# 7

# *CP Violation*

The charge conjugation ($C$), the parity ($P$), and the time-reversal ($T$) are the three important discrete symmetries in nature. These symmetries are exactly conserved in the electromagnetic, the strong and the gravitational interactions, although they are violated in the weak interaction. It is also known that the consistency of the field theory requires that the product $CPT$ should be conserved in any theory. Another product $CP$ plays a special role in particle physics and cosmology.

In the weak interaction, $CP$ is not conserved, and in 1964 $CP$ violation was observed [22] in the $K^\circ - \overline{K^\circ}$ oscillation. During the past few years, evidence of $CP$ violation has been observed in the $K$-decays [93] as well as in the $B$-decays [94]. Although $CP$ violation has not been observed in the strong interaction, theory predicts a large $CP$ violation which needs to be suppressed; this is known as the strong $CP$ problem. There could be $CP$ violation in the leptonic sector, which may have its origin in the Majorana nature of the neutrinos. Since $CP$ violation implies an asymmetry between particles and antiparticles, the fact that there is more matter in the universe compared with antimatter requires $CP$ violation in the early universe.

In the standard model it is possible to accommodate a $CP$ phase that can explain all the observed $CP$ violation in the $K$-system as well as in the $B$-system. But any extensions of the standard model could provide newer sources of $CP$ violation. The strong $CP$ problem requires a new dynamical origin of $CP$ violation that predicts a new particle, the *axion*. There are additional sources of $CP$ violation in the leptonic sector if neutrinos are Majorana particles. The $CP$ violation required to explain the baryon asymmetry of the universe could have a completely different origin.

## 7.1 *CP* Violation in the Quark Sector

As discussed in section 5.4, when the up-quark and the down-quark mass matrices are diagonalized, one complex phase in the charge–current interaction cannot be removed. This complex phase manifests itself as $CP$ violation in decays of heavy mesons. $CP$ violation was first observed in $K - K^\circ$ oscillations and recently it has been observed in the $K$-decays and also in the $B$-meson system. In this section we review some salient features of $CP$ violation in the $K$-system as well as in the $B$-system.

### *K*-System

*CP* violation was first observed in the $K^\circ - \overline{K^\circ}$ oscillations, in which the state $K^\circ$ ($d\bar{s}$, with strangeness number $+1$) oscillates into the state $\overline{K^\circ}$ ($\bar{d}s$ with strangeness number $-1$). Thus, strangeness is changed by two units in the process and the *CP* violating phase enters in the mass matrix of the $K^\circ - \overline{K^\circ}$ system. This is called the indirect *CP* violation, which is different from the *CP* violation in the *K*-decays, where strangeness is changed by one unit and is called the direct *CP* violation. In the standard model, there is one *CP* phase in the CKM quark mixing matrix, which can explain both the direct as well as the indirect *CP* violations. However, different extensions of the standard model could contain new sources of *CP* violation that may have different contributions to the two *CP* violations.

Since the $K^\circ - \overline{K^\circ}$ states are formed in strong interactions, their initial strangeness number is known. However, they decay through weak interactions, in which strangeness is not conserved. This can lead to a mixing of these states

$$K^\circ \to \pi^+\pi^- \to \overline{K^\circ}.$$

The dominant contribution to this $K^\circ - \overline{K^\circ}$ mixing comes from a box diagram (see figure 7.1) in the standard model.

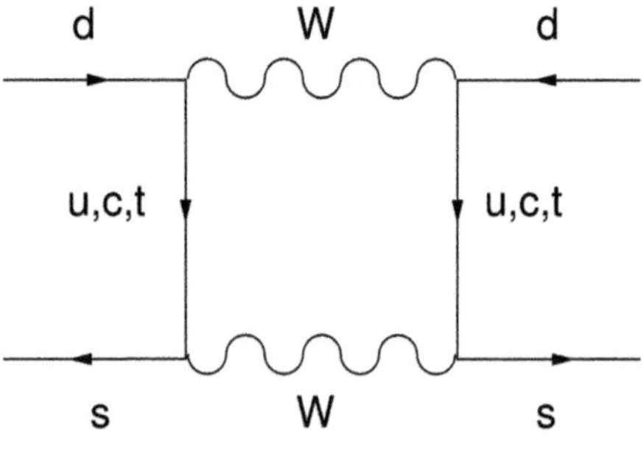


**FIGURE 7.1**
Box diagram contributing to $K^\circ - \overline{K^\circ}$ oscillations.

We shall now explain how *CP* violation can enter in the $K^\circ - \overline{K^\circ}$ oscillations. We first adopt a phase convention

$$CP|K^\circ\rangle = |\overline{K^\circ}\rangle \quad \text{and} \quad CP|\overline{K^\circ}\rangle = |K^\circ\rangle,$$

so the states with *CP* eigenvalues $\pm 1$ can be written as

$$|K_1^\circ\rangle = \frac{1}{\sqrt{2}}[|K^\circ\rangle + |\overline{K^\circ}\rangle] \tag{7.1}$$

$$|K_2^\circ\rangle = \frac{1}{\sqrt{2}}[|K^\circ\rangle - |\overline{K^\circ}\rangle]. \tag{7.2}$$

These states can decay through weak interactions into the states containing pions, which have odd parity $P|\pi^{\pm},\pi^0\rangle = -|\pi^{\pm},\pi^0\rangle$. The pions transform under charge conjugation as $C|\pi^{\pm},\pi^0\rangle = |\pi^{\mp},\pi^0\rangle$, and hence, a two pion state ($\pi^0\pi^0$ or $\pi^+\pi^-$) has positive *CP* eigenvalue. A $CP = +1$ state $K_1^\circ$ will thus decay into two pions, while the $CP = -1$ state $K_2^\circ$ will decay into three pions and more complex decay modes if *CP* is conserved. $K_2^\circ$ will have much shorter half-life than $K_1^\circ$, since the phase space consideration will suppress the three-body decays compared with the two-body decays. There will also be a small mass difference between the two states $K^\circ - \overline{K^\circ}$ , which has been observed, due to the different intermediate states to which they can make virtual transitions.

We shall now write the time development of the $K^\circ - \overline{K^\circ}$ system, which is determined by an effective Hamiltonian containing the masses ($M_{ij}$) and the decay widths ($\Gamma_{ij}$) of these states, in the basis $\{K^\circ,\ \overline{K^\circ}\ \}$:

$$H_{ij} = M_{ij} - \frac{i}{2}\Gamma_{ij}. \tag{7.3}$$

The mass matrix and the decay width are constrained by the Hermiticity, *CPT* and *CP* conservation:

$$\begin{aligned} M_{ij} &= M_{ji}^*, & \Gamma_{ij} &= \Gamma_{ji}^* \\ M_{11} &= M_{22} & \Gamma_{11} &= \Gamma_{22} \\ M_{12} &= M_{21} & \Gamma_{12} &= \Gamma_{21}. \end{aligned}$$

Since we are interested in the *CP* violation, we consider both $M$ and $\Gamma$ to be complex. Then the physical eigenstates can be obtained by diagonalizing the Hamiltonian, which can be parameterized by a measure of the *CP* violation $\varepsilon$ and is given by

$$|K_S\rangle = \frac{1}{\sqrt{2(1+|\varepsilon|^2)}}\,[(1+\varepsilon)|K^\circ\rangle + (1-\varepsilon)|\overline{K^\circ}\rangle] \tag{7.4}$$

$$|K_L\rangle = \frac{1}{\sqrt{2(1+|\varepsilon|^2)}}\,[(1+\varepsilon)|K^\circ\rangle - (1-\varepsilon)|\overline{K^\circ}\rangle]. \tag{7.5}$$

The *CP* violating parameter $\varepsilon$ enters in the mass matrix of the $K^\circ - \overline{K^\circ}$ system and is given by

$$\varepsilon = \frac{\sqrt{H_{12}} - \sqrt{H_{21}}}{\sqrt{H_{12}} + \sqrt{H_{21}}} = \frac{i\mathrm{Im}M_{12} + \frac{1}{2}\mathrm{Im}\Gamma_{12}}{(M_L - M_S) - \frac{i}{2}(\Gamma_L - \Gamma_S)}. \tag{7.6}$$

The *CP* violation also introduces nonorthogonality between the states $K_S$ and $K_L$:

$$\langle K_S|K_L\rangle = \frac{2\ \mathrm{Re}\ \varepsilon}{1+|\varepsilon|^2}. \tag{7.7}$$

The states $K_1^\circ$ and $K_2^\circ$ have definite *CP* eigenstates with eigenvalues $\pm 1$ and are orthogonal to each other. If there is no *CP* violation, the state $K_L$ will become identical to $K_2^\circ$ and can decay into only three pions or more complex states. If there is *CP* violation, $K_L$ will be able to decay into two pions. Thus, the measurement of the branching ratio

$$R = \frac{(K_L \to \pi^+\pi^-)}{(K_L \to \text{all charged modes})} = (2.0 \pm 0.4) \times 10^{-3}$$

established *CP* violation in the $K^\circ - \overline{K^\circ}$ oscillations [22]. This implies that the measure of *CP* violation amounts to $|\varepsilon| = 2.3 \times 10^{-3}$, which has been improved and has been reducing the errors drastically since then.

To understand all the physical observables and determine the amount of *CP* violation, we need one more *CP* violating parameter $\varepsilon'$, which is the *CP* violation in the decay amplitudes. The dominant contribution to this new *CP* violating parameter $\varepsilon'$ comes from the gluon-mediated penguin diagram of figure 7.2.

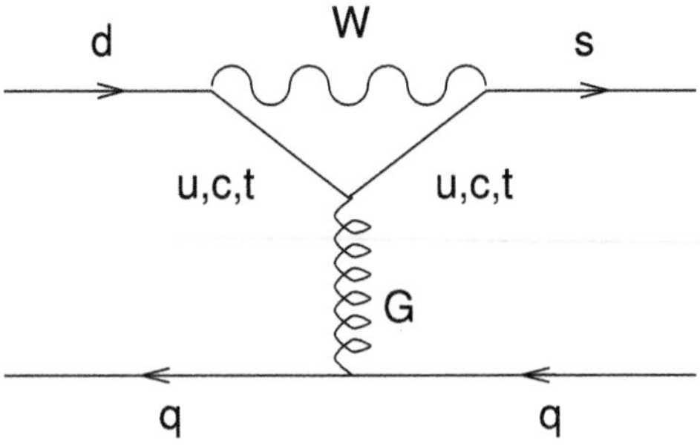


**FIGURE 7.2**

Penguin diagram contributing to the *CP* violating parameter $\frac{\varepsilon'}{\varepsilon}$ in *K*-decays

Let us first parameterize the $2\pi$ decay amplitudes in terms of the final states that are characterized by two isospin states $I = 0$ and $I = 2$, as

$$A(K^\circ \to 2\pi, I) = A_I \exp^{i\delta_I} \quad \text{and} \quad A(\overline{K^\circ} \to 2\pi, I) = \bar{A}_I \exp^{i\delta_I}. \tag{7.8}$$

Only the difference in the phase shift $\Delta = \delta_2 - \delta_0$ appears in the observable, since the phase shift $\delta_I$ is produced by the strong final state interaction of the pions.

*CPT* implies that we can write the complex quantities $A_I$ and $\bar{A}_I$ as

$$A_I = |A_I| \exp^{i\theta_I} \quad \text{and} \quad \bar{A}_I = -|A_I| \exp^{-i\theta_I}, \tag{7.9}$$

where $\theta_I$ is a phase of weak origin. In the Wu and Yang phase convention [95], the amplitude $A_0$ is real and we can now express

$$\varepsilon' = \frac{i}{\sqrt{2}} \left| \frac{A_2}{A_0} \right| \sin\theta_2 \exp^{i\Delta} \tag{7.10}$$

$$\omega = \left| \frac{A_2}{A_0} \right| \cos\theta_2 \exp^{i\Delta}, \tag{7.11}$$

where $\omega$ is not related to $CP$ violation.

The $CP$ violating quantities that are directly related to the physical observables can now be defined as

$$\eta_{+-} = |\eta_{+-}|\exp^{i\phi_{+-}} = \frac{A(K_L \to \pi^+\pi^-)}{A(K_S \to \pi^+\pi^-)} \tag{7.12}$$

$$\eta_{00} = |\eta_{00}|\exp^{i\phi_{00}} = \frac{A(K_L \to \pi^0\pi^0)}{A(K_S \to \pi^0\pi^0)} \tag{7.13}$$

$$\delta = \frac{\Gamma(K_L \to \pi^-\ell^+\nu) - \Gamma(K_L \to \pi^+\ell^-\bar{\nu})}{\Gamma(K_L \to \pi^-\ell^+\nu) + \Gamma(K_L \to \pi^+\ell^-\bar{\nu})}, \tag{7.14}$$

which are related to the $CP$ violating parameters $\varepsilon$ and $\varepsilon'$ and the parameter $\omega$ through the relations

$$\eta_{+-} = \varepsilon + \frac{\varepsilon'}{1+\frac{\omega}{\sqrt{2}}} \tag{7.15}$$

$$\eta_{00} = \varepsilon - 2\frac{\varepsilon'}{1-\sqrt{2}\omega} \tag{7.16}$$

$$\delta = \frac{2\mathrm{Re}\varepsilon}{1+|\varepsilon|^2}. \tag{7.17}$$

These quantities have been measured and found to be nonvanishing, confirming $CP$ violation in both $K^\circ - \overline{K^\circ}$ oscillations as well as in $K$-decays. The above discussions are also valid for other neutral meson systems such as $B^\circ - \overline{B^\circ}$ or $D^\circ - \overline{D^\circ}$, although the short lifetime of the heavy mesons complicates the analysis much more.

The experimental values of the $K_L$ and $K_S$ mass differences and the decay width differences are fairly precise [44]

$$\Delta M_K \equiv M_L - M_S = 3.48 \times 10^{-15}\ \mathrm{GeV}$$
$$\Delta\Gamma_K \equiv \Gamma_L - \Gamma_S = -7.37 \times 10^{-15}\ \mathrm{GeV}. \tag{7.18}$$

Taking these numbers it is possible to arrive at an empirical relation within about 5% error,

$$\Delta M_K = -\frac{1}{2}\Delta\Gamma_K. \tag{7.19}$$

The interference of the $K_L$ and $K_S$ states allows a measurement of the $CP$ violating parameter $\varepsilon$ from the measurement of the parameters

$$|\eta_{+-}| = (2.276 \pm 0.017) \times 10^{-3}$$
$$\phi_{+-} = (43.3 \pm 0.5)^\circ$$

$$|\eta_{00}| = (2.262 \pm 0.017) \times 10^{-3}$$
$$\phi_{00} = (43.2 \pm 1.0)^\circ$$

$$\delta = (3.27 \pm 0.12) \times 10^{-3}. \tag{7.20}$$

The value of $\varepsilon$ comes out to be [44]

$$|\varepsilon| = (2.271 \pm 0.017) \times 10^{-3}$$
$$\phi(\varepsilon) \approx \tan^{-1}\frac{2(M_L - M_S)}{\Gamma_L - \Gamma_S} = (43.49 \pm 0.08)^{\circ}. \tag{7.21}$$

The phase of $\varepsilon_K$ is obtained by applying *CPT* and unitarity.

Using the experimental input of $\Gamma_{12} \approx 0$, we simplify the relations

$$\Delta M_K = 2\mathrm{Re}M_{12},$$

$$\Delta\Gamma_K = 2\mathrm{Re}\Gamma_{12},$$

$$|\varepsilon| = \frac{1}{\sqrt{2}}\frac{\mathrm{Im}M_{12}}{\Delta M_K} = \frac{1}{2\sqrt{2}}\frac{\mathrm{Im}M_{12}}{\mathrm{Re}M_{12}}.$$

The *CP* violating observables can also be simplified

$$\eta_{+-} \approx \varepsilon + \varepsilon', \quad \eta_{00} \approx \varepsilon - 2\varepsilon', \quad \delta \approx 2\mathrm{Re}\ \varepsilon. \tag{7.22}$$

The measurements of the *CP* violating parameters then give [93]

$$\mathrm{Re}(\frac{\varepsilon'}{\varepsilon}) = \frac{\varepsilon'}{\varepsilon} = (2.1 \pm 0.5) \times 10^{-3}$$
$$\phi(\varepsilon') = \delta_2 - \delta_0 + \frac{\pi}{4} \approx (48 \pm 4)^{\circ}. \tag{7.23}$$

The predictions of the standard model for the *CP* parameters are so far consistent with all measurements and there are no indications of new physics coming from these measurements [96].

### *B*-System

During the past few years the *B*-physics has made much progress. The *CP* violation in the *B*-system has also been established in the *B*-factories. The new generation of *B*-factory experiments intensified activities in *B*-physics. One of the main expectations from these experiments is to discover whether there are any signals of new physics beyond the standard model. Earlier experiments at LEP and SLD collected $b\bar{b}$ events operating at the *Z*-resonance. But in the current experiments *B*-decays are studied near their production threshold at the $\Upsilon(4S)$ resonance [44]. Thus, heavier states such as $B_s^{\circ}(s\bar{b})$ or $B_c^{+}(c\bar{b})$ and *b*-flavored baryons are not produced. In the $\Upsilon(4S)$ decays only equal amounts of $B^{\circ}(d\bar{b})\overline{B^{\circ}}(\bar{d}b)$ and $B^{+}(u\bar{b})B^{-}(\bar{u}b)$ pairs are expected.

The short half-life of the *B*-mesons makes the analysis of *B*-system different from that of the *K*-system. In addition, there are two neutral $B^{\circ} - \overline{B^{\circ}}$ meson systems, $B_d^{\circ}$ $(d\bar{b})$ and $B_s^{\circ}$ $(s\bar{b})$. For the heavy quarks the calculation of the hadronic matrix elements is also different.

Let us consider the evolution of the flavor state $B^{\circ}$ or $\overline{B^{\circ}}$. The effective Hamiltonian in the basis $[B^{\circ} \ \ \overline{B^{\circ}}]$

$$H = M - \frac{i}{2}\Gamma$$

will govern the evolution of these states $B^\circ$ or $\overline{B^\circ}$. Here $M$ and $\Gamma$ are the mass and decay width. Similar to the $K$-system, we shall write the eigenstates of this effective Hamiltonian

$$|B_\pm\rangle = p|B^\circ\rangle \pm q|\overline{B^\circ}\rangle, \tag{7.24}$$

where

$$\frac{q}{p} = \frac{1-\varepsilon}{1+\varepsilon} = \sqrt{\frac{M_{12}^* - \frac{i}{2}\Gamma_{12}^*}{M_{12} - \frac{i}{2}\Gamma_{12}}}. \tag{7.25}$$

These states $|B_\pm\rangle$ are not orthogonal. If $CP$ is conserved, these states will have definite $CP$ eigenvalues $CP|B_\pm\rangle = \pm|B_\pm\rangle$, where we considered the convention $CP|B^\circ\rangle = |\overline{B^\circ}\rangle$ and Re $(q/p) > 0$.

We now consider the time development of any general state, which is a superposition of physical states. At a given time $t$, it is given by

$$\begin{aligned}
|\psi(t)\rangle &= C_1 \exp^{-i(M_1 - i\Gamma_1/2)t}|B_+\rangle + C_2 \exp^{-i(M_2 - i\Gamma_2/2)t}|B_-\rangle \\
&\propto \left(C_1 \exp^{-i(M_1 - i\Gamma_1/2)t} + C_2 \exp^{-i(M_2 - i\Gamma_2/2)t}\right)|B^\circ\rangle \\
&\quad + \frac{q}{p}\left(C_1 \exp^{-i(M_1 - i\Gamma_1/2)t} - C_2 \exp^{-i(M_2 - i\Gamma_2/2)t}\right)|\overline{B^\circ}\rangle.
\end{aligned}$$

In the two cases $C_1 = \pm C_2 = 1$, which corresponds to the initial states to be pure $B^\circ$ and $\overline{B^\circ}$ states, the time evolutions are given by

$$\begin{aligned}
|B^\circ(t)\rangle &= g_+(t)|B^\circ\rangle + \frac{q}{p} g_-(t)|\overline{B^\circ}\rangle \\
|\overline{B^\circ}(t)\rangle &= g_+(t)|\overline{B^\circ}\rangle + \frac{q}{p} g_-(t)|B^\circ\rangle,
\end{aligned} \tag{7.26}$$

where

$$g_\pm(t) = \frac{1}{2}\left(\exp^{-i(M_1 - i\Gamma_1/2)t} \pm \exp^{-i(M_2 - i\Gamma_2/2)t}\right). \tag{7.27}$$

This gives the time-dependent probabilities for the flavor oscillations.

In the $B^\circ - \overline{B^\circ}$ oscillations, decays to flavor specific final states are studied. We write the decay amplitudes

$$A_f = \langle f|H|B^\circ\rangle \quad \text{and} \quad \bar{A}_{\bar{f}} = \langle \bar{f}|H|\overline{B^\circ}\rangle \tag{7.28}$$

for the $B^\circ$ decay into a final state $f$ and $\overline{B^\circ}$ decays into $\bar{f}$, respectively. Although the instantaneous decay amplitudes $A_{\bar{f}} = \langle \bar{f}|H|B^\circ\rangle$ and $\bar{A}_f = \langle f|H|\overline{B^\circ}\rangle$ vanish, the $B^\circ - \overline{B^\circ}$ mixing allows $B^\circ$ to decay into $\bar{f}$ and $\overline{B^\circ}$ to decay into $f$. We thus consider the time-integrated mixing probability [44]

$$\chi_f^{B^\circ \to \overline{B^\circ}} = \frac{\int_0^\infty |\langle \bar{f}|H|B^\circ(t)\rangle|^2\, dt}{\int_0^\infty |\langle \bar{f}|H|B^\circ(t)\rangle|^2 + \int_0^\infty |\langle f|H|B^\circ(t)\rangle|^2}. \tag{7.29}$$

If there is no $CP$ violation, a simplified expression follows

$$\chi_f^{B^\circ \to \overline{B^\circ}} = \chi_f^{\overline{B^\circ} \to B^\circ} = \chi = \frac{x^2 + y^2}{2(x^2+1)}, \tag{7.30}$$

where

$$x = \frac{\Delta M}{\Gamma}, \quad y = \frac{\Delta\Gamma}{2\Gamma} \tag{7.31}$$

and $\Delta M$ and $\Delta\Gamma$ are given by

$$\begin{aligned} \Delta M &= -2\mathrm{Re}[\frac{q}{p}(M_{12} - \frac{i}{2}\Gamma_{12})] \approx -2|M_{12}| \\ \Delta\Gamma &= -4\mathrm{Im}[\frac{q}{p}(M_{12} - \frac{i}{2}\Gamma_{12})] \approx 2|\Gamma_{12}|\cos\zeta, \end{aligned} \tag{7.32}$$

where $\Gamma_{12}/M_{12} = r\,\exp^{i\zeta}$. Since $r \sim 10^{-3}$, we neglect higher order terms. Experiments confirm the $B^\circ - \overline{B^\circ}$ oscillation [97], and the amount of mixing is given by

$$\chi_d = 0.174 \pm 0.009\,. \tag{7.33}$$

However, flavor specific final states cannot be studied to look for $CP$ violation in the $B$-system [98].

It is convenient to look for some common nonleptonic decay channels of $B^\circ$ and $\overline{B^\circ}$, which are also $CP$ eigenstates, say, $CP|f_\pm\rangle = \pm|f_\pm\rangle$. Thus, if $CP$ is conserved, a $CP$ positive state $B^\circ$ can decay only into $|f_+\rangle$. If we now start with $N_+(N_-)$ numbers of $B_+(B_-)$ mesons, after a time $t$, there will be a different combination of $B^\circ$ and $\overline{B^\circ}$ mesons, denoted as $B(t)$, whose decay rate into the state $|f_+\rangle$ will be

$$\Gamma(B(t) \to f_+) = \frac{N_+}{N_+ + N_-}\Gamma(B_+(t) \to f_+). \tag{7.34}$$

However, when there is $CP$ violation, the time evolution of this decay rate will be different from pure exponential.

Since both $B^\circ$ and $\overline{B^\circ}$ can decay into a final state $|f_+\rangle$ when there is $CP$ violation, the time-dependent decay rates are given by [44]

$$\begin{aligned} &\Gamma(B^\circ(t) \to f) \propto \\ &|A_f|^2\left\{|g_+(t)|^2 + |g_-(t)|^2\,|\xi(f)|^2 + 2\mathrm{Re}\left[g_-(t)g_+^*(t)\xi(f)\right]\right\} \\ &\Gamma(\overline{B^\circ}(t) \to f) \propto \\ &\exp^{-\Gamma_1 t}|\bar{A}_f|^2\left\{|g_+(t)|^2 + |g_-(t)|^2\,|\xi(f)|^{-2} + 2\mathrm{Re}\left[\frac{g_-(t)g_+^*(t)}{\xi(f)}\right]\right\}, \end{aligned} \tag{7.35}$$

where $\xi(f) = (q/p)\,(\bar{A}_f/A_f)$. $\Delta\Gamma/\Gamma$ is significant only for $B_s$ decays, but it can be neglected for the $B_d$ decays. We can, thus, define an asymmetry in the wrong-sign semi-leptonic $B$-decays as a measure of indirect $CP$ violation as

$$\begin{aligned} a_{SL}(B) &= \frac{\Gamma(\overline{B}(t) \to \ell^+X) - \Gamma(B(t) \to \ell^-X)}{\Gamma(\overline{B}(t) \to \ell^+X) + \Gamma(B(t) \to \ell^-X)} \\ &= \frac{|p/q|^2 - |q/p|^2}{|p/q|^2 + |q/p|^2} = r_B \sin\zeta_B. \end{aligned} \tag{7.36}$$

Standard model predicts a tiny asymmetry, so any large asymmetry would imply new physics beyond the standard model.

Since $|\bar{A}_{\bar{f}}| \neq |A_f|$, measurement of the rate asymmetries for *CP* conjugate decays can tell us about direct *CP* violation. If the final state $f$ for two amplitudes $(A_i)$, with different strong phases $(\delta_i)$ coming from final state interaction effects and different weak phases $(\xi_i)$, which contribute to the same decay, has different flavor content than its *CP* conjugate, the asymmetry is given by

$$a = \frac{2A_1A_2\sin(\xi_1-\xi_2)\sin(\delta_1-\delta_2)}{A_1^2+A_2^2+2A_1A_2\cos(\xi_1-\xi_2)\cos(\delta_1-\delta_2)}. \tag{7.37}$$

Any difference between the *CP* asymmetries for *B*-decays to flavor eigenstates can also allow measurement of direct *CP* violation, provided the weak decay phase in $\bar{A}_f/A_f$ is not cancelled by the mixing weak phase in $q/p$.

Measurement of the decays of $B^\circ$ and $\overline{B^\circ}$ to *CP* eigenstates provides the golden mode for observing *CP* violation. When the initial states are $B^\circ$ or $\overline{B^\circ}$ only, the time-dependent decay rates to the same final state $f$ with definite *CP* eigenvalue will be different and the corresponding *CP* asymmetry is given by [44]

$$a_f(t) = \frac{\Gamma(\overline{B^\circ}(t)\to f)-\Gamma(B^\circ(t)\to f)}{\Gamma(\overline{B^\circ}(t)\to f)+\Gamma(B^\circ(t)\to f)}. \tag{7.38}$$

For the final state $f=(c\bar{c})K^\circ$ with *CP* eigenvalue $\eta_f$, the asymmetry becomes

$$a_f(t) = -\eta_f \sin 2\phi_1 \sin \Delta M t, \tag{7.39}$$

for nonvanishing $A_f$ and $\bar{A}_f$ and when weak mixing phase is different from the weak decay phase. $\phi_1$ is one of the three internal angles of the *unitary triangle*, defined as

$$\phi_1 = \pi - \arg\left(\frac{-V_{tb}^*V_{td}}{-V_{cb}^*V_{cd}}\right).$$

This mode has the advantage that the decays of the final state $(c\bar{c})K^\circ$ have very little contribution from direct *CP* violation and there is no ambiguity due to strong interactions [98]. This reduces the uncertainty in interpreting the data and has recently been studied at the KEKB asymmetric-energy $e^+e^-$ collider by the Belle detector. They reported the first evidence for *CP* violation in the *B*-system from an analysis of the asymmetry in the distribution of the time intervals between the two $B^\circ$ meson decay points

$$\sin 2\phi_1 = 0.99 \pm 0.14(\text{stat}) \pm 0.06(\text{syst}). \tag{7.40}$$

The large positive value for $\sin 2\phi_1$ is consistent with the *CP* violation coming from the CKM phase in the standard model.

## 7.2 Strong $CP$ Problem

After discussing the question of $CP$ violation in $K$- and $B$-decays, which have been observed and could be explained consistently in the context of standard model, we now come to another question of $CP$ violation in the standard model, the strong $CP$ problem. While we try to study the weak decays to find if this can provide us any information of physics beyond the standard model, in the strong $CP$ problem we try to solve a more theoretical problem. The strong interaction predicts a large $CP$ violation within the context of standard model, which has not been observed, and hence, we should find some underlying theoretical justification to explain the smallness of the prediction for $CP$ violation.

To understand the problem, let us consider the limit when the $u$ and $d$ quarks are massless. The QCD Lagrangian then has a chiral symmetry, $U(2)_L \times U(2)_R$, which is spontaneously broken resulting in four massless Nambu–Goldstone bosons, $\eta$ meson and the pions. If we now introduce nonvanishing quark masses, they will break the chiral symmetry explicitly, and hence, the Nambu–Goldstone bosons will pick up small masses satisfying the relation $m_\eta \leq \sqrt{3}\, m_\pi$, which is inconsistent with the actual masses of these mesons [99]. It was then pointed out [100] that the QCD Lagrangian is invariant under a $U(1)_A$ transformation

$$q_i \to \exp^{i\alpha\gamma_5} q_i \quad \text{and} \quad m_j \to \exp^{-2i\alpha} m_j,$$

which is broken by the axial anomaly of the $U(1)_A$ current and the instanton effect, which gives a mass to the $\eta$ meson even for massless quarks. So, the pions are the only Nambu–Goldstone bosons in the problem, which remains light after the quarks are given masses.

Although the axial anomaly of the $U(1)_A$ current solves this problem, it makes the QCD vacuum complex [101]. This $CP$ violating term in the QCD Lagrangian is parameterized by a phase $\Theta$, which can take any values between 0 and $\pi$, but its experimental bound is $\Theta \leq 10^{-9}$. From naturalness consideration we should expect the value of $\Theta$ to be of the order of 1, so the smallness of $\Theta$ becomes a problem, which is called the *strong CP problem*.

There are suggestions to solve this problem, but none of the suggestions could be verified so far. The most popular and interesting solution for the problem makes the $\Theta$ parameter of QCD vanish dynamically through the Peccei–Quinn mechanism [102]. A global $U(1)$ symmetry, called *the Peccei–Quinn symmetry*, ensures that the Lagrangian conserves $CP$ and $\Theta = 0$. This symmetry is then broken spontaneously at a very high scale and the corresponding Nambu–Goldstone boson, called *the axion*, picks up small mass due to instanton effects. The most interesting aspect of this theory is that it predicts the new particle axion, which is being detected.

## *CP* Violation in Strong Interactions

The strong coupling constant of QCD makes the nonperturbative effects important. Among other interesting effects, the transition matrix elements split up into a family of disconnected sectors due to some static field configurations [101]. These disconnected sectors are topologically distinct and labelled by a topological quantum number $n$, called the winding number. In a temporal gauge $A_a^0 = 0$, these distinct vacuum states $|n\rangle$ are specified by demanding that at spatial infinity the gauge fields should vanish. The vacuum states are labelled by the winding numbers $n$, which are topological numbers arising due to instanton effects, defined by

$$n = \frac{g^2}{32\pi^2} \int d^4x F_{\mu\nu}^a \tilde{F}^{\mu\nu a}, \tag{7.41}$$

where

$$F_{\mu\nu}^a = \partial_\mu A_\nu^a - \partial_\nu A_\mu^a + g f^{abc} A_{\mu b} A_{\nu c}$$
$$\tilde{F}_{\mu\nu}^a = \frac{1}{2} \varepsilon_{\mu\nu\alpha\beta} F^{\alpha\beta a}.$$

$n$ is an integer number for any pure gauge, characterizing the QCD vacuum. Since for an instanton $n = 1$, it is possible to transform $|n\rangle \to |n+1\rangle$ in the presence of an instanton. Any gauge transformations cannot take back a generated field to a zero field smoothly. When any gauge transformation takes a generated field to a definite vacuum state $|n\rangle$, the corresponding gauge fields are denoted as $A_n^i$. Except for quantum tunnelling due to instanton effects, no gauge transformations can take one vacuum to another.

The $\Theta$ vacua are gauge invariant vacuum states of QCD, which are constructed as

$$|\Theta\rangle = \sum_n \exp^{-in\Theta} |n\rangle. \tag{7.42}$$

Different $\Theta$ correspond to distinct theories, which are not connected by any gauge transformations.

A Green's function in QCD gives the vacuum to vacuum transition in the presence of an external source. In the temporal gauge this gives the transition between the pure gauge configurations $A_n^i \to A_m^i$ at $t = \infty$. In addition to the usual QCD vacuum, this Green's function will have an additional phase $\exp^{i\nu\Theta}$ coming due to the transition $A_n^i \to A_m^i$, where winding number is changed by $\nu = m - n$ in a particular $\Theta$ vacuum. This additional phase is associated with an additional term in the QCD Lagrangian

$$\mathscr{L}_\Theta = \frac{g^2}{32\pi^2} \, \Theta \, F_{\mu\nu}^a \tilde{F}^{\mu\nu a}. \tag{7.43}$$

This gives a new source of *CP* violation in QCD. There is no *CP* violation for $\Theta = 0, \pi$, and from current algebra consideration $\Theta = \pi$ is not allowed [101].

This term also has its origin in the $U(1)_A$ axial current

$$J_5^\mu = \sum_{i=1}^{N_f} \bar{q}_i \gamma^\mu \gamma_5 q_i, \tag{7.44}$$

which is classically conserved for massless quarks and is generated by a conserved charge

$$Q_5 = \int d^3x J_5^0. \tag{7.45}$$

The anomaly introduces an one-loop divergent contribution

$$\partial_\mu J_5^\mu = 2N_f \left( \frac{g^2}{32\pi^2} F_{\mu\nu}^a \tilde{F}^{\mu\nu a} \right) = 2N_f \left( \frac{g^2}{32\pi^2} \partial^\mu K_\mu \right), \tag{7.46}$$

where $N_f$ is the number of flavors and

$$K_\mu = \frac{1}{2} \varepsilon_{\mu\alpha\beta\gamma} A_a^\alpha [F_a^{\beta\gamma} - \frac{g}{3} \varepsilon_{abc} A_b^\beta A_c^\gamma].$$

The conserved current has an additional contribution from the anomaly term

$$\tilde{J}_5^\mu = J_5^\mu - 2N_f \left( \frac{g^2}{32\pi^2} K^\mu \right). \tag{7.47}$$

The associated charge $\tilde{Q}_5 = \int d^3x \tilde{J}_5^0$ is time independent, but not gauge invariant. Any gauge transformation, where winding number is changed by 1, transforms $\tilde{Q}_5 \to \tilde{Q}_5 + 2N_f$. Thus, a chiral rotation changes the $\Theta$ vacuum to

$$\exp^{i\alpha\tilde{Q}_5} |\Theta\rangle = |\Theta + 2N_f \alpha\rangle. \tag{7.48}$$

In short, the anomaly made the chiral current divergent. Absence of any associated conserved charge implies there is no Nambu–Goldstone boson, and hence, the $\eta$ meson becomes massive. But the new conserved current is no longer gauge invariant. This generates the disconnected sectors of the vacuum and, hence, introduces the $\Theta$ term in the QCD Lagrangian.

If the quarks are massless, a chiral transformation can take one $\Theta$ vacuum to another, and hence, it is possible to choose a chiral rotation that takes to $\Theta = 0$. However, since quarks are not massless, this cannot be the solution to this strong *CP* problem.

The *CP* violation in the weak interactions coming from a complex phase in the quark mass matrices can also contribute to the $\Theta$ term of the QCD Lagrangian. In general the quark mass matrices could be complex. A chiral rotation to the quark fields transforms the $\Theta$ vacuum by an amount

$$\Theta \to \bar{\Theta} = \Theta + \mathrm{Arg}[\det M].$$

Thus, any chiral rotation can make one of the two terms vanishing but cannot make both of them vanish simultaneously. When quarks are not massless, it will not be possible to apply any chiral rotations that can make the $\Theta$ term vanish.

Let us now consider the effect of the $\Theta$ term in the QCD Lagrangian. For simplicity we consider a one generation example, where the $\Theta$ term can be explicitly given by the expression

$$\mathscr{L}_\Theta = i \frac{m_u m_d \sin\Theta}{\sqrt{m_u^2 + m_d^2 + 2m_u m_d \cos\Theta}} (\bar{u}\gamma_5 u + \bar{d}\gamma_5 d). \tag{7.49}$$

This predicts a nonvanishing electric dipole moment of the neutron in the range of [103]

$$4 \times 10^{-17}\ \bar{\Theta} e.cm < d_n < 2 \times 10^{-15}\ \bar{\Theta} e.cm,$$

which is much larger than the present experimental limit [44]

$$d_n < 0.63 \times 10^{-25} e.cm,$$

unless the $\Theta$ parameter is very small,

$$\bar{\Theta} = \Theta + \text{Arg}[\det M] < 10^{-10}. \tag{7.50}$$

Why $\Theta$ is so small and not of the order of 1 is the strong *CP* problem. Standard model does not provide any solution to this problem. We shall now discuss some extensions of the standard model that can provide a solution to this problem.

## 7.3 Peccei–Quinn Symmetry

The strong *CP* problem has its origin in the topological structure of the QCD vacuum. The nonperturbative effects predict a large *CP* violation originating from the QCD Lagrangian, but the experimental constraints require that this parameter be extremely small. Higher order corrections will again predict large corrections, so the counter terms have to be fine tuned to many decimal points to keep this parameter small to all orders in perturbation theory and explain the experimental constraints. This fine tuning is unnatural and known as the strong *CP* problem.

The first possibility one considers is that at least one of the quarks is massless, say, the *u*-quark. The different $\Theta$ vacua states are now connected by chiral rotations, and hence, it is possible to consider a $\Theta$ vacuum which solves the $U(1)_A$ problem. Since all the vacuum states are equivalent, there is no *CP* violation. But none of the quarks are massless, and hence, one starts with a massless *u*-quark and tries to generate an effective *u* quark mass through instanton effects [104]. Although this solution is definitely a reasonable possibility, this does not explain why the *u*-quark mass is vanishing to start with.

In another possible solution to the strong *CP* problem, the theory allows a very small arg$[\det M]$, estimates the $\bar{\Theta}$ term, and shows that it is small. But the constraints from the *CP* violation in the *K*- and the *B*-system do not allow a small enough strong *CP* term, as required.

The most popular solution to the strong *CP* problem is the one suggested by Peccei and Quinn [102], which provides a dynamical solution. One introduces a global $U(1)_{PQ}$ symmetry under which the quark mass terms are forbidden, but they come from the *vev* of the Higgs fields. The chiral transformation of the Peccei–Quinn $U(1)_{PQ}$ global symmetry ensures that $\bar{\Theta}$ vanishes dynamically. The *vev* of the Higgs fields break the Peccei–Quinn symmetry spontaneously, but the axial anomaly also breaks this symmetry explicitly, giving a small mass to the corresponding Nambu–Goldstone boson, the axion [105].

Under the Peccei–Quinn $U(1)_{PQ}$ symmetry, the left-handed and right-handed quarks carry different charges, so the mass terms are forbidden. However, the transformations of the Higgs scalar ensure that the Yukawa couplings, which can give masses to the quarks after the Higgs scalars acquire *vevs*, are allowed. The chiral nature of the symmetry will introduce a new anomaly term, given by

$$\mathscr{L}_{PQ} = \frac{a}{v_a}\xi\frac{g^2}{32\pi^2}F^a_{\mu\nu}\tilde{F}^{\mu\nu a}, \tag{7.51}$$

where $a$ is the axion field, the scale of the $U(1)_{PQ}$ breaking is $v_a$, and $\xi$ is a parameter which depends on how the $PQ$ charges are assigned to the fermions. This anomaly term provides a potential for the axion field and gives it a vacuum expectation value

$$\langle a\rangle = \langle\bar{\Theta}|a|\bar{\Theta}\rangle = -\bar{\Theta}\frac{1}{\xi}\,v_a. \tag{7.52}$$

The physical axion can then be written as $a_{phys} = a - \langle a\rangle$, so the axion eliminates the unwanted $\bar{\Theta}$ term in the Lagrangian and replaces it with the physical axion field

$$\begin{aligned}\mathscr{L}_{eff} &= \bar{\Theta}\frac{g^2}{32\pi^2}F^a_{\mu\nu}\tilde{F}^{\mu\nu a} + \frac{a}{v_a}\xi\frac{g^2}{32\pi^2}F^a_{\mu\nu}\tilde{F}^{\mu\nu a}\\ &= \frac{a_{phys}}{v_a}\xi\frac{g^2}{32\pi^2}F^a_{\mu\nu}\tilde{F}^{\mu\nu a}.\end{aligned} \tag{7.53}$$

The axion is the Nambu–Goldstone boson corresponding to the global $U(1)_{PQ}$ symmetry breaking, and hence, the effective Lagrangian contains only derivative couplings

$$\mathscr{L}_a = -\frac{1}{2}\partial_\mu a\partial^\mu a + \mathscr{L}(\partial_\mu a, \psi), \tag{7.54}$$

where the second term gives the interactions of the axions with the fermions, which are determined by the structure of the $U(1)_{PQ}$ symmetry. The vacuum expectation value of the axion conspires to cancel the $\bar{\Theta}$ term exactly, and the axion gets a small mass with its coupling determined by the scales of symmetry breaking.

**Models of the Axion**

The minimal extension of the standard model to accommodate a chiral symmetry requires at least two Higgs doublets $\phi_1$ and $\phi_2$ carrying the same $U(1)_{PQ}$ charges. The transformations of the quarks, the leptons and the Higgs scalars under the $U(1)_{PQ}$ Peccei–Quinn symmetry can be defined as

$$\{q,\ell\}_L \to \exp^{-i\alpha}\{q,\ell\}_L$$

$$\{u,d,e\}_R \to \exp^{i\alpha}\{u,d,e\}_R$$

$$\phi_{1,2} \to \exp^{-2i\alpha}\phi_{1,2}.$$

The usual Yukawa couplings that can give masses to the quarks and leptons after the Higgs doublets $\phi_1$ and $\phi_2$ acquire *vevs* are now given by

$$\mathscr{L}_Y = h^u \bar{q}_L u_R \phi_2 + h^d \bar{q}_L d_R \phi_1 + h^e \bar{\ell}_L e_R \phi_1 + H.c. \tag{7.55}$$

The *vevs* of the Higgs doublets $\langle \phi_i \rangle = v_i/\sqrt{2}$ break the $U(1)_{PQ}$ symmetry at a scale $v = \sqrt{v_1^2 + v_2^2}$. These Higgs scalars also break the electroweak symmetry and give masses to the quarks and leptons. In the simplest model the scale of $U(1)_{PQ}$ symmetry breaking scale is the same as the electroweak symmetry breaking scale. Since the axion couples to matter directly, these are called the visible axion models.

Corresponding to the two Higgs doublets, there are now four physical Higgs scalars and four Nambu–Goldstone modes. Three of the Nambu–Goldstone modes will give masses to the $W^\pm, Z$ bosons and the remaining one becomes the axion field

$$a = \frac{1}{v}(v_1 \mathrm{Im}\phi_1^0 - v_2 \mathrm{Im}\phi_2^0).$$

Since the $U(1)_{PQ}$ symmetry is spontaneously broken, the corresponding Nambu–Goldstone boson, the axion, will be massless. But the QCD gluon anomaly will break this symmetry explicitly, and hence, the axion will become a pseudo-Nambu–Goldstone boson with a small mass, given by

$$m_a = \frac{f_\pi m_\pi}{v} N_g \left( \frac{v_1}{v_2} + \frac{v_2}{v_1} \right) \frac{\sqrt{m_u m_d}}{m_u + m_d}, \tag{7.56}$$

where $f_\pi$ is the pion decay constant.

In the visible axion models, when the axion mass is less than the electron mass, the axions can decay into two photons. The nonderivative interaction of the axion to the electromagnetic field comes from loops containing the right-handed fermions

$$\mathscr{L}_{a\gamma\gamma} = \frac{a_{phys}}{f_\pi} \frac{m_a}{m_\pi} \sqrt{\frac{m_u}{m_d}} \frac{e^2}{16\pi^2} F_{\mathrm{em}}^{\mu\nu} \tilde{F}_{\mathrm{em}\ \mu\nu}, \tag{7.57}$$

and hence, the axion becomes long-lived. When the axion mass is more than the electron mass, it can decay rapidly into two electrons and become short-lived

$$\tau(a \to e^+ e^-) = \frac{8\pi v^2 v_2^2}{m_e^2 v_1^2 \sqrt{m_a^2 - 4m_e^2}}. \tag{7.58}$$

But experiments ruled out both these possibilities. The strongest constraint comes from the nonobservation of the $K$-decays to axions [106]

$$Br(K^+ \to \pi^+ + nothing) \leq 3.8 \times 10^{-8}, \tag{7.59}$$

where (*nothing*) in the decay products includes long-lived axions, which would escape detection. The quarkonium decay $Q\bar{Q} \to a\gamma$ measures the coupling of axions to heavy quarks [107]. Nonobservation of such processes [44, 108]

$$Br(\Upsilon \to a\gamma) < 3 \times 10^{-4}$$
$$Br(J/\Psi \to a\gamma) < 1.4 \times 10^{-5}$$

rules out the long-lived axions. The short-lived axions are ruled out from a measurement of the process

$$Br(\pi^+ \to ae^+\nu_e) < 10^{-10}. \tag{7.60}$$

This rules out the short-lived axions that couple to matter directly.

In the visible axion model the Peccei–Quinn symmetry was broken along with the electroweak symmetry and the axion couples to matter directly. Since this model is ruled out, an extension of the model was considered, where the scale of $U(1)_{PQ}$ symmetry breaking is much higher. The couplings of the axion to the ordinary matter are then suppressed by the ratio of the *vevs* of the electroweak symmetry breaking scale to the $U(1)_{PQ}$ symmetry breaking scale. Such weak coupling of the axion would allow it to evade existing experimental searches, and hence, this invisible axion model is allowed by present experiments.

In the simplest extension [109] of the visible axion model a standard model singlet scalar $\sigma$ is added, which transforms under the $U(1)_{PQ}$ as $\sigma \to \exp^{2i\alpha}\sigma$. This field interacts with the fermions only through its mixing with the usual Higgs scalars

$$V(\phi_1,\phi_2,\sigma) = V(\phi_1,\phi_2) + V(\sigma) + (a|\phi_1|^2 + b|\phi_2|^2)|\sigma|^2 + c\phi_1\phi_2\sigma^2.$$

The $U(1)_{PQ}$ symmetry is broken by the *vev* of the singlet Higgs $\langle\sigma\rangle = v_\sigma$ at a scale much higher than the electroweak symmetry breaking scale $v_a = v_\sigma \gg v$.

The axion field is now a combination of the fields $\mathrm{Im}\phi_1^0$, $\mathrm{Im}\phi_2^0$, and $\mathrm{Im}\sigma$ and is given by

$$a = \frac{1}{v(v^2 v_\sigma^2 + 4v_1^2 v_2^2)^{1/2}}[2v_1 v_2(v_1 \mathrm{Im}\phi_1^0 + v_2 \mathrm{Im}\phi_2^0) - v^2 v_\sigma \mathrm{Im}\sigma]. \tag{7.61}$$

In the limit of very large $v_\sigma$, the mass of the axion becomes

$$m_a \approx \frac{f_\pi m_\pi}{v_\sigma} \sim \frac{(0.2\ \mathrm{GeV})^2}{v_\sigma}, \tag{7.62}$$

and the axion decay constant becomes $f_a \sim v_\sigma/2$. This makes the axion very light with derivative coupling to ordinary matter, which is suppressed by $v/v_\sigma$. Axions can now decay into only two photons.

The main criterion of the Peccei–Quinn mechanism is that the fermions with nontrivial transformation under $SU(3)_c$ should have chiral transformation under the Peccei–Quinn symmetry, so the symmetry is broken explicitly by the $SU(3)$ anomaly. This criterion is exploited in another version of the invisible axion model where new heavy quarks were introduced [110]. The quarks, the leptons and the usual Higgs doublets are singlets under the $U(1)_{PQ}$ group. Only the new heavy quarks and the singlet Higgs scalar $\sigma$ have nontrivial transformation under $U(1)_{PQ}$. $SU(3)_c$ anomaly cancellation require both left- and right-handed heavy quarks $X_L$ and $X_R$. Under the Peccei–Quinn symmetry the masses of the heavy fermions should not be invariant, which is satisfied by the transformation

$$X_{[L,R]} \to \exp^{i\gamma_5\alpha} X_{[L,R]}. \tag{7.63}$$

A singlet, transforming under this $U(1)_{PQ}$ symmetry as

$$\sigma \to e^{-2i\alpha}\,\sigma,$$

can then have the required Yukawa coupling $\bar{X}_L X_R \sigma$, which can give mass to the heavy quarks when the singlet $\sigma$ acquires a *vev*, $\langle\sigma\rangle = v_\sigma$.

Since the heavy quarks transform nontrivially under $SU(3)_c$, a $\gamma_5$ rotation of the heavy quark fields can contribute to the QCD axial anomaly and change the $\bar{\Theta}$ term by an amount $\alpha$. Since anomaly does not depend on the mass of the particles, the new heavy quark fields $X_{L,R}$ can make the $\bar{\Theta}$ term vanish dynamically. There will be terms in the scalar potential that would mix the singlet Higgs with the Higgs doublet, and the axion will be given by

$$a = \frac{1}{(v^2 + v_\sigma^2)^{1/2}}[v\mathrm{Im}\phi^0 + v_\sigma \mathrm{Im}\sigma], \tag{7.64}$$

where $v$ is the *vev* of the standard model Higgs doublet $\phi$, and $v_a = v_\sigma$ is the $U(1)_{PQ}$ symmetry breaking scale. The axion mass is given by

$$m_a = \frac{f_\pi m_\pi}{2v_a}\frac{\sqrt{m_u m_d}}{m_u + m_d}, \tag{7.65}$$

with the decay constant $f_a = v_a/2$. The axion now decays into two photons with a very long lifetime. There is no direct coupling of the axion with the electrons, except a tiny coupling at the one-loop level through its coupling to photons. A highly suppressed coupling of the axion to ordinary matter follows from its couplings with pions and $\eta$ mesons.

In another variant of the invisible axion model, the role of the heavy quarks is replaced by the gluinos in supersymmetric theories, which are heavy colored fermions. Replacing the $\mu$ term in the superpotential, the $R$-parity could be uplifted to a global $U(1)_R$ symmetry which is then identified with the $U(1)_{PQ}$ symmetry. This symmetry is then broken at a large scale without breaking supersymmetry, solving the strong *CP* problem [111].

There is another solution in $N = 1$ supergravity models [112], where the $R$-parity in supersymmetric models is considered as the Peccei–Quinn symmetry. In this theory $R$-symmetry is broken by the soft terms in supergravity theories, and hence, there is no axion. There are also interesting solutions to the strong *CP* problem originating from superstring theories. The pseudoscalar moduli fields in $D = 10$ superstring theories contain a pseudoscalar field in four dimensions. This 4-dimensional pseudoscalar field could become the axion and the Yang–Mills Chern–Simmons form could introduce an anomaly term and remove the $\bar{\Theta}$ term from the QCD Lagrangian [113]. In this case the heavy scale is the Planck scale or the compactification scale, and hence, the axion is superlight with a large decay constant. This model has another problem with a new source of the $\Theta$ term coming from the requirement of a hidden sector confining force for supersymmetry breaking.

**Status of Axion Search**

The most natural solution to the strong *CP* problem is the Peccei–Quinn mechanism, which predicts a new light particle, the axion, with very weak coupling to ordinary matter. This is the invisible axion model, in which the scale of Peccei–Quinn symmetry breaking is very high and the coupling of the axion is suppressed by this large scale. Although this new particle has not yet been discovered, the allowed range of parameters is highly constrained from astrophysical and cosmological considerations and also from laboratory measurements.

The effective couplings that are relevant for the invisible axion searches are its couplings with electrons $g_{ae^+e^-}$, nucleons $g_{aNN}$, and photons $g_{a\gamma\gamma}$. Although these quantities are model dependent, they can be expressed in terms of some known parameters and the inverse of the Peccei–Quinn symmetry breaking scale $v_a$, which also determines the only free parameter of the theory, the mass of the axion

$$m_a = \frac{\sqrt{z}}{1+z} N \frac{f_\pi m_\pi}{v_a}, \tag{7.66}$$

where $N$ is the *color-anomaly* of the Peccei–Quinn symmetry, which is model dependent and $z = m_u/m_d$. Thus, all the astrophysical and cosmological bounds could be translated to a bound on the axion mass.

The emission of axions could affect the stellar dynamics, which constrains the mass of the axion. In models, in which electrons couple to axions directly through derivative couplings [109], the axion emission occurs through the Compton axion production, $\gamma + e^- \to a + e^-$, and the axion bremsstrahlung, $e^- + Z \to e^- + Z + a$. On the other hand, in hadronic models [110] where the axions couple to the electrons very weakly, the main axion production mechanism is through the Primakoff process: $\gamma + Z \;\; (or\; e^-) \to a + Z \;\; (or\; e^-)$. So the emission rate is substantially reduced, weakening the bound on axion mass.

The cooling of stars due to axion emission gives a bound on the axion mass to be around 1 eV [114], while for the hadronic axions it is about 20 eV. These bounds are applicable up to an axion mass of about 10 keV, since the production of such axions will be suppressed at higher energies. The suppression of helium ignition due to the axions in low-mass red giants [115] gives a constraint on the axion mass to be less than $10^{-2}$ eV and for the hadronic axions around 2 eV. Nonobservation of shortening of the neutrino emission from the supernovae SN1987a rules out any axion mass within the range of 2 eV to $10^{-2}$ eV [116]. These bounds translate to a lower bound on the scale of the Peccei–Quinn symmetry breaking of $v_a > 10^9$ GeV.

We now discuss the upper bound on $v_a$ coming from the cosmological considerations. Below the Peccei–Quinn symmetry breaking scale $T = v_a$, the axion field takes some arbitrary value and appears as a massless Nambu–Goldstone particle, but after the QCD phase transition it again picks up a mass

$$m_a(T) = 0.1\, m_a(T=0) \left( \frac{\lambda_{QCD}}{T} \right)^{3.7}. \tag{7.67}$$

The *vev* will now adjust and start oscillating to its final dynamical value corresponding to $\bar{\Theta} = 0$. The upper bound on $v_a$ comes from the requirement that the energy density stored in the oscillating axion field should not be larger than the energy density of the universe [117].

The *vev* of the axion field satisfies the equation of motion

$$\frac{d^2\langle a\rangle}{dt^2} + 3H(t)\frac{d\langle a\rangle}{dt} + m_a^2(t)\langle a\rangle = 0, \tag{7.68}$$

where $H$ is the Hubble constant and $t \sim T^{-2}$. In a fast expanding universe $H(t) \gg m_a(t)$, and the expectation value of $\langle a\rangle$ remains approximately constant. But when the mass attains equilibrium $m_a(t) \gg H(t)$ and changes adiabatically $(dm_a/dt \ll m_a^2(t))$, the value of $\langle a\rangle$ will oscillate around its true expectation value corresponding to the vanishing $\Theta$ parameter and is given by

$$\langle a\rangle \approx \frac{A}{\sqrt{m_a(t)R^3(t)}} \cos m_a(t)t, \tag{7.69}$$

where $A$ is a constant. The energy stored in this axion field

$$\rho_a(t) = \frac{1}{2}m_a^2(t)\langle a\rangle^2 + \frac{1}{2}\left(\frac{d\langle a\rangle}{dt}\right)^2 = \frac{1}{2}\frac{A^2 m_a(t)}{R^3(t)} \tag{7.70}$$

should not be more than the total energy of the universe

$$\rho_a \leq \rho_{crit}.$$

The axion density today $\rho_a(t)$ can be estimated in terms of the density $\rho_a(t_{osc})$ at the time $t_{osc}$ when $\langle a\rangle$ started oscillating

$$m_a(t_{osc}) \sim \Lambda_{QCD}^2/M_{Pl}$$

and

$$\rho_a(t_{osc}) \sim \frac{1}{2}m_a^2(t_{osc})(C\, v_a)^2,$$

where $C$ is some number $O(1)$. The fact that the axion density should not overclose the universe gives a lower bound on the mass of the axion to be around $10^{-5}$ eV [117]. Combining this estimate of the cosmological bound with the astrophysical bound, we are left with a small range of symmetry breaking scale

$$10^9 \text{ GeV} < v_a < 10^{12} \text{ GeV} \tag{7.71}$$

and, hence, a small window of axion mass in the range of $10^{-2}$ eV $> m_a > 10^{-5}$ eV.

Several experiments are going on to detect the axions with mass in this allowed range. In a novel detection mechanism virtual photons are produced in a strong magnetic field, which should interact with the axion in the magnetic field releasing a single mono-energetic photon [118]. There are also experiments where axions are first produced in the laboratory through the axion–photon mixing [119], and then they are detected. The axion searches have not provided any positive information so far. However, if axions are detected, it will be a very important result for our understanding of the physics beyond the standard model.

## 7.4 *CP* Violation in the Leptonic Sector

The *CP* violation in the leptonic sector is different from the quark sector, mainly because of the Majorana nature of the neutrinos. Since we expect the neutrinos to be Majorana particles to explain why the neutrino masses are so much lower than the charged fermion masses, the freedom of phase rotation of the right-handed fermions is lost. In the quark sector it is possible to absorb some of the phases by the right-handed fields. On the other hand, there are new Majorana *CP* phases contributing to the neutrinos.

In general all the Yukawa couplings could be complex giving rise to *CP* violation. This will make the elements of the charged lepton mass matrix be complex. The Majorana mass matrices of the left-handed and right-handed neutrinos and the Dirac mass matrix of the neutrinos could also be complex. It is possible to make the charged lepton mass matrix real and diagonal, particularly because of the freedom of transforming the right-handed charged leptons without affecting other observable parameters of the standard model. For the neutrino sector, we can integrate out the effects of the heavy right-handed neutrinos and work in the basis of the effective light neutrinos, which are dominantly left-handed neutrinos with very little or no admixtures of right-handed neutrinos. In the rest of this section we shall denote these effective light neutrinos, which are also the flavor eigenstates, by the states $\nu_\alpha, \alpha = e, \mu, \tau$.

We restrict our discussion to three generations, since the cosmological, astrophysical and laboratory constraints favor three generations of neutrinos at present. Without referring to the origin of the neutrino mass, we shall only introduce a neutrino mass term in the standard model. We start with the Lagrangian containing the charged current interaction, charged lepton masses and the neutrino Majorana mass terms

$$\mathscr{L} = \frac{g}{\sqrt{2}} \bar{l}_{mL}\, \gamma^\mu\, \nu_{mL}\, W_\mu^- - \bar{l}_{mL}\, M_{lmn}\, l_{nR} - \nu_{mL}^T\, C^{-1}\, M_{\nu mn}\, \nu_{nL}. \tag{7.72}$$

The charged lepton mass matrix can be diagonalized by a bi-unitary transformation $E^\dagger_{\alpha mR} M_{lmn} E_{n\beta L} = M_{l\alpha} \delta_{\alpha\beta}$, where $\alpha, \beta = e, \mu, \tau$ are the physical states and $m, n = 1, 2, 3$ are the states where the charged leptons are not diagonal. Thus, it is convenient to work in the basis in which the charged lepton mass matrix is real and diagonal. In this basis the neutrinos can be identified with the corresponding physical charged leptons and are called the flavor basis.

In the flavor basis the charged current interactions and the lepton masses can be written as

$$\mathscr{L} = \frac{g}{\sqrt{2}} \bar{l}_{\alpha L}\, \gamma^\mu\, \nu_{\alpha L}\, W_\mu^- - \bar{l}_{\alpha L}\, M_{l\alpha}^{\text{diag}}\, l_{\alpha R} - \nu_{\alpha L}^T\, C^{-1}\, M_{\nu\alpha\beta}\, \nu_{\beta L}. \tag{7.73}$$

The flavor neutrino eigenstates $\nu_{\alpha L}$ are given by

$$\nu_{\alpha L} = E^\dagger_{\alpha m L} \nu_{mL}. \tag{7.74}$$

The symmetric neutrino mass matrix can be diagonalized by a single unitary matrix $U_{\alpha i}$ through

$$U^T_{i\alpha} M_{\nu\alpha\beta} U_{\beta j} = K_i^2 M_{\nu i} \delta_{ij}, \tag{7.75}$$

where $M_{\nu i}$ is a real diagonal matrix and $K$ is a diagonal phase matrix. The unitary transformation $U_{\alpha i}$ relates the flavor neutrino states to the neutrino mass eigenstates (equation (6.68))

$$\nu_{\alpha L} = \sum_i U_{\alpha i} \nu_{iL}. \tag{7.76}$$

The mixing matrix $U_{\alpha i}$ is called the *PMNS* (Pontecorvo–Maki–Nakagawa–Sakata) mixing matrix [67].

If there is *CP* violation, the Yukawa couplings will be complex. Some of these phases could be removed by redefining the fermions, but some of the phases will remain and show up in physical processes. To find out which phases cannot be removed, it is convenient to construct some rephasing invariant quantities [120, 121], in the same way the Jarlskog invariants were constructed for the quark CKM mixing matrix (see equation (5.52)).

Any phase transformation of the physical neutrinos $\nu_i \longrightarrow \exp^{i\delta_i} \nu_i$ should not affect the charged current interactions and the Majorana mass matrix, and hence, the mixing matrix $U$ and the phase matrix $K$ should transform as

$$\{U_{\alpha i}, K_i\} \longrightarrow \exp^{-i\delta_i} \{U_{\alpha i}, K_i\}.$$

This will allow us to remove the overall phase of $K$ leaving two independent phases for three generations. For the mixing matrix $U$, all other phases except for one could be eliminated with this transformation for three generations. Appropriate rephasing of the neutrinos may make $K$ real and have all three phases in $U$, but it will not be possible to reduce the numbers of phases to less than 3 by any rephasing.

The rephasing of the neutrinos transforms both $U$ and $K$, but the physical processes are independent of the rephasing of the neutrinos. Thus, $U$ and $K$ must enter in any physical observables in the rephasing invariant combinations

$$\begin{aligned} s_{\alpha ij} &= \mathrm{Im}[U_{\alpha i}\, U^*_{\alpha j}\, K^*_i\, K_j] \\ t_{\alpha i\beta j} &= \mathrm{Im}[U_{\alpha i}\, U_{\beta j}\, U^*_{\alpha j}\, U^*_{\beta i}]. \end{aligned} \tag{7.77}$$

All of the $s_{\alpha ij}$ and $t_{\alpha i\beta j}$ are not independent, for example,

$$t_{\alpha i\beta j} = s_{\alpha ij} \cdot s_{\beta ji}. \tag{7.78}$$

We now look for a minimal set of these invariant measures of *CP* violation. For three generations, if none of the elements of the mixing matrix vanishes, one possibility is to work with the three independent $s_{\alpha ij}$. But a more useful set of minimal independent rephasing invariant quantities is

$$\{J_{CP} = t_{\alpha i13};\ J_1 = s_{113};\ J_2 = s_{123}\}. \tag{7.79}$$

If any of the elements of $U$ vanish, this argument fails. For example, if $U_{13} = 0$, all these $J$'s will vanish, although $s_{112}$ and $s_{123}$ will still remain nonvanishing.

This choice of the minimal set $\{J_{CP}, J_1, J_2\}$ has the advantage that it interprets $J_{CP}$ to be equivalent to the Jarlskog invariant in the quark sector. $J_{CP}$ can appear in all processes, but the two *CP* violating measures $J_1$ and $J_2$ enter only in the lepton-number violating interactions. However, some combinations of $J_1$ and $J_2$ may enter in some higher order lepton-number conserving processes as well. If there are loops containing lepton-number violating processes, although the final states are *CP* conserving, there may exist intermediate lepton-number violating vertices where $J_1$ and $J_2$ will appear. In this case the *CP* violating phases $J_1$ and $J_2$ may contribute to the amplitude of the *CP* conserving processes [122].

Let us now consider the case of neutrino oscillations. If a neutrino of one flavor oscillates into another flavor of neutrino, lepton number is not violated. These are the conventional neutrino oscillations we usually discuss. If $J_{CP}$ is nonvanishing, these oscillations will be affected. It is also possible that a neutrino will oscillate into an antineutrino, violating lepton number. This is the neutrino oscillation that was proposed originally by Pontecorvo. In this case *CP* violation can come from $J_1$, $J_2$, or $J_{CP}$. However, this type of neutrino oscillation is not possible to observe, since the amplitude for the process is highly suppressed from the phase–space consideration.

We shall now discuss only the *CP* violation in the lepton-number conserving neutrino oscillations. We use the convention that the probability of any neutrino oscillation from one flavor $\alpha$ to another flavor $\beta$ is denoted by $P_{\nu_\alpha \to \nu_\beta}$. The *CPT* invariance relates the transition probabilities [123]

$$P_{\nu_\alpha \to \nu_\beta} = P_{\bar{\nu}_\beta \to \bar{\nu}_\alpha}. \tag{7.80}$$

It is then convenient to define a measure of *CP* violation

$$\Delta^{CP}_{e\mu} = P_{\bar{\nu}_e \to \bar{\nu}_\mu} - P_{\nu_e \to \nu_\mu} = P_{\nu_\mu \to \nu_e} - P_{\nu_e \to \nu_\mu}, \tag{7.81}$$

in terms of the difference between transition probabilities of *CP* conjugate channels. Unitarity implies that for three generation there is only one such measure

$$\Delta^{CP} = \Delta^{CP}_{e\mu} = \Delta^{CP}_{\mu\tau} = \Delta^{CP}_{e\tau}, \tag{7.82}$$

which is proportional to $J_{CP}$. The *CP* measures $J_1$ and $J_2$ cannot enter in these expressions for lepton-number conserving processes.

In case of lepton-number violating neutrino oscillations, the *CP* measures are defined as

$$\Delta^{CP}_{\alpha\beta}(\Delta L = 2) = P_{\bar{\nu}_\alpha \to \nu_\beta} - P_{\nu_\alpha \to \bar{\nu}_\beta}. \tag{7.83}$$

The different measures $\Delta^{CP}_{e\mu}(\Delta L = 2)$, $\Delta^{CP}_{\mu\tau}(\Delta L = 2)$ and $\Delta^{CP}_{e\tau}(\Delta L = 2)$ will depend on all the *CP* violating measures.

Let us now consider a particular parmetrization of the mixing matrix $U$ for three generations including all the three *CP* violating phases:

$$U = \begin{pmatrix} c_1c_3 & s_1c_3e_\eta & s_3e_{\rho\phi} \\ (-s_1c_2 - c_1s_2s_3e_\phi)e_{-\eta} & c_1c_2 - s_1s_2s_3e_\phi & s_2c_3e_{\rho\eta} \\ (s_1s_2 - c_1c_2s_3e_\phi)e_{-\rho} & (-c_1s_2 - s_1c_2s_3e_\phi)e_{\eta\rho} & c_2c_3 \end{pmatrix}, \tag{7.84}$$

where $c_i = \cos\theta_i$, $s_i = \sin\theta_i$, $e_\alpha = \exp^{i\alpha}$, $e_{\alpha\beta} = \exp^{i(\alpha-\beta)}$, and $\alpha, \beta = \eta, \delta, \phi$ are the three *CP* violating phases. The corresponding three *CP* violating measures are

$$J_{CP} = -c_1 c_2 c_3^2 s_1 s_2 s_3 \sin\phi \tag{7.85}$$

$$J_1 = -c_1 c_3 s_3 \sin(\rho - \phi) \tag{7.86}$$

$$J_2 = -s_1 c_3 \sin(\rho - \phi - \eta)\,. \tag{7.87}$$

For a special case, $\rho = \phi$ and $\eta = 0$, both $J_1$ and $J_2$ vanishes, but there can be *CP* violation in any processes. On the other hand, for $\phi = 0$ the *CP* violating measure $J_{CP}$ vanishes, implying there is no *CP* violation in any lepton-number conserving processes like neutrino oscillations. We can rewrite this parameterization by defining the masses of the neutrinos with Majorana phases

$$\tilde{m}_1 = m_1 \qquad \tilde{m}_2 = \exp^{2i\eta} m_2 \qquad \tilde{m}_3 = \exp^{2i(\rho-\phi)} m_3 \tag{7.88}$$

and then write the mixing matrix $U$ with one complex phase $\phi$ like the mixing matrix in the quark sector

$$U = \begin{pmatrix} c_1 c_3 & s_1 c_3 & s_3 e_{-\phi} \\ -s_1 c_2 - c_1 s_2 s_3 e_\phi & c_1 c_2 - s_1 s_2 s_3 e_\phi & s_2 c_3 \\ s_1 s_2 - c_1 c_2 s_3 e_\phi & -c_1 s_2 - s_1 c_2 s_3 e_\phi & c_2 c_3 \end{pmatrix}. \tag{7.89}$$

The *CP* invariant measures remain the same in both cases.

There is another major difference between the *CP* violation in the quark sector and the lepton sector. In case of leptons, in the flavor basis it is possible to find a measure of the *CP* violation only in terms of the elements of the effective light neutrino Majorana mass matrix [121]. This is possible because the mass matrix can be diagonalized by only one mixing matrix and, hence, contains all the information available in the mixing matrix. Any rephasing of the charged leptons $E \to XE$, where X is the phase transformation to the charged leptons, can introduce phases in the mixing matrix $U \to X^* U$. This transformation can be interpreted as a transformation to the mass matrix, $M \to X^* M X^*$.

Under the rephasing of the charged leptons and the neutrinos the effective light neutrino mass matrix transforms as

$$\begin{aligned} \nu_\alpha &\to e^{i\delta_\alpha} \nu_\alpha \\ \ell_\alpha &\to e^{i\eta_\alpha} \ell_\alpha \\ M_{\nu\alpha\beta} &\to e^{i(\delta_\alpha + \delta_\beta - \eta_\alpha - \eta_\beta)} M_{\nu\alpha\beta}\,. \end{aligned} \tag{7.90}$$

While the mixing matrix is unitary, the mass matrix is symmetric. Writing the elements of the mass matrix $M_\nu$ as $m_{\alpha\beta}$, we shall construct rephasing invariant measures in terms of $m_{\alpha\beta}$.

Any rephasing invariant quadratic terms are of the form $m^*_{\alpha\beta} m_{\alpha\beta} = |m_{\alpha\beta}|^2$ and, hence, real. Thus, the simplest rephasing invariant measure we can consider is quartic terms

$$\mathcal{I}_{\alpha\beta\eta\phi} = m_{\alpha\beta} m_{\eta\phi} m^*_{\alpha\phi} m^*_{\eta\beta}\,. \tag{7.91}$$

Any $n \times n$ symmetric matrix has $n(n+1)/2$ independent phases. But by appropriate rephasing, $n$ independent phases can be removed leaving $n(n-1)/2$ independent phases.

To find the minimal set of the rephasing invariants, we make use of the transitive and conjugation properties

$$\mathscr{I}_{\alpha\beta\rho\phi}\mathscr{I}_{\rho\beta\eta\phi} = |m_{\rho\beta}m_{\rho\phi}|^2\mathscr{I}_{\alpha\beta\eta\phi}$$
$$\mathscr{I}_{\alpha\beta\eta\rho}\mathscr{I}_{\alpha\rho\eta\phi} = |m_{\alpha\rho}m_{\eta\rho}|^2\mathscr{I}_{\alpha\beta\eta\phi}$$

and

$$\mathscr{I}_{\alpha\beta\eta\phi} = \mathscr{I}^*_{\alpha\phi\eta\beta} = \mathscr{I}_{\eta\phi\alpha\beta} = \mathscr{I}^*_{\eta\beta\alpha\phi}\,. \tag{7.92}$$

Using these relations we can express any of the measures in terms of the invariants $\mathscr{I}_{\eta\phi\alpha\alpha}$ as

$$\mathscr{I}_{\rho\beta\eta\phi} = \frac{\mathscr{I}_{\rho\beta\alpha\alpha}\mathscr{I}_{\eta\phi\alpha\alpha}\mathscr{I}^*_{\phi\rho\alpha\alpha}\mathscr{I}^*_{\eta\beta\alpha\alpha}}{|m_{\alpha\alpha}|^4|m_{\rho\alpha}m_{\beta\alpha}m_{\eta\alpha}m_{\phi\alpha}|^2}\,, \tag{7.93}$$

where $\rho,\beta,\eta,\phi \neq \alpha$ and $\alpha = 1,2,...,n$, where $n$ is the number of generations. We can also express any quartics of the form $\mathscr{I}_{\rho\phi\alpha\alpha}$ in terms of quartics such as $\mathscr{I}_{\beta\beta\alpha\alpha}$ as

$$\text{Im}\,[\mathscr{I}_{\rho\phi\alpha\alpha}] = -\text{Im}\,[\mathscr{I}_{\rho\alpha\alpha\phi}] = -\frac{\text{Im}[\mathscr{I}_{\rho\rho\alpha\alpha}\cdot\mathscr{I}_{\alpha\alpha\phi\phi}\cdot\mathscr{I}_{\rho\rho\phi\phi}]}{\text{Re}\,[\mathscr{I}_{\rho\alpha\alpha\phi}]\,(|m_{\rho\rho}|^2\,|m_{\phi\phi}|^2)}. \tag{7.94}$$

When there are texture zeroes, this analysis has to be extended to include some other quartics. In the absence of any zeroes, the *CP* violating measures can be defined as

$$I_{\alpha\beta} = \text{Im}\,[\mathscr{I}_{\alpha\alpha\beta\beta}] = \text{Im}\,[m_{\alpha\alpha}m_{\beta\beta}m^*_{\alpha\beta}m^*_{\beta\alpha}], \qquad (\alpha < \beta) \tag{7.95}$$

For a given neutrino mass matrix, we can infer if there is *CP* violation by just observing it. Since

$$I_{\alpha\beta} = I_{\beta\alpha} \quad \text{and} \quad I_{\alpha\alpha} = 0,$$

we can conclude that there are $n(n-1)/2$ independent measures for $n$ generations.

The rephasing invariants $I_{\alpha\beta}$ gives the total numbers of *CP* violating measures, including both Dirac and Majorana type phases. If we now define the mass-squared matrix

$$\tilde{M} = (M_\nu^\dagger\, M_\nu) = (M_\nu\, M_\nu^\dagger)^*, \tag{7.96}$$

with $\tilde{m}_{\alpha\beta}$ as its element, then

$$J_{ijk} = \text{Im}\left[\tilde{m}_{ij}\,\tilde{m}_{jk}\,\tilde{m}_{ki}\right] \qquad (i \neq j \neq k) \tag{7.97}$$

is the measure of *CP* violation that tells us if any lepton number conserving processes can have *CP* violation. This quantity $J_{ijk}$ vanishes when $J_{CP}$ vanishes, and in this case, there is no *CP* violation in lepton-number conserving processes such as neutrino oscillations.

# Part III

# Grand Unification and Supersymmetry

# 8

# *Grand Unified Theory*

The standard model of the strong and the electroweak interactions is a spontaneously broken gauge theory with at least 18 parameters. During the electroweak phase transition, the Higgs doublet $\phi$ acquires a *vev* breaking the symmetry group

$$SU(3)_c \times SU(2)_L \times U(1)_Y \xrightarrow{m_W} SU(3)_c \times U(1)_Q .$$

The gauge coupling constants corresponding to the three groups, $g_3, g_2$ and $g_Y$ are three parameters of the theory. Other parameters include the six quark masses, three mixing angles, and one *CP* violating phase in the quark mixing matrix appearing in the charged currents, three charged lepton masses and the mass and the quartic coupling constant of the Higgs scalar. When the standard model is extended to include neutrino masses, the number of free parameters becomes 27. The additional parameters are three neutrino masses, three mixing angles, and one *CP* phase in the leptonic sector. If neutrinos are Majorana particles, there will be two additional Majorana *CP* phases.

One natural extension of the standard model is to consider a grand unified theory, in which all three groups will be unified [124]. There will be only one unified gauge group with only one coupling constant [125]. At some higher energy, which is the scale of unification $M_U$, the grand unified group will break down to the standard model

$$\mathcal{G}_U \xrightarrow{M_U} SU(3)_c \times SU(2)_L \times U(1)_Y .$$

At the unification scale, the gauge coupling constant of the unified group will be the same as the gauge coupling constants of the standard model. As we come down to lower energies, the three gauge coupling constants evolve differently, and we arrive at the standard model near the electroweak symmetry breaking scale. This gauge coupling evolution determines the scale of unification.

Another motivation for grand unification is to treat the quarks and leptons in the same footing at higher energies by putting them in the same representation of the unification group. In the first attempt of unification, lepton was considered as a fourth color in a partially unified $SU(4)_c \times SU(2)_L \times SU(2)_R$ gauge theory. This quark–lepton unification implies baryon and lepton number violation and, hence, predicts proton decay [59]. The quark–lepton unified theory breaks down to the standard model through Higgs mechanism at some high scale $M_X$. Another interesting symmetry at higher energies could be the left–right symmetric theory with the gauge group $SU(3)_c \times SU(2)_L \times SU(2)_R \times U(1)_{B-L}$. In the left–right symmetric models, parity could be made an exact symmetry. Thus, spontaneous breaking of the left–right symmetry could also be associated with spontaneous breaking of parity [59, 60]. Another interesting feature of the left–right symmetric models is that the difference between the baryon and lepton number becomes a gauge symmetry. Thus, neutrino masses or the matter–antimatter asymmetry in the universe could also be related to the left–right symmetry breaking.

There are several possible grand unified theories depending on the unification gauge group and the symmetry breaking pattern with differing predictions. While some of the possibilities are ruled out by present experiments, none of the predictions has been verified so far. One major uncertainty with all grand unified theories is that the scale of unification is close to the Planck scale when gravity becomes strong. Thus, nonrenormalizable interactions coming from gravity can modify many of the predictions. However, general features of the grand unified theories remain unaffected.

## 8.1 $SU(5)$ Grand Unified Theories

Our main purpose in constructing a grand unified theory is to unify the semi-simple gauge group of the standard model $SU(3)_c \times SU(2)_L \times U(1)_Y$ into a simple gauge group $\mathcal{G}$, so the theory contains only one gauge coupling constant $g_U$ corresponding to the unified gauge group. At the unification scale ($M_U$) we have

$$g_3(M_U) = g_2(M_U) = g_1(M_U) = g_U(M_U).$$

The rank of the standard model gauge group is 4, so the unification group $\mathcal{G}$ must have at least 4 diagonal generators. The simplest grand unified theory with the rank of the unification group $\mathcal{G}$ to be more than 4, which can accommodate all the known fermions, is based on a gauge group $SU(5)$ [124].

In the standard model the first generation contains 15 fermions, the left-handed up and down quarks of three flavors ($u_L^{1,2,3}$ and $d_L^{1,2,3}$), the right-handed up and down quarks of three flavors ($u_R^{1,2,3}$ and $d_R^{1,2,3}$), the left-handed neutrinos ($\nu_{eL}$), and the left-handed and right-handed electrons ($e_L, e_R$). All of these particles have to be put in some representations of the group $SU(5)$.

Since the unification group $\mathcal{G}$ must commute with the Lorentz group, the left-handed and right-handed particles should belong to different representations of the group. In any grand unified theory, it is thus convenient to include the left-handed particles ($\psi_L$) and the $CP$ conjugates of the right-handed particles, which are the left-handed antiparticles (${\psi^c}_L = {\psi_R}^c$) in any particular representation $\mathcal{R}$, so the conjugate representation $\bar{\mathcal{R}}$ contains the $CP$ conjugates of the left-handed particles, which are the right-handed antiparticles (${\psi^c}_R = {\psi_L}^c$) and the right-handed particles ($\psi_R$). In case the left-handed fermions belong to $\mathcal{R} = \mathcal{R}_1 + \mathcal{R}_2$, the right-handed fermions will belong to $\bar{\mathcal{R}} = \bar{\mathcal{R}}_1 + \bar{\mathcal{R}}_2$. We shall now specify the representation $\mathcal{R}$ containing all the fermions of one generation. The second and third generation fermions will also belong to another representation $\mathscr{R}$ and have similar interactions. The *CPT* theorem implies that the Hermitian conjugate terms will contain the interactions of $\bar{\mathscr{R}}$, the same as that of $\mathscr{R}$, except when *CP* is violated. If there is *CP* violation, the interactions of $\mathscr{R}$ and $\bar{\mathscr{R}}$ will differ by a phase. For the discussions of the grand unified theory, we shall not include the *CP* violation and, hence, shall not discuss the Hermitian conjugate terms separately.

We first identify the representations of the grand unified theory that can include all the known fermions:

$$q_L = \begin{pmatrix} u \\ d \end{pmatrix}_L \equiv (3,2,\frac{1}{6}); \quad {u^c}_L \equiv (\bar{3},1,-\frac{2}{3}); \quad {d^c}_L \equiv (\bar{3},1,\frac{1}{3});$$

$$\ell_L = \begin{pmatrix} \nu \\ e^- \end{pmatrix}_L \equiv (1,2,-\frac{1}{2}); \quad {e^c}_L = {e^+}_L \equiv (1,1,1).$$

The transformation properties are given under the standard model gauge group, with the normalization that the electric charge is given by

$$Q = T_{3L} + Y.$$

The representations of the unified group $\mathscr{R}$ should be free of anomaly for consistency. In any $SU(n)$, $n > 3$ group, all complex representations have anomaly, so one considers two representations $\mathscr{R}_1 + \mathscr{R}_2$ with equal and opposite anomalies. It was found that the fundamental representation 5-plet and antisymmetric rank 2 tensor representation 10-plet of $SU(5)$ contribute equally to the anomaly and all the fermions could be accommodated in a combination of the two representations $\bar{5}+10$, which is anomaly free.

Since we want to embed the $SU(3)_c \times SU(2)_L \times U(1)_Y$ group in the $SU(5)$, before assigning the fermions to any representations we first try to see the decompositions of a few representations of $SU(5)$ under its $SU(3)_c \times SU(2)_L \times U(1)_Y$ subgroup:

$$\begin{aligned} \bar{5} &= (\bar{3},1,1/3) + (1,2,-1/2) \\ 10 &= (3,2,1/6) + (\bar{3},1,-2/3) + (1,1,1) \\ 24 &= (8,1,0) + (1,3,0) + (1,1,0) + (3,2,-5/6) + (\bar{3},2,5/6). \end{aligned}$$

From these decompositions, we can identify which fermions belong to $\bar{5}$ and 10, which are

$$\psi_{\bar{5}L} = \begin{pmatrix} d_1^c \\ d_2^c \\ d_3^c \\ \nu_e \\ e^- \end{pmatrix}_L \quad \psi_{10L} = \begin{pmatrix} 0 & u_3^c & -u_2^c & -u^1 & -d^1 \\ -u_3^c & 0 & u_1^c & -u^2 & -d^2 \\ u_2^c & -u_1^c & 0 & -u^3 & -d^3 \\ u^1 & u^2 & u^3 & 0 & -e^+ \\ d^1 & d^2 & d^3 & e^+ & 0 \end{pmatrix}_L . \quad (8.1)$$

The right-handed fermions and antifermions will belong to the 5 and $\overline{10}$ representations of $SU(5)$:

$$\psi_{5R} = \begin{pmatrix} d_1 \\ d_2 \\ d_3 \\ e^+ \\ -\nu_e^c \end{pmatrix}_R \quad \psi_{\overline{10}R} = \begin{pmatrix} 0 & u_3 & -u_2 & -u_1^c & -d_1^c \\ -u_3 & 0 & u_1 & -u_2^c & -d_2^c \\ u_2 & -u_1 & 0 & -u_3^c & -d_3^c \\ u_1^c & u_2^c & u_3^c & 0 & -e^- \\ d_1^c & d_2^c & d_3^c & e^- & 0 \end{pmatrix}_R . \tag{8.2}$$

All fermions can be accommodated in the anomaly free $\bar{5}+10$ representations of $SU(5)$. Right-handed neutrino $\nu_{eR}$ and its $CP$ conjugate $\nu^c_{eL}$ have not been included in these representations.

The gauge bosons of the grand unified group $SU(5)$ belong to the adjoint 24-representation, which are the 8 gluons $(G^a_\mu, a = 1...8)$ generating the group $SU(3)_c$, 3 gauge bosons $(W^i_\mu, i = 1,2,3)$ for the group $SU(2)_L$ and the gauge boson corresponding to the hypercharge $(B_\mu)$ generating the $U(1)_Y$. The remaining gauge bosons mix the quarks with the leptons and antiquarks and are denoted by

$$\mathscr{X}^\mu_{\alpha i} \equiv \begin{pmatrix} X^\mu_\alpha \\ Y^\mu_\alpha \end{pmatrix} \quad \text{and} \quad \mathscr{X}^{c\mu}_{\alpha i} \equiv \begin{pmatrix} X^{c\mu}_\alpha \\ Y^{c\mu}_\alpha \end{pmatrix}, \tag{8.3}$$

where $\alpha = 1,2,3$ are the $SU(3)_c$ and $i = 1,2$ are the $SU(2)_L$ indices. The gauge bosons $X_\mu$ and $Y_\mu$ carry color quantum numbers and belong to a doublet of $SU(2)_L$ with electric charges $-1/3$ and $-4/3$, respectively.

For convenience, we express the 24 gauge bosons of the $SU(5)$ grand unified theory $(\mathscr{A}^m_\mu, m = 1...24)$ in a matrix form:

$$\mathscr{A} = \begin{pmatrix} G_1^D & G^{12} & G^{13} & X_1^c & Y_1^c \\ G^{21} & G_2^D & G^{23} & X_2^c & Y_2^c \\ G^{31} & G^{32} & G_3^D & X_3^c & Y_3^c \\ X_1 & X_2 & X_3 & G_4^D & W^+ \\ Y_1 & Y_2 & Y_3 & W^- & G_5^D \end{pmatrix}, \tag{8.4}$$

with the diagonal elements

$$\begin{aligned} G_1^D &= \frac{1}{2}G_3 + \frac{1}{2\sqrt{3}}G_8 - \frac{B}{\sqrt{15}} \\ G_2^D &= -\frac{1}{2}G_3 + \frac{1}{2\sqrt{3}}G_8 - \frac{B}{\sqrt{15}} \\ G_3^D &= -\frac{1}{\sqrt{3}}G_8 - \frac{B}{\sqrt{15}} \\ G_4^D &= \frac{1}{2}W_3 - \frac{3B}{2\sqrt{15}} \\ G_5^D &= -\frac{1}{2}W_3 - \frac{3B}{2\sqrt{15}}. \end{aligned}$$

The four independent diagonal generators of $SU(5)$, $G_3, G_8, W_3$ and $B$, correspond to the two diagonal generators of $SU(3)_c$ and one each of $SU(2)_L$ and $U(1)_Y$. The gauge bosons $\mathscr{X}_\mu$ and $\mathscr{X}_\mu^c$ are the raising and lowering operators connecting the leptons (antileptons) with the antiquarks (quarks) or the quarks with the antiquarks. Their interactions are shown diagrammatically in figure 8.1, which can lead to new processes such as proton decay. This is prediction of any quark–lepton unification.

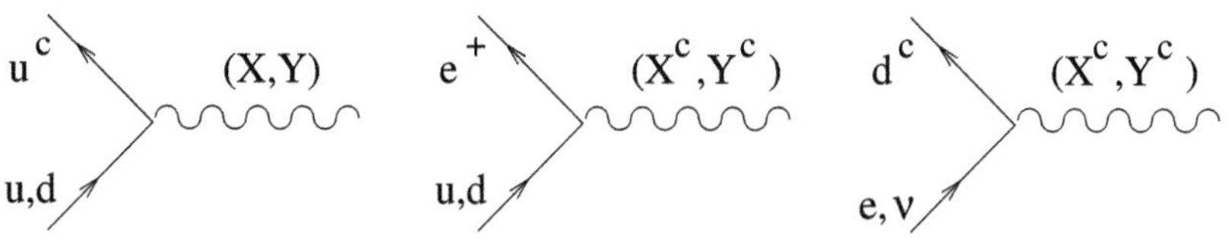


**FIGURE 8.1**
Interactions of $X_\mu$ and $Y_\mu$ gauge bosons in $SU(5)$ grand unified theory.

## 8.2 Particle Spectrum

The choice of Higgs scalar and the components that acquire *vev* determine the phenomenology in any grand unified theory. A conventional choice of symmetry breaking pattern in $SU(5)$ grand unified theory can be written as

$$\begin{aligned} SU(5) &\xrightarrow{M_U} SU(3)_c \times SU(2)_L \times U(1)_Y \\ &\xrightarrow{M_W} SU(3)_c \times U(1)_Q \,. \end{aligned} \tag{8.5}$$

At the unification scale $M_U$, the symmetry breaking does not change the rank of the group, and this symmetry breaking could be achieved by a Higgs scalar in the adjoint representation $\Sigma$ $\{24\}$.

The scalar potential for the field $\Sigma$, which acquires a vacuum expectation value (*vev*), can be written as

$$V[\Sigma] = -\frac{1}{2}m_\Sigma^2 \mathrm{Tr}\,\Sigma^2 + \frac{1}{4}a(\mathrm{Tr}\,\Sigma^2)^2 + \frac{1}{2}b\mathrm{Tr}\,\Sigma^4, \tag{8.6}$$

where $m_\Sigma^2 > 0$. The expected symmetry breaking pattern requires that the standard model singlet component of the 24-plet Higgs scalar acquires a *vev*

$$<\Sigma>= v_0 \begin{pmatrix} 1 & 0 & 0 & 0 & 0 \\ 0 & 1 & 0 & 0 & 0 \\ 0 & 0 & 1 & 0 & 0 \\ 0 & 0 & 0 & -\frac{3}{2} & 0 \\ 0 & 0 & 0 & 0 & -\frac{3}{2} \end{pmatrix} . \tag{8.7}$$

This *vev* commutes with the generators of $SU(3)$ that act on the first three elements and commutes with the generators of $SU(2)$ that act on the last two elements and one $U(1)$ subgroup of the $SU(5)$. Gauge bosons corresponding to these three groups remain massless, and the remaining gauge bosons become superheavy with mass $M_X \approx M_U$. From the quantum numbers of the different states, it is possible to identify the three groups with the standard model gauge groups. The normalization of the group $U(1)$ has to be suitably defined to identify it with the $U(1)_Y$.

The electroweak symmetry breaking requires a standard model Higgs doublet field, which can now belong to a 5-plet of $SU(5)$, $\phi\ \{5\}$, whose scalar potential can be written as

$$V[\phi] = -\frac{1}{2}m_\phi^2 \phi^\dagger \phi + \frac{1}{4}\lambda(\phi^\dagger \phi)^2. \tag{8.8}$$

$\langle m_\phi^2 \rangle > 0$ ensures a *vev* of this field, given by

$$<\phi>= \frac{v}{\sqrt{2}} \begin{pmatrix} 0 \\ 0 \\ 0 \\ 0 \\ 1 \end{pmatrix}. \tag{8.9}$$

Similar to the standard model, this Higgs scalar also gives masses to the fermions.

The usual mass term for the fermions comes from Yukawa coupling

$$\mathscr{L}_Y = f_{ij} \overline{\psi_L} \psi_R \phi + H.c. \tag{8.10}$$

When the scalar field $\phi$ acquires a *vev*, this gives the fermion masses

$$m_{ij} \overline{\psi_L} \psi_R \phi + H.c. \quad \text{with } m_{ij} = f_{ij} \langle \phi \rangle.$$

Since the left-handed and right-handed fermions belong to the conjugate representations $\Psi$ and $\bar{\Psi}$, respectively, in grand unified theories, it is convenient to use the following form of the Yukawa coupling for the fermion masses

$$\mathscr{L}_Y = f_{ij} \psi_L^T C^{-1} {\psi^c}_L \phi + H.c. \tag{8.11}$$

where $C$ is the charge conjugation matrix acting on the spinor space. Since the left-handed particles $\psi_L$ and their $CP$ conjugate left-handed antiparticles ${\psi^c}_L$ belong to the same representations $\Psi$ of any grand unified theories, the Yukawa coupling terms can be written as

$$\mathscr{L}_Y = f_{ij} \Psi_L^T C^{-1} \Psi_L \Phi + H.c. \tag{8.12}$$

where $C$ acts on both the spinor space as well as on the grand unified group space. Although this form of the Yukawa coupling appears to be similar to the Majorana mass terms, since both left-handed particles and antiparticles belong to the same representation $\Psi$, this Yukawa coupling includes both the Dirac and Majorana mass terms, if any.

The Yukawa couplings of the left-handed fermions $\psi_{\bar{5}L}$ and $\psi_{10L}$ with the Higgs scalars are given by

$$\mathscr{L}_{Yuk} = h^1_{ab} \psi^T_{\bar{5}L} C^{-1} \psi_{10L} \phi^\dagger + h^2_{ab} \psi^T_{10L} C^{-1} \psi_{10L} \phi + H.c. \tag{8.13}$$

The Hermitian conjugate terms will include all the right-handed particles and their *CP* conjugate states. Only the 5-plet of Higgs scalar can enter in the Yukawa couplings with the fermions. The first term gives masses to the down-quarks and the charged-leptons when the field $\phi$ acquires a *vev*, while the second term gives mass to the up-quarks

$$M_d = h^1_{ab} \langle \phi \rangle, \qquad M_e = h^1_{ab} \langle \phi \rangle, \qquad \text{and} \qquad M_u = h^2_{ab} \langle \phi \rangle. \tag{8.14}$$

These give mass relations at the grand unification scale $M_U$, which is then extrapolated down to the electroweak symmetry breaking scale $M_W$ using the renormalization group equations for Yukawa couplings.

If these are the only Yukawa couplings in the theory, then neutrinos are massless. So, to explain the tiny neutrino mass, one can introduce an $SU(5)$ singlet fermion $S$ and allow the couplings

$$\mathscr{L}_Y = h^S_{ab} \psi_{\bar{5}L} S \phi + M_S S S + H.c. \tag{8.15}$$

This will give a small see-saw mass to the neutrinos, ${h^S_{ab}}^2 \langle \phi \rangle^2 / M_S$. Another possibility for giving neutrino masses is to introduce a 15-plet ($\supset (1,3,-2)$ under $SU(3)_c \times SU(2)_L \times U(1)_Y$) of Higgs scalar $\xi$, which allows the Yukawa coupling

$$\mathscr{L}_Y = f_{ab} \psi^T_{\bar{5}L} C^{-1} \psi_{\bar{5}L} \xi^\dagger + H.c. \tag{8.16}$$

If the scalar potential contains a term $\mu \xi \phi \phi$, then the scalar $\xi$ can acquire a very tiny *vev*

$$\langle \xi \rangle = \mu \frac{\langle \phi \rangle^2}{m_\xi^2},$$

where both $\mu$ and $m_\xi$ are very large, of the order of the grand unification scale, so $\langle \xi \rangle$ is of the order of eV. This can, in turn, give the required neutrino masses $m_{\nu ab} = f_{ab} \langle \xi \rangle$.

Although the minimal $SU(5)$ grand unified theory is very simple and elegant, many predictions of the model are in contradiction with experiments. The neutrino mass problem is one of them. Another problem is the charged fermion mass relations. From the form of the charged fermion masses we notice that $m_b = m_\tau$, $m_\mu = m_s$ and $m_e = m_d$ at the unification scale $M_U$. We now extrapolate these relations to low energy using the renormalization group equation for the evolution of the fermion mass operators

$$m_f(m_W) = m_f(M_U) \prod_{i=1}^{3} \left( \frac{\alpha_i(m_W)}{\alpha_i(M_U)} \right)^{(\gamma_i / b_i)}, \tag{8.17}$$

where $\gamma_i = 3(N^2 - 1)/2N$ is the gamma function for the group $SU(N)$ with $i = 3, 2, 1$ corresponding to the groups $SU(3)_c$, $SU(2)_L$ and $U(1)_Y$, respectively, and we shall define the beta functions later in equation (8.37). Although the low energy prediction for the ratio $m_b/m_\tau$ is consistent, the conditions for the first and second generations $m_\mu = m_s$ and $m_e = m_d$ at the scale $M_U$ are badly in contradiction with low-energy mass ratios.

The main reason for the fermion mass problem is that the *vevs* of $\phi$ do not distinguish between quarks and leptons. An interesting mechanism to solve this problem includes a 45 Higgs scalar ($\chi_k^{ij}$, antisymmetric in $i, j$) [126]. It is then possible to give *vev* to the component

$$\langle \chi_k^{i5} \rangle = \frac{v_{45}}{\sqrt{2}}(\delta_k^i - 4\delta^{i4}\delta_{k4})$$

of $\chi_k^{ij}$, which gives unequal contributions for the quarks and the leptons. Including this field the additional Yukawa interactions are given by

$$\mathscr{L}_{Yuk} = f_{ab}^1 \psi_{\bar{5}L}\psi_{10L}\chi^\dagger + f_{ab}^2 \psi_{10L}\psi_{10L}\chi + H.c. \tag{8.18}$$

Taking the *vevs* of $\phi$ and $\chi$, discussed above, the down quark and charged lepton masses become

$$\sqrt{2}M_d = h_{ab}^1 v + f_{ab}^1 v_{45}, \quad \text{and} \quad \sqrt{2}M_\ell = h_{ab}^1 v - 3f_{ab}^1 v_{45}, \tag{8.19}$$

where $a, b = 1, 2, 3$ are generation indices. When only $f_{22}, h_{12}$ and $h_{33}$ are nonvanishing, the down quark and charged lepton mass matrices will be given by

$$\sqrt{2}M_d = \begin{pmatrix} 0 & h_{12}^1 v & 0 \\ h_{12}^1 v & f_{22}^1 v_{45} & 0 \\ 0 & 0 & h_{33}^1 v \end{pmatrix}, \quad \sqrt{2}M_\ell = \begin{pmatrix} 0 & h_{12}^1 v & 0 \\ h_{12}^1 v & -3f_{22}^1 v_{45} & 0 \\ 0 & 0 & h_{33}^1 v \end{pmatrix}, \tag{8.20}$$

which can be diagonalized to get the relations

$$m_b = m_\tau, \quad m_\mu = 3m_s \quad \text{and} \quad 3m_e = m_d \tag{8.21}$$

at the grand unification scale. When extrapolated to low energies for comparison with experimental numbers, these relations turns out to be very good. The factors 3 enter, where they should be, which comes from the choice of the *vev* of 45-plet. Since $\chi_k^{i5}$ is traceless and antisymmetry requires $i \neq 5$, the quarks get equal contributions and the leptons get a contribution equal to the number of quarks, which is 3. This nice mechanism can be implemented in other grand unified theories such as $SO(10)$ and the factor will always come out to be 3, because there are 3 colors of quarks.

## 8.3 Proton Decay

Unification of the standard model gauge groups to a grand unified theory implies unification of quarks and leptons as well. There will then be new interactions between the quarks and the leptons, to be mediated by new gauge bosons at the grand unification scale. This is a generic feature of any grand unified theory, which leads to proton decay and baryon number violation. In $SU(5)$ grand unified theory the interactions of these gauge bosons $\mathscr{X}_\mu$ and $\mathscr{X}^c_\mu$ are given by

$$\begin{aligned}\mathscr{L} = & \, g_5 \bar{D}^c_{\alpha L} \gamma_\mu \Psi_{iL} \mathscr{X}^{c\mu}_{\alpha i} + g_5 \bar{E}^c_L \gamma_\mu Q_{\alpha i L} \mathscr{X}^{c\mu}_{\alpha i} \\ & + g_5 \bar{U}^c_{\alpha L} \gamma_\mu Q_{\beta i L} \mathscr{X}^\mu_{\gamma i} \varepsilon_{\alpha\beta\gamma} + H.c., \end{aligned} \tag{8.22}$$

where $g_5$ is the $SU(5)$ gauge coupling constant and $\varepsilon_{\alpha\beta\gamma}$ is a totally antisymmetric $SU(3)_c$ tensor with $\varepsilon_{123} = 1$. In general, the new interactions between the quarks and leptons are mediated by three types of gauge bosons; *leptoquarks* interact with a quark and a lepton, *diquarks* interact with a quark and an antiquark, and *dileptons* ineract with a lepton and an antilepton. In the $SU(5)$ grand unified theory, dileptons are absent, while the leptoquarks and diquarks are the same fields and mediate proton decay, as shown in figure 8.2.

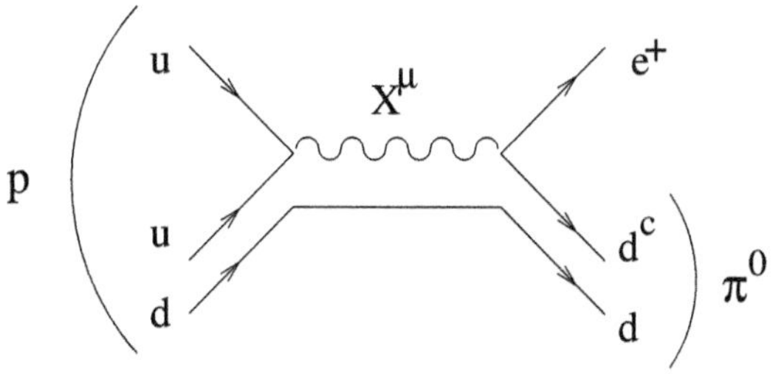


**FIGURE 8.2**
Example of gauge bosons mediating proton decay $p \to e^+\pi^\circ$.

The new gauge bosons $\mathscr{X}_\mu$ belong to $SU(5)/SU(3)_c \times SU(2)_L \times U(1)_Y$ and are superheavy with mass of the order of the unification scale ($M_X \approx M_U$). Since these heavy bosons appear in the propagator, the amplitude for proton decay of the above form ($p \to e^+\pi^0$) will be given by

$$\mathscr{A}(p \to e^+\pi^0) = \frac{\alpha_5}{M_X^2}, \tag{8.23}$$

where the $SU(5)$ gauge coupling constant $\alpha_5$ is given by

$$\alpha_5 = \frac{g_5^2}{4\pi}. \tag{8.24}$$

The proton lifetime is then given by

$$\tau_p \approx \frac{M_X^4}{\alpha_5^2 m_p^5}, \tag{8.25}$$

where $m_p$ is the mass of the proton. The unification scale and the coupling constant at the unification scale can be obtained from an analysis of the evolution of the gauge coupling constants using the renormalization group equations and applying the boundary condition of coupling constant unification. We shall discuss this in the next section. For the minimal $SU(5)$ grand unified theory, the prediction for the proton lifetime (for the decay mode $p \to e^+\pi^0$) comes out to be [125]

$$10^{30} \text{ years} < \tau_p < 10^{31} \text{ years}, \tag{8.26}$$

which is ruled out by the present experimental limit of $\tau_p > 1.6 \times 10^{33}$ years on the proton lifetime [127]. However, some extensions such as inclusion of gravity effects or the supersymmetric version of the model may still be consistent, so we shall continue with the discussions with $SU(5)$ grand unified theory.

In order to study different aspects of proton decay, including various decay modes and the choice of Higgs scalars, we need to know more about the grand unified theories and the interactions of the Higgs scalars. There is also an elegant formalism to study these details starting from our knowledge of the standard model. It is possible to construct effective higher-dimensional operators, consistent with the standard model, which can provide us some important information about the proton decays [54] without any knowledge about the grand unified theories. We only assume that the origin of these operators lies in the grand unified theories, and they are obtained after integrating out all the heavy degrees of freedom in the theory.

A baryon-number violating operator must contain quarks (denoted by $Q$ here) and the $SU(3)_c$ invariance with nonzero baryon number would require at least three quarks ($QQQ$). Lorentz invariance allows only even number of fermions, so the simplest form of the operators is $QQQL$ or $QQQL^c$, where $L$ represents a lepton. These nonrenormalizable four-fermion effective operators have mass-dimension six and hence they must have effective coupling constant suppressed by a factor of $M^{-2}$, where $M$ is the heavy mass scale in the theory of the order of the grand unification scale $M \sim M_U$. The proton lifetime, thus, comes out to be

$$\tau_p \approx \text{Const.}\ \frac{M^4}{m_p^5}, \tag{8.27}$$

which is similar to the one obtained in the $SU(5)$ model.

There are six possible form of the dimension-six baryon-number violating operators, which are allowed by the standard model and Lorentz invariance. They are given by [54]

$$\mathcal{O}^1_{abcd} = (\bar{D}^c_{a\alpha R} U_{b\beta R})(\bar{Q}^c_{ci\gamma L} \Psi_{djL}) \varepsilon_{\alpha\beta\gamma} \varepsilon_{ij}$$

$$\mathcal{O}^2_{abcd} = (\bar{Q}^c_{ai\alpha L} Q_{bj\beta L})(\bar{U}^c_{c\gamma R} E_{dR}) \varepsilon_{\alpha\beta\gamma} \varepsilon_{ij}$$

$$\mathcal{O}^3_{abcd} = (\bar{Q}^c_{ai\alpha L} Q_{bj\beta L})(\bar{Q}^c_{ck\gamma L} \Psi_{dlL}) \varepsilon_{\alpha\beta\gamma} \varepsilon_{ij} \varepsilon_{kl}$$

$$\mathcal{O}^4_{abcd} = (\bar{Q}^c_{ai\alpha L} Q_{bj\beta L})(\bar{Q}^c_{ck\gamma L} \Psi_{dlL}) \varepsilon_{\alpha\beta\gamma} (\tau^m \varepsilon)_{ij} (\tau^m \varepsilon)_{kl}$$

$$\mathcal{O}^5_{abcd} = (\bar{D}^c_{a\alpha R} U_{b\beta R})(\bar{U}^c_{c\gamma R} E_{dR})\varepsilon_{\alpha\beta\gamma}$$

$$\mathcal{O}^6_{abcd} = (\bar{U}^c_{a\alpha R} U_{b\beta R})(\bar{D}^c_{c\gamma R} E_{dR})\varepsilon_{\alpha\beta\gamma}. \tag{8.28}$$

Here, $\alpha, \beta, \gamma = 1,2,3$ are $SU(3)_c$ indices; $i, j = 1,2$ are $SU(2)_L$ doublet indices; $m = 1,2,3$ is a $SU(2)_L$ triplet index; and $a,b,c,d = 1,2,3$ are generation indices. We exploioted the Fierz transformations to transform the vector and tensor operators into these forms.

All of these operators are of the form $QQQL$, and hence, $B-L$ is conserved, since quarks carry baryon number $1/3$ and leptons carry lepton number $L = 1$. To the lowest order only the $B-L$ conserving baryon number violation will be allowed in all grand unified theories. The $B-L$ violating baryon number violating operators are of the form $QQQL\phi$, where $\phi$ is the standard model Higgs doublet scalar. This operator will have an additional suppression factor of $\langle\phi\rangle/M$ and the decay mode of the proton will be suppressed by about 25 orders of magnitude. These operators will not allow processes where an $s$-quark state is created, implying that processes such as $p \to e^+\overline{K}^0$ or $n \to e^+\overline{K}^-$ with $\Delta S = \Delta B$ are forbidden. These operators also relate several decay channels of the proton.

At this stage we have not included any intermediate scale in the theory, which is valid for the $SU(5)$ grand unified theory. But in grand unified theories with larger gauge groups such as $SO(10)$, there are intermediate symmetry breaking scales, which may allow some of the higher-dimensional operators to dominate over the dimension-6 operators. For example, an operator of the form $QQQL\Delta$ may be suppressed by $M_I^{-3}$, where $M_I$ is some intermediate symmetry breaking scale and it is also possible that $\langle\Delta\rangle \sim M_I$. There are no such processes, but $SO(10)$ grand unified theory allows $B-L$ violating neutron–antineutron oscillations [61] and $B-L$ violating three lepton decay modes of proton [128] that are comparable with $B-L$ conserving proton decays [129].

## 8.4 Coupling Constant Unification

In grand unified theories the strong, the weak and the electromagnetic interactions are unified into a single unified theory. So, the three gauge coupling constants $g_3, g_2$ and $g_1$, corresponding to the groups $SU(3)_c$, $SU(2)_L$ and $U(1)_Y$, also should get unified at the unification scale $M_U$, and we should have

$$g_3(M_U) = g_2(M_U) = g_1(M_U) = g_5(M_U). \tag{8.29}$$

If we now use the renormalization group equations to study the evolution of the different gauge coupling constants, it will be possible to estimate the unification scale $M_U$ and the gauge coupling constant at the unification scale $g_5$.

When the symmetry breaking takes place at the scale of grand unification, the standard model gauge group emerges from the unified $SU(5)$ group. This requires normalization of the generators of the different groups in the same way:

$$\text{Tr}\,[T_i T_j] = N\delta_{ij}, \tag{8.30}$$

where N is a constant, which has to be the same for the unification group and all of its subgroups. We choose $N = 1/2$ for the fundamental representation of any $SU(n)$ group and then normalize the $U(1)$ subgroup by comparing its normalization with any complete representations of $SU(5)$.

Consider the $\bar{5}$ representation of $SU(5)$, which contains the down antiquarks ${d^c}_L$ with hypercharge $Y = 1/3$ and the lepton doublets containing $e_L$ and $\nu_L$ with $Y = -1/2$. The $U(1)_Y$ normalization $Y_N$ must be the same as the normalization of $\bar{5}$, which is $N = \frac{1}{2}$. Writing the normalizations of the components of $\bar{5}$

$$\text{Tr}\,(Y^2) = \left[3 \times Y_{d^c}^2 + \times Y_e^2 + Y_\nu^2\right] = \left[3 \times \frac{Y_N^2}{9} + 2 \times \frac{Y_N^2}{4}\right] = \frac{5}{6}\,Y_N^2, \tag{8.31}$$

we can get the $U(1)_Y$ normalization to be

$$Y_N = \sqrt{\frac{3}{5}}. \tag{8.32}$$

The corresponding normalization of the $U(1)_Y$ coupling constant becomes

$$g'^2 = \frac{3}{5} g_1^2. \tag{8.33}$$

At the unification scale the weak mixing angle then becomes

$$\sin^2\theta_w = \frac{g'^2}{g_2^2 + g'^2} = \frac{3}{8}. \tag{8.34}$$

We shall now demonstrate how the different gauge coupling constants evolve with energy and how we arrive at the gauge coupling unification.

We start with the experimental values of $\sin^2\theta_W$ and $\alpha_s$ at the electroweak scale $m_W$, obtained from the measurements of the Z-mass and Z-width at LEP and also the jet cross-sections and energy–energy correlations. Using the fine structure constant at the electroweak scale $\alpha_{em}(m_W) = 1/127.9$, along with the values of $\sin^2\theta(m_W) = .2334$ and $\alpha_3(m_W) = .118$ from the high precision LEP data, we can write the values of the three coupling constants at $m_W$ as

$$\begin{aligned}
\alpha_1^{-1}(m_W) &\equiv \frac{3}{5}\alpha^{-1}(m_W)\cos^2\theta_w(m_W) = 58.83 \\
\alpha_2^{-1}(m_W) &\equiv \alpha^{-1}(m_W)\sin^2\theta_w(m_W) = 29.85 \\
\alpha_3^{-1}(m_W) &\equiv \alpha_s^{-1}(m_W) = 8.47,
\end{aligned} \tag{8.35}$$

where $\alpha_i = g_i^2/4\pi$. The evolution of these coupling constants with energy is governed by the renormalizable group equations

$$\mu \frac{dg_i}{d\mu} = \beta_i(g_i) = b_i \frac{g_i^3}{16\pi^2}, \tag{8.36}$$

where $g_i, i = 3, 2, 1$ are the gauge coupling constants for the groups $SU(3)_c$, $SU(2)_L$ and $U(1)_Y$, respectively. The one-loop beta functions are given by

$$\beta_i(g_i) = \frac{g_i^3}{16\pi^2}\left[-\frac{11}{3}\mathrm{Tr}\,[T_a^2] + \frac{2}{3}\sum_f \mathrm{Tr}\,[T_f^2] + \frac{1}{6}\sum_s \mathrm{Tr}\,[T_s^2]\right], \tag{8.37}$$

where $T$'s are generators of different representations of the groups, to which the gauge bosons, fermions and scalars belong. The quadratic Casimir invariants Tr $[T_i^2]$ are given by equation (8.30). The gauge boson contribution vanishes for any $U(1)$ group, but for an $SU(n)$ group it is Tr $[T_a^2] = n$ since the gauge bosons belong to the adjoint representation of the group.

There are four quarks, the left-handed and right-handed up and down quarks, which belong to the fundamental triplet representation of $SU(3)_c$. For the fundamental representations of any $SU(n)$ group, Tr $[T_iT_i] = 1/2$. Since the leptons do not contribute to the $SU(3)_c$ beta function, the total contribution of the fermions to the $SU(3)_c$ beta function is 2. Similarly, for the $SU(2)_L$ contribution to the beta function, we consider all the $SU(2)_L$ doublets, which are three colored left-handed quarks and the left-handed leptons. Since the contributions of the fundamental doublet representation is $1/2$, the total $SU(2)_L$ contribution to the beta function is also 2. If we now consider the normalization of the $U(1)_Y$ group as discussed above, the total contribution again comes out to be 2.

If we include the contribution of the $n_h$ numbers of standard model doublet Higgs scalars to the beta function, which are 0, $n_h$ and $3n_h/5$ for the groups $SU(3)_c$, $SU(2)_L$ and $U(1)_Y$, we can write the solutions to the renormalizable group equations in the one-loop approximation as

$$\begin{aligned}
\alpha_3^{-1}(M_U^2) &= \alpha_3^{-1}(m_W^2) + \frac{(11 - \frac{4}{3}n_g)}{4\pi}\log\left(\frac{M_U^2}{m_W^2}\right) \\
\alpha_2^{-1}(M_U^2) &= \alpha_2^{-1}(m_W^2) + \frac{(\frac{22}{3} - \frac{4}{3}n_g - \frac{1}{6}n_h)}{4\pi}\log\left(\frac{M_U^2}{m_W^2}\right) \\
\alpha_1^{-1}(M_U^2) &= \alpha_1^{-1}(m_W^2) + \frac{(-\frac{4}{3}n_g - \frac{1}{10}n_h)}{4\pi}\log\left(\frac{M_U^2}{m_W^2}\right),
\end{aligned} \tag{8.38}$$

where $n_g = 3$ is the number of generations of fermions.

The evolution of the three gauge coupling constants $\alpha_1^{-1}, \alpha_2^{-1}$ and $\alpha_3^{-1}$ are plotted as a function of energy in figure 8.3. They do not meet at a point, even including the errors in the measurements of the coupling constants. This result is valid even when the two-loop contributions are included and hence there is no grand unification

[130]. This rules out the minimal $SU(5)$ grand unified theory as well as any other grand unified theory in which there are no intermediate symmetry breaking scales and there are no extra particles. Before the precision measurements at LEP, the uncertainty in the coupling constants allowed unification [125] for the $SU(5)$ group at a scale $6 \times 10^{14}$ GeV predicting a proton lifetime of $10^{30}$ years, which has also been ruled out by the present experimental bounds on proton lifetime.

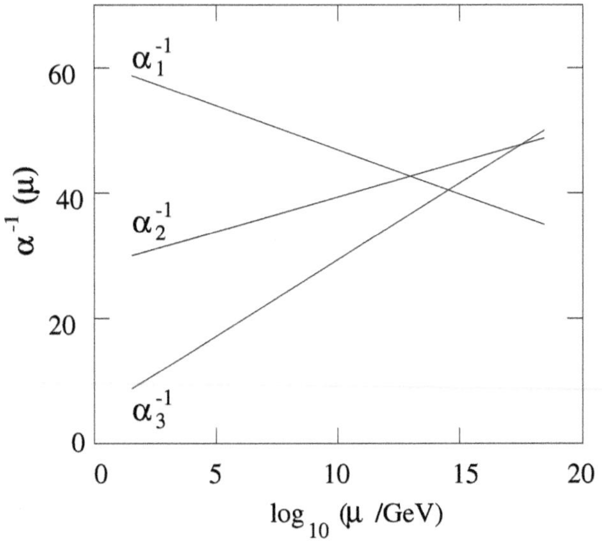


**FIGURE 8.3**
Evolution of the gauge coupling constants in the $SU(5)$ grand unified theory. In the minimal model they do not meet at a point.

If there are intermediate symmetry breaking scales or new particles with masses above the electroweak scale, the gauge coupling constants meet at a point and the unification scale comes out to be usually around $m_U \sim 10^{16}$ GeV. The gauge coupling unification, neutrino mass, fermion mass relations and proton decay problem can also be solved when gravity effects are included or some new physics originating at the Planck scale is considered.

### Gravity Effects

From the coupling constant unification it is possible to predict the unification scale, which comes out to be very close to the Planck scale, where gravity becomes strong. So, although we are extrapolating our low-energy theory to the unification scale, the results may get modified due to effects of gravitational interaction, which could be very large since gravity allows nonrenormalizable interactions. There are some phenomenological studies to show how some nonrenormalizable interactions, which can be induced by gravity, could solve almost all the problems of $SU(5)$ grand unified theory [131, 132, 133]. Since the nonrenormalizable terms are not calculable, the parameters in this analysis are arbitrary. The main idea behind this analysis is to see if, for some range of the parameters, these interactions can make the $SU(5)$ grand unified theory consistent with the gauge coupling unification [131].

Since the nonrenormalizable terms cannot be derived from our present understanding of quantum gravity, a phenomenological Lagrangian is considered:

$$\mathscr{L} = \mathscr{L}_0 + \sum_{n=1} \mathscr{L}^{(n)}, \tag{8.39}$$

where $\mathscr{L}_0$ is the usual renormalizable $SU(5)$ Lagrangian and effects of nonrenormalizable terms of different order are considered in the second term. All of these terms should be invariant under the group $SU(5)$, so the possible dimension-5 and dimension-6 terms can be written as

$$\mathscr{L}^{(1)} = -\frac{1}{2}\frac{\eta^{(1)}}{M_{Pl}}\mathrm{Tr}(F_{\mu\nu}\Sigma F^{\mu\nu}) \tag{8.40}$$

$$\begin{aligned}\mathscr{L}^{(2)} = -\frac{1}{2}\frac{1}{M_{Pl}^2}\Big[&\eta_a^{(2)}\{\mathrm{Tr}(F_{\mu\nu}\Sigma^2F^{\mu\nu}) + \mathrm{Tr}(F_{\mu\nu}\Sigma F^{\mu\nu}\Sigma)\}\\ &+ \eta_b^{(2)}\mathrm{Tr}(\Sigma^2)\mathrm{Tr}(F_{\mu\nu}F^{\mu\nu}) + \eta_c^{(2)}\mathrm{Tr}(F^{\mu\nu}\Sigma)\mathrm{Tr}(F_{\mu\nu}\Sigma)\Big],\end{aligned} \tag{8.41}$$

where $\eta^{(n)}$ specify the couplings of the higher-dimensional operators and $\Sigma$ is the 24-plet of Higgs scalars whose *vev* breaks $SU(5)$.

Once we replace the scalar field $\sigma$ by its *vev*, there will be effective dimension-4 terms that break $SU(5)$ but invariant under $SU(3)_c \times SU(2)_L \times U(1)_Y$. The resulting Lagrangian becomes

$$\begin{aligned}\mathscr{L}_{321} = &-\frac{1}{2}(1+\varepsilon_3)\mathrm{Tr}(F_{\mu\nu}^{(3)}F^{(3)\mu\nu}) - \frac{1}{2}(1+\varepsilon_2)\mathrm{Tr}(F_{\mu\nu}^{(2)}F^{(2)\mu\nu})\\ &-\frac{1}{2}(1+\varepsilon_1)\mathrm{Tr}(F_{\mu\nu}^{(1)}F^{(1)\mu\nu}).\end{aligned} \tag{8.42}$$

This is the usual $SU(3)_c \times SU(2)_L \times U(1)_Y$ invariant Lagrangian, except the physical gauge fields are now scaled by constant factors $(1+\varepsilon_i)$ $i = 3,2,1$. If we now define the physical gauge fields including the scaling factors

$$A_i' = A_i(1+\varepsilon_i)^{\frac{1}{2}}, \tag{8.43}$$

we recover the usual $SU(3)_c \times SU(2)_L \times U(1)_Y$ invariant Lagrangian with modified coupling constants

$$g_i^2(M_U) = \bar{g}_i^2(M_U)(1+\varepsilon_i)^{-1}. \tag{8.44}$$

We measure the coupling constants at low energy $g_i$, and they then evolve with energy and, get scaled $\bar{g}_i$ by the nonrenormalizable interactions near the unification scale. The boundary condition at the unification scale now reads

$$\bar{g}_3^2 = \bar{g}_2^2 = \bar{g}_1^2 = g_0^2. \tag{8.45}$$

So the coupling constants that are unified are not the same as the low-energy constants we measure. When we evolved the low energy gauge couplings, even if they do not meet at a point, that may not be interpreted as lack of unification.

It is possible to include the dimension-5 and dimension-6 terms and achieve unification at a scale as high as $10^{17}$ GeV. Thus, the problem of gauge unification and proton decay are simultaneously solved in the minimal $SU(5)$ grand unified theory in the presence of gravity induced terms [132]. The fermion mass relations as well as the neutrino mass problem can also be solved if gravity effects are included [133]. Consider the fermion mass problem where it is possible to have an effective gravity induced term $(1/M_{Pl})\bar{\psi}_L\psi_R\phi\Sigma$, which contains an effective 45-plet ($\subset 5\times 24$) of Higgs scalar and can solve the fermion mass relations. The small Majorana neutrino mass can come from an effective term $\phi\phi/M_{Pl}$. The quantum gravity effects can, in principle, cure all the problems of the minimal $SU(5)$ grand unified theory. On the other hand, predictions of any grand unified theory could get modified due to the gravity effects.

## 8.5 $SO(10)$ GUT

The grand unified theory based on the gauge group $SO(10)$ has several new interesting features [134] compared with the $SU(5)$ grand unified theory. An $SO(10)$ group is a rank-5 group, and there is one additional diagonal generator, which allows some interesting intermediate symmetry breaking stages before the group reduces to the standard model gauge group $SU(3)_c \times SU(2)_L \times U(1)_Y$. One of the most interesting intermediate symmetries is the left–right symmetry, in which parity is broken spontaneously [59, 60]. The difference between the baryon and lepton number $B-L$ is also a gauge symmetry [129] in the theory and hence $B-L$ is also spontaneously broken. As a result, there are new $B-L$ violating processes, which may have observable predictions such as neutron–antineutron oscillations, neutrinoless double beta decay or $B-L$ violating proton decay modes, making the model even more interesting. Tiny neutrino masses are naturally explained and gauge coupling constants meet at a point for some range of the intermediate symmetry breaking scale.

The $SO(10)$ group can descend to the standard model through different intermediate symmetries, and depending on the choice of Higgs scalars, many versions of the theory are possible. If the group $SO(10)$ breaks down to the standard model directly at the scale of grand unification, the model will inherit some of the problems of the $SU(5)$ grand unified theory. We shall discuss only some salient features of the $SO(10)$ grand unified theory, with one or two intermediate symmetry breaking scales.

### The Model

The $SO(10)$ group is an orthogonal group with a 10-dimensional fundamental or vector representation. The generators of the group are $10\times 10$ antisymmetric matrices of dimension 45. There is a 16-dimensional spinor representation with opposite helicity compared with its conjugate $\overline{16}$ spinor representation. All the known fermions that belong to $\bar{5}+10$ of $SU(5)$ and the $CP$ conjugate of the right-handed neutrino that could be accommodated in a singlet of $SU(5)$, can be accommodated in the spinor representation of $SO(10)$:

$$\Psi_L = \begin{pmatrix} u_1 & u_2 & u_3 & \nu_e \\ d_1 & d_2 & d_3 & e^- \end{pmatrix}_L + \begin{pmatrix} u_1^c & u_2^c & u_3^c & \nu_e^c \\ d_1^c & d_2^c & d_3^c & e^+ \end{pmatrix}_L . \tag{8.46}$$

The right-handed fermions can then be identified with the components of a $\overline{16}$ representation

$$\overline{\Psi}_R = \begin{pmatrix} u_1^c & u_2^c & u_3^c & \nu_e^c \\ d_1^c & d_2^c & d_3^c & e^+ \end{pmatrix}_R + \begin{pmatrix} u_1 & u_2 & u_3 & \nu_e \\ d_1 & d_2 & d_3 & e^- \end{pmatrix}_R . \tag{8.47}$$

We have written this particular form with a motivation. The leptons can now be identified as a fourth color of quarks after including the right-handed neutrino, which means that we can extend the group $SU(3)_c\times SU(2)_L\times U(1)_Y \equiv \mathscr{G}[3_c-2_L-1_Y]$ to $SU(4)_c\times SU(2)_L\times SU(2)_R \equiv \mathscr{G}[4_c-2_L-2_R]$ and have quark–lepton unification [59]. This partially unified group $\mathscr{G}[4_c-2_L-2_R]$ has rank 5 and is the maximal subgroup of $SO(10)$. The decomposition of the 16-spinor containing the fermions under the group $\mathscr{G}[4_c-2_L-2_R]$ can be written as

$$16 = (4,2,1)+(\bar{4},1,2) \quad \text{and} \quad \overline{16} = (\bar{4},2,1)+(4,1,2). \tag{8.48}$$

One of the popular and interesting intermediate symmetry breaking chain of $SO(10)$ group is via the partially unified group and then the left–right symmetric group

$$\begin{aligned}
SO(10) &\xrightarrow{M_U} SU(4)_c\times SU(2)_L\times SU(2)_R \qquad \mathscr{G}[4_c-2_L-2_R] \\
&\xrightarrow{M_1} SU(3)_c\times SU(2)_L\times SU(2)_R\times U(1)_{(B-L)} \\
&\qquad\qquad\qquad\qquad \mathscr{G}[3_c-2_L-2_R-1_{(B-L)}] \\
&\xrightarrow{M_R} SU(3)_c\times SU(2)_L\times U(1)_Y . \qquad \mathscr{G}[3_c-2_L-1_Y] \\
&\xrightarrow{m_W} SU(3)_c\times U(1)_Q \qquad\qquad \mathscr{G}[3_c-1_Q] .
\end{aligned}$$

The partially unified group

$$\mathscr{G}[4_c-2_L-2_R] \equiv G_{PS}$$

is also called the Pati–Salam group and we denote the left–right symmetric group as

$$\mathscr{G}[3_c-2_L-2_R-1_{(B-L)}] \equiv G_{LR}$$

and the standard model symmetry group as

$$\mathcal{G}[3_c - 2_L - 1_Y] \equiv G_{std} \quad \text{and} \quad \mathcal{G}[3_c - 1_Q] \equiv G_Q.$$

The $U(1)$ subgroup of $SU(4)_c$ has been identified with the $B-L$ quantum number [129], which is true for all the known fermions. But for scalars or some new exotic fermions this may not be true. However for the present discussion this identification is consistent.

$B-L$ is a gauge symmetry in the conventional left–right symmetric model. In these models the hypercharge and the electric charge are related to the $B-L$ and the $SU(2)_R$ quantum number by the relation

$$Q = T_{3L} + T_{3R} + \frac{B-L}{2} = T_{3L} + Y. \tag{8.49}$$

Thus, any $B-L$ violating process can occur only after the $B-L$ symmetry is spontaneously broken.

The gauge bosons belong to the adjoint 45-dimensional antisymmetric representation of $SO(10)$, which transform under the left–right symmetric model $G_{LR}$ as

$$\begin{aligned} 45 = {} & (8,1,1,0) + (1,3,1,0) + (1,1,3,0) + (1,1,1,0) \\ & + (3,2,2,-\frac{2}{3}) + (\bar{3},2,2,\frac{2}{3}) + (3,1,1,\frac{4}{3}) + (\bar{3},1,1,-\frac{4}{3}) \,. \end{aligned} \tag{8.50}$$

The gauge bosons corresponding to the first line are the ones that generate the left–right symmetry group and remain massless until the left–right symmetry breaking scale. The remaining fields given in the second line, which are the leptoquark and diquark gauge bosons, become massive at the scale of grand unification. These gauge bosons can mediate proton decay as in the $SU(5)$ grand unified theory. However, in all these interactions $B-L$ is conserved, since it is a gauge symmetry and broken only after the left–right symmetry breaking. During the left–right symmetry breaking, the charged components $W_R^{\pm}$ of the gauge bosons $(1,1,3,0)$ and a combination of the neutral component $W_R^3$ of $(1,1,3,0)$ and $W_{BL}$ corresponding to $(1,1,1,0)$ become massive, leaving the standard model gauge bosons massless.

### Symmetry Breaking and Fermion Masses

As we discussed earlier, in any grand unified theory the Higgs scalars are very important in deciding the symmetry breaking pattern, giving mass to the fermions, evolution of the gauge coupling constants, and mediating exotic processes through their interactions. In the $SO(10)$ grand unified theory several Higgs scalars are required, since there are now intermediate symmetry breaking scales. Moreover, the rank of the group has to be reduced, so it can descend to the standard model.

There are many combinations of the Higgs scalars that can allow the different symmetry breaking chains. Each of these combinations can have different phenomenological implications. In addition to the Higgs scalars that are required for any symmetry breaking, there could be additional scalars, which only contribute in solving some problems such as fermion mass relations, neutrino mass, or baryon number nonconservation. In supersymmetric theories there are more possibilities and some Higgs scalars are required to allow symmetry breaking without breaking supersymmetry. We shall discuss a few of these issues and concentrate on the generic features of the $SO(10)$ grand unified theory.

We shall consider the symmetry breaking chain

$$SO(10) \xrightarrow{M_U} G_{LR} \xrightarrow{M_{LR}} G_{std} \xrightarrow{M_W} G_Q .$$

We have not included the Pati–Salam subgroup in the discussion, since the renormalization group analysis of the gauge coupling evolution does not allow the $G_{PS}$ to be much lower than the unification scale. Moreover, the $G_{PS}$ intermediate symmetry does not allow the new phenomenology that we plan to discuss. However, to identify different components of the scalars, we shall write the decomposition of the various fields in terms of their decomposition under the group $G_{PS}$. We start with the decomposition of some of the scalar fields:

$$\begin{aligned}
210 &= (1,1,1)+(6,2,2)+(15,3,1)+(15,1,3) \\
&\quad +(15,1,1)+(10,2,2)+(\overline{10},2,2) \\
54 &= (1,1,1)+(1,3,3)+(20',1,1)+(6,2,2) \\
45 &= (1,3,1)+(1,1,3)+(15,1,1)+(6,2,2) \\
126 &= (15,2,2)+(10,1,3)+(\overline{10},3,1)+(6,1,1) \\
120 &= (1,2,2)+(10,1,1)+(\overline{10},1,1)+((6,1,3) \\
&\quad +(6,3,1)+(15,2,2) \\
16 &= (4,2,1)+(\overline{4},1,2) \\
10 &= (6,1,1)+(1,2,2).
\end{aligned} \tag{8.51}$$

The $SO(10)$ breaking to the left–right symmetric group $G_{LR}$ could be achieved by a 210-dimensional field or a combination of the fields $45+54$. We shall consider only the scalar belonging to the 210 representation, when two of its components $(1,1,1)$ and $(15,1,1)$ acquire *vev*. Another alternative of this symmetry breaking is when the $(1,1,1)$ component of 210 or 45 acquires *vev* and breaks $SO(10) \to G_{PS}$ and subsequently the $(15,1,1)$ of 210 or 54 acquires *vev* and breaks $G_{PS} \to G_{LR}$.

The next stage of symmetry breaking is more interesting from phenomenological point of view. The left–right symmetry breaking can take place when the Higgs scalar belongs to a 126-dimensional representation or a 16-dimensional representation of $SO(10)$. The studies with these two choices of the Higgs scalar have widely different consequences. The next stage of symmetry breaking requires a Higgs scalar 10,120,126 or a 16 of $SO(10)$. Since the fermion masses come from the *vev* of these scalars, different choices of the Higgs would give different fermion mass relations.

In the simplest version of the $SO(10)$ grand unified theory the left–right symmetry breaking also breaks the parity spontaneously. This is an interesting feature of the theory that explains the origin of the low-energy parity violation. In the scalar sector, parity could be broken even before the left–right symmetry is broken. When $SO(10)$

is broken to $G_{PS}$ by the *vev* of $(1,1,1)$ belonging to a 45 or a 210 of $SO(10)$, parity is also broken since the field is odd under parity [135]. This can then cause mass splitting between scalars that are related to each other under $SU(2)_L \leftrightarrow SU(2)_R$. This is referred to as *D*-parity to distinguish it from the parity of the Lorentz group.

Another interesting feature associated with the left–right symmetric model is that the $B-L$ symmetry is gauged. Thus, $B-L$ violating interactions can appear only after this symmetry is broken spontaneously. Since $B-L$ symmetry breaking is required for giving neutrino masses and explaining the baryon asymmetry of the universe, different choices of Higgs scalars for the $B-L$ breaking can lead to different consequences. The fermion mass relations can also be affected by the $B-L$ symmetry breaking. Although the left–right symmetry could be broken with the $B-L$ symmetry, it is also possible to break the left–right symmetry at a much higher scale than the $B-L$ symmetry breaking scale. It is also possible to associate the *D*-parity violating scale with the neutrino masses, decoupling it from the $B-L$ symmetry breaking.

In the popular version of the $SO(10)$ grand unified theory, a $\overline{126}$ representation of $SO(10)$ breaks the left–right symmetry along with the $U(1)_{B-L}$. There is an $SU(2)_R$ triplet Higgs scalar transforming as $\Delta_R \equiv (1,1,3,-2)$ under $G_{LR}$, whose neutral component acquires a *vev*. Left-right *D*-parity then implies the existence of a left-handed triplet Higgs scalar $\Delta_L \equiv (1,3,1,-2)$ under $G_{LR}$ belonging to a 126 representation of $SO(10)$. For a choice of parameters in the scalar potential, it is possible to give *vev* to $\Delta_R$ but not to $\Delta_L$. However, after the electroweak symmetry breaking, $\Delta_L$ will receive an induced *vev*, which is extremely small.

At the left–right symmetry breaking scale $M_R$, we have

$$G_{LR} \xrightarrow{\langle \Delta_R \rangle} G_{std},$$

so $\langle \Delta_R \rangle \sim M_R$. The electroweak symmetry is then broken by the bi-doublet Higgs scalar $\Phi_{10} \equiv (1,2,2,0)$ belonging to a 10-plet of $SO(10)$, which is doublet under both $SU(2)_L$ and $SU(2)_R$. This field also gives masses to the fermions. At the electroweak symmetry breaking scale $M_W$, we thus have

$$G_{std} \xrightarrow{\langle \Phi_{10} \rangle} G_Q.$$

As in the $SU(5)$ grand unified theory, this scalar gives the wrong fermion mass relations. Therefore, we need another component of 126-plet Higgs, which is also a bi-doublet $\Phi_{126} \equiv (15,2,2,0)$, to contribute to the fermion masses. The 15-plet of $SU(4)_c$ contains a component, which is a singlet under $SU(3)_c \times U(1)_{B-L}$ and has the form diag$[1,1,1,-3]$, which can then give the correct fermion mass relations. The minimization of the scalar potential would now give us the relation

$$\langle \Delta_L \rangle = \text{const.} \frac{\langle \Phi_{10} \rangle^2}{\langle \Delta_R \rangle}, \tag{8.52}$$

where the constant (const.) includes parameters in the scalar potential.

The Yukawa couplings that give masses to the fermions can now be written as

$$\mathscr{L}_Y = f_{ab}\Psi_L\Psi_L\Phi_{10} + \tilde{f}_{ab}\Psi_L\Psi_L\Phi^\dagger_{126} + f_{Lab}\Psi_L\Psi_L\Delta^\dagger_L + f_{Rab}\Psi_L\Psi_L\Delta_R + H.c., \quad (8.53)$$

where $a, b = 1, 2, 3$ are the generation indices. As previously, $\Psi_L$ contains both the left-handed particles and antiparticles, and hence, the first two terms and their Hermitian conjugates give Dirac masses to all the fermions. The third term can only give Majorana masses to the left-handed neutrinos with a tiny *vev* of $\Delta_L$, while the fourth term gives large Majorana masses to the right-handed neutrinos. The light neutrino masses are given by

$$m_\nu = f_L\langle\Delta_L\rangle - \frac{f^2\langle\Phi_{10}\rangle^2}{f_R\langle\Delta_R\rangle}. \quad (8.54)$$

Both the contributions are of similar order of magnitude. These Higgs fields are listed in table 8.1.

**TABLE 8.1**
The Higgs scalars contributing to fermion masses and their transformations under the left–right symmetric group $G_{LR}$ and the standard model symmetry group $G_{std}$.

| Scalar fields | Transforms under $\mathscr{G}[3_c - 2_L - 2_R - 1_{(B-L)}]$ | Transforms under $\mathscr{G}[3_c - 2_L - 1_Y]$ |
|---|---|---|
| $\Delta_R$ | $(1,1,3,-2)$ | $(1,1,0)$ |
| $\Phi_{10}$ | $(1,2,2,0)$ | $(1,2,\pm\frac{1}{2})$ |
| $\Phi_{126}$ | $(1,2,2,0)$ | $(1,2,\pm\frac{1}{2})$ |
| $\Delta_L$ | $(1,3,1,-2)$ | $(1,3,-1)$ |

In another class of models there are no triplet Higgs scalars $\Delta_L$ or $\Delta_R$ belonging to a 126-plet of $SO(10)$. Although the main purpose of these models is to have smaller representations of Higgs scalars, these models have different consequences. In these models the electroweak symmetry breaking as well as the left–right symmetry breaking could be implemented by only 16-plet and $\overline{16}$-plet of Higgs scalars. For charged fermion masses one may include a 10-plet of Higgs scalar and give *vev* to the bi-doublet, but there are also models where the fermion masses as well as the Majorana neutrino masses come from similar dimension-5 operators from physics near the scale of grand unification. The simplest of these models breaks the left–right symmetry with a 16-plet field $\chi_R \equiv (1,1,2,-1)$ and breaks the electroweak symmetry with a 10-plet field $\Phi_{10} \equiv (1,2,2,0)$. The left–right symmetry requires a field $\chi_L \equiv (1,2,1,-1)$, which acquires a *vev* at the electroweak symmetry breaking scale. An $SO(10)$ singlet fermion $S$ has to be included, so the Majorana mass terms of $S$ can generate the see-saw Majorana masses to the left-handed and right-handed neutrinos. The fermion mass relations can also be explained by the higher order radiative contributions in these models.

### $B-L$ Violating Interactions

We shall now discuss some of the consequences of spontaneous $B-L$ violation in the $SO(10)$ grand unified theories. We shall first consider the triplet Higgs models, where the left–right symmetry $G_{LR}$ is broken by the Higgs scalar $\Delta_R \equiv (1,1,3,-2)$. The *vev* of $\Delta_R$ can then mediate $B-L$ violating processes such as neutron–antineutron oscillations [61] or three-lepton decays of the proton [128] and give Majorana masses to the right-handed neutrinos. We shall demonstrate how processes such as neutron–antineutron oscillations take place in the $SO(10)$ grand unified theory.

The leptoquark and diquark gauge bosons can take part in the baryon-number violating processes, but they have masses of the order of $M_U$, and their dominant contribution will be to $B-L$ conserving proton decays. In $SO(10)$ grand unified theory the left–right symmetry breaking Higgs scalar $\Delta_R^\dagger$ belongs to 126 of $SO(10)$. Although the component $\Delta_R$ only acquires *vev*, several other components of the same representation 126 remain light, which includes the diquarks $\Delta_{qq}$ and the leptoquarks $\Delta_{lq}$. Since the mixing between $\Delta_{qq}$ and $\Delta lq$ is strongly suppressed, these scalars will not mediate proton decay. If we now represent a 126 representation by $\Delta$, then a quartic coupling of $\Delta$ can allow $B-L$ violating processes, when $\Delta_R$ acquires *vev*.

Consider a coupling $\Delta_{qq}^3\Delta_R^\dagger$, where the diquarks take right-handed quarks to right-handed antiquarks. When $\Delta_R$ acquire a *vev*, the effective operator $\Delta_{qq}^3$ would take three quarks into three antiquarks leading to neutron–antineutron oscillations. We can write the relevant interactions as

$$\mathscr{L} = f\Delta_{qq}^3\Delta_R + f_q\overline{Q_R{}^c}Q_R\Delta_{qq} + f_R\overline{L_R{}^c}L_R\Delta_R, \tag{8.55}$$

where we have adopted the convention that

$$Q_R = \begin{pmatrix} U_R \\ D_R \end{pmatrix} \quad \text{and} \quad L_R = \begin{pmatrix} N_R \\ E_R \end{pmatrix}, \tag{8.56}$$

and $N_R$ are the right-handed neutrinos. From these interactions it is now possible to allow a transition from a neutron to an antineutron as shown in figure 8.4. The amplitude for this neutron–antineutron oscillation is given by [136]

$$\delta m_n = A\frac{ff_q^3 v_R}{m_{\Delta_{qq}}^6}, \tag{8.57}$$

where $A \sim 10^{-4}$ is a correction factor due to renormalization [137, 138], wave function at the origin [138, 139] and nuclear effects [140].

The neutron–antineutron oscillation requires an effective six quark dimension-9 effective operator $QQQQQQ$. The amplitude for the process has a suppression factor $M^{-5}$, whereas the lowest-dimensional proton decay operator has dimension-6 and is suppressed by $M^{-2}$. However, in the case of proton decay the observable is proton lifetime, which has a suppression of $M^{-4}$, whereas in case of neutron–antineutron oscillation the observable is the oscillation amplitude, which has a suppression $M^{-5}$. In addition, since the mass scales involved in the two cases are different, in some models the neutron–antineutron oscillation may be more probable than the proton decay in spite of the additional suppression coming from nuclear effects.

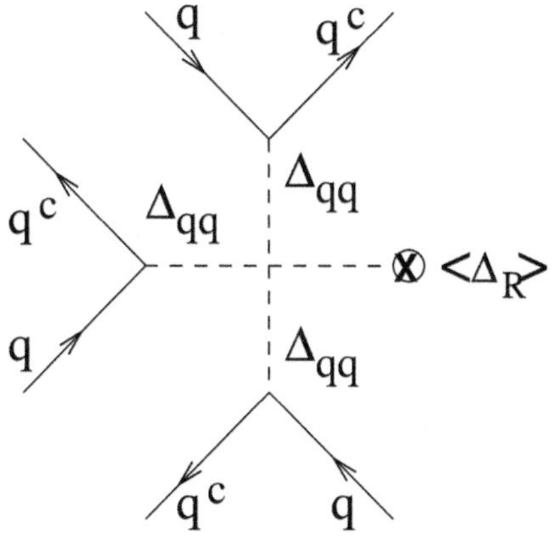


**FIGURE 8.4**
Diagram for the $n-\bar{n}$ oscillation mediated by the diquark Higgs scalars $\Delta_{qq}$.

According to the *CPT* theorem the masses of the neutron and the antineutron should be the same and we can write the mass matrix in the basis $[|n> \quad |\bar{n}>]$ as

$$M_n = \begin{pmatrix} m & \delta m \\ \delta m & m \end{pmatrix}. \tag{8.58}$$

Then the two physical states become

$$n_{\pm} = \frac{1}{\sqrt{2}}(|n> \pm |\bar{n}>), \tag{8.59}$$

with masses

$$m_{\pm} = m \pm \delta m. \tag{8.60}$$

If we start with a pure neutron beam, then after a time $t$ there will be a finite probability of finding an antineutron

$$P_{n\to\bar{n}}(t) = \sin^2 \delta m\, t. \tag{8.61}$$

For small mass difference, the oscillation period may be defined as

$$\tau_{n\bar{n}} = \frac{1}{\delta m}, \tag{8.62}$$

so the probability of finding the antineutron in the pure neutron beam becomes

$$P_{n\to\bar{n}}(t) = \left(\frac{t}{\tau_{n\bar{n}}}\right). \tag{8.63}$$

Experiments could not detect any antineutrons in a neutron beam at a distance, from which a bound on the neutron–antineutron oscillation period [141] is established $\tau_{n\bar{n}} > 0.86 \times 10^8$ secs, corresponding to a baryon-number violating lifetime of a physical neutron to be $\tau_+ > 10^{32}$ years [127].

### Gauge Coupling Unification

We demonstrated that in $SU(5)$ grand unified theory, coupling constant unification is not possible. We also commented that the result is valid for any grand unified theory with no intermediate symmetry breaking scales. We shall now show how the problem is solved and the coupling constant unification can be achieved in $SO(10)$ grand unified theory with intermediate symmetry breaking scales. Consider the symmetry breaking pattern

$$SO(10) \xrightarrow{M_U} G_{PS} \xrightarrow{M_I} G_{LR} \xrightarrow{M_R} G_{std} \xrightarrow{M_W} G_Q\,.$$

To study the evolution of the gauge coupling constants, we shall apply boundary conditions at the different symmetry breaking scales and then evolve the couplings to the next symmetry breaking scales.

The contribution of the Higgs scalars can be large when there are intermediate symmetry breaking scales, since several Higgs scalars take part in the symmetry breaking and giving fermion masses. Moreover the Higgs representations are also large compared with $SU(5)$. In considering the Higgs contributions to the renormalization group equation, we shall assume the *decoupling theorem*. Following this theorem, we shall assume that the components of any Higgs scalars that are not protected by any symmetry are heavy.

Let us start with a symmetry group $G$ and consider the symmetry breaking pattern

$$G \xrightarrow{M_1} G_1 \xrightarrow{M_2} G_2. \tag{8.64}$$

Consider a representation $\mathscr{R}$ of $G$, which decomposes under the different subgroups $G_1$ and $G_2$ as

$$\mathscr{R} \to \sum_n \mathscr{R}_n \to \sum_m \left[\sum_n \mathscr{R}_{nm}\right]. \tag{8.65}$$

Any representation $\mathscr{R}_i$ of $G_1$ can be decomposed under $G_2$ as $\sum_m \mathscr{R}_{im}$. According to the decoupling theorem, if any component $\mathscr{R}_{kl} \subset \mathscr{R}_k \subset \mathscr{R}$ acquires a *vev* at the scale $M_2$, then until $M_1$, only the component $\mathscr{R}_k$ will contribute to the evolution of the gauge coupling constants, and above $M_1$, the entire representation $\mathscr{R}$ will contribute. After the component $\mathscr{R}_{kl}$ of $G_2$ acquires a *vev* at a scale $M_2$, the $G_1$ symmetry above $M_2$ ensures that all the components of the representation $\mathscr{R}_k$ of $G_2$ have mass less than $M_2$. Similarly, above the scale $M_1$ the symmetry group is $G$ and it implies that all components of the representation $\mathscr{R}$ should have mass less than the symmetry breaking scale $M_1$.

We shall now start with the values of the couplings at the electroweak scale $m_W$, whose central values are

$$\begin{aligned}\alpha_1(m_W) &= 0.01699\\ \alpha_2(m_W) &= 0.0335\\ \alpha_3(m_W) &= 0.118.\end{aligned} \tag{8.66}$$

We shall evolve these couplings to the left–right symmetry breaking scale $M_R$ and then apply the boundary conditions at $M_R$

$$\begin{aligned}\alpha_{1Y}^{-1}(M_R) &= \frac{3}{5}\alpha_{2R}^{-1}(M_R) + \frac{2}{5}\alpha_{1(B-L)}^{-1}(M_R)\\ \alpha_{2L}^{-1}(M_R) &= \alpha_{2R}^{-1}(M_R).\end{aligned} \tag{8.67}$$

For simplicity we assume $M_I = M_U$, so the next matching condition is at the unification scale $M_U$. The evolution equations then become

$$\begin{aligned}\alpha_{1Y}^{-1}(m_W) &= \alpha_{10}^{-1}(M_U) + \left(\frac{6}{5}b_{2R} + \frac{4}{5}b_4\right)M_{U1}\\ &\quad + \left(\frac{6}{5}b_{2R} + \frac{4}{5}b_{1(B-L)}\right)M_{1R} + 2b_{1Y}M_{RW}\\ \alpha_{2L}^{-1}(m_W) &= \alpha_{10}^{-1}(M_U) + 2b_{2L}M_{U1} + 2b_{2L}M_{1R} + 2b_{2L}^{ew}M_{RW}\\ \alpha_{3c}^{-1}(m_W) &= \alpha_{10}^{-1}(M_U) + 2b_4M_{U1} + 2b_{3c}M_{1R} + 2b_{3c}M_{RW},\end{aligned} \tag{8.68}$$

where $M_{ij} \equiv 4\pi\ \ln(M_i/M_j)$ and the beta function coefficients, $b_i$'s, are given as

$$\begin{aligned}b_{1(B-L)} &= \frac{1}{(4\pi)^2}\frac{3}{2};\ b_{1Y} = \frac{1}{(4\pi)^2}\frac{2}{5};\ b_{2L} = b_{2R} = -\frac{1}{(4\pi)^2}\frac{20}{3};\\ b_{2L}^{ew} &= -\frac{1}{(4\pi)^2}\frac{41}{6};\ b_3 = -\frac{1}{(4\pi)^2}11.\end{aligned} \tag{8.69}$$

Here we have taken the number of fermion families $n_f = 3$. The gauge coupling constants now meet at a point as long as we consider $M_R > 10^{11}$ GeV. This scale of left–right symmetry breaking is also interesting from the point of view of neutrino masses and generating a baryon asymmetry of the universe. We shall elaborate on these points at a later stage. Now depending on the intermediate scale $M_I$, the unification scale can be $M_U > 10^{15}$ GeV. The proton lifetime also comes out to be consistent with present experiments. Proper fermion mass relations require the existence of a 126-plet of Higgs field, which is also present in the theory. Most of the inconsistencies of the minimal $SU(5)$ grand unified theory are solved in the $SO(10)$ grand unified theory.

The popular version of the $SO(10)$ grand unified theory is the one with supersymmetry above the symmetry breaking scale. Another advantage of the supersymmetric models is that it is possible to calculate the evolution of the fermion masses. Among supersymmetric models, one interesting class of models is the minimal supersymmetric $SO(10)$ grand unified theory, in which the number of parameters is chosen to be minimal, although it was subsequently proved to be inconsistent with the coupling constant unification.

Nonsupersymmetric extensions of the $SO(10)$ grand unified theory have also been studied with some motivations. The baryon asymmetry of the universe adds more constraint to the models. One of the main motivations of the grand unified theories was to explain the baryon asymmetry of the universe. Later it was found that the baryon asymmetry generated in the grand unified theories were $B-L$ conserving, which is washed out by the sphaleron transition before the electroweak phase transition. At present the most popular explanation of the baryon asymmetry of the universe comes from the lepton number violation. The $B-L$ violating interactions that generate the neutrino masses can also generate a lepton asymmetry, which gets converted to a baryon asymmetry of the universe before the electroweak phase transition by the sphaleron transition.

We discussed some generic features of the grand unified theories and it appears that some of the grand unified theories can solve all the problems. But once we start probing them in details, more and more problems start creeping up. One of the most important constraints comes from the gauge coupling unification. Other constraints include the fermion mass relations, neutrino masses, proton decay, doublet–triplet splitting, baryon asymmetry of the universe, gravitino problem, and more. The arbitrariness of the symmetry breaking patterns are also of concern. Some new variants of the grand unified theories have also been studied in the recent times. The orbifold grand unified theories were proposed to solve the doublet–triplet splitting problem. The models with large extra dimensions opened up the possibility of grand unification at the TeV scale. Some of these issues will be discussed later.

# 9

# *Supersymmetry*

Supersymmetry is a symmetry that relates a fermion to a boson, so any representation of a *supermultiplet* must contain both fermions and bosons and all particles will have their supersymmetric counterparts which are called the *superpartners*. A fermion will have *scalar superpartners* (usually called *sfermions*, e.g., the scalar superpartner of the quarks are called *squarks*) while a scalar or a vector will have *fermionic superpartner* (usually the superpartner of a boson is referred to as *bosino*, e.g., *Higgsino* and *gauginos* are fermionic superpartners of Higgs bosons and gauge bosons respectively). The electron and its scalar superpartner, the *selectron*, will belong to the same supermultiplet. In a supersymmetric theory, all interactions will have their supersymmetric counterterms, obtained by interchanging $particles \leftrightarrow superparticles$. However, because we have not observed any superparticles so far, supersymmetry must be a broken symmetry and the superparticles will have masses of the order of supersymmetry breaking scale, if they exist.

Supersymmetry established itself as one of the most popular extensions of the standard model because it provides a solution to the gauge-hierarchy problem and allows unification of the space–time symmetry with the internal symmetry. In addition, it offers an extremely rich phenomenology due to the presence of all the superparticles that are yet to be observed. The lightest superparticle could be the most natural candidate for dark matter. The aesthetic beauty of relating a fermion to a boson in a supersymmetric theory makes it even more attractive. We hope that at higher energies supersymmetry exists, and we should be able to see signals of supersymmetry in the next generation accelerators. In this chapter I shall present a brief introduction to supersymmetry.

## 9.1 Why Supersymmetry

The standard model of the strong, the weak and the electromagnetic interactions is highly successful, but definitely not the ultimate theory. Among other things, it does not appear to have three coupling constants to explain the theory. Grand unified theories are proposed, in which the gauge coupling constants get unified at a very high scale. Beyond a large scale of grand unification there is only one gauge coupling constant corresponding to the grand unified group, and there are several interesting predictions of these theories that make them highly attractive. However, the grand unified theories suffer from a serious gauge hierarchy problem. The scale of grand unification is about 14 orders of magnitude higher than the electroweak symmetry breaking scale. The interactions of the Higgs scalars that break the grand unified group at the grand unification scale will interact with the electroweak symmetry breaking scalars and will introduce quadratic divergent corrections to the light particles. To keep the electroweak symmetry breaking scale low, counterterms have to be introduced that can cancel these divergent corrections to 14 decimal points. Fine tuning of parameters to all orders is unnatural technically and this is the gauge hierarchy problem.

Consider an $SU(5)$ grand unified theory, where the $SU(5)$ symmetry is broken by a 24-plet Higgs field $\Sigma$ and the electroweak symmetry is broken by a 5-plet Higgs field $\phi$. The scalar potential can be written as

$$V = V[\Sigma] + V[\phi] + V[\Sigma, \phi], \tag{9.1}$$

where the first two terms contain the usual quadratic and quartic terms

$$\begin{aligned} V[\Sigma] &= -\frac{1}{2} m_\Sigma^2 \mathrm{Tr}\, \Sigma^2 + \frac{1}{4} a (\mathrm{Tr}\, \Sigma^2)^2 + \frac{1}{2} b \mathrm{Tr}\, \Sigma^4 \\ V[\phi] &= -\frac{1}{2} m_\phi^2 \phi^\dagger \phi + \frac{1}{4} \lambda (\phi^\dagger \phi)^2. \end{aligned} \tag{9.2}$$

In addition there are the dangerous cross terms

$$V[\Sigma, \phi] = \alpha \phi^\dagger \phi \; \mathrm{Tr}\, \Sigma^2 + \beta \phi^\dagger \Sigma^2 \phi. \tag{9.3}$$

If we now write the Higgs scalar fields around their *vevs*, $\langle \Sigma \rangle = v_0$ and $\langle \phi \rangle = v$, the cross terms will contribute to the masses of $\phi$ of the order of $v_0$. If we introduce counterterms to eliminate these cross terms, they will again be generated at all orders of perturbation through loop diagrams. So, we need to introduce counterterms at all orders.

Let us take the complete potential and get the condition for the minimum by substituting the *vevs* of the scalar fields

$$m_\Sigma^2 = \frac{1}{2}(15a + 7b) v_0^2 + (\alpha + \frac{3}{10}\beta) v^2 \tag{9.4}$$

$$m_\phi^2 = \frac{1}{2}\lambda v^2 + (15\alpha + \frac{9}{2}\beta) v_0^2. \tag{9.5}$$

The *vev* of the 5-plet Higgs scalar $v$ will receive contributions proportional to the *vev* of the 24-plet Higgs scalar $v_0$. So, to prevent such large corrections we have to fine tune parameters to all orders of perturbation with increasing precision, which is technically unnatural. Any theory with two widely different mass scales, originating from the *vevs* of some scalar fields, always requires unnatural fine tuning of parameters at all orders of perturbation theory. This is known as the *gauge hierarchy problem* and is present in all grand unified theories.

Let us consider the problem from a different approach. If there are a massless fermion $\psi$ and a Higgs scalar $\phi$ with mass $m_h$, the most general Lagrangian can be written as

$$\mathscr{L}_\phi = i\overline{\psi}(\gamma^\mu \partial_\mu)\psi + \partial_\mu \phi^\dagger \partial_\mu \phi - m_s^2 \phi^\dagger \phi + \lambda_s (\phi^\dagger \phi)^2 - (\frac{\lambda_F}{2}\overline{\psi}\phi\psi + H.c.). \quad (9.6)$$

When the Higgs scalar $\phi$ acquires a vacuum expectation value ($vev$) $\langle \phi \rangle = v$, the fermion gets a mass $m_F = \lambda_F v$. The interaction term will generate the one-loop divergent Higgs self-energy diagram of figure 9.1, which can be eliminated by the renormalization of the fermion mass. The correction to the fermion mass is given by

$$\delta m_F = -\frac{3\lambda_F^2 m_F}{64\pi^2} \log\left(\frac{\Lambda^2}{m_F^2}\right), \quad (9.7)$$

where $\Lambda$ is a momentum space cut-off and is some large scale in the theory; in the present case it will be the unification scale. This correction vanishes for $m_F \to 0$, and hence, it is technically natural, because the Lagrangian is invariant under a chiral symmetry,

$$\psi_{[L,R]} \to \exp^{i\theta_{[L,R]}} \psi_{[L,R]}, \quad (9.8)$$

in the limit of $m_F \to 0$. If we set $\lambda_F = 0$, the chiral symmetry will protect the mass term to all orders of perturbation theory, and it will be possible to have the mass terms only when this symmetry is broken. All corrections to the mass are proportional to $\lambda_F$ and to $m_F$.

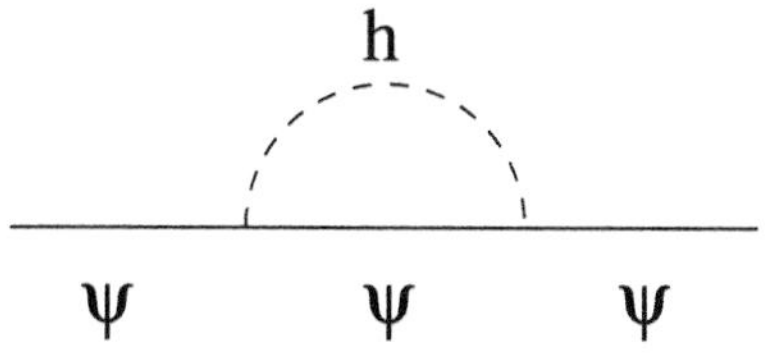


**FIGURE 9.1**
Self-energy diagram of fermions ($\psi$) with scalars ($h$) in the loop

Let us consider the divergence of this diagram. The loop integral gives $p^4$, while the fermion and the scalar propagators give $p^{-3}$, but by inserting a fermion mass $m_F$ in the numerator the diagram becomes logarithmically divergent. This is not true for the fermion loop of figure 9.2, which has $p^4$ coming from the loop integral, $p^{-2}$ from the fermion propagators, making the diagram quadratically divergent. So the corrections to the Higgs scalar masses are not proportional to the scalar masses and there is no symmetry that can protect the scalar masses.

The quadratically divergent one-loop contributions to the Higgs scalar mass, coming from the fermion loop of figure 9.2, is given by

$$\delta m_h^2 = -\frac{\lambda_F^2}{8\pi^2}\Lambda^2. \quad (9.9)$$

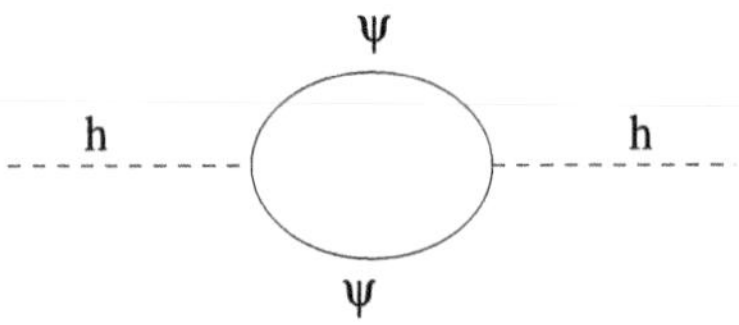


**FIGURE 9.2**
Self-energy diagram of scalars ($h$) with fermions ($\psi$) in the loop

This renormalizes the Higgs scalar mass, but it is not proportional to $m_h$. Even if we start with a massless Higgs scalar, the corrections will be of the order of the large scale in the theory, which is the scale of grand unification in the present case. The only way to keep the mass of the Higgs scalar light is by adding counter terms

$$m_h^2 = m_{h,0}^2 + \delta m_h^2 + \text{counter terms}, \tag{9.10}$$

and adjusting parameters to an extremely high degree of accuracy, so the divergent contributions are cancelled. To keep the mass of $h$ at around 100 GeV when the cut-off scale or the scale of grand unification is around $10^{15}$ GeV, the counter terms have to be fine tuned to 13 decimal points at the one-loop level, while at the two-loop level the fine tuning will be to 26 decimal points. To all orders of perturbation theory counter terms have to be included and fine tuning of parameters have to be performed. This fine tuning of parameters is required whenever there are two different scales in the theory, which are generated by the Higgs *vevs*, as in the grand unified theories.

One of the main motivations of supersymmetry is that quadratic divergences are absent. Although the fine tuning of parameters is required at the tree level, there are no loop corrections that may require any fine tuning. This is because the scalars and the fermions in the loop contribute quadratic divergences with opposite sign and similar form, so they cancel in the limit of equal masses of fermions and scalars in the loop. Since supersymmetry [142], i.e., a $boson \leftrightarrow fermion$ symmetry, ensures that the masses of the bosons and the fermions are identical naturally, there is no quadratic divergence [143]. Thus, in the limit of exact supersymmetry, there are no divergent radiative corrections that can destabilize any hierarchical symmetry breaking pattern. However, we have not observed the superpartners of ordinary particles with same mass, so even if supersymmetry exists in nature, it must be broken before the electroweak symmetry breaking. The mass difference between the particles and their superpartners will be of the order of supersymmetry breaking scale. The corrections from the radiative corrections will also be of the same order. Thus, to keep the Higgs mass at less than 1 TeV, supersymmetry must also break around the 1 TeV.

In addition to solving the gauge hierarchy problem by means of cancellation of the quadratic divergent corrections, supersymmetry beautifies any theory by making it invariant under a transformation between fermions and bosons. Another feature of supersymmetry is that it allows unification of the Poincaré group with the internal symmetry groups. The Poincaré group of space–time symmetry or the rotations and translations in four-dimensional Minkowski space can at most be a direct product of some internal symmetry group according to the Coleman–Mandula theorem [144]. This forbids a true unification of the space–time symmetry with internal symmetry.

The generators of the Poincaré group satisfy commutation relations, which in turn restricts the field operators to satisfying only commutation relations. The supersymmetry algebra is generated by anticommuting generators that relate a fermion to a boson, and the Coleman–Mandula theorem is no longer valid. The Coleman–Mandula theorem was generalized by Haag, Lopuzanski and Sohnius to include the anticommuting generators [145], which then allows a true unification of the Poincaré group with the internal symmetry groups in a supersymmetric theory.

In addition to providing a solution to the gauge hierarchy problem and allowing unification of the space–time symmetry with internal symmetries, we now believe that the correct quantum theory of gravity is supersymmetric. The superpartners and their interactions predict interesting phenomenology in the next generation accelerators, which are added attractions of supersymmetry. There are also many cosmological consequences of supersymmetry including its prediction for a natural candidate of cold dark matter.

## 9.2 Formalism

Supersymmetry algebra is generated by anticommuting generators with spin $1/2$. As a result, the generators of supersymmetry change the spin of the particle by a half-integer, and a fermion changes over to a boson and vice versa. Invariance under supersymmetry implies invariance of the theory under the exchange of fermions with bosons. Let us denote a supersymmetry generator by Weyl spinor operators $Q_\alpha$ and $\bar{Q}_{\dot{\alpha}}$, which are left-chiral and right-chiral, respectively. These operators satisfy the commutation relations with the energy-momentum four-vector $P^\mu$

$$[P^\mu, \bar{Q}_{\dot{\alpha}}] = [P^\mu, Q_\alpha] = [P^2, Q_\alpha] = [P^2, \bar{Q}_{\dot{\alpha}}] = 0, \tag{9.11}$$

and they anticommutes with themselves

$$\{Q_\alpha, Q_\beta\} = \{\bar{Q}_{\dot{\alpha}}, \bar{Q}_{\dot{\beta}}\} = 0. \tag{9.12}$$

Since the left-chiral operator $Q_\alpha$ belongs to the $(1/2, 0)$ representation of the $SU(2)_L \times SU(2)_R \subset O(4)$ and $Q_{\dot{\alpha}}$ belongs to the $(0, 1/2)$ representations, their anticommutation should belongs to the $(1/2, 1/2)$ representation, which is a four-vector. Thus, the anticommutation of left-chiral operator and right-chiral operator is related to the energy-momentum four-vector

$$\{Q_\alpha, \bar{Q}_{\dot{\beta}}\} = 2\sigma^\mu{}_{\alpha\dot{\beta}} P_\mu. \tag{9.13}$$

Another equivalent approach to constructing the supersymmetry algebra is to work with Majorana spinors $Q_M$, which are related to the Weyl spinors $Q_\alpha$ and satisfy the anticommutation relation,

$$\{Q_M, \bar{Q}_M\} = 2\gamma^\mu P_\mu, \tag{9.14}$$

where $\bar{Q}_M = Q_M^\dagger \gamma_0$.

In a supersymmetric theory the vacuum state must satisfy

$$Q_\alpha |vac>= 0, \tag{9.15}$$

which immediately implies that the vacuum energy must vanish $< 0|P_0|0 >= 0$. The energy of any supersymmetric nonvacuum state is positive definite and contains equal numbers of fermionic and bosonic states with equal mass.

We now concentrate on the simplest supersymmetric theory, the $N = 1$ supersymmetry, which is generated by only one operator $Q$. The lowest representation of $N = 1$ supersymmetry, the chiral representation, contains a scalar and a fermion that transform to each other when operated by the supersymmetry generator $Q$:

$$\text{fermion} \overset{Q}{\longleftrightarrow} \text{boson}. \tag{9.16}$$

The next higher supersymmetric representation, the vector representation, contains the gauge bosons and their fermionic superpartners, called the gauginos. The spin-2 graviton is associated with its superpartner spin-3/2 gravitino. All the superparticles are massive and their mass depends on the details of the supersymmetry breaking mechanism.

Let us now demonstrate a supersymmetric theory containing a massless scalar $\phi$ and a massless fermion $\psi$. If we now apply supersymmetry transformation generated by the $N = 1$ supersymmetry generator $Q$, the algebra does not close and the action does not remain invariant. Since the Weyl fermion $\psi$ has 2 complex degrees of freedom, whereas the complex scalar $\phi$ has only one complex degree of freedom, we need one more complex degree of freedom to match the off-shell total numbers of degrees of freedom. We introduce a complex scalar auxiliary field $F$, so we can write the supersymmetric transformations of the different fields and check that the algebra closes.

To verify the invariance of the theory under supersymmetry transformations, we start with the Lagrangian

$$\mathscr{L} = (\partial_\mu \phi^*)(\partial^\mu \phi) + i\bar{\psi}\bar{\sigma}^\mu \partial_\mu \psi + F^* F, \tag{9.17}$$

and then utilize the equation of motion for the auxiliary field to eliminate it, which is a trivial one in this case

$$F = 0. \tag{9.18}$$

We now write the transformations of the scalar, fermion and auxiliary fields:

$$\begin{aligned} \delta_\xi \phi &= \sqrt{2}\xi\psi \\ \delta_\xi \psi &= i\sqrt{2}\sigma^\mu \bar{\xi} \partial_\mu \phi + \sqrt{2}\xi F \\ \delta_\xi F &= \frac{i}{\sqrt{2}} \bar{\xi} \bar{\sigma}^\mu \partial_\mu \psi \, , \end{aligned} \tag{9.19}$$

where $\xi$ is a Grassmann variable and parameterizes the supersymmetry transformation. The Lagrangian then transforms to a total derivative, and the action becomes invariant. We shall now proceed with superfield formalism to construct a supersymmetry theory after presenting the spinor notations required for the formalism.

**Spinor Notations**

The Weyl representation was formulated to study the massless solutions of the Dirac equation. In this representation the gamma-matrices are defined as

$$\gamma^\mu = \begin{pmatrix} 0 & \sigma^\mu \\ \bar{\sigma}^\mu & 0 \end{pmatrix} \quad (\mu = 0, 1, 2, 3), \tag{9.20}$$

where

$$\sigma^\mu \equiv (I_2, \sigma_i); \;\; \bar{\sigma}^\mu \equiv (I_2, -\sigma_i) = \sigma_\mu \quad (i = 1, 2, 3)\ , \tag{9.21}$$

and $I_2$ is a $2 \times 2$ unit matrix, while $\sigma_i$, $i = 1, 2, 3$ are the Pauli matrices. The main advantage of the Weyl representation is that the $\gamma_5$ matrix is diagonal,

$$\gamma_5 = i\gamma^0\gamma^1\gamma^2\gamma^3 = \begin{pmatrix} -I_2 & 0 \\ 0 & I_2 \end{pmatrix}. \tag{9.22}$$

Thus, a 4-dimensional spinor representation of the group $O(4)$ will have the first two rows belonging to the representation $(1/2, 0)$ of the subgroup $SU(2)_L \times SU(2)_R$, while the last two rows belong to $(0, 1/2)$. In other words, a Dirac spinor $\Psi_D$, which is a 4-spinor of $O(4)$, can be decomposed into two Weyl spinors as

$$\Psi_D = \Psi_L + \Psi_R, \tag{9.23}$$

where

$$\begin{aligned} \Psi_L &= \frac{1}{2}(1 - \gamma_5)\Psi_D = \begin{pmatrix} \psi_\alpha \\ 0 \end{pmatrix} \\ \Psi_R &= \frac{1}{2}(1 + \gamma_5)\Psi_D = \begin{pmatrix} 0 \\ \bar{\chi}^{\dot{\alpha}} \end{pmatrix}. \end{aligned} \tag{9.24}$$

The $\psi_\alpha, \alpha = 1, 2$ and $\chi^{\dot{\alpha}}, \dot{\alpha} = 1, 2$ are two-component spinors, belonging to the $(1/2, 0)$ and $(0, 1/2)$ representations of $SU(2)_L \times SU(2)_R$, respectively.

In the Weyl basis, the definition of the charge conjugation operator remains the same but the form changes

$$\Psi_D^c = C\bar{\Psi}_D^T = -i\gamma^0\gamma^2\bar{\Psi}_D^T = \begin{pmatrix} \chi_\alpha \\ \bar{\psi}^{\dot{\alpha}} \end{pmatrix}. \tag{9.25}$$

Here we define

$$\begin{aligned} \bar{\psi}^{\dot{\alpha}} &\equiv \varepsilon^{\dot{\alpha}\dot{\beta}}\bar{\psi}_{\dot{\beta}} \equiv \varepsilon^{\dot{\alpha}\dot{\beta}}(\psi_\beta)^* \\ \chi_\alpha &\equiv \varepsilon_{\alpha\beta}\chi^\beta \equiv (\bar{\chi}^{\dot{\alpha}})^*, \end{aligned} \tag{9.26}$$

with the antisymmetric tensors given by

$$\varepsilon_{\alpha\beta} = \varepsilon_{\dot{\alpha}\dot{\beta}} = i\sigma^2 = \begin{pmatrix} 0 & 1 \\ -1 & 0 \end{pmatrix}$$

$$\varepsilon^{\alpha\beta} = \varepsilon^{\dot{\alpha}\dot{\beta}} = -i\sigma^2 = \begin{pmatrix} 0 & -1 \\ 1 & 0 \end{pmatrix}. \tag{9.27}$$

We define a Majorana spinor, which is its own charge conjugate, satisfying

$$\Psi_M^c = \Psi_M \Longrightarrow \psi_\alpha = \chi_\alpha; \; \bar{\chi}^{\dot{\alpha}} = \bar{\psi}^{\dot{\alpha}}, \tag{9.28}$$

and hence, it has four real components. We can construct a Majorana spinor from a Weyl spinor as

$$\Psi_M = \begin{pmatrix} \psi_\alpha \\ \bar{\psi}^{\dot{\alpha}} \end{pmatrix}. \tag{9.29}$$

In the Weyl basis we now have the indices $\alpha, \beta, \cdots$ for the representations of the group $SU(2)_L$, which are called the left-chiral fields. There are also indices $\dot{\alpha}, \dot{\beta}, \cdots$ for the right-chiral fields that belong to the representations of the group $SU(2)_R$. For the purpose of simplification we define an abbreviated notation including summation over indices. In this notation the scalar bilinears are defined as

$$\psi\chi = \chi^\alpha \psi_\alpha = \psi^\alpha \chi_\alpha = -\chi_\alpha \psi^\alpha = \chi\psi$$

$$\bar{\chi}\bar{\psi} = \bar{\chi}_{\dot{\alpha}} \bar{\psi}^{\dot{\alpha}} = \bar{\psi}_{\dot{\alpha}} \bar{\chi}^{\dot{\alpha}} = -\bar{\chi}^{\dot{\alpha}} \bar{\psi}_{\dot{\alpha}} = \bar{\psi}\bar{\chi}. \tag{9.30}$$

The four-vector bilinears can be written as

$$\bar{\chi}_{\dot{\alpha}} (\bar{\sigma}^\mu)^{\dot{\alpha}\alpha} \psi_\alpha = -\psi^\alpha (\sigma^\mu)_{\alpha\dot{\alpha}} \bar{\chi}^{\dot{\alpha}} = -\psi\sigma^\mu \bar{\chi} = \bar{\chi}\sigma^\mu \psi. \tag{9.31}$$

Similarly, a tensor may be defined as

$$\chi\sigma^{\mu\nu}\psi = \chi^\alpha (\sigma^{\mu\nu})_\alpha{}^\beta \psi_\beta = -\psi\sigma^{\mu\nu}\chi. \tag{9.32}$$

In this notation we can write the possible bilinears formed by two Dirac spinors

$$\Psi = \begin{pmatrix} \psi_\alpha \\ \bar{\chi}^{\dot{\alpha}} \end{pmatrix} \quad \text{and} \quad \Phi = \begin{pmatrix} \phi_\alpha \\ \bar{\eta}^{\dot{\alpha}} \end{pmatrix} \tag{9.33}$$

in the following five forms:

$$\bar{\Psi}\Phi = \bar{\psi}\bar{\eta} + \chi\phi = (\bar{\Phi}\Psi)^\dagger$$

$$\bar{\Psi}\gamma_5\Phi = \bar{\psi}\bar{\eta} - \chi\phi = -(\bar{\Phi}\gamma_5\Psi)^\dagger$$

$$\bar{\Psi}\gamma^\mu\Phi = \chi\sigma^\mu\bar{\eta} + \bar{\psi}\bar{\sigma}^\mu\phi = (\bar{\Phi}\gamma^\mu\Psi)^\dagger$$

$$\bar{\Psi}\gamma^\mu\gamma_5\Phi = \chi\sigma^\mu\bar{\eta} - \bar{\psi}\bar{\sigma}^\mu\phi = (\bar{\Phi}\gamma^\mu\gamma_5\Psi)^\dagger$$

$$\bar{\Psi}\Sigma^{\mu\nu}\Phi = i\bar{\psi}\bar{\sigma}^{\mu\nu}\bar{\eta} + i\chi\sigma^{\mu\nu}\phi = (\bar{\Phi}\Sigma^{\mu\nu}\Psi)^\dagger. \tag{9.34}$$

The five forms are the scalar, pseudo-scalar, vector, axial-vector and tensor.

Starting with the two Majorana spinors

$$\Psi_M = \begin{pmatrix} \psi_\alpha \\ \bar{\psi}^{\dot{\alpha}} \end{pmatrix} \quad \text{and} \quad \Phi_M = \begin{pmatrix} \phi_\alpha \\ \bar{\phi}^{\dot{\alpha}} \end{pmatrix}, \tag{9.35}$$

we write the five bilinears:

$$\begin{aligned}
\bar{\Psi}_M\Phi_M &= \bar{\psi}\bar{\phi} + \psi\phi = \bar{\Phi}_M\Psi_M = (\bar{\Psi}_M\Phi_M)^\dagger \\
\bar{\Psi}_M\gamma_5\Phi_M &= \bar{\psi}\bar{\phi} - \psi\phi = \bar{\Phi}_M\gamma_5\Psi_M = -(\bar{\Psi}_M\gamma_5\Phi_M)^\dagger \\
\bar{\Psi}_M\gamma^\mu\Phi_M &= \psi\sigma^\mu\bar{\phi} + \bar{\psi}\bar{\sigma}^\mu\phi = -\bar{\Phi}_M\gamma^\mu\Psi_M = -(\bar{\Psi}_M\gamma^\mu\Phi_M)^\dagger \\
\bar{\Psi}_M\gamma^\mu\gamma_5\Phi_M &= \psi\sigma^\mu\bar{\phi} - \bar{\psi}\bar{\sigma}^\mu\phi = \bar{\Phi}_M\gamma^\mu\gamma_5\Psi_M = (\bar{\Psi}_M\gamma^\mu\gamma_5\Phi_M)^\dagger \\
\bar{\Psi}_M\Sigma^{\mu\nu}\Phi_M &= i\bar{\psi}\bar{\sigma}^{\mu\nu}\bar{\phi} + i\psi\sigma^{\mu\nu}\phi = -\bar{\Phi}_M\Sigma^{\mu\nu}\Psi_M \\
&= -(\bar{\Psi}_M\Sigma^{\mu\nu}\Phi_M)^\dagger.
\end{aligned} \tag{9.36}$$

With this brief introduction we shall now proceed to construct superspace forlamisn.

**Superspace Formalism**

It is highly convenient to develop superspace formalism, in order to construct a supersymmetric theory. The first ingredient for this purpose is the introduction of Grassmann parameters. In the Weyl representation two independent constant anticommuting Grassmann numbers $\theta_\alpha$ $(\alpha = 1,2)$ and $\bar{\theta}_{\dot{\alpha}}$, $(\dot{\alpha} = \dot{1},\dot{2})$ belong to the $(1/2,0)$ and $(0,1/2)$ representations of the group $SU(2)_L \times SU(2)_R \subset O(4)$. In the Majorana representation we need to introduce the anticommuting Grassmann numbers $\varepsilon_a$ $(a = 1,2,3,4)$, where $\varepsilon$ satisfies the Majorana condition. We shall now continue with the Weyl representation.

The anticommutation relations of the Grassmann numbers are given by

$$\{\theta_\alpha, \theta_\beta\} = \{\bar{\theta}_{\dot{\alpha}}, \bar{\theta}_{\dot{\beta}\}} = \{\bar{\theta}_{\dot{\alpha}}, \theta_\beta\} = 0\,. \tag{9.37}$$

The only possible bilinears that can be formed with these Grassmann variables are

$$\begin{aligned}
\theta\theta &= \theta_\alpha\theta^\alpha = \theta_\alpha\theta_\beta\varepsilon_{\alpha\beta} \\
\bar{\theta}\bar{\theta} &= \bar{\theta}_{\dot{\alpha}}\bar{\theta}^{\dot{\alpha}} = \bar{\theta}_{\dot{\alpha}}\bar{\theta}_{\dot{\beta}}\varepsilon_{\dot{\alpha}\dot{\beta}} \\
\bar{\theta}\sigma^\mu\theta &= -\theta\sigma^\mu\bar{\theta} = \bar{\theta}_{\dot{\alpha}}(\bar{\sigma}^\mu)^{\dot{\alpha}\beta}\theta_\beta = -\theta^\alpha(\sigma^\mu)_{\alpha\dot{\beta}}\bar{\theta}^{\dot{\beta}}\,.
\end{aligned} \tag{9.38}$$

There are no cubic or higher products of $\theta$ or $\bar{\theta}$,

$$\theta\theta\theta = \theta\theta\theta\theta = \cdots = \bar{\theta}\bar{\theta}\bar{\theta} = \bar{\theta}\bar{\theta}\bar{\theta}\bar{\theta} = \cdots = 0.$$

The only cubic and quadratic terms possible are

$$(\theta\theta)\bar{\theta}_{\dot{\alpha}}, \quad (\bar{\theta}\bar{\theta})\theta_\alpha \quad \text{and} \quad (\theta\theta)(\bar{\theta}\bar{\theta}).$$

We complete this discussion by defining the derivatives with respect to the Grassmann variables, where $\partial/\partial\theta^\alpha$ satisfies

$$\frac{\partial}{\partial\theta^\alpha}\theta^\beta = \delta_\alpha{}^\beta; \quad \frac{\partial}{\partial\theta_\alpha}\theta_\beta = \delta^\alpha{}_\beta; \quad \frac{\partial}{\partial\theta^\beta}\theta_\alpha = \varepsilon_{\alpha\beta}. \tag{9.39}$$

The derivative $\partial/\partial\bar{\theta}_{\dot{\alpha}}$ satisfies analogous formulas.

A superspace is constructed with the variables $(x^\mu, \theta, \bar{\theta})$, so a superfield can be defined as $S(x^\mu, \theta, \bar{\theta})$. The action of the generators,

$$\begin{aligned} P_\mu &= i\partial_\mu \\ iQ_\alpha &= \frac{\partial}{\partial\theta^\alpha} - i\sigma^\mu_{\alpha\dot{\alpha}}\bar{\theta}^{\dot{\alpha}}\partial_\mu \\ i\bar{Q}_{\dot{\alpha}} &= -\frac{\partial}{\partial\bar{\theta}^{\dot{\alpha}}} + i\theta^\alpha\sigma^\mu_{\alpha\dot{\alpha}}\partial_\mu, \end{aligned} \tag{9.40}$$

then gives the supersymmetry algebra.

We would now like to classify the different superfields and find the irreducible representations of the superalgebra. We start with defining the fermionic derivatives or the covariant derivatives for superfields

$$\begin{aligned} D_\alpha &= \frac{\partial}{\partial\theta^\alpha} + i\sigma^\mu_{\alpha\dot{\alpha}}\bar{\theta}^{\dot{\alpha}}\partial_\mu \\ \bar{D}_{\dot{\alpha}} &= -\frac{\partial}{\partial\bar{\theta}^{\dot{\alpha}}} - i\theta^\alpha\sigma^\mu_{\alpha\dot{\alpha}}\partial_\mu, \end{aligned} \tag{9.41}$$

which satisfy the anticommutation relations

$$\{D_\alpha, D_\beta\} = \{\bar{D}_{\dot{\alpha}}, \bar{D}_{\dot{\beta}}\} = 0 \quad \text{and} \quad \{D_\alpha, \bar{D}_{\dot{\alpha}}\} = 2i\sigma^\mu_{\alpha\dot{\alpha}}\partial_\mu. \tag{9.42}$$

The fermioninc derivatives also anticommute with the generators of the supersymmetry algebra

$$\{D_\alpha, Q_\beta\} = \{\bar{D}_{\dot{\alpha}}, Q_\beta\} = \{D_\alpha, \bar{Q}_{\dot{\beta}}\} = \{\bar{D}_{\dot{\alpha}}, \bar{Q}_{\dot{\beta}}\} = 0. \tag{9.43}$$

Thus, these covariant derivatives are invariant under supersymmetry transformations.

The covariant derivatives can project out superfields into two chiral representations. We define the left-chiral superfields satisfying

$$\bar{D}_{\dot{\alpha}}\Phi = 0 \tag{9.44}$$

and the right-chiral superfields satisfying

$$D_\alpha\Phi = 0. \tag{9.45}$$

This becomes transparent when we make the transformations of the variable

$$y^\mu = x^\mu + i\theta\sigma^\mu\bar{\theta}, \tag{9.46}$$

which satisfies

$$\bar{D}_{\dot{\alpha}} y^\mu = 0. \tag{9.47}$$

It terms of the new coordinates, any superfields satisfying the condition for left-chiral superfield of equation (9.44) will have explicit dependence only in $\theta$, while the superfields satisfying equation (9.45) can be expressed only in terms of $\bar{\theta}$.

We can expand any left-chiral superfield satisfying equation 9.44 in powers of $\theta$ and $y^\mu$ as

$$\Phi(y^\mu, \theta) = \phi(y) + \sqrt{2}\theta\psi(y) + \theta\theta F(y)\ , \tag{9.48}$$

while we can expand a right-chiral superfield that satisfies equation 9.45 in powers of $\bar{\theta}$ and $y^\mu$ as

$$\Phi^\dagger(y^\mu, \bar{\theta}) = \phi(y) + \sqrt{2}\bar{\theta}\psi(y) + \bar{\theta}\bar{\theta}F(y). \tag{9.49}$$

The right-chiral superfields are called the conjugate superfields. The left-handed particles and antiparticles belong to the left-chiral superfields, while the right-handed particles and antiparticles belong to the right-chiral superfields.

The supersymmetric transformations of the component fields $\phi$, $\psi$ and $F$, given in equation (9.19), can be retrieved by defining the infinitesimal supersymmetry transformation of the superfield $\Phi$ as

$$\Phi \to \Phi + \delta\Phi = \Phi + i(\xi Q + \bar{\xi}\bar{Q})\Phi. \tag{9.50}$$

In terms of the superfields we can write the Lagrangian that is invariant under the supersymmetry transformation as

$$\mathcal{L}_{susy} = \sum_i [\Phi_i^\dagger \Phi_i]_D + ([W(\Phi)]_F + H.c.). \tag{9.51}$$

The second term is called the $F$-term and contains only left-chiral superfields. The superpotential $W(\Phi)$ can be written in terms of $1, \theta$ and $\theta\theta$. Only the coefficient of $\theta\theta$ transforms into total derivative under a supersymmetry transformation and this coefficient is invariant under supersymmetry. This coefficient of $\theta\theta$ in the superpotential $W(\Phi)$ is denoted as the $F$-term, $[W(\Phi)]_F$. The $F$-term in expansion of the left-chiral superfields contains only the left-handed particles and antiparticles. The right-handed particles and antiparticles are included in the Hermitian conjugate term, which is the coefficient of $\bar{\theta}\bar{\theta}$ in the superpotential $W(\Phi^\dagger)$ with the right-chiral superfields. The first term contains products of left-chiral and right-chiral superfields, so it contains all powers of $\theta$ and $\bar{\theta}$. The $D$-term, which is the coefficient of $\theta\theta\bar{\theta}\bar{\theta}$, is invariant under supersymmetry transformation. All the kinetic energy terms are included in the $D$-term.

We now demonstrate with an example with superfields $\Phi_i$. We write the most general superpotential

$$W(\Phi) = \frac{1}{2} m_{ij}\Phi_i\Phi_j + \frac{1}{3}\lambda_{ijk}\Phi_i\Phi_j\Phi_k. \tag{9.52}$$

Taking the $D$-term of $\sum_i \Phi_i^\dagger \Phi_i$ and the $F$-term of the superpotential we can write the Lagrangian as

$$\begin{aligned}\mathscr{L} &= \partial_\mu \phi_i^\dagger \partial^\mu \phi_i + i\bar{\psi}_i \sigma^\mu \partial_\mu \psi_i + F_i^\dagger F_i \\ &\quad + m_{ij}\phi_i F_j - \frac{1}{2} m_{ij} \psi_i \psi_j \\ &\quad + \lambda_{ijk}\phi_i \phi_j F_k - \lambda_{ijk} \psi_i \psi_j \phi_k + H.c. \end{aligned} \tag{9.53}$$

The first line comes from the $D$-terms and the rest from $F$-terms. We use the equations of motion for the auxiliary fields $F_i$:

$$F_i^\dagger = - \left. \frac{\partial W(\phi)}{\partial \phi_i} \right|_F = -m_{ij}\phi_j - \lambda_{ijk}\phi_j \phi_k \tag{9.54}$$

and eliminate them to write the Lagrangian in terms of physical fields as

$$\mathscr{L} = \mathscr{L}_{KE} + \mathscr{L}_{Yuk} + V(\phi_i), \tag{9.55}$$

where the kinetic energy, Yukawa interactions including the fermion mass terms, and the scalar potential are given by

$$\begin{aligned}\mathscr{L}_{KE} &= \partial_\mu \phi_i^\dagger \partial^\mu \phi_i + i\bar{\psi}_i \sigma^\mu \partial_\mu \psi_i \\ \mathscr{L}_{Yuk} &= -\frac{1}{2} m_{ij} \psi_i \psi_j - \lambda_{ijk} \psi_i \psi_j \phi_k + H.c.. \\ V(\phi_i) &= F_i^\dagger F_i = \left| - \left[ \frac{\partial W(\phi)}{\partial \phi_i} \right]_F \right|^2 = \left| -m_{ij}\phi_j - \lambda_{ijk}\phi_j\phi_k \right|^2 . \end{aligned} \tag{9.56}$$

The right-handed fermions and their antiparticles are included in the Hermitian conjugate term. The equality of the fermion and scalar masses is manifest in the construction and vanishing of the quadratically divergent radiative corrections is guaranteed.

Any internal symmetry breaking can be implemented in supersymmetric theories by giving vacuum expectation values to the scalar partners of any superfield. However, as long as the $F$-terms and $D$-terms do not acquire any *vevs*, supersymmetry remains unbroken. Consider one example of internal symmetry breaking without breaking supersymmetry. Let us start with two superfields $X$ and $Y$, whose interactions are represented by the superpotential

$$W(X,Y) = fX(Y^2 - M^2), \tag{9.57}$$

where $f$ and $M$ are two parameters in the theory with mass dimensions 0 and 1, respectively. In general, it is possible to start with a superpotential of the form we require in a supersymmetric theory, because there are no radiative corrections that can generate any new terms. We may need to fine tune some parameters at the tree level, but since there are no higher order corrections, we do not require any further fine tuning.

The $F$-terms corresponding to this superpotential are

$$F_X = \left[\frac{\partial W}{\partial \phi_X}\right]_F = f(\phi_Y^2 - M^2) \quad \text{and} \quad F_Y = 2f\phi_X\phi_Y,$$

which vanish at the minima preserving supersymmetry when the *vevs* of the scalar components of $X$ and $Y$ are given by

$$\langle \phi_X \rangle = 0 \qquad \langle \phi_Y \rangle = M. \tag{9.58}$$

The nonvanishing of the *vev* of $Y$ can now break any internal symmetry at the scale $M$ if $Y$ transforms nontrivially under the internal group.

## 9.3 Gauge Theory

We shall now develop supersymmetric theories with local internal symmetries. Since our ultimate goal is to construct supersymmetric standard model at energies above the electroweak symmetry breaking scale, we shall discuss both Abelian as well as non-Abelian gauge theories.

### Abelian Gauge Theory

So far we have been working with chiral superfields, which include a scalar and a fermion. Introduction of local symmetries would require covariant derivatives including the vector fields, which should replace the ordinary derivatives in the kinetic energy term. We should introduce vector superfields, which should be linked with the $D$-term and depend on both $\theta$ and $\bar{\theta}$. A vector superfield $V \equiv V(x, \theta, \bar{\theta})$ can be written as

$$\begin{aligned}
V = {} & C(x) + i\theta\chi(x) - i\bar{\theta}\bar{\chi}(x) + \frac{i}{2}[\theta\theta G(x) - \bar{\theta}\bar{\theta}G^*(x)] \\
& + \theta\sigma^\mu\bar{\theta}V_\mu(x) + i\theta\theta\bar{\theta}[\bar{\lambda}(x) + \frac{i}{2}\bar{\sigma}^\mu\partial_\mu\chi(x)] \\
& - i\bar{\theta}\bar{\theta}\theta[\lambda(x) + \frac{i}{2}\sigma^\mu\partial_\mu\bar{\chi}(x)] \\
& + \frac{1}{2}\theta\theta\bar{\theta}\bar{\theta}[D(x) - \frac{1}{2}\partial_\mu\partial^\mu C(x)],
\end{aligned} \tag{9.59}$$

where $V_\mu$ is the vector field; $G = M + iN$ and $C, M, N, D$ are real scalar fields and $\chi, \lambda$ are Weyl spinors.

To construct an Abelian gauge theory, we start with two complex chiral superfields $X_\pm$ with the $U(1)$ charges $\pm 1$, which transform under the $U(1)$ gauge symmetry as

$$X_\pm \to X'_\pm = \exp^{\mp[2iq\Lambda(x)]} X_\pm. \tag{9.60}$$

We introduce a vector superfield $V(x,\theta,\bar{\theta})$, which transforms under the $U(1)$ gauge symmetry as

$$V \to V' = V + i(\Lambda - \Lambda^\dagger). \tag{9.61}$$

In order to write the gauge kinetic energy term we define a supersymmetric field strength, which is a chiral superfield, constructed from this vector superfield as

$$W_\alpha = \bar{D}^2 D_\alpha V\,, \tag{9.62}$$

so

$$\bar{D}_{\dot{\alpha}} W_\alpha = 0. \tag{9.63}$$

This chiral superfield can be expanded as

$$W_\alpha(y,\theta) = 4i\lambda_\alpha(y) + 4D(y)\theta_\alpha + 2i(\sigma^{\mu\nu}\theta)_\alpha F_{\mu\nu}(y) + 4\theta\theta(\sigma^\mu\partial_\mu\bar{\lambda})_\alpha, \tag{9.64}$$

where $\lambda$ is the gaugino and $F_{\mu\nu}$ is the field strength tensor. The gaugino can be expressed as a four-component Majorana spinor:

$$\Lambda_M = \begin{pmatrix} \lambda_\alpha \\ \bar{\lambda}^{\dot{\alpha}} \end{pmatrix}. \tag{9.65}$$

We can then write the gauge-invariant Lagrangian in terms of the superfields as

$$\begin{aligned}\mathscr{L} = &\frac{1}{32}[W^\alpha W_\alpha]_F + [X_+^\dagger \exp^{2qV} X_+ + X_-^\dagger \exp^{-2qV} X_-]_D \\ &+m[X_+X_-]_F + H.c.\end{aligned} \tag{9.66}$$

The first term contains the kinetic energy for the gauge boson and the gaugino, the second term gives the interaction of the gauge superfields with the fermions, and the scalar potential is given in the second line. Only a mass term is allowed by the gauge symmetry for the chiral superfields $X_\pm$, which combine the two Majorana spinors to form a Dirac spinor

$$\Psi = \begin{pmatrix} \psi_{X_+\alpha} \\ \bar{\psi}^{\dot{\alpha}}_{X_-} \end{pmatrix}. \tag{9.67}$$

The physical fields in this theory includes the Dirac spinor $\Psi$, the corresponding scalars $\phi_{X_+}$ and $\phi_{X_-}$, the vector boson $V_\mu$, and the gaugino $\Lambda_M$. Eliminating the auxiliary fields, the Lagrangian becomes

$$\begin{aligned}\mathscr{L} = &\, i\bar{\Psi}\gamma^\mu D_\mu \Psi - m\bar{\Psi}\Psi \\ &+(D_\mu\phi_{X_+})^\dagger(D^\mu\phi_{X_+}) + (D_\mu\phi_{X_-^\dagger})^\dagger(D^\mu\phi_{X_-^\dagger}) \\ &-m^2(\phi_{X_+}^\dagger\phi_{X_+} + \phi_{X_-}^\dagger\phi_{X_-}) - \frac{q^2}{2}(\phi_{X_+}^\dagger\phi_{X_+} - \phi_{X_-}^\dagger\phi_{X_-})^2 \\ &+\frac{q}{\sqrt{2}}[\bar{\Lambda}_M\Psi(\phi_{X_+}^\dagger + \phi_{X_-}) - \bar{\Lambda}_M i\gamma_5\Psi(\phi_{X_+}^\dagger - \phi_{X_-}) \\ &-\bar{\Psi}\Lambda_M(\phi_{X_+} + \phi_{X_-}^\dagger) - \bar{\Psi}i\gamma_5\Lambda_M(\phi_{X_+} - \phi_{X_-}^\dagger)] \\ &-\frac{1}{4}V_{\mu\nu}V^{\mu\nu} + \frac{i}{2}\bar{\Lambda}_M\gamma^\mu\partial_\mu\Lambda_M.\end{aligned} \tag{9.68}$$

The covariant derivative is given by

$$D^\mu = \partial^\mu + iqV^\mu. \tag{9.69}$$

The interactions in this theory are similar to the nonsupersymmetric case, except the new interactions involving the superparticles. These new interactions can be obtained simply by replacing any two of the particles at any vertex by their corresponding superpartners.

**Non-Abelian Gauge Theory**

We generalize the supersymmetric gauge theory to include non-Abelian gauge group. Any chiral superfield $\Phi$ transforms as

$$\Phi \rightarrow \Phi' = \exp^{-2igT^a\Lambda^a}\Phi, \tag{9.70}$$

where $T^a, a = 1...N$ are the $N$ generators of the non-Abelian gauge group. Gauge invariance would now require $N$ vector superfields $V^a$ and corresponding $N$ Majorana gauginos. The gauge bosons will interact with two fermions or two scalars, but the gauginos can interact with only a fermion and a scalar.

Corresponding to any fermions or any scalars in a nonsupersymmetric theory, we now have to include chiral superfields $\Phi_i$. The superpotential will be a function of all of these fields. Depending on the transformations of the fields under the non-Abelian gauge group and any other discrete symmetries in the theory, the superpotential will be determined. The superpotential will contain all of the left-chiral superfields, while its Hermitian conjugate will contain all of the right-chiral superfields. The Yukawa couplings and the scalar potential will then be contained in

$$\mathscr{L}_{int} = [W(\Phi_i) + H.c.]_F. \tag{9.71}$$

The scalar potential is obtained by eliminating the auxiliary fields. The superpotential contains terms of mass-dimension 3 only. The $F$-term implies an integration $\int d^2\theta$ or $\int d^2\bar{\theta}$, which adds one mass-dimension. However, the scalar quartic interactions can enter in the scalar potential.

The spinor field strength superfield can now be defined as

$$\begin{aligned} W^a_\alpha &= \bar{D}^2 D_\alpha V^a + igf^{abc}\bar{D}^2(D_\alpha V^b)V^c \\ &= 4i\lambda^a_\alpha + [4\delta^\beta_\alpha D^a(y) + 2i(\sigma^\mu\bar{\sigma}^\nu)_\alpha{}^\beta V^a_{\mu\nu}(y)]\theta_\beta \\ &\quad +4\theta^2\sigma^\mu_{\alpha\dot{\alpha}}\mathscr{D}_\mu\bar{\lambda}^{a\dot{\alpha}}(y). \end{aligned} \tag{9.72}$$

The field-strength tensor corresponding to the gauge bosons is given by

$$V^a_{\mu\nu} = \partial_\mu V^a_\nu - \partial_\nu V^a_\mu - igf^{abc}V^b_\mu V^c_\nu, \tag{9.73}$$

where $f^{abc}$ is the structure function of the group, which defines the commutation relations of the generators $[T^a, T^b] = if^{abc}T^c$. The covariant derivative can then be defined as

$$\mathscr{D}_\mu\bar{\lambda}^{a\dot{\alpha}} = \partial_\mu\bar{\lambda}^{a\dot{\alpha}} - gf^{abc}V^b_\mu\bar{\lambda}^{c\dot{\alpha}}. \tag{9.74}$$

The gauge invariant supersymmetric Lagrangian will now contain the kinetic energy terms of the vector superfields

$$\begin{aligned}\mathscr{L}_V &= \frac{1}{64}\left[(W^{a\alpha}W^a_\alpha)+(W^a{}_\alpha{}^\dagger)(W^{a\alpha\dagger})\right]_F \\ &= -\frac{1}{4}V^a_{\mu\nu}V^{a\mu\nu}+i\lambda^a\sigma^\mu\mathscr{D}_\mu\bar{\lambda}^a+\frac{1}{2}D^aD^a \end{aligned} \tag{9.75}$$

and the $D$-term containing the interactions of the chiral superfields with the gauge superfields

$$\begin{aligned}\mathscr{L}_D &= [\Phi^\dagger \exp^{2gT^aV^a_\mu}\Phi]_D \\ &= (\mathscr{D}_\mu\phi)^\dagger(\mathscr{D}^\mu\phi)+i\psi\sigma^\mu\mathscr{D}^\dagger_\mu\bar{\psi}+F^\dagger F \\ &\quad +i\sqrt{2}g(\phi^\dagger T^a\lambda^a\psi-\bar{\psi}T^a\bar{\lambda}^a\phi)+g\phi^\dagger T^aD^a\phi. \end{aligned} \tag{9.76}$$

The gauge covariant derivative is given by

$$\mathscr{D}_\mu = \partial_\mu + igT^aV^a_\mu. \tag{9.77}$$

Finally we eliminate the auxiliary fields by using their equations of motion, given by

$$F_i^\dagger = -\frac{\partial W}{\partial \phi_i} \quad \text{and} \quad D^a = -\sum_i g\phi_i^\dagger T_i^a\phi_i, \tag{9.78}$$

where $T_i^a$ is the generator of the representations of $\phi_i$ of the group. The effective scalar potential now becomes

$$V(\phi_i) = \sum_i \left|\frac{\partial W}{\partial \phi_i}\right|^2 + \frac{1}{2}g^2\sum_a\left(\sum_i \phi_i^\dagger T_i^a\phi_i\right)^2. \tag{9.79}$$

There is an additional contribution coming from the $D$-term in this case.

It is possible to break the gauge symmetry without breaking supersymmetry in the same way we discussed earlier. Consider the superpotential

$$W(X,Y) = fX(Y^2-M^2), \tag{9.80}$$

where $X$ is a singlet under the gauge group but $Y$ transforms nontrivially. The minimization of this superpotential will lead to vanishing of the $F$-terms preserving supersymmetry with nonvanishing *vev* of $Y$, which will break the group. This will provide mass to the gauge bosons. Since supersymmetry is unbroken, the gauginos will also get mass equal to the mass of the gauge bosons. After supersymmetry breaking the masses of the gauge bosons and the gauginos will differ and the mass difference will be determined by the supersymmetry breaking mechanism and the scale.

There are now chiral superfields corresponding to the fermions as well as to the scalars. To distinguish the fermions from scalars, in some cases a discrete matter parity is imposed to distinguish the two types of chiral superfields. A schematic chart of the superfields corresponding to the ordinary particles of the standard model is depicted in figure 9.3.

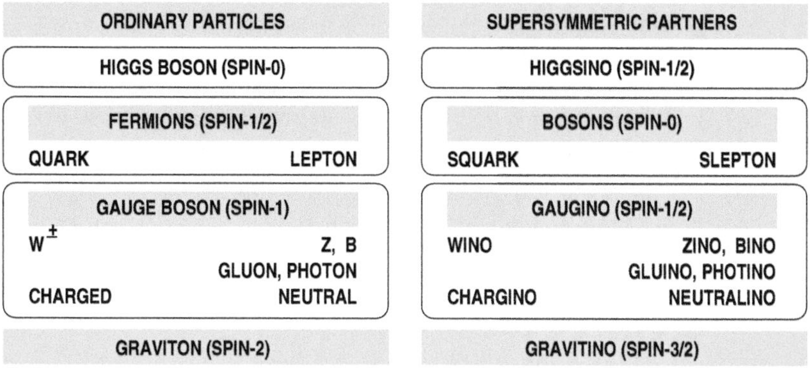


**FIGURE 9.3**
Ordinary particles and their superpartners in supersymmetric theories.

## 9.4 Supersymmetry Breaking

Although supersymmetry solves several theoretical problems and we expect it should be there, so far we have not seen it. So, even if there is supersymmetry in nature, it should be broken at energies higher than the electroweak symmetry breaking scale. On the other hand, the solution to the gauge hierarchy problem would dictate that supersymmetry should be broken near the electroweak symmetry breaking scale. Therefore, we expect that supersymmetry should be broken at around a few TeV.

The supersymmetry breaking mechanism determines the spectrum of particles after the breaking. However, there are several possible mechanisms of supersymmetry breaking, which introduces uncertainty in the low-energy predictions of supersymmetry. In the following we present two generic supersymmetry breaking mechanisms.

### Spontaneous Supersymmetry Breaking

Any spontaneous symmetry breaking implies that the symmetry of the theory is not respected by the vacuum. But the vacuum should respect the Lorentz invariance, which implies that only a scalar field can acquire a vacuum expectation value (*vev*). When a scalar field acquires a *vev*, the vacuum energy can still vanish in a supersymmetric theory, if both the $F$-term and the $D$-term vanish. However, if either the $F$-term or the $D$-term becomes nonvanishing after any scalar field acquires *vev*, the vacuum energy becomes nonvanishing and supersymmetry is broken spontaneously. When supersymmetry is spontaneously broken, there will be a Nambu–Goldstone fermion corresponding to the fermionic supersymmetry generator. This massless fermion is a problem in any theory. However, in supergravity theories, when the supersymmetry generators become local, this Nambu–Goldstone fermion combines with the gravitino and the gravitino becomes massive.

We shall present simple examples where the $F$-term and the $D$-term can become nonvanishing after any scalar field acquires *vev*. We discussed spontaneous symmetry breaking in supersymmetric theories, where some scalars can acquire *vev*, breaking the internal symmetry group but preserving supersymmetry. We now consider an example, which will not allow any minima that correspond to the vanishing of all the $F$-terms, and supersymmetry will be spontaneously broken after the scalars acquire *vev*. This is known as the *O'Raifeartaigh mechanism* [146]. Let us consider three superfields $A, B, C$, which could even be singlets under any internal gauge symmetry. If the superpotential is given by

$$W(A,B,C) = mAB + fC(B^2 - \mu^2), \tag{9.81}$$

where $f, g$ are couplings and $m$ and $\mu$ are some mass parameters, the $F$-terms are given by

$$\begin{aligned} F_A^\dagger &= \frac{\partial W}{\partial A} = mB \\ F_B^\dagger &= \frac{\partial W}{\partial B} = mA + 2fCB \\ F_C^\dagger &= \frac{\partial W}{\partial C} = f(B^2 - \mu^2). \end{aligned} \tag{9.82}$$

It is clear that all the $F$-terms cannot be made to vanish simultaneously. Since the scalar potential $V = \sum_i |F_i|^2$ vanishes only when all the $F$-terms vanish, the minimum energy now corresponds to nonvanishing of the potential. Thus, supersymmetry is broken spontaneously with this superpotential.

The minima of the potential correspond to a nonvanishing $F$-term, which is $\sqrt{f}\,\mu$ and determines the scale of supersymmetry breaking. The corresponding Nambu–Goldstone fermion, $\psi_C$, remains massless and there is a mass splitting among the bosonic and fermionic components of $B$. The mass splitting is of the order of supersymmetry breaking scale.

To demonstrate the $D$-term supersymmetry breaking [147], consider a $U(1)$ gauge theory with chiral superfields $\Phi_i$ with charge $e$. Any term proportional to the vector superfield ($V$) may be added to the Lagrangian without breaking the gauge symmetry or Lorentz invariance

$$\mathscr{L} = \frac{1}{32}(W^a W_a)_F + (\Phi^\dagger \exp^{2eV} \Phi)_D + \xi(V)_D, \tag{9.83}$$

where $\xi$ is a constant. The $D$-term is now given by

$$D = -(\xi + e\phi^\dagger \phi). \tag{9.84}$$

When the two terms have opposite sign, i.e., $\xi e < 0$, the minimum of the potential will correspond to $\langle \phi \rangle \neq 0$ but the $D$-term vanishes, whereas for $\xi e > 0$ the minimum would correspond to $< \phi >= 0$ and $V = D^2/2$. In the latter case the gauge symmetry remains unbroken but breaks supersymmetry spontaneously, and the mass splitting between particles and their superpartners becomes $\sqrt{e\xi}$.

Although both the $F$- and $D$-term spontaneous supersymmetry breaking looks attractive and allows mass-splitting between the particles and their superpartners, it is not possible to get a phenomenologically acceptable mass spectrum. One of the two bosonic superpartners of any fermions always comes out to be lighter than the fermions. Since we have not observed any superpartners of the known fermions in nature, we need to make these bosonic superpartners heavier than the electroweak symmetry breaking scale. Because of this, both the spontaneous supersymmetry mechanisms have been abandoned in their simplest form. We now expect that the supersymmetry breaking will come from some hidden sector and at low energy it will appear as some kind of explicit symmetry breaking by soft terms.

### Explicit Supersymmetry Breaking by Soft Terms

One solution to the problem of light bosonic superpartner is to introduce soft mass terms for the superpartners, which breaks supersymmetry explicitly. Although in the low-energy theory we introduce these explicit supersymmetry breaking soft terms, we expect that these terms will originate from some definite supersymmetry breaking mechanism at some high scale. In fact, in some simple $N = 1$ supergravity theory, such soft supersymmetry breaking terms do emerge after the supersymmetry breaking in the hidden sector and are communicated to the observable sector through gravity interactions. In any explicit symmetry breaking the main concern is the renormalizability of the original theory, which is protected if the explicit symmetry breaking terms are all soft terms with mass dimension less than 4. If the symmetry breaking terms have dimension 4 or more, then the higher order corrections will introduce divergent terms which will destroy the symmetries of the original theory. However, if the symmetry breaking terms are only the soft terms, the renormalizability of the original theory will not be lost.

The mass terms for the scalars, the gauginos, and the bilinear and trilinear scalar interactions could be the possible soft terms that break supersymmetry:

$$\begin{aligned}\mathscr{L}_{soft} &= -\frac{1}{2}M^a_\lambda \lambda^a \lambda^a - \frac{1}{2}(m^2)_{ij}\phi_i\phi_j^* \\ &\quad -\frac{1}{2}(BM)_{ij}\phi_i\phi_j - \frac{1}{6}(Ay)_{ijk}\phi_i\phi_j\phi_k + H.c.\end{aligned} \tag{9.85}$$

The bilinear and the trilinear terms are referred to as $A$- and $B$-terms, respectively. The mass parameters in the soft supersymmetry breaking terms could be of the order of the supersymmetry breaking scale. Although the soft terms could introduce mass terms for the gauginos, the mass terms for the gauge bosons are not allowed by requirement of gauge invariance. We hope that a complete theory will also provide an understanding of the origin of these soft terms. But for phenomenological studies we shall include all possible soft terms allowed by the symmetries of the theory.

In the next chapter we shall make use of the supersymmetric formalism presented so far and construct supersymmetric models. We shall first supersymmetrize the standard model and then add all possible supersymmetry breaking soft terms, which becomes the so-called minimal supersymmetric standard model with the gauge group $SU(3)_c \times SU(2)_L \times U(1)_Y$. Below the electroweak symmetry breaking, this theory gives us the nonsupersymmetric standard model. We then consider possible supersymmetry breaking mechanisms, which may restrict the allowed soft terms and reduce the numbers of parameters in the minimal supersymmetric standard model. In these phenomenologically interesting models, supersymmetry breaking scale is taken to be around a few TeV, so the electroweak symmetry breaking scalar mass is protected. This will also allow these models to be tested in the next generation accelerators. The lightest supersymmetric particle could also become the dark matter candidate.

# 10

# *Supersymmetric Models*

Supersymmetry provides a solution to the gauge hierarchy problem and promises a unified theory of Poincaré group with the internal symmetry groups. It is also a beautiful theory by itself, since it unifies the concept of fermions and bosons and keeps them in the same supermultiplets. Since supersymmetry has not been observed in nature, it must be broken at some higher energies, if it exists. So, the phenomenology of supersymmetry will depend strongly on the supersymmetry breaking mechanism, while to see any signature of supersymmetry we need to consider a supersymmetry breaking mechanism. To overcome this problem, some supersymmetry breaking mechanism has been proposed from theoretical considerations and then the predictions are generalized to study their phenomenological consequences. For example, supergravity induced supersymmetry breaking gives us the soft supersymmetry breaking terms. But there is no unique theory of supergravity that can be used to calculate the possible soft supersymmetry breaking terms at low energy. So, in a phenomenological model of supergravity induced supersymmetry breaking, all possible soft terms will be included with proper constraints.

With this general concept of supersymmetric models, we shall now discuss a few phenomenological models of supersymmetry and their consequences. We start with the simplest supersymmetric extension of the standard model of particle physics, called the minimal supersymmetric standard model (MSSM). These phenomenological models of supersymmetry have several interesting predictions for the next generation accelerators.

## 10.1 Minimal Supersymmetric Standard Model

The minimal supersymmetric standard model, also referred to as MSSM, is an extension of the standard model of electroweak theory, where all particles and interactions are made supersymmetric and all possible supersymmetry breaking soft terms, whose origin is not specified, are included. The uncertainty in determining the soft terms comes from our lack of knowledge about the supersymmetry breaking mechanism. *R*-parity invariance is imposed to eliminate fast baryon and lepton-number violating terms. This ensures that at any vertex, superpartners of the standard model particles should appear in pairs. We shall discuss the *R*-parity and some of the symmetry breaking mechanisms at the end.

In this chapter, we follow the conventions:

> We use the same notation for the superfields and the usual particles. In the superpotential any field would correspond to a superfield, while in the scalar potential or usual interaction terms the same field would correspond to the usual particles of the standard model;
>
> The superpartners are denoted by a tilde, for example, $\widetilde{W}^{\pm}$ for the fermionic superpartner of the charged gauge bosons $W^{\pm}$.

**Particle Contents**

The gauge symmetry of the minimal supersymmetric standard model is the same as the standard model gauge group $SU(3)_c \times SU(2)_L \times U(1)_Y$. So there are massless gauge bosons, namely, 8 gluons corresponding to the group $SU(3)_c$, 3 gauge bosons corresponding to the group $SU(2)_L$, and one generator of the group $U(1)_Y$. Since these gauge bosons belong to vector superfields, there will also be the corresponding supersymmetric partners, the gauginos. They are 8 gluinos, 2 charged Winos, and 2 neutralinos. These neutralinos are now neutral fermions, which mix with the neutral components of the Higgs scalars. As a result, they cannot be treated independently of the superpartners of the neutral Higgs bosons. There are two Higgs scalars, $H_1$ and $H_2$, which transform under $SU(3)_c \times SU(2)_L \times U(1)_Y$ as $(1,2,-1/2)$ and $(1,2,1/2)$, respectively. The two Higgs scalars are required, since the superpartners of the Higgs scalars are fermions and contribute to the axial anomaly. So, anomaly cancellation requires the presence of both scalars. The two Higgs scalars are also required for giving mass to the up and the down quarks. Unlike the nonsupersymmetric theories, one Higgs scalar alone cannot give mass to both the up and the down quarks. So there will be four neutralinos which are combinations of two gauginos and two Higgsinos.

The supersymmetry breaking scale is assumed to be around the TeV scale to ensure that the electroweak symmetry breaking scale is not destabilized by the quadratic divergences coming from some high scale. So, at the time of electroweak symmetry breaking, this model appears to be similar to the nonsupersymmetric model. However, the couplings are now restricted by the requirement of supersymmetry. For example, the same Higgs field cannot couple to the up and the down quarks even after the symmetry breaking, and we need at least two Higgs scalars to give masses to the up and the down quarks. Soft supersymmetry breaking terms are added to give masses to the gauginos, while keeping the gauge bosons massless, since the supersymmetry breaking can introduce mass splitting of the order of supersymmetry breaking. Such soft supersymmetry breaking terms may have their origin from supergravity, in which case they are naturally of the order of the supersymmetry breaking scale. Other allowed soft terms are also included in the model.

We include the quarks and the leptons of three generations in the minimal supersymmetric standard model. Since they are components of supersymmetry multiplets, the scalar superpartners squarks and sleptons will also be present in the theory. They transform under the gauge group $SU(3)_c \times SU(2)_L \times U(1)_Y$ in the same way as the quarks and leptons. The sfermions transform as scalars under the Lorentz group and handedness does not make any sense. But in supersymmetric models the left-handed fermions $\psi_L$ are included in left-chiral superfields $\Psi_L$, while the right-handed fermions $\psi_R$ are replaced by their $CP$ conjugates ${\psi^c}_L$ and belong to the conjugate superfield $\Psi^c_L$. To distinguish the partner of $\psi_L$ from the partner of ${\psi^c}_R$, we use the subscripts $[L,R]$ for the scalar partners and denote them as $\phi_L \subset \Psi_L$ and $\phi_R \subset \Psi^c_R$, although $\phi_L$ and $\phi_R$ transform in the same way under the Lorentz group.

The supersymmetry breaking soft terms that gives masses to the sfermions and are proportional to the supersymmetry breaking scale are also included; hence, the sfermions will have masses around the same scale. The hierarchy in the masses of the sfermions is not large, since they all have masses within the range of a few GeV to a few TeV. As a result, if supersymmetry is observed, we expect to see a proliferation of new particles given by all of these new superparticles and their couplings to ordinary matter in the planned next generation accelerators at LHC and ILC. At present these soft terms are arbitrary, and we are not in a position to predict these masses. But if they are observed in experiments, we can study their phenomenology.

The fermion masses come from the couplings of the fermion superfields with the Higgs scalars. The superpotential containing the Yukawa couplings of the up and the down quarks can be written as

$$W = h_d \, QH_1D^c + h_u QH_2U^c + H.c. \tag{10.1}$$

The $F$-term of this superpotential will give the Yukawa couplings which can give masses to the up and the down quarks after the Higgs scalars $H_1$ and $H_2$ acquire *vevs*. The conjugate of the Higgs superfields ${H_1}^\dagger$ and ${H_2}^\dagger$ are right-chiral superfields and will enter in the Hermitian conjugate part of this superpotential. Both Higgs scalars ${H_1}^\dagger$ and ${H_2}^\dagger$ are required for anomaly cancellation. All particles of the minimal supersymmetric standard model are listed in table 10.1.

### Interactions in the MSSM

The minimal supersymmetric standard model contains the usual quarks and leptons, gauge bosons corresponding to the standard model gauge group, and two Higgs scalars. The superpartners of all these particles are also present, although the mass splitting between the ordinary particles and the superpartners is of the order of supersymmetry breaking scale and coming from the soft terms.

The supersymmetric Lagrangian also contains all the interactions of the standard model with some restrictions coming from supersymmetry. The main constraints are in the scalar potential. We write the supersymmetric Lagrangian in terms of the superfields, which will contain all the interactions [148].

To write the gauge interactions, we first define the spinor field-strength tensors for the three groups as

$$\begin{aligned} G^a_\alpha &= \bar{D}^2 D_\alpha G^a + i g_s f^{abc} \bar{D}^2 (D_\alpha G^b) G^c \\ W^i_\alpha &= \bar{D}^2 D_\alpha W^i + i g \varepsilon^{ijk} \bar{D}^2 (D_\alpha W^j) W^k \\ B_\alpha &= \bar{D}^2 D_\alpha B, \end{aligned} \tag{10.2}$$

**TABLE 10.1**
Chiral and vector superfields in the minimal supersymmetric standard model, their components, and transformations under the standard model gauge groups $G_{std} \equiv SU(3)_c, SU(2)_L$ and $U(1)_Y$.

| Chiral superfields | Component fields | Component superfields | $G_{std}$ quantum numbers |
|---|---|---|---|
| $Q$ | $\begin{pmatrix} u_L \\ d_L \end{pmatrix} = Q_L$ | $\begin{pmatrix} \tilde{u}_L \\ \tilde{d}_L \end{pmatrix} = \tilde{Q}$ | $(3, 2, 1/6)$ |
| $U^c$ | $u_L^c$ | $\tilde{u}_L^c$ | $(\bar{3}, 1, -2/3)$ |
| $D^c$ | $d_L^c$ | $\tilde{d}_L^c$ | $(\bar{3}, 1, 1/3)$ |
| $L$ | $\begin{pmatrix} \nu_{eL} \\ e_L^- \end{pmatrix} = L_L$ | $\begin{pmatrix} \tilde{\nu}_{eL} \\ \tilde{e}_L^- \end{pmatrix} = \tilde{L}$ | $(1, 2, -1/2)$ |
| $E^c$ | $e_L^+$ | $\tilde{e}_L^+$ | $(1, 1, 1)$ |
| $H_2$ | $\begin{pmatrix} H_2^+ \\ H_2^0 \end{pmatrix} = H_2$ | $\begin{pmatrix} \tilde{H}_2^+ \\ \tilde{H}_2^0 \end{pmatrix} = \tilde{H}_2$ | $(1, 2, 1/2)$ |
| $H_1$ | $\begin{pmatrix} H_1^0 \\ H_1^- \end{pmatrix} = H_1$ | $\begin{pmatrix} \tilde{H}_1^0 \\ \tilde{H}_1^- \end{pmatrix} = \tilde{H}_1$ | $(1, 2, -1/2)$ |
| **Vector superfields** | **Gauge bosons** | **Gauginos** | **$G_{std}$ quantum numbers** |
| $G^a$ | $G^a$ | $\tilde{G}^a$ | $(8, 1, 0)$ |
| $W^i$ | $\begin{pmatrix} W^{\pm} \\ W^3 \end{pmatrix} = W^i$ | $\begin{pmatrix} \tilde{W}^{\pm} \\ \tilde{W}^3 \end{pmatrix}$ | $(1, 3, 0)$ |
| $B$ | $B$ | $\tilde{B}$ | $(1, 1, 0)$ |

where $f^{abc}$ is the structure function of the $SU(3)_c$ group and for the $SU(2)_L$ group the structure function is the totally antisymmetric tensor $\varepsilon_{ijk}$ with $\varepsilon_{123} = 1$. The gauge interaction of the vector superfields and their kinetic energies would now be given by

$$\mathscr{L}_V = \frac{1}{64}\left[G^{i\alpha}G^i_\alpha + G^{i\,\dagger}_\alpha G^{i\alpha\dagger} + W^{i\alpha}W^i_\alpha + W^{i\,\dagger}_\alpha W^{i\alpha\dagger} + 2B^\alpha B_\alpha\right]_F. \tag{10.3}$$

The interactions of the gauge multiplets with the chiral superfields are contained in the $D$-term where both the chiral superfields and the conjugate superfields will enter in addition to the vector superfields. The $D$-term in the Lagrangian is given by

$$\begin{aligned}\mathscr{L}_D = \Bigg[ & Q^\dagger \exp^{2\left(g_s T^p G^p + g\sigma^m W^m + \frac{g'}{6}B\right)} Q \\ & + U^{c\dagger} \exp^{2\left(g_s T^p G^p - \frac{2g'}{3}B\right)} U^c + D^{c\dagger} \exp^{2\left(g_s T^p G^p + \frac{g'}{3}B\right)} D^c \\ & + L^\dagger \exp^{2\left(g\sigma^m W^m - \frac{g'}{2}B\right)} L + E^{c\dagger} \exp^{2(g'B)} E^c \\ & + H_2^\dagger \exp^{2\left(g\sigma^m W^m + \frac{g'}{2}B\right)} H_2 + H_1^\dagger \exp^{2\left(g\sigma^m W^m - \frac{g'}{2}B\right)} H_1 \Bigg]_D \quad . \qquad (10.4)\end{aligned}$$

Here, $T^p$ and $\sigma^m$ are the generators of the groups $SU(3)_c$ and $SU(2)_L$, respectively and we have suppressed the generation and group indices in the different chiral fields. This Lagrangian will include all the interactions of the gauge bosons and the gauginos with the fermions and the scalars.

The superpotential contains the scalar interactions, Yukawa couplings, and the interactions of the superpartners of the chiral superfields

$$W_F = \left[h_u Q H_2 U^c + h_d Q H_1 D^c + h_e L H_1 E^c + \mu H_2 H_1\right]_F . \qquad (10.5)$$

Generation indices are suppressed. The $F$-term of the superpotential includes the Yukawa coupling terms. When the Higgs fields acquire *vevs* they give masses to the fermions through these terms.

The most general potential includes more terms, which violate baryon and lepton numbers very strongly [149]. To avoid these catastrophic terms in the minimal supersymmetric standard model, an additional discrete symmetry is imposed, which is known as the $R$-parity and defined as [150, 151]

$$R = (-1)^{3(B-L)+2S} \quad \text{or} \quad R = (-1)^{3B+L+2S}, \qquad (10.6)$$

where $S$ is the spin of the particle and $B$ and $L$ are the baryon and lepton numbers. Both the definitions have the same consequences. $R$-parity implies that all ordinary particles will have even $R$-parity, while the superpartners will have odd $R$-parity. The $R$-parity conservation then implies

(i) The renormalizable baryon and lepton-number violating terms are absent.

(ii) In any vertex, superparticles will enter in pairs. So, when a superparticle decays, the decay products contain at least one superparticle.

(iii) The lightest of the superparticles cannot decay and must be absolutely stable. This lightest supersymmetric particle (LSP) could be a dark matter candidate. The LSP is neutral and color singlet and it interacts with other particles very weakly.

In the minimal supersymmetric standard model $R$-parity conservation simplifies the analysis substantially. In some extensions of the minimal supersymmetric standard model $R$-parity violation is considered, which will be discussed later.

We write all the possible soft supersymmetry breaking terms that are allowed by the standard model gauge symmetry and $R$-parity:

$$\begin{aligned}
-\mathscr{L}_{soft} &= m_1^2|H_1|^2 + m_2^2|H_2|^2 - B\mu(H_1H_2 + H.c.) \\
&+ M_{\tilde{Q}}^2(\tilde{u}_L^*\tilde{u}_L + \tilde{d}_L^*\tilde{d}_L) + M_{\tilde{u}}^2\tilde{u}_R^*\tilde{u}_R + M_{\tilde{d}}^2\tilde{d}_R^*\tilde{d}_R \\
&+ M_{\tilde{L}}^2(\tilde{e}_L^*\tilde{e}_L + \tilde{\nu}_L^*\tilde{\nu}_L) + M_{\tilde{e}}^2\tilde{e}_R^*\tilde{e}_R \\
&+ \frac{1}{2}\left[M_3\bar{\tilde{g}}\tilde{g} + M_2\bar{\tilde{\omega}}_i\tilde{\omega}_i + M_1\bar{\tilde{b}}\tilde{b}\right] \\
&+ \frac{g}{\sqrt{2}M_W}\left[\frac{M_d}{\cos\beta}A_dH_1\tilde{Q}\tilde{d}_R^* + \frac{M_u}{\sin\beta}A_uH_2Q\tilde{u}_R^*\right. \\
&+ \left.\frac{M_e}{\cos\beta}A_eH_1\tilde{L}\tilde{e}_R^* + H.c.\right],
\end{aligned} \tag{10.7}$$

where the angle $\tan\beta = v_2/v_1$ is the ratio of the *vevs* of the two Higgs doublets $H_1$ and $H_2$. The soft terms contain the masses of the superpartners of the ordinary fermions and gauge bosons, the trilinear A-terms that allow mixing of the scalar partners of the left-handed and right-handed fermions when the Higgs scalars acquire *vev* and the B-terms that mix the scalar components of the two Higgs doublets. Our lack of knowledge about the origin of the soft terms increases the unknown parameters in the model. In spite of that, the minimal supersymmetric standard model provides us with very rich phenomenology.

**Higgs Potential and Electroweak Symmetry Breaking**

We can now write the complete Lagrangian for the minimal supersymmetric standard model as

$$\mathscr{L} = \mathscr{L}_V + \mathscr{L}_D + \mathscr{L}_F + \mathscr{L}_{soft}, \tag{10.8}$$

where the different parts of the Lagrangian are given in equations (10.3), (10.4), (10.5) and (10.7). The different terms give the gauge interactions, the interactions of the gauge bosons with the fermions and scalars and also the kinetic energy terms for the superfields, the Yukawa interactions and the scalar interactions, and the supersymmetry breaking soft terms, which also provide the mass splitting between the ordinary particles and their superpartners.

Let us now write the scalar potential containing the masses and self interactions of the Higgs scalars

$$\begin{aligned}
V &= \frac{g^2}{2}|H_1^*H_2|^2 + \frac{g^2+g'^2}{8}\left(|H_2|^2 - |H_1|^2\right)^2 \\
&+ |\mu|^2\left(|H_1|^2 + |H_2|^2\right) \\
&+ m_1^2|H_1|^2 + m_2^2|H_2|^2 + B\mu H_1H_2 + H.c.
\end{aligned} \tag{10.9}$$

The terms in the first line with couplings $g$ and $g'$ come from the $D$-terms corresponding to the $SU(2)_L$ and $U(1)_Y$ gauge groups, and the terms in the second line are the $F$-terms. The terms in the last line are the soft supersymmetry breaking masses and the bilinear B-term. In terms of the $SU(2)_L$ components, these terms can be written as

$$\begin{aligned} |H_1|^2 &= |H_1^0|^2 + |H_1^-|^2 \\ |H_2|^2 &= |H_2^+|^2 + |H_2^0|^2 \\ H_1 H_2 &= H_1^0 H_2^0 + H_1^- H_2^+ \, . \end{aligned} \tag{10.10}$$

The sneutrino can also mix with the Higgs scalars and can even acquire *vev* in some cases, which is constrained by phenomenology [151].

The parameters of the potential are more restrictive than in the standard model. In the standard model with two Higgs doublets, there are six free parameters and a phase, whereas for the minimal supersymmetric standard model there are only three independent combinations of parameters

$$|\mu|^2 + m_1^2, \quad |\mu|^2 + m_2^2, \text{ and } \mu B, \tag{10.11}$$

and there is no $CP$ phase in $V$. When these parameters satisfy the relations

$$\begin{aligned} (\mu B)^2 &> (|\mu|^2 + m_1^2)(|\mu|^2 + m_2^2) \\ |\mu|^2 + \frac{m_1^2 + m_2^2}{2} &> |\mu B|, \end{aligned} \tag{10.12}$$

the Higgs fields acquire *vevs*

$$< H_{1,2}^0 >= v_{1,2}, \tag{10.13}$$

where $v_{1,2}$ can be given in terms of the three independent parameters mentioned above and can be chosen to be positive. This breaks the electroweak symmetry, and the $W$ bosons get a mass

$$M_W^2 = \frac{g^2}{2}(v_1^2 + v_2^2). \tag{10.14}$$

The experimental value of the $W$ boson mass eliminates one free parameter.

The three components of the Higgs scalars will become the longitudinal modes of the massive gauge bosons $W^{\pm}, Z$ leaving five other degrees of freedom of the two complex Higgs doublets $H_1$ and $H_2$. These five components are

(i) a $CP$-odd neutral pseudoscalar Higgs boson $A$,

(ii) two $CP$-even Higgs fields $h$ and $H$, with $M_h < M_H$ and

(iii) the charged Higgs bosons $H^{\pm}$.

The masses of the physical Higgs scalars are given by

$$\begin{aligned}
M_A^2 &= \frac{2 \mid \mu B \mid}{\sin 2\beta} \\
M_{h,H}^2 &= \frac{1}{2}\left\{M_A^2 + M_Z^2 \mp \left((M_A^2 + M_Z^2)^2 - 4M_Z^2 M_A^2 \cos^2 2\beta\right)^{1/2}\right\} \\
M_{H^\pm}^2 &= M_W^2 + M_A^2,
\end{aligned} \tag{10.15}$$

where the angle $\beta$ is defined as

$$\tan\beta = \frac{v_2}{v_1}. \tag{10.16}$$

The angle $\beta$ lies in the range $0 \leq \beta \leq \pi/2$ because $v_{1,2}$ are positive. In terms of the mass eigenstates the Higgs fields are given by

$$\begin{aligned}
H_1 &= \begin{pmatrix} v_1 + \frac{1}{\sqrt{2}}(H\cos\alpha - h\sin\alpha + iA\sin\beta) \\ H^- \sin\beta \end{pmatrix} \\
H_2 &= \begin{pmatrix} H^+ \cos\beta \\ v_2 + \frac{1}{\sqrt{2}}(H\sin\alpha + h\cos\alpha + iA\cos\beta) \end{pmatrix},
\end{aligned} \tag{10.17}$$

where the Higgs mixing angle $\alpha$ is given by

$$\tan 2\alpha = \tan 2\beta \left[\frac{M_H^2 + M_h^2}{M_A^2 - m_Z^2}\right]. \tag{10.18}$$

We can now choose two of the parameters as free parameters and determine all others only in terms of these two parameters. In one convention $\tan\beta$ and $M_A$ are considered as the free parameters and all other quantities are expressed in terms of these two parameters.

These expressions can be combined to predict

$$M_{H^+} > M_W, \quad M_H > M_Z, \quad M_h < M_A, \quad \text{and} \quad M_h < M_Z|\cos 2\beta|. \tag{10.19}$$

These relations are tree-level predictions and loop corrections will modify them. For example, loop corrections make the lightest Higgs $h$ heavier than the $Z$ boson because of the corrections from the top and the squarks. However, this does not change the predictions too drastically. Including all corrections and taking a large $\tan\beta$, the limit on the lightest of the neutral Higgs comes out to be [152]

$$M_h < 130 GeV, \tag{10.20}$$

which is definitely within the reach of LHC.

The Higgs sector is described by two parameters $M_h$ and $\tan\beta$, although a certain region of this parameter space has been ruled out by experiments. Experimental bound on the supersymmetric Higgs mass [153] for all values of $\tan\beta$ is less stringent than the nonsupersymmetric Higgs bound due to suppression of their couplings to the vector bosons. In some models this bound becomes weaker since there are invisible decay modes such as $h, A \to \tilde{\chi}_1^0 \tilde{\chi}_1^0$ with significant branching ratios.

One of the main aims of LHC is to look for the Higgs scalars. Since the mass of the lightest Higgs scalar in the supersymmetric models is constrained to be lower than 135 GeV, it is expected that LHC will be able to find the Higgs. One can complicate the model and go beyond the minimal supersymmetric standard model, but in any extensions the lightest neutral Higgs cannot be made heavier than 150 GeV, which is also within the reach of LHC. If the neutral Higgs is not seen at this energy then the supersymmetric models will be in trouble.

**Spectrum of Superparticles**

The masses of the fermions come from the Yukawa terms. One can now invert the problem to get the Yukawa couplings in equation (10.5) in terms of the fermion masses,

$$h_{e,d} = \frac{g m_{e,d}}{\sqrt{2} M_W \cos\beta} \qquad h_u = \frac{g m_u}{\sqrt{2} M_W \sin\beta}, \tag{10.21}$$

where $g$ is the $SU(2)_L$ gauge coupling constant. The Higgs boson couplings to fermions are now determined in terms of the two parameters $M_A$ and $\tan\beta$, which leaves only $\mu$ as a free parameter in the superpotential.

The gauge coupling constants give the interactions of the Higgs bosons to the gauge bosons. To lowest order, the neutral Higgs boson couplings to fermions are specified in terms of the two variables $M_A$ and $\tan\beta$:

$$\begin{aligned} \mathscr{L} = & -\frac{g m_u}{2 M_W \sin\beta}(\cos\alpha\, \overline{u}uh + \sin\alpha\, \overline{u}uH + \cos\beta\, \overline{u}\gamma_5 uA) \\ & -\frac{g m_d}{2 M_W \cos\beta}(\sin\alpha\, \overline{d}dh + \cos\alpha\, \overline{d}dH + \sin\beta\, \overline{d}\gamma_5 dA) \\ & -\frac{g m_e}{2 M_W \cos\beta}(\sin\alpha\, \overline{e}eh + \cos\alpha\, \overline{e}eH + \sin\beta\, \overline{e}\gamma_5 eA). \end{aligned} \tag{10.22}$$

The dependence on the squark masses and the mixing parameters cannot be neglected when radiative corrections are included. For large $M_A$ (of the order of 300GeV or higher), the Higgs couplings become close to the nonsupersymmetric standard model couplings.

The masses of the superpartners come mainly from the soft mass terms, which depend on the scale of supersymmetry breaking. The trilinear $A$-terms give the mixing between the scalar partners of the left-handed and right-handed fermions. To see the spectrum of the superpartners, we can write the mass matrices for the scalar partners of the up and the down quarks and the charged leptons in the basis $[\tilde{f}_L \ \tilde{f}_R]$,

$$M_{\tilde{u}}^2 = \begin{pmatrix} M_{\tilde{Q}}^2 + m_u^2 + m_{1u}^2 & m_u(A_u - \mu\cot\beta) \\ m_u(A_u - \mu\cot\beta) & M_{\tilde{u}}^2 + m_u^2 + m_{2u}^2 \end{pmatrix}$$

$$M_{\tilde{d}}^2 = \begin{pmatrix} M_{\tilde{Q}}^2 + m_d^2 + m_{1d}^2 & m_d(A_d - \mu\tan\beta) \\ m_d(A_d - \mu\tan\beta) & M_{\tilde{d}}^2 + m_d^2 + m_{2d}^2 \end{pmatrix}$$

$$M_{\tilde{e}}^2 = \begin{pmatrix} M_{\tilde{L}}^2 + m_e^2 + m_{1e}^2 & m_e(A_e - \mu\tan\beta) \\ m_e(A_e - \mu\tan\beta) & M_{\tilde{e}}^2 + m_e^2 + m_{2e}^2 \end{pmatrix}, \tag{10.23}$$

where we defined

$$\begin{aligned} m_{1a}^2 &= M_Z^2 (T_{3L}^a - Q^a \sin^2\theta_W)\cos 2\beta \\ m_{2a}^2 &= Q^a M_Z^2 \sin^2\theta_W \cos 2\beta \qquad (a = u,d,e), \end{aligned} \tag{10.24}$$

with $T_{3L}^a$ and $Q^a$ as the $SU(2)_L$ quantum numbers and electric charges of $a = u_L, d_L$ and $e_L$. The various mixings depend on the soft parameters $A_i$ and $\tan\beta$. For large $\tan\beta$ the mixings in the down quark sector and the charged lepton sector are large. Depending on the parameters $A_i$, when the mixing is large the lightest squark will be a stop. The requirement that the stop mass does not become negative, and hence the stop acquire a *vev*, dictates a condition $|A_t| < M_{\tilde{Q},\tilde{u}}$.

The superpartners of the charged $SU(2)_L$ gauge bosons $W^\pm$ and the physical charged Higgs $H^\pm$ are the four charged Majorana fermions called *the charginos*. They mix with each other and the mass term for these charginos is given by

$$M_{ch} = -\frac{1}{2}\begin{pmatrix}\tilde{W}^- & \tilde{H}^-\end{pmatrix}\begin{pmatrix} m_2 & \sqrt{2}M_W\cos\beta \\ \sqrt{2}M_W\sin\beta & \mu \end{pmatrix}\begin{pmatrix}\tilde{W}^+ \\ \tilde{H}^+\end{pmatrix}. \tag{10.25}$$

A bi-unitary transformation (similar to the quark masses) can diagonalize this matrix to give two physical Dirac charginos with masses

$$\begin{aligned} m_{\tilde{\chi}_{1,2}^\pm}^2 = \frac{1}{2}\Big[ & m_2^2 + 2M_W^2 + \mu^2 \mp \big[(m_2^2 - \mu^2)^2 + 4M_W^4\cos^2 2\beta \\ & + 4M_W^2(m_2^2 + \mu^2 + 2m_2\mu\sin^2\beta)\big]^{\frac{1}{2}}\Big]. \end{aligned} \tag{10.26}$$

The usual convention is $M_{\chi_1} < M_{\chi_2}$.

The chargino mass vanishes for $\mu \to 0$. So, to keep the charginos heavier than the electroweak symmetry breaking scale, we need $\mu \neq 0$. On the other hand the Z-boson mass is related to the $\mu$-parameter through

$$m_Z^2 = 2\left[\frac{M_h^2 - M_H^2\tan^2\beta}{\tan^2\beta - 1}\right] - 2\mu^2. \tag{10.27}$$

The correct Z mass needs a cancellation of the term containing the Higgs mass with the $\mu$ term quite accurately. This fine tuning of the parameter $\mu$ is called the *$\mu$-problem*.

Finally we shall discuss the mass spectrum of the neutral fermions. There are four of them: the fermion superpartners of the neutral gauge bosons $B$ and $W^3$ denoted by $\tilde{b}$ and $\tilde{\omega}^3$, respectively and the fermion superpartners of the Higgs bosons, $\tilde{h}_1^0$ and $\tilde{h}_2^0$. All four mix with each other and the mass matrix is given in the basis $[\tilde{b}\ \ \tilde{\omega}\ \ \tilde{h}_1^0\ \ \tilde{h}_2^0]$

by

$$M_{\tilde{\chi}_i^0} = \begin{pmatrix} m_1 & 0 & -M_Z c_\beta s_W & M_Z s_\beta s_W \\ 0 & m_2 & M_Z c_\beta c_W & -M_Z s_\beta c_W \\ -M_Z c_\beta s_W & M_Z c_\beta c_W & 0 & -\mu \\ M_Z s_\beta s_W & -M_Z s_\beta c_W & -\mu & 0 \end{pmatrix}, \tag{10.28}$$

where $c_\beta = \cos\beta, s_\beta = \sin\beta, c_W = \cos\theta_W$, $s_W = \sin\theta_W$ and $\theta_W$ is the electroweak mixing angle. The physical masses and eigenstates are obtained by diagonalizing this mass matrix. The mass eigenstates are not the fermionic partners of the usual photon and Z. The photino state

$$\tilde{\gamma} = \tilde{W}^3 \sin\theta_W + \tilde{B}\cos\theta_W \tag{10.29}$$

becomes a mass eigenstate only when $m_1 = m_2$. For the physical neutralino mass eigenstates, the usual convention is $M_{\tilde{\chi}_1^0} < M_{\tilde{\chi}_2^0} < M_{\tilde{\chi}_3^0} < M_{\tilde{\chi}_4^0}$. The lightest of the four neutralinos, $\tilde{\chi}_1^0$, is usually assumed to be the lightest supersymmetric particle (LSP). All other particles would decay into the LSP. Since supersymmetric particles enter in pairs in any interaction due to the $R$-parity conservation, the LSP cannot decay into any other particles. Thus, the LSP remains stable and could be a candidate for the dark matter in the universe. The couplings of the LSP have to be very weak.

In any detector, the supersymmetric particles will produce a cascade of decays. But any such cascade should involve the LSP, since that is the LSP to which all other supersymmetric particles can decay. On the other hand, the weak coupling of the LSP will not allow it to be detected inside the detector. Thus, one of the most prominent signatures of a supersymmetric particle would be missing energy from the undetected LSP. Supersymmetric particles would produce a cascade of decays, but the final state would consist of leptons + jets + $E_T$(missing) [154]. The missing LSP will not allow the reconstruction of the masses of the supersymmetric particles completely, but still from a combination of signatures much information about supersymmetry will become available. The Majorana properties of the gluinos will generate some like-sign dileptons [155]. Three-lepton production is another typical signature of the minimal supersymmetric standard model. So far nonobservation of any supersymmetry signal at LEP, SLC and Tevatron has only restricted the parameter space, which includes direct searches as well as indirect searches including the precision measurements.

## 10.2 Supersymmetry Breaking Models

The minimal supersymmetric standard model does not specify the origin of the soft supersymmetry breaking terms and, hence, adds all possible soft terms in the Lagrangian. This makes the total numbers of free parameters 124. This sometimes makes the minimal supersymmetric standard model phenomenologically inconvenient [156]. This problem is partially solved when one tries to derive the soft terms from some other theory. For example, when supersymmetry is broken in the hidden sector of any supergravity theory, it is possible to obtain the soft masses. In this case the mass of the gravitino sets the mass scale in the theory and all mass parameters get related to the same mass scale at some high energy scale of grand unification. Although it is not possible to derive the complete theory from any supergravity theory, a phenomenological theory may be constructed in which the constraints that could come from the supergravity models on the soft terms of the minimal supersymmetric standard model are imposed. This reduces the numbers of free parameters and makes the theory more predictable in this supergravity inspired model or in the constrained minimal supersymmetric standard model. In other words, one assumes some *hidden world* where supersymmetry is broken, which then *induces the soft terms*. From the nature of the hidden world and how the supersymmetry breaking information is communicated to the ordinary world, we get some constraints on the minimal supersymmetric standard model parameters.

**Supergravity-Inspired Models**

The most popular scenario of supersymmetry breaking comes from the supergravity models. We shall discuss in the next chapter some formal aspects of the supergravity theory and how the soft terms originate in these theories. The requirement of unification of several parameters constrains some of the parameters of the minimal supersymmetric standard model. These theories are called the constrained minimal supersymmetric standard model (CMSSM) or the *supergravity-inspired* minimal supersymmetric standard model or sometimes mSUGRA model [157].

In the supergravity-inspired models, one conjectures that it is natural that some of the parameters were unified at very large scale without specifying the actual model. For example, along with the gauge coupling unification, one assumes that the gaugino masses ($M_i$) were also unified at the scale of grand unification ($M_U$),

$$M_i(M_U) = m_{1/2}. \tag{10.30}$$

The renormalization group equations then predict how they descend to low energy

$$\frac{dM_i}{dt} = -b_i \alpha_i M_i / 4\pi, \tag{10.31}$$

where $b_i$ are the usual beta functions and $\alpha_i = \frac{g_i^2}{4\pi}$ is related to the gauge coupling constants. The evolution of the parameters then gives the low-energy masses of the gauge fermions to be

$$\begin{aligned} M_2(M_W) &= \left( \frac{\alpha(M_W)}{\alpha_s(M_W) \sin^2\theta_W(M_W)} \right) M_3(M_W) \\ M_1(M_W) &= \frac{5}{3} \tan^2\theta_W(M_W) M_2(M_W), \end{aligned} \tag{10.32}$$

which makes the gluino the heaviest of the gaugino masses.

It is also assumed that there is a common scalar mass at the unification scale $(M_U)$,

$$m_1^2 = m_2^2 = M_{\tilde{Q}}^2 = M_{\tilde{u}}^2 = M_{\tilde{d}}^2 = M_{\tilde{L}}^2 = M_{\tilde{e}}^2 = m_0^2, \tag{10.33}$$

which gives $M_{h,H}^2(M_U) = m_0^2 + \mu^2$. Since the evolution of the squarks has some contribution from the $SU(3)_c$ gauge bosons, the squarks become heavier than the sleptons at low energy,

$$M_{\tilde{q}}^2(M_W) > M_{\tilde{\ell}}(M_W). \tag{10.34}$$

In this picture it is assumed that the supersymmetry will be broken at some intermediate scale in the hidden sector, which will be communicated to the observable sector through gravity interactions [158]. Since the hidden sector and how the supersymmetry breaking is communicated cannot distinguish the components of any multiplet of the gauge groups, any contribution coming from the hidden sector should be independent of the gauge transformation properties of the fields. In $N = 1$ supergravity models supersymmetry is broken at some intermediate scale $M_S \sim m \sim 10^{11}$ GeV and all the soft terms are generated with the same mass, which is the gravitino mass $m_{3/2}$ (see section 11.3)

$$m_{3/2} \sim M_{soft} \sim \frac{M_S^2}{M_{Pl}} \sim 1TeV. \tag{10.35}$$

All A-parameters are also equal at the unification scale, independent of the fields. This assumption leads to only 5 input parameters at the grand unification scale $M_U$ [155], which are, $m_0, m_{1/2}, A_0, \mu$ and the Higgs mixing parameter $B$. At low energy we then have only a few parameters from which the $Z$ mass restricts $|\mu B|$ and makes $\tan\beta$ a free parameter instead of $B$, so we have the new set of parameters

$$m_0, m_{1/2}, A_0, \tan\beta, \text{sign}(\mu). \tag{10.36}$$

One variation of this model assumes nonuniversal scalar masses at the unification scale, changing the low-energy phenomenology of the model. Theoretical considerations also suggest that the universality is not generic [159].

**Gauge-Mediated Supersymmetry Breaking**

There is a variant of the supergravity-inspired model called the *gauge-mediated supersymmetry breaking* model. Supersymmetry is again broken in the hidden sector, but the connection between the hidden sector and the low-energy phenomenology is now through a messenger chiral superfield, which transforms under the $SU(3)_c \times SU(2)_L \times U(1)_Y$ group nontrivially. When supersymmetry is broken by a gauge singlet superfield in the hidden sector, the messenger gauge nonsinglet chiral superfield obtains a mass $\Lambda$, and the supersymmetry breaking effect is communicated to the minimal supersymmetric standard model particles through gauge interaction [160]. The mass parameter $\Lambda$ is chosen to be of the order of

$$\Lambda \sim 10^4 - 10^5 GeV, \tag{10.37}$$

so the soft terms in the Lagrangian are of the order of a few TeV. The messenger gauge nonsinglet chiral superfield couples to the ordinary particles, which then communicate the supersymmetry breaking effect through the soft terms.

One-loop contributions give masses to the gauginos,

$$M_i \sim \frac{\alpha_i}{4\pi}\Lambda, \tag{10.38}$$

and scale in the same way as the corresponding gauge coupling constants at low energy. The scalar superpartners of the minimal supersymmetric standard model obtain mass at the two-loop level from diagrams involving the gauge fields and the messenger fields. The scalar masses are

$$\tilde{M}^2 \sim \left(\frac{\Lambda}{4\pi}\right)^2 \left\{\alpha_s^2 C_3 + \alpha_2^2 C_2 + \alpha_1^2 C_1\right\}, \tag{10.39}$$

where $C_i$ are the quadratic Casimir operators of the gauge groups $SU(3)_c \times SU(2)_L \times U(1)_Y$.

One of the problems with the minimal supersymmetric standard model is that the mass splitting of the squarks and sleptons does not allow GIM type cancellation and gives large flavor-changing neutral current (FCNC). However, by assuming that the masses of the squarks and sleptons are almost degenerate, the FCNC problem is reduced. In the gauge-mediated supersymmetric model the near degeneracy of the squarks and the sleptons comes out naturally, which reduces the FCNC problem somewhat. The $\mu$ and $B$ parameters are model dependent and are not predicted in this scenario.

One characteristic feature of the gauge-mediated supersymmetry breaking is that it predicts a very tiny gravitino mass [160, 161]

$$m_{3/2} \sim \frac{\Lambda^2}{M_{Pl}} \sim 10^{-10} GeV\,, \tag{10.40}$$

which is the LSP. All other supersymmetric particles will first decay into the next-to-lightest supersymmetric particle (NLSP), which is heavier than the LSP but lighter than all other superparticles, since the coupling with the gravitino is small. This NLSP then decays into the LSP (gravitino) very weakly resulting in a very definite signature of the gauge-mediated supersymmetry breaking models. Since the NLSP can eventually decay, it can also be a charged particle. The most likely candidates for NLSP are $\tilde{\chi}_1^0$ and $\tilde{\tau}_R^{\pm}$. For both of these NLSPs, the decays contain a gravitino

$$\tilde{\chi}_1^0 \to \gamma\, \tilde{G}_{3/2} \text{ and } \tilde{\chi}_1^0 \to Z\, \tilde{G}_{3/2} \text{ or } \tilde{\tau}_R^{\pm} \to \tau^{\pm}\, \tilde{G}_{3/2}. \tag{10.41}$$

Since the gravitino cannot be detected, the energy carried away by the gravitino gives a distinctive signal for supersymmetry of missing transverse momentum with model-dependent lifetimes and phenomenologies.

The minimal gauge-mediated supersymmetry breaking scenario has fewer parameters compared with the supergravity inspired models, but the gauge-mediated models are somewhat less compelling compared with the supergravity-inspired models.

The minimal gauge-mediated supersymmetry breaking model predicts an upper limit on the lightest Higgs mass $M_h < 124$ GeV for a top quark mass of $m_t = 175$ GeV [161]. But this minimal model is by no means complete, and even the more complicated models are not fully satisfactory. Finally the arbitrariness in the messenger sector of the gauge-mediated supersymmetry breaking makes it less attractive.

### Anomaly-Mediated Supersymmetry Breaking

We shall now consider another supersymmetry breaking mechanism with different phenomenology, which has its origin in the superstring theory [162]. Superstring theories are defined in 10 dimensions. The extra six dimensions are then compactified to get the low-energy effective 4-dimensional theory. In some of the compactification scheme it is possible that the resulting 4-dimensional effective theory has some low-energy anomalous $U(1)$ gauge group. Since the original theory is anomaly free, the effective theory also should be free of anomalies. To have anomaly cancellation in the effective theory, one needs to introduce an additional complex chiral superfield, whose scalar partner is the dilaton field, which should then acquire a *vev* for consistency. This *vev* of the dilaton field then introduces a $D$-type term, which in turn breaks supersymmetry. This supersymmetry breaking also introduces the soft mass terms. One drawback of the model is that it predicts too low gaugino mass. However, since the A and the B terms are also likely to be small in this model, this may not be so bad from a phenomenological point of view. However, this model has not been studied in detail so far.

Another related supersymmetry breaking mechanism is based on the supergravity model [163]. It has been observed that a supergravity-coupled gauge theory has a conformal invariance in the absence of mass terms. When the parameters are evolved with energy using renormalization group equations, the renormalization process will break the conformal symmetry by introducing the mass scale in the theory. This will break the conformal symmetry generating conformal anomaly in the theory leading to soft terms with a very definite pattern. In fact, all soft terms become proportional to the gravitino mass. So, the gravitino mass needs to be around 10 to 100 TeV. The sfermion masses come out to be nearly degenerate as in the gauge-mediated supersymmetry breaking, which helps to solve the FCNC problem. But there is one serious problem with this scenario, namely, that it leads to tachyonic mass for the color singlet superpartners. This negative mass squared term for the sleptons will induce a *vev* to the sleptons, which is unacceptable. Any further consideration of this scenario will depend on solving this issue. One attempt toward this end is to extend the minimal supersymmetric standard model by an extra $U(1)$, where the Fayet–Illiopoulos $D$-terms solve the negative-mass squared problem [164].

## 10.3 *R*-Parity Violation

Although we have not seen any signals for supersymmetry, we expect that the supersymmetric particles have masses close to the electroweak symmetry breaking scale. To ensure that we do not miss any signals for supersymmetry, there are attempts to study some extensions of the minimal supersymmetric standard model, which may have some testable predictions. One such extension is the $R$-parity violating models.

The renormalizable baryon and lepton-number violating couplings have been prevented in the minimal supersymmetric standard model by imposing $R$-parity [149]

$$R = (-1)^{3(B-L)+2S} \quad \text{or} \quad R = (-1)^{3B+L+2S}. \tag{10.42}$$

Otherwise the theory would mediate unacceptably fast proton decay. This implies that at any vertex there have to be at least two supersymmetric particles. Thus, decays of any supersymmetric particle should have one more supersymmetric particle, and the LSP becomes stable.

Although it is a sufficient condition for suppressing proton decay to have $R$-parity conservation, this is not necessary. From phenomenological considerations $R$-parity violating models were considered, in which either B or L, but not both, is conserved to prevent proton decay. Since proton decay requires violation of both baryon and lepton numbers, we do not require both B and L conservation to prevent fast proton decay. These $R$-parity violating theories with either B or L violation have several predictions which are different from the minimal supersymmetric standard model.

In the $R$-parity violating extension of the minimal supersymmetric standard model, one starts with the Lagrangian

$$\mathscr{L} = \mathscr{L}_{MSSM} + [W_R]_F, \tag{10.43}$$

where the second term contains the new $R$-parity violating interactions [149, 150] and is given by the superpotential,

$$\begin{aligned} W_R &= W_R^L + W_R^B \\ W_R^B &= \frac{1}{2}\lambda''_{ijk} U_i^c D_j^c \bar{D}_k^c \\ W_R^L &= \frac{1}{2}\lambda_{ijk} L_i L_j \bar{E}_k^c + \lambda'_{ijk} L_i Q_j \bar{D}_k^c + \mu_i L_i H_2, \end{aligned} \tag{10.44}$$

where $i, j, k$ are flavor indices. The first part of these interactions $W_R^B$ contains only one term that violates baryon number but conserves lepton number. This term is antisymmetric in $j, k$, and hence, $\lambda''_{ijk}$ has nine components.

The second part of the $R$-parity violating interactions $W_R^L$ contains three terms that violate lepton number but conserve baryon number. The first of these terms is antisymmetric in $i, j$. $\lambda_{ijk}$ has nine components, while $\lambda'_{ijk}$ has 27 components. The last term of $W_R^L$ is the bilinear term, which can be absorbed in the $\mu$-term in the superpotential of the minimal supersymmetric standard model by suitable transformations of the superfield $H_1$. But removing this term at some energy does not ensure its removal at all energies. Moreover, one should ensure that this does not lead to a nonvanishing *vev* of the sneutrino, $< \tilde{\nu}_i > \neq 0$. In general, this bilinear term will be present in most cases of $R$-parity violation. So, counting these three bilinear couplings, in total 48 $R$-parity violating couplings are allowed, 9 $B$ violating terms in $W_R^B$ and 39 $L$ violating terms in $W_R^L$. Since there is no motivation for $B$ violation, while $L$ violation is required for neutrino masses, in most $R$-parity violating models one assume, $W_R^B$ is absent and only $W_R^L$ is nonvanishing.

Any $R$-parity violating models have the characteristics that they allow the trilinear $R$-parity violating terms in which there is only one supersymmetric particle. Thus, a supersymmetric particle will be allowed to decay into two ordinary particles. This would mean that the LSP can decay into ordinary particles and there is no stable LSP. This will change the phenomenology of supersymmetric particles. The search strategy for supersymmetry will also change accordingly. While the $R$-parity conserving scenarios would find a missing $p_T$ due to the energy carried by the LSP, the $R$-parity violating scenarios may have, for example, dilepton signals as more definite indication of supersymmetry.

The bilinear terms in the models of $R$-parity violation give rise to a mixing of the neutralinos with the neutrinos. In the simplest scenario with one of the neutrinos, say, $\tau$ neutrino, mixing with the neutralinos, the LSP can decay through its mixing with the $\nu_\tau$ to ordinary particles. This also enriches the phenomenology of supersymmetry [165, 166]. Moreover, this lepton-number violating mixing of the neutralinos with the neutrinos can contribute to the Majorana neutrino masses.

The parameters in the $R$-parity violating interactions can be constrained [167] by the processes that these interactions allow, such as the baryon and lepton-number violating processes including proton decay, neutron–antineutron oscillations and Majorana neutrino masses. There are also constraints coming from charged current universality, $e - \mu - \tau$ universality, $\nu_e - e$ scattering, atomic parity violation, $\nu_e$ deep inelastic scattering, quark mixing, $K^+ \to \pi^+ \nu \bar{\nu}$ or $D^0 - \bar{D}^0$ mixing, $\tau$-decays, $D$-decays and LEP precision measurements. Astrophysical constraints can also be severe for some parameters. Since there are no predictions for the $R$-parity violating interactions, using these bounds for the $R$-parity violating parameters, possible observable signals are studied.

## 10.4 Supersymmetry and $L$ Violation

There could be some new sources of lepton number violation in supersymmetric models. The minimal supersymmetric standard model extends the standard model, so there are no right-handed neutrinos and no lepton number violation, and hence, neutrinos are massless. However, any of the extensions of the standard model that give a Majorana mass to the neutrinos can be implemented into some extensions of the minimal supersymmetric standard model. The effective lepton-number violating operator in supersymmetric theories is similar to that of the nonsupersymmetric theories

$$\mathcal{O}_L = \frac{f_{ij}}{M_L} \left[L_i L_j H_2 H_2\right]_F . \tag{10.45}$$

The $F$-term makes $\mathcal{O}_L$ a dimension-5 operator, so it is suppressed by some heavy lepton-number violating scale $M_L$. This operator can be realized by either introducing right-handed neutrinos or triplet Higgs scalar superfields, as discussed for the nonsupersymmetric theories. The radiative Zee-type model would now require four Higgs doublets and a charged scalar superfield. Except for some phenomenological details, these models share the same features as the nonsupersymmetric models. In supersymmetric models lepton number violation could come from some new lepton-number violating soft supersymmetry breaking terms [168]. In one such scenario it induces a *vev* of the sneutrinos [168]. This can allow mass splitting among the scalar components giving rise to sneutrino–antisneutrino oscillations and contribute to the Majorana masses of the neutrinos.

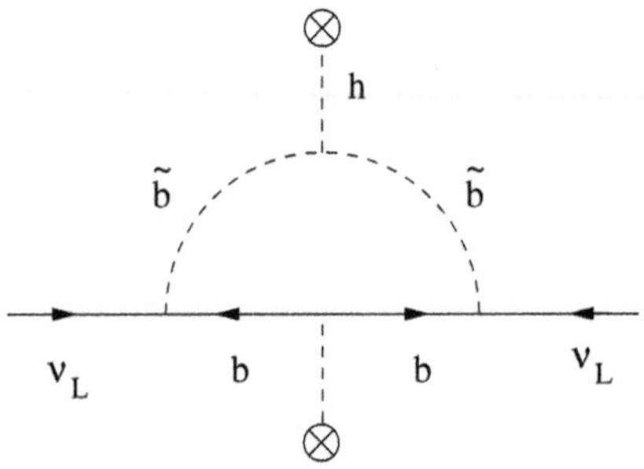


**FIGURE 10.1**
One-loop diagram generating a neutrino mass in $R$-parity violating models.

When $R$-parity is violated through the lepton-number violating terms, which conserves baryon number, there can be several new mechanisms of giving neutrino masses. The $R$-parity violating superpotential containing the lepton-number violating interactions $W_R^L$ is given in equation (10.44). All of these terms can then generate neutrino masses through one-loop diagrams. As an example, we present one such radiative diagram in figure 10.1).

We shall now discuss the constraints on the various $R$-parity violating parameters in $W_R^L$ coming from lepton-number violating considerations. These terms can contribute to the neutrino masses, so from the present mass-squared differences between different neutrino flavors, as obtained from the atmospheric and solar neutrino experiments and also the laboratory experiments, we can constrain these parameters [166]

$$\lambda' \sim 10^{-4} \tag{10.46}$$

with the supersymmetry breaking scale to be around 200 GeV. Similarly, the requirement for $\lambda$ is also of the same order of magnitude to explain the neutrino masses. These general constraints will hardly allow direct detection of these $R$-parity violating interactions, although a more rigorous analysis may still allow some interesting phenomenology.

The bilinears and the lepton-number violating soft terms in the potential together allow *vevs* for the sneutrinos. This will then allow mixing of the neutralinos with the neutrinos, which in turn will introduce a mass to the neutrinos. Given the present constraints on the parameters, these contributions from the bilinears alone are not enough to explain all neutrino experiments, and one needs to consider the trilinears together with the bilinears. The origin of the bilinears could be from the usual holomorphic terms, or from the nonholomorphic terms. When all of these terms are taken together, these models explain the neutrino masses and also generate a baryon asymmetry of the universe. However, none of the simple models with only one source of lepton number violation could explain all the present results in neutrino masses and mixing [165]. These models also fail to solve the problem of baryon asymmetry of the universe.

In addition to the neutrino oscillation experiments, the present bound on the neutrinoless double beta decay ($0\nu\beta\beta$ decay) can also constrain several lepton-number nonconserving parameters of the supersymmetric theories. The lepton number violation coming from soft supersymmetry breaking, which gives rise to sneutrino–antisneutrino oscillations, contributes to the amplitude of the $0\nu\beta\beta$ decay. These supersymmetry breaking soft terms will then be constrained by the present bound on the $0\nu\beta\beta$ decay [168], and the bound on the effective lepton-number violating MSSM parameter is $\eta^{susy} < 1\times 10^{-8}(m_{susy}/100\text{ GeV})^5$. Assuming all superparticles to have a common mass $m_{susy} \sim 100$ GeV, this amounts to a bound for the Majorana-like $B-L$ violating sneutrino mass

$$\tilde{m}_M < (2-11)\text{ GeV}. \tag{10.47}$$

This lepton-number violating mass $\tilde{m}_M$ corresponds to the mass difference between a sneutrino and an antisneutrino.

All the $R$-parity violating interactions given by $W_R^L$ in equation (10.44) would imply Majorana masses to the neutrinos and contribute to the neutrinoless double beta decay. As a result these couplings are constrained by the present bound on the neutrinoless double beta decay [169]. For the first generation this bound is

$$\lambda'_{111} < 3.3\times 10^{-4}\left(\frac{m_{\tilde{q}}}{100\text{ GeV}}\right)^2\left(\frac{m_{\tilde{g}}}{100\text{ GeV}}\right)^{1/2}, \tag{10.48}$$

assuming equal mass for the up-squark and the down-squark. If both $R$-parity violating interaction and the neutrino mass contribute to the neutrinoless double beta decay, we get bounds on the products of two couplings

$$\lambda'_{112}\lambda'_{121} < 1.1\times 10^{-6}\left(\frac{\Lambda_{susy}}{100\text{ GeV}}\right)^3$$

$$\lambda'_{113}\lambda'_{131} < 3.8 \times 10^{-8} \left( \frac{\Lambda_{susy}}{100 \text{ GeV}} \right)^3 \tag{10.49}$$

for supersymmetric mass parameters of the order of 100 GeV.

Indirect constraints on the $R$-parity violating couplings come from the $K^\circ - \overline{K^\circ}$ oscillations, $B^\circ - \overline{B^\circ}$ mixing, $D^\circ - \overline{D^\circ}$ mixing, or lepton flavor changing processes. Some of the bounds are even stronger, such as the bound from $K^\circ - \overline{K^\circ}$ oscillations $\lambda'_{i12}\lambda'_{i21} < 1 \times 10^{-9}$ or the bound from $B^\circ - \overline{B^\circ}$ mixing $\lambda'_{i13}\lambda'_{i31} < 8 \times 10^{-8}$, both for $m_{\tilde{e}} = 100$ GeV.

It is also possible to constraint the elements of the neutrino mass matrix from a combined fit to the neutrino oscillation results, the bound on the neutrinoless double beta decay, and WMAP constraints on the total mass of the three neutrinos. This, in turn, can give bounds on the products of the standard couplings [169, 170]

$$\begin{aligned} \lambda'_{i33}\lambda'_{i'33} &< 3.6 \times 10^{-8}, \quad \lambda'_{i32}\lambda'_{i'23} < 8.9 \times 10^{-7}, \\ \lambda'_{i22}\lambda'_{i'22} &< 2.2 \times 10^{-5}, \quad \lambda'_{i33}\lambda'_{i33} < 6.3 \times 10^{-7}, \\ \lambda'_{i32}\lambda'_{i23} &< 1.1 \times 10^{-5}, \quad \lambda'_{i22}\lambda'_{i22} < 1.7 \times 10^{-4}. \end{aligned} \tag{10.50}$$

For this representative set of constraints the masses of the scalar superpartners are assumed to be around 100 GeV. In the hierarchical scenario the bounds on the $R$-parity violating couplings would be stronger than in the degenerate case.

## 10.5 Grand Unified Theories

Supersymmetry was originally proposed as an interesting symmetry of nature that relates fermions to bosons and unifies space–time symmetry with internal symmetries. However, it became popular only when it could provide a solution to the gauge hierarchy problem of the grand unified theories. There are two widely different mass scales separated by the great desert in any grand unified theories. To protect the light Higgs scalars against quadratic divergences coming from the radiative corrections, severe fine tuning is required at all orders. This problem is solved in supersymmetric theories, where the divergent diagrams are cancelled by equivalent diagrams with supersymmetric particles. Thus, it is natural to construct a supersymmetric version of the grand unified theories.

Let us first consider a supersymmetric version of the $SU(5)$ grand unified theory. We first incorporate the fermions in chiral superfields, and the gauge bosons are included in vector superfields belonging to the adjoint 24-dimensional representation of $SU(5)$. Thus, the two chiral superfields, $\psi$ and $T$, belonging to the $\bar{5}$ and 10-dimensional representations of $SU(5)$, respectively, will contain the usual fermions and their scalar superpartners. The vector superfield will contain the 24 gauge bosons and the gauginos.

We consider the Higgs scalars belonging to a 24-dimensional scalar superfield $\Sigma$, which breaks the $SU(5)$ group. The adjoint representation is the minimal choice for the purpose, since we do not want to break the rank of the group. For the electroweak symmetry breaking and for giving masses to the up and the down quarks and the leptons, we introduce two Higgs doublets, $H_1$, belonging to the representation $\bar{5}$ and $H_2$, belonging to the representation 5 of $SU(5)$. The two Higgs doublets are also required to cancel chiral anomaly, since the fermions in the scalar superfields contribute to anomaly.

Once the particle spectrum is determined, the gauge interactions and the kinetic energy terms are determined. We shall now write the superpotential that can give us the required symmetry breaking

$$
\begin{aligned}
SU(5) \quad &\xrightarrow{\Sigma\{24\}} \quad SU(3)_c \times SU(2)_L \times U(1)_Y \\
&\xrightarrow{H_1\{5\},H_2\{\bar{5}\}} \quad SU(3)_c \times U(1)_Q
\end{aligned}
$$

and give masses to the fermions. To give TeV scale masses to the superpartners coming from supersymmetry breaking, soft supersymmetry breaking terms are introduced as in the minimal supersymmetric standard model. We shall not discuss them here.

One possible choice of the superpotential containing the Higgs scalars is

$$
\begin{aligned}
W = {} & h_u T T H_2 + h_d T \psi H_1 + \lambda_2 X (Tr\Sigma^2 - M_\Sigma^2) \\
& + \lambda_1 H_2 H_1 \Sigma + \mu H_1 H_2, 
\end{aligned}
\qquad (10.51)
$$

where we have introduced one singlet chiral field $X$ to write this superpotential. We assume that all supersymmetry breaking effect should be included in the soft terms, so the minima of this superpotential should not break supersymmetry at the unification scale. The vanishing of the $F$-terms

$$
\begin{aligned}
F_X &= \frac{\partial W}{\partial X} = Tr\Sigma^2 - M_\Sigma^2 = 0 \\
F_\Sigma &= \frac{\partial W}{\partial \Sigma} = 2\lambda_2 X \Sigma
\end{aligned}
$$

now gives *vevs* of the fields

$$
<\Sigma> = M_\Sigma \text{diag}(1 \quad 1 \quad 1 \quad -3/2 \quad -3/2) \quad \text{and} \quad <X> = 0. \qquad (10.52)
$$

There are two other degenerate vacua in which different components of $\Sigma$ acquire *vevs*, which will break $SU(5) \to SU(4) \times U(1)$ or preserve $SU(5)$. However, we assume that supergravity effects will remove this degeneracy and $SU(3)_c \times SU(2)_L \times U(1)_Y$ will become the absolute minimum. Although we presented one particular superpotential for demonstration, there could be many more possibilities for the superpotential. Moreover, to write a particular superpotential one needs to impose certain discrete symmetries, all of these complicate the theory. In addition, there are now new problems in the theory which require newer solutions. Some of the problems of the nonsupersymmetric grand unified theories, such as the fermion mass problem or the neutrino mass problem, are also present in the supersymmetric models. Again gravitational corrections could also solve some of these problems. There is another problem, called the doublet–triplet splitting problem, which we shall mention below. The $\mu$-problem is also difficult to solve in supersymmetric grand unified theories. At present there is no standard supersymmetric $SU(5)$ grand unified theory, which is fully consistent and can provide solution to all of these problems. But all of them have a generic structure, which is similar.

The *vevs* of the scalars $H_1$ and $H_2$ break the electroweak symmetry. These fields also give masses to the fermions. In the effective $SU(3)_c \times SU(2)_L \times U(1)_Y$, invariant superpotential for the fermion masses will be exactly the same as the minimal supersymmetric standard model, and hence, we shall not write them here. However, the triplet components of these scalars can in general mediate very fast proton decay. This requires that there has to be some mechanism that makes the $SU(3)_c$ triplet components as heavy as the grand unification scale, while keeping the doublet component light. This is the so-called doublet–triplet splitting problem of the grand unified theories.

Similar to the minimal supersymmetric standard model we shall impose $R$-parity to forbid fast proton decay by forbidding the renormalizable baryon-number violating terms. The proton decay now comes from a dimension-5 operator,

$$\mathcal{O} = \frac{f}{M_U}\,[QQQL]_F\,, \tag{10.53}$$

where $M_U$ is the unification scale and $f$ is the effective coupling constant. The scalar superfields have mass dimension 1 and the $F$-term adds one more mass dimension. This gives us an interaction vertex, violating baryon and lepton numbers, in which two fermions and two scalars enter. Although the operator is suppressed by only one power of $M_U$, this is still acceptable since the coupling constants are now the Yukawa couplings and could be smaller than the gauge couplings and the unification scale is also large. This operator implies that the decay modes of the proton are different from those of the nonsupersymmetric case. In this case the dominant proton decay mode is $\mu^+ K^0$, which violates strangeness number. This comes because of the symmetry property of the operator.

We shall now discuss the evolution of the gauge coupling constants in supersymmetric theories. All of the superpartners of the ordinary particles contribute to the renormalization group equation. This changes the coefficients of the beta function in the following way. In the nonsupersymmetric case a gauge boson contribution to the beta function comes with a coefficient $-11/3$, while a Weyl or Majorana fermion contributes an amount $2/3$ and a scalar an amount $1/6$ (see equation (8.36)). In the supersymmetric case a vector multiplet has a gauge boson, which contributes $-11/3$ and the corresponding gaugino, which is a Majorana fermion, contributes $2/3$. So, a vector multiplet contributes an amount $-3$ to the beta function. For a chiral multiplet there is a Majorana fermion contributing $2/3$ and the corresponding complex scalar superpartner contributing $1/3$. So for a chiral superfield, the contribution is

1, whether it contains a fermion or a scalar. Thus, in the supersymmetric case the renormalization group equations and the one-loop beta functions are given by

$$\mu \frac{dg_i}{d\mu} = \beta_i(g_i) = b_i \frac{g_i^3}{16\pi^2}, \tag{10.54}$$

with the beta functions

$$\beta_i(g_i) = \frac{g_i^3}{16\pi^2}\left[-3\text{Tr}\,[T_a^2] + \sum_f \text{Tr}\,[T_f^2] + \sum_s \text{Tr}\,[T_s^2]\right]. \tag{10.55}$$

The quadratic Casimir invariants have been discussed in section 8.4. In the standard model, the fermion contributions are 2 for all the groups and Tr $[T_a^2] = N$ for $SU(N)$ and 0 for $U(1)$.

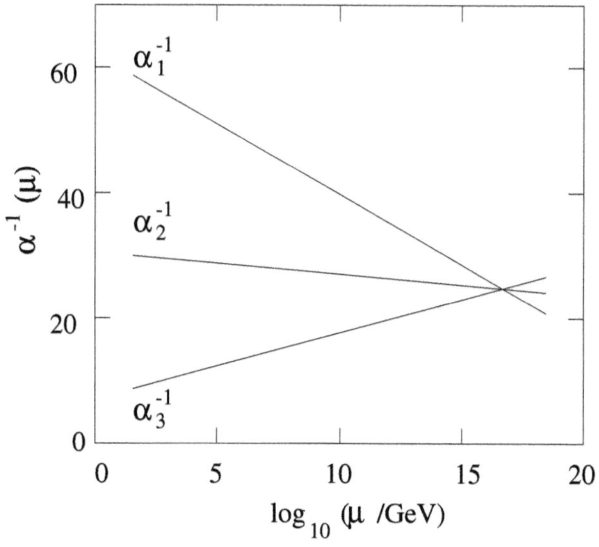


**FIGURE 10.2**
Gauge coupling unification in the supersymmetric $SU(5)$ grand unified theory.

At the one-loop level the $SU(3)_c \times SU(2)_L \times U(1)_Y$ gauge coupling constants in the supersymmetric $SU(5)$ grand unified theory will evolve as

$$\alpha_3^{-1}(m_X^2) = \alpha_3^{-1}(m_W^2) + \frac{(9-2n_g)}{4\pi}\log\left(\frac{m_X^2}{m_W^2}\right)$$

$$\alpha_2^{-1}(m_X^2) = \alpha_2^{-1}(m_W^2) + \frac{(6-2n_g-\frac{1}{2}n_h)}{4\pi}\log\left(\frac{m_X^2}{m_W^2}\right)$$

$$\alpha_1^{-1}(m_X^2) = \alpha_1^{-1}(m_W^2) + \frac{(-2n_g - \frac{3}{10}n_h)}{4\pi} \log\left(\frac{m_X^2}{m_W^2}\right), \tag{10.56}$$

where $n_g = 3$ is the number of generations of fermions. We have included the effect of $n_h$ Higgs doublets. Starting with the gauge coupling constants at the electroweak scale, as obtained from the precision measurements, we evolve them to the grand unification scale and find that all three meet at a point around $10^{16}$ GeV, as shown in figure 10.2. In practice, one considers the uncertainty in the measured couplings and then varies the supersymmetry breaking scale, below which the superparticles will not contribute to the renormalization group equation. The best fit that corresponds to coupling constant unification comes for the supersymmetry breaking scale to be around a TeV [130, 171]. The present values of the precision results again disfavor the single stage unification.

The supersymmetric $SO(10)$ grand unified theories with some intermediate symmetry breaking scales are thus more likely unified theory, which can have solution to many of the problems present in the minimal $SU(5)$ grand unified theory. Some of the problems that could be addressed naturally in an $SO(10)$ grand unified theory are correct neutrino mass, solution to strong $CP$ problem, left–right symmetry breaking, leptogenesis, gravitino problem and inflation. Gauge coupling unification also prefers theories with intermediate symmetry breaking scale.

Supersymmetric theories are more interesting than the nonsupersymmetric grand unified theories, but $N = 1$ supergravity models are even more interesting. Starting with the Higgs masses to be $m_0^2 + \mu^2$ and the squarks and sleptons masses to be $m_0$ at the unification scale $M_U$, if we evolve downward to the electroweak scale, we find one very interesting phenomenon. The squark and the slepton masses remain positive, but the lighter Higgs mass-squared becomes negative $M_h^2 < 0$ around the electroweak scale. This will imply a *vev* for the Higgs scalar, which then breaks the electroweak symmetry and gives masses to the fermions. Thus, the negative mass-squared term required for the Higgs mechanism in the standard model comes out naturally in supergravity theories [172].

Starting from the unification scale, as we evolve the top quark mass with energy it reaches a fixed point value, beyond which it stops evolving. This means that the final value of the top quark mass is independent of the initial values of the couplings at the unification scale and the observed top quark mass is predicted for $\tan\beta \sim 1 - 3$.

Supersymmetric theories have many advantages, but there are also some disadvantages. The superpartners and the constraints in the couplings sometimes create new problems. But considering all aspects we now believe that supersymmetry should be present at high energies. In addition to the phenomenological considerations discussed in this chapter, it may play an important role in unifying the gauge interactions with gravity.

# Part IV

# Extra Dimensions

# 11

# *Extended Gravity*

The gravitational interaction plays the most important role in the evolution of the universe although this is the weakest of all the forces. However, at very high energies gravity should become as strong as any other interaction. Since the strong, the weak and the electromagnetic interactions are gauge theories and could be unified in a grand unified theory, we envisage an ultimate theory at higher energies in which gravity will be unified with all other interactions. At higher energies we expect our theory to be supersymmetric to solve the gauge hierarchy problem and allow unification of Poincaré group with the internal symmetry groups. So, we must also formulate a supersymmetric theory of gravity, which is the supergravity.

The strong, the weak and the electromagnetic interactions could be explained by theories arising from symmetries in some internal degree of freedom, whereas gravity originates from space–time symmetry. So the obvious choice would be to think of considering the internal degrees of freedom that give rise to the different gauge interactions to be some new space dimensions. This concept of extending gravity to unify it with other gauge interactions at higher dimensional theories makes us think that there are extra space dimensions beyond the four space–time dimensions we experience at present. Of course in all these theories, one needs to explain why the extra dimensions have not yet shown up in any of our experiments. There are also some new ideas leading to accessibility of the extra dimensions in the next generation accelerators. In the next few chapters we shall review the basics and mention some of these recent developments. Although it will not be possible to go into any details of these theories, we shall try to convey the excitements in the field.

## 11.1 Gravity in Four Dimensions

All matter attracts each other through gravitational interaction, and hence, we experience gravity in our daily life more than any other force. The next common interaction is the electromagnetic interaction, which is the interaction among charged particles. Both interactions follow an inverse square law, but the origin of the two forces are completely different. While the electromagnetic interaction originates from some internal symmetry, the gravitational interaction originates from the space–time symmetry.

Unlike the electromagnetic interaction, the gravitational interaction is mediated by spin-2 boson, the graviton, and hence, the theory is not renormalizable. Although we do not understand the gravitational interaction at the quantum level, we have a well-established classical theory of gravity, which is the *general theory of relativity*. We hope that at very high energy we shall be able to embed the general theory of relativity in a consistent renormalizable quantum theory of gravity. At present the superstring theory is emerging as the most consistent theory of quantum gravity, which also promises to unify gravity with other interactions. We start here with the general theory of relativity, which has been an extremely successful theory as far as our present observations are concerned.

**General Theory of Relativity**

The general theory of relativity is based on the hypothesis that any massive object leaves an imprint in the space–time geometry around the object and any theory of gravity must deal with this geometry of space–time. The equivalence between the gravitational interaction of the object and the space–time geometry it created is formulated in the *equivalence principle*, which is the foundation stone of the general theory of relativity. The equivalence principle originates from the observation of equivalence between the gravitational mass and the inertial mass.

The *equivalence principle* identifies the mass of a particle we measure on the surface of the Earth with the mass of the particle that enters in the gravitational interaction between Earth and that body. This equivalence between the inertial mass and the gravitational mass can be stated in another form. Inside a freely falling lift it will not be possible to experience the gravitational force under which the lift is falling. This implies that any test body will appear to be massless inside the freely falling lift, which is the locally accelerated space–time frame. Thus, we can replace any gravitational interaction locally by an equivalent accelerated frame of reference over a small interval of time. This is the essence of the *weak equivalence principle*.

This can be generalized to include effects of other interactions in the *strong equivalence principle*, which states that any physical interaction other than the gravity behaves in a locally inertial frame as if gravitation were absent. Since gravity is the only interaction that can be understood in terms of space–time geometry, the Maxwell's equations or any other interactions must remain the same in any locally inertial frame.

To compare the weak and the strong equivalence principle, let us come back to the example of the freely falling lift. According to the weak equivalence principle, the curvature created by the mass of the Earth will not be felt by any test body inside a freely falling lift. That is, any measurement of the mass of the test body will not reveal any difference between the gravitational interaction of the Earth and the accelerating frame of the lift. The strong equivalence principle states that not only any measurement of the mass, which is experienced only by gravity, but any other measurements that involve other kinds of interactions will also fail to reveal any difference between the locally accelerating frame and the gravitational interaction of the Earth.

The equivalence principle will imply that we can study the gravitational interaction by studying an equivalent geometry of the space–time. To study any other interactions in the presence of gravity, we can simply study these interactions in the curved background space–time. The background curvature or the metric is determined by the background gravitational interaction. Let us consider an example of bending of light. If there is a massive object between two space–time points, the light will not travel in straight line while going from one of these points to the other. The Maxwell's equations will find the background metric to be curved because of the massive object, and hence, light will travel along the curved space–time.

In the absence of matter, two points in space–time are connected by straight lines. But if there are some massive objects in the neighbourhood, those will introduce curvature to the background space–time metric. The effects of any massive object or any gravitational interaction can be studied by replacing the object with an equivalent curvature in space–time.

As an example, consider a flat wire mesh whose end-points are fixed at the boundary. The wires are stretched along straight lines. If a massive ball is placed at a point, the wires will become curved (see figure 11.1). If we now place any test particle on the wire mesh, the test particle will role down toward the massive ball. If we take out the massive ball, but by pressing keep the curvature of the wire mesh same, the test particle will not see any difference and will be attracted toward the point where the massive object was placed.

In a similar way, any massive object makes the space–time around the object curved. If a test particle is placed near the massive object, it will experience the gravitational attraction of the massive object. According to the equivalence principle, the test particle cannot distinguish whether a massive object is attracting it or the background metric is curved causing it to be pulled in that direction. With this discussion of the equivalence principle we now proceed to develop the general theory of relativity.

Before we explain how to relate the gravitational interactions to the curvature of space–time, we shall first develop some mathematical tools to study the geometry of space–time. We shall introduce the notion of covariant transformations in ordinary space–time, starting with some discussions on local Lorentz transformations and general coordinate transformations.

In a flat inertial space, the distance between two space–time points is given by

$$ds^2 = dt^2 - dx^2 - dy^2 - dz^2 = g_{\mu\nu} dx^\mu dx^\nu. \tag{11.1}$$

From now on we shall imply summation over repeated indices, unless otherwise specified. The coefficient $g_{\mu\nu}$ is a function of the space–time coordinate $x^\mu$ and is called the metric. The metric $g_{\mu\nu}$ takes the simple form diag $[1, -1, -1, -1]$ for inertial flat space, but in a curved space–time the metric contains all the information about the geometry of space–time. As we shall discuss, all the information about any gravitational interaction will be contained in the metric.

We now introduce quantities that are invariant under any coordinate transformations. The distance between two points on the surface of a sphere and the distance

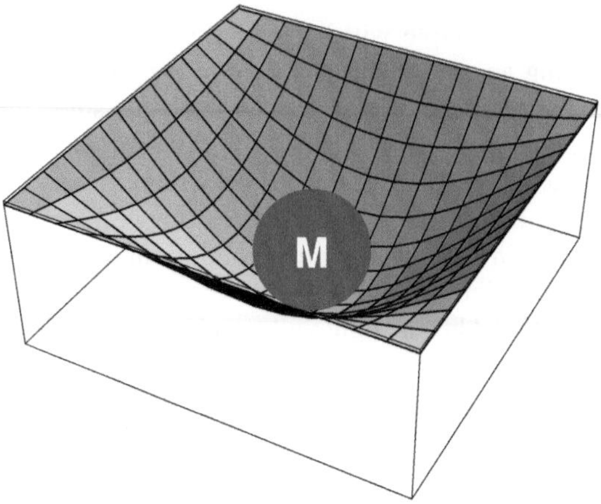


**FIGURE 11.1**
Curvature on a wire mesh in the presence of a massive object M.

between two point in a flat space but written in spherical coordinate system appear to be similar. In other words, whether $g_{\mu\nu}$ correspond to the coordinate system in which we are expressing the distance or to the curvature in space has to be distinguished. We, thus, introduce invariant quantities that remain invariant under any coordinate transformations.

Any scalars remain invariant under any coordinate transformation

$$\phi(x^\mu) = \phi[x^\mu(x'^\nu)] = \phi'(x'^\nu). \tag{11.2}$$

Let us consider the $G_{22} = SU(2) \times SU(2)$ subgroup of the group $O(4)$, which is locally isomorphic to the Lorentz group. A scalar transforms under $G_{22}$ as (0,0). The mass of a particle is a scalar quantity. Under any Lorentz transformation, the form of the function may change, but its value remains the same.

A *contravariant* and a *covariant vector* transform as

$$\begin{aligned} A'^\mu &= \frac{\partial x'^\mu}{\partial x^\nu} A^\nu \\ A'_\mu &= \frac{\partial x^\nu}{\partial x'^\mu} A_\nu, \end{aligned} \tag{11.3}$$

respectively. Any vector transforms under the group $G_{22}$ as $(1/2,1/2)$, while a fermion transforms as $(1/2,0)+(0,1/2)$. The difference between a contravariant and a covariant vector is that a tangent to a curve is a contravariant vector while the normal to any hypersurface is a covariant vector. This concept of a contravariant and a covariant vector can be generalized to a tensor. The metric tensor $g_{\mu\nu}$ is a covariant tensor of rank 2, while $g^{\mu\nu}$ is a contravariant tensor of rank 2 and $g^\mu_\nu$ is a mixed tensor of rank 2. A metric tensor is symmetric, but in general any tensor could be symmetric $S_{\mu\nu} = S_{\nu\mu}$ or antisymmetric $A_{\mu\nu} = -A_{\nu\mu}$. When any two of the indices are repeated, sum over these indices are implied. When an upper index is contracted with a lower index, the rank of the tensor is reduced by two.

Although a derivative of a scalar transforms as a vector,

$$\frac{\partial \phi'}{\partial x'^{\mu}} = \frac{\partial x^{\nu}}{\partial x'^{\mu}} \frac{\partial \phi}{\partial x^{\nu}}, \tag{11.4}$$

a derivative of a vector does not transform like a tensor. It transforms as

$$\frac{\partial B'_{\mu}}{\partial x'^{\nu}} = \frac{\partial x^{\rho}}{\partial x'^{\mu}} \frac{\partial x^{\sigma}}{\partial x'^{\nu}} \frac{\partial B_{\rho}}{\partial x^{\sigma}} + \frac{\partial^2 x^{\rho}}{\partial x'^{\nu} \partial x'^{\mu}} B_{\rho}. \tag{11.5}$$

This is because the derivative of a vector depends on how a vector is transported from one point to another. As shown in figure 11.2, if a unit vector is taken from a point $A$ to a point $B$ without changing its direction, it is called parallel transport. When the points $A$ and $B$ are connected by a straight line, the unit vector remains a unit vector even after the parallel transport. But when the points $A$, $B$ and $C$ lie along a curved line, the unit vector ceases to be a unit vector after the parallel transport.

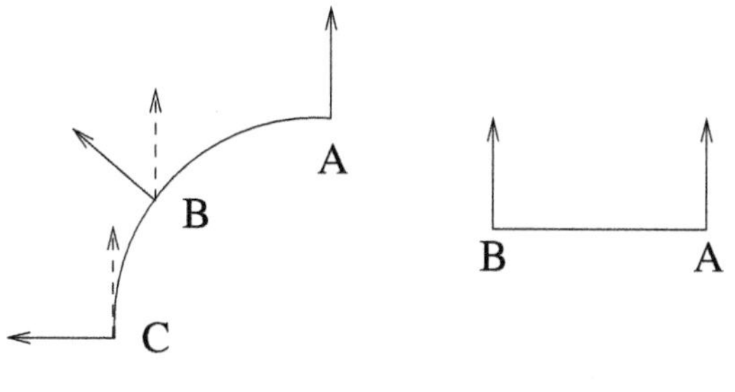


**FIGURE 11.2**
A unit vector at the point A has been parallel transported to the points B and C (dotted lines) along the curved line and compared with the unit vectors at these points. When the unit vector is parallel transported to B along a straight line, it remains a unit vector.

After the parallel transport the deviation from the unit vector is given by the additional term in the above transformation of the derivative of a vector. To represent this change we introduce the *Christoffel symbol* $\Gamma^{\rho}_{\mu\nu}$, which denotes the affine connection between the two space–time points. With this change we can then define the covariant derivative as

$$B_{\mu;\nu} = \frac{\partial B_{\mu}}{\partial x^{\nu}} - \Gamma^{\rho}_{\mu\nu} B_{\rho} = B_{\mu,\nu} - \Gamma^{\rho}_{\mu\nu} B_{\rho}, \tag{11.6}$$

where $B_{\mu,\nu} = \frac{\partial B_{\mu}}{\partial x^{\nu}}$. The *covariant derivative* of any vector now behaves like a tensor under any general coordinate transformation. This is ensured by the transformation of the Christoffel symbol:

$$\Gamma'^{\mu}_{\nu\rho} = \frac{\partial x'^{\mu}}{\partial x^{\delta}} \frac{\partial x^{\sigma}}{\partial x'^{\nu}} \frac{\partial x^{\pi}}{\partial x'^{\rho}} \Gamma^{\delta}_{\sigma\pi} + \frac{\partial^2 x^{\pi}}{\partial x^{\nu} \partial x^{\rho}} \frac{\partial x'^{\mu}}{\partial x^{\pi}}. \tag{11.7}$$

The covariant derivatives of a contravariant vector and the covariant derivative of the metric tensor are then given by

$$\begin{aligned} A^{\mu}_{;\nu} &= \frac{\partial A^{\mu}}{\partial x^{\nu}} + \Gamma^{\mu}_{\rho\nu} A^{\rho} = A^{\mu}_{,\nu} + \Gamma^{\mu}_{\rho\nu} A^{\rho}, \\ g_{\mu\nu;\rho} &= \frac{\partial g_{\mu\nu}}{\partial x^{\rho}} - \Gamma^{\sigma}_{\mu\rho} g_{\sigma\nu} - \Gamma^{\sigma}_{\nu\rho} g_{\mu\sigma}. \end{aligned} \tag{11.8}$$

We can simplify these relations by working in a *Riemannian space*, in which it is possible to have a locally inertial frame at any given point. Here we have

$$g_{\mu\nu;\rho} \equiv 0. \tag{11.9}$$

The Christoffel symbol can now be expressed in terms of the metric tensor and of ordinary derivatives of the metric tensor as

$$\Gamma^{\rho}_{\mu\nu} = \frac{1}{2} g^{\rho\sigma} \left( g_{\sigma\mu,\nu} + g_{\sigma\nu,\mu} - g_{\mu\nu,\sigma} \right). \tag{11.10}$$

The advantage of working in a Riemannian space is that it is always possible to go to a locally inertial reference frame. This means that we can always go to a coordinate system in the neighborhood of any point $P$, in which all the Christoffel symbols vanish. At the point $P$ the metric is given by the metric of a flat *Minkowski space*

$$\eta_{mn} = \mathrm{diag}\,(+1, -1, -1, -1). \tag{11.11}$$

We now define another useful quantity, the *vierbeins* $e_n{}^{\mu}$, which relates the unit vectors in the local inertial coordinate system with the unit vectors of the general coordinate system. The metrics in the two coordinate systems are related by the vierbeins through the relation

$$g_{\mu\nu} e_m{}^{\mu} e_n{}^{\nu} = \eta_{mn}. \tag{11.12}$$

The inverse vierbeins $e^m{}_{\nu}$ is defined with the property

$$e_m{}^{\mu} e^m{}_{\nu} = \delta^{\mu}_{\nu}, \tag{11.13}$$

so the inverse coordinate transformations become possible

$$g_{\mu\nu} = e^m{}_{\mu} e^n{}_{\nu} \eta_{mn}. \tag{11.14}$$

All the information of a metric is, thus, contained in the vierbeins.

The vierbeins connect the general coordinate system to the locally flat inertial coordinates, so they carry indices from both the coordinate system. We have to define a new covariant derivative for the vierbeins

$$D_{\mu} e^m{}_{\nu} = e^m{}_{\nu,\mu} - \Gamma^{\lambda}_{\mu\nu} e^m{}_{\lambda} + \omega_{\mu}{}^m{}_n e^n{}_{\nu}. \tag{11.15}$$

The covariant derivative now contain the spin-connection $\omega_\mu{}^m{}_n$, which is a vector under the general coordinate transformations and also associated with the local Minkowski space.

These spin-connections have some similarity with the gauge fields in gauge theories ($A^i_\mu$). A gauge boson is a vector field that carries an index of the internal group. For the spin-connection, the internal group is substituted by the local Lorentz group, and hence, the indices $m,n$ of the spin-connection replaces the group index $i$ of the gauge fields. We, thus, define a tensor under the general coordinate transformations, analogous to the field-strength tensor of gauge theories

$$R_{\mu\nu}{}^{mn} = \partial_\mu \omega_\nu{}^{mn} - \partial_\nu \omega_\mu{}^{mn} + \omega_\mu{}^m{}_k \omega_\nu{}^{kn} - \omega_\nu{}^m{}_k \omega_\mu{}^{kn}, \tag{11.16}$$

where $\omega_\nu{}^{mn}\eta_{kn} = \omega_\nu{}^m{}_k$. This tensor is antisymmetric in $\mu,\nu$, so the symmetric Christoffel symbols $\Gamma^\lambda_{\mu\nu} = \Gamma^\lambda_{\nu\mu}$ do not enter into this definition.

Various contractions of this tensor allow us to define some important tensors:

(a) *Riemann–Christoffel curvature tensor:*

$$R_{\mu\nu\rho\sigma} = R_{\mu\nu}{}^{mn} e_{m\rho} e_{n\sigma}, \tag{11.17}$$

satisfying the symmetry relations

$$R_{\mu\nu\lambda\rho} = -R_{\nu\mu\lambda\rho} = -R_{\mu\nu\rho\lambda} = R_{\lambda\rho\mu\nu} \tag{11.18}$$

and

$$R_{\mu\nu\lambda\rho} + R_{\mu\rho\nu\lambda} + R_{\mu\lambda\rho\nu} = 0. \tag{11.19}$$

The curvature tensor also satisfies the *Bianchi identity*

$$R_{\mu\nu\lambda\rho;\sigma} + R_{\mu\nu\sigma\lambda;\rho} + R_{\mu\nu\rho\sigma;\lambda} = 0. \tag{11.20}$$

The important characteristics of the Riemann–Christoffel tensor is that it vanishes everywhere when the space–time is flat.

(b) *Ricci tensor:*

$$R_{\mu\nu} = R_{\lambda\nu}{}^{mn} e_n{}^\lambda e_{m\mu} = g^{\lambda\rho} R_{\lambda\mu\nu\rho} = R^\lambda{}_{\mu\nu\lambda}$$

$$= \partial_\nu \Gamma^\lambda_{\mu\lambda} - \partial_\lambda \Gamma^\lambda_{\mu\nu} - \Gamma^\sigma_{\mu\nu}\Gamma^\lambda_{\sigma\lambda} + \Gamma^\lambda_{\mu\sigma}\Gamma^\sigma_{\nu\lambda}\,. \tag{11.21}$$

This is the only possible rank-2 tensor that could be constructed from the Riemann–Christoffel tensor.

(c) *Scalar curvature:*

$$R = R_{\mu\nu}{}^{mn} e_m{}^\nu e_n{}^\mu = g^{\mu\nu} R_{\mu\nu} = R^\mu{}_\mu\,. \tag{11.22}$$

The scalar curvature appears in the equation of motion of the metric tensor.

(d) *Einstein tensor:*

$$G_{\mu\nu} = R_{\mu\nu} - \frac{1}{2} g_{\mu\nu} R. \tag{11.23}$$

The Bianchi identity implies vanishing of the covariant derivative of the Einstein tensor

$$G_{\mu\nu;\lambda} = 0 \tag{11.24}$$

implying the Einstein tensor is divergence free.

**Einstein Equation**

The Einstein equation determines the metric and is given by the vanishing of the Einstein tensor [173]

$$G_{\mu\nu} = R_{\mu\nu} - \frac{1}{2} g_{\mu\nu} R = 0. \tag{11.25}$$

The corresponding Einstein action is given by [174]

$$S^{(E)} = \int d^4x \, \mathscr{L}^{(E)} = -\int d^4x \, \frac{eR}{2\kappa^2}, \tag{11.26}$$

where

$$e = \det(e^m{}_\lambda) = [\det(e_m{}^\lambda)]^{-1} = [-\det(g_{\mu\nu})]^{1/2}, \tag{11.27}$$

and $\kappa^2 = 8\pi G_N$, where $G_N = 6.67259(85) \times 10^{-11} \mathrm{m^3 kg^{-1} s^{-2}}$ is the Newton's gravitational constant.

In this form of the Einstein equation we have not included matter, which needs an additional term proportional to the energy-momentum tensor $T_{\mu\nu}$. This will act as a source in the Einstein equation. One more term could be included in the Einstein equation: a constant may be added to the scalar curvature. This term has important cosmological consequences, and hence, this constant $\Lambda$ is called the *cosmological constant.* Including all the terms the final form of the Einstein equation in the presence of matter and cosmological constant is given by

$$G_{\mu\nu} + \Lambda g_{\mu\nu} = R_{\mu\nu} - \frac{1}{2} g_{\mu\nu} R + \Lambda g_{\mu\nu} = -8\pi G_N T_{\mu\nu}. \tag{11.28}$$

This is the most general equation that determines the metric or the geometry of space–time in the presence of matter. The cosmological constant term does not affect the gravitational interactions between two objects and hence this term was not included in the original equation. Later Einstein included this term to find a static solution of the universe. Although the observation of expanding universe ruled out the possibility of static universe, we may still need the cosmological constant. Present observations require about 70% of the total matter in the universe in the form of dark energy or cosmological constant.

**Schwarzschild Solution**

To understand the implications of the Einstein equation, we need to find exact solutions with some ansatz. One of the interesting solutions of the Einstein equation, the *Schwarzschild solution* [175], is used for many practical purposes. The space–time geometry outside a spherical distribution of matter of mass $M$ is given in the form of a line element

$$ds^2 = \left(1 - \frac{2G_N M}{R}\right) dt^2 - \left(1 - \frac{2G_N M}{R}\right)^{-1} dR^2 - R^2(d\theta^2 + \sin^2\theta\, d\phi^2). \quad (11.29)$$

This line element determines the metric and hence the geometry of space–time outside the mass distribution.

This solution provides us with several interesting aspects of the general theory of relativity. The characteristic scale of the general theory of relativity is given by the *Schwarzschild radius* $R_s = 2G_N M$, which is an important parameter in the theory. The Schwarzschild radius $R_s$ shows a singularity in the solution for any given mass. For any compact object with radius smaller than $R_s$, the Schwarzschild solution shows that near the surface of such object the geometry is strongly curved and even photons will be trapped inside such objects. Such objects are called *black holes*.

Another implication of the Schwarzschild radius is that it determines the scale of quantum gravity. When the Schwarzschild radius becomes comparable to the Compton wavelength of $\lambda \sim \frac{1}{M}$, quantum effects cannot be neglected. This is called the *Planck scale*, when the characteristic scales of the quantum theory ($\lambda$) and the general theory of relativity ($R$) become comparable,

$$M_{Pl} \sim \frac{1}{\sqrt{G_N}} \sim 10^{19}\ \text{GeV}.$$

The corresponding length, the Planck length, and the time, the Planck time, are defined as

$$\begin{aligned} L_{Pl} &\sim \sqrt{G_N} \sim 10^{-33}\ \text{cm} \\ t_{Pl} &\sim \sqrt{G_N} \sim 5 \times 10^{-44}\ \text{s}. \end{aligned}$$

To probe a distance scale smaller than the Planck length, the required kinetic energy of the test particle should be greater than the Planck scale. At this scale, gravity becomes strong and quantum uncertainty comes into the picture. This introduces a restriction in our study of the early universe. Our lack of understanding of quantum gravity does not allow us to understand the evolution of the universe before the Planck time.

**Tests of General Relativity**

The general theory of relativity have been successfully tested by many experiments. The simplest prediction is the bending of light. Since light propagates along a straight line, if there is curvature in space due to a massive body, light will travel along the curved trajectory. This has been verified with differing accuracy at different times [176, 177].

Newtonian mechanics also predicts a bending of light in a gravitational field. The angle of deflection comes out to be

$$\Delta\phi_N \approx 2/e \approx 2GM/c^2 r_m,$$

where $e$ is the eccentricity of the deflected path. For the light grazing the sun, this gives $\phi_N = 0.875''$. On the other hand, for the light grazing the sun the general theory of relativity predicted the deflection of light to be $\Delta\phi = 1.75''$. Since light follows the null geodesics, special theory of relativity does not predict any deflection of light. During total eclipse of the sun, a measurement of positions of about 400 stars gives an average deflection of light $\phi = 1.9''$.

The gravitational red-shift of spectral lines measurement also supports the general theory of relativity. Another important observation in support of the general theory of relativity is the precession of the perihelion of Mercury [178]. According to the Newtonian gravity, Mercury follows an elliptical path around the sun. But due to other planets and stars, the orbit precesses. The Newtonian mechanics predicted the precession of the perihelion (when the distance between the sun and Mercury is minimum) of Mercury to be $\Delta\phi_N = (5557.62 \pm 0.20)''$ while the observed value is $\Delta\phi_{obs} = (5600.73 \pm 0.40)''$. The discrepancy of $\Delta\phi_{dis} = (43.11 \pm 0.45)''$ could be explained by the general theory of relativity, which predicted $\Delta\phi_{dis}^{gtr} = 43.03''$ for the discrepancy.

### Gravitational Radiation

The general theory of relativity describes the the gravitational interaction, and the solution of the Einstein equation (equation (11.28)) gives the metric or the geometry of space–time that contains the information about the gravitational interaction. The traceless symmetric metric tensor $g_{\mu\nu}$ also includes the gravitational radiation, but this is not transparent in the Einstein action as the quadratic kinetic energy term is not present explicitly. For this purpose we shall now work with the action for linearized Einstein gravity [179]. We first make the weak field approximation, which states that the space is almost flat and the metric contains small fluctuations around the Minkowski metric $\eta_{\mu\nu}$. This is expressed as

$$g_{\mu\nu} = \eta_{\mu\nu} + k\, h_{\mu\nu}, \tag{11.30}$$

where $|h_{\mu\nu}| \ll 1$ and the factor $k$ indicates that the mass dimension of $h_{\mu\nu}$ is 1. For convenience we set $k = 1$. In one formulation, quantization of gravity makes similar weak field approximation with the Minkowski metric as the vacuum and the perturbation $h_{\mu\nu}$ corresponding to the quantum field mediating gravitational interaction–*the graviton*.

Since the perturbation $h_{\mu\nu}$ is small, we keep only first order terms in $h_{\mu\nu}$. This will allow us to contract or raise or lower indices using $\eta_{\mu\nu}$. The linearized Einstein action can then be given by

$$S_L^{(E)} = -\frac{1}{2}\int d^4x(R^L_{\mu\nu} - \frac{1}{2}\eta_{\mu\nu}R^L)h^{\mu\nu}, \tag{11.31}$$

with the linearized Ricci tensor and the linearized curvature scalar given by

$$\begin{aligned} R^L_{\mu\nu} &= \frac{1}{2}\left(\frac{\partial^2 h_{\mu\nu}}{\partial x^\lambda \partial x_\lambda} - \frac{\partial^2 h^\lambda{}_\nu}{\partial x^\mu \partial x^\lambda} - \frac{\partial^2 h^\lambda{}_\mu}{\partial x^\nu \partial x^\lambda} + \frac{\partial^2 h^\lambda{}_\lambda}{\partial x^\mu \partial x^\nu}\right) \\ R^L &= \eta_{\mu\nu}R^L_{\mu\nu}, \end{aligned} \tag{11.32}$$

where we retained terms only up to first order in $h_{\mu\nu}$. The linearized Einstein equation including the cosmological constant and the energy-momentum tensor ($T_{\mu\nu}$) for the matter fields then becomes

$$R^L_{\mu\nu} - \frac{1}{2}\eta_{\mu\nu}R^L + \Lambda\eta_{\mu\nu} = -8\pi G_N T_{\mu\nu}. \tag{11.33}$$

Since $T_{\mu\nu}$ is independent of $h_{\mu\nu}$ to first order, it satisfies the condition

$$\frac{\partial}{\partial x^\mu}T^\mu{}_\nu = 0. \tag{11.34}$$

We now demand that general coordinate transformations of any weak field should leave the field weak, so after the general coordinate transformation the metric should remain close to the Minkowski metric. This restricts the allowed general coordinate transformations to the form

$$x^\mu \to x'^\mu = x^\mu + \varepsilon^\mu(x), \tag{11.35}$$

where $O(\partial\varepsilon^\mu/\partial x^\nu) \lesssim h_{\mu\nu}$. This ensures that after the coordinate transformation, the new fluctuation

$$h'_{\mu\nu} = h_{\mu\nu} - \frac{\partial \varepsilon_\mu}{\partial x^\nu} - \frac{\partial \varepsilon_\nu}{\partial x^\mu} \tag{11.36}$$

is also a solution, where $\varepsilon_\mu = \varepsilon^\nu\eta_{\mu\nu}$ are some small arbitrary functions of $x^\mu$.

We now work in the harmonic coordinate system utilizing the gauge freedom of the linearized Einstein equation (11.33), which is defined as

$$g^{\mu\nu}\Gamma^\lambda_{\mu\nu} = 0. \tag{11.37}$$

The field equation then becomes

$$\frac{\partial^2 h_{\mu\nu}}{\partial x^\lambda \partial x_\lambda} = -16\pi G_N(T_{\mu\nu} - \frac{1}{2}\eta_{\mu\nu}T^\lambda{}_\lambda) = -16\pi G_N S_{\mu\nu}, \tag{11.38}$$

where $S_{\mu\nu}$ is the source for the gravitational field $h_{\mu\nu}$ and the solution is a retarded potential

$$h_{\mu\nu}(\vec{x},t) = -4\pi G_N \int d^3\vec{x}' \frac{S_{\mu\nu}(\vec{x}', t - |\vec{x} - \vec{x}'|)}{|\vec{x} - \vec{x}'|}. \tag{11.39}$$

This is the gravitational radiation produced by the source $S_{\mu\nu}$, having two longitudinal degrees of freedom like the photon, corresponding to the helicity $\pm 2$. The quantum of the gravitational radiation, the graviton, can interact with itself due to the nonlinearity in the Einstein equation making the general theory of relativity more complex than any gauge theories.

If there is any radiation coming from infinity, that will satisfy the equations

$$\frac{\partial^2 h_{\mu\nu}}{\partial x^\lambda \partial x_\lambda} = 0$$
$$\frac{\partial}{\partial x^\mu} h^\mu{}_\nu = \frac{1}{2} \frac{\partial}{\partial x^\nu} h^\mu{}_\mu \tag{11.40}$$

and can be given by the plane wave solution

$$h_{\mu\nu}(x) = e_{\mu\nu} \exp^{ik_\lambda x^\lambda} + e^*_{\mu\nu} \exp^{-ik_\lambda x^\lambda}, \tag{11.41}$$

where the symmetric matrix $e_{\mu\nu}$ is the polarization tensor. This solution will add to the retarded potential due to the source $S_{\mu\nu}$. The energy and momentum carried by this plane wave coming from infinity is

$$t_{\mu\nu} = \frac{1}{8\pi G_N} [R^{(2)}_{\mu\nu} - \eta_{\mu\nu} \eta^{\lambda\rho} R^{(2)}_{\lambda\rho}], \tag{11.42}$$

where $R^{(2)}_{\mu\nu}$ is the second order term in the Ricci tensor in $h^2_{\mu\nu}$. The first order terms in $h^2_{\mu\nu}$ cancel out since they satisfy the Einstein equation.

The *Mach's principle* seeded the idea of gravitational radiation by stating that the motions of local bodies are not independent of the influence from the rest of the universe [180]. Analogy of gravitation and electromagnetism then implies that the acceleration of the distant cosmic objects should produce gravitational waves that will affect the motion of local matter. The Einstein equation demonstrates this gravitational radiation mathematically and point out that this radiation carry energy. Detection of gravitational radiation has been going on since its inception, but so far we have not been able to detect the radiation. Earlier experiments tried to use small mechanical or hydrodynamical resonant quadrupole antennas with natural mode of free oscillation, while the present generations of detectors take into account possible sources of the radiation and make use of the advancements in space technology.

## 11.2 Gravity in Higher Dimensions

Unification of the gravitational interaction with the strong, weak and electromagnetic interactions is the ultimate goal we have been aspiring. In this direction the first consistent theory was the *Kaluza–Klein* theory [181]. In the Kaluza–Klein theory the space–time dimensions have been extended to 5-dimensions with the expectation that the metric in 5-dimensions should contain the 4-dimensional metric as well as the gauge boson of the electromagnetic interaction. Since we know that there are only four noncompact dimensions for consistency, the fifth dimension has been assumed to be compact. The radius of compactification of the fifth dimension is assumed to be extremely small, so our present energy probes will not be able to see this extra dimensions. In this 5-dimensional theory of the general theory of relativity, the Einstein action describes all the interactions. After compactification the effective theory will have the Einstein equation in 4-dimensions and the extra degrees of freedom of the 5-dimensional metric will contain the photon or the gauge field corresponding to the $U(1)_Q$ gauge theory in 4-dimensions, which is the theory of electromagnetic interaction. Higher-dimensional theories are required to accommodate the non-Abelian gauge interactions.

The Kaluza–Klein theory was not considered seriously for many years. However, now we cannot think of unifying gravity with other interactions without extending our theory to include new extra dimensions. Thus, the Kaluza–Klein theory now forms the basis of many recent developments in the field.

**Kaluza–Klein Theory**

In the Kaluza–Klein theory [181] one starts with the theory of gravity in more than three space and one time dimensions. The simplest case is a five-dimensional space–time, where the fifth dimension is assumed to be a compact extra space dimension. Since we have not seen the fifth dimension, we assume that it is compact with a very small radius. The distance to which we could probe with our present energy accelerators is larger than the radius of the compact fifth dimension and that is why we have not been able to resolve the fifth dimension. In most theories we expect the radius of the compact dimensions to be close to the Planck length of $l \sim G_N \sim 10^{-35} m$, so we may resolve them only when we reach the Planck energy scale of $M_{Pl} \sim 10^{19}$ GeV.

The energy scale corresponding to the radius of the compact dimensions is also called the scale of compactification. At energies below the compactification scale, it will not be possible to resolve the extra dimensions and hence they will only appear as points. However, the scaling invariance along the extra dimensions will manifest itself as gauge invariance in some internal symmetry space. In general, when we work in a $(4+n)$-dimensional space, it correspond to a $M^4 \times \mathscr{K}$ space, where $M^4$ is the usual 4-dimensional Minkowski space–time and $\mathscr{K}$ is the compactified $n$-dimensional space [182]. In the present example, $\mathscr{K}$ is a one-dimensional space denoted by the coordinate $y$ and the coordinate of the usual 4-dimensional Minkowski space is represented by $x$. Since the fifth dimension is compact, the coordinate of the fifth dimension must satisfy the boundary condition

$$y = y + 2\pi r, \tag{11.43}$$

where $r$ is the radius of compactification, which gives the size of the compact space. Although the simplest form of the compact space is a circle with radius $r$, in general the shape of the compact space is irrelevant. As long as this boundary condition is satisfied, the space is compact and its size is specified by $r$.

This boundary condition will imply that any scalar field $\phi(x)$ should satisfy

$$\phi(x,y) = \phi(x,y+2\pi r), \tag{11.44}$$

which would then allow a Fourier series expansion of the field

$$\phi(x,y) = \sum_{n=-\infty}^{\infty} \phi_n(x)\exp^{iny/r}. \tag{11.45}$$

The fact that the $\phi_n(x)\exp^{iny/r}$ should be single-valued under $y \to y+2\pi r$ implies that the $y$ component of the momentum of any state $\phi_n$ will be quantized and is of the order of $\sim O(|n|/r)$. In 4-dimensions these states will represent physical states or quantized excitations with momentum $O(|n|/r)$.

The physical spectrum of particles will now contain a zero-mode $\phi_0$, which corresponds to $n = 0$, and remain massless at this stage. All other states will be very massive with mass of the order of $1/r \sim M_c$, where $M_c$ is the compactification scale. In the conventional theories this is taken to be around the Planck scale $M_{Pl} \sim 10^{19}$ GeV. Higher order excitations will have mass in multiples of the Planck mass. Thus, all the particles we experience at present energies will have corresponding infinite towers of Kaluza–Klein excitations above the compactification scale of $M_c$ if the theory originates from any higher-dimensional theories.

In any realistic Kaluza–Klein theory, the zero modes must contain the standard model fermions, gauge bosons and the graviton. We shall now demonstrate that the 5-dimensional theory of gravity contains both the 4-dimensional graviton and a 4-dimensional gauge boson corresponding to quantum electrodynamics. Let us start with a line element in 5 dimensions:

$$ds^2 = g_{MN}dx^M dx^N = g_{\mu\nu}dx^\mu dx^\nu - (dy - \kappa A_\mu(x)dx^\mu)^2, \tag{11.46}$$

where $g_{MN}, M, N = 0,1,2,3,4$ is the 5-dimensional metric; $g_{\mu\nu}, \mu, \nu = 0,1,2,3$ is the 4-dimensional metric; and $A_\mu$ is a four vector. Any coordinate transformation in the $y$ direction will now appear as a gauge transformation that leaves the line element invariant

$$\begin{aligned} x^\mu &\to x^\mu \\ y &\to y + \Lambda(x) \\ g_{\mu\nu} &\to g_{\mu\nu} \\ A_\mu(x) &\to A_\mu(x) + \frac{1}{\kappa}\partial_\mu \Lambda. \end{aligned} \tag{11.47}$$

Let us now include a complex scalar field ($\Phi$) with mass $\mu$. We assume that the field satisfies the constraint

$$\partial_y \Phi = \frac{i}{r}\Phi. \tag{11.48}$$

As we shall see, this would imply that the field behaves as a charged scalar, where the charge is given in units of the couplings in the fifth dimension.

The Lagrangian for the field is now given by

$$\mathscr{L}_\Phi = e\left[g^{\mu\nu}D_\mu\Phi_n D_\nu\Phi_n - (\mu^2 + \frac{n^2}{r^2})|\Phi_n|^2\right], \tag{11.49}$$

where we define the gauge covariant derivative as

$$D_\mu = \partial_\mu - i\frac{\kappa}{r}A_\mu. \tag{11.50}$$

As mentioned earlier, the electromagnetic interaction comes out from the fifth dimension. Comparing with the usual gauge transformations of the quantum electrodynamics, we can see that the electric charge is given by

$$Q_n = \frac{\kappa}{r}, \tag{11.51}$$

and the coordinate transformations in the fifth dimension become equivalent to a $U(1)$ gauge transformation in 4-dimensions.

The mass of the different Kaluza–Klein states are given by

$$m_n^2 \sim \mu^2 + \left(\frac{n}{r}\right)^2. \tag{11.52}$$

Here $\mu \ll 1/r$ is the mass parameter and gives the mass of the zero mode. If we start with $\mu = 0$, we find that the zero mode remains massless, while all other Kaluza–Klein excitations have picked up mass in multiples of $n/r$. For any fermions, if the zero modes belong to any chiral representations, only then can they remain massless and protected by the chiral symmetry. Otherwise all the left-handed and right-handed components would combine to make them massive with mass of the order of $O(1/r)$. We now discuss the zero modes of the gauge fields and the metric.

We first assume

$$\frac{\partial g_{MN}}{\partial y} = 0 \tag{11.53}$$

to get the zero modes of the metric and their field equations in 4-dimensions. The Fourier transform of the 5-dimensional metric $g_{MN}$ can be written as

$$g_{MN}(x,y) = \sum_n g_{MN}^{(n)}(x)\exp^{iny/r}. \tag{11.54}$$

Since the 4-dimensional metric $g_{\mu\nu}$ and the 4-dimensional gauge boson $A_\mu$ should remain massless, they should belong to the zero mode of the 5-dimensional metric $g_{MN}^{(0)}$. We parameterize the zero mode of the 5-dimensional metric $g_{MN}^{(0)}$ in terms of $g_{\mu\nu}$, $A_\mu$ and a scalar field $\phi$ as

$$g_{MN}^{(0)} = \phi^{-1/3}\begin{pmatrix} g_{\mu\nu} + \phi A_\mu A_\nu & \phi A_\mu \\ \phi A_\nu & \phi \end{pmatrix}. \tag{11.55}$$

The field $\phi$ is called a *dilaton field*, since it appears as a scaling parameter in the fifth dimension. Any higher-dimensional theory of gravity contains the dilaton field, which may have some interesting phenomenological consequence as well.

If we now start with the Einstein's action in 5-dimensions

$$S^{(5)}_{(E)} = -\frac{1}{\kappa_5^2} \int d^4x dy e^{(5)} R^{(5)}, \tag{11.56}$$

in 4-dimensions the zero modes contain the 4-dimensional Einstein's action, field equations of the gauge fields, and the kinetic energy term of the dilaton field

$$S^{(5)}_{(E)} = S_{(E)} - \frac{e}{2\kappa^2} \int d^4x \left[ \frac{1}{4} \phi F_{\mu\nu} F^{\mu\nu} + \frac{1}{6\phi^2} \partial^\mu \phi \partial_\mu \phi \right]. \tag{11.57}$$

Here $F_{\mu\nu}$ is the field-strength tensor and $\kappa$ is the four-dimensional gravitational constant

$$\begin{aligned} F_{\mu\nu} &= \partial_\mu A_\nu - \partial_\nu A_\mu \\ \kappa^2 &= \frac{\kappa_5^2}{(2\pi r)}. \end{aligned} \tag{11.58}$$

The zero mode gauge fields $A_\mu$ remain massless due to the gauge invariance. Thus, after integrating out all the heavy degrees of freedom from the theory, at low energies we find the 4-dimensional metric satisfying the Einstein equation, the massless gauge bosons satisfying the field equations and maintaining gauge invariance of electrodynamics, and the scalar dilaton field satisfying the Klein–Gordon equation. In addition, if there are zero modes of scalars and fermions, they satisfy the usual field equations.

**Non-Abelian Gauge Theory in Higher Dimension**

In the simplest Kaluza–Klein theory the general theory of relativity is extended to a 5-dimensional theory. The extra space dimension is then assumed to be compact so 5-dimensional metric can be expanded in terms of its harmonics. The zero mode of the metric then contains the regular massless graviton, or the 4-dimensional metric satisfying the Einstein equation. In addition the metric contains the vector bosons of a $U(1)$ gauge theory and a scalar field, the dilaton. Any translation along the fifth dimension appears as a $U(1)$ gauge transformation in 4-dimensions.

The next step in this approach is to consider more than 5-dimensions, so the compactification leads to larger gauge groups, including non-Abelian gauge theories. When the zero modes of the higher-dimensional metric are decomposed in terms of the zero modes of the four-dimensional metric, the zero modes of the remaining components become the gauge fields in four dimensions. For any non-Abelian gauge theory, the gauge fields $A^a_\mu, a = 1, \cdots, n$ require $4m$ components of the higher-dimensional metric, where $m$ is determined by the isometry group, and the metric of the compactified space depends on $n$ as well. This will be possible in a $4+m$-dimensional space.

Let us consider a $D = 4 + d$-dimensional space $\mathscr{M}^4 \otimes \mathscr{K}$, where $\mathscr{K}$ is the compact $d$-dimensional space and $\mathscr{M}^4$ is the 4-dimensional Minkowski space. If the radius of compactification is so small that an observer cannot resolve it, the observer will find the space around him to be locally a Minkowski space $\mathscr{M}^4$ and given by a flat metric $\eta_{\mu\nu}$, $\mu, \nu = 0, 1, 2, 3$. He will not know the existence of the extra dimensions. To ensure the validity of the equivalence principle, any transformations in the local coordinate system should not affect the extra dimensions. On the other hand, any fluctuations in the extra dimensions should not perturb our locally flat space $\mathscr{M}^4$. This means that $\mathscr{K}$ is orthogonal to $\mathscr{M}^4$ and is a compact $d$-dimensional manifold given by a metric $g_{\alpha\beta}(y)$, where the indices $\alpha, \beta = 4, \cdots, D$ run over the $d$-dimensional space coordinates $y$ ensuring causality. We can, thus, write the metric of the $D$-dimensional space as

$$g_{MN} = \begin{pmatrix} \eta_{\mu\nu}(x) & 0 \\ 0 & g_{\alpha\beta}(y) \end{pmatrix}, \tag{11.59}$$

where $M, N$ are the $D = 4 + d$-dimensional indices in the space $\mathscr{M}^4 \otimes \mathscr{K}$.

We now identify any transformations along the extra dimensions with the gauge transformations in four dimensions, which will be restricted by the metric on $\mathscr{K}$. Given a particular manifold $\mathscr{K}$ in the $d$-dimensional space, the symmetry group for the gauge transformations that can be generated in 4-dimensions is called the *isometry group* of the manifold $\mathscr{K}$. The isometry group is the group of transformations of the coordinates $y^\alpha$, generated by the *Killing vectors* $K_a^\alpha(y)$ of the manifold $\mathscr{K}$, that leave the metric of $\mathscr{K}$ unchanged.

Consider a group of transformations generated by the Killing vectors $K_a^\alpha(y)$

$$y^\alpha \to y'^\alpha = y^\alpha + \sum_{a=1}^{n_k} \varepsilon^a(x) K_a^\alpha(y), \tag{11.60}$$

where $n_k < d(d+1)$ is the dimension of the isometry group of $\mathscr{K}$. The Killing vectors satisfy the Killing equations

$$\nabla^\alpha K_a^\beta + \nabla^\beta K_a^\alpha = 0. \tag{11.61}$$

The $D$-dimensional metric can then be parameterized as

$$g_{MN} = \begin{pmatrix} g_{\mu\nu} + g_{\alpha\beta} K_a^\alpha A_\mu^a K_b^\beta A_\nu^b & g_{\alpha\beta} K_a^\alpha A_\mu^a \\ g_{\alpha\beta} K_a^\beta A_\nu^a & g_{\alpha\beta} \end{pmatrix}. \tag{11.62}$$

Since any transformations generated by the Killing vectors should leave the metric of the manifold invariant, the gauge bosons must transform as

$$A_\mu^a \to A_\mu^a + \partial_\mu \varepsilon^a(x). \tag{11.63}$$

These gauge bosons mediate the gauge interactions in 4-dimensions.

If we now try to get the standard model gauge groups from compactification of some extra dimensions, then the isometry group of the required manifold $\mathscr{K}$ must contain the $SU(3)_c \times SU(2)_L \times U(1)_Y$ gauge group. The lowest-dimensional manifold that can have such isometry group to accommodate the standard model is a 7-dimensional manifold [183]. Thus, we need at least an 11-dimensional space to accommodate the standard model gauge group. However, when we try to include the quarks and leptons in any 11-dimensional theory, we encounter new problems.

In any odd-dimensional space, all fermion representations are vector-like, and after compactification they all acquires a mass of the order of the scale of compactification, which is around the Planck scale. It can be proved in a general way that in odd dimensions it is not possible to have any zero modes and there are no massless fermions. This makes it difficult to construct any 11-dimensional Kaluza–Klein theory, which can accommodate the standard model gauge group and the chiral fermions that can give us the particle spectrum of the standard model.

Although the 11-dimensional Kaluza–Klein theory could not accommodate the standard model, it demonstrated that higher-dimensional theories can treat gravity and other gauge interactions in the same way. One natural extension of the theory is to supersymmetrize the theory. In a supersymmetric theory of gravity, the *supergravity*, all the gauge and scalar fields will have their fermionic superpartners, which can remain massless after compactification. With this motivation, theories of supergravity were constructed in four and higher dimensions. We shall now discuss the simplest case of $N = 1$ supergravity theory in four dimensions.

## 11.3 Supergravity

We discussed supersymmetry in chapter 9, where we did not consider the parameters of supersymmetry transformation to be local. We also restricted ourselves to $N = 1$ supersymmetry generated by one spinorial operator $Q$. The minimal supersymmetric standard model, which is also based on the $N = 1$ global supersymmetry, was then discussed in chapter 10. However, if we intend to combine gravity with supersymmetry, then we have to consider *local supersymmetry* or *supergravity*, where the coefficients of the generators in the supersymmetry transformations are space–time dependent [158].

We shall discuss only $N = 1$ supergravity generated by one spinorial operator $Q$. The generators of supersymmetry now vary from point to point, generate local coordinate system at different space–time points, and should include the general theory of relativity. We first supersymmetrize the Einstein equation (equation (11.28)) of the general theory of relativity. We include the fermionic superpartner $\Psi_\mu$ of the spin-2 graviton, which is described by a traceless symmetric tensor $g_{\mu\nu}$. The kinetic energy for this spin-3/2 gravitino or the Rarita–Schwinger field $\Psi_\mu$, which is a spinor carrying a vector index, satisfies the Rarita–Schwinger Lagrangian [184],

$$\mathscr{L}_{RS} = -\frac{1}{2}\varepsilon^{\mu\nu\rho\sigma}\bar{\Psi}_\mu \gamma_5 \gamma_\nu \tilde{D}_\rho \Psi_\sigma, \tag{11.64}$$

where $\varepsilon^{\mu\nu\rho\sigma}$ is the totally antisymmetric tensor, with $\varepsilon^{0123} = 1$. The covariant derivative $(\tilde{D}_\mu)$ includes terms quadratic in the Rarita–Schwinger fields and is defined by

$$\tilde{D}_\mu = \partial_\mu - \frac{i}{4}\tilde{\omega}_{\mu mn}\sigma_{mn}, \tag{11.65}$$

where

$$\tilde{\omega}_{\mu mn} = \omega_{\mu mn} + \frac{i\kappa^2}{4}(\bar{\Psi}_\mu\gamma_m\Psi_n + \bar{\Psi}_m\gamma_\mu\Psi_n - \bar{\Psi}_\mu\gamma_n\Psi_m). \tag{11.66}$$

To get the field equations of the gravitons, we expand the metric $g_{\mu\nu}$ around a flat metric $\eta_{\mu\nu}$ as

$$g_{\mu\nu} = \eta_{\mu\nu} + \kappa h_{\mu\nu}, \tag{11.67}$$

where $\kappa$ makes the mass dimension of $h_{\mu\nu}$ to be 1, so it can describe the bosonic field graviton. This expansion will linearize Einstein gravity [179], the Einstein action is now replaced by the linearized Einstein action, and the vierbein is expressed in terms of the graviton $h_{\mu\nu}$.

The Lagrangian corresponding to this linearized Einstein action is now given by

$$\mathscr{L}_{(E)} = -\frac{eR}{2\kappa^2} = -\frac{e}{\kappa^2}R_{\mu\nu}{}^{mn}e_m{}^\nu e_n{}^\mu. \tag{11.68}$$

When this Lagrangian is combined with the Rarita–Schwinger Lagrangian of equation (11.64)

$$\mathscr{L}_{sugra} = \mathscr{L}_{RS} + \mathscr{L}_{(E)}, \tag{11.69}$$

it becomes invariant under the on-shell local supersymmetry transformations,

$$\delta_\xi\Psi_\mu = \frac{1}{\kappa}D_\mu\xi(x) \tag{11.70}$$

$$\delta_\xi e^m{}_\mu = -\frac{i}{2}\kappa\bar{\xi}(x)\gamma^m\Psi_\mu. \tag{11.71}$$

The spinorial field $\xi(x)$ generates local supersymmetry.

We now introduce the auxiliary fields to close the algebra off-shell since the degrees of freedom for the fields $e^m{}_\mu$ and $\Psi_\mu$ are different. These fields are: a *scalar* $S$, a *pseudoscalar* $P$ and an *axial vector* $A_\mu$. These auxiliary fields will be eliminated by the equations of motion. We can then define the complete Lagrangian as

$$\mathscr{L} = \mathscr{L}_{RS} + \mathscr{L}_{(E)} - \frac{e}{3}(S^2 + P^2 - A_m^2), \tag{11.72}$$

where $A_m = e^m{}_\mu A_\mu$. This completes the discussion on the supersymmetric gravity multiplet.

To include the matter multiplet [158, 185], we start with the Wess–Zumino model (equation (9.17)) and write the on-shell supersymmetric Lagrangian of a scalar $\phi$ and a fermion $\psi$, given by

$$\mathscr{L} = (\partial_\mu\phi^*)(\partial^\mu\phi) + i\bar{\psi}\bar{\sigma}^\mu\partial_\mu\psi. \tag{11.73}$$

Under local supersymmetry transformation this Lagrangian is not invariant and there is one additional term proportional to $\partial_\mu \bar{\xi}(x)$. This can be compensated by including the Rarita–Schwinger field in combination with the energy-momentum tensor $T^{\mu\nu}$. But this is still not complete and the graviton field has to be included to compensate the transformation of the Rarita–Schwinger field. After including all of these terms, the final Lagrangian can be made invariant under local supersymmetry transformation, which is given by

$$\begin{aligned}\mathscr{L} = & -\frac{1}{2\kappa^2}eR - \frac{1}{2}\varepsilon^{\mu\nu\rho\sigma}\bar{\Psi}_\mu\gamma_5\gamma_\nu\tilde{D}_\rho\Psi_\sigma + e(\partial_\mu\phi^*)(\partial^\mu\phi) \\ & +\frac{i}{2}e\bar{\Psi}\gamma^\mu D_\mu\Psi - \frac{\kappa}{2}e\bar{\Psi}_\mu\phi(A - i\gamma_5 B)\gamma^\mu\Psi \\ & -\frac{i\kappa^2}{4}\varepsilon^{\mu\nu\rho\sigma}\bar{\Psi}_\mu\gamma_\nu\Psi_\rho A\overleftrightarrow{D}_\sigma B - \frac{\kappa^2}{4}e\bar{\Psi}\gamma_5\gamma^\mu\Psi A\overleftrightarrow{D}_\mu B \\ & +\text{four fermion quartic interaction terms,}\end{aligned} \tag{11.74}$$

where the complex scalar $\phi$ has been written in terms of two real scalars,

$$\phi = \frac{1}{\sqrt{2}}(A + iB). \tag{11.75}$$

This is the local supersymmetric Lagrangian corresponding to the Wess–Zumino model.

We can now follow the same procedure to construct a Lagrangian for the chiral ($\Phi$) and vector ($V$) superfields, which is given by

$$\mathscr{L} = \mathscr{L}_B + \mathscr{L}' + \mathscr{L}_F, \tag{11.76}$$

where the first term includes the interactions of the scalar fields and the kinetic energy terms of the scalars and the gauge bosons; the second term includes the kinetic energy terms for the Rarita–Schwinger spinor, the fermions, the gauginos and some nonrenormalizable terms; and the last term contains fermion Yukawa couplings, gaugino couplings and all other remaining nonrenormalizable interactions.

For an explicit form of the Lagrangian, we first define the Kähler potential

$$G(\phi^*, \phi) = J(\phi^*, \phi) + \ln|W|^2, \tag{11.77}$$

where

$$J(\phi^*, \phi) = -3\ln(-K/3). \tag{11.78}$$

Here $K \equiv K(\Phi^\dagger \exp^{2gV}, \Phi)$ is an arbitrary function of the chiral superfields ($\Phi$) and their conjugates ($\Phi^\dagger$) and appears in the global supersymmetric Lagrangian. Since gravity can induce nonrenormalizable terms, the global supersymmetric Lagrangian is not necessarily renormalizable in any supergravity theory and is given by

$$\mathscr{L}_G = [K(\Phi^\dagger \exp^{2gV}, \Phi)]_D + [W(\Phi)]_F + [f_{ab}(\Phi)W_a^\alpha W_{\alpha b}]_F + H.c., \tag{11.79}$$

where $W_{a\alpha} = \bar{D}^2 D_\alpha V^a$ is the *gauge field strength superfield* and $f_{ab}(\Phi) = f^R + i f^I$ is an arbitrary function of the superfields (with $f^R_{ab} = \text{Re}f_{ab}; f^I_{ab} = \text{Im}f_{ab}$). Both $K$ and

$f$ can now include nonrenormalizable terms. Interactions of the gauge bosons with the chiral fields enter through the terms involving the vector superfield $V$.

Before we write the complete Lagrangian, we define the derivatives of the Kähler potential as

$$\begin{aligned} G^i &= \frac{\partial G}{\partial \phi_i} \\ G_i &= \frac{\partial G}{\partial \phi^{i*}} \\ G^i_j &= \frac{\partial^2 G}{\partial \phi_i \partial \phi^{j*}} \end{aligned} \tag{11.80}$$

and define the inverse

$$(G^{-1})^i_j G^j_k = \delta^i_k. \tag{11.81}$$

In the unit of $\kappa = 1$ (which means all masses are in units of the Planck mass), the different terms in the Lagrangian are then given by

$$\begin{aligned} \frac{\mathscr{L}_B}{e} &= -\frac{1}{2}R + G^i_j D_\mu \phi_i D^\mu \phi^{j*} + \exp^G \left(3 - G_i (G^{-1})^i_j G^j\right) \\ &\quad -\frac{1}{4} f^R_{ab} F_{a\mu\nu} F^{\mu\nu}_b + \frac{i}{4} f^I_{ab} F_{a\mu\nu} \tilde{F}^{\mu\nu}_b \\ &\quad -\frac{g^2}{2} (\mathrm{Re} f^{-1}_{ab}) G^i (T_a)_{ij} \phi_j G^k (T_b)_{kl} \phi_l. \end{aligned} \tag{11.82}$$

$$\begin{aligned} \frac{\mathscr{L}'}{e} &= \frac{1}{2e} \varepsilon^{\mu\nu\sigma\rho} \bar{\Psi}_\mu \gamma_5 \gamma_\nu \tilde{D}_\rho \Psi_\sigma + \frac{1}{4e} \bar{\Psi}_\mu \gamma_\nu \Psi_\rho (G^i D^{\mu\nu\rho} \phi_i - G_i D^{\mu\nu\rho} \phi^{i*}) \\ &\quad + \left( \frac{i}{2} G^i_j \bar{\Psi}_{iL} \gamma^\mu D_\mu \Psi^j_L + \frac{i}{2} \bar{\Psi}_{iL} \gamma^\mu D_\mu \phi_j \Psi_{kL} (-G^{ij}_k + \frac{1}{2} G^i_k G^j) \right. \\ &\quad \left. + \frac{1}{\sqrt{2}} G^i_j \bar{\Psi}_{\mu L} \gamma^\nu D_\nu \phi^{i*} \gamma^\mu \Psi_{jR} \right) - \frac{1}{2} \frac{\partial f_{ab}}{\partial \phi_i} \bar{\Psi}_{iR} \sigma^{\mu\nu} F_{a\mu\nu} \lambda_{bL} \\ &\quad + \frac{1}{4} f^R_{ab} (\bar{\lambda}_a \gamma^\mu D_\mu \lambda_b + \bar{\lambda}_a \gamma^\mu \sigma^{\nu\rho} \Psi_\mu F_{b\nu\rho} + G^i D_\mu \phi_i \bar{\lambda}_{aL} \gamma^\mu \lambda_{bL}) \\ &\quad - \frac{i}{8} f^I_{ab} D_\mu (e \bar{\lambda}_a \gamma_5 \gamma^\mu \lambda_b) + H.c. \end{aligned} \tag{11.83}$$

$$\begin{aligned} \frac{\mathscr{L}_F}{e} &= \frac{1}{2} \exp^{G/2} \left( i \bar{\Psi}_\mu \sigma^{\mu\nu} \Psi_\nu - (G^{ij} + G^i G^j - G^{ij}_k (G^{-1})^k_l G^l) \bar{\Psi}_{iL} \Psi_{jR} \right. \\ &\quad \left. + i\sqrt{2} G^i \bar{\Psi}_{\mu L} \gamma_\mu \psi_{iL} + \frac{1}{2} \frac{\partial f^*_{ab}}{\partial \phi^{j*}} (G^{-1})^j_k G^k \lambda_a \lambda_b \right) \\ &\quad - \frac{i}{2} g \left( G^i (T_a)_{ij} \phi_j \bar{\Psi}_{\mu L} \gamma^\mu \lambda_{aL} - 4 G^i_j (T_a)_{ik} \phi_k \bar{\lambda}_{aR} \Psi_{iL} \right. \\ &\quad \left. + \mathrm{Re} f^{-1}_{ab} \frac{\partial f_{bc}}{\partial \phi_k} G^i (T_a)_{ij} \phi_j \bar{\psi}_{kR} \lambda_{cL} \right) + H.c. \\ &\quad + \text{four fermion quartic interaction terms,} \end{aligned} \tag{11.84}$$

where $\psi_L$ and $\psi_R$ are the left and right chiral components of the Majorana spinors and we defined $\varepsilon^{\alpha\beta\gamma\delta}D_\delta = D^{\alpha\beta\gamma}$. Although we presented the Lagrangian for completeness, it is too complicated to explain every term in details.

The global supersymmetric Lagrangian $\mathscr{L}_G$ contains most of the low-energy interactions. When supersymmetry is broken in the hidden sector, supersymmetry breaking soft terms are introduced into the Lagrangian (see chapter 10) leaving the low-energy theory renormalizable. This will ensure that there are no quadratic divergent corrections even after supersymmetry breaking.

Although supersymmetry may be broken in the hidden sector at some intermediate symmetry breaking scale, the scale of soft terms ($\mu_{susy}$) is of the order of TeV. As a result the mass splitting among the superpartners is of the order of the $\mu_{susy}$ and any low-energy supersymmetry breaking effect will be of the same order of magnitude. In any supergravity theory, when supersymmetry is broken the gravitino gets a mass,

$$m_{3/2} = \exp^{G/2}. \tag{11.85}$$

This sets the effective supersymmetry breaking scale at low-energy $\mu_{susy}$, and all soft terms have the same mass parameter $\mu_{susy}$ [158].

We shall now explain how the superHiggs mechanism breaks local supersymmetry giving us the low-energy effective theory [158]. Consider a simple form of the Kähler potential

$$G(\phi_i, \phi_i^*) = \phi^i \phi_i^* + \ln |W(\phi_i)|^2. \tag{11.86}$$

The potential contained in the Lagrangian 11.82

$$V = \exp^G [G_i (G^{-1})^i_j G^j - 3] \tag{11.87}$$

then becomes

$$V = \exp^{\phi^i \phi_i^*} \left[ \left| \frac{\partial W}{\partial \phi^i} + \phi_i^* W \right|^2 - 3|W|^2 \right]. \tag{11.88}$$

Since we are working in units of $\kappa = 1$, all masses are in units of Planck mass. Any terms with mass dimension greater than 4 will be suppressed by powers of the Planck mass. Neglecting these small terms this potential will reduce to the global supersymmetric potential $V = |\partial W / \partial \phi^i|^2$.

The hidden sector supersymmetry breaking through the superHiggs mechanism introduces a chiral multiplet $z$, which is governed by the Polonyi potential [186]

$$W(z) = m^2 (z + \beta) = m^2 z'. \tag{11.89}$$

This potential breaks supersymmetry since $V = |\partial W / \partial z|^2 \neq 0$. To study the effect of this supersymmetry breaking in the hidden sector on the ordinary matter, we consider a superpotential involving the hidden sector field $z$ and the matter field $\phi_i$ to be of the form

$$W(z, \phi^i) = W_h(z) + W_o(\phi^i).$$

This gives the potential

$$V = \exp^{(|z|^2+|\phi_i|^2)}\left\{m^4\left(\left|1+z'z^*\right|^2-3|z'|^2\right)+\left|\frac{\partial W_o}{\partial \phi^i}+\phi_i^* W_o\right|^2-3|W_o|^2\right\}. \tag{11.90}$$

If we now assume that the vanishing of the cosmological constant before supersymmetry breaking in the observable sector, the minimum of $V$ must occur at $V = 0$. This determines the vacuum expectation values $z = a$, $W_o = bm^2$, and

$$m_{3/2} = \exp^{G/2} = |W_o|\exp^{z^2/2} = bm^2\exp^{a^2/2}. \tag{11.91}$$

For $W_o = 0$, when ordinary matter is not included, $a = \sqrt{3}-1$, $b = 1$, $\beta = 2-\sqrt{3}$, and $m \approx (m_{3/2}M_{Pl})^{1/2} \approx 10^{11}$ GeV for $m_{3/2} \sim 1$ TeV.

To get the low-energy contributions we neglect all terms that are suppressed by the Planck scale and then substituting the *vevs* of the hidden sector field $z$ and hidden sector superpotential $W_h$, we get

$$V_{eff} = Pm^4 + \left|\frac{\partial \widehat{W}_o}{\partial \phi^i}\right|^2 + V_{soft}, \tag{11.92}$$

where $P = [(1+ab)^2-3b^2]\exp^{a^2}$ and $\widehat{W}_o = \exp^{a^2/2}W_o$. The cosmological constant vanishes (not counting the contribution of the electroweak symmetry breaking scale) for $P = 0$. The second term is the scalar potential coming from the global supersymmetric theory, and the third term contains all the low-energy soft supersymmetry breaking terms

$$V_{soft} = m_{3/2}^2|\phi^i|^2 + m_{3/2}\left(\frac{\partial \widehat{W}_o}{\partial \phi^i}\phi^i + (a^2+\frac{a}{b}-3)\widehat{W}_o + H.c.\right). \tag{11.93}$$

The soft terms contain only one mass parameter $\mu_{susy} = m_{3/2}$. In this model of supergravity, supersymmetry is broken in the hidden sector by the Polonyi potential, which is communicated to the observable sector through supergravity interactions at the Planck scale. Thus, the soft term has suppression by the Planck scale. At present there does not exist any supersymmetry breaking mechanism in the observable sector that is phenomenologically consistent. So, all the models that are studied at present break supersymmetry in the hidden sector and the predictions of these models depend on how the supersymmetry breaking in the hidden sector is mediated to the observable sector. In all of these models low-energy theory contains only the soft supersymmetry breaking contributions.

## 11.4 Extended Supergravity

The gravitational interactions we encounter at present energies could be explained by the general theory of relativity. All other known interactions could be explained by the standard model, which is a gauge theory. One natural extension of gravity to include the standard model interactions is the Kaluza–Klein theory, which is the theory of gravity at higher dimensions. In the Kaluza–Klein theory, after compactification of the extra dimensions, the higher-dimensional metric contains the usual 4-dimensional metric and also the gauge fields that can generate the gauge interactions. Any translation in the extra dimensions would appear as gauge transformations in 4-dimensions. The isometry group of the compact manifold determines the gauge groups in 4-dimensions. To accommodate the standard model gauge group we require at least 11-dimensional space. However, in 11-dimensional space it is not possible to have massless fermions after compactification because all representations are vectorial. It is also extremely difficult to reproduce the standard model particle spectrum starting from any 11-dimensional Kaluza–Klein theory.

The next step in this direction was to extend gravity to make it supersymmetric. We start with a $N = 1$ local supersymmetric theory or a $N = 1$ supergravity theory. All representations contain both fermions and bosons, and hence, accommodating fermions is no longer a problem in this theory. All the soft terms have the same origin, and gravitino mass is the only mass parameter that determines all supersymmetry breaking effects. However, all the interactions are put in by hand and there is no geometric origin of the interactions.

Considering the problems of the higher-dimensional Kaluza–Klein theory and 4-dimensional $N = 1$ supergravity theory, one can now think of two possible extensions of these theories. One possibility is to have a $N = 1$ supergravity theory in 11-dimensions, while the other alternative is to have $N = n$ supersymmetry in 4-dimensions. As we shall discuss later, both approaches lead to the same result. But for the present let us discuss the two possibilities separately.

In an $N = 1$ supergravity theory in 11-dimensions, we start with only the graviton supermultiplet. The fields in this theory are the components of the gravity supermultiplet, which are the symmetric tensor $g_{MN}$ with 44 degrees of freedom, Rarita–Schwinger spinor $\psi_M$ with 128 degrees of freedom and an antisymmetric tensor $A_{MNP}$ with 84 degrees of freedom. The 4-dimensional massless fermions can come from the Rarita–Schwinger spinor fields in 11-dimensions. This theory has several interesting features including the possibility of spontaneous compactification of $D = 11$ to $\mathscr{M}^4 \otimes \mathscr{K}$. While the standard model gauge group $SU(3)_c \times SU(2)_L \times U(1)_Y$ can be extracted from the isometry group of a manifold if the manifold has at least 7-dimensions, another study of the extended supersymmetry reveals that for more than 7 compact extra dimensions, the theory becomes inconsistent. Thus, the 11-dimensional supergravity theory becomes a unique choice for higher dimensional supersymmetric theory of gravity.

We will discuss in detail the $N = 1$ supersymmetry, which is generated by one spinorial operator $Q_\alpha$ and its conjugate $Q_{\dot{\alpha}}$. In general there could be more generators of supersymmetry $Q^a_\alpha, a = 1, \cdots, n$ and their conjugates $Q^a_{\dot{\alpha}}, a = 1, \cdots, n$ and accordingly $N = n$ supersymmetric theories may be constructed by these generators. Let us now consider an $N = 2$ supersymmetric theory generated by two generators $Q^1$ and $Q^2$ and their conjugates. These operators satisfy the algebra

$$\begin{aligned} \{Q^a_\alpha, Q^b_\beta\} &= \varepsilon_{ab}\varepsilon_{\alpha\beta} Z \\ \{Q^a_\alpha, \bar{Q}^b_{\dot{\beta}}\} &= \delta_{ab}(\sigma^\mu)_{\alpha\dot{\beta}} P_\mu, \end{aligned} \tag{11.94}$$

where $Z$ is called the central charge and commutes with $Q^a_\alpha$, $\bar{Q}^a_{\dot{\alpha}}$ and $P_\mu$. We shall discuss the simple case with $Z = 0$.

The lowest-dimensional multiplet of the $N = 2$ supersymmetry will contain more particles than the lowest-dimensional multiplet of an $N = 1$ supersymmetry. Consider a multiplet of an $N = 2$ supersymmetry. There will be a left-chiral $N = 1$ supermultiplet ($\Phi$) and its conjugate right-chiral $N = 1$ supermultiplet ($\bar{\Phi}$), both belonging to the same representation of the $N = 2$ supermultiplet. If the components of these superfields are written as

$$\Phi \equiv \begin{pmatrix} \phi \\ \psi_L \end{pmatrix} \quad \text{and} \quad \bar{\Phi} \equiv \begin{pmatrix} \bar{\phi} \\ \psi_R \end{pmatrix},$$

then the generator of the $N = 1$ supersymmetry $Q^1$ will transform these fields as

$$\phi \overset{Q^1}{\leftrightarrow} \psi_L, \quad \text{and} \quad \bar{\phi} \overset{Q^1}{\leftrightarrow} \psi_R. \tag{11.95}$$

In the $N = 2$ supersymmetry the second spinor operator $Q^2$ will act on the other combinations

$$\phi \overset{Q^2}{\leftrightarrow} \psi_R, \quad \text{and} \quad \bar{\phi} \overset{Q^2}{\leftrightarrow} \psi_L, \tag{11.96}$$

so the $N = 1$ superfields $\Phi$ and $\bar{\Phi}$ become components of the same $N = 2$ scalar superfield.

The lowest representation of the $N = 2$ supersymmetry contains both the left-chiral and right-chiral superfields of the $N = 1$ supersymmetry. A scalar superfield of the $N = 2$ supersymmetry contains both the left-handed fermions and the right-handed fermions and their corresponding scalars, and there are no chiral multiplets. A vector superfield of the $N = 2$ supersymmetry contains a vector and a scalar superfield of the $N = 1$ supersymmetry. While the components of the vector and the scalar superfields are related by $Q^1$, the second supersymmetry generator of the $N = 2$ supersymmetry $Q^2$ relates the components of the vector superfield to the components of the scalar superfield. Thus, the vector superfield of the $N = 2$ supersymmetry contains a vector, two fermions and a scalar. The representations of the $N = 2$ supersymmetry are depicted in figure 11.3.

The next higher supersymmetry is the $N = 4$ supersymmetry, which is generated by four spinor operators, $Q^a_\alpha, a = 1, 2, 3, 4$, and their conjugates. Let us consider $N = 2$ superfields generated by the operators $Q^1$ and $Q^2$. A vector and a scalar $N = 2$

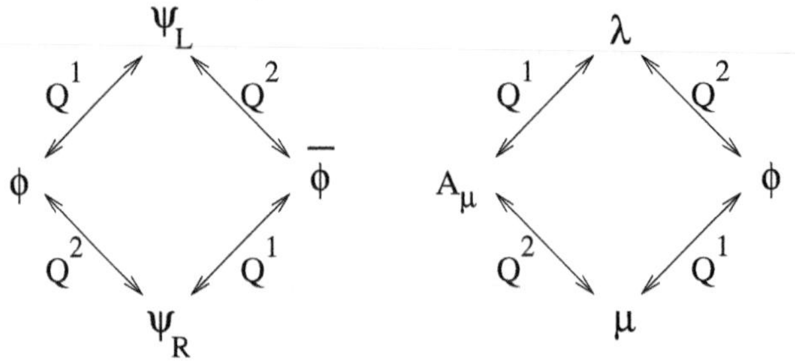


**FIGURE 11.3**
$N=2$ scalar (left) and vector (right) supermultiplets. $Q^1$ and $Q^2$ are the $N=2$ supersymmetry generators.

superfield will now become related by the other two generators $Q^3$ and $Q^4$. The lowest representation of the $N=4$ supersymmetry will be represented by a $N=4$ superfield, whose components are

$$\begin{pmatrix} N=4 \\ superfield \end{pmatrix} \equiv \begin{pmatrix} \begin{pmatrix} N=2\ vector \\ superfield \end{pmatrix} \\ \\ \begin{pmatrix} N=2\ scalar \\ superfield \end{pmatrix} \end{pmatrix} \equiv \begin{pmatrix} \begin{pmatrix} N=1\ vector \\ superfield \\ \\ N=1\ scalar \\ superfield \end{pmatrix} \\ \\ \begin{pmatrix} N=1\ left\ chiral \\ superfield \\ \\ N=1\ right\ chiral \\ superfield \end{pmatrix} \end{pmatrix}.$$

The minimal representation of the $N=4$ supersymmetry contains the vectors, fermions and scalars.

As we increase the number of supersymmetry generators, the minimal supermultiplets get larger containing higher spin particles. The minimal $N=1$ supermultiplet contains particles with spins 0 and 1/2, whereas $N=4$ supermultiplet contains particles with spins 0, 1/2 and 1. Similarly, the minimal $N=8$ supermultiplet would contain particles with spins 0, 1/2, 1, 3/2 and 2. Since it is not possible to construct consistent theories with spin higher than 2, one can set a restriction on the number of supersymmetry generators to be $N \leq 8$.

Let us now consider $N=1$ supersymmetry in 5-dimensions. Any spinor in 5-dimensions belongs to the 8-dimensional representation of $O(5)$. In $O(4)$, the projection operators decompose this 8-dimensional representation of $O(5)$ into $4+\bar{4}$ of

$O(4)$. After compactification of the 5-dimensional space, the spinor operator of $N = 1$ supersymmetry in 5-dimensions becomes two spinor operators in 4-dimensions generating $N = 2$ supersymmetry. This implies that there is one-to-one correspondence between a 4-dimensional $N = 2$ supersymmetry and a 5-dimensional $N = 1$ supersymmetry. In a more general way it has been shown that an 11-dimensional $N = 1$ supergravity theory has one-to-one correspondence with a 4-dimensional $N = 8$ supergravity theory [187]. Since $N > 8$ supersymmetry is not consistent, $D = 11$ is the highest-dimensional consistent supersymmetric theory. On the other hand, to accommodate the standard model gauge group, $D = 11$ is the lowest-dimensional theory. This makes the 11-dimensional $N = 1$ supergravity theory unique.

As we mentioned earlier, one of the drawbacks of the 11-dimensional theory is that it does not allow any chiral representations. The simplest modification to this theory is to consider a 10-dimensional theory, which allows chiral representations. However, by going to 10-dimensions we have to relax the condition of obtaining the standard model gauge groups from the metric itself. The isometry group of the 10-dimensional theory does not allow the standard model gauge groups to emerge from the 10-dimensional metric alone. In 10-dimensions, the graviton multiplet now contains the metric $g_{MN}$ with 35 degrees of freedom, the 56-component left-handed Majorana Rarita–Schwinger spinor $\Psi_M$, the 28-component antisymmetric tensor $B_{MN}$, the right-handed Majorana spinors $\lambda$ with 8 degrees of freedom, and the scalar dilaton field $\phi$. The vector multiplet in 4-dimensions contains only 8 gauge fields and 8 left-handed spinors, and the standard model gauge group cannot be included. This problem has been addressed by introducing chiral and vector fields in the 10-dimensional $N = 1$ supergravity theory. The vector fields in 10-dimensional $N = 1$ supergravity theory will contain both the gauge bosons and the massless chiral fermions in 4-dimensions, which can reproduce the standard model gauge group and also the standard model particle spectrum.

Although this prescription to include the 10-dimensional gauge bosons in the theory solves the problem of accommodating the standard model, the theory now contains 10-dimensional gauge as well as gravitational anomalies. The hexagon loops with fermions in the loop and gravitons at all the six external legs give the divergent gravitational anomalies, while the gauge fermions in the loop introduce the gauge anomalies. So, although the theory is consistent at the classical level, one-loop quantum corrections will make it inconsistent. This problem of anomaly could be solved in the superstring theory [188].

In a 10-dimensional superstring theory with the gauge group $E_8 \times E_8$ or a gauge group $SO(32)$, the gauge and the gravitational anomalies cancel. After compactification in certain manifolds with $SU(3)$ holonomy, these theories contain all known fermions as the zero modes of the d=10 superstring theory [189]. Thus, the theory reproduces the standard model interactions and the phenomenologically consistent particle spectrum [190] and in some cases the Yukawa couplings could also be calculated [191]. In addition to these phenomenological consistencies, the superstring theory has many virtues. In recent times the duality conjectures and brane solutions provided us many new possibilities. However, the most important feature of the superstring theory is that it is now emerging as a successful theory of quantum gravity.

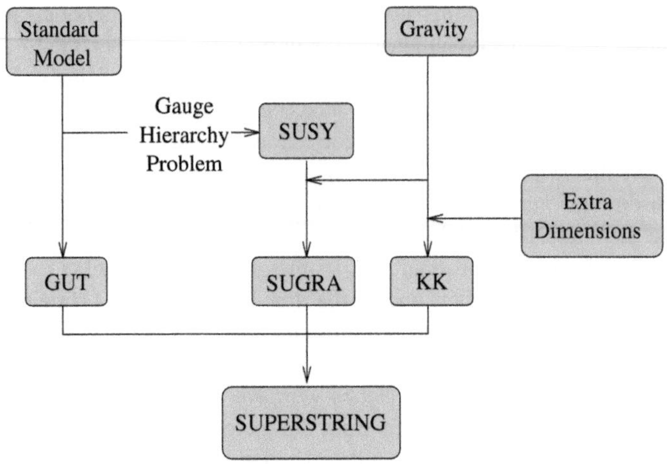


**FIGURE 11.4**
Development of the superstring theory.

In summary, we started with the gauge theories of the standard model and the general theory of relativity in 4-dimensions. To unify gauge interactions with gravity, higher-dimensional Kaluza–Klein theory was developed. Unification of all the gauge interactions could be achieved in the grand unified theories. The gauge hierarchy problem of the grand unified theories could be solved by making the theory supersymmetric, which also allows us to unify the Poincaré group with the internal symmetry groups. Gravity could be incorporated in these theories by making supersymmetry transformations local, which is the supergravity theory. All of these concepts of grand unification, Kaluza–Klein theory and supergravity could then be consistently incorporated into the 10-dimensional superstring theory (see figure 11.4).

# 12

# *Introduction to Strings*

There are four fundamental interactions in nature: the strong, weak, electromagnetic and gravitational interactions. The first three of these interactions are explained as gauge theories while the gravitational interaction is explained by the general theory of relativity, classically. The superstring theory promises to be a unified theory of all of these interactions at very high energy, when all the interactions become strong. It is also emerging as a consistent quantum theory of gravity, which would ensure that the ultimate unified theory is also a quantum theory of all the interactions.

The string theory replaces the point particles by objects, extended along one internal direction. Depending on the boundary conditions at the end-points of the strings, there will be different vibrational modes of these strings. If the string tension is too high about the Planck scale, the higher excited states cannot be created with our present available energy. The infinite towers of particles corresponding to every particle we see are similar to the Kaluza–Klein excited states and decouple from our low-energy world. Only the different massless zero modes of vibration will become available to us as almost point particles. Let us draw an analogy with the musical instruments for purpose of illustration. There are infinite towers of octaves. The first diatonic scale may be identified with the massless zero modes, while the octaves will correspond to excitations with mass of the order of the Planck scale. The different tones in the diatonic scale will correspond to different particles we see at low energy.

Consider an ordinary string that is moving around us with its different points representing points in some internal degree of freedom. The two end-points should satisfy certain boundary conditions. In an open string the end-points are fixed like the usual musical instruments, while for a closed string the two end-points are identified with each other. String theory can be consistently developed only in higher dimensions that are also compactified near the Planck scale. At our present energies, all string excitations as well as the Kaluza–Klein excitations will also disappear and we shall be left with only a 4-dimensional theory with the standard model interactions.

Superstring theory is developing very fast and it is not possible to give any details in this short chapter. So the idea of presenting this chapter is to provide familiarity with some basic aspects of the theory, which will allow the reader to follow what is going on in the subject, the phenomenological applications including the recent developments of models with extra dimensions, and some cosmological consequences.

## 12.1 Bosonic Strings

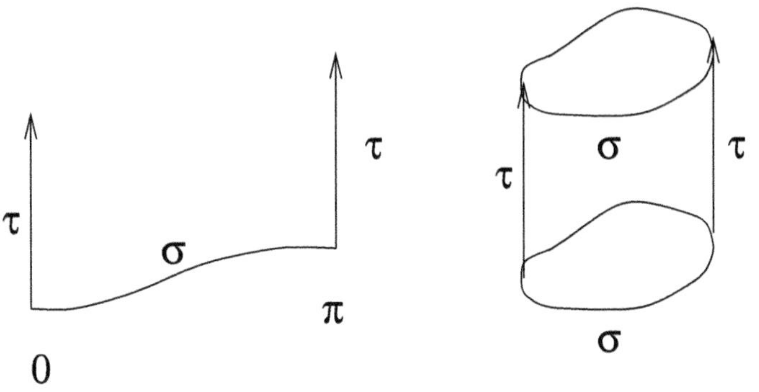


**FIGURE 12.1**
Open and closed strings.

The trajectory of a point particle in $d$ dimensions can be specified by $X^\mu(\tau), \mu = 0, 1, ...(d-1)$ at different times $\tau$. If the particle has some extensions along an internal degree of freedom given by $\sigma$ at every point of time, which is the case for a string, we should specify $X^\mu(\tau, \sigma)$ for every $\tau$ and at different space-like points of $\sigma$. It is convenient to take the range of $\sigma$ to be

$$0 \leq \sigma \leq \pi. \tag{12.1}$$

Then the boundary condition for a closed string will be

$$X^\mu(\tau, \sigma + \pi) = X^\mu(\tau, \sigma), \tag{12.2}$$

whereas for an open string it will be (see figure 12.1)

$$\frac{\partial X^\mu(\tau, \sigma = 0)}{\partial \sigma} = \frac{\partial X^\mu(\tau, \sigma = \pi)}{\partial \sigma} = 0. \tag{12.3}$$

We can now start from an action

$$S = -\frac{T}{2} \int_{\tau_1}^{\tau_2} d\tau \int_0^\pi d\sigma (-\det h)^{1/2} h^{\alpha\beta} \eta^{\mu\nu} \partial_\alpha X_\mu \partial_\beta X_\nu, \tag{12.4}$$

and obtain the Euler–Lagrange equations for the closed and open strings applying the corresponding boundary conditions. We have adopted the notation that $\eta_{\mu\nu}, \mu, \nu = 0, 1, ...(d-1)$ is the flat $d$-dimensional space–time metric with signature $(+ - - - ...)$ and $h_{\alpha\beta}(\tau\sigma), \alpha, \beta = 0, 1$ is the world sheet metric with signature $(+-)$. Here $T$ is the string tension, which characterizes the strength of interaction.

Under the reparametrization $(\tau,\sigma) \to (\tau',\sigma')$ this action remains invariant and we can exploit this freedom to choose our gauge to be a conformal gauge and then use the Lorentz invariance and scale invariance to make $h_{\alpha\beta} = \eta_{\alpha\beta}$, where $\eta_{\alpha\beta} = \mathrm{diag}(1, -1)$. We can write the equation of motion by varying the action with respect to $X^\mu$ and $h_{\alpha\beta}$ and applying boundary conditions to eliminate the surface terms. In this covariant gauge the equation of motion becomes

$$\partial_\sigma \partial^\sigma X^\mu = \left( \frac{\partial^2}{\partial \tau^2} - \frac{\partial^2}{\partial \sigma^2} \right) X^\mu = 0. \tag{12.5}$$

In addition we have to satisfy a constraint equation given by the vanishing of the energy-momentum tensor of the scalar fields $X^\mu$. In the conformal gauge the components of the constraint equations are

$$\begin{aligned} T_{00} = T_{11} &= -\frac{1}{2}(\partial_\tau X^\mu \partial_\tau X_\mu + \partial_\sigma X^\mu \partial_\sigma X_\mu) = 0 \\ T_{01} = T_{10} &= -\partial_\tau X^\mu \partial_\sigma X_\mu = 0, \end{aligned} \tag{12.6}$$

where the indices 0 and 1 refer to $\tau$ and $\sigma$, respectively.

The general solutions for these wave equations can be expanded in terms of oscillator coefficients, which can then be quantized for two different boundary conditions of the closed and the open strings. The solution for the closed string can be separated into two parts:

*left-mover*: where all terms depend on $\tau + \sigma$

*right-mover*: where all terms depend on $\tau - \sigma$.

The general solution then contains both the left-movers and the right-movers

$$X^\mu(\tau,\sigma) = X_R^\mu(\tau-\sigma) + X_L^\mu(\tau+\sigma), \tag{12.7}$$

where

$$\begin{aligned} X_R^\mu(\tau-\sigma) &= \frac{1}{2}x^\mu + \frac{1}{2}l^2 p^\mu(\tau-\sigma) + \frac{i}{2}l \sum_{n\neq 0} \frac{1}{n}\alpha_n^\mu \exp^{-2in(\tau-\sigma)} \\ X_L^\mu(\tau+\sigma) &= \frac{1}{2}x^\mu + \frac{1}{2}l^2 p^\mu(\tau+\sigma) + \frac{i}{2}l \sum_{n\neq 0} \frac{1}{n}\tilde{\alpha}_n^\mu \exp^{-2in(\tau+\sigma)} \end{aligned} \tag{12.8}$$

and $l$ is the fundamental length scale, defined in terms of the string tension $T$ by $l = (\pi T)^{-1/2}$.

The commutation relation for the center-of-mass coordinates ($x^\mu$) and the momentum ($p^\mu$) of the string is given by

$$[x^\mu, p^\nu] = -i\eta^{\mu\nu}. \tag{12.9}$$

Similarly, commutation relation satisfied by the oscillators are

$$\begin{aligned} [\alpha_m^\mu, \alpha_n^\nu] &= -m\delta_{m+n,0}\eta^{\mu\nu} \\ [\tilde{\alpha}_m^\mu, \tilde{\alpha}_n^\nu] &= -m\delta_{m+n,0}\eta^{\mu\nu} \\ [\alpha_m^\mu, \tilde{\alpha}_n^\nu] &= 0. \end{aligned} \tag{12.10}$$

The Hamiltonian in the covariant gauge

$$H = -\frac{T}{2}\int_0^{\pi} d\sigma(\partial_\tau X^\mu \partial_\tau X_\mu + \partial_\sigma X^\mu \partial_\sigma X_\mu) \tag{12.11}$$

may now be expressed in terms of the oscillator coefficients in the mode expansion as

$$H = -\frac{1}{2}\sum_{n\neq 0}(\alpha^\mu_{-n}\alpha_{\mu n} + \tilde{\alpha}^\mu_{-n}\tilde{\alpha}_{\mu n}) - \frac{l^2}{2}p^\mu p_\mu. \tag{12.12}$$

We shall now discuss the constraint equations.

The independent constraints equations can be written by defining the new combinations

$$T_{\pm\pm} \equiv \frac{1}{2}(T_{00} \pm T_{01}) = -\frac{1}{4}(\partial_\tau X^\mu \pm \partial_\sigma X^\mu)(\partial_\tau X_\mu \pm \partial_\sigma X_\mu) = 0, \tag{12.13}$$

where $T_{++}$ depends only on $X^\mu_L(\tau+\sigma)$, while $T_{--}$ depends only on $X^\mu_L(\tau-\sigma)$. If we now Fourier transform these quantities, we can write the *Virasoro operators* as

$$L_m = \frac{T}{2}\int_0^{\pi} d\sigma \exp^{2im(\tau-\sigma)} T_{--} = -\frac{1}{2}\sum_{n=-\infty}^{\infty}\alpha^\mu_{m-n}\alpha_{\mu n}$$

$$\tilde{L}_m = \frac{T}{2}\int_0^{\pi} d\sigma \exp^{2im(\tau+\sigma)} T_{++} = -\frac{1}{2}\sum_{n=-\infty}^{\infty}\tilde{\alpha}^\mu_{m-n}\tilde{\alpha}_{\mu n},$$

where we defined $\alpha^\mu_0 = \tilde{\alpha}^\mu_0 = (l/2)p^\mu$.

Acting on the physical states, these constraint equations give the conditions

$$L_m|\phi\rangle = \tilde{L}_m|\phi\rangle = 0 \qquad m > 0, \tag{12.14}$$

and for $m = 0$, these conditions become

$$L_0|\phi\rangle = 0 \Longrightarrow \left[-\frac{1}{2}\alpha^\mu_0\alpha_{\mu 0} - \sum_{n=1}^{\infty}\alpha^\mu_{-n}\alpha_{\mu n}\right]|\phi\rangle = a|\phi\rangle$$

$$\tilde{L}_0|\phi\rangle = 0 \Longrightarrow \left[-\frac{1}{2}\tilde{\alpha}^\mu_0\tilde{\alpha}_{\mu 0} - \sum_{n=1}^{\infty}\tilde{\alpha}^\mu_{-n}\tilde{\alpha}_{\mu n}\right]|\phi\rangle = a|\phi\rangle. \tag{12.15}$$

The constant $a$ is infinite, but proper renormalization will allow us to choose $a = 1$.

The algebra satisfied by the Virasoro operators $L_m$ is known as the *Virasoro algebra*

$$[L_m, L_n] = (m-n)L_{m+n} + b(m)\delta_{m+n,0}, \tag{12.16}$$

where the constant $b(m)$ is called the conformal anomaly and is obtained by normal-ordering when $m+n=0$. Similarly the left-mover part of the Virasoro algebra comes out to be exactly similar with $L_m$ replaced by $\tilde{L}_m$. The algebra is completed by

$$[L_m, \tilde{L}_n] = 0. \tag{12.17}$$

The *conformal anomaly* comes out to be $b(m) = (D/12)m(m^2-1)$. This theory becomes consistent in $D = 26$ and $a = 1$.

The Hamiltonian is now given by the Virasoro operator $L_0$

$$H = L_0,$$

and the masses of the physical states are then obtained from the condition $L_0|\phi\rangle = 0$ and are given in units of $l = 1$ by

$$\begin{aligned}
\frac{1}{4}M_R^2 &= -\sum_{n=1}^{\infty} \alpha^{\mu}_{-n}\alpha_{\mu n} - a \\
\frac{1}{4}M_L^2 &= -\sum_{n=1}^{\infty} \tilde{\alpha}^{\mu}_{-n}\tilde{\alpha}_{\mu n} - a \\
M^2 &= M_L^2 + M_R^2 = 2M_L^2 = 2M_R^2 .
\end{aligned} \tag{12.18}$$

The masses of the different states receive equal contributions from the left-movers and the right-movers because $L_0|\phi\rangle = \tilde{L}_0|\phi\rangle$.

For the open string the general solution to the wave equation is given by

$$X^{\mu}(\tau,\sigma) = x^{\mu} + l^2 p^{\mu}(\tau) + il\sum_{n\neq 0}\frac{1}{n}\alpha^{\mu}_{n}\exp^{-in\tau}\cos(n\sigma). \tag{12.19}$$

In this case the decomposition into left-movers or right-movers is not justified. The nonzero commutators are now given by

$$\begin{aligned}
[x^{\mu}, p^{\nu}] &= -i\eta^{\mu\nu} \\
[\alpha^{\mu}_{m}, \alpha^{\nu}_{n}] &= -m\delta_{m+n,0}\eta^{\mu\nu}.
\end{aligned} \tag{12.20}$$

The Hamiltonian now takes the form,

$$H = -\frac{1}{2}\sum_{n\neq 0}\alpha^{\mu}_{-n}\alpha_{\mu n} - \frac{l^2}{2}p^{\mu}p_{\mu}. \tag{12.21}$$

We can now construct the Virasoro algebra from the equation of constraints, following the same prescription as in the case of closed strings.

Proceeding in the same way we can now write the masses of the physical states in units of $l = 1$, which are

$$\frac{1}{2}M^2 = -\sum_{n=1}^{\infty}\alpha^{\mu}_{-n}\alpha_{\mu n} - a. \tag{12.22}$$

As previously the infinite contribution $a$ is chosen to be 1, after proper renormalization. Exploiting the remaining freedom, we work in the light cone gauge. In this gauge it is possible to find an expression for $a$, which after regularization gives

$$a = \frac{d-2}{24}. \tag{12.23}$$

For the choice of $a = 1$, we then get $d = 26$. In a more rigorous treatment it can be shown that the consistency of the quantum theory also requires that the bosonic strings are formulated in 26-dimensions only. In any other dimensions the conformal invariance and the Lorentz invariance will be destroyed by an anomaly in the quantum theory [188].

## 12.2 Superstrings

In the superstring theory, we extend the bosonic string theory to include the fermions and then ensure supersymmetry in the world sheet. Similar to the $d$-dimensional scalar field $X^\mu$ we include a $d$ component two-dimensional Majorana spinor field $\Psi^\mu$ in the action. Observing that the scalar $X^\mu$ couples to two-dimensional gravity, we expect that the spinor $\Psi^\mu$ would form a world-sheet supersymmetry that can then couple to two-dimensional supergravity. This is possible with the Lagrangian

$$S = -\frac{1}{2\pi}\int d\tau d\sigma(-\det h)^{1/2}(h^{\alpha\beta}\partial_\alpha X^\mu \partial_\beta X_\mu + i\bar{\Psi}^\mu \rho^\alpha \partial_\alpha \Psi_\mu), \tag{12.24}$$

where $\alpha = 0, 1$ are the world-sheet coordinates. The world-sheet gamma matrices are given by

$$\rho^0 = \begin{pmatrix} 0 & -i \\ i & 0 \end{pmatrix} \quad \rho^1 = \begin{pmatrix} 0 & i \\ i & 0 \end{pmatrix} \quad \text{and} \quad \{\rho^\alpha, \rho^\beta\} = 2\eta^{\alpha\beta} I, \tag{12.25}$$

where $\eta^{\alpha\beta} = \text{diag}(+1, \ -1)$ and $I$ is the $2 \times 2$ unit matrix. This Lagrangian possesses a world-sheet supersymmetry, generated by

$$\begin{aligned} \delta X^\mu &= \bar{\xi}\Psi^\mu \\ \delta \Psi^\mu &= -i\rho^\alpha \partial_\alpha X^\mu \xi, \end{aligned} \tag{12.26}$$

but there is no space–time supersymmetry. However, in the final consistent theory when the problem of tachyonic states is solved by applying projections of the states, the theory possesses space–time supersymmetry.

To include gravity we now make the supersymmetry transformation local $\xi \equiv \xi(\tau, \sigma)$. The algebra will close only after we include the supergravity gravitino $\chi^\alpha$ and the graviton $h_{\alpha\beta}$ fields, define their transformation properties, and include their action in the Lagrangian. This world-sheet supergravity theory can be consistently defined only in $d = 10$, satisfying the criterion of cancellation of the conformal anomaly.

We proceed in the same way as in the bosonic strings. Making use of the conformal invariance and reparametrization invariance to choose $h_{\alpha\beta} = \eta_{\alpha\beta}$, we can write the Euler–Lagrange equations as

$$\begin{aligned} \partial_\alpha \partial^\alpha X^\mu &= 0 \\ i\rho^\alpha \partial_\alpha \Psi^\mu &= 0. \end{aligned} \tag{12.27}$$

We now give the boundary conditions for the open and closed strings independently. For the open strings the boundary conditions for the fermions are

$$\begin{aligned}\Psi_L^\mu(\tau,\sigma=0) &= \Psi_R^\mu(\tau,\sigma=0)\\ \Psi_L^\mu(\tau,\sigma=\pi) &= \pm\Psi_R^\mu(\tau,\sigma=\pi),\end{aligned} \tag{12.28}$$

where we have written

$$\Psi^\mu = \begin{pmatrix}\Psi_R^\mu(\tau-\sigma)\\ \Psi_L^\mu(\tau+\sigma)\end{pmatrix}. \tag{12.29}$$

The components satisfy the equations

$$\begin{aligned}\left(\frac{\partial}{\partial\tau}+\frac{\partial}{\partial\sigma}\right)\Psi_R^\mu &= 0\\ \left(\frac{\partial}{\partial\tau}-\frac{\partial}{\partial\sigma}\right)\Psi_L^\mu &= 0.\end{aligned} \tag{12.30}$$

These boundary conditions ensure that the surface terms vanish to give us the equations of motion. In addition we have to consider the constraints coming from the vanishing of the energy-momentum tensor $T_{\alpha\beta}$ for $X^\mu$ and $\Psi^\mu$ and the world-sheet supercurrent $J^\alpha$, where

$$\begin{aligned}T_{\alpha\beta} &= -\partial_\alpha X^\mu\partial_\beta X_\mu - \frac{i}{4}\bar{\Psi}^\mu(\rho_\alpha\partial_\beta+\rho_\beta\partial_\alpha)\Psi_\mu\\ &\quad+\frac{\eta_{\alpha\beta}}{2}(\partial^\gamma X^\mu\partial_\gamma X_\mu+\frac{i}{2}\bar{\Psi}^\mu\rho^\gamma\partial_\gamma\Psi_\mu)\\ J^\alpha &= \frac{1}{2}\,\rho^\beta\rho^\alpha\Psi^\mu\partial_\beta X_\mu.\end{aligned} \tag{12.31}$$

Similar to the Virasoro algebra, the Fourier transforms of certain combinations of these quantities define the super-Virasoro algebra.

The boundary conditions for the closed strings can now be defined as

$$\Psi_{L,R}^\mu(\tau,\sigma+\pi) = \pm\Psi_{L,R}^\mu(\tau,\sigma). \tag{12.32}$$

We can now choose the boundary conditions for the left-movers and right-movers independently. For both left-movers and right-movers we can have two choices of periodic and antiperiodic boundary conditions. The periodic boundary condition, given by $+$, is called the Ramond boundary conditions [192] (denoted by $R$) and the antiperiodic boundary condition, given by $-$, is called the Neveu–Schwarz boundary conditions [193] (denoted by $NS$).

We shall now concentrate mostly on the closed strings, which are phenomenologically more promising. However, since some recent developments with brane solutions make use of the open strings, we shall also mention the open strings in the beginning. The mode expansions for the left- and the right-movers of the open strings are not independent of each other. Both the left-movers and the right-movers should be either periodic ($R$) or antiperiodic ($NS$) over a range 0 to $\pi$ of $\sigma$. For the periodic ($R$) and antiperiodic ($NS$) boundary conditions, the mode expansions for the open strings are given by

$$\Psi^{\mu}_{\substack{R\\L}}(R) = \frac{1}{\sqrt{2}}\sum_n d^{\mu}_n \exp^{-in(\tau\mp\sigma)}$$
$$\Psi^{\mu}_{\substack{R\\L}}(NS) = \frac{1}{\sqrt{2}}\sum_r b^{\mu}_r \exp^{-ir(\tau\mp\sigma)}, \qquad (12.33)$$

where $n = 1,2,\ldots$ are integers and $r = 1/2, 3/2, \ldots$ are half-integers. The anticommutation relations of the fermions now give the quantization conditions

$$\{d^{\mu}_m, d^{\nu}_n\} = -\delta_{m+n,0}\eta^{\mu\nu}$$
$$\{b^{\mu}_r, b^{\nu}_s\} = -\delta_{r+s,0}\eta^{\mu\nu}, \qquad (12.34)$$

where $m,n$ are integers and $r,s$ are half-integers. The superstring Hamiltonian for the Ramond ($R$) and Neveu–Schwarz ($NS$) boundary conditions can now be written as

$$H(R) = -\frac{1}{2}p^{\mu}p_{\mu} - \frac{1}{2}\sum_{n\neq 0}\alpha^{\mu}_{-n}\alpha_{\mu n} - \frac{1}{2}\sum_n n d^{\mu}_{-n} d_{\mu n}$$
$$H(NS) = -\frac{1}{2}p^{\mu}p_{\mu} - \frac{1}{2}\sum_{n\neq 0}\alpha^{\mu}_{-n}\alpha_{\mu n} - \frac{1}{2}\sum_r r b^{\mu}_{-r} b_{\mu r}. \qquad (12.35)$$

Finally we can write the mass-squared operators

$$\frac{1}{2}M^2(R) = \sum_{n=1}^{\infty} n d^{\mu}_{-n} d_{n\mu} - a_R$$
$$\frac{1}{2}M^2(NS) = \sum_{r=1/2}^{\infty} r b^{\mu}_{-r} b_{r\mu} - a_{NS}, \qquad (12.36)$$

which contain the bosonic part and the fermionic part for the $R$ and the $NS$ boundary conditions. The constant $a_R$ vanishes for the $R$ sector because there is exact cancellation of the fermionic and the bosonic contributions. For the $NS$ sector, it comes out to be $a_{NS} = (d-2)/16$. For the cancellation of conformal anomaly and the consistency of the quantum theory, the superstring theory can be defined only in $d = 10$, so we get $a_{NS} = 1/2$.

The left-movers and the right-movers for a closed string are distinct, and we have to write the mode expansion for each of them separately for the periodic and antiperiodic boundary conditions, which are

$$\Psi^{\mu}_{R}(R) = \sum_n d^{\mu}_n \exp^{-2in(\tau-\sigma)}$$
$$\Psi^{\mu}_{R}(NS) = \sum_{r=n+1/2} b^{\mu}_r \exp^{-2ir(\tau-\sigma)}$$
$$\Psi^{\mu}_{L}(R) = \sum_n \tilde{d}^{\mu}_n \exp^{-2in(\tau+\sigma)}$$
$$\Psi^{\mu}_{L}(NS) = \sum_{r=n+1/2} \tilde{b}^{\mu}_r \exp^{-2ir(\tau+\sigma)}, \qquad (12.37)$$

where $n$ are integers and $r$ are half-integers. The anticommutation relations of the fermionic oscillators for the left-movers and right-movers are given by

$$\begin{aligned}\{d_m^\mu, d_n^\nu\} &= \{\tilde{d}_m^\mu, \tilde{d}_n^\nu\} = -\delta_{m+n,0}\eta^{\mu\nu}\\ \{b_r^\mu, b_s^\nu\} &= \{\tilde{b}_r^\mu, \tilde{b}_s^\nu\} = -\delta_{r+s,0}\eta^{\mu\nu}.\end{aligned} \tag{12.38}$$

The Hamiltonians for the Ramond and the Neveu–Schwarz boundary conditions can be given in terms of the fermionic oscillators of the mode expansions by

$$\begin{aligned}H(R) &= -\frac{1}{2}p^\mu p_\mu - \sum_{n\neq 0}(\alpha_{-n}^\mu \alpha_{\mu n} + \tilde{\alpha}_{-n}^\mu \tilde{\alpha}_{\mu n})\\ &\qquad - \sum_n n(d_{-n}^\mu d_{\mu n} + \tilde{d}_{-n}^\mu \tilde{d}_{\mu n})\\ H(NS) &= -\frac{1}{2}p^\mu p_\mu - \sum_{r\neq 0}(\alpha_{-r}^\mu \alpha_{\mu r} + \tilde{\alpha}_{-r}^\mu \tilde{\alpha}_{\mu r})\\ &\qquad - \sum_r r(b_{-r}^\mu b_{\mu r} + \tilde{b}_{-r}^\mu \tilde{b}_{\mu r}).\end{aligned} \tag{12.39}$$

Vanishing of the energy-momentum tensor $T^{\alpha\beta}$ and the world sheet supercurrent $J^\alpha$ will then give the constraint equations and the mass spectrum. These constraint equations can be written in the covariant gauge as

$$T_{00} = T_{11} = 0, \quad T_{10} = T_{01} = 0, \quad J^0 = 0 \quad \text{and} \quad J^1 = 0.$$

The 0 and 1 in the subscript correspond to the $\sigma$ and $\tau$ directions. We can now write the independent constraint equations as

$$\begin{aligned}T_{\pm\pm} &= \frac{1}{2}(T_{00} \pm T_{01}) = -\partial_\pm X^\mu_{\substack{L\\R}} \partial_\pm X_{\mu \substack{L\\R}} - \frac{i}{2}\Psi^\mu_{\substack{L\\R}} \partial_\pm \Psi_{\mu \substack{L\\R}} = 0\\ J_\pm &= \Psi^\mu_{\substack{L\\R}} \partial_\pm X_{\mu \substack{L\\R}} = 0,\end{aligned} \tag{12.40}$$

with $\partial_\pm = (\partial_\tau \pm \partial_\sigma)/2$.

The super-Virasoro operators can now be defined by taking the Fourier transforms of $T_{\mp\mp}$ as

$$\begin{aligned}L_m &= \frac{1}{2\pi}\int_0^\pi d\sigma \, \exp^{2im(\tau-\sigma)} \, T_{--} \qquad m \neq 0\\ \tilde{L}_m &= \frac{1}{2\pi}\int_0^\pi d\sigma \, \exp^{2im(\tau+\sigma)} \, T_{++} \qquad m \neq 0.\end{aligned} \tag{12.41}$$

For the $R$ sector we can Fourier transform $J_\mp$ and define

$$\begin{aligned}F_m &= \frac{1}{2\pi}\int_0^\pi d\sigma \, \exp^{2im(\tau-\sigma)} \, J_- \qquad m \neq 0\\ \tilde{F}_m &= \frac{1}{2\pi}\int_0^\pi d\sigma \, \exp^{2im(\tau+\sigma)} \, J_+ \qquad m \neq 0,\end{aligned} \tag{12.42}$$

where $m$ is an integer. Similarly we define for the $NS$ sector

$$\begin{aligned} G_r &= \frac{1}{2\pi}\int_0^{\pi} d\sigma \exp^{2ir(\tau-\sigma)} J_- \\ \tilde{G}_r &= \frac{1}{2\pi}\int_0^{\pi} d\sigma \exp^{2ir(\tau+\sigma)} J_+, \end{aligned} \tag{12.43}$$

where $r$ is half-integer.

These quantities can now be expressed in terms of the oscillators of the mode expansion for the periodic and antiperiodic boundary conditions separately. For the right-mover fermionic degrees of freedom, these operators are given for the $NS$ and $R$ sectors by

$$\begin{aligned} L_m(NS) &= -\frac{1}{2}\sum_n \alpha^{\mu}_{m-n}\alpha_{\mu n} + \frac{1}{2}\sum_r (\frac{m}{2}-r) b^{\mu}_{m-r} b_{\mu r} \quad m \neq 0 \\ L_m(R) &= -\frac{1}{2}\sum_n \alpha^{\mu}_{m-n}\alpha_{\mu n} + \frac{1}{2}\sum_n (\frac{m}{2}-n) d^{\mu}_{m-n} d_{\mu n} \quad m \neq 0 \\ G_r(NS) &= -\frac{1}{2}\sum_n b^{\mu}_{r-n}\alpha_{\mu n} \\ F_m(R) &= -\frac{1}{2}\sum_n d^{\mu}_{m-n}\alpha_{\mu n}, \end{aligned} \tag{12.44}$$

where $r$ is half integer and $n$ is integer. For the left-movers these quantities are obtained by replacing all of the operators with their corresponding *tilde* operators, including the oscillators of the mode expansion.

For $m = 0$ we define for the right-movers and the left-movers

$$L_0 = \frac{1}{2\pi}\int_0^{\pi} d\sigma\, T_{--} \quad \text{and} \quad \tilde{L}_0 = \frac{1}{2\pi}\int_0^{\pi} d\sigma\, T_{++}. \tag{12.45}$$

For the right-movers these operators can be written in terms of the mode expansion oscillators in the $NS$ and $R$ sectors as

$$\begin{aligned} L_0(NS) &= -\frac{1}{2}\sum_n \alpha^{\mu}_{-n}\alpha_{\mu n} - \frac{1}{2}\sum_r r b^{\mu}_{-r} b_{\mu r} \\ L_0(R) &= -\frac{1}{2}\sum_n \alpha^{\mu}_{-n}\alpha_{\mu n} - \frac{1}{2}\sum_n n d^{\mu}_{-n} d_{\mu n}. \end{aligned} \tag{12.46}$$

In the same way we can write these operators for the left-movers by replacing the operators and the oscillators with their corresponding *tilde* operators and oscillators.

The super-Virasoro algebra, given by the commutation relations of the Virasoro operators, for the right-movers in the $NS$ sector is given by

$$\begin{aligned} [L_m, L_n] &= (m-n)L_{m+n} + \frac{D}{8}(m^3 - m)\delta_{m+n,0} \\ [L_m, G_r] &= (\frac{m}{2} - r)G_{m+r} \\ \{G_r, G_s\} &= 2L_{r+s} + \frac{D}{8}(r^2 - \frac{1}{4})\delta_{r+s,0}. \end{aligned} \tag{12.47}$$

For the right-movers in the $R$ sector it is

$$
\begin{aligned}
[L_m, L_n] &= (m-n)L_{m+n} + \frac{D}{8}m^3\delta_{m+n,0} \\
[L_m, F_n] &= (\frac{m}{2}-n)F_{m+n} \\
\{F_m, F_n\} &= 2L_{m+n} + \frac{D}{8}m^2\delta_{m+n,0}.
\end{aligned}
\tag{12.48}
$$

We are not writing the super-Virasoro algebra for the left-movers, which can be obtained by replacing the operators by the *tilde* operators.

We can now write the Hamiltonian as

$$H = 2(L_0 + \tilde{L}_0). \tag{12.49}$$

The physical states for both the $NS$ and $R$ sectors should now satisfy the constraint equations

$$L_m|\phi\rangle = 0 \quad \text{for } (m>0) \text{ and } \quad L_0|\phi\rangle = 0. \tag{12.50}$$

The $NS$ and $R$ sectors should also satisfy further constraints

$$G_r|\phi\rangle = 0 \quad (r>0) \qquad F_m|\phi\rangle = 0 \quad (m>0). \tag{12.51}$$

There exist similar constraints for the left-movers.

The mass-squared operators turn out to be the same for the left-movers and for the right-movers. From these constraints we can write the mass-squared operators as

$$M^2 = M_L^2 + M_R^2 = 2M_L^2 = 2M_R^2. \tag{12.52}$$

For the Ramond and Neveu–Schwarz boundary conditions, the mass-squared operators for the left-movers and right-movers are given by

$$
\begin{aligned}
\frac{1}{4}M_R^2(NS) &= -\sum_{n=1}^{\infty}\alpha^{\mu}_{-n}\alpha_{\mu n} - \sum_{r=1/2}^{\infty} r b^{\mu}_{-r}b_{r\mu} - a_{NS} \\
\frac{1}{4}M_R^2(R) &= -\sum_{n=1}^{\infty}\alpha^{\mu}_{-n}\alpha_{\mu n} - \sum_{n=1}^{\infty} n d^{\mu}_{-n}d_{n\mu} - a_R \\
\frac{1}{4}M_L^2(NS) &= -\sum_{n=1}^{\infty}\tilde{\alpha}^{\mu}_{-n}\tilde{\alpha}_{\mu n} - \sum_{r=1/2}^{\infty} r \tilde{b}^{\mu}_{-r}\tilde{b}_{r\mu} - a_{NS} \\
\frac{1}{4}M_L^2(R) &= -\sum_{n=1}^{\infty}\tilde{\alpha}^{\mu}_{-n}\tilde{\alpha}_{\mu n} - \sum_{n=1}^{\infty} n \tilde{d}^{\mu}_{-n}\tilde{d}_{n\mu} - a_R\,.
\end{aligned}
\tag{12.53}
$$

Similar to the open strings the consistency condition for the cancellation of the conformal anomaly requires $d = 10$ and the constants then become $a_R = 0$ and $a_{NS} = 1/2$.

In the covariant gauge, where the world-sheet gravitino has been gauged away, it is still possible to eliminate some more degrees of freedom utilizing the freedom to reparameterize the world-sheet and local Weyl scaling. This will finally give us the physical states. For this purpose we work in the light-cone gauge, where we choose

$$X^+(\tau,\sigma) = \frac{1}{\sqrt{2}}(X^0 + X^{D-1}) = x^+ + p^+\tau, \tag{12.54}$$

where $x^+$ and $p^+$ are constants. For the fermionic fields we consider the world-sheet supersymmetry transformation, which allows us to choose

$$\Psi^+ = \frac{1}{\sqrt{2}}(\Psi^0 + \Psi^{D-1}) = 0. \tag{12.55}$$

The field

$$X^-(\tau,\sigma) = \frac{1}{\sqrt{2}}(X^0 - X^{D-1}) \quad \text{and} \quad \Psi^- = \frac{1}{\sqrt{2}}(\Psi^0 - \Psi^{D-1})$$

can then be expressed in terms of the transverse degrees of freedom $X^i$ and $\Psi^i$, $i = 1,2,...,(D-2)$, leaving only the $X^i$ and $\Psi^i$ as the physical degrees of freedom.

We now consider only the physical fields to write the mass-squared operators. In terms of the mode expansion parameters of the physical fields the mass-squared operators are now given by

$$\begin{aligned}
M^2 &= M_R^2 + M_L^2 = 2M_L^2 = 2M_R^2 \\
\frac{1}{4}M_R^2(R) &= \sum_{n=1}^{\infty} \alpha_{-n}^i \alpha_n^i + \sum_{n=1}^{\infty} n d_{-n}^i d_n^i \\
\frac{1}{4}M_L^2(R) &= \sum_{n=1}^{\infty} \tilde{\alpha}_{-n}^i \tilde{\alpha}_n^i + \sum_{n=1}^{\infty} n \tilde{d}_{-n}^i \tilde{d}_n^i \\
\frac{1}{4}M_R^2(NS) &= \sum_{n=1}^{\infty} \alpha_{-n}^i \alpha_n^i + \sum_{r=1/2}^{\infty} r b_{-r}^i b_r^i - \frac{1}{2} \\
\frac{1}{4}M_L^2(NS) &= \sum_{n=1}^{\infty} \tilde{\alpha}_{-n}^i \tilde{\alpha}_n^i + \sum_{r=1/2}^{\infty} r \tilde{b}_{-r}^i \tilde{b}_r^i - \frac{1}{2},
\end{aligned} \tag{12.56}$$

where $i = 1,2,...8$. As previously, $R$ and $NS$ correspond to the Ramond and Neveu–Schwarz boundary conditions.

The physical states in the light-cone gauge are now given by $\alpha_{-n}^i d_{-n}^i|0>_R$ or $\alpha_{-n}^i b_{-r}^i|0>_R$ for the right-movers, multiplied by $\tilde{\alpha}_{-n}^i \tilde{d}_{-n}^i|0>_L$ or $\tilde{\alpha}_{-n}^i \tilde{b}_{-r}^i|0>_L$ for the left-movers. Combining the left-movers and right-movers, there are now four types of states:

$$\begin{aligned}
(R)-(R): &\quad \alpha_{-n}^i d_{-n}^i|0>_R \tilde{\alpha}_{-n}^i \tilde{d}_{-n}^i|0>_L \\
(R)-(NS): &\quad \alpha_{-n}^i d_{-n}^i|0>_R \tilde{\alpha}_{-n}^i \tilde{b}_{-r}^i|0>_L \\
(NS)-(R): &\quad \alpha_{-n}^i b_{-r}^i|0>_R \tilde{\alpha}_{-n}^i \tilde{d}_{-n}^i|0>_L \\
(NS)-(NS): &\quad \alpha_{-n}^i b_{-r}^i|0>_R \tilde{\alpha}_{-n}^i \tilde{b}_{-r}^i|0>_L .
\end{aligned} \tag{12.57}$$

The ground state for the $(R)-(R)$ case is a massless boson state, while for the $(NS)-(NS)$ case the ground state is a tachyon and original supersymmetry is broken. The ground states for the mixed states have massless spinors.

The $(NS)-(NS)$ Neveu–Schwarz sector ground state now becomes tachyonic, since $M_L^2 = M_R^2 = -1/2$. We can now apply the GSO projection (due to Gliozzi, Scherk and Olive [194]) to the states of the theory, remove the problem of tachyonic states, and simultaneously restore $d = 10$ supersymmetry. The GSO projection operators are given by

$$
\begin{aligned}
P_{NS} &= \frac{(1+(-1)^{F_{NS}+1})}{2} \\
P_R &= \frac{(1+\eta(-1)^{F_R+1})}{2},
\end{aligned} \tag{12.58}
$$

where the number operators for the left- and right-movers in the Neveu–Schwarz and the Ramond sectors are given by,

$$
\begin{aligned}
F_{NS}^L &= \sum_{r=1/2}^{\infty} b_{-r}^i b_r^i \\
F_{NS}^R &= \sum_{r=1/2}^{\infty} \tilde{b}_{-r}^i \tilde{b}_r^i \\
F_R^L &= \sum_{n=1}^{\infty} d_{-n}^i d_n^i \\
F_R^R &= \sum_{n=1}^{\infty} \tilde{d}_{-n}^i \tilde{d}_n^i.
\end{aligned} \tag{12.59}
$$

In terms of the GSO projections there are now three types of string theories, which are classified as:

***Type IIA:*** This is a closed string theory and has opposite values of $\eta$ for the left- and right-movers in the definition of the GSO projection operator for the Ramond sector $P_R$. The theory has $N = 2$ supersymmetry and the ground state contains an $N = 2$ supermultiplet. There are no chiral multiplets. The massless particles include an $N = 2$ supergravity multiplet at 10 dimensions.

***Type IIB:*** This is also a closed string theory, but has the same value of $\eta$ for the left- and right-movers in the definition of the GSO projection operator $P_R$. The theory has $N = 2$ supersymmetry, but the massless states contain chiral multiplets.

***Type I:*** This contains both closed and open strings and have some similarity with type IIB. However, in this case the string is not orientable, i.e., it is symmetrical under the exchange $\sigma \to -\sigma$. In this case the theory has only $N = 1$ supersymmetry and the massless states contain chiral multiplets and the graviton.

The vector multiplets come from the closed and open string sectors, respectively. Perturbation is more complicated in this case, since it is unoriented and breakable.

None of these three types of the string theories can accommodate any internal gauge groups in addition to the multiplets they contain. Since the $N = 1$ and $N = 2$ supergravity multiplets in 10-dimensions are insufficient to include the particle spectrum of the standard model as well as the standard model gauge group, attempt is made to construct heterotic strings [195]. In the heterotic strings it will be possible to include a gauge group at 10-dimensions.

## 12.3 Heterotic String

The *heterotic string* theory uses the right-movers of type II superstring theory and the left-movers of closed bosonic strings [195]. The right-movers then become 10-dimensional, whereas the left-movers become 26-dimensional. To match the two dimensions, the 16-extra dimensions of the left-movers are compactified, which gives the stringy gauge fields and generates the non-Abelian gauge group. Once the new gauge symmetry is generated, the superpartners of the corresponding gauge bosons, the gauginos, will participate in the gauge and gravitational anomaly. The cancellation of the gauge and the gravitational anomalies then restricts the gauge group to either an $SO(32)$ or an $E_8 \times E_8$. The latter group is found to be phenomenoloically more attractive [190].

### Compactification of Closed Bosonic Strings

We shall first discuss about some relevant features of compactification of closed bosonic strings and then come back to the construction of a heterotic string theory. We start with a closed bosonic string and compactify one of the dimensions on a circle of radius $R$ along the direction $\tilde{d} = 25$. The mode expansions of the left-movers and the right-movers of the usual noncompact 25 dimensions $X^\mu, \mu = 0, 1, ..24$ are

$$X_R^\mu(\tau - \sigma) = \frac{1}{2}x^\mu + \frac{1}{2}p^\mu(\tau - \sigma) + \frac{i}{2}l \sum_{n\neq 0} \frac{1}{n}\alpha_n^\mu \exp^{-2in(\tau-\sigma)}$$
$$X_L^\mu(\tau + \sigma) = \frac{1}{2}x^\mu + \frac{1}{2}p^\mu(\tau + \sigma) + \frac{i}{2}l \sum_{n\neq 0} \frac{1}{n}\tilde{\alpha}_n^\mu \exp^{-2in(\tau+\sigma)}. \qquad (12.60)$$

For the compactified dimension $\tilde{d} = 25$, the compactification condition

$$x^{\tilde{d}} = x^{\tilde{d}} + 2\pi R \qquad (12.61)$$

will modify the boundary condition and introduce two integers, one corresponding to the Kaluza–Klein mode expansion and the second corresponding to winding number of the string around the compact dimension.

The compactification condition will constrain the center of mass momentum $p^{\tilde{d}}$ by the requirement that $\exp^{(ip^{\tilde{d}}x^{\tilde{d}})}$ should be single-valued, implying

$$p^{\tilde{d}} = m/R \tag{12.62}$$

for any integer $m$. The momentum is now quantized and the integer $m$ defines the Kaluza–Klein modes in the theory. Since the strings can wind around the compact dimensions any number of times, in string compactification a new quantization results from the winding of the strings. If the strings wind around the extra dimensions $n$ times without changing the boundary conditions, the periodic boundary condition of a closed string modifies to

$$X^{\tilde{d}}(\tau,\sigma+\pi) = X^{\tilde{d}}(\tau,\sigma) + 2\pi Rn. \tag{12.63}$$

The boundary conditions will modify the mode expansion of the compactified dimensions along left-movers and right-movers to

$$\begin{aligned} X_R^{\tilde{d}}(\tau-\sigma) &= x_R^{\tilde{d}} + p_R^{\tilde{d}}(\tau-\sigma) + \frac{i}{2}l\sum_{n\neq 0}\frac{1}{n}\alpha_n^{\tilde{d}}\exp^{-2in(\tau-\sigma)} \\ X_L^{\tilde{d}}(\tau+\sigma) &= x_L^{\tilde{d}} + p_L^{\tilde{d}}(\tau+\sigma) + \frac{i}{2}l\sum_{n\neq 0}\frac{1}{n}\tilde{\alpha}_n^{\tilde{d}}\exp^{-2in(\tau+\sigma)}, \end{aligned} \tag{12.64}$$

where

$$\begin{aligned} x^{\tilde{d}} &= x_L^{\tilde{d}} + x_R^{\tilde{d}} \\ p_R^{\tilde{d}} &= \frac{1}{2}(p^{\tilde{d}} - 2nR) \\ p_L^{\tilde{d}} &= \frac{1}{2}(p^{\tilde{d}} + 2nR). \end{aligned} \tag{12.65}$$

The momentum now depends on the Kaluza–Klein state $m$ with momentum $p^{\tilde{d}}$ and also on the string winding number $n$.

The masses of the physical states can now be obtained by working in the light-cone gauge defined by the equation (12.54). Since $\tilde{d} = 25$ is the compactified dimension, we define

$$X^{\pm} = (X^0 \pm X^{24}), \tag{12.66}$$

which gives us the physical degrees of freedom to be $X^i, i = 1,2,...,23$. From the mass shell condition

$$M^2 = -\sum_{\mu=0}^{24} p_\mu p^\mu, \tag{12.67}$$

and the constraints (see equation (12.13)) $T_{++} = 0$ and $T_{--} = 0$, we finally obtain the mass spectrum for the physical states

$$\begin{aligned} M_R^2 &= 4\left(\frac{m}{2r} - \frac{n}{r}\right)^2 + 8N - 8 \\ M_L^2 &= 4\left(\frac{m}{2r} + \frac{n}{r}\right)^2 + 8\tilde{N} - 8, \end{aligned} \tag{12.68}$$

where

$$N = \sum_{n=1}^{\infty} (\alpha^i_{-n}\alpha_{in} + \alpha^{\tilde{d}}_{-n}\alpha_{\tilde{d}n})$$
$$\tilde{N} = \sum_{n=1}^{\infty} (\tilde{\alpha}^i_{-n}\tilde{\alpha}_{in} + \tilde{\alpha}^{\tilde{d}}_{-n}\tilde{\alpha}_{\tilde{d}n}). \tag{12.69}$$

The level matching condition $M_L^2 = M_R^2$ then implies that $N - \tilde{N} = mn$. This condition has interesting implications for the duality between the numbers $n$ and $m$. The mass-squared operator now becomes

$$\frac{1}{8}M^2 = \frac{n^2R^2}{2} + \frac{m^2}{8R^2} + \frac{N}{2} + \frac{\tilde{N}}{2} - 1. \tag{12.70}$$

The integer $m$ denotes the Kaluza–Klein state and gives the momentum in the compactified dimension, while the integer $n$ is the winding number of the compactified manifold and its contribution to mass is given in the first term. In ordinary Kaluza–Klein theories there are the momenta of the compactified dimensions, but there is no analogue of the first term arising from string winding. We can now denote these two types of superheavy states as the string excited states and the Kaluza–Klein excited states.

The 25-dimensional theory, resulting after the compactification of one of the dimensions in a circle, has the physical states: the graviton; the dilaton; the 25-dimensional rank-2 tensor, $\alpha^i_{-1}|0>_R \tilde{\alpha}^j_{-1}|0>_L$, which can be decomposed to the trace; the traceless symmetric tensor; and the antisymmetric part. In addition to these states, the compactification now provides new massless vector particles

$$V^i = \alpha^i_{-1}|0>_R \tilde{\alpha}^{\tilde{d}}_{-1}|0>_L \quad \text{and} \quad \tilde{V}^i = \alpha^{\tilde{d}}_{-1}|0>_R \tilde{\alpha}^i_{-1}|0>_L\,. \tag{12.71}$$

They correspond to gauge fields of a $U(1) \times U(1)$ gauge symmetry. In the Kaluza–Klein compactification, there are massless vector states, where one index belongs to the compactified dimension ($g_{5\mu} = A_\mu$). In string theory there are now two such states, $(V^i \pm \tilde{V}^i)/\sqrt{2}$, and the evolved gauge group is larger than the Kaluza–Klein compactification because of the topological property associated with the winding number $n$ of the compactified dimension.

The target space also has a scalar $\alpha^{\tilde{d}}_{-1}|0>_R \tilde{\alpha}^{\tilde{d}}_{-1}|0>_L$. In addition there are all the excited states with Kaluza–Klein and winding numbers $m$ and $n$. For some special values of $R$, some of these states with nonzero $m$ and $n$ can also remain massless satisfying the mass shell condition $M_L^2 = M_R^2 = 0$. Consider, for example, the special case with

$$R^2 = \frac{1}{2} = \alpha'. \tag{12.72}$$

In earlier equations we set $l = (\pi T){-1/2} = \sqrt{2\alpha'} = 1$. This is one of the most interesting solutions for the massless particles. If we interchange $n \to m$ and $R \to \alpha'/R$, the spectrum of particles remains the same and is called *T-duality*. We shall come back to duality in the last section of this chapter. Corresponding to the different values of $m, n, N$ and $\tilde{N}$, there are now six massless vectors and nine massless scalars. We include the states with vanishing winding number and Kaluza–Klein number in this counting. The six massless vector states now combine into an $SU(2) \times SU(2)$ gauge field whereas the scalars belong to a $(\mathbf{3}, \mathbf{3})$ representation of this group. In this special case of $R^2 = 1/2$, the gauge symmetry has been enhanced to a $SU(2) \times SU(2)$ gauge symmetry, without changing the rank of the group.

### Construction of Heterotic Strings

We shall now construct the heterotic strings, whose right-movers are the ordinary closed superstrings. The bosonic degrees of freedom $X^\mu_R(\tau - \sigma)$ and the fermionic degrees of freedom $\Psi^\mu_R(\tau - \sigma)$ of the right-movers satisfy the usual mode expansions

$$\begin{aligned} X^\mu_R(\tau-\sigma) &= \frac{1}{2}x^\mu + \frac{1}{2}p^\mu(\tau-\sigma) + \frac{i}{2}l \sum_{n\neq 0} \frac{1}{n}\alpha^\mu_n \exp^{-2in(\tau-\sigma)} \\ \Psi^\mu_R(R) &= \sum_n d^\mu_n \exp^{-2in(\tau-\sigma)} \\ \Psi^\mu_R(NS) &= \sum_r b^\mu_r \exp^{-2ir(\tau-\sigma)}, \end{aligned} \tag{12.73}$$

where $n$ is integers and $r$ is half-integers and $\mu = 0, 1, ..., 9$. $(R)$ and $(NS)$ correspond to the Ramond (periodic) and the Neveu–Schwarz (antiperiodic) boundary conditions, respectively.

The left-movers are the bosonic strings, which are formulated in 26 dimensions. The first 10 dimensions ($\mu = 0, 1, ..., 9$) have their right-mover superstring counterparts and the usual mode expansion

$$X^\mu_L(\tau+\sigma) = \frac{1}{2}x^\mu + \frac{1}{2}p^\mu(\tau+\sigma) + \frac{i}{2}l \sum_{n\neq 0} \frac{1}{n}\tilde{\alpha}^\mu_n \exp^{-2in(\tau+\sigma)}. \tag{12.74}$$

The coefficients satisfy the usual commutator and anticommutator relations and

$$[x^\mu, p^\nu] = -i\eta^{\mu\nu}. \tag{12.75}$$

For the remaining 16 dimensions ($I = 1, 2, ..., 16$) there is no corresponding right-mover counterpart. We, thus, compactify these 16 dimensions to get the resulting 10-dimensional heterotic strings. We shall expand these bosonic operators $X^I_L(\tau + \sigma)$ as

$$X^I_L(\tau+\sigma) = \frac{1}{2}x^I_L + \frac{1}{2}p^I_L(\tau+\sigma) + \frac{i}{2}l \sum_{n\neq 0} \frac{1}{n}\tilde{\alpha}^\mu_n \exp^{-2in(\tau+\sigma)}. \tag{12.76}$$

In this case, there are no corresponding right-movers, and hence, there is a factor of $1/2$ in the commutation relation,

$$[x^I_L, p^I_L] = -\frac{i}{2}\eta^{IJ}. \tag{12.77}$$

We compactify the 16-dimensional space ($I = 1,2,...,16$) on a torus by identifying

$$x_L^I = x_L^I + \sqrt{2}\pi \sum_{a=1}^{16} n_a R_a e_a^I, \tag{12.78}$$

where $e_a^I, a = 1,2,...,16$ defines the basis vectors of the lattice $\Gamma$, which defines the 16-dimensional torus with radii $R_a$, whose length is chosen to be $\sqrt{2}$. The single-valuedness of the momentum corresponding to the compactification condition introduces 16 integer numbers $m_a$ that specify the Kaluza–Klein states with momenta

$$p_L^I = \frac{1}{\sqrt{2}} \sum_{a=1}^{16} \frac{m_a}{R_a} e^{*I}_{a}, \tag{12.79}$$

where the dual $e^{*I}_a$ of the basis vector $e_a^I$ is defined by

$$e_a^I e^{*I}_b = \delta_{ab}. \tag{12.80}$$

In the light-cone gauge the mass-squared operator is now given by

$$M^2 = M_R^2 + M_L^2, \tag{12.81}$$

where

$$\frac{1}{4} M_R^2 = N \quad \text{and} \quad \frac{1}{4} M_L^2 = \frac{1}{2} \sum_{I=1}^{16} (p_L^I)^2 + \tilde{N} - 1, \tag{12.82}$$

with

$$\begin{aligned} N(R) &= \sum_{n=1}^{\infty} \alpha_{-n}^i \alpha_n^i + \sum_{n=1}^{\infty} n d_{-n}^i d_n^i \\ N(NS) &= \sum_{n=1}^{\infty} \alpha_{-n}^i \alpha_n^i + \sum_{r=1/2}^{\infty} r b_{-r}^i b_r^i - \frac{1}{2} \\ \tilde{N} &= \sum_{n=1}^{\infty} \tilde{\alpha}_{-n}^i \tilde{\alpha}_n^i + \sum_{n=1}^{\infty} \tilde{\alpha}_{-n}^I \tilde{\alpha}_n^I. \end{aligned} \tag{12.83}$$

Here, the sum over $i$ extends from 1 to 8, while that over $I$ extends from 1 to 16. The physical states satisfy the condition $M_R^2 = M_L^2$. The massless states correspond to $M_R^2 = M_L^2 = 0$, which includes the vector bosons generating a gauge symmetry at $d = 10$. The modular invariance of the strings ensures cancellation of gauge and gravitational anomaly, which dictates two possible solutions in 16 dimensions with $M_L^2 = 0$. These two solutions correspond to the gauge groups $SO(32)$ and $E_8 \times E_8$. There are no tachyonic states in this case, so there is no necessity to enforce GSO projections. The massless states now contain the superstring right-movers in the $NS$ and $R$ sectors and bosonic string left-movers with $M_R^2 = 0$ and $M_L^2 = 0$; they are

$$\begin{aligned} (NS): &\quad b_{-1/2}^i |0>_R \; \tilde{\alpha}_{-1}^j |0>_L \qquad i,j = 1,2,...,8 \\ (R): &\quad |0>_R \; \tilde{\alpha}_{-1}^j |0>_L \qquad j = 1,2,...,8, \end{aligned} \tag{12.84}$$

which includes a massless graviton, a scalar field (dilaton), and an antisymmetric tensor in the (*NS*) sector and a gravitino and an eight-component ten-dimensional spinor. Together they form a complete 10-dimensional $N = 1$ supergravity multiplet. In addition, there are now the gauge bosons corresponding to the groups $SO(32)$ or $E_8 \times E_8$ and the gauginos coming from the compactification.

## 12.4 Superstring Phenomenology

The goal of superstring theory is to unify gauge interactions with the gravitational interaction. Since these interactions will become strong at large scale, the unified theory should also be the theory of quantum gravity, should incorporate supersymmetry to solve the gauge hierarchy problem and unify Poincaré group with internal gauge groups, and should incorporate the standard model gauge group and the particle spectrum. Among the five possible superstring theories, namely, type I, type IIA, type IIB, heterotic strings with gauge group $SO(32)$, and heterotic strings with gauge group $E_8 \times E_8$, the $E_8 \times E_8$ heterotic string emerges as the most promising candidate to accommodate all these requirements. When the extra 6 dimensions are compactified in the Calabi–Yau manifold, the residual low-energy 4-dimensional theory could retain the $N = 1$ supersymmetry and the standard-model gauge group $SU(3)_c \times SU(2)_L \times U(1)_Y$ with proper low-energy particle spectrum [189, 196]. It is also possible to calculate the Yukawa couplings in some of these models [197], which can at least confirm the consistency of the theory, although at this stage there may not be any predictions.

The $N = 1$ supergravity Lagrangian in 10 dimensions with the physical particle spectrum of the 10-dimensional heterotic string with the gauge group $E_8 \times E_8$, as presented in table 12.1, will be almost equivalent to the heterotic string Lagrangian and will have the same low-energy limit. The next step is to obtain the 4-dimensional theory by compactifying the extra 6 dimensions, which preserves the required low-energy aspects of the theory, such as unbroken supersymmetry and the low-energy gauge group $SU(3)_c \times SU(2)_L \times U(1)_Y$ with its particle content. We should first find some six-dimensional manifold, which is orthogonal to our locally flat Minkowski space and satisfy the following criteria:

(i) The geometry of the 10-dimensional space should be $\mathscr{M}^4 \times \mathscr{K}$, where $\mathscr{M}^4$ is the 4-dimensional flat Minkowski space–time and $\mathscr{K}$ is a 6-dimensional compact space.

(ii) $N = 1$ supersymmetry should be unbroken in four dimensions at the time of compactification. We should break supersymmetry around the electroweak scale through some other mechanism so the quadratic divergences are prevented until the electroweak symmetry breaking scale.

**TABLE 12.1**
Particle spectrum of a $d = 10$ heterotic string theory with the gauge group $E_8 \times E_8$. $M, N = 0, 1, ..., 9$ are 10-dimensional space–time index, and $\alpha$ is the index for the group $E_8 \times E_8$.

| State Vectors | Fields | Particles |
|---|---|---|
| $b^i_{-1/2}\|0>_R\ \tilde{\alpha}^j_{-1}\|0>_L$ | $h_{MN}$ | graviton |
| | $B_{MN}$ | antisymmetric rank-2 tensor |
| | $\phi$ | dilaton |
| $\|0>_R\ \tilde{\alpha}^j_{-1}\|0>_L$ | $\Psi_M$ | gravitino |
| | $\lambda$ | dilatino |
| $b^i_{-1/2}\|0>_R\ \tilde{\alpha}^I_{-1}\|0>_L$ | $A^\alpha_M$ | gauge boson |
| $\|0>_R\ \tilde{\alpha}^I_{-1}\|0>_L$ | $\chi^\alpha$ | gaugino |

(iii) The standard model gauge group and the known particle spectrum should be reproduced after the compactification.

**Compactification on Calabi–Yau Manifold**

These requirements are not as simple as they appear. Once we assume that the space is of the form $\mathcal{M}^4 \times \mathcal{K}$, the six-dimensional manifold $\mathcal{K}$ becomes highly constrained. The metric on $\mathcal{K}$ should be Hermitian with respect to the almost complex structure, whose integrability requires the structure to be complex and Kähler. The condition for space–time supersymmetry is most restrictive. It requires that the space should be Ricci-flat, which can be ensured by the existence of a spinor, which should be a triplet under the $SU(3)$ holonomy group. In six dimensions the spin connection is a priori a $O(6)$ gauge field, and hence, the spinor is a 4-spinor. The condition of supersymmetry has a trivial solution that the spin connection vanishes everywhere leading to a flat noncompact space, which is undesirable. The other nontrivial solution requires that one component of the spinor gets a value, which breaks the group $SU(4) \to SU(3)$. Then the tangent space should have an $SU(3)$ symmetry, and the spinor transforms as a triplet under this $SU(3)$ holonomy group. The existence of such covariant spinor makes the space *Ricci-flat*.

The simple physical conditions we required, could be satisfied only by the Calabi–Yau manifolds, which are the complex Kähler Ricci-flat manifolds with an $SU(3)$ holonomy. These conditions are based on field theory considerations, but later a consistent string theory has also been constructed on the Calabi–Yau manifold. Another advantage of the compactification in the Calabi–Yau manifold is that the anomaly cancellation gives a condition that the spin connection has to be identified with the gauge connection. This is done by associating the $SU(3)$ holonomy group with the $SU(3)$ subgroup of one of the $E_8$ gauge group of the $E_8 \times E_8$ heterotic strings [189]. This breaks the gauge group $E_8$ spontaneously to $E_6$, which contains the right particle spectrum of the standard model and has already been considered as a possible grand unified group [198]. There are also some discrete symmetries of the manifold, which ensure that the group $E_6$ is further broken during compactification to one of its subgroups, leaving the correct particle spectrum.

Compactification on a Calabi–Yau manifold thus gives chiral fermion in the fundamental representation of the group $E_6$ automatically. The number of generations of chiral representations is determined by the *Euler characteristics* $\chi$, which is a topological quantity of the manifold. The Euler characteristics are related to the number of superfields $n_{27}$ in the fundamental representation 27 and the number of superfields $n_{\overline{27}}$ in the conjugate representation $\overline{27}$ by the relation

$$\chi = n_{27} - n_{\overline{27}}. \tag{12.85}$$

There will be $n_{\overline{27}}$ numbers of $27 + \overline{27}$ superfields, which will be the Higgs scalars. Thus, the required numbers of fermions and scalars come out from the geometry of the compact space. Since the compactification breaks the $E_6$ gauge group also, only the 27 and $\overline{27}$ Higgs scalars are required for any symmetry breaking or giving masses to the fermions, which are also present in the theory.

The particle content of the 10-dimensional $N = 1$ supergravity theory originating from superstring theory includes the massless graviton, gravitino, dilaton, dilatino and an antisymmetric rank-2 tensor. In addition there are the gauge fields $A_M^\alpha$ and their superpartners, the gauginos $\chi^\alpha$ in 10 dimensions that generate the group $E_8 \times E_8$. After compactification some of the components of these gauge superfields can only contribute to the chiral zero modes in 4 dimensions. In the following we shall assume $\alpha$ to be the index for only one of the $E_8$ group and the other $E_8$ index remains as a hidden sector index, which could contribute to low-energy supersymmetry breaking.

The gauginos and the gauge bosons belong to the 248-dimensional representation of the group $E_8$. Under its maximal subgroup $SU(3) \times E_6$, these fields transform as

$$248 = (8,1) + (1,78) + (3,27) + (\bar{3},\overline{27}). \tag{12.86}$$

On the other hand, the 10-dimensional vector $V_M$ has a decomposition

$$V_M = V_\mu + V_a + V_{\bar{a}}, \tag{12.87}$$

where $M = 0,1,...,9$ is the index of the 10 dimensions, while $\mu = 0,1,2,3$ denotes the four dimensions. Under the $SU(3)$ holonomy group, the compact 6 dimensions now split into $6 = 3 + \bar{3}$, which corresponds to the indices $a = 1,2,3$ and $\bar{a} = 1,2,3$. So, their transformation is governed by the spin connection of the Calabi–Yau manifold. When these $SU(3)$ spin connections are identified with the gauge connections of the $SU(3)$ subgroup of $E_8$, we contract the index $a$ with the $SU(3)$ index in the triplet representation. If we now write the $E_8$ index as $\alpha \equiv (\eta, m)$, where $\eta$ is an $E_6$ index and $m$ is an $SU(3)$ index, then the 10-dimensional vectors would contain

$$V_M^\alpha \supset V_\mu^{(\eta_{78}, m_1)} \oplus V_a^{(\eta_{27}, m_3)} \delta_{m_3}^a \oplus V_{\bar{a}}^{(\eta_{\overline{27}}, m_{\bar{3}})} \delta_{m_{\bar{3}}}^{\bar{a}}. \tag{12.88}$$

The first term represents the gauge bosons of the group $E_6$ in 4-dimensions, while the second and the third terms are the scalar superfields in 4-dimensions, containing all the fermions and Higgs scalars. Thus, for the field $A_M^\alpha$, when we identify these two $SU(3)$ groups, the 27 and $\overline{27}$ representations acquire the topological property that their numbers get related to some topological quantum numbers of the manifold, which are the *Hodge numbers*. The low-lying states become blind to the second $E_8$ group, and hence, it does not affect any low-energy phenomenology. The gauginos corresponding to this second $E_8$ group can form condensates at very low energy and can break supersymmetry in the hidden sector. For our discussions of the observable sector phenomenology, we shall ignore this second $E_8$ group.

**Three-Generation Superstring Model**

Without going into the details, we shall present an example of a Calabi–Yau manifold, which gives three generations of fermions [196]. We demonstrate the discrete symmetries in such manifold using algebraic geometry methods, which can restrict some of the couplings of the fermions and make the theory consistent with low-energy phenomenology [197].

Let us consider a complex space $\mathscr{C}^{n+1}$, described by $(n+1)$ complex numbers $(z_0, z_1, \ldots, z_n)$ with the origin removed $\mathscr{C}^{n+1} - \{0\}$. A complex projective space $\mathscr{C}\mathscr{P}^n$ of dimension $n$ is defined by associating a scaling invariance or a projection, $(z_0, z_1, ..., z_n) = \lambda(z_0, z_1, ..., z_n)$ on $\mathscr{C}^{n+1} - \{0\}$, where $\lambda$ is any nonzero complex number. The coordinates $(z_0, z_1, ..., z_n)$ are called homogeneous coordinates of $\mathscr{C}\mathscr{P}^n$, which is a complex Kähler manifold. A Calabi–Yau manifold is a hypersurface on a $\mathscr{C}\mathscr{P}^3 \times \mathscr{C}\mathscr{P}^3$ space with the homogeneous coordinates $x_i$ and $y_i$, $i = 0, 1, 2, 3$, which is defined by the zeros of the polynomials

$$\begin{aligned} &\sum x_i^3 + a_1 x_0 x_1 x_2 + a_2 x_0 x_1 x_3 + a_3 x_0 x_2 x_3 + a_4 x_1 x_2 x_3 = 0 \\ &\sum y_i^3 + b_1 y_0 y_1 y_2 + b_2 y_0 y_1 y_3 + b_3 y_0 y_2 y_3 + b_4 y_1 y_2 y_3 = 0 \\ &\sum c_{ij} x_i y_j = 0, \end{aligned} \tag{12.89}$$

with $c_{00} = 1$. These are the most general polynomials and have 23 parameters $a_i, b_i, c_{ij}$, which define the metric of the Calabi–Yau space. There are now 23 independent superfields belonging to the 27 representations, which can be represented by the 23 independent monomials, as given in table 12.2.

We shall now act on this space by a $\mathscr{Z}_3$ discrete symmetry

$$\begin{aligned} \mathscr{Z}_3 \, : \; & (x_0, x_1, x_2, x_3) \times (y_0, y_1, y_2, y_3) \\ & \longrightarrow (x_0, \alpha^2 x_1, \alpha x_2, \alpha x_3) \times (y_0, \alpha y_1, \alpha^2 y_2, \alpha^2 y_3), \end{aligned} \tag{12.90}$$

where $\alpha = \exp[2\pi i/3]$, so $\alpha^3 = 1$. After the action of the discrete symmetry, this Calabi–Yau space has an Euler characteristic 3 and gives a three-generation model.

**TABLE 12.2**
Monomial representation of the 23 independent 27-dimensional superfields of $E_6$, cataloged according to their $\mathscr{G} \equiv \mathscr{Z}_3$ transformation properties.

| $\mathscr{G}(1)$ | $\mathscr{G}(\alpha)$ | $\mathscr{G}(\alpha^2)$ |
|---|---|---|
| $\lambda_1 \equiv x_0x_1x_2$ | $q_1 \equiv x_1x_2x_3$ | $Q_1 \equiv x_0x_2x_3$ |
| $\lambda_2 \equiv x_0x_1x_3$ | $q_2 \equiv y_0y_2y_3$ | $Q_2 \equiv y_1y_2y_3$ |
| $\lambda_3 \equiv y_0y_1y_2$ | $q_3 \equiv x_0y_1$ | $Q_3 \equiv x_0y_2$ |
| $\lambda_4 \equiv y_0y_1y_3$ | $q_4 \equiv x_1y_2$ | $Q_4 \equiv x_0y_3$ |
| $\lambda_5 \equiv x_1y_1$ | $q_5 \equiv x_1y_3$ | $Q_5 \equiv x_1y_0$ |
| $\lambda_6 \equiv x_2y_2$ | $q_6 \equiv x_2y_0$ | $Q_6 \equiv x_2y_1$ |
| $\lambda_7 \equiv x_3y_3$ | $q_7 \equiv x_3y_0$ | $Q_7 \equiv x_3y_1$ |
| $\lambda_8 \equiv x_2y_3$ | | |
| $\lambda_9 \equiv x_3y_2$ | | |

This discrete $\mathscr{G} \equiv \mathscr{Z}_3$ group will also act on the gauge group $E_6$, which will then be broken through flux breaking mechanism [199] to $SU(3)_c \times SU(3)_L \times SU(3)_R$, under which the 27 superfields transform as

$$27 = (1,3,\bar{3}) + (\bar{3},1,\bar{3}) + (3,3,1). \tag{12.91}$$

The first 9 superfields are the leptons, while the second 9 superfields are the antiquarks and the last 9 superfields are the quarks. The embedding of $\mathscr{G}$ on $E_6$ is given by

$$U_g = (1) \otimes \alpha(1) \otimes \alpha(1), \tag{12.92}$$

where $(1)$ is a $3 \times 3$ unit matrix. This means that the leptons transform under $\bar{\mathscr{G}}$ as $\alpha \cdot \bar{\alpha} = 1$, while antiquarks and quarks transform under $\bar{\mathscr{G}}$ as $\bar{\alpha} = \alpha^2$ and $\alpha$, respectively. We can thus identify the $\lambda$'s with the leptons, $q$'s with the quarks, and $Q$'s with the antiquarks.

After identifying the particles with the algebraic varieties, it is possible to calculate the Yukawa couplings using the correspondence between the Calabi–Yau compactification and compactifications on tensor products of minimal $N = 2$ superconformal field theories [191]. A simpler group theoretic approach to find the discrete symmetries that can restrict the couplings utilizes another topological quantum number, the intersection number of the manifold. In terms of the algebraic varieties, any symmetry group, which commutes with the discrete group $\mathscr{G} \equiv \mathscr{Z}^3$ will then be able to generate a discrete symmetry of the theory. These discrete symmetries can then restrict various couplings of the fermion superfields, which determine the low-energy phenomenology. These calculations could establish the consistency of the theory, although there are no unique phenomenological predictions of the superstring theory.

A large class of low-energy phenomenology of the superstring theory means only superstring-inspired phenomenology [190]. In the superstring-inspired phenomenology, a phenomenological model is considered taking the particle spectrum from the superstring theory. One assumes that the theory contains fermions belonging only to the 27-plet representation of $E_6$ and the Higgs scalars belonging to the superfields in the 27 or $\overline{27}$ representations. The Lagrangian is the $N = 1$ supergravity Lagrangian with an arbitrary superpotential, constructed based on the problem under consideration. For the soft terms one needs to assume string-motivated supersymmetry breaking mechanisms.

It is also possible to construct four-dimensional superstring models. Here one does not have to first take the field theoretical limit and then compactify. The compactification takes place at the string level, which makes these theories more complicated. Moreover, heterotic strings are no longer the unique possibility. The string scale could also be lower than the Planck scale. The intersecting branes open up another new aspect of the superstring phenomenology. All these newer developments are changing the concept of superstring phenomenology very fast [200]. Whether or not superstring theory can predict our low-energy world, the fact that it can accommodate the standard model consistently with all the known fermions and their interactions without any contradiction is already an achievement. This also supports our belief that superstring theory can be the theory of all the basic interactions. Which solution is preferred is difficult to decide from our present knowledge, but it is established that the superstring theory can provide consistent low-energy phenomenology.

## 12.5 Duality and Branes

The superstring theory has been emerging as the most consistent theory of quantum gravity, which can also unify gauge interactions with gravitational interaction and have built-in supersymmetry. The low-energy standard model gauge group and the standard model particle spectrum also could be accommodated in the theory. The heterotic strings with $E_8 \times E_8$ gauge group in 10-dimensions with the extra six dimensions compactified on a three-generation Calabi–Yau manifold appeared to be the only possibility that could give us the low-energy phenomenology in 4-dimensions consistently. But with newer developments this uniqueness is lost and we observe many interesting results.

There is a general prescription for compactifying the extra dimensions by replacing six of the target space dimensions with a conformal field theory with the desired charge. Compactification on a Calabi–Yau manifold with $SU(3)$ holonomy in the field theory limit is only one of the solutions of this more general prescription. The coordinates of the compact Calabi–Yau space will then belong to a conformal field theory with a consistent central charge. Techniques such as $F$-theory provide new classes of 4-dimensional $N = 1$ string vacua [201]. These developments with newer solutions for the low-energy world are making the superstring theory appear less unique, but more interesting [200]. Some recent advances in the field are reverting our idea about nonuniqueness of the string vacua and giving indications that the different low-energy theories may emerge from one ultimate theory [202].

Many of the newer developments in the field started when the string theory went through a major breakthrough following the proposition of the duality conjectures. These conjectures relate a theory with strong couplings to a theory with weak couplings, which allows us to study some strongly coupled theories at least partially. One of the significant results from such developments is the calculation of the black hole entropy. While other approaches to the problem end up with a puzzle, string theory could approach the problem from a microscopic point of view and provide a partial solution.

Another interesting outcome of the duality conjecture is the brane solutions, which now allow us to construct string theory at a scale as low as a few TeV. These theories have many interesting features, some of which will be mentioned in the next two chapters. The brane solutions also have other implications as well in connection with the studies of duality conjectures and construction of new phenomenological models.

**Information Loss Puzzle**

The string theory met with triumph in explaining the information loss puzzle in black holes [203]. Since other competing theories of quantum gravity could not provide any explanation for this puzzle, string theory is now accepted as the most consistent theory of quantum gravity.

Black holes are classical solutions, in which the matter density and the space–time curvature is so high that when any particle or even light reaches a certain distance near the black hole, called the event horizon, it is pulled inside increasing the size of the black hole. Since nothing can come out of the black hole once it is inside the event horizon, it can be considered as a perfect blackbody. The black holes could be formed by collapse of matter under its own gravitational pull or could be primordial.

When quantum effects are considered, the event horizon cannot be sharply defined. Thus, the spread of the wave function of a particle falling inside the event horizon allows the particle to leave from near the event horizon. Hawking radiation [204] can now be viewed as pair creation near the event horizon, where some particles may combine with the ensemble of particles. The remaining particles are pulled back into the black hole. Black holes could as well evaporate radiating particles that are near its event horizon. In quantum theory it is expected that the black holes emit thermal blackbody radiation and carry entropy. Consider an ensemble of particles in a pure state near the event horizon of a black hole. Hawking radiation will imply that the state leaving the black hole will carry some information about the initial state, and the wave functions of the particles that combine with the ensemble will make it a mixed state. Since the Hawking radiation is thermal, although the ensemble started as pure state, most of the information about the initial state is lost.

Any consistent quantum theory of gravity should explain this information loss puzzle from a microscopic description of the radiation from the black holes. Among several other competing theories, only the string theory could at least partially solve this problem. In a special class of black holes, the string theory could count the number of quantum states of black holes and relate this number with the entropy of the black holes. If $N$ is the degeneracy of states with a certain quantum number of these black holes, then the Bekenstein–Hawking entropy of a black hole is given by

$$S_{BH} = \ln\,(N). \tag{12.93}$$

It is also possible in this microscopic approach of the string theories to calculate the Hawking radiation from these black holes due to quantum scattering processes inside the black hole [203].

The classical theory of gravity is explained by the general theory of relativity. Although the theory has been extremely successful at low energies, the theory is nonrenormalizable since it is described by a spin-2 particle, the graviton. In field theory, the graviton exchange in the loop will have infinite contributions, which cannot be removed by ordinary renormalization prescriptions. These quantum corrections, coming from the loop diagrams, cannot be dealt with any techniques of ordinary point-particle field theory. On the other hand the quantum string theory has no ultraviolet divergences, and since the particle spectrum of string theory contains a massless spin-2 particle which can be interpreted as the graviton, it appears to be a consistent theory of quantum gravity. The solution to the information loss puzzle now puts string theory ahead of all other approaches to a consistent theory of quantum gravity.

### String Duality

One of the major results in the string theory is the duality conjecture [205], which establishes an equivalence between two or more apparently distinct string theories. This has many consequences in string theory as well as several fascinating phenomenological applications. The duality conjectures started with reducing the nonuniqueness of the string theory. We mentioned the five different superstring theories: type I superstring theory, type IIA superstring theory, type IIB superstring theory, heterotic string theory with gauge group $E_8 \times E_8$ and heterotic string theory with gauge group $SO(32)$. These theories are defined in 10 dimensions and have consistent weak-coupling perturbation expansions, so they can be described by $N = 1$ supergravity theory in 10-dimensions.

The new era in string theory or the *second string revolution* started when some of these theories could be shown to be equivalent by the duality symmetries of the string theories. These duality symmetries were proposed as conjectures, which provided many mathematical identities. When these mathematical identities could be eventually proved through rigorous calculations, these conjectures were established. The main problem of proving these conjectures directly is that these duality conjectures show equivalence between a weakly coupled string theory to a strongly coupled string theory and we do not have any tools to calculate anything in the strongly coupled string theories.

Before we discuss any of the recent developments, let us explain the duality concept that was already known, the *T-duality*. Consider a 10-dimensional string theory in which one of the dimensions has been compactified on a circle, so the space is now $R^9 \times S^1$. We now consider two different types of string theories with compactification radius $R_1$ and $R_2$ in this compactified space $R^9 \times S^1$, with the condition,

$$R_1 R_2 = l_s^2, \tag{12.94}$$

where $l_s$ is the fundamental string length scale, with tension

$$T = \frac{1}{2\pi l_s^2} = 2\pi m_s^2, \tag{12.95}$$

where $m_s$ is the fundamental string mass scale. The T-duality implies that the two types of string theories under consideration are identical, if for $R_1 \to 0$, the dual theory decompactifies and $R_2 \to \infty$. While compactifying the bosonic strings we encountered two quantum numbers: string winding number $n$, which denotes the number of times a string can wind the compactified dimension and the Kaluza–Klein mode $m$, which comes from the single-valuedness of $\exp^{ipx}$ requiring $p = m/R$. Both these quantum numbers contribute to the mass-squared operator

$$\frac{1}{8}M^2 = \frac{N}{2} + \frac{\tilde{N}}{2} - 1 + \frac{n^2R^2}{2} + \frac{m^2}{8R^2}. \tag{12.96}$$

Under T-duality, the last two terms exchange their roles [206]. If we interchange the string winding number ($n$) with the quantum number associated with the quantized Kaluza–Klein momentum modes ($m$) and simultaneously invert the compactification radius $R \to 1/R$, the spectrum remains invariant.

In superstring theories, type IIA and type IIB are T-dual. So, if the nonchiral type IIA theory is compactified on $S^1$ with almost vanishing radius to get a 9-dimensional theory, it becomes equivalent to a chiral type IIB theory in 10 dimensions. The radius of the IIA theory is equivalent to the *vev* of a scalar field. When this *vev* vanishes, the scalar field appears as one of the components of the 10-dimensional metric tensor. Thus, these two theories are not independent and are two limiting points in a continuous moduli space of quantum vacua in 10 dimensions. Similarly, the two heterotic strings are also T-dual.

To elaborate this let us consider a type II superstring theory in ten dimensions and then compactify its tenth dimension on a circle $S^1$. Then consider the T-duality transformation

$$m \leftrightarrow n \qquad R \leftrightarrow \frac{\alpha'}{R},$$

where the string coupling constant is given by

$$\alpha' = \frac{1}{2\pi T} = \frac{1}{2}l^2 = \frac{1}{2}.$$

This can be achieved for type II strings by assigning $X^9 = X^9_L - X^9_R$ instead of $X^9 = X^9_L + X^9_R$ and reversing the sign of the right-movers but still combining the left-movers with the right-movers. The corresponding changes in the fermionic sector, following the same prescription for the T-duality, would take us from a type IIA superstring to a type IIB superstring. Thus, a type IIA superstring compactified on a circle with radius $R$ is equivalent to a type IIB superstring compactified on a circle with radius $\alpha'/R$. This is consistent with the observation that when type IIA and type IIB superstrings are compactified on a circle, the mass spectra of these two theories come out to be the same.

We shall now discuss a new class of duality, where one conjectures equivalence of two different string theories [207] and the duality maps an elementary particle to a composite particle depending on which string theory we use. This duality conjecture could not be proved but has undergone many consistency checks and provided several major important results. It maps a closed-string excitation to an open-string excitation and also maps a strongly coupled theory to a weakly coupled theory through $g\tilde{g} = 1$. A perturbation expansion in one type of string theory contains information about the nonperturbative results of the other that describes the full quantum string theory. Our lack of knowledge about calculations in a strongly coupled theory makes it difficult to prove the conjecture more rigorously. In simple language this establishes a duality between gravity and field theory in the context of branes, which is also known as AdS/CFT correspondence. This conjecture establishes an equivalence between a type IIB string theory on $S^5 \times AdS_5$ and an $N = 4$ superconformal SU(N) gauge theory in 4-dimensions [207, 208].

We first define a few terms and then discuss the duality conjecture. An $S^5$ is a 5-dimensional sphere embedded in a 6-dimensional space with its radius defined by $R^2 = \sum_{i=1}^{6} y_i^2$. We can define an antideSitter space ($AdS_{p+2}$) of dimension $p+2$ as a hypersurface on a $2+p+1$-dimensional flat space with signature $(--+++)$, whose size is given by

$$R^2 = -X_{-1}^2 - X_0^2 + \sum_{i=1}^{p+1} X_\alpha^2. \tag{12.97}$$

Solving this equation one can get the metric in the $AdS_{p+2}$ space as

$$ds^2 = \frac{U^2}{R^2}\,\eta_{ij}dx^i dx^j + R^2\frac{dU^2}{U^2}, \tag{12.98}$$

where $U = X_{-1} + X_{p+1}$ and $x^i = RX_i/U \quad (i = 0,..p)$ parameterize the hypersurface satisfying the condition $X_{-1} - X_{p+1} = x^2 U/R^2 + R^2/U$ with $x^2 = \eta_{ij}x^i x^j$. All 10 dimensions in $S^5 \times AdS_5$ are assumed to have the same radius $R$. A type IIB string theory in $S^5 \times AdS_5$ is related to an $N = 4$ superconformal SU(N) Yang–Mills theory. An $N = 4$ superconformal field theory has four supersymmetric generators, which relate four Majorana fermions and six bosons with each of the gauge bosons. There is only one gauge coupling constant ($g_Y$) for the entire SU(N) gauge group and all of their $N = 4$ supersymmetric partners. In addition there is one more parameter, which is the vacuum angle $\theta$.

According to the Maldacena conjecture, the physical Green's functions in type IIB string theory on $S^5 \times AdS_5$, which are the physical excitations on the flat 4-dimensional boundary of $AdS_5$, have one to one correspondence with the correlation functions of gauge-invariant operators in $N = 4$ supersymmetric SU(N) gauge theory. The parameters of the type IIB string theory, string coupling constant $g_s$, inverse of the string tension $\alpha'$, and the *vev* of the *Ramond–Ramond scalar a*, get related to the parameters of the $N = 4$ supersymmetric SU(N) gauge theory $g_Y$, $\theta$ and $N$, through the relations

$$g_s = g_Y^2, \qquad a = \theta \quad \text{and} \quad \frac{R}{\sqrt{\alpha'}} = (4\pi N g_Y^2)^{1/4}. \tag{12.99}$$

This conjecture has many interesting consequences, such as the derivation of the holographic principle [209], which states that there is one degree of freedom per Planck area on the boundary in a consistent theory of quantum gravity. The fundamental degrees of freedom reside only in the boundary and there is no fundamental degree of freedom in the interior.

There are other dualities, which relate the different types of superstring theories. The type I string theories and the 10-dimensional heterotic $SO(32)$ string theory are conjectured to be dual; the heterotic $SO(32)$ string theory compactified on a four-dimensional torus is conjectured to be dual to a four-dimensional type IIA string theory compactified on K3; and type IIB is conjectured to be self dual. These dualities seem to imply that all of the string theories are related to each other. It is our limitations that we can do calculations with the five string theories only in the weak-coupling limit. These theories are related to each other by duality and all of them are dual to one strongly coupled string theory, the $M$-theory. All these equivalences seem to indicate that there is an ultimate theory, referred to as the $U$-theory, and if we have the full understanding of the strong-coupling limits of the ultimate theory, we can derive all these string theories in the weak-coupling limit from the ultimate theory (see figure 12.2).

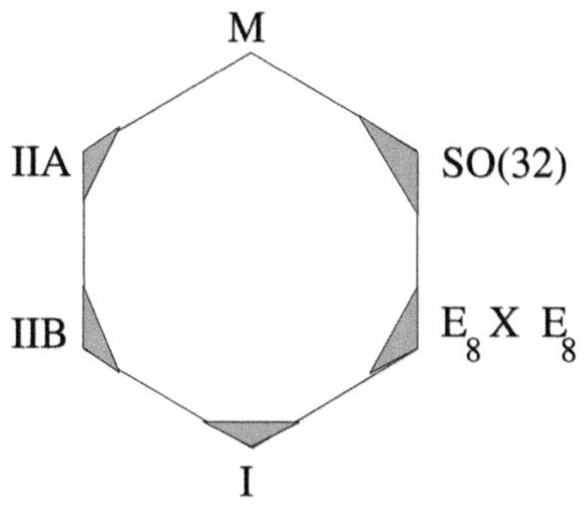


**FIGURE 12.2**
The perturbative regions of different string theories are shown in shaded areas, which are related to each other and to the $M$-theory.

At present the only information we have about the nonperturbative region of the moduli space, except for the duality conjectures, is about the $M$-theory, which is a well-defined quantum theory residing in 11 dimensions. In the low-energy limit it reduces to an 11-dimensional $N = 1$ supergravity theory, which allows us to get some insight into $M$-theory. Other string theories approach the $M$-theory in the strong-coupling limit. The strong-coupling limits of all the string theories in 10 dimensions become 11-dimensional, since the coupling constant manifests itself as an extra dimension in the strong-coupling limit. As a result, the 11-dimensional $M$-theory does not have any coupling constant and until now there does not exist any systematic procedure to do calculations in $M$-theory, except for the low-energy limit.

The field equations of the 11-dimensional supergravity describe a supermembrane (or M2-brane), where the particles have two space-like internal extensions and they are defined on a (2+1)-dimensional world-sheet in 11 dimensions. When one of the world-sheet space-like dimensions wraps around one space–time dimension, it is also compactified and gives the type IIA superstring world-volume action in 10 dimensions. Thus, type IIA superstring theory is an M2-brane of $M$-theory with one of its dimensions wrapped around the circular spatial dimension. Truncation to zero modes is not required. The matrix theory is another attempt to give a nonperturbative description of the $M$-theory. In the infinite-momentum limit, the $M$-theory is described by a quantum system with $N \times N$ matrices as the fundamental degrees of freedom. The appropriate correlation functions then give the scattering amplitudes of the $M$-theory in the limit $N \to \infty$, which reproduce the low-energy 11-dimensional $N = 1$ supergravity results.

We now understand the five weakly coupled superstring theories at 10 dimensions, which are not independent and also related to the strongly coupled $M$-theory in 11 dimensions, whose low-energy limit correspond to the 11-dimensional $N = 1$ supergravity theory. However, we expect that all these theories should emerge from yet another ultimate theory or the $U$-theory, which is also a strongly coupled theory in 11 dimensions. There are attempts to understand the $U$-theory by studying the five weakly coupled theories as well as the strongly coupled $M$-theory. Although we are yet to understand the $U$-theory, all these studies start giving us hope for an ultimate theory of everything.

### Branes

Brane solutions have several interesting consequences in string theory. They are useful in deriving gauge theory results from string theories [210]. Observations from branes lead to several duality conjectures. The branes solutions allow lowering of the Planck scale, which leads to interesting low-energy phenomenology. Brane solutions also allow new phenomenological models, in which only gravity propagates along the extra dimensions while the gauge interactions are localized on a boundary brane (which will be discussed in the next chapter) [211, 212].

Branes are static classical solutions in string theories [213]. A $p$-brane denotes a static configuration which extends along $p$ spatial directions and is localized in all other directions. Such solutions are invariant under translation in the transverse directions. Strings are equivalent to 1-brane, membranes are 2-brane, particles are 0-brane, and a $p$-brane is described by a $(p+1)$-dimensional gauge field theory.

Two kinds of boundary conditions are frequently used for the open strings. We constructed open strings with the Neumann boundary condition

$$\frac{\partial X^\mu(\tau,\sigma=0)}{\partial\sigma}=\frac{\partial X^\mu(\tau,\sigma=\pi)}{\partial\sigma}=0, \tag{12.100}$$

which states that the variations of the coordinates $X^\mu$ are arbitrary at the boundary and no momentum can flow out of the end of the open string. The other one is the Dirichlet-boundary condition

$$X^a(\tau,\sigma=0)=X^a(\tau,\sigma=\pi)=c^a, \tag{12.101}$$

where $c^a$ is a constant vector. The boundaries are now frozen, so the variation of the coordinates vanishes although the momentum can now leave the boundaries.

The $p$-branes, which are extended solitons in string theory, are dynamical objects and can move independently of the size of the $d$-dimensional compact manifold in which they reside. The $(p+1)$-coordinates $X^i$, $i=0,1,2,...,p$ now satisfy the Neumann boundary conditions, so no momentum can leave the boundaries along these $p+1$ dimensions. The remaining coordinates $X^a$, $a=p+1,...,d-1$ satisfy the Dirichlet boundary conditions, so momentum can flow out of these boundaries to a $(d-p-1)$-dimensional surface, called the $D$-brane and extend along the $p+1$ dimensions, whose positions are fixed at the end of the open strings.

We now discuss these solutions for superstrings. We have to impose opposite boundary conditions for the Neumann and Dirichlet directions to maintain space–time supersymmetry. In our convention the Neumann condition is

$$\begin{aligned}\Psi^i_L(\tau,\sigma=0) &= \Psi^i_R(\tau,\sigma=0)\\ \Psi^i_L(\tau,\sigma=\pi) &= \pm\Psi^i_R(\tau,\sigma=\pi),\end{aligned} \tag{12.102}$$

so the Dirichlet condition would be

$$\begin{aligned}\Psi^a_L(\tau,\sigma=0) &= -\Psi^a_R(\tau,\sigma=0)\\ \Psi^a_L(\tau,\sigma=\pi) &= \pm\Psi^a_R(\tau,\sigma=\pi).\end{aligned} \tag{12.103}$$

The advantage of this convention is that using a doubling trick it is possible to relate the Dirichlet and Neumann boundary conditions to the periodic and antiperiodic boundary conditions for closed strings. We can then classify the open strings in the same way as the $R$ or $NS$ types. When both the ends satisfy Dirichlet [Neumann] conditions, the corresponding bosonic string is called DD [NN] and is related to a periodic boundary condition. In this case, for the $NS$ (antiperiodic fermionic) sector we take the boundary conditions at $\sigma=\pi$ to be

$$\Psi^i_L=-\Psi^i_R \quad\text{and}\quad \Psi^a_L=\Psi^a_R, \tag{12.104}$$

and for the $R$ (periodic fermionic) sector we take

$$\Psi^i_L=\Psi^i_R \quad\text{and}\quad \Psi^a_L=-\Psi^a_R. \tag{12.105}$$

When the opposite ends of the bosonic string satisfy different boundary conditions DN [ND], it is related to the antiperiodic boundary condition in the bosonic sector. In this case the fermionic sector is periodic in the $NS$ sector and antiperiodic in the $R$ sector.

Let us now discuss the case of single $D$-brane, which is DD in the transverse direction and NN in the longitudinal direction. Then $X^\mu$ is always periodic, and $\Psi^\mu$ is periodic in the $R$ sector and antiperiodic in the $NS$ sector. We can now write the Neumann boundary conditions in the bosonic sector in terms of the mode expansion oscillators as

$$\alpha_n^i = \tilde{\alpha}_n^i,$$

and those for the Dirichlet direction are

$$x^a = c^a, \quad p^a = 0, \quad \alpha_n^a = -\tilde{\alpha}_n^a.$$

So the bosonic modes are given by

$$\begin{aligned} X^i &= x^i + p^i\tau + i\sum_{n\neq 0}\frac{1}{n}\alpha_n^i \exp^{-in\tau} \cos n\sigma \\ X^a &= c^a - \sum_{n\neq 0}\frac{1}{n}\alpha_n^a \exp^{-in\tau} \sin n\sigma. \end{aligned} \tag{12.106}$$

The mode expansion for the fermionic sector is given by

$$\begin{aligned} \Psi^i_{\substack{R\\L}}(R) &= \frac{1}{\sqrt{2}}\sum_n d_n^i \exp^{-in(\tau\mp\sigma)} \\ \Psi^a_{\substack{R\\L}}(R) &= \pm\frac{1}{\sqrt{2}}\sum_n d_n^a \exp^{-in(\tau\mp\sigma)} \\ \Psi^i_{\substack{R\\L}}(NS) &= \frac{1}{\sqrt{2}}\sum_r b_r^i \exp^{-ir(\tau\mp\sigma)} \\ \Psi^a_{\substack{R\\L}}(NS) &= \pm\frac{1}{\sqrt{2}}\sum_r b_r^a \exp^{-ir(\tau\mp\sigma)}, \end{aligned} \tag{12.107}$$

where $n = 1, 2, \ldots$ are integers and $r = 1/2, 3/2, \ldots$ are half-integers. The commutation relations of the bosonic oscillators and the anticommutation relations of the fermions are the same as previously.

The left-movers are not independent of the right-movers. We work in the light cone gauge and eliminate two directions, the time-like Neumann direction and one space-like Neumann direction $x^1$. The mass-squared operators for the $R$ and the $NS$ sectors now become

$$\begin{aligned} \frac{1}{2}M^2(R) &= \sum_{n=1}^{\infty}\alpha_{-n}^i\alpha_n^i + \sum_{n=1}^{\infty} n d_{-n}^\mu d_{n\mu} - a_R \\ \frac{1}{2}M^2(NS) &= \sum_{n=1}^{\infty}\alpha_{-n}^i\alpha_n^i + \sum_{r=1/2}^{\infty} r b_{-r}^\mu b_{r\mu} - a_{NS}. \end{aligned} \tag{12.108}$$

The physical GSO invariant states in the $NS$ sector are

$$\alpha^i_{-n} b^i_{-r} |0\rangle \,, \qquad \alpha^a_{-n} b^a_{-r} |0\rangle, \tag{12.109}$$

and similarly the states for the $R$ sector are

$$\alpha^i_{-n} d^i_{-n} |0\rangle \,, \qquad \alpha^a_{-n} d^a_{-n} |0\rangle. \tag{12.110}$$

The quantum consistency allows us to define these superstrings in a target space of $d = 10$ and the normal-ordering constants turn out to be $a_R = 0$ and $a_{NS} = (d-2)/16 = 1/2$. The lowest GSO invariant states in both the sectors turn out to be massless.

The T-duality interchanges the role of string winding number $n$ and the Kaluza–Klein mode $m$. But for open string the T-duality interchanges a DD string with an NN string. This is because the DD string can have definite winding number but not the Kaluza–Klein mode, while the NN string can have Kaluza–Klein mode expansion but not definite winding number. For DD strings we can have definite winding number since the end-points are fixed, and they cannot unwind themselves to change the winding number. But it is not possible to have nontrivial momentum modes since the $D$-branes at the end can absorb any momentum carried by the strings in the compact directions. On the other hand for an NN string definite winding number is not possible since the end points can move freely and can unwind themselves. But quantized Kaluza–Klein mode expansion is possible since no momentum can flow out of the end-points.

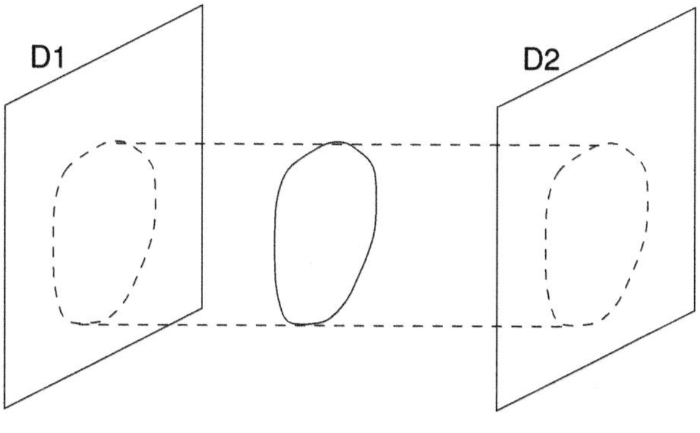


**FIGURE 12.3**
Closed string mediating interaction between two $D$-branes, D1 and D2 (see text)

It is possible to have a stack of $D$-branes, in which case non-Abelian field theories may emerge. Now there will be ND directions which have to be even numbered to have target space supersymmetry. Since GSO projection is possible only for multiples of 4 dimensions, a lower-dimensional $D$-brane could exist inside a 4-

or 8-dimensional $D$-brane. It is also possible to construct sets of parallel Dp-branes; which have $p$ space-like and one time-like dimensions. Consider two parallel Dp-branes, the separation in the transverse direction can now take any values. There are now two strings stretching between the two branes with opposite orientation, so a $U(1) \times U(1)$ gauge symmetry would emerge, each $U(1)$ residing on each of the branes. The interaction of the $D$-branes takes place through exchange of closed type IIA (B) strings (see figure 12.3). This allows a mapping of a Dp-brane with even p to a Dq-brane with odd q under T-duality, which implies that under T-duality along one circle type IIA interchanges with type IIB strings.

# 13

# *Extra Dimensions and Low-Scale Gravity*

We are familiar with the four space–time dimensions, which are noncompact. But from theoretical considerations we believe that there could be more than four space–time dimensions. The simplest possibility is to consider the extra dimensions to be compact with very small radius, so only the probes with energy close to the Planck energy ($M_{Pl} \sim 10^{19}$ GeV) could see these extra dimensions. Then the Newtonian gravity in our effective four-dimensional space–time would remain unaffected. When the dimensions higher than four are compactified, the components of the metric corresponding to the extra compact dimensions will become gauge fields in four dimensions. Any translation along the extra dimensions would then appear as gauge transformations in four dimensions. The graviton in higher dimensions would decompose into the graviton and gauge bosons in four dimensions (see section 11.2).

With the advent of the brane solutions and duality, a new scenario of low effective Planck scale opened up. According to this possibility only gravity propagates in the extra dimensions, while all other interactions are confined in the four dimensions. As a result, gravity could be very strong in the extra dimensions, although due to small overlap of the extra dimensions with our brane, we experience very weak gravitational interaction. At higher energies all of the dimensions will open up and gravity will become strong in all the dimensions. The possibility of models with low Planck scale opens up new venues for extra dimensions. Several new scenarios have been proposed since then, some of which have rich phenomenological implications, while others have some shortcomings. But all of these models have their own merits, which are worth investigating in details.

We shall now discuss a few selected topics in this new area of research. It will not be possible to include all the interesting topics and there will be bias in selecting these topics, but I hope this presentation will at least provide some feeling about these developments and motivate the readers to follow the developments in more details from the references. Some related exciting works will be reviewed in the next chapter. Since this field of research is still developing many of the results may be changed, but I hope the basic concepts that I try to emphasize here will remain useful.

## 13.1 Large Extra Dimensions

Any higher-dimensional theories must give us four space–time dimensions with the standard model interactions at energies below the electroweak symmetry breaking scale. Any Kaluza–Klein excited states, which enter the theory after compactification of the extra dimensions, should decouple from the low-energy theory. Thus, the compactification scale, determined by the inverse radius of compactification, must be much heavier than the electroweak scale. This implies that the radius of compactification should be much smaller than the nuclear size. Thus, the possibility of large extra dimension [211], with the radius of compactification of the order of mm, became so drastically different from the usual higher-dimensional theories.

The main difference between the models of large extra dimensions and the conventional higher-dimensional theories is that the large extra dimensions are blind to the all interactions except gravity. Thus, although the extra dimensions are large, the gauge interactions are not affected by their presence. Only gravitational interaction will have some effect because of these large extra dimensions, which we shall discuss later.

Another key issue for considering a large extra dimension [211] is that the fundamental scale of gravity need not be the Planck scale, $M_{Pl} \sim 10^{19}$ GeV. Gravitational interaction can become strong even at a few TeV in theories with extra dimensions. In a $(4+d)$-dimensional space, gravity becomes strong at an energy $M_*$ of about a few TeV, which is the fundamental scale in the theory. Below this scale ordinary matter fields start decoupling from the extra dimensions and get confined to only four space–time dimensions, although gravity continues to propagate along all dimensions. As a result the coupling of gravity with ordinary matter is suppressed by the large volume of the extra dimensions. Thus, the effective Planck scale $M_{Pl}$ that we see today is due to the fact that the four-dimensional effective coupling to gravity is suppressed by the volume of the extra dimensions at present energies.

### Branes

The construction of theories of large extra dimensions or low-scale gravity relies on the brane solutions in string theory (see section 12.5). The branes are static classical solutions in string theories [213]. A $p$-brane extends along $p$ spatial directions and is localized in all other directions. The $p+1$ coordinates now satisfy the Neumann boundary conditions, so no momentum can flow out at the end-points and the particles residing in this $p$-brane are explained by the $(p+1)$-dimensional gauge field theory.

If a $p$-brane is embedded in a $d$-dimensional space, the remaining $(d-p-1)$ coordinates satisfy the Dirichlet boundary condition so the end-points are frozen although momentum can leave the boundary (see section 12.5). These directions constitute the *bulk* of the entire space–time. The $p$-branes are sitting only at the boundaries (a $p+1$-dimensional wall with thickness $M_*^{-1}$). The Neumann boundary conditions would imply that no momentum can flow out of the boundaries of these walls, so the particles are confined in these branes. However, the Dirichlet condition would allow momentum flow along the boundaries of the bulk, which are absorbed by the $p$-branes at the boundary. So, particles moving in the bulk can enter or leave these $p$-branes at the boundary. Moreover, such solutions are invariant under translation in the transverse directions, so any particles residing in the $p$-branes cannot feel the extra dimensions.

As an example consider a higher-dimensional theory, in which ordinary standard model particles are residing on a 3-brane at one of the boundaries. These particles in the 3-brane are described by the usual 4-dimensional gauge field theory. Only gravity resides in the $n$-dimensional bulk, and the strength of gravitational interaction in the bulk is very high. In addition, there could be other particles also in the bulk, which do not have interactions with the standard model gauge bosons. The standard model particles can experience gravity only when the gravitons enter our 3-brane, so the gravity coupling in our 3-brane will be suppressed by the total volume of the bulk. Essentially the interaction will be determined by the probability of the gravitons to be in the $p$-brane. As a result, although the actual gravitational interaction strength in the bulk is strong, the effective gravitational strength that we feel today in our 3-brane is weak.

At higher energies, all the dimensions were treated equally by all interactions and even the particles confined in our 3-brane would have felt strong gravitational interactions. So, the scale $M_*$ at which gravity becomes strong and quantum gravity becomes effective, is the fundamental scale in the theory and can be as low as a few TeV. At this scale the string action would treat the gravity and the ordinary particles equally, which is now the string scale. Even the compactification scale has to be of the same order of magnitude, implying that all the new physics we mentioned in the context of higher dimensions would open up at this scale. The new Kaluza–Klein excited states would make the evolution of the gauge coupling constants very fast, and grand unification would also take place around this scale [214].

**Low Planck Scale**

We start with the simplest Kaluza–Klein theory in 5 dimensions, whose extra one dimension is compactified on a circle (see section 11.2). After compactification, the four-dimensional effective gravitational coupling constant $\kappa$ is related to the 5-dimensional gravitational constant $\kappa_5$ by

$$\kappa^2 = \frac{\kappa_5^2}{2\pi R}, \tag{13.1}$$

where $R$ is the radius of compactification of the fifth dimension. In a more general case, when the space is a $(4+d)$-dimensional pseudo-Euclidean space, $E^{4+d} = M_4 \times \mathscr{K}$ ($\mathscr{K}$ is a $d$-dimensional compact space), the effective four-dimensional gravitational constant is given by

$$\kappa^2 = \frac{\kappa_{4+d}^2}{V_d}, \tag{13.2}$$

where $V_d$ is the volume of the space of extra dimensions and $\kappa_{4+d}$ is the gravitational coupling constant in $E^{4+d}$. So the smallness of the effective four-dimensional gravitational constant could be due to a large volume (or the radius of compactification) of the extra dimensions.

Let us now consider a $4+d$-dimensional space, the $\mathscr{M}^4 \times \mathscr{K}$ space. The fundamental scale $M_*$ gives the gravitational potential between two test particles with masses $m_1$ and $m_2$ in $\mathscr{K}$ separated by a distance $r \ll R$,

$$V(r) \sim \frac{m_1 m_2}{M_*^{d+2}} \frac{1}{r^{d+1}}. \tag{13.3}$$

When these test particles are separated by large distance $r \gg R$, the gravitational flux cannot penetrate the extra dimensions and hence the potential would be

$$V(r) \sim \frac{m_1 m_2}{M_*^{d+2} R^d} \frac{1}{r}. \tag{13.4}$$

In our four-dimensional world, the effective potential would be

$$V(r) \sim \frac{m_1 m_2}{M_{Pl}^2} \frac{1}{r}, \tag{13.5}$$

which relates the true Planck scale $M_*$ in the higher-dimensional theories, which is the fundamental scale in the theory and the effective Planck scale $M_{Pl}$ of our 4-dimensional space $\mathscr{M}^4$ through

$$M_{Pl}^2 = M_*^{2+d} R^d. \tag{13.6}$$

Using the observed Planck scale in 4 dimensions, we can get an estimate of the radius of compactification of the extra dimensions to be

$$R \sim M_*^{-1} \left( \frac{M_{Pl}}{M_*} \right)^{2/d} \sim 10^{(32-17d)/d} \times \left( \frac{1\ TeV}{M_*} \right)^{(d+2)/2} \text{cm}. \tag{13.7}$$

We now assume that the fundamental scale is of the order of a few TeV from phenomenological considerations. If we now demand that $M_*$ is of the order of TeV, then $d = 1$ is not allowed since it requires a deviation from the Newtonian gravity at a distance $R \sim 10^{13}$ cm, which is about the distance scale of the solar system.

Since the large extra dimensions imply deviation from Newton's law, improved experiments were performed to verify Newton's law to shorter distances. The new experiments have now verified Newton's law down to a distance of 0.2 mm [215]. This means that if the fundamental scale is in the TeV range, any value of $d \geq 2$ is allowed. In fact, any value of $d \geq 2$ with TeV scale new physics will imply deviation from Newton's law at $1 - 10\ \mu$m. In all earlier theories the compactification scale was assumed to be the same as the Planck scale, but now it is possible to make $M_*$ to be around a few TeV scale, making it experimentally testable.

Once the possibility of a fundamental scale of TeV with large extra dimensions with compactified radius $R \sim 1 - 10\ \mu$m is considered, the problem comes with the ordinary particles. When the extra dimensions are compactified to a radius of $R$, it would mean that all particles in these extra dimensions would have the Kaluza–Klein excited states with mass $\sim n/R$, where $n$ is some integer number and $n = 0$ corresponds to the zero modes. All the Kaluza–Klein excited states will have the same quantum numbers as the zero mode. In the usual Kaluza–Klein compactification only the zero modes could remain light and all the excited modes have masses of the order of the Planck scale. Now we have a compactification radius $R \sim 1 - 10\ \mu$m, which corresponds to a scale of about a few eV. So, we shall get the infinite tower of Kaluza–Klein excited states corresponding to each of the particles in the extra dimensions with mass starting at eV. For gravitons this is not a problem, rather the Kaluza–Klein states of the graviton make gravity strong at high energy that may have interesting phenomenology. But for ordinary particles this is a problem. Since none of the Kaluza–Klein excited states corresponding to the ordinary quarks, leptons and gauge bosons have been observed so far, the only possibility is that the ordinary particles do not see the extra dimensions. The standard model particles and the $SU(3)_c \times SU(2)_L \times U(1)_Y$ gauge interactions should be confined to our 4-dimensional world. Although such solutions are provided by branes, let us consider here the possibility of a classical solution providing us with localized matter [216].

**Localized Matter**

Let us consider a real scalar field $\phi_c$ in 5 dimensions. There is a classical kink solution $\phi_c(y)$ depending on its fifth coordinate $y$, whose asymptotic forms are

$$\phi_c(y \to +\infty) = +v \quad \text{and} \quad \phi_c(y \to -\infty) = -v. \tag{13.8}$$

This can be achieved starting with an action where the scalar potential $V(\phi)$ has two degenerate minima at $\phi = \pm v$,

$$S_\phi = \int \mathrm{d}^4x\mathrm{d}y \left[\frac{1}{2}(\partial_A \phi_c)^2 - V(\phi_c)\right] = 0. \tag{13.9}$$

$A \equiv \{x, y\}$ are the coordinates of the five dimensions. $\phi_c$ now describes a domain wall separating two classical vacua along the y-direction, which breaks translational invariance in this direction.

The action for a fermion $\Psi$ in this model can be written as

$$S_\Psi = \int \mathrm{d}^4x\mathrm{d}y \left(i\bar{\Psi}\Gamma^A \partial_A \Psi - h\phi_c \bar{\Psi}\Psi\right). \tag{13.10}$$

$\Gamma^A \equiv \{\gamma^\mu, -i\gamma^5\}$ are five-dimensional gamma matrices. In the domain wall background the Dirac equation becomes

$$\left[i\Gamma^A \partial_A - h\phi_c(y)\right]\Psi = 0. \tag{13.11}$$

The Poincaré invariance along 4 dimensions is now unbroken and there is a zero mode of the fermion characterized by

$$\gamma^\mu p_\mu \Psi_0 = 0 \quad \text{and} \quad \gamma^5 \partial_5 \Psi_0 = h\phi_c(y)\Psi_0. \tag{13.12}$$

In four dimensions this fermion is left-handed $\gamma_5\Psi_0 = \Psi_0$ and has the form

$$\Psi_0 = \exp\left[-\int_0^y dy' h\phi_c(y')\right]\psi_L(p), \tag{13.13}$$

where $\psi_L(p)$ satisfies the 4-dimensional Weyl equation.

This fermion is massless at the domain wall (near $y = 0$) and is localized near $y = 0$. At large $|y|$ it decays exponentially, $\Psi_0 \propto \exp[-m_5|y|\ ]$. At the scalar vacuum $\phi = \pm v$, the 5-dimensional fermion acquires a mass $m_5 = hv$, and hence, there are continuum states starting with masses $m_5$ in the 5 dimensions which are not bound to the domain wall and could escape to $|y| = \infty$.

At low energy compared with $v$, the zero mode fermions are confined to 4 dimensions and they form the ordinary matter around us. At energies above $v$, the zero modes could produce the massive modes ($m_5$) of the continuum and particles could escape to the fifth dimension. This mimics the brane world, where the fermions are confined to the 3-brane and only at higher energies do the extra dimensions open up.

This construction could be generalized to higher dimensions by considering topological defects such as Abrikosov–Nielsen–Olsen vortex or 't Hooft–Polyakov monopole. In some cases the index theorem ensures the zero modes. However it is very difficult to localize the gauge fields. In one mechanism [217] a gauge theory has been considered which is in confinement phase in the bulk but not in the brane. Then the electric field of a charge residing on the brane will not penetrate the bulk. On the other hand, all states propagating in the bulk are heavy, and light particles carrying gauge charges are bound to the brane.

We have demonstrated the possibility of confining the ordinary particles in the 3-brane and allowing gravitons to propagate in the bulk. The gravity coupling in the bulk is not so weak, but in 4 dimensions it is suppressed by the volume of extra dimensions and would appear to be weak. There are all the Kaluza–Klein excited modes of the graviton in the bulk, all of which couple weakly with the ordinary matter. The long range gravitational force is due to the massless graviton exchange. The fundamental scale $M_*$ is now about a TeV, when ordinary particles could transfer momentum to the bulk. In addition to the gravitons, there could be singlet scalars residing in the bulk.

### Four-Dimensional Interactions

Let us now consider the interactions in our 4-dimensional world. For simplicity we assume that the 3-brane is located at $y = 0$. For 4-dimensional fermions $\psi_i(x)$ and scalars $\phi_a(x)$, the action will be simply

$$S_\psi = \int \mathrm{d}^4x\mathrm{d}y \mathscr{L}(\psi_i, \phi_a)\delta(y) = \int \mathrm{d}^4x \mathscr{L}\left[\psi_i(x), \phi_a(x)\right]. \tag{13.14}$$

If we now consider a singlet scalar $\chi(x,y)$ in the bulk, the Yukawa interaction of this field with the fermions in the 3-brane will be given by

$$S_{int} = \int \mathrm{d}^4x\mathrm{d}y \frac{f_{ij}}{\sqrt{M_*}} \bar{\psi}_i(x)\psi_j(x)\chi(x, y=0)\delta(y). \tag{13.15}$$

Here the Yukawa coupling $f_{ij}$ is dimensionless. If the mass dimension of a field in 4 dimensions is $D_4$, then its mass dimension in $d$ dimensions would be $D_4 + d/2$. Because of this the above 5-dimensional interaction term contains the mass parameter ($M_*$), which is the fundamental scale in the theory.

Since the extra dimensions are compact with large radius $R$, the bulk field can be expressed as the zero mode $\chi_0$ and Kaluza–Klein excited modes $\chi_n$ with masses $n/R$. In four dimensions the effective interactions with the zero modes and the Kaluza–Klein excited modes of $\chi$ will then be given by

$$S_{int} \approx \int d^4x \frac{M_*}{M_{Pl}} f_{ij} \bar{\psi}_i \psi_j \left( \chi_0 + \sqrt{2} \sum_{n=1}^{\infty} \chi_n \right). \tag{13.16}$$

Thus, any coupling between the bulk fields and the ordinary matter is suppressed by a factor $M_*/M_{Pl}$.

**Bulk Matter**

A generic features of any higher-dimensional theories is that the higher-dimensional graviton must include the 4-dimensional graviton and a 4-dimensional scalar field, the radion. The higher-dimensional gravitons would give us the zero mode of the graviton $G^0_{\mu\nu}$, the Kaluza–Klein excited modes of the graviton $G^n_{\mu\nu}$, $(n > 0)$ with masses $n/R$, and zero mode and Kaluza–Klein excited modes of a scalar field, the radion $b^n$, $(n \geq 0)$. Their couplings with ordinary matter in 4 dimensions reduce to

$$\mathscr{L} = -\frac{1}{M_{Pl}} \sum_n \left( G^n_{\mu\nu} - \frac{\kappa}{3} b^n \eta_{\mu\nu} \right) T_{\mu\nu}. \tag{13.17}$$

Here $\kappa$ is a parameter of order 1. The zero mode of the radion $b^0$ usually picks up a mass from the stabilization mechanism [218]. The interaction with the zero mode of the graviton is generally responsible for the gravitational interactions, and the Newton constant in our world remains the same, $G_N \sim M_{Pl}^{-2}$ to leading order. Due to the effects of the excited gravitons there is a small correction at short distances, which is being tested in experiments.

It is possible to introduce singlet fermions in the bulk. The free action for the massless field in 5 dimensions will contain the free action for the 4-dimensional left- and right-chiral fermions as well as the Dirac mass term, making all the Kaluza–Klein states massive with mass $n/R$. However, one can add a Majorana mass term in 5 dimensions in the action forbidding the Dirac mass term using an orbifold compactification. In this case it is possible to have chiral fermions in four dimensions. However, if gauge fields are introduced in the bulk, the Kaluza–Klein modes become massive leaving the zero mode massless in 4 dimensions. But these gauge fields should not be the gauge fields of the standard model, otherwise their coupling to ordinary matter will be highly suppressed.

Let us now summarize the salient features of the model. Ordinary particles are confined in the 3-brane at the boundary of the main higher-dimensional bulk. All interactions in the 3-brane are described by usual 4-dimensional gauge field theory. Only gravity propagates in the bulk. The gravity coupling to ordinary matter is suppressed by the volume of the extra dimensions, and we find it weak. The extra dimensions are compactified with very large radius, and the Kaluza–Klein modes of the graviton get masses of the order of eV. Near the fundamental scale of about TeV, all dimensions become equal and gravity and ordinary matter are treated similarly. All excited modes now show up, gravity becomes strong in all directions, and all matter can propagate along any directions.

## 13.2 Phenomenology of Large Extra Dimensions

The models with extra dimensions that can allow low scale strong gravity can be broadly classified into two classes, the models with large extra dimensions or those with small extra dimensions. In the large extra dimensions the standard model interactions are confined in a 3-brane and only gravity propagates in the bulk, whereas in models with small extra dimensions, the geometry is warped. Since the phenomenology of the models with large extra dimensions is comparatively easy to explain, we shall restrict ourselves to the phenomenological studies of the large extra dimensions, except for some comments on small extra dimensions.

There are many fascinating predictions of these models, but in constructing a model, many new problems creep up due to the lack of any high scales. Explanation of these new issues requires new physics from the extra dimensions. All these new ideas for model building and new phenomenological predictions, assuming that the new fundamental scale is of the order of TeV, started a new era in high energy physics phenomenology. We shall try to give a brief introduction to this vast new field.

Among the major predictions of these models with extra dimensions, the gravitational force law should change from $1/r^2$ to $1/r^4$ for distances $\sim O(10^{-4} - 1)$ mm. For any hard collisions of energy greater than the fundamental scale $E > M_*$, some of the momentum could be carried away by the bulk particles in the extra dimensions and escape from our 4-dimensional world. However, no quantum numbers could be lost, since the standard model particles reside only in our 4 dimensions. The new particles and the large extra dimensions will modify cosmological predictions. Since there are no large scales in the theory, solutions to the neutrino mass problem, strong *CP* problem, and other problems require new physics from extra dimensions.

When supersymmetry is broken by compactification at the TeV scale, the particle spectrum and the evolution of the gauge and Yukawa couplings are much different. The squarks and sleptons are naturally an order of magnitude lighter than gauginos. The sparticle spectrum now depends on only two new parameters. The Higgsino mass is automatically generated when supersymmetry is broken. All this new phenomenology promises new testable signatures in the next generation accelerators along with the signatures of new Kaluza–Klein states.

The models of extra dimensions will be most attractive if the fundamental scale turns out to be close to the TeV scale [219]. In that case all the new physics we have been mentioning could be accessible to the next generation accelerators. Since gravity will become strong at this scale, we can expect many new signals that could be beyond our imagination at present. However, so far there is no experimental evidence that supports this idea, and hence, in reality it could be possible that the fundamental scale is several orders of magnitude higher than TeV and we will not see anything in the next generation laboratory experiments. So, none of the consequences we shall be discussing can be truly considered as predictions of these models, but they are the possibilities one should look for to test if the fundamental scale is really of the order of TeV.

All of these new rich phenomenological and astrophysical consequences have made this new idea of TeV scale extra dimensions extremely attractive. If signatures of any of these predictions such as deviation from Newton's law at a distance of $\mu$m or signatures of new particles in the colliders are seen, then we have to understand physics at the TeV scale in a different way than we see now.

**Deviation from Newton's Law**

The first prediction of the models with large extra dimensions is deviation from Newtonian gravity at short distances. Until recently Newton's law was tested up to a distance of about a mm. But if gravity propagates in the bulk with strong coupling and the fundamental scale is about TeV, then for $d = 2$, a deviation of the inverse squared law of gravity was predicted at the submillimeter distances. So new experiments were performed to test gravity in the mm range. These experiments now improved the bound to a distance of about 0.2 mm.

**Missing Energy**

At low energy there are no new particles in the models with extra dimensions. But at higher energies all the Kaluza–Klein excited states of the ordinary particles will become accessible in the colliders. But even at lower energies there are some distinct signals of the models with extra dimensions.

In all models of extra dimensions, the gravitons and some scalars propagate in the bulk. In our 3-brane or the 4-dimensional world, there will be all the Kaluza–Klein excited states of the graviton with masses given by the inverse radius of the extra dimensions. All these gravitons can, in principle, interact with the ordinary particles giving rise to new signals. When these excited gravitons interact with ordinary particles, they can carry some transverse momentum of the ordinary particles. The probability of such processes increases as we approach the fundamental scale at which gravity becomes strong. So, the most distinct signals of these models with extra dimensions are to emit Kaluza–Klein gravitons into the bulk which we visualize as missing energy.

Each of the Kaluza–Klein gravitons interact with matter in our world with 4-dimensional gravitational strength. If there are $d$ extra dimensions with large radius

$R$, then $(\sqrt{s}R)^d$ Kaluza–Klein excited gravitons will be produced during a collision of center of mass energy $\sqrt{s}$. When we write the interaction of the gravitons with ordinary matter in 4 dimensions, the integration over the extra dimensions gives a volume factor suppression, which makes these couplings suppressed as $(s/M_{Pl}^2)$. The total effect will be a missing energy of the order of

$$\frac{(\sqrt{s}R)^d}{M_{Pl}^2} \sim \frac{1}{s}\left(\frac{\sqrt{s}}{M_*}\right)^{d+2}. \tag{13.18}$$

Thus, even though the coupling of every Kaluza–Klein state with matter is weak, the total emission rate of Kaluza–Klein gravitons increases due to the large number of states. The produced Kaluza–Klein gravitons will appear as missing energy and there will be processes such as

$$\begin{aligned} e^+\, e^- \ &\rightarrow \gamma + \text{missing energy} \\ q\,\bar{q} \ &\rightarrow jet + \text{missing energy} \\ \text{gluons} + \text{gluons} \ &\rightarrow \text{missing energy}. \end{aligned} \tag{13.19}$$

The cross-section for the $e^+\, e^-$ process can be estimated as

$$\sigma \sim \frac{\alpha}{M_{Pl}^2}\,(ER)^d \sim \frac{\alpha}{E^2}\left(\frac{E}{M_*}\right)^{d+2}. \tag{13.20}$$

Hence, the cross-section increases rapidly with energy, and near the fundamental scale $M_*$, it becomes comparable to the electromagnetic cross-section. Detection of these processes at LHC, ILC as well as in Tevatron could probe the fundamental scale up to several TeV. There are also the indirect signals of extra dimensional origin of some specific model, such as low-energy grand unification or existence of dileptons in the TeV range and simultaneously lepton number violation.

**Supernova**

At present the most severe constraints for the fundamental scale come from supernovae of about 30 TeV. The productions of the Kaluza–Klein gravitons will carry away energy from inside the stars and sun, which will cause fast cooling and change the evolution of the stars and sun. This gives strong bound on the fundamental scale of any models with extra dimensions.

There are other comparable constraints coming from cosmology since all the Kaluza–Klein states of the graviton will be produced at high temperature in the early universe which could change the evolution of the universe or the present composition of the universe. The evolution of the universe also gets modified in these models. We shall discuss the astrophysical constraints and cosmological issues in more detail in chapter 18.

## 13.3 TeV Scale GUTs

Since there are no large scales in the theories with large extra dimension, the concept of grand unification has to be changed. Apparently the new Kaluza–Klein excited states change the evolution of the gauge coupling constants and allow unification of the strong, the weak and the electromagnetic interactions at the TeV scale.

In conventional theories the gauge coupling constants evolve with energy following a logarithmic behavior. All the known fermions contribute to this evolution through loop diagrams. The gauge coupling constants $\alpha_3$, $\alpha_2$ and $\alpha_1$ corresponding to the groups $SU(3)_c$, $SU(2)_L$ and $U(1)_Y$ now evolve in different ways, but all three gauge couplings meet at a point. In the conventional grand unified theories the gauge coupling constants are unified at around $M_U \sim 10^{16}$ GeV, which is the unification scale. Above this energy all the interactions will be explained by one unified gauge symmetry. All fermions would then belong to some representations of this grand unified symmetry group. The fact that the quarks and leptons belong to the same representation of the group would imply that some of these new gauge bosons will mediate proton decay. Again the present limit on the proton lifetime requires the unification scale to be about the same value as predicted by the gauge coupling unification. At the unification scale the unified gauge group is broken to its low-energy subgroup by the Higgs mechanism. Some Higgs scalars belonging to the representations of the unified group acquire a *vev* at the unification scale and break the group. This Higgs has a mass and *vev* of the order of $M_U$. In the scalar potential, these heavy Higgs scalars will have mixing with the light Higgs scalar, which breaks the electroweak symmetry. The *vev* of the heavy Higgs would then induce large mass to the light Higgs, which has to be protected by means of fine tuning at all orders of perturbation theory, giving rise to the gauge hierarchy problem.

Theories with large extra dimensions do not require any large scale for grand unification. Above the fundamental scale of a few TeV, new physics takes over and grand unification is possible in these theories with extra dimensions slightly above the fundamental scale and there is no hierarchy problem [214]. This means that all the nice features of grand unification would now be accessible to the next generation accelerators. At the fundamental scale all the Kaluza–Klein excited modes start influencing the evolution of the gauge coupling constants. Above some scale $\mu_0$ the Kaluza–Klein excited states of the ordinary particles would contribute to the gauge coupling running. This would modify the evolution of the gauge coupling constants and they would start evolving much faster. The evolution of the gauge coupling will then follow a power law behavior, instead of the usual logarithmic behavior. As a result of the fast evolution of the gauge coupling constants above the TeV scale, the grand unification will be achieved within tens of TeV.

Above the fundamental scale, all the Kaluza–Klein excitations of the ordinary particles will be produced and will not be localized in the brane. So they will be able to propagate along the extra dimensions. As a result all the dimensions will open up to the standard model particles around the fundamental scale. Since supersymmetric models become nonrenormalizable at dimensions higher than four, the nice logarithmic behavior of the evolution of the gauge coupling constant will be lost. The fundamental scale $M_*$ will become the ultraviolet cut-off scale for the higher-dimensional supersymmetric standard model. So, for a consistent grand unification, the gauge coupling constants have to evolve fast enough beyond the scale $\mu_0$ by the influence of the Kaluza–Klein states, so unification can be achieved below $M_*$. As we shall see, this is indeed the case and all three gauge coupling constants meet at a point leading to grand unification just below the fundamental scale $M_*$ of tens of TeV for $\mu_0$ to be around TeV. So the gauge hierarchy problem is solved. The proton decay then becomes a problem, which we shall discuss shortly.

We demonstrate a simple case of gauge coupling unification with $d$ extra dimensions. Although this result is true for $d = 1, 2$, it may be extended to higher dimensions without loss of generality. We assume all the dimensions to have the same radius $R = \mu_0^{-1}$, but this can also be generalized. All the standard model particles will now have the Kaluza–Klein excited states with mass

$$m_n^2 = m_0^2 + \sum_{i=1}^{d} \frac{n_i^2}{R^2}. \tag{13.21}$$

Since the evolution of the gauge coupling constants will be studied from the $Z$ boson mass, the zero mode masses of the ordinary particles can be neglected in the following considerations.

Let us first consider only the gauge and Higgs boson contributions. At energies above $\mu_0$, all the Kaluza–Klein towers of particles will show up and then the states can be identified with representations of $N = 2$ supersymmetry. At each of the Kaluza–Klein mass levels $n$, the particle content of the supersymmetric standard model is augmented by an $N = 2$ supermultiplet. An $N = 2$ vector supermultiplet appears for each of the gauge bosons ($A_\mu$) and an $N = 2$ hyper-multiplet for each of the two Higgs multiplets ($H_1$ and $H_2$). Consider an $N = 2$ supersymmetry generated by two spinorial generators $Q_1$ and $Q_2$. In an $N = 2$ hyper-multiplet of the $n$th Kaluza–Klein mode, they relate the two scalars to left- and right-chiral spinors

$$H_1^n \overset{Q^1}{\leftrightarrow} \psi_L^n, \quad \text{and} \quad H_2^n \overset{Q^1}{\leftrightarrow} \psi_R^n$$

$$H_1^n \overset{Q^2}{\leftrightarrow} \psi_R^n, \quad \text{and} \quad H_2^n \overset{Q^2}{\leftrightarrow} \psi_L^n.$$

This is the minimal representation of $N = 2$ supersymmetry which is also a vectorial (contains both left- and right-chiral fermions with same quantum numbers) representation. So, if we include fermions in our simple discussion, there have to be *mirror fermions* in the theory. We assume that the brane solution is constructed in such a way that only chiral fermions come out as zero modes, and we neglect effects of Kaluza–Klein towers for the fermions. For the vector multiplet the transformations are

$$A_\mu^n \overset{Q^1}{\leftrightarrow} \lambda^n, \quad \text{and} \quad \phi^n \overset{Q^1}{\leftrightarrow} \chi^n$$

$$A_\mu^n \overset{Q^2}{\leftrightarrow} \chi^n, \quad \text{and} \quad \phi^n \overset{Q^2}{\leftrightarrow} \lambda^n.$$

For each gauge group, there will be gauge bosons corresponding to the adjoint representation of the group (see also section 11.4).

The one-loop evolution of the gauge coupling in the ordinary supersymmetric theory is given by

$$\alpha_i^{-1}(\mu) = \alpha_i^{-1}(M_Z) - \frac{b_i}{2\pi} \ln \frac{\mu}{M_Z}, \tag{13.22}$$

where the one-loop $\beta$-function coefficients in supersymmetric standard model are $\{b_1, b_2, b_3\} = \{33/5, 1, -3\}$. Beyond the scale $\mu_0$, when the Kaluza–Klein states start entering into the picture, the evolution of the gauge coupling constant is modified due to the additional $N = 2$ supersymmetric states and is given by

$$\alpha_i^{-1}(\Lambda) = \alpha_i^{-1}(\mu_0) - \frac{b_i - \tilde{b}_i}{2\pi} \ln \frac{\Lambda}{\mu_0} - \frac{\tilde{b}_i X_d}{2\pi d} \left[ \left( \frac{\Lambda}{\mu_0} \right)^d - 1 \right], \tag{13.23}$$

where $\tilde{b}_i$ are the new beta function and $X_d$ is a numerical factor, essentially coming from the phase space factor in $d$-dimensions which comes from the $d$ dimensional integrations and is given by

$$X_d = \frac{2\pi^{d/2}}{d\Gamma(d/2)}. \tag{13.24}$$

$\Gamma$ is the Euler gamma function ($X_0 = 1$, $X_1 = 2$, $X_3 = 4\pi/3$ and so on). We also assume that below $\mu_0$, there are no Kaluza–Klein states and the couplings evolve logarithmically, while immediately above $\mu_0$ all the Kaluza–Klein states contribute. Then we match the two solutions at $\mu = \mu_0$ to obtain

$$\alpha_i^{-1}(\Lambda) = \alpha_i^{-1}(M_Z) - \frac{b_i}{2\pi} \ln \frac{\Lambda}{M_Z} + \frac{\tilde{b}_i}{2\pi} \ln \frac{\Lambda}{\mu_0} - \frac{\tilde{b}_i X_d}{2\pi d} \left[ \left( \frac{\Lambda}{\mu_0} \right)^d - 1 \right]. \tag{13.25}$$

This power law behavior now makes the evolution very fast above the scale $\mu_0$ allowing unification below the fundamental scale $M_*$. In other words, the consistency of unification still allows a very low fundamental scale of about tens of TeV with $\mu_0$ of the order of TeV (as shown in figure 13.1) [214].

It is not obvious that by including the Kaluza–Klein states the theory would automatically guarantee unification. Considering that single stage unification does not work in the case of $SU(5)$ grand unified theory, it is very interesting to find that in case of extra dimensions even a single stage unification is possible. The unification coupling constant comes out to be smaller than the unification coupling constant in ordinary grand unified theory, and hence, perturbative calculation is still valid in spite of including the contributions due to all the Kaluza–Klein particles.

**Fermion Masses and Proton Decay**

The low scale grand unification in models with extra dimensions is highly attractive, but it introduces a few problems as well. Two imminent questions are the fermion mass hierarchy and the proton decay. In grand unified theories the fermion masses

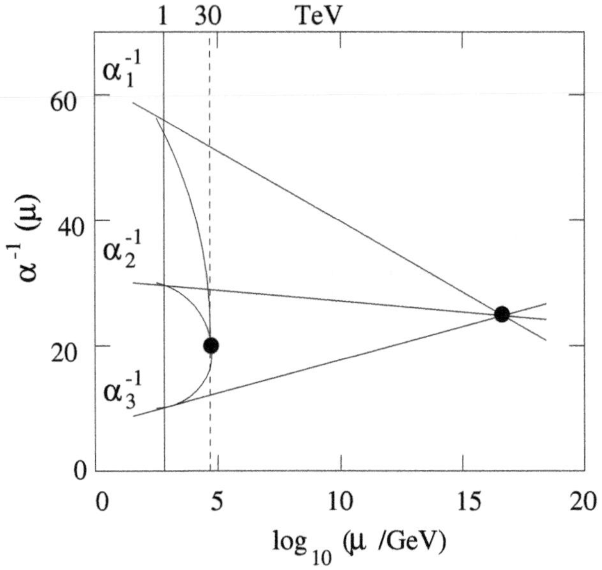


**FIGURE 13.1**
Unification of coupling constants in models with large extra dimensions at a TeV. The excited modes make the coupling constants evolve very quickly above 1 TeV so they get unified at say, 30 TeV.

originate from the Yukawa interactions of the fermions belonging to the representations of the grand unification group. Thus, all components of any multiplet get similar contributions at the grand unification scale. As they evolve down to low energy, the Yukawa couplings are scaled down following the renormalization group equations. The large unification scale again helps to get the large fermion mass hierarchy, since the Yukawa couplings evolve linearly with $\log \mu$. Because there is no large scale in theories with extra dimensions, we have to find a new solution to this problem using some features of extra dimensions. It has been found that similar to the gauge coupling evolution, even in this case the Yukawa couplings evolve exponentially instead of linearly with $\log \mu$. Thus, the power law behavior allows a large fermion mass hierarchy even with TeV scale unification [214]. In case of proton decay, in the conventional grand unified theories the large unification scale gives strong suppression, which is not possible now. So, new mechanism to solve the proton decay problem has to be invoked.

Within the usual minimal supersymmetric standard model the Yukawa couplings ($Y_F$) are related to the mass of the corresponding fermion ($m_F$) by

$$m_F = y_F \times v_i, \tag{13.26}$$

where $v_{1,2}$ are the *vevs* of the neutral components of the Higgs scalars $H_{1,2}$, where

$H_1$ couples to the down quark and the leptons and $H_2$ couples to the up quark. These *vevs* are related to the parameters $v = \sqrt{v_1^2 + v_2^2} \approx 174$ GeV and $\tan\beta = v_2/v_1$. These couplings evolve with energy as

$$\frac{d}{d\ln\mu}\alpha_F^{-1}(\mu) = -\frac{b_F(\mu)}{2\pi}, \tag{13.27}$$

where we define $\alpha_F = y_F^2/4\pi$ and the one-loop beta function coefficients $b_F(\mu)$ now depend on the gauge coupling as well as on the Yukawa coupling $b_F(\mu) \equiv b_F(\alpha_i, \alpha_F)$ and they change with energy.

In the presence of the Kaluza–Klein excited states, the Yukawa coupling constants evolve according to the general power law form

$$\alpha_F^{-1}(\Lambda) = \alpha_F^{-1}(\mu_0) - \frac{c_F}{2\pi}\frac{X_{\Delta_F}}{\Delta_F}\left[\left(\frac{\Lambda}{\mu_0}\right)^{\Delta_F} - 1\right], \tag{13.28}$$

where $\Delta_F = (n_F + 1)d$, $R \sim \mu_0^{-1}$ is the radius of the extra dimensions, $c_F \sim \Lambda^{2n_F}(1 + ...)$ is a dimensionful coefficient, and $X_{\Delta F} \sim 2\pi^{\Delta_F/2}/\Delta_F\Gamma(\Delta_F/2)$.

The Yukawa coupling constants now evolve from their low-energy value logarithmically until the scale $\mu_0$. From the scale $\mu_0$, all the Kaluza–Klein states start contributing to the loop diagrams controlling the evolution of the Yukawa couplings. All of the $N = 2$ supersymmetric Kaluza–Klein excited states will then run the coupling constants too fast, so soon all the Yukawa couplings approach a simultaneous Landau pole at which $\alpha_F^{-1} \to 0$ (see figure 13.2). All the Yukawa couplings approach the Landau pole in a flavor-dependent manner, and on the way they get unified. If we now assume that they get unified before the Landau pole and then evolve them downwards starting from this unification point, the low-energy physical Yukawa couplings would be given by

$$\alpha_F^{-1}(\mu_0) = \frac{X_{\Delta_F}}{2\pi\Delta_F}\left(\frac{\Lambda}{\mu_0}\right)^{2n_F}\left[\left(\frac{\Lambda}{\mu_0}\right)^{\Delta_F} - 1\right] \approx \frac{X_{\Delta_F}}{2\pi\Delta_F}\left(\frac{\Lambda}{\mu_0}\right)^{\Delta_F+2n_F}. \tag{13.29}$$

Thus, even for $M_* \sim 10\mu_0$ we can achieve Yukawa coupling unification, starting with a large fermion mass hierarchy through the power law behavior. However, this does not explain why the fermion mass hierarchy exists at low scale.

There are a few suggestions made to solve this problem. These suggestions require new mechanisms from extra dimensions. In the *distant breaking* mechanism to solve the fermion mass hierarchy [220] (we shall discuss this mechanism in detail later in the context of neutrinos), one introduces flavor symmetry in our brane to distinguish the different generations. This symmetry is conserved in our brane, but it is badly broken in another brane at a large distance. There is a bulk scalar field, which transforms nontrivially under the flavor symmetry. Symmetry breaking in another brane acts as a source to this bulk scalar, and it carries this symmetry breaking information to our brane. When this bulk scalar interacts with the fermions in our world, the shinned value of the bulk scalar mediates this symmetry breaking very

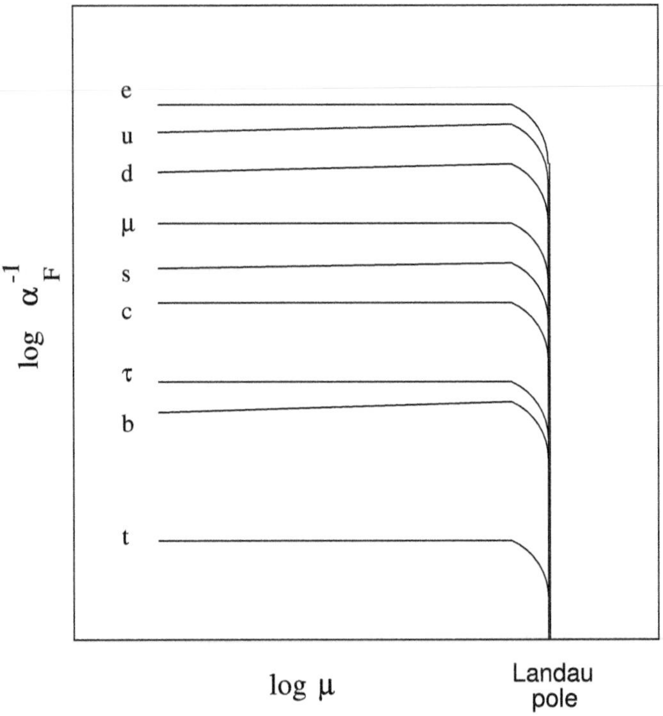


**FIGURE 13.2**
Yukawa coupling evolution toward the Landau pole in models with large extra dimensions.

weakly. The effective amount of symmetry breaking in our world will depend on the profile of the bulk scalar and the separation between the two branes. There could be several distant branes breaking the symmetry by differing amounts, giving rise to hierarchical fermion masses.

### Thick Wall Scenario

There is another interesting mechanism, which originates from new inputs in the extra dimensions and assumes that our 3-brane has a width [221]. The thickness of the brane may be considered to be around $\Delta y = (\text{TeV})^{-1}$. If the 3-brane in which all matter is localized has this finite thickness, then it is possible that different fermions are localized in different points in the higher dimensions within this thick wall. The Higgs fields and the gauge fields are uniformly spread over the entire thick wall, but the fermions are localized with an exponential profile.

Let us consider a 5-dimensional model in which the brane thickness is $\Delta y$ along the $y$-direction. Within the thickness $\Delta y$ let us consider two points $y_1$ and $y_2$ and assume that two 4-dimensional fermion fields $\psi_1(y = y_1)$ and $\psi_2(y = y_2)$ are localized in two different points in the fifth direction. The gauge and the Higgs fields are free to move anywhere within the thickness $\Delta y$, but these fermions are not. We assume that the fermions have a Gaussian profile around the point where they are confined (see figure 13.3). If a particle ($\psi$) is confined at a point $y = 0$, the Gaussian profile would be given by

$$f_\psi(y) = \frac{\mu^{1/2}}{(\pi/2)^{1/4}} \exp^{-\mu^2 y^2}, \tag{13.30}$$

where $\mu$ is some constant, which depends on the scalar field vacuum. So, at a distance r, the wave function falls off exponentially.

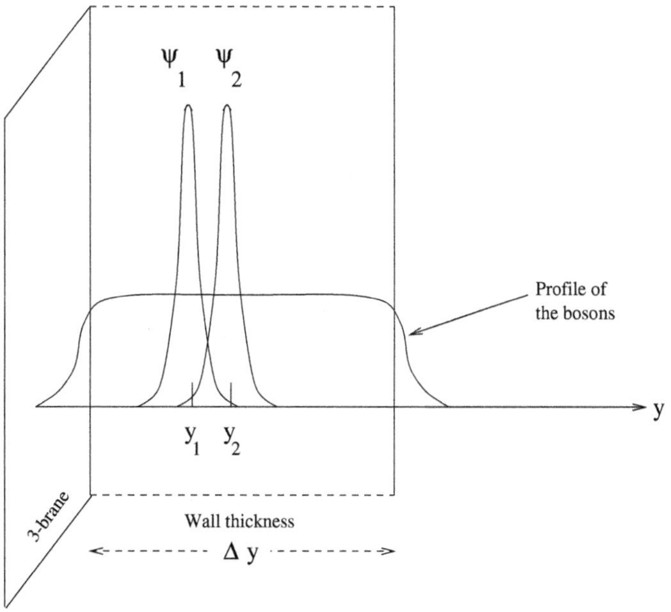


**FIGURE 13.3**
Bosons have a constant profile in a thick wall scenario with thickness $\Delta y$, while the fermions $\psi_1$ and $\psi_2$ have sharp Gaussian profiles at the points $y_1$ and $y_2$.

Now consider an interaction of the two fermions through a 4-dimensional Higgs scalar $\phi$. The Yukawa coupling of the two fermions with the Higgs will be given by

$$S = \int d^4x \kappa \phi(x) \psi_1(x) \psi_2(x) \int \mathrm{d}y f_{\psi_1}(y) f_{\psi_2}(y). \tag{13.31}$$

The integration in the $y$ coordinate gives the overlap of the two fermions at any point $y$, which will depend on the separation of the two points $y_1 = 0$ (say) and $y_2 = r$, at which the two fermions are localized. This integral will give an exponential suppression

$$\int \mathrm{d}y f_{\psi_1}(y) f_{\psi_2}(y) = \exp^{-\mu^2 r^2}. \tag{13.32}$$

This is a generic feature of this mechanism of localizing fermions at different points. For $\mu r \sim 1$ this gives no suppression, which is required for the top mass, while for $\mu r \sim 5$ this gives a suppression of about $10^{-6}$, which can explain the electron mass.

The low-scale grand unification eliminates the gauge hierarchy problem, since the ratio of the Planck scale (or the grand unification scale) to the electroweak scale is only about 2 to 3 orders of magnitude. This will apparently imply that there is no need for supersymmetry. However, the original theory is most likely the superstring theory for consistency with quantum gravity and the brane solution, which is supersymmetric. Because of this most of the analysis is done in the context of the supersymmetric standard model.

**Proton Decay**

We now turn to the question of proton decay. In the ordinary grand unified theories the scale of unification is required to be above about $10^{16}$ GeV to explain the present experimental limit on the proton lifetime. In theories with extra dimensions the required unified value of the gauge coupling constant is slightly lower. Since the gauge coupling enters in the proton lifetime as fourth power, this would mean that the proton lifetime would give a constraint of $10^{14}$ GeV. So, a unification scale of about TeV is too low to explain the proton decay.

A solution to this problem should come from some new mechanisms in the extra dimensions. There are several suggestions made to solve this problem. In one of them the higher-dimensional theory is compactified on a $\mathscr{Z}_2$ orbifold. The higher-dimensional fields $\Phi(x)$ are then decomposed into even and odd functions of these extra coordinates $\Phi_\pm(x)$ [214]. The grand unified scale is close to the fundamental scale, but the scale of compactification $\mu_0$ is one order of magnitude smaller. So, we assume that at the time of compactification we can classify the particles belonging to the minimal supersymmetric standard model as $\Psi_{MSSM}$ and all the new particles coming from the grand unification can be represented by $\Psi_{GUT}$. It is then clear that all the new physics including proton decay are caused by the particles belonging to $\Psi_{GUT}$.

We now assume that the particles $\Psi_{GUT}$ are odd with respect to the compactified coordinates $y_i$, so after compactification we do not see them at low energy. They become the bulk particles and their interactions with the ordinary matter is suppressed by the effective Planck scale $M_{Pl}$. The usual minimal supersymmetric standard model particles $\Psi_{MSSM}$ are even with respect to the $y_i$ coordinates, and the zero modes of these particles remain massless at the compactification scale. These are all the particles present in our 3-brane at low energy. They pick up mass at the time of electroweak symmetry breaking. This mechanism also ensures the doublet–triplet splitting mechanism, since the triplet now belongs to $\Psi_{GUT}$ and becomes heavy naturally, while the doublets belong to $\Psi_{MSSM}$ and remain light.

Another interesting proposition to suppress proton decay comes from the idea of delocalizing the fermions [221]. As we discussed while explaining the fermion mass hierarchy, this mechanism localizes different fermions at different points in the thick wall of our brane. The gauge and Higgs bosons can propagate anywhere in this thick wall, which is again confined to a point in the fifth dimension.

Let us now consider the operator for proton decay. In any proton decay operator baryon number is violated by 1, which requires presence of three quarks. For Lorentz invariance even numbers of fermions have to be present, so the simplest operator is a four-fermion operator with three quarks and one lepton. Let us denote it by $QQQL$, where these fields are all 5-dimensional (the corresponding operator in 4 dimensions will be $qqq\ell$). Now assume that all the quarks are localized at $y = 0$ and all leptons are localized at $y = r$. Then the proton decay operator will be

$$S \sim \frac{1}{M^3} \int d^5x(QQQL). \tag{13.33}$$

In 4 dimensions, in terms of the zero modes quark and lepton fields, this will become

$$S \sim \frac{1}{M^2} \int \mathrm{d}^4x\lambda\ (qqq\ell), \tag{13.34}$$

where $\lambda \sim \exp[-3\mu^2 r^2/4]$ comes from the overlap of the wave functions of quarks and leptons along the $y$ direction. For a separation of $\mu r \sim 10$ we obtain $\lambda \sim 10^{-33}$ which can safely explain the present limit on the proton lifetime in a theory with TeV scale grand unification.

In the earlier explanation, it was assumed that all the new particles responsible for proton decay have to be heavier than the compactification scale and they become bulk fields. So, the leptoquarks and diquarks will both reside in the bulk. But in this case, localizing quarks and leptons in different points in extra dimensions implies that diquarks and leptoquarks both can be there at low energy residing in our 3-brane. But the leptoquark coupling to quarks and leptons will have an exponential wave function overlap suppression. So, in this scenario it may be possible to see diquarks in the next generation colliders, but leptoquarks should not be observed.

## 13.4 Neutrino Masses and Strong *CP* Problem

The neutrino masses are the only evidence we have about physics beyond the standard model. This tiny mass is naturally explained by introducing a large lepton-number violating scale. On the other hand the strong *CP* problem can be dynamically explained by introducing a new Peccei–Quinn symmetry at some high scale. Since there is no large scale in models of extra dimensions, these two problems have to be explained with some new physics. We shall now discuss some of these proposals.

### Neutrino Masses

In the standard model neutrinos are massless, but recent experiments confirm a nonzero mass of the neutrinos. This requires new physics beyond the standard model.

One simple way to understand the smallness of the neutrino mass is to write an effective operator for the neutrino masses in terms of the standard model particles. Since there are no right-handed neutrinos in the standard model, one can only write an operator for a Majorana mass. There is only one such effective operator, which has dimension-5 and is given by [54]

$$\mathscr{L}_{Maj} = \frac{f_{eff}}{M}(\nu\phi^\circ - e\phi^+)^2, \tag{13.35}$$

where $M$ is some large mass indicating the scale of lepton number violation and $f_{eff}$ is the effective coupling constant, which is determined by the details of the model.

This effective operator could be realized in different extensions of the standard model. The most popular ones are to extend the standard model with right-handed neutrinos $N_R$ or triplet Higgs scalars $\xi$. In models with right-handed neutrinos, there is a Dirac mass term coming from the *vev* of the usual Higgs doublet and there is a large Majorana mass term of the $N_R$. The smallness of the Majorana mass is naturally explained by considering the lepton-number violating scale to be of the order of $10^{10}$ GeV. In the triplet Higgs models the scalar potential contains a lepton-number violating coupling of $\xi$ with a large mass scale of about $10^{10}$ GeV that induces a small *vev* to $\xi$, which in turn gives the tiny Majorana mass to the neutrinos naturally.

In models with extra dimensions there is no scale above the fundamental scale of the theory, which is only tens of TeV. In the above operator, if we take the *vev* of $\phi^\circ$ to be around 100 GeV, then for $M \sim 10$ TeV we require an effective coupling constant of $f_{eff} \sim 10^{-10}$, which is highly unnatural.

**Bulk Singlet**

In models with large extra dimensions it is possible to have a tiny Dirac neutrino mass exploiting the special features of the extra dimensions. In addition to the standard model particles in our 3-brane and graviton in bulk, we now introduce right-handed neutrinos in the bulk, which are singlet fermions [222, 223]. Since the bulk field interacts with particles in the 3-brane very weakly (suppressed by the volume of the extra dimensions), we can write a Dirac mass of the neutrinos, which is very small. It is also possible to generate a Majorana mass but the mass will be highly suppressed in that case.

As an example consider a 5-dimensional theory with a right-handed neutrino $N_R$ in the bulk, which is a singlet under all gauge interactions. Writing the $8 \times 8$ gamma matrices of the 5 dimensions in the Weyl basis as

$$\Gamma^\mu = \begin{pmatrix} 0 & \sigma^\mu \\ \bar{\sigma}^\mu & 0 \end{pmatrix} \quad \text{and} \quad \Gamma^5 = \begin{pmatrix} i & 0 \\ 0 & -i \end{pmatrix}, \tag{13.36}$$

where $\sigma^\mu = \{\gamma^\circ, \gamma^i\}$ and $\bar{\sigma}^\mu = \{\gamma^\circ, -\gamma^i\}$, we can decompose the Dirac spinor as

$$\Psi \equiv \begin{pmatrix} N_R \\ N^c{}_R \end{pmatrix}. \tag{13.37}$$

In 5 dimensions we can write the Dirac mass as $\bar{\Psi}\Psi$ and the Majorana masses as $\Psi^T C_5 \Psi$ with the five-dimensional charge conjugation matrix $C_5 = \gamma^0\gamma^2\gamma^5$.

To get the interactions of this bulk fermion we expand its components using Fourier transformation as

$$N_R(x,y) = \sum_n \frac{1}{\sqrt{2\pi R}} N_{Rn}(x) \exp^{iny/R}, \tag{13.38}$$

where $N_{Rn}$ are the Kaluza–Klein excited states of the right-handed neutrinos with quantized momentum. Integrating over the extra dimensions the four-dimensional action now contains the kinetic energy terms for the four-dimensional fields and an interaction term of the fields $N_R$, $\ell_L$ and $\phi$. This Yukawa interaction or the mass term for the neutrinos is given by

$$\begin{aligned} S_{int} &= \int d^4x \, \frac{\kappa}{\sqrt{M_*}} \, \bar{\ell}(x) N_R(x, y=0) \phi^\dagger(x) \\ &= \int d^4x \, \frac{\kappa v}{\sqrt{2\pi M_* R}} \, \bar{\nu}_L(x) \sum_n N_{Rn}(x), \end{aligned} \tag{13.39}$$

where $v = \langle \phi^0 \rangle$ and $\kappa$ is a dimensionless coefficient. For the lowest Kaluza–Klein state of the bulk fermion ($n = 0$), the Dirac mass term becomes

$$m_{\nu 0}^D = \frac{\kappa v}{\sqrt{2\pi M_* R}}. \tag{13.40}$$

So, for $m_{\nu 0}^D \ll 1/R$ the Kaluza–Klein modes with $n \neq 0$ will be irrelevant at very low energies and the 4-dimensional theory contains a Dirac neutrino with mass $m_{\nu 0}^D$ and Kaluza–Klein states.

In $d$ extra dimensions this expression for the neutrino mass generalizes to

$$m_{\nu 0}^D = \frac{\kappa v}{\sqrt{V_d M_*^d}} = \frac{\kappa v M_*}{M_{Pl}}. \tag{13.41}$$

$V_d = M_{Pl}^2 / M_*^{d+2}$ is the volume of the $d$ extra dimensions. Keeping the effects of the Kaluza–Klein states, the mass matrix now generalizes to a form, which can be written in the basis $[\nu_L, N_{1R}, N_{2R}, \ldots]$ as

$$\mathcal{M}_\nu = \begin{pmatrix} m_{\nu 0}^D & 0 & 0 & . & . & . \\ m_{\nu 0}^D & 1/R & 0 & . & . & . \\ m_{\nu 0}^D & 0 & 2/R & . & . & . \\ . & . & . & . & & \\ . & . & . & . & . & \\ . & . & . & . & & . \end{pmatrix}. \tag{13.42}$$

The lowest eigenvalue now gets corrected by a factor and the corrected neutrino mass becomes

$$m_\nu^D = m_{\nu 0}^D \left[ 1 - \sum_n \left( \frac{m_\nu^D R}{n} \right)^2 \right]^{1/2}. \tag{13.43}$$

This sum is well behaved for $d > 2$, which is also required for phenomenological considerations.

Thus, making the singlet sterile neutrino to be a bulk fermion, we could suppress the neutrino Dirac mass. For $M_* \sim 1$ TeV, the neutrino mass comes out to be $m_\nu^D = 10^{-4}\ \kappa$ eV, which is too low. A trivial solution is to make $M_* \sim 1000$ TeV and keep $\kappa \sim 1$ for the heaviest neutrino. In this scenario it is possible to include a lepton-number violating Majorana mass term for the right-handed neutrinos in 5 dimensions. But this will give a light Majorana neutrino with mass of the order of $10^{-11}$ eV.

The phenomenology of the right-handed neutrino in the bulk has been extensively studied [223]. Since the right-handed neutrinos are bulk matter, they will also have all the Kaluza–Klein excited states such as the graviton. There will be an infinite tower of states starting at a scale of about eV. Since they interact with ordinary matter very weakly, this is consistent with all present constraints. However, the existence of the Kaluza–Klein states gives rise to some new features of the model, which are mixing of neutrinos of different flavors with the Kaluza–Klein states. They induce neutrino oscillations which are distinct and future experiments may distinguish them from the usual 4-dimensional mechanisms of neutrino mass.

There is another suggestion to adjust the suppression factor without increasing the fundamental scale or the 5-dimensional Yukawa coupling constant $\kappa$. Here one assumes that the right-handed neutrino does not reside on the entire $d$-dimensional bulk space. Instead it can move on an $m$-dimensional subspace. This can arise if our brane is situated at the intersection of two or more branes, of which at least one brane has $m+3$ spatial dimensions. Without going into the details of how such a scenario may arise, let us consider the effect of such construction. Now the Kaluza–Klein mode expansion of this field is

$$N_R(x,y) = \frac{1}{\sqrt{V_m}} \sum_{\vec{l}} N_{R\vec{l}}(x) \exp\left[\frac{-2\pi i\, \vec{l}\cdot\vec{y}}{(V_m)^{1/m}}\right]. \tag{13.44}$$

Using this Fourier expansion of the field and integrating over the extra dimensions we get the Dirac neutrino mass to be

$$m_\nu^D = kv\left[1 - \sum_n \left(\frac{m_\nu^D R}{n}\right)^2\right]^{1/2} \left(\frac{M_*}{M_{Pl}}\right)^{m/d}. \tag{13.45}$$

In this case the suppression factor is $\sqrt{V_m M_*^m} \sim \sqrt{(RM_*)^m}$. Using the relation $M_{Pl}^2 = R^d M_*^{d+2}$, the suppression factor becomes $(M_*/M_{Pl})^{m/d}$. Thus, with suitable choice of $m$ and $d$, it is possible to get the required neutrino masses.

### Triplet Higgs

In models with extra dimensions it is possible to get naturally small neutrino masses with triplet Higgs scalars in the distant breaking mechanism, which has some interesting phenomenological consequences [222, 224]. In this scenario lepton number

$L$ is assumed to be conserved in our brane at $\mathscr{P}(y=0)$, but broken spontaneously by the *vev* of a scalar $\eta$ $(L=2)$ in another brane $\mathscr{P}'(y=y_*)$ at a distance $r=|y_*|$ away from our brane in the extra dimension. This field $\eta$ couples to a bulk scalar $\chi(x,y)$ $(L=-2)$, through the interaction [224]

$$\mathscr{S}_{other} = \int_{\mathscr{P}'} \mathrm{d}^4x' \, \mu^2 \, \eta(x')\chi(x', y=y_*), \tag{13.46}$$

where $\mu$ is a mass parameter. This will induce an effective *vev* to $\chi$ breaking lepton number in the bulk, which in turn induces a tiny lepton number violation in our brane through $\langle\chi\rangle$:

$$\langle\chi(x,y=0)\rangle = \Delta_d(r)\langle\eta(x,y=y_*)\rangle, \tag{13.47}$$

where $\langle\eta\rangle$ acts as a point source and $\Delta_d(r)$ is the Yukawa potential in $d$ transverse dimensions. For $m_\chi r \ll 1$, and $d>2$, the Yukawa potential takes an interesting form so the asymptotic form of the profile of $\chi$ becomes

$$\langle\chi\rangle \approx \frac{\Gamma(\frac{d-2}{2})}{4\pi^{\frac{d}{2}}} \frac{M_*}{(M_* r)^{d-2}}, \tag{13.48}$$

which is the amount of lepton number violation in our world and is suitably small for large $r$. Let us now introduce a triplet Higgs scalar [224] in our brane $\xi$, whose quantum numbers are defined by its interaction

$$S_\xi = \int \mathrm{d}^4x f_{ij} \xi(x)\ell_i(x)\ell_j(x), \tag{13.49}$$

then the shinned value of $\langle\chi\rangle$ will induce explicit lepton number violation in our brane

$$\begin{aligned}\mathscr{S}_\chi &= h \int_{\mathscr{P}} \mathrm{d}^4x \, \xi^\dagger(x)\phi(x)\phi(x)\chi(x,y=0) \\ &= h\,\langle\chi\rangle \int_{\mathscr{P}} \mathrm{d}^4x \, \xi^\dagger(x)\phi(x)\phi(x).\end{aligned} \tag{13.50}$$

This lepton number violation will then give a neutrino mass

$$(\mathscr{M}_\nu)_{ij} \approx \frac{\Gamma(\frac{d-2}{2})}{2\pi^{\frac{d}{2}}} \, h \, f_{ij} \, \frac{\langle\phi\rangle^2 M_*}{m_\xi^2} \left(\frac{M_*}{M_{Pl}}\right)^{(2-4/d)}. \tag{13.51}$$

For $d=3$, $M_* \sim 2$ TeV, $m_\xi \lesssim 1$ TeV and $h \sim f_{ij} \sim 0.5$, we get $(\mathscr{M}_\nu)_{33} \sim 0.24$ eV, which is of the right magnitude.

Explanation of the present experiments on neutrino mass requires that the triplet Higgs mass has to be around TeV. Thus, this model predicts a triplet Higgs scalar in the detectable range [224]. Since the same-sign dilepton signals of the triplet Higgs are very clean, they should be observed in LHC or Tevatron or ILC if this is the true mechanism of neutrino mass. Again the branching fractions of $\xi^{++}$ into same-sign charged leptons $l_i^+ l_j^+$ determine directly the Yukawa coupling $|f_{ij}|$, which in turn gives us the elements of the neutrino mass matrix modulo an overall scale factor. This is a unique feature of this model. If this happens to be the model of neutrino mass, then we can get all the information about the neutrino mass from collider experiments.

There are a few other suggestions to explain the smallness of neutrino masses without invoking a large lepton-number violating scale [222, 225]. While these models have some interesting features, they do not involve any new mechanisms of the extra dimensions. The model with bulk singlet right-handed neutrino can be implemented in theories with warped compactification (small extra dimensions) [226]. Although it has some differences in its construction, the basic idea is similar to that of a bulk singlet mechanism of the large extra dimensions. However, the model has no new predictions at low energy. It is difficult to implement the other models of neutrino masses in models with small extra dimensions.

### Strong *CP* Problem

The solution to the $U(1)_A$ problem through breaking of the chiral symmetry introduced *CP* violation in the QCD Lagrangian. The coefficient $\Theta$ in the *CP* violating term

$$\mathscr{L}_{\rm eff} = \mathscr{L}_{\rm QCD} + \bar{\Theta}\frac{g^2}{32\pi^2}F_a^{\mu\nu}\tilde{F}_{\mu\nu a} \qquad (13.52)$$

is experimentally constrained to be very small. To explain why this number is so small naturally, a dynamical mechanism has been proposed which predicts a new light particle, the axion.

The mass of the axion $m_a$ is constrained to be very low from astrophysical and cosmological considerations. The mass is given by the inverse of its decay constant $f_a$. In conventional theories the decay constant turns out to be the scale $v_a$ at which the Peccei–Quinn symmetry is broken and the bound on the axion mass implies $10^9$ GeV $< f_a < 10^{12}$ GeV. It is a typical characteristics of any pseudo-Nambu–Goldstone boson that its coupling is determined by the *vevs* of the Higgs scalars in the theory and there is no extra parameter which can be tuned to suppress its coupling to matter. A large scale is required so its coupling to matter can be suppressed.

In theories with extra dimensions there is no large scale in our world. So, the global $U(1)$ Peccei–Quinn symmetry cannot be broken at some high scale. If the $U(1)$ symmetry is broken at a scale below the fundamental scale, the axion coupling to matter would be very strong which is ruled out. Then the natural possibility would be to suppress the axion coupling to matter with some mechanisms involving extra dimensions.

The obvious choice to suppress the axion coupling to matter would be to make the axions a bulk field [227]. Then similar to the suppression of the coupling of matter to gravity, the axion coupling will be suppressed by the ratio of the fundamental scale to the Planck scale in 5 dimensions $(M_*/M_{Pl})$. If the symmetry breaking scale is around $M_*$, the decay constant would be $\sim O(M_{Pl})$. This suppression of the coupling constant can be viewed as effectively raising the scale of $v_a$ to the Planck scale, which is again very large. The axion coupling will then be too small and fail to satisfy the cosmological constraint. Following the discussions for neutrino mass, the natural choice is then to consider the axion to be a bulk field only along some ($m$-dimensions) of the directions and not all ($d$-dimensions) the extra dimensions, so the coupling to matter has a suppression $(M_*/M_{Pl}^{m/d})$. This will make the axion coupling to be the right amount.

Let us now discuss how a higher-dimensional Peccei–Quinn mechanism works. Consider an example of a 5-dimensional space in which a complex scalar field $\phi$ transforms nontrivially under the $U(1)_{PQ}$ Peccei–Quinn symmetry. The *vev* of the scalar breaks this symmetry at around the fundamental scale $\langle\phi\rangle = v_a = \hat{f}_a/\sqrt{2} \sim M_*$. The Nambu–Goldstone boson, the axion $a$, corresponding to the $U(1)_{PQ}$ symmetry breaking would translate under the $U(1)_{PQ}$ symmetry making it massless. We now assume that our 3-brane is located at $y = 0$ and the field $a$ interacts with the gluons in our brane at the boundary, which gives a small mass to the axion. The effective action of the 5-dimensional axion is given by

$$\mathscr{S}_{\text{eff}} = \int d^4x\, dy \left[ \frac{1}{2} M_* \, \partial_A a \partial^A a + \frac{\xi}{\hat{f}_a} \frac{g^2}{32\pi^2} a F_a^{\mu\nu} \tilde{F}_{\mu\nu a} \, \delta(y) \right] , \tag{13.53}$$

where $\xi$ is a model dependent parameter.

We now Fourier expand the axion field assuming an orbifold compactification on a $Z_2$ under which the axion is assumed to be symmetric. In terms of the Kaluza–Klein modes the axion decomposes as

$$a(x,y) = \frac{1}{\sqrt{V}} \sum_{n=0}^{\infty} a_n(x) \cos\left(\frac{ny}{R}\right), \tag{13.54}$$

where $V = 2\pi R$ is the volume of the extra dimension. Now only the zero mode translates under the $U(1)_{PQ}$ symmetry and remains massless in the absence of the anomaly term, but the other Kaluza–Klein modes become massive. This can be seen from the effective 4-dimensional effective action

$$\begin{aligned} \mathscr{L}_{\text{eff}} = & \frac{1}{2} \sum_{n=0}^{\infty} (\partial_\mu a_n)^2 - \frac{1}{2} \sum_{n=1}^{\infty} \frac{n^2}{R^2} a_n^2 \\ & + \frac{\xi}{f_a} \frac{g^2}{32\pi^2} \left( a_0 + \sum_{n=1}^{\infty} \sqrt{2} a_n \right) F_a^{\mu\nu} \tilde{F}_{\mu\nu a} , \end{aligned} \tag{13.55}$$

where we defined the new effective decay constant as $f_a = (VM_*)^{1/2} \hat{f}_a$. For $d$ extra dimensions this becomes

$$f_a = \sqrt{V_d M_*^d} \hat{f}_a = \frac{M_{Pl}}{M_*} \hat{f}_a. \tag{13.56}$$

The Kaluza–Klein excited modes will have masses proportional to the inverse of the radius of compactification $m_{an} \sim n/R$ and only the zero mode becomes the true axion, which picks up a tiny effective mass from the anomaly term, proportional to the inverse of the decay constant. Since $f_a \sim M_{Pl}$ is very large, an approximate form of the axion can be given by

$$m_a = \frac{\sqrt{2} f_\pi m_\pi}{f_a} \sim \frac{(0.3 \text{ GeV})^2}{f_a}. \tag{13.57}$$

This expression is strictly valid in the limit $R \to 0$. For large $R$, the effect of the Kaluza–Klein modes is felt by the zero mode by decreasing its effective mass. For very large $R$, the mass of the zero mode becomes $m_a \sim R^{-1}$ and the mass of the axion gets bounded by the inverse radius of compactification.

The axion coupling to ordinary matter comes from the

$$\mathscr{S}_{a\psi\psi} \sim \int d^4x \frac{1}{f_a} (\partial_\mu a|_{y=0}) (\bar{\psi} \gamma^\mu \gamma^5 \psi) . \tag{13.58}$$

In terms of Kaluza–Klein components this becomes

$$\mathscr{S}_{a\psi\psi} \sim \frac{1}{f_a} \int d^4x (\partial_\mu a_0 + \sqrt{2} \sum_{n=1}^{\infty} \partial_\mu a_n) (\bar{\psi} \gamma^\mu \gamma^5 \psi). \tag{13.59}$$

Although only the zero mode could have a derivative coupling with matter, now all the Kaluza–Klein modes have derivative coupling with matter suppressed by the effective decay constant $f_a$.

The effective decay constant $f_a \sim M_{Pl}$ makes the axion mass unacceptably small in this scenario. As we mentioned earlier, if we now consider a scenario in which the axion propagates in the $m$-dimensional subspace of the $d$-dimensional bulk, the effective decay constant becomes

$$f_a = \sqrt{V_m M_*^m} \hat{f}_a = \left( \frac{M_{Pl}}{M_*} \right)^{m/d} \hat{f}_a. \tag{13.60}$$

Suitable choice of $m$ and $d$ then makes the axion mass and, hence, the decay constant consistent with present constraints.

In this scenario there are now new constraints on the axion coupling coming from the coupling of matter with the Kaluza–Klein states with equal strength. The main constraints are from cosmological considerations [228]. There are constraints from the consideration of evolution of stars, mainly supernovae. This gives the strongest lower bound. Then there are bounds coming from big-bang nucleosynthesis, overclosure of the universe, cosmic microwave background radiation, and diffused photon background. We shall discuss these details in chapter 18.

In another possible higher-dimensional solution to the axion mass problem, a $U(1)_{PQ}$ symmetry is introduced in our brane, but broken in another brane at a distance [229]. This distant breaking of $U(1)_{PQ}$ then induces a small coupling of axion to ordinary matter. Here one introduces heavy quarks with another anomalous low-energy $U(1)$ gauge symmetry (expecting that this symmetry will be anomaly free at high energy). The details of the mechanism are similar to the distant breaking mechanism for neutrino masses.

There is a somewhat different suggestion to have quasi-localized gluons in higher dimensions [230]. The vacuum structure then turns out to be trivial. At large distances the gluons would then behave as higher-dimensional, which does not support finite action instantonic configurations. Then the $\Theta$ term vanishes on higher dimensions and there is no *CP* problem in such scenario.

## 13.5 Warped Extra Dimensions

So far all the discussions we made about higher dimensions are based on the fact that our 4-dimensional metric does not depend on the space–time points in the extra dimensions. We usually consider higher-dimensional theories based on the fact that the higher-dimensional metric is factorizable and can be written as a direct product of the noncompact four space–time dimensions and the compact space of extra dimensions. If the space–time geometry is warped, i.e., the 4-dimensional metric depends on its position in the extra dimension, then it is possible to construct a consistent theory of small extra dimensions that can allow a TeV scale strong gravity [231].

Several interesting features of the theories with large extra dimensions are shared by theories with the small extra dimensions or the Randall–Sundrum model. This theory with small extra dimensions promises a true solution to the gauge hierarchy problem. In theories with large extra dimension, although the large Planck scale is not present, there exists a very small scale corresponding to the distance of a few microns, at which the law of gravity is modified. Since the extra dimensions are compactified and the radius of compactification is of the order of a few microns, there will be Kaluza–Klein excited states with masses starting from eV, corresponding to all the particles in the bulk. Hence, in the bulk there is another scale of eV in addition to the fundamental scale of a few TeV. For example, when the (4+d)-dimensional Planck scale or the fundamental scale $M_*$ is related to the effective four-dimensional Planck scale $M_{Pl}$ by the volume of the compact space $V_d$,

$$M_{Pl}^2 = M_*^{d+2} V_d, \tag{13.61}$$

the corresponding compactification scale

$$\mu_c \sim \left(\frac{1}{V_d}\right)^{1/d} \tag{13.62}$$

is much smaller than the electroweak scale. Although there is no hierarchy between the Planck scale of $M_*$ to the electroweak scale in theories with large extra dimensions, a large hierarchy between the electroweak scale and the compactification scale of order eV still exists.

In theories with small extra dimensions, the warped geometry relates the Planck scale in our brane to the exponentially large Planck scale in another brane. So the new parameters appear only in the exponent, eliminating any hierarchy of scales. The metric is not factorizable in this case, rather the four-dimensional metric is multiplied by a warp factor, which is a rapidly changing function of an additional dimension. Let us consider an example of such warped geometry, provided by a nonfactorizable metric that respects the four-dimensional Poincaré invariance,

$$ds^2 = \exp^{-2kr_c|\phi|}\,\eta_{\mu\nu}\,dx^\mu\,dx^\nu + r_c^2\,d\phi^2, \tag{13.63}$$

where $k$ is a scale of the order of the Planck scale and $x^\mu$ are the four-dimensional coordinates for constant $\phi$. The extra dimension is considered to be an $S^1/Z_2$ orbifold, whose coordinate is $y = r_c\,\phi$, where $r_c$ is the radius of a compact fifth dimension and the angular coordinate $\phi$ has the periodicity in the range $-\pi \le \phi \le \pi$. $\phi$ may be rescaled to absorb $r_c$, but that will alter the periodicity of $\phi$.

The $S^1/Z_2$ orbifold may be viewed as a circle $S^1$, whose two sides are identified with each other by the $Z_2$ mapping. If every point in the upper half of the circle is identified with some points in the lower half of the circle, we get only a half-circle. If some current was flowing through the circle, then at each point in this half-circle the current flowing in two directions will cancel each other, except for the two end-points. These two end-points then become the two singular fixed points in this orbifold. The $Z_2$ relates the points $(x,\phi)$ and $(x,-\phi)$, so there are now two 3-branes extending along the $x^\mu$ directions and localized at the two fixed points in the fifth dimension. We call the fixed point at $\phi = \pi$ the visible brane and the other at $\phi = 0$ the hidden sector brane. Both the branes couple to the purely four-dimensional components of the bulk metric

$$\begin{aligned} g_{\mu\nu}^{vis} &= G_{\mu\nu}(x^\mu, \phi = \pi) \\ g_{\mu\nu}^{hid} &= G_{\mu\nu}(x^\mu, \phi = 0), \end{aligned} \tag{13.64}$$

where $G_{MN}$ $(M,N = \mu, 4)$ is the five-dimensional metric. The curvature parameter $k$ relates the Planck scale in the hidden sector brane $M_{Pl}$ to the fundamental scale in our visible brane $M_*$ through the exponential warp factor of the five dimensional spacetime,

$$M_{Pl}^2 = \frac{M_*^3}{k}\left[1 - \exp^{-2kr_c\pi}\right]. \tag{13.65}$$

The bulk space connecting the visible sector brane and the hidden sector brane is a slice of $AdS_5$ space. The standard model particles are confined in our brane, while gravity propagates in the bulk.

**Four-Dimensional Interactions**

We can now define an effective metric $\overline{g}_{\mu\nu}(x)$ and express the metric in the visible and hidden sector branes in terms of this effective metric as

$$\begin{aligned} g_{\mu\nu}^{vis} &= \exp^{-2\sigma}\overline{g}_{\mu\nu} \\ g_{\mu\nu}^{hid} &= \overline{g}_{\mu\nu}, \end{aligned} \tag{13.66}$$

where $\sigma = kr_c\pi$. The effective four-dimensional action can now be written in terms of this effective metric $\overline{g}_{\mu\nu}(x)$. The total action now splits into

$$S = S_{bulk} + S_{vis} + S_{hid}, \tag{13.67}$$

where the bulk action now contains the 5-dimensional gravity including the cosmological constant term

$$S_{bulk} = \int d^4x \int_{-\pi}^{\pi} d\phi \sqrt{-G}(2M_*^3 R - \Lambda). \tag{13.68}$$

By properly normalizing the fields we can determine the physical masses. Consider a Higgs field $H$ containing one symmetry breaking mass parameter $v_0$,

$$S_{vis} \supset \int d^4x \sqrt{-g_{vis}} \, \{g_{vis}^{\mu\nu} D_\mu H^\dagger D_\nu H - \lambda(|H|^2 - v_0^2)^2\}. \tag{13.69}$$

We then write the action in terms of the effective metric $\overline{g}_{\mu\nu}(x)$ and renormalize the wave functions as

$$\begin{aligned}
\psi_{vis}(\phi = \pi) &\to \exp^{3\sigma/2} \psi_{vis}(\phi = \pi) \\
H_{vis}(\phi = \pi) &\to \exp^{\sigma} H_{vis}(\phi = \pi) \\
\psi_{hid}(\phi = 0) &\to \psi_{hid}(\phi = 0) \\
H_{hid}(\phi = 0) &\to H_{hid}(\phi = 0)
\end{aligned} \tag{13.70}$$

and obtain

$$S_{eff} = \int \mathrm{d}^4x \sqrt{-\overline{g}} \left\{ \overline{g}_{\mu\nu} D_\mu H^\dagger D_\nu H - \lambda \left( |H|^2 - \exp^{-2\sigma} v_0^2 \right)^2 \right\}. \tag{13.71}$$

The mass scale (in general, any mass scale) in the visible sector 3-brane will correspond to a physical mass

$$\{v_0, m_0\}_{phys} = \exp^{-\sigma} \{v_0, m_0\}. \tag{13.72}$$

Since all operators get rescaled by their four-dimensional conformal weight, the effective metric appears in the four-dimensional Einstein action. Hence, all mass scales in the visible sector brane will be rescaled to the physical mass. In other words, the Planck mass scale $M_{Pl}$ in the distant brane now scales down to the $M_*$ scale in our world (see figure 13.4).

By considering $\exp^{\sigma}$ to be of the order of $10^{15}$, this mechanism produces TeV scale physics. In other words, for $kr_c \approx 10$, all the mass parameters in the theory $v_0, m_0, k, M$, and $\mu_c \sim O(1/r_c)$ become of the same order of magnitude, which is close to the effective Planck scale of $10^3$ GeV.

Similar to the models of large extra dimensions, in our brane the physical mass scales in the theory are of the TeV scale. All the Kaluza–Klein modes and their couplings are also determined by the TeV scale physics. This gives rise to a distinct and rich phenomenology for the next generation accelerators. For an observer in the

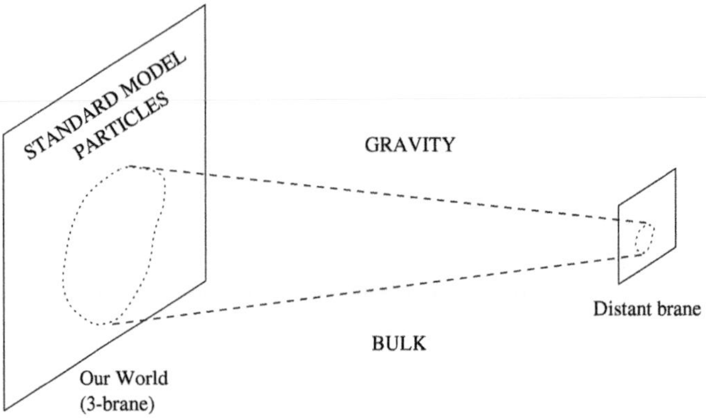


**FIGURE 13.4**
The length scales $M_{Pl}^{-1}$ in the hidden sector at the 3-brane in one boundary are scaled up exponentially to $M_*^{-1}$ in the visible sector in our 3-brane at the other boundary.

visible 3-brane, the TeV scale will become the fundamental scale and the effective Planck scale in the hidden sector brane will appear as derived scale. The large Planck scale and the weakness of the gravitational interaction arise because of the small overlap of the graviton wave function with our brane in the fifth dimension. Since the graviton resides mostly in the bulk, the wave function overlaps with our 3-brane at the boundary of the fifth dimension is very small.

**Radion Stability**

The 5-dimensional metric $G_{MN}$ now contains the 4-dimensional metric $g_{\mu\nu}$ satisfying the 4-dimensional effective action

$$S_{eff} = M_{Pl}^2 \int d^4x \sqrt{-g} R^{(4)}(g), \tag{13.73}$$

where $R^{(4)}(g)$ is the 4-dimensional scalar curvature. In addition it contains another physical field, which is generated by the fluctuation of $G_{44}$, called the radion $b_N$, which remains massless. Its *vev* $r_c$ is highly unstable. Stabilization of the radion, therefore, becomes an important issue for theories with small extra dimensions.

In one proposal [218] to stabilize the radion a scalar field $\Phi$ is introduced in the bulk. In the Randall–Sundrum background metric its action in the bulk is given by

$$S_{bulk} = \frac{1}{2} \int d^4x \int_{-\pi}^{\pi} d\phi \, \sqrt{-G} \left( G^{MN} \partial_M \Phi \partial_N \Phi - m^2 \Phi^2 \right). \tag{13.74}$$

The action of the scalar field in the visible and hidden branes is given by

$$\begin{aligned} S_{hid} &= -\int d^4x \sqrt{-g^{hid}} \lambda_1 (\Phi^2 - v_1^2)^2 \\ S_{vis} &= -\int d^4x \sqrt{-g^{vis}} \lambda_2 (\Phi^2 - v_2^2)^2, \end{aligned} \tag{13.75}$$

where $v_i$ and $\lambda_i$ are parameters with nonzero mass dimensions.

We then make an ansatz that $\Phi$ depends only on the coordinates of the extra dimension and assume that the $\lambda_i$ are large enough to give the *vev* of the scalar fields at the two branes

$$\Phi(0) = v_1 \quad \text{and} \quad \Phi(\pi) = v_2. \tag{13.76}$$

With another assumption $\varepsilon = m^2/4k \ll 1$, we can write the potential for the distance of the two branes as

$$\begin{aligned} V(r_c) = k\varepsilon v_1^2 + 4k\exp^{-4kr_c\pi}\left(v_2 - v_1\exp^{-\varepsilon kr_c\pi}\right)^2\left(1+\frac{\varepsilon}{4}\right) \\ -k\varepsilon v_1\exp^{-(4+\varepsilon)kr_c\pi}\left(2v_2 - v_1\exp^{-\varepsilon kr_c\pi}\right), \end{aligned} \tag{13.77}$$

which has a minimum at

$$kr_c = \frac{4k^2}{\pi m^2}\log\left(\frac{v_1}{v_2}\right). \tag{13.78}$$

The expression for the stable distance $r_c$ between the branes in the presence of the bulk scalar field $\Phi$ shows that there is no necessity for any fine tuning to get $kr_c \sim 10$, which gives the fundamental scale to be around TeV to get an effective Planck scale in 4 dimensions.

One very important feature of the Randall–Sundrum model comes from the duality conjecture between warped $AdS_5$ geometries with boundary branes and strongly coupled (broken) conformal field theories coupled to dynamical gravity. For every conformal field theory (CFT) operator there is a corresponding bulk field. Given any boundary condition on these bulk fields at the four-dimensional boundary of the $AdS$ slice, there is a unique solution of the string effective action (which includes gravity). Introduction of the Planck brane allows a nice holographic interpretation of this duality, which allows us to study the phenomenology of this scenario in a consistent manner.

Although the hierarchy problem is solved and it gives us very rich phenomenology, several new problems emerge in these models of extra dimensions. For example, the question of proton decay becomes very important. Grand unification could be achieved in these models below the fundamental scale of TeV, which makes the proton lifetime too small. Several mechanisms have been suggested to solve this problem. Two other problems need special attention in theories with extra dimensions. Since the smallness of the neutrino mass and the axion coupling is attributed to a large scale and there is no large scale in these theories, solutions to these problems require new approaches.

The theories with large and small extra dimensions are just two special cases of a large class of theories, which bring us promise of new physics. There are other constructions of brane worlds which can trap gravity in a very large number of extra dimensions. But the important point that these models of extra dimensions are making is that the gravity can propagate in some dimensions, which are not accessible to the standard model particles. This will mean that the strength of gravitational interaction can become strong at fairly low scale, which is the fundamental scale in the theory. The new low Planck scale depends on various unknown parameters and cannot be determined by any theory. The Planck scale could be as low as TeV or this could be as high as the effective Planck scale of $10^{19}$ GeV. We hope this new scale to be in the TeV range, so all the new physics will become accessible to the next generation experiments. This will allow us to probe all the new exciting physics of models of extra dimensions and even study quantum gravity in the next few years. However, if the scale is much higher, we may not observe any new signatures. In any case, the possibility of observing new phenomenology in the TeV range is highly welcome, since the next generation detectors have to be designed keeping these new signatures in mind. The future experiments can only discriminate among the several possible theories at the TeV scale and lead us to the next stage.

# 14

# *Novelties with Extra Dimensions*

The possibility of new physics coming from extra space dimensions opened up many new directions in this field. Most of the higher dimensional theories consider only four noncompact space–time dimensions and the remaining compact space dimensions. However, it has been demonstrated that the extra space dimensions could also be noncompact in some cases. For example, the extra dimensions can be noncompact if the geometry is warped and the radius of curvature of the extra dimensions are small enough. It is also possible to construct models with extra dimensions, where the extra dimensions do not play any crucial role with respect to gravity. The extra dimensions are constructed in an interval, and the boundary points in this interval provide new physics. If the end points are singular, one may construct grand unified theories in such orbifolds with several interesting features. The boundary conditions at the end-points can also be used to break gauge symmetries spontaneously. This could be an alternative to the Higgs mechanism of spontaneous symmetry breaking. The boundary conditions may also break supersymmetry giving rise to interesting supersymmetric particle spectrum. If the ordinary particles are allowed to propagate along the interval, it can lead to newer phenomenological implications.

## 14.1 Noncompact Warped Dimensions

There are evidences that we live in four noncompact dimensions. Newton's law has been established to a very high degree of accuracy, which is valid for four dimensions. Although there are only four noncompact dimensions, it is possible to have several compact extra dimensions. Even for compact higher dimensions the radius of curvature should be very small. Otherwise, since gravity propagates along all the dimensions, at short distances deviation from Newton's $1/r^2$ law should be observed. If the usual particles can also propagate in the extra dimensions, then there will be even stronger constraints. For $n$ extra compact dimensions, the Planck scale in higher dimensions $M_*$ will be reduced compared with the Planck scale in four dimensions $M_{Pl}$ by the volume $V_n$ of the extra dimensions $M_*^{2n} = M_{Pl}^2/V_n$. However, these general comments would change in case of warped geometry.

If the metric of our four space–time dimensions depends on its position in the extra dimensions, then the metric is called warped. The general comments about the extra dimensions are based on the fact that the metric is factorizable and the space of extra dimensions can be written as a direct product of the usual noncompact four space–time dimensions. If the space–time geometry is warped, then it is possible to construct a consistent noncompact $4+n$-dimensional theory of gravity [232]. In this case the Planck scale in higher dimensions will be related to the four-dimensional Planck scale by the curvature of the extra dimensions. Due to the warped geometry, the extra dimensions can have a localized bound state of the higher-dimensional graviton. Consider the wave equation for small gravitational fluctuations (see section 11.1) in the presence of a nontrivial potential $V$ due to the curvature, which acts like a source term, given by

$$[\partial_\mu \partial^\mu - d_j d^j + V(z_j)]\, h(x^\mu, z_j) = 0. \tag{14.1}$$

Here the coordinates are $x^\mu$ for the four dimensions and $z_j$ for the extra dimensions. The higher-dimensional gravitational fluctuations $h(x^\mu, z_j)$ can be written in terms of the four-dimensional Kaluza–Klein states $\psi(z)$ with fixed wave function in the extra dimensions

$$h(x^\mu, z_j) = \exp^{ip\cdot x} \psi(z).$$

The masses $m$ of the four-dimensional Kaluza–Klein states is given by the eigenvalues of

$$[-d_j d^j + V(z)]\, \psi(z) = -m^2\, \psi(z)\,, \tag{14.2}$$

where $p^2 = m^2$. If the background preserves Poincaré invariance, then there is a normalizable zero mode state representing the four-dimensional graviton, in addition to all its Kaluza–Klein excited modes. In case of factorizable geometry, there is a gap in the spectrum of the excited modes, which determines the scale at which deviation from usual four-dimensional gravity is expected. In the present case although the excited states have a continuum with no gap, the nontrivial potential $V(z)$ implies a single bound state corresponding to a massless four-dimensional graviton, whose wave function is centered around the 3-brane containing the standard model particles. This reproduces four-dimensional gravity predictions at low energy and long distances, in spite of the presence of the continuum of excited gravitons.

We present here the original model [232] which is a five-dimensional space–time containing a single 3-brane with a positive tension. The standard model particles are confined in the 3-brane. A regulator brane (denoted by *brane*$'$) at a distance of $y_0 = \pi r_c$ from the 3-brane of interest is introduced for checking consistency, which will ultimately be taken to infinity (taking the limit $y_0 \to \infty$) and be removed from the physical setup. The two branes, the standard model 3-brane and the regulator brane, constitute the boundaries of a finite fifth dimension. The action is given by

$$\begin{aligned} S &= S_{grav} + S_{brane} + S_{brane'}\,, \\ S_{grav} &= \int d^4x \int dy \sqrt{-G}(2M_*^3 R - \Lambda) \\ S_{brane} &= \int d^4x \sqrt{-g_{brane}}(V_{brane} + L_{brane})\,, \end{aligned} \tag{14.3}$$

where $G_{MN}$ $(M,N=\mu,4;\ \mu=0,1,2,3)$ is the five-dimensional metric, $R$ is the five-dimensional Ricci scalar, and $M_*$ is the fundamental scale for gravity in five dimensions.

A solution to the Einstein's equations which respects four-dimensional Poinaré invariance may be obtained when the boundary and bulk cosmological terms are related by

$$V_{brane} = -V_{brane'} = 24M^3k, \quad \text{and} \quad \Lambda = -24M^3k^2,$$

with $k$ to be the curvature parameter that would define the effective Planck scale in four dimensions. The solution is a slice of the symmetric space, $AdS_5$, and is given by

$$ds^2 = \exp^{-2k|y|}\ \eta_{\mu\nu}\ dx^\mu\ dx^\nu + dy^2, \tag{14.4}$$

where $0 \leq y \leq y_0$ is the coordinate of the fifth dimension.

The four-dimensional graviton zero mode follows from this equation and is described by an effective action

$$S_{eff} \subset \int d^4x \int_0^{\pi r_c} dy\ 2M_*^3 r_c \exp^{-2k|y|} \bar{R}, \tag{14.5}$$

where $\bar{R}$ denotes the four-dimensional Ricci scalar made of the four-dimensional metric $\bar{g}_{\mu\nu}(x)$. From this the effective Planck scale in four dimensions can be related to the fundamental scale through the relation

$$M_{Pl}^2 = \frac{M_*^3}{k}\left[1 - \exp^{-2ky_0}\right], \tag{14.6}$$

so there is a well-defined effective Planck scale even in the limit $y_0 \longrightarrow \infty$. This allows the removal of the regulator brane from the setup.

The next step is to study the spectrum of general linearized fluctuations

$$G_{MN} = \exp^{-2k|y|}\ \eta_{\mu\nu} + h_{\mu\nu}(x,y)$$

and check for consistency with four-dimensional gravity. This can be done by a Kaluza–Klein reduction through separation of variable $h(x,y) = \exp^{ip\cdot x}\psi(y)$ (where $p^2 = m^2$) and obtaining a solution to the linearized equation of motion expanded about the warped metric

$$\left[-\frac{m^2}{2}\exp^{2k|y|} - \frac{1}{2}\partial_y^2 - 2k\delta(y) + 2k^2\right]\ \psi(y) = 0. \tag{14.7}$$

Only the even functions of $y$ are relevant for the boundary condition with the regulator brane taken to infinity. The $\mu\nu$ indices are eliminated by working in the gauge $\partial^\mu h_{\mu\nu} = h_\mu^\mu = 0$, where $\mu\nu$ indices are the same for all the terms.

To understand the concept without going into details, we make a change of variables

$$z = \frac{\text{sgn}(y)}{k}\left(\exp^{k|y|} - 1\right);\ \ \hat{\psi}(z) = \psi(y)\exp^{k|y|/2};\ \ \hat{h}(x,z) = h(x,y)\exp^{k|y|/2}$$

to obtain

$$\left[-\frac{1}{2}\partial_z^2 + V(z)\right] \hat{\psi}(z) = m^2 \hat{\psi}(z), \tag{14.8}$$

where

$$V(z) = \frac{15\,k^2}{8(k|z|+1)^2} - \frac{3k}{2}\delta(z). \tag{14.9}$$

The general shape of this potential tells us that the delta function supports a single normalizable bound state mode, which is the massless graviton reproducing gravity in four dimensions. The remaining eigenstates correspond to a continuum of the excited Kaluza–Klein states. Since the potential falls off to zero for $z_0 \to \infty$, there is no gap and the continuum of excited states has all possible $m^2 > 0$. Tuning of the cosmological terms ensure that the bound state mode corresponds to the massless graviton. Taking only even functions of $z$ and the limit $z_0 \to \infty$, only a semi-infinite extra dimension is achieved. This can be extended to an infinite extra dimension by allowing both the even and odd functions of $z$.

In the infinite extra dimension limit the effective nonrelativistic gravitational potential between two particles with mass $m_1$ and $m_2$ on the standard model brane at $z = 0$ is the static potential generated by exchange of the zero-mode and continuum excited mode propagators. It is given by

$$\begin{aligned} V(r) &\sim G_N \frac{m_1 m_2}{r} + \int_0^\infty \frac{dm}{k} G_N \frac{m_1 m_2 \exp^{-mr}}{r} \frac{m}{k} \\ &\sim G_N \frac{m_1 m_2}{r}\left(1 + \frac{1}{r^2 k^2}\right). \end{aligned} \tag{14.10}$$

The leading term produces an effective four-dimensional gravity with usual Newtonian potential, which is due to the massless graviton or the bound state zero-mode. The sum over excited continuum states adds an correction, extremely suppressed by Planck scale squared ($k$ is of the order of Planck scale). The production of the continuum modes from the brane at $z = 0$ will also be suppressed due to the wave function suppression there.

In this model the tension in the standard model brane is fine tuned, which is equivalent to the fine tuning of the cosmological constant problem, since the bound state graviton mode is determined by the brane tension and the bulk cosmological constant. Planck scale is the same as the fundamental scale of the higher-dimensional theory. The excited continuum of states thus decouples from the low-energy physics, but is required for the consistency of the theory. At this stage this theory is not complete and may not have significant phenomenological implications, but the basic idea of noncompact extra dimensions with warped space–time geometry opens up new possibilities.

## 14.2 Orbifold Grand Unified Theories

Although the idea of grand unified theory (GUT) is fascinating and has several interesting predictios, we are yet to find a model of grand unified theory that is free of all problems. The gauge hierarchy problem, doublet–triplet splitting, and predictions of fermion masses and flavor mixing, are few of the generic problems. Supersymmetry solves the guage hierarchy problem and in fact, that is the main motivation for taking supersymmetry seriously. There are other complications of supersymmetric grand unified theories, but because of many other interesting features of supersymmetry, we study supersymmetry independent of any grand unified theories.

The orbifold grand unified theories were started to solve the doublet–triplet splitting problem. But the beauty of the models made them important on their own spirit. The orbifold GUTs are constructed in higher dimensions and compactification of the extra dimensions on singular orbifolds breaks the grand unified groups down to low-energy subgroups by the boundary conditions. Gravity does not play any special role in extra dimensions. The symmetries of the orbifold and their action on the group space make the models predictive. The lack of arbitrariness of these models is the best feature of the orbifold GUTs.

Supersymmetric orbifold grand unified theories have been constructed based on the gauge groups $SU(5)$ and $SO(10)$ with one and two extra dimensions, respectively. For the purpose of demonstration of the basic idea, we shall restrict ourselves to only $SU(5)$ GUTs in 5 dimensions. The $SO(10)$ GUTs in 6 dimensions have the added constraints coming from the requirement of quadrangle anomaly cancellation. The supersymmetric partners of the Higgs scalars enter the anomaly diagrams and the choice of Higgs representations is highly constrained.

We present here a supersymmetric $SU(5)$ orbifold GUT in 5 dimensions [233]. The first part of the construction is to describe the nature of the 5-dimensional space–time. The 5-dimensions are taken to be a factorizable manifold $M_4 \times K$, where $M_4$ is the usual four-dimensional Minkowski space with coordinates $x_\mu$, $\mu = 0,1,2,3$, the extra dimensions are compactified on an orbifold $K$, and the radius of compactification $R$ gives the grand unification scale $R^{-1} = M_c < M_{GUT} = M_U$. In this example, $K$ is considered to be an orbifold $S^1/(Z_2 \times Z_2')$, where $S^1$ is a circle of radius $R$, which has been mod out by the discrete symmetry $Z_2 \times Z_2'$. Denoting the coordinate of the circle $S^1$ in the fifth dimension by $y = x_4$, modding out by $Z_2$ transformations, implies the equivalence relation

$$\mathscr{P} : y \to -y.$$

To visualize it, consider a circular loop made by a string. Assume some current is flowing along the loop in one direction. Now shrink the loop to a line by superimposing the upper half circle onto the lower half circle. Then on any given point on the line there will be currents flowing in opposite directions cancelling each other, except for the two end-points. These two end-points will then be singular. For completeness, we further mod this out by the second $Z_2'$ which acts as

$$\mathscr{P}' : y' \to -y',$$

with $y' = y + \pi R/2$. With this we complete the construction of the orbifold in the fifth dimension, which is essentially a line interval $y \in [0, \pi R/2]$ with two fixed points at the boundaries at $y = 0$ and $y = \pi R/2 \equiv \ell$. The fixed points at the boundaries will have 4-dimensional 3 branes $\mathscr{O}$ and $\mathscr{O}'$, respectively.

The basic assumption of the orbifold GUTs is that the fields propagating along the fifth dimension or the bulk should also possess the two discrete symmetries $Z_2$ and $Z_2'$. In other words, corresponding to the action of the discrete groups $Z_2 \times Z_2'$ on the circle $S^1$, there will be equivalent $Z_2 \times Z_2'$ parity transformations on the fields propagating in the bulk. Any generic field in the bulk, $\Phi(x_\mu, y)$, will then be acted upon by the action of the two $Z_2$ and $Z_2'$ parities $\mathscr{P}$ and $\mathscr{P}'$ as

$$\begin{aligned} \mathscr{P} &: \Phi(x_\mu, y) \to \Phi(x_\mu, -y) = P_\Phi\, \Phi(x_\mu, y) \\ \mathscr{P}' &: \Phi(x_\mu, y') \to \Phi(x_\mu, -y') = P'_\Phi\, \Phi(x_\mu, y'). \end{aligned} \tag{14.11}$$

The parities $\mathscr{P}$ and $\mathscr{P}'$ give eigenvalues $\pm 1$. In general, there is no definite rule to decide the parities of any field. Depending on the requirement of the model, one needs to assign the parities of any field. However, once the parities of any field are assigned, the interactions and the profile of the field will be determined.

To get the profile of the various fields, we write the mode expansions of the fields $\Phi_{\pm\pm}(x_\mu, y)$ with eigenvalues $\{\mathscr{P}, \mathscr{P}'\} \equiv \{\pm, \pm\}$:

$$\begin{aligned} \Phi_{++}(x_\mu, y) &= \sqrt{\frac{2}{\pi R}} \sum_{n=0}^{\infty} \Phi_{++}^{(2n)}(x_\mu) \cos\frac{2ny}{R}, \\ \Phi_{+-}(x_\mu, y) &= \sqrt{\frac{2}{\pi R}} \sum_{n=0}^{\infty} \Phi_{+-}^{(2n+1)}(x_\mu) \cos\frac{(2n+1)y}{R}, \\ \Phi_{-+}(x_\mu, y) &= \sqrt{\frac{2}{\pi R}} \sum_{n=0}^{\infty} \Phi_{-+}^{(2n+1)}(x_\mu) \cos\frac{(2n+1)y}{R}, \\ \Phi_{--}(x_\mu, y) &= \sqrt{\frac{2}{\pi R}} \sum_{n=0}^{\infty} \Phi_{--}^{(2n+2)}(x_\mu) \cos\frac{(2n+2)y}{R}. \end{aligned} \tag{14.12}$$

Only the 4D Kaluza–Klein field with eigenvalues $++$ can have a massless zero mode. The fields $\Phi_{++}$ and $\Phi_{+-}$ can be nonvanishing at the brane $\mathscr{O}$ at $y = 0$, while the fields $\Phi_{++}$ and $\Phi_{-+}$ can be nonvanishing at the brane $\mathscr{O}'$ at $y = \ell$.

In 5D the local Lorentz group is O(5). The Weyl projection operator $\gamma_5$ is part of O(5) and hence both the left-chiral and right-chiral fields of 4D belong to the same representation of any 5D field. The $N = 1$ supersymmetry in 5D will thus contain 8 real supercharges, which in 4D will imply an $N = 2$ supersymmetry. For any realistic orbifold grand unified theory, the parity assignment corresponding to the discrete symmetries $\mathscr{P}$ and $\mathscr{P}'$ should reduce $N = 2$ supersymmetry to $N = 1$ supersymmetry in 4D and also break the $SU(5)$ symmetry to the standard model gauge group

$SU(3)_c \times SU(2)_L \times U(1)_Y$. This can be achieved by the parity assignments

$$\mathscr{P} = \text{diag}\,\{+1,+1,+1,+1,+1\}$$
$$\text{and} \quad \mathscr{P}' = \text{diag}\,\{-1,-1,-1,+1,+1\}, \tag{14.13}$$

where these matrix representations of $\mathscr{P}$ and $\mathscr{P}'$ act on the fundamental representation of $SU(5)$. Thus, all $SU(5)$ components of any multiplet will have the same parity under $\mathscr{P}$, while the components of the $SU(5)$ multiplets that are invariant under the standard model $SU(3)_c \times SU(2)_L \times U(1)_Y$ (denoted by the index $a$) will have opposite parity compared with the fields belonging to the coset space $SU(5)/SU(3)_c \times SU(2)_L \times U(1)_Y$ (denoted by the index $\hat{a}$) under $\mathscr{P}'$. A coset space $G/H$ means the elements of the group space $G$, which are not elements of $H$.

The vector multiplet of $N=2$ supersymmetry contains a vector supermultiplet $V_a$ and a scalar supermultiplet $\Sigma_a$ of $N=1$ supersymmetry. The parity assignments are inputs in orbifold GUTs, which determine the matter contents. One convenient choice for the parity operator $\mathscr{P}$ is even for the vector multiplets and odd for the scalar multiplets. The $(\mathscr{P}, \mathscr{P}')$ assignments for the vector and scalar multiplets are then given by

$$V^a \equiv (+,+), \quad V^{\hat{a}} \equiv (+,-), \quad \Sigma^{\hat{a}} \equiv (-,+), \quad \Sigma^a \equiv (-,-).$$

Thus, only $V^a$ will have massless zero modes, and hence, at low energies only the minimal supersymmetric standard model gauge bosons are present in the theory.

Although the 4-dimensional theory has only the standard model gauge symmetry, the 5-dimensional theory has a larger position-dependent gauge symmetry. The gauge transformation parameters $\xi_a$ and $\xi_{\hat{a}}$ now have the Kaluza–Klein expansion

$$\xi_a(x_\mu, y) = \sum_{n=0}^{\infty} \xi_a^n(x_\mu) \cos\frac{2ny}{R}$$
$$\xi_{\hat{a}}(x_\mu, y) = \sum_{n=0}^{\infty} \xi_{\hat{a}}^n(x_\mu) \cos\frac{(2n+1)y}{R}. \tag{14.14}$$

At $y = \pi R/2$, all the modes of the gauge transformation parameter $\xi_{\hat{a}}^n$ vanish. Thus, in the brane $\mathscr{O}'$ at $y = \ell$, only the standard model symmetry is present, whereas in all other points in the fifth dimension all the gauge parameters are nonvanishing and possess the $SU(5)$ symmetry. For the gauge kinetic terms this implies

$$S = \int \mathrm{d}^4x dy \left[\frac{1}{g_5^2}F^2 + \delta(y)\frac{1}{\tilde{g}^2}F^2 + \delta(y-\ell)\frac{1}{\tilde{g}_a^2}F_a^2\right], \tag{14.15}$$

where $F_a$ correspond to the field strengths for the standard model gauge groups ($a = 1,2,3$) and $F$ represents the field strengh for the entire $SU(5)$ gauge symmetry. The first and second terms are $SU(5)$ invariant bulk and $\mathscr{O}$ brane gauge kinetic energies, while the third term represents standard model gauge kinetic energy localized at $y = \ell$. The standard model gauge couplings $g_a$ are obtained by integrating over the extra dimensions

$$\frac{1}{g_a^2} = \frac{\ell}{g_5^2} + \frac{1}{\tilde{g}_a^2}, \tag{14.16}$$

where the contribution of $\tilde{g}$ has been absorbed into a shift of $\tilde{g}_a$. The standard model gauge couplings $g_a$ now depend on the localized kinetic operators $\tilde{g}_a$, making them nonuniversal at the large cut-off scale. However, this problem can be removed by making the extra dimensions large so that the nonuniversal contribution is suppressed by the volume factor.

Next we introduce the Higgs scalars in the model. The usual Higgs doublet which breaks the electroweak symmetry belongs to a **5** of $SU(5)$. Supersymmetry then requires both **5** and $\overline{\mathbf{5}}$ representations of the $SU(5)$ Higgs fields. The Higgs fields could be introduced in the bulk and then impose parity $\mathscr{P}'$ to ensure that only the standard model doublets have the zero modes and can remain massless, while the triplet components become heavy naturally and there is no doublet–triplet splitting problem. This was the main motivation to consider orbifold GUTs initially. One may also introduce the Higgs fields in the standard model brane at $y = \ell$. Then only the standard model representations will appear in the theory and one may consider only the Higgs doublets in the theory. There are no Higgs triplets in this model, which also solves the doublet–triplet splitting problem.

The fermions can now be introduced in the bulk or in the $SU(5)$ invariant brane at $y = 0$ to preserve the $SU(5)$ understanding of matter quantum numbers. If they are in the brane, then they appear as supermultiplets in **10** + **5** representations. But, if they are in the bulk, they appear in $N = 2$ hypermultiplets. However, the parities will allow only one $N = 1$ supermultiplet to be the massless zero mode. Gauge coupling unification in these theories requires the compactification scale $M_c \sim 10^{15}$ GeV to be slightly less than the grand unification scale $M_U \sim 2 \times 10^{16}$ GeV. The Kaluza–Klein states contribute to the evolution of the gauge coupling between these two scales.

## 14.3 Split Supersymmetry

Supersymmetry, a symmetry between fermions and bosons, is a beautiful symmetry to exist in nature. For constructing a consistent theory of gravity, the superstring theory, supersymmetry plays a crucial role. However, a more popular reason for introducing supersymmetry is to solve the gauge hierarchy problem. This implies that the supersymmetry breaking scale should not be higher than TeV. Supersymmetry also provides solution to the gauge coupling unification.

Recently it has been argued that the naturalness may not be a problem, rather we must get used to it. We have noticed a strong naturalness problem with the cosmological constant. The observed value of the cosmological constant requires new dynamics at the scale of $10^{-3}$ eV, which should not be affected by any physics at much higher energies, for example, the electroweak symmetry breaking scale or even the Planck scale.

If we now extrapolate our lack of understanding of this naturalness question to the Higgs sector, we may not require supersymmetry near the TeV scale. There could be supersymmetry at much higher scale so all the scalar superpartners are heavy. However, the gauginos and Higgsinos remain light so the gauge coupling unification is not disturbed. The model also predicts a dark matter candidate. This new class of models with supersymmetry breaking scale to be very large is called the split supersymmetry [234, 235].

Split supersymmetric models may not be as appealing as the regular supersymmetric models, but they are a new possibility with distinct phenomenological predictions, which could be verified in the next generation accelerator experiments. The essential features of any split supersymmetric models include predictions for a dark matter candidate, which is usually a gravitino LSP. The standard model gauge coupling constants evolve to a single unified gauge coupling at the grand unification scale. It is assumed that grand unification into a simple group occurs without any intermediate symmetry breaking scale and the gauginos and the Higgsinos have masses of the order of 100 GeV. In split supersymmetric models all scalar fermions have masses of the order of some intermediate symmetry breaking scale. The flavor changing neutral current problem of the low-energy supersymmetry and the proton decay problems are absent in all split supersymmetric models. Observable electron dipole moments are one of the generic predictions of these models, which could be tested in the next generation of experiments. The particle spectrum required for split supersymmetry can be obtained in some models originating from extra dimensions. We shall illustrate the basic structure of these models with one example of an orbifold grand unified theory [236]. However, the mechanism is generic in nature and could be applicable to any 5-dimensional space with one of the dimensions an interval, and the effective Planck scale around the grand unification scale [234].

Consider a supersymmetric $SU(5)$ orbifold GUT in 5 dimensions with $N = 1$ supersymmetry, the same as the one discussed in the previous section. The standard model particles are localized at the $\mathcal{O}'$ brane at $y = \ell$, where $SU(5)$ is broken to the standard model. The quarks and the leptons remain massless in this brane. Supersymmetry is broken by Scherk–Schwarz mechanism in extra dimensions [237]. The scalar radius moduli $T$ acquire kinetic terms only through gravity couplings. The $F$-component of this radion chiral superfield $T = r + \theta^2$ in the $SU(5)$ invariant brane at $\mathcal{O}$ then breaks supersymmetry [238] in the background making the minimum of the potential negative. Fine tuning the $F$-component of the chiral superfield $X$ in the $\mathcal{O}'$ brane, the vacuum energy is made to vanish. The coupling of this field $X$ makes all scalar superpartners heavy, while the gauginos and the Higgsinos receive only 100 GeV order anomaly mediated masses.

We add a constant superpotential

$$W = cM_5^3,$$

localized in the $SU(5)$ invariant brane at $\mathcal{O}$ and write the tree-level effective Lagrangian with $T$ and the conformal compensator field $\phi = 1 + \theta^2 F_\phi$

$$L = \int d^4\theta \phi^\dagger \phi K + \int d^2\theta \phi^3 W + H.c., \tag{14.17}$$

where the Kähler potential $K = M_5^3(T+T^\dagger)$. The resulting scalar potential

$$V = M_5^3\left(r|F_\phi|^2 + F_T^* F_\phi + 3cF_\phi + h.c\right) \tag{14.18}$$

leads to supersymmetry breaking minimum with vanishing potential

$$F_\phi = 0 \quad \text{and} \quad F_T = -3c.$$

The gravitino then receives a mass $m_{3/2} = 1/r$ (assuming $c = 1$) by combining with the fermionic partner of $T$.

The one-loop corrections make $F_\phi$ nonvanishing

$$F_\phi \sim \frac{1}{16\pi^2}\frac{1}{M_5^3 r^4} \ll m_{3/2}$$

with a negative potential at the minima. As discussed earlier, we now add a chiral superfield $X$ at the standard model brane $\mathcal{O}'$ and write the $R$-parity invariant superpotential at $\mathcal{O}'$ as

$$W = m^2 X = \frac{1}{4\pi r^2}X. \tag{14.19}$$

We assigned an $R$-symmetry charge 2 to the superfield $X$. The Kähler potential at $\mathcal{O}'$ is then given by

$$K = X^\dagger X - \frac{(X^\dagger X)^2}{M_5^2} + \cdots. \tag{14.20}$$

The minimum of the potential then breaks supersymmetry with

$$|F_\phi|^2 = m^4 = \frac{1}{16\pi^2 r^4} \tag{14.21}$$

and a vanishing cosmological constant.

The Kähler potential and the superpotential then gives the mass spectrum of all the fields. This leads to masses of

$$\left.\begin{array}{ll} scalar\ component\ of\ X & m_X^2 \sim m^4/M_5^2 \\ fermionic\ component\ of\ X & m_{\psi_X} \sim m^4/M_5^3 \\ \text{vev}\ for\ X & \langle X\rangle \sim m^2/M_5 \end{array}\right\} \tag{14.22}$$

and $F_X \sim m^2$, where $m = M_5^3/M_{Pl}^2$. Contact interactions on the standard model brane or anomaly or gravitationally induced effective operators

$$\int d^4\theta \frac{1}{M_5^2}X^\dagger X Q^\dagger Q, \quad \int d^2\theta \frac{m^2 X}{M_5^2}WW \quad \text{and} \quad \int d^4\theta \frac{X^\dagger X}{M_5^2}WW,$$

leads to the mass spectrum:

$$\left.\begin{array}{ll} scalar\ superpartners\ of\ fermions & m_S \sim \frac{|F_X|}{M_5} \sim \frac{M_5^5}{M_{Pl}^4} \\ gaugino\ masses & M_i \sim \frac{|F_X|^2}{M_5^3} \sim \frac{M_5^9}{M_{Pl}^8} \end{array}\right\} \tag{14.23}$$

We then assign a vanishing $R$-charge to $H_u$ and $H_d$ to prevent terms such as $M_5 H_u H_d$. The leading order terms contributing to $\mu B$ and $\mu$ are

$$\int d^4\theta \frac{m^2 X^\dagger}{M_5^3} H_u H_d, \quad \int d^2\theta \frac{m^2 X}{M_5^2} H_u H_d \quad \text{and} \quad \int d^4\theta \frac{X^\dagger X}{M_5^2} H_u H_d,$$

This gives us Higgsino masses similar to those of the gauginos with

$$\mu B \sim \frac{|F_X|^2}{M_5^2} \sim m_S^2 \quad \text{and} \quad \mu \sim M_i. \tag{14.24}$$

In the present model, the natural scale is the grand unification scale, which is of the order of $M_5 \sim M_G \sim 3 \times 10^{13}$ GeV. Assuming usual flat space relationship $rm_5^3 = M_{Pl}^2$, we get the different mass scales:

$$\left.\begin{array}{ll} \textit{Gravitino mass} & m_{3/2} \sim 10^{13} GeV \\ \textit{Supersymmetry breaking scale} & m_S \sim 10^9 GeV \\ \textit{Masses of superpartners of fermions} & m_{\tilde{f}} \sim m_S \sim 10^9 GeV \\ \textit{Gaugino and Higgsino masses} & M_i \sim 100 \text{ GeV} \end{array}\right\} \tag{14.25}$$

The split supersymmetry spectrum comes out of this orbifold GUT model. Gauge coupling unification is verified and the model predicts an LSP dark matter candidate. Since the standard model fermions are now localized to the $\mathcal{O}'$ brane, where the $SU(5)/SU(3)_c \times SU(2)_L \times U(1)_Y$ multiplets are absent, there are no $SU(3)_c$ triplet Higgs scalars, which could cause fast proton decay. Proton decay is naturally suppressed, and fermion masses come from the usual standard model Yukawa couplings.

Split supersymmetric model is a phenomenological model with distinct predictions that the superpartners of the quarks and leptons are heavy, but the gauginos and Higgsinos are light. The gauge coupling unification is assured and LSP dark matter is there. However, there exist some models of extra dimensions, including the models of orbifold grand unified theories, which can predict this spectrum consistently. If superpartners of the fermions are not detected in the next generation accelerators, then this possibility will become more attractive.

## 14.4 Higgsless Models

The standard model of electroweak interactions is based on the Higgs mechanism, but we have yet to find the Higgs scalar. In fact, we have yet to find any fundamental scalar particle until now. So, the discovery of the Higgs particle will establish that spin-0 scalar particles exist in nature and the Higgs mechanism of spontaneous symmetry breaking will be established, which makes the discovery of the Higgs particle so important. The precision measurements of the parameters of the standard model provided bounds on the mass of the Higgs particle. There is a stronger bound on the Higgs mass in the supersymmetric theories. If the Higgs particle is heavier than about 135 GeV, then that may rule out the minimal supersymmetric standard model. But in case we do not find the Higgs particle at all within the allowed mass range, it will pose a serious problem.

With a view to overcoming this problem several alternatives have been attempted. The technicolor theory appeared promising in the beginning, but the precision measurements of the electroweak parameters ruled out this possibility. In another possibility one adds a second Higgs scalar, which does not acquire any *vev*. For some choice of parameters this scenario pushes up the mass of the Higgs scalar. Another interesting possibility is the little Higgs model. These are essentially extensions of the composite Higgs models, in which a new strong force binds the more fundamental constituents of the Higgs field. Inconsistency with the precision electroweak measurements is avoided by making the Higgs scalars pseudo-Nambu–Goldstone bosons and then incorporating collective symmetry breaking mechanism, which enforces the cancellation of the quadratic divergent contributions to the Higgs mass by introducing new particles. These new particles may have phenomenological predictions. However, at present the most promising alternative to the Higgs mechanism seems to be the Higgsless model, which is a new scenario that may emerge in models with extra dimensions.

The basic idea behind the Higgsless model is to break the gauge symmetry spontaneously and give masses to the fermions by applying specific boundary conditions on an extra dimension in an interval [239, 240, 241]. We start with a model with one extra dimension in an interval, which means that the extra dimension is neither extended to infinity nor compact. Consider a circle and identify all points of the circle on the left with a corresponding point in the right. This straight line with singularity at the end-points represents an extra dimensions in an interval. The boundary conditions on the singular end-points can then break any gauge symmetry in this interval.

This is an interesting idea, but there is no completely consistent theory based on this idea. There is also an argument that these theories may not satisfy the unitarity constraint [242]. But considering the importance of an alternative for the Higgs mechanism, we expect that the problems of the Higgsless model can eventually be removed.

Let us consider a 5-dimensional theory, where the fifth dimension is an interval starting at 0 and ending at $\pi R$ [239]. The end-points have singularity like the orbifold models. We start with a scalar field $\phi$ in the bulk and write its action

$$S = \int d^4x \int_0^{\pi R} dy \left( \frac{1}{2}\partial^N \phi \partial_N \phi - V(\phi) + \frac{1}{2}\delta(y) M_1^2 \phi^2 + \frac{1}{2}\delta(y-\pi R) M_2^2 \phi^2 \right) . \qquad (14.26)$$

The two end-points have the masses for the scalar field. We now impose that the variation of the equation vanishes to get a consistent set of boundary conditions. The bulk equation of motion is obtained by the variation of this equation and then integrating by parts

$$\partial^N \partial_N \phi + \frac{\partial V}{\partial \phi} = 0. \tag{14.27}$$

For the boundary conditions there will be contributions from the boundary pieces

$$\begin{aligned} \delta\phi(\partial_5\phi + M_1^2\phi)\big|_{y=0} &= 0 \\ \delta\phi(\partial_5\phi - M_2^2\phi)\big|_{y=\pi R} &= 0. \end{aligned} \tag{14.28}$$

These two equations are then solved using the consistent set of boundary conditions that respects 4-dimensional Lorentz invariance at $y = 0, \pi R$:

$$(i) \quad (\partial_5\phi \pm M_i^2\phi)\big|_{y=0,\pi R} = 0 \tag{14.29}$$

$$(ii) \quad \phi\big|_{y=0,\pi R} = const. \tag{14.30}$$

The former is a mixed boundary condition, and for $M_i = 0$, it reduces to the Neumann boundary condition. The latter boundary condition becomes the Dirichlet boundary condition for $const. = 0$ (see section 12.5 for discussions on these boundary conditions). Depending on the problem under consideration, the different boundary conditions may be used.

Let us now include the gauge bosons in the problem. We denote the 5-dimensional gauge bosons as $A^a_M$, $M = 0,1,2,3,4$, where $a$ is the index for the internal gauge group in the adjoint representation. This field will decompose to a 4-dimensional gauge boson $A^a_\mu$ and a scalar $A^a_5$. In the $R_\xi$ gauge, the mixing between these fields $A^a_\mu$ with $A^a_5$ is eliminated by gauge fixing and the action becomes

$$\begin{aligned} S = \int d^4x \int_0^{\pi R} dy \bigg( &-\frac{1}{4} F^a_{\mu\nu}\, F^{a\mu\nu} - \frac{1}{2} F^a_{5\nu}\, F^{a5\nu} \\ &- \frac{1}{2\xi} (\partial_\mu\, A^{a\mu} - \xi\, \partial_5\, A^a_5)^2 \bigg). \end{aligned} \tag{14.31}$$

$\xi \to \infty$ will take it to unitary gauge, making all $A^a_5$ mode massive. Since the Kaluza–Klein modes of $A^a_5$ become unphysical, we eliminate them. They become the longitudinal modes of the gauge bosons.

Similar to the scalar field, we can write the usual bulk equations of motion by integrating by parts and taking the variation of this action

$$\begin{aligned} \partial_M F^{aM\nu} - g_5 f^{abc} F^{bM\nu} A^c_M + \frac{1}{\xi} \partial^\nu \partial^\sigma A^a_\sigma - \partial^\nu \partial_5 A^a_5 &= 0, \\ \partial^\sigma F^a_{\sigma 5} - g_5 f^{abc} F^b_{\sigma 5} A^{c\sigma} + \partial_5 \partial_\sigma A^{a\sigma} - \xi\, \partial_5^2 A^a_5 &= 0. \end{aligned} \tag{14.32}$$

The boundary conditions are now given by

$$\begin{aligned} F^a_{\nu 5}\, \delta A^{a\nu}\big|_{y=0,\pi R} &= 0 \\ (\partial_\sigma A^{a\sigma} - \xi \partial_5 A^a_5)\, \delta A^a_5\big|_{y=0,\pi R} &= 0, \end{aligned} \tag{14.33}$$

where $f^{abc}$ is the structure function for the gauge group and $g_5$ is the 5-dimensional gauge coupling constant.

Maintaining the Lorentz invariance in 4-dimensions, the different choices of the boundary conditions can be written as

$$(i) \quad A^a_\mu\big|_{y=0,\pi R} = 0; \quad A^a_5\big|_{y=0,\pi R} = const. \tag{14.34}$$

$$(ii) \quad A^a_\mu\big|_{y=0,\pi R} = 0; \quad \partial_5 A^a_5\big|_{y=0,\pi R} = 0 \tag{14.35}$$

$$(iii) \quad \partial_5 A^a_\mu\big|_{y=0,\pi R} = 0; \quad A^a_5\big|_{y=0,\pi R} = const. \tag{14.36}$$

There can be varieties of boundary conditions, which would depend on the problem under consideration. If there are localized scalars on the boundary, the boundary conditions on the gauge bosons can also be mixed, Neumann or Dirichlet. The boundary conditions for different gauge directions could also be different.

We shall now demonstrate how some combinations of the boundary conditions could be used to break the gauge symmetry. We shall consider a variant of the standard model without the Higgs scalar. Our main goal is to achieve the electroweak symmetry breaking $SU(3)_c \times SU(2)_L \times U(1)_Y \to SU(3)_c \times U(1)_Q$ and give masses to the fermions from the suitable boundary conditions in the fifth dimension without giving a *vev* to the Higgs scalar.

We start with a 5-dimensional space, in which the fifth dimension is compactified in an interval in a warped Randall–Sundrum background [232]. We shall consider a left–right symmetric group in the bulk $SU(3)_c \times O(4) \times U(1)_{B-L} \equiv SU(3)_c \times SU(2)_L \times SU(2)_R \times U(1)_{B-L}$ and assume that the metric is conformally flat

$$ds^2 = \left(\frac{R}{z}\right)^2 (\eta_{\mu\nu} dx^\mu dx^\nu - dz^2). \tag{14.37}$$

The coordinate of the fifth dimension extends over the interval corresponding to range of mass scales TeV$\to M_{Pl}$, which is the interval $[R, R']$, where $R \sim 1/M_{Pl}$ and $R' \sim \text{TeV}^{-1}$.

We now apply the following boundary conditions in the branes at $z = R$ and $z = R'$

$$\text{at } z = R': \begin{cases} \partial_z A^{+a}_\mu = 0, \ A^{-a}_\mu = 0, \ \partial_z B_\mu = 0, \\ A^{+a}_5 = 0, \ \partial_z A^{-a}_5 = 0, \ B_5 = 0, \end{cases} \tag{14.38}$$

$$\text{at } z = R: \begin{cases} \partial_z (g_5 B_\mu + \tilde{g}_5 A^{R3}_\mu) = 0, \ A^{R1,2}_\mu = 0, \ \partial_z A^{La}_\mu = 0, \\ \tilde{g}_5 B_\mu - g_5 A^{R3}_\mu = 0, \ A^{L,Ra}_5 = 0, \ B_5 = 0, \end{cases} \tag{14.39}$$

where we defined $A^{\pm a}_M = (A^{La}_M \pm A^{Ra}_M)/\sqrt{2}$. These boundary conditions break the group $O(4) \to SU(2)_D$ in the brane at $z = R'$, while in the brane at $z = R$ the group breaks down to $SU(2)_R \times U(1)_{B-L} \to U(1)_Y$.

In the brane at $z = R'$, the symmetry breaking $O(4) \to SU(2)_D$ could equivalently be achieved by giving a large *vev* to a Higgs field in the boundary. These scalars would then decouple from the gauge boson scattering. The Higgs scalar transforming as $(2,2,0)$ under the group $SU(2)_L \times SU(2)_R \times U(1)_{B-L}$ could be used for this symmetry breaking. This will imply that a combination of the $SU(2)_L$ and $SU(2)_R$ gauge bosons will become massive in this brane. In the similar way, in the brane at $z = R$, the symmetry breaking $SU(2)_R \times U(1)_{B-L} \to U(1)_Y$ could be achieved by the *vev* of the Higgs scalar $(1,2,1/2)$ in the boundary. This will mean that the charged gauge bosons and one combination of the neutral gauge boson will become heavy in this brane.

When we match both the boundary conditions at $z = R$ and $z = R'$, all the four charged gauge bosons and the two combinations of the neutral gauge bosons of the group $SU(2)_L \times SU(2)_R \times U(1)_{B-L}$ will become massive, leaving only the photon to be massless in the bulk. The matter fields in the bulk will then experience gauge interaction mediated only by the photon. Thus, the symmetry breakings $O(4) \times U(1)_{B-L} \to SU(2)_D \times U(1)_{B-L}$ at $z = R'$ and $SU(2)_L \times SU(2)_R \times U(1)_{B-L} \to SU(2)_L \times U(1)_Y$ at $z = R$ will leave only the $U(1)_Q$ symmetry unbroken in the bulk so the matter field in the bulk experiences the electromagnetic interaction only.

These boundary conditions will then give the Kaluza–Klein mode expansions

$$B_\mu = g_5 a_0 \gamma_\mu(x) + \sum_{k=1}^{\infty} \psi_k^{(B)}(z) Z_\mu^{(k)}(x), \tag{14.40}$$

$$A_\mu^{L,R3} = \tilde{g}_5 a_0 \gamma_\mu(x) + \sum_{k=1}^{\infty} \psi_k^{(L,R3)}(z) Z_\mu^{(k)}(x), \tag{14.41}$$

$$A_\mu^{L,R\pm} = \sum_{k=1}^{\infty} \psi_k^{(L,R\pm)}(z) W_\mu^{(k)\pm}(x). \tag{14.42}$$

From this expansion it is clear that only the 4-dimensional photon field $\gamma_\mu(x)$ has a flat wave function and could remain massless. All other fields become massive. The lowest states in the Kaluza–Klein towers are the fields $W_\mu^{(k)\pm}$ and $Z_\mu^{(k)}$ with masses

$$M_W^2 = \frac{1}{R'^2 \log(R'/R)} \tag{14.43}$$

$$M_Z^2 = \frac{g_5^2 + 2\tilde{g}_5^2}{g_5^2 + \tilde{g}_5^2} \frac{1}{R'^2 \log(R'/R)}. \tag{14.44}$$

These fields correspond to the usual $W^\pm$ and $Z$ bosons of the electroweak symmetry. For $R^{-1} = 10^{19}$ GeV we require $R'^{-1} = 500$ GeV for the correct masses of the $W^\pm$ and $Z$ gauge bosons. This will then imply that the first Kaluza–Klein excitations of these bosons will have masses of around 1.2 TeV and the next ones will have masses of around 1.9 TeV.

The gauge coupling constants of the standard model are now given in terms of the 5-dimensional gauge coupling constants and the extensions of the extra dimension by the relations

$$g^2 = \frac{g_5^2}{R \log(R'/R)},$$

$$e^2 = \frac{g_5^2 \tilde{g}_5^2}{(g_5^2 + 2\tilde{g}_5^2) R \log(R'/R)}, \tag{14.45}$$

and the $\rho$ parameter remains consistent with experiments

$$\rho = \frac{M_W^2}{M_Z^2 \cos^2 \theta_w} \approx 1.$$

The other parameters of the standard model are also found to be consistent. The chiral fermions will only have zero modes in the warped space in an interval, but for the fermion masses suitable boundary conditions for the different fermions have to be introduced. Although none of these parameters come out as predictions of the model, it is possible to find a consistent choice of the boundary conditions that can explain the quark and lepton masses and predict the masses of their resonances [241]. If this mechanism is realized in nature, the Kaluza–Klein excitations of the gauge bosons should be observed in the next generation accelerators.

# Part V

# Astroparticle Physics

# 15

# *Introduction to Cosmology*

In cosmology we study the evolution of the universe starting with some simple principles, known as the cosmological principles. We try to relate astrophysical observations with the cosmological models that are developed based on the cosmological principles and the earlier observations and improve the model to explain newer phenomena. With many new inputs from satellite based observations and also highly sophisticated ground based experiments, the cosmological parameters have been determined with much higher accuracy, and we have a much better understanding of the evolution of our universe.

Out of the many possible cosmological models we have now converged to the standard model of cosmology. Although the standard model does not explain all the phenomena and we need to extend the model, it remains the starting point for our understanding of the evolution of the universe. Most of the alternate cosmological models have been ruled out by experiments.

To understand some aspects of cosmology, we need inputs from particle physics. In the standard model of cosmology the universe started from a big-bang and continued to expand. If we extrapolate backwards, the universe was very dense in the past and the average energy per particle was much higher. To understand the evolution of the universe in that epoch, an understanding of the particle interactions at that energy is required. Our present observations could have some information imprinted in them about the particle interactions when the universe was hotter and the average energy of the particles were much higher than the reach of our present day accelerators.

In this chapter we shall present a brief introduction to cosmology keeping in mind the interplay between particle physics and cosmology, which we shall discuss in the subsequent chapters. Most of the materials in this chapter are taken from one of the books or reviews we consulted. Since this is a separate subject on its own, we tried to include only the materials that are needed for the completeness of the present book.

We shall use the following units for length in this chapter:
*Light year*, the distance travelled by light in one year;
*Parsec* (pc), a distance of around 3.26 light years, at which a star would have a parallax of one second of arc, where parallax is the apparent change in position of a nearby star due to orbital motion of Earth around the sun;
*Megaparsec* (Mpc), $10^6$ pc $\sim 3.09 \times 10^{24}$ cm;
*Astronomical Unit* (AU), the average distance between Earth and the sun ($\sim 1.5 \times 10^8$ km).

## 15.1 The Standard Model of Cosmology

If we want to make a precise statement about the universe, we can only state what we see around us at present. But we want to know much more than that by generalizing our present observations in a model and then extrapolating it to earlier times and larger distances. Finally we want to test the predictions of the model from further observations. Our knowledge of the universe is so limited that the simplest models may fail to explain many phenomena, and we may have to extend the model in many ways. So, the first task is to have a base structure, which can then be improved continuously with newer observations.

From the several alternative cosmological models proposed in the beginning to explain our universe, the hot big-bang cosmology has become the most successful model and is now called the standard model of cosmology. As expected, this model is extended continuously with newer ingredients, but it gives the base structure for our understanding of the universe.

The big-bang cosmology should not be interpreted as a model of the universe that started from a point with a big-bang, as if everything was created from some singularity. It only means that if we extrapolate our present understanding of the universe backward in time without any change, then the universe would have started from a singularity at some point of time, which we denote as the beginning of the universe and refer to any event in the universe with respect to this time. For example the age of our present universe refers to this time in the past. However, it should be kept in mind that the universe could have gone through different phases and the evolution of the universe could have a different history than what we think at present. Particularly at very early time when the universe was too dense and the average energy was close to the Planck mass, the evolution of the universe was governed by quantum gravity. Without any knowledge of quantum gravity, we cannot talk about the evolution of the universe before that time. So, we can at most start our discussions from that point of time.

Given the success of the standard model of cosmology and its extensions, we shall not discuss the alternate cosmological models. The essential ingredients for the standard model of cosmology are the following:

*Cosmological Principle:* It is a postulate that the universe is homogeneous and isotropic. The basic idea originates from Mach's principle that there is no preferred frame of reference in the universe. Any object in the universe is moving under the influence of all other objects in the universe. Theoretically this may sound quite logical, but from an observational point of view this is a bold step. At the scale of our galaxy, it would appear that our Milky Way galaxy is an isolated island in the universe and there is inhomogeneity all around. Only at a very large scale of, say, Hubble length $\sim$ 4000 Mpc, does the universe appear to be homogeneous. The small inhomogeneity may be attributed to some kind of perturbation.

*Expanding Universe:* It has been established from observations that the universe is expanding in all directions at present. Every galaxy is moving away from any other galaxies. The velocity of recession of any two objects is proportional to the mean distance ($\ell$) between the two objects

$$\frac{d\ell}{dt} = H_0\ell. \tag{15.1}$$

This is known as Hubble's law, and the constant of proportionality $H_0$ is known as Hubble constant, which gives the present expansion rate. The Hubble constant is time dependent, and its variation with time tells us if the universe is accelerating or decelerating. We shall denote any cosmological parameters at present time with a subscript 0.

*General Theory of Relativity:* The evolution of the universe is governed by the general theory of relativity. All observed phenomena are found to be consistent with the explanations of the general theory of relativity. So, we extrapolate the theory to all times, with a word of caution that the theory will fail to explain the evolution of the universe in the quantum regime, when gravity was too strong.

*Cosmic Background Radiation:* As we go back in time the universe shrunk to smaller size and become denser and hotter. At some very early time the universe would have been dominated mostly by thermal blackbody radiation. When matter decoupled from radiation, this thermal radiation remained as cosmic background radiation with the information about thermal history imprinted on them.

With these assumption we try to get a global picture of the evolution of the universe and then understand the small deviations and fluctuations as some additional effects, which may contain more information about the early universe. If we do not consider the small scale inhomogeneity and anisotropy of our universe, then the dynamics of our universe may be understood by the symmetric solution of the general theory of relativity.

The expansion of the universe may be assumed to be uniform as a first order approximation. This will give a simple picture of the universe–that it started from a singular point and evolved continuously. We already know that this picture is not correct, but with this picture we can extrapolate our knowledge to earlier times and try to find out which relic signals may contain some information that can lead us to a better understanding of our past. The most significant result it points to is the existence of cosmic microwave background radiation.

If we assume that the simplest picture of the universe we are talking about was almost the exact picture of the universe except for some small deviations, then we can smoothly extrapolate our knowledge to earlier times. This extrapolation of our simple notion of an expanding universe will lead us to a singular point from which the universe started. This is referred to as the origin of our universe, the so-called big-bang. We refer to this time as the origin of time and refer to any event in the universe with respect to this time. This will also allow us to define the age of our universe.

Immediately after the big-bang the universe was extremely hot and the average energy per particle was very high. Slowly the universe cooled down and the average energy came down continuously. After the atoms were formed matter decoupled from radiation and the universe became matter-dominated. The thermal blackbody radiation then constituted the cosmic background radiation. As the universe continued to expand, the wavelength of this background radiation also got stretched by the same scale factor. So, the thermal background blackbody gamma-radiation with very high energy of the early universe would become the $3^\circ$ K microwave radiation at present. Existence of this microwave background radiation confirmed this simplistic picture as the backbone of our cosmological model.

When we include the effect of matter in this discussion the uniform expansion can lead to three different possible fates of the universe. If the total density of the universe is below a certain density, called the critical density, then the universe will continue to expand forever. If the total density of the universe is above the critical density, then the universe will eventually stop expanding and start collapsing again. However, several considerations suggest that the most desirable value of the density of the universe is the same as the critical density, in which case the expansion will slow down but will never start collapsing again. In the presence of cosmological constant these dynamics will get modified. If there is positive cosmological constant, which has been observed, the universe will accelerate even when the density of the universe is equal to the critical density.

With this simple picture of the universe we can write a brief history of the universe. However, we have already evidenced many deviations from this picture, some of which can be considered as small fluctuations to this basic structure, while other deviations require drastic extensions and modifications of the model. For completeness we shall note some of the major deviations during this narration of the history of the universe.

### Brief History

The universe started with the big-bang at all points in space, not a single singular point, and every point started moving away from each other uniformly. It was like the surface of a balloon; as one blows a balloon, every point moves away from each other. We can define a temperature of the universe, which corresponds to the average energy per particle at any given time. We can also represent a particular time of the universe by the available energy per particle at that time. When the universe cooled down to a temperature corresponding to the Planck scale of $M_{Pl} \sim 10^{19}$ GeV, gravity started becoming weak. Before this time the evolution of the universe was governed by quantum gravity. Due to our lack of knowledge of quantum gravity, we cannot extrapolate our results to any time earlier than this. This would mean that there might not have been any big-bang to start with and the universe could have been evolving in a completely different manner. So, we begin all our discussions from the Planck scale.

**TABLE 15.1**
Some of the possible important events during the evolution of the universe

| Events | Energy in GeV | time in secs/yr | temp in °K |
|---|---|---|---|
| BIG-BANG | $\infty$ | 0 | $\infty$ |
| Planck scale: End of quantum gravity era | $10^{19}$ | $10^{-44}s$ | $10^{32}$ |
| Grand unification | $10^{15}$ | $10^{-37}s$ | $10^{28}$ |
| LR symmetry, PQ symmetry, L-violation, Supersymmetry, Extra dimensions, Baryogenesis, Inflation | | | |
| Electroweak symmetry breaking | $10^2$ | $10^{-10}s$ | $10^{15}$ |
| QCD chiral symmetry breaking and color confinement | $10^{-1}$ | $10^{-5}s$ | $10^{12}$ |
| Nucleosynthesis | $10^{-4}$ | $100s$ | $10^9$ |
| Matter dominates over radiation<br>Formation of atoms<br>Decoupling of matter and radiation | $2 \times 10^{-10}$ | $3 \times 10^5 y$ | 3000 |
| Formation of large scale structures | $10^{-12}$ | $10^9 y$ | 15 |
| PRESENT UNIVERSE | $2.3 \times 10^{-13}$ | $13.7 \times 10^9 y$ | 2.73 |

As the universe continued expanding it went through many phase transitions corresponding to the gauge symmetry breaking in particle physics. The first one was immediately after the Planck scale, the phase transition corresponding to the grand unification. The exotic particles giving rise to proton decay and other new phenomena decoupled below this temperature of grand unification phase transition. This means that these particles decayed into lighter particles, but the energy available to the lighter particles was not large enough to recreate these particles again. So, the number density of these particles reduced drastically. The next important landmark in the history of the universe was the temperature corresponding to around $10^{10}$ GeV. It is likely that lepton number violation took place around this energy, which in turn generated a small asymmetry in the number density of matter and antimatter. Around this time several other important transitions might have taken place, such as the left–right symmetry breaking, spontaneous parity violation, Peccei–Quinn symmetry breaking, supersymmetry breaking in the hidden sector, or split supersymmetric partners of matter decoupled. If there are dimensions higher than four, the extra dimensions should have decoupled from our four-dimensional world around this time. Another very important event, which took place around this time, is inflation. We now believe that the universe went through a phase of very rapid expansion, known as the inflation, when all prehistory would have been erased. The scale of inflation is expected to be as high as the GUT scale, but the reheating temperature is much lower. In most supersymmetric models the scale is expected to have been as low as $10^6$ GeV to avoid the overproduction of gravitinos.

After the electroweak phase transition at around 100 GeV, the massive gauge bosons $W^{\pm}, Z$ decoupled because the inverse decay rates became very small. The strength of weak interaction also started decreasing rapidly and became inversely proportional to the square of the $W^{\pm}, Z$ gauge boson masses. This made the neutrinos very weakly interacting. At energies around 1 MeV, the neutrino interaction rate became smaller than the expansion rate of the universe, so the neutrino could no longer remain in equilibrium with the thermal bath and, hence, decoupled.

The next major event in the history of the universe was nucleosynthesis. After the *QCD* chiral symmetry breaking at around 100 to 300 GeV, baryons and mesons were formed. The primordial light nuclei formed after this at around 1 MeV. The most attractive feature of nucleosynthesis is that the prediction of primordial light nuclei fits very well with observations. So, any modifications of the present picture of our cosmos should retain the predictions of nucleosynthesis.

After $3 \times 10^5$ yrs, the universe became matter-dominated and atoms started forming. As the electron density reduced due to formation of atoms, interactions of photons became much less and radiation decoupled from matter. This radiation became the cosmic microwave background radiation that we see today. The radiation is uniformly distributed in all directions, but contains small fluctuations. This anisotropy of the cosmic microwave provides us much important information about our early universe and also about structure formation.

One of the most important modifications of the simple model of our cosmos is the addition of the cosmological constant or dark energy. All the observations including the anisotropy of the cosmic microwave background radiation now requires that our universe is dominated by some background energy which acts like matter with negative pressure, so although the matter density of our universe is same as the critical density, the universe is accelerating and will expand forever. Explanation of this dark energy is very difficult. A large fraction of the matter is again in the form of dark matter. Some of the suggestions to solve these problems require drastic modifications of the model of cosmology which can change the fate of our universe at a later time. Therefore, although we have a simple structure of our universe that is beautiful and can explain many things, it is far from complete.

## 15.2 Evolution of the Universe

We shall develop our understanding of the standard cosmological model in steps, beginning with the cosmological principle. The isotropy and homogeneity of space and time refer only to very large scale. But theoretically we assume that at any scale the universe is isotropic and homogeneous and the small anisotropy and inhomogeneity are due to some perturbation, and we try to find the cause of this perturbation. In slightly technical language this means that the effect of the rest of the universe on any particular object is the same, which is denoted by the almost flat background metric. Since the gravitational interaction of all objects in the universe on any particular test body can be described by the curvature of space in the neighbourhood of the test body, we include the effect of gravitational interactions of all the objects in the universe in the background metric. The cosmological principle then states that the background metric is isotropic and homogeneous. In this chapter we shall refer to section 11.1 for our concept of space–time and the gravitational interaction.

### Geometry

We shall now develop the background metric satisfying the cosmological principle starting from our geometrical understanding of space–time. Let us consider an invariant line element between two space–time points

$$ds^2 = g_{\mu\nu} dx^\mu dx^\nu.$$

This line element tells us how two space–time points are connected when the geometry of space connecting the two points is described by the metric tensor $g_{\mu\nu}$. In the general theory of relativity the coordinates have no absolute significance, only the line elements between two points are a relevant quantity of interest. This ensures Mach's principle that there is no preferred coordinate system. The geometry of space at any space–time point, represented by the metric tensor $g_{\mu\nu}$, determines the effect of gravitational interactions on a test object at the given space–time point. The cosmological principle thus implies a background metric that is isotropic and homogeneous.

For any comoving observer, there is no spatial separation. Hence, the proper time interval ($dx^0 = dt$ corresponding to $dx^i = 0, i = 1,2,3$) will be the invariant interval connecting two neighbouring events $ds^2 = g_{00} dt^2$ for the comoving observer implying $g_{00} = 1$. In a homogeneous and isotropic universe, synchronization of clocks should be possible irrespective of the position of the clock, which implies $g_{0i} = 0$. The line element can then be written as

$$ds^2 = dt^2 + g_{ij} dx^i dx^j = dt^2 - dl^2, \tag{15.2}$$

where $dl^2$ is the *proper distance* between two points, which is the distance between the two points at any instant of time $t$.

We discussed the example of a balloon while explaining the expansion of the universe uniformly, in which all points are moving away from each other. A circle is a one-dimensional space, embedded in a two-dimensional space with the constraint that the distance from the origin to any point is fixed. This satisfies the condition that as the radius of the circle increases, all points on the circle move away from each other. A two-dimensional extension of this picture is a sphere, which can be constructed by taking all points in a 3-dimensional surface which are at a fixed distance away from the origin. We generalize this to a 3-sphere, which can be constructed in a 4-dimensional space with coordinates $(x, y, z, w)$. All points on the sphere will be at a fixed distance away from the origin, satisfying the condition

$$R^2 = x^2 + y^2 + z^2 + w^2,$$

where $R$ is a constant radius. This constraint equation will allow us to express the fourth coordinate as

$$w^2 = R^2 - (x^2 + y^2 + z^2). \tag{15.3}$$

If we now express the coordinates $(x, y, z)$ to a polar coordinate system $(r, \theta, \phi)$, so

$$x = r \sin\theta \cos\phi; \quad y = r \sin\theta \sin\phi; \quad y = r \cos\theta,$$

we can express

$$dw = \frac{r\,dr}{w} = \frac{r\,dr}{(R^2 - r^2)^{1/2}}.$$

This will allow us to write the spatial line element on the surface of this three-sphere as

$$\begin{aligned} dl^2 &= dx^2 + dy^2 + dz^2 + dw^2 \\ &= \frac{dr^2}{1 - r^2/R^2} + r^2 d\theta^2 + r^2 \sin^2\theta d\phi^2. \end{aligned} \tag{15.4}$$

It is sometimes convenient to write the spatial line element in a different form as

$$dl^2 = R^2[d\chi^2 + \sin^2\chi d\Omega], \tag{15.5}$$

where we change the coordinates to

$$r = R \sin\chi$$

and write the angular part as

$$d\Omega = d\theta^2 + \sin^2\theta d\phi^2.$$

This form of the equation shows explicit dependence of the line element on the radius of the sphere.

**Scale Factor**

The next important ingredient for the understanding of the evolution of the universe is the *scale factor*. We now demand the proper physical distance $dl$ between two comoving points on the 3-sphere scales with time. From our earlier discussions we noticed that as the balloon inflates the distance between any two points increases. For any sphere, the distance between any two points on its surface increases as the radius of the sphere increases. From the last expression we can thus conclude that homogeneity and isotropy of space allows the scaling of the line element only in the form

$$dl^2 = a(t)^2\, g^0_{ij} dx^i dx^j.$$

The invariant 4-dimensional line element representing a homogeneous and isotropic universe, in which the distance between any two comoving points is scaling as $l(t) = a(t)$, can then be written as

$$\begin{aligned} ds^2 &= dt^2 - a(t)^2 dl^2 \\ &= dt^2 - a^2(t)\left\{\frac{dr^2}{1-kr^2} + r^2 d\theta^2 + r^2 \sin^2\theta d\phi^2\right\}, \end{aligned} \tag{15.6}$$

where $k = 1/R^2$ is a constant. This is the maximally symmetric Robertson–Walker metric for a homogeneous and isotropic spatial section [243].

In our expanding universe the scale factor $a(t)$ may be related to an observable, the red-shift of any receding object, which is the change in wavelengths of spectral lines. If light comes from a distant source (S) at $(r,t)$, then the light started from the source at an earlier time $(t)$ corresponding to a scale factor $a(t)$, which is different from the scale factor $a(t_0)$ corresponding to the time $(t_0)$ when the light reaches the observer $O$ at $(0,t_0)$. Thus, measurement of the red-shift of the distant object can give us the scale factor of an earlier time.

Let the wavelength of the light emitted by $S$ be $\lambda$, so the time difference between the emission of two crests is $\delta t = \lambda$. Let us assume that the first crest reached us at $t_0$ and the second crest reached us at $t_0 + \delta t_0$, so the wavelength we shall measure for this light is $\lambda_0 = \delta t_0$. For the light traveling along the geodesics $ds^2 = 0$, assuming $d\phi = d\theta = 0$, we can write from equation (15.6)

$$\int_{t_0}^{t} \frac{dt}{a(t)} = \int_{t_0+\delta t_0}^{t+\delta t} \frac{dt}{a(t)} = \int_0^r \frac{dr'}{(1-kr'^2)^{1/2}}.$$

If we now assume that the scale factor has not changed within the short period $\delta t_0$ or $\delta t$, we can get the relation

$$\frac{\delta t}{a(t)} = \frac{\delta t_0}{a(t_0)} \qquad \Longrightarrow \qquad \frac{\lambda}{\lambda_0} = \frac{a(t)}{a(t_0)}.$$

This allows us to define the red-shift $z$ as

$$1 + z = \frac{\lambda_0}{\lambda} = \frac{a(t_0)}{a(t)}, \tag{15.7}$$

which is related to the change in wavelength of any light which has travelled a long distance from a time when the universe was smaller in size with scale factor $a(t)$ to our present time when the scale factor is $a(t_0)$.

The rate of expansion of the universe is given by the Hubble parameter $H$ (see equation (15.1)). The proper distance between two points $d(t)$, which is the distance corresponding to $dt = 0$, given by

$$d(t) = a(t) \int_0^r \frac{dr'}{(1-kr'^2)^{1/2}},$$

and the velocity at which these two points are receding from each other, given by

$$v(t) = \dot{a}(t) \int_0^r \frac{dr'}{(1-kr'^2)^{1/2}},$$

will give us the expression for the Hubble constant at any given time

$$H = \frac{v(t)}{d(t)} = \frac{\dot{a}(t)}{a(t)}. \tag{15.8}$$

The present value of the Hubble constant is denoted by $H_0 = \dot{a}(t_0)/a(t_0)$. Usually the Hubble constant is expressed as a dimensionless quantity by defining

$$h = \frac{H_0}{100\ km\ s^{-1}\ Mpc^{-1}}, \tag{15.9}$$

and the present best fit value of the parameter is [44]

$$h = 0.73^{+0.3}_{-0.4}. \tag{15.10}$$

Another related parameter of importance is the deceleration parameter, which gives the rate of change of the Hubble constant, defined as

$$q = -\frac{\ddot{a}(t)}{\dot{a}(t)^2} a(t) = -\frac{1}{H^2} \frac{\ddot{a}(t)}{a(t)}. \tag{15.11}$$

The present value of the deceleration parameter is denoted by $q_0$. We follow the convention that the present values of any parameters will have a subscript 0.

The curvature parameter $k$ has important physical significance and gives the topological nature of space. When $k$ is positive, the space is similar to a sphere with positive curvature. Since it is related to the radius of the sphere in the four-dimensional space in which our 3-sphere is embedded, and is a constant parameter, we can always normalize it to $k = 1$. The spatial line element is given by

$$dl^2 = d\chi^2 + \sin^2 \chi d\Omega\,. \tag{15.12}$$

This corresponds to a closed universe. The spatial metric corresponds to a three-sphere.

In the case of negative curvature, i.e., $k$ is negative, we can make the substitution

$$R \to iR, \quad \chi \to i\chi$$

and normalize to $k = -1$ and write the line element as

$$dl^2 = d\chi^2 + \sinh^2 \chi d\Omega \,. \tag{15.13}$$

The hyperbolic plane is unbounded with infinite volume implying that the universe is open.

The other topologically interesting case is the flat universe corresponding to zero curvature $k = 0$. At any instant of time the metric is equal to the Minkowski space:

$$dl^2 = dr^2 - r^2 d\Omega \,. \tag{15.14}$$

Any line element is the same as that of a Minkowski space times the scale factor. We shall now see how the universe evolved for the different values of $k = \pm 1, 0$.

To understand how the scale factor evolves with time, we start with Einstein's equation (equation (11.28)) [173]

$$R_{\mu\nu} - \frac{1}{2} g_{\mu\nu} \mathscr{R} + \Lambda g_{\mu\nu} = -8\pi G T_{\mu\nu}. \tag{15.15}$$

We include the cosmological constant $\Lambda$ in the equation, which can have only cosmological consequences but no consequence on gravitational interactions between two objects. Initially this term was introduced by Einstein to construct a static model of the universe, but later he removed the term because expansion of the universe was established. However, a most general form of the equation should include the cosmological constant. At present, observations require a nonvanishing cosmological constant, so we shall include its effect in our discussions.

### Equation of State

In addition to the geometry of space–time we need to know the form of matter distribution, given by the stress–energy tensor $T_{\mu\nu}$, to understand the development of the metric with time. The homogeneity and isotropy of space simplifies this problem tremendously. Consistency with the symmetries of the metric dictates that the stress–energy tensor must be diagonal, and isotropy implies the space-components are equal. A simple form of the stress–energy tensor, consistent with these requirements can be taken to be of the form

$$T^{\mu\nu} = \text{diag}[\rho, -p, -p, -p], \tag{15.16}$$

where both the energy density $\rho(t)$ and the pressure $p(t)$ of this perfect fluid is time-dependent. The time component of the conservation of stress–energy ($T^{\mu\nu}_{;\nu} = 0$) gives

$$d(\rho a^3) = -p d(a^3), \tag{15.17}$$

which states that the change in energy in a comoving volume is given by the negative of pressure times the change in volume.

The equation of state can be written as

$$p = \omega \rho. \tag{15.18}$$

We consider a simple form where $\omega$ is independent of time. The energy density will then evolve as

$$\rho \propto a^{-3(1+\omega)}. \tag{15.19}$$

Different values of $\omega$ correspond to

| | | |
|---|---|---|
| $(i)$ Matter dominated universe : | $p = 0$ | $\rho \propto a^{-3}$ |
| $(ii)$ Radiation dominated universe : | $p = \rho/3$ | $\rho \propto a^{-4}$ |
| $(iii)$ Vacuum energy dominated universe : | $p = -\rho$ | $\rho \propto const.$ |

In general, for ordinary matter $\omega$ lies in the *Zeldovich interval* of $0 \leq \omega \leq 1$. However, in some models it could be as low as $-1/3$ for matter and lower than $-1$ for vacuum energy. We shall come back to this discussion again in the context of cosmological constant.

**Evolution of the Scale Factor**

To find out the evolution of the scale factor, we start with the nonvanishing components of the Ricci tensor ($R_{ij}$) and the Ricci scalar ($\mathscr{R}$) for the Robertson–Walker metric

$$\begin{aligned} R_{00} &= -3\frac{\ddot{a}}{a} \\ R_{ij} &= -\left(\frac{\ddot{a}}{a} + \frac{2\dot{a}^2}{a^2} + \frac{2k}{a^2}\right) g_{ij} \\ \mathscr{R} &= -6\left(\frac{\ddot{a}}{a} + \frac{\dot{a}^2}{a^2} + \frac{k}{a^2}\right). \end{aligned} \tag{15.20}$$

The field equations governing the evolution of the scale factor $a(t)$ are then given by

$$\frac{\dot{a}^2}{a^2} + \frac{k}{a^2} = \frac{8\pi G}{3}\rho + \frac{\Lambda}{3} = \frac{8\pi G}{3}(\rho + \rho_v) \tag{15.21}$$

$$2\frac{\ddot{a}}{a} + \frac{\dot{a}^2}{a^2} + \frac{k}{a^2} = -8\pi G\, p + \Lambda = 8\pi G\,(-p + \rho_v). \tag{15.22}$$

These two equations are called the Einstein–Friedmann–Lemaitre equations [244]. Here we defined $\rho_v = \Lambda/(8\pi G)$, which could be nonvanishing when any field acquires a nonvanishing vacuum expectation value $T_{\mu\nu}^{vac} = \rho_v g_{\mu\nu}$ or if there is a nonvanishing cosmological constant $\Lambda$ in Einstein's equation.

We may rewrite equation (15.21) in terms of the Hubble parameter, defined in equation (15.8), as

$$\frac{k}{H^2a^2} = \frac{8\pi G}{3H^2}(\rho + \rho_v) - 1 = \Omega - 1, \tag{15.23}$$

where $\Omega = \Omega_m + \Omega_\Lambda$; $\Omega_m = \rho/\rho_c$; $\Omega_\Lambda = \rho_v/\rho_c$ and we define the critical density $\rho_c$ as

$$\rho_c = \frac{3H^2}{8\pi G}. \tag{15.24}$$

Since $H^2a^2 \geq 0$, we can infer

$$k = +1 \Rightarrow \Omega > 1$$
$$k = 0 \Rightarrow \Omega = 1$$
$$k = -1 \Rightarrow \Omega < 1.$$

Thus, the universe is closed if the density of the universe is greater than the critical density of the universe, flat if the density if same as the critical density, and open if the density is less than the critical density.

Although it is difficult to get an explicit form of the scale factor, we can express $a(t)$ in some special cases in terms of some parameter related to time $t$. For this discussion we shall assume a vanishing cosmological constant. For $k = +1$ (and hence $\Omega_0 > 1$) and a matter-dominated universe, we can express $a(t)$ in terms of a parameter $\theta$ related to time as

$$a(t) = a_0(1 - \cos\theta)\frac{\Omega_0}{2(\Omega_0 - 1)}. \tag{15.25}$$

This gives the evolution of $a(t)$ for a closed universe. At $t \equiv \theta = 0$, the universe starts from $a(0) = 0$ and expands to a maximum value at $\theta = \pi$ and then collapse, back to a point ($a(t_{col}) = 0$) again at a later time $t_{col} \equiv \theta = 2\pi$. However, for the same matter-dominated universe with $k = -1$ (and hence $\Omega < 1$) the time development is given in terms of $\psi = -i\theta$ as

$$a(t) = a_0(\cosh\psi - 1)\frac{\Omega_0}{2(1 - \Omega_0)^{3/2}}. \tag{15.26}$$

The universe expands forever with increasing scale factor. $k = 0$ is the limiting case when the universe expands forever without any acceleration. The same result will also be valid for a radiation-dominated universe. However, if we include contributions from cosmological constant, this conclusion will change which we shall discuss later. The best fit for the present density of the universe is [44]

$$\Omega_0 = 1.003^{+0.013}_{-0.017}. \tag{15.27}$$

Thus, the total density (including contributions from cosmological constant) is almost same as the critical density

$$\rho_{c0} = 6.8 \times 10^{-27}\ \text{kg m}^{-3}.$$

If the universe is open or closed $\Omega_0 \neq 1$, then from equation (15.23) it is implied that near the Planck scale the universe started with $|\Omega - 1| \ll 10^{-50}$. Then, a natural choice would be a flat universe $\Omega = 1$ and $k = 0$, so it was flat at all times.

Combining the two Einstein–Friedmann–Lemaitre equations (15.21) and (15.22) we can write an expression for the acceleration

$$\frac{\ddot{a}}{a} = -\frac{4\pi G}{3}(\rho + 3p) + \frac{\Lambda}{3}. \tag{15.28}$$

In the absence of the cosmological constant, this equation states that the present expansion ($\dot{a} > 0$) of the universe must have started from $a = 0$, unless the sign of $(\rho + 3p)$ changed sometime in the past. This simple extrapolation leads to the concept of the singular big-bang corresponding to $a = 0$. We denote the corresponding time as $t = 0$ and mark any event in the history of the universe with reference to this time. In other words we normalize the model with $a(t = 0) = 0$. If $\Lambda$ contributes, the origin of time may shift depending on the dynamics.

With reference to this origin of time, the present time defines the *age of the universe* $t_0$. The age of the universe depends directly on the Hubble constant, the deceleration parameter and other parameters of the model. For the simple cases of flat universe, it can be expressed as $t_0 \sim CH_0^{-1}$ where $C = 2/3, 1/2, 1$ for matter-dominated, radiation-dominated or vacuum energy-dominated universes, respectively. For $\Omega = \Omega_m + \Omega_\Lambda = 1$, it is possible to get an analytical expression for the age of the universe:

$$H_0 t_0 = \frac{2}{3\sqrt{\Omega_\Lambda}} \ln\left(\frac{1 + \sqrt{\Omega_\Lambda}}{\sqrt{1 - \Omega_\Lambda}}\right). \tag{15.29}$$

The present best fit value of the age of the universe is [44]

$$t_0 = 13.7^{+0.1}_{-0.2}\ \text{Gyr}. \tag{15.30}$$

The improved accuracy of this value assumes a cold dark matter model.

### Cosmological Constant

One of the most important findings in cosmology is the nonvanishing cosmological constant. The first clue for the nonvanishing cosmological constant came from supernova 1a observations and subsequently it was confirmed with results from studies of rich clusters of galaxies and cosmic microwave background power spectrum, which measured temperature anisotropy as a function of angular scale. A best fit for the cosmological constant and matter density is

$$\Omega_\Lambda = 0.62 \pm 0.16 \qquad \text{and} \qquad \Omega_m = 0.24 \pm 0.10. \tag{15.31}$$

The different components of the total matter density are [44]

$$\begin{aligned} \text{Matter Density} \quad \Omega_m h^2 &= 0.127^{+0.007}_{-0.009} \\ \text{Baryon Density} \quad \Omega_b h^2 &= 0.0223^{+0.0007}_{-0.0009} \\ \text{Radiation Density} \quad \Omega_r h^2 &= 2.47 \times 10^{-5}. \end{aligned}$$

Thus, the cosmological constant constitutes about 70% of the total density of the universe, another 25% of the density is in the form of dark matter, and only about 5% is the baryonic matter.

The positive and nonvanishing cosmological constant is difficult to explain from any theory. Writing in a different unit the value for the cosmological constant is

$$\Lambda \sim 0.6 \times 10^{-56} \text{ cm}^{-2}.$$

This is too small when we look at it from a different point of view. Consider the electroweak phase transition, in which we start with a scalar potential

$$V(\phi) = -\frac{1}{2}\mu^2 \phi^\dagger \phi + \frac{\lambda}{4}(\phi^\dagger \phi)^2.$$

For $\mu^2, \lambda > 0$, the field $\phi$ gets a vacuum expectation value $\langle \phi \rangle = \sqrt{\mu^2/\lambda}$ and the cosmological constant then corresponds to the minimum of the energy

$$\Lambda_\phi = -4\pi G\, \mu^2 |\langle \phi \rangle|^2 \sim -2 \times 10^{-4} \text{ cm}^{-2}.$$

This is too large compared with the present value of the cosmological constant. If we consider the grand unified theory phase transition or the Planck scale, the situation will be much worse. However, if there is supersymmetry in nature, then the cosmological constant will remain vanishing as long as supersymmetry is unbroken. Supersymmetry breaking scale will be the relevant scale for the cosmological constant, which is also much higher. Several solutions to the cosmological constant problem has been discussed in the literature, which we shall discuss again in chapter 17. Some of these solutions require a different equation of state ($p = \omega \rho$) for the dark energy. Present observations give the best fit value for the equation of state parameter to be

$$\omega = -0.98 \pm 0.12. \tag{15.32}$$

This value is consistent with the dark energy solution of the cosmological constant, which requires $\omega = -1$.

The presence of the cosmological constant changes the evolution of the universe. From equation (15.21) it is clear that the cosmological constant adds up to the total density of the universe as far as expansion of the universe is concerned. Equation 15.22 shows the difference between the ordinary matter and cosmological constant for the evolution of the universe. A positive cosmological constant or the dark energy produces negative pressure for the expansion of the universe. This can be understood better by writing the Poisson equation for the Newtonian potential $\phi$:

$$\nabla^2 \phi = 4\pi G \rho - \Lambda. \tag{15.33}$$

Solving for $\phi$ for a spherical mass distribution $M$, one can write the force on a test body with mass $m$ at a distance $r$ as

$$\vec{F} = -\frac{GMm}{r^3}\vec{r} + \frac{1}{3}\Lambda m \vec{r}.$$

Thus, the cosmological constant $\Lambda$ produces a repulsive force on any test body.

Einstein introduced the cosmological constant to obtain a static universe, which could be achieved with $k = 1$ and a positive cosmological constant

$$\Lambda_E = 4\pi G\rho = \frac{k}{a(t)^2} \Longrightarrow \ddot{a}(t) = \dot{a}(t) = 0.$$

This model was soon ruled out by the observation that the universe is expanding.

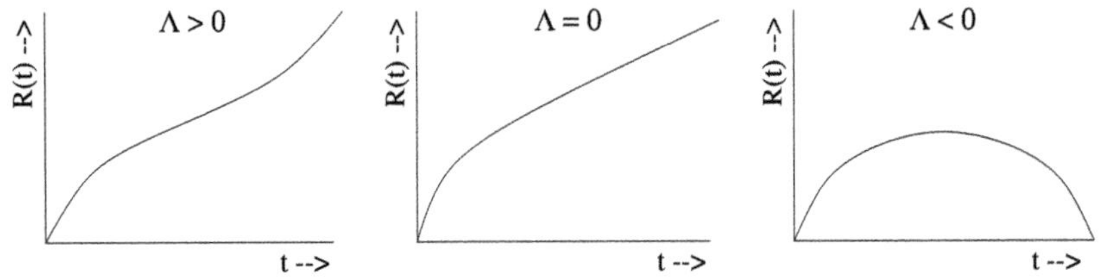


**FIGURE 15.1**
Evolution of the universe for vanishing curvature ($k = 0$) and different cosmological constant.

For a flat universe ($k = 0$), the universe expands without any acceleration if there is no cosmological constant. In the presence of a cosmological constant, the evolution of the flat universe is shown in figure 15.1. Since the observations confirmed a positive cosmological constant and the universe most likely has zero curvature, the universe is expanding and accelerating at present. In this case the present evolution of the scale factor can be analytically expressed as

$$a(t) = a_0 \left\{ \frac{1}{2\Omega_\Lambda} \left[\cosh\left(\sqrt{3\Lambda}\, t\right) - 1\right] \right\}^{1/3}, \tag{15.34}$$

normalized to $a(0) = 0$. It shows that asymptotically the universe grows exponentially for large $t$

$$a(t) \sim \exp[H_\Lambda t] \sim \exp\left[ \left(\frac{\Lambda}{3}\right)^{1/2} t \right], \tag{15.35}$$

where $H_\Lambda^2 = \Lambda/3$ is the Hubble constant in the *de Sitter phase*.

We now sum up the discussions in this section. We start with a homogeneous and isotropic universe described by the Robertson–Walker metric. The universe is expanding and the expansion is governed by the Hubble constant. The total matter density (including the dark matter and the dark energy) is same as the critical density of the universe and the universe is flat. The cosmological constant is nonvanishing and positive, so the universe is flat and accelerating.

## 15.3 Primordial Nucleosynthesis

In the early universe, we try to extrapolate our particle physics models to get some information. But we could not verify most of these results. The earliest time from which we could get some experimental inputs is the time of formation of nuclei, when the average energy per particle was around 1 MeV. From our knowledge of the present abundances of elements we could extrapolate them to the time of primordial nucleosynthesis, when these elements had synthesized. This allows us to verify the theoretical predictions for the elements, which is one of the major triumphs of the standard model of cosmology.

### Equilibrium Thermodynamics

We shall first introduce some thermodynamic quantities. Although in the early universe the expansion rate of the universe was too high for any reaction to come into equilibrium, at the time of nucleosynthesis the universe may be described by equilibrium thermodynamics. In equilibrium, the number density ($n$), energy density ($\rho$), and pressure ($p$) of a dilute, weakly interacting gas of $g$ internal degrees of freedom is given by

$$\begin{aligned} n &= \frac{g}{(2\pi)^3} \int f(\vec{p})\, d^3p \\ \rho &= \frac{g}{(2\pi)^3} \int E(\vec{p})\, f(\vec{p})\, d^3p \\ p &= \frac{g}{(2\pi)^3} \int \frac{|\vec{p}|^2}{3E}\, f(\vec{p})\, d^3p, \end{aligned} \tag{15.36}$$

where $E^2 = |\vec{p}|^2 + m^2$ and $f(\vec{p})$ is the phase space distribution function,

$$[f(\vec{p})]_{fermion} = \frac{1}{\exp^{(E-\mu)/T} + 1} \quad \text{and} \quad [f(\vec{p})]_{boson} = \frac{1}{\exp^{(E-\mu)/T} - 1}, \tag{15.37}$$

and $\mu$ is the chemical potential. In a reaction $i + j \leftrightarrow k + l$ in chemical equilibrium, the chemical potentials of these particles satisfy

$$\mu_i + \mu_j = \mu_k + \mu_l. \tag{15.38}$$

In our convention, the Boltzmann constant $k_B = 1$, so $1^\circ\text{K} = 0.86 \times 10^{-13}\text{GeV}$.

The number densities of relativistic particles (with mass $m \ll T$) and nonrelativistic particles (with mass $m \gg T$) are

$$\begin{aligned} n_R &= C\frac{\zeta(3)}{\pi^2} g T^3 \\ n_{NR} &= g\left(\frac{mT}{2\pi}\right)^{3/2} \exp^{-(m-\mu)/T}, \end{aligned} \tag{15.39}$$

where $C = 1$ for bosons, $C = 3/4$ for fermions, and the zeta function of 3 is $\zeta(3) = 1.202...$ Thus, in equilibrium the total energy density and pressure will have contributions mostly from the relativistic species, since the number density of the nonrelativistic species is exponentially suppressed. Neglecting the contributions of the nonrelativistic species, we can then write a simplified expression for the energy density and pressure of all the particles as

$$\rho_R = \frac{\pi^2}{30} g_* T^4$$
$$p_R = \frac{\rho_R}{3} = \frac{\pi^2}{90} g_* T^4, \tag{15.40}$$

where the effective number of massless degrees of freedom is given by the contributions from both fermions and bosons

$$g_* = \sum_i \frac{1+C}{2} g_i \frac{T_i^4}{T^4}, \tag{15.41}$$

where the summation extends over all the species, including both fermions and bosons. We assume that $g_i$ numbers of species are at a temperature $T_i$. Here $T$ is the photon temperature and all the species with masses $m_i \ll T$ contribute. Around 100 GeV, the effective number of degrees of freedom is $g_* \sim 100$, while at temperatures less than 1 MeV, it is $g_* \sim 3.36$.

Another important quantity is the entropy of the universe $S$ per comoving volume, given by

$$S = \frac{a^3(\rho + p)}{T} = a^3\, s, \tag{15.42}$$

where the entropy density $s$ is dominated by the relativistic particles and is given by

$$s = \frac{2\pi^2}{45} g_{*s} T^3, \tag{15.43}$$

where

$$g_{*s} = \sum_i \frac{1+C}{2} g_i \frac{T_i^3}{T^3},$$

and the photon number density is related to entropy as $s = 1.8 g_{*s} n_\gamma$. For the present value we can write $s/n_\gamma = 1.8 g_{*s}(\text{today}) = 7.04$. Conservation of energy implies that the entropy per comoving volume remains constant in thermal equilibrium,

$$S = sa^3 = \frac{2\pi^2}{45} g_* a^3 T^3 = \text{constant} \implies a \sim T^{-1}. \tag{15.44}$$

Thus, during adiabatic expansion, the universe cools down.

With this definition of entropy we can now define the number density per comoving volume $N \equiv n/s$, which remains constant in equilibrium. For example, the baryon number density per comoving volume

$$\frac{n_B}{s} = \frac{n_b - n_{\bar{b}}}{s} \tag{15.45}$$

remains unchanged if there is no baryon-number violating interactions. If there is baryon-number violating interaction in equilibrium, the baryon number density vanishes. Any reaction can reach equilibrium, when the reaction rate $\Gamma(T)$ at any temperature $T$ is faster than the expansion rate of the universe

$$\Gamma(T) > H(T) = 1.66\sqrt{g_*}\frac{T^2}{M_{Pl}}, \tag{15.46}$$

where $M_{Pl}$ is the Planck mass. Although this out-of-equilibrium condition can give us a rough estimate, in actual problem the departure from equilibrium can be understood by solving the Boltzmann equations.

**Synthesis of Elements**

At around 1 sec after the big-bang or about 1 sec after gravity started becoming weak, synthesis of light elements, such as H, D, $^3$He, $^4$He and $^7$Li, started, when the temperature was about $10^{10}$ K and the average kinetic energy was about 1 MeV [245]. Although proton, neutron and light elements had formed at that time, they were nonrelativistic. The number density of such nonrelativistic particles in thermal equilibrium with mass number $A$ is given by

$$n_A = g_A \left(\frac{m_A T}{2\pi}\right)^{3/2} \exp^{(\mu_A - m_A)/T}, \tag{15.47}$$

where $g_A$ is the statistical factor. The chemical potential $\mu_A$ of the light elements $A$ with charge $Z$ is related to that of the proton $\mu_p$ and the neutron $\mu_n$ in chemical equilibrium by

$$\mu_A = Z\mu_p + (A - Z)\mu_n. \tag{15.48}$$

With the knowledge of the chemical potential of the species $A$, we can now write the chemical potential and hence the number density of the species $A$ in terms of the number densities of the proton and neutron as

$$n_A = g_A A^{3/2} 2^{-A} \left(\frac{2\pi}{m_N T}\right)^{3(A-1)/2} n_p^Z n_n^{A-Z} \exp^{B_A/T}, \tag{15.49}$$

where $B_A$ is the binding energy of the species $A$ and is defined as

$$B_A = Zm_p + (A - Z)m_n - m_A. \tag{15.50}$$

Binding energies ($B_A$ in MeV) and statistical factors ($g_A$) of some of the light nuclei $^AZ$ are $^2$H [2.22,3], $^3$H [6.92,2], $^3$He [7.72,2], $^4$He [28.3,1], $^{12}$C [92.2,1], written as $^AZ$ [$B_A$,$g_A$].

**Light Element Abundances**

The primordial abundances of any nuclei are difficult to estimate. It is convenient to define the relative mass fraction of the different species of nuclei $A$ as

$$X_A = \frac{An_A}{n_B} = \frac{n_A A}{n_n + n_p + \sum_i (An_A)_i}, \tag{15.51}$$

with $\sum_i X_i = 1$. Using the expression for the number density of the particles, the mass fraction for a species $A(Z)$ can then be written as

$$X_A = \frac{1}{2} g_A \left[ \frac{\eta \zeta(3)}{\sqrt{\pi}} \left( \frac{2T}{m_N} \right)^{3/2} \right]^{A-1} A^{5/2} X_p^Z X_n^{A-Z} \exp^{B_A/T}, \qquad (15.52)$$

where the ratio of number of baryons to number of photons $\eta = n_b/n_\gamma$ is an unknown parameter in the model. This is a measure of the excess of baryons over antibaryons in the universe and plays a crucial role in the evolution of the universe.

Theoretical estimates of the light element abundances are possible for the time of primordial nucleosynthesis at a time $t \sim 180$ s. However, actual measurements are possible from a much later epoch, which introduce some uncertainty due to the steller nucleosynthesis when some light elements are ejected and some heavy elements are formed. High resolution spectra in astrophysical sites with low metal contents are extrapolated to zero-metallicity to reduce errors. Taking care of theoretical uncertainties and statistical and systematic errors, the relative mass fraction of $^4$He, denoted by $Y = 2n_n/(n_n + n_p)$, comes out to be [44]

$$Y = 0.249 \pm 0.009. \qquad (15.53)$$

The main source of error in $^4$He abundance is its steller production. On the other hand, for deuterium abundance, the main error comes from its depletion at temperatures above $6 \times 10^5$ K through the reaction $p$+D$\rightarrow$ $^3$He+$\gamma$. The present limit on deuterium abundance is [44]

$$\frac{D}{H} = (2.78 \pm 0.29)105. \qquad (15.54)$$

It is difficult to measure $^3$He abundance with any reasonable accuracy. The production of $^3$He from deuterium makes it more uncertain, and hence, this data is not used to extract any information from nucleosynthesis. The $^7$Li abundance is even more difficult to determine and its average value is [44]

$$\frac{^7\text{Li}}{H} = (1.7 \pm 0.02^{+1.1}_{-0.0}) \times 10^{-10}. \qquad (15.55)$$

We shall now compare these light element abundances with the theoretical predictions.

### Nucleosynthesis Predictions

The observational evidences for the abundances of light elements spread over 10 orders of magnitude with very little uncertainty. One of the major successes of the big-bang cosmology is to predict this large hierarchy in the element abundances. Therefore, any modification of the standard model of cosmology has to ensure that this prediction is not jeopardized. For example, if there are more than four dimensions, the extra dimensions should decouple from our present four-dimensional world before the time of nucleosynthesis.

The weak interaction processes

$$p + e^- \leftrightarrow n + \nu_e, \qquad p + \bar{\nu}_e \leftrightarrow n + e^+, \qquad n \leftrightarrow p + e^- + \bar{\nu}_e$$

keep the protons and neutrons in thermal equilibrium at around 10 MeV ($10^{-2}$ s). Any interaction continues to be in thermal equilibrium at any time $t$, if the interaction at that time $\Gamma(t)$ takes place at a faster rate than the expansion rate of the universe, which is given by the Hubble parameter

$$H(T) = 1.66\sqrt{g_*}\frac{T^2}{M_{Pl}}. \tag{15.56}$$

$T$ is the temperature at time $t$, and $g_*$ is the number of relativistic degrees of freedom available at that time. As the temperature comes down with time, the interaction rate slows down faster than the Hubble parameter. When the interaction rate becomes slower than the expansion rate of the universe, the particles move apart from each other before they can interact with each other, and hence, the interaction goes out-of-equilibrium. For the weak interaction, the departure from equilibrium starts at around a temperature that satisfies

$$\frac{\Gamma}{H} \approx \left(\frac{T}{0.8\ \text{MeV}}\right)^3 < 1, \tag{15.57}$$

which is about 0.8 MeV. Below this temperature the universe expands faster than the weak interaction rate.

When the neutrons and the protons are in equilibrium, their relative abundance is given by (neglecting chemical potentials)

$$\frac{n_n}{n_p} = \exp^{-\Delta m/kT}, \tag{15.58}$$

where $\Delta m = m_n - m_p$. This abundance continues until they reach the temperature 0.8 MeV, when their interaction rate goes out-of-equilibrium and the neutron to proton ratio becomes

$$\frac{n_n}{n_p} = \exp^{-\Delta m/kT_f} \simeq \frac{1}{6}. \tag{15.59}$$

At this time since the weak interaction rate becomes slower than the expansion rate of the universe, the neutron to proton conversion freezes out and the neutron to proton ratio becomes almost constant. Since neutrinos interact only through weak interaction, the neutrinos decouple around the same time and the electrons and the positrons annihilate. At a little lower temperature of $0.3 - 0.1$ MeV, corresponding to about 1 to 3 minutes after the big-bang, formation of helium starts via the reactions

$$\begin{aligned}
n + p &\leftrightarrow D + \gamma \\
D + D &\leftrightarrow {}^4\text{He} + \gamma \\
D + p &\leftrightarrow {}^3\text{He} + \gamma \\
D + n &\leftrightarrow {}^3\text{H} + \gamma,
\end{aligned}$$

which give an amount of helium abundance of

$$Y = \frac{2n_n}{n_n + n_p}. \tag{15.60}$$

The parameters involved in the prediction for the helium abundance are the fraction of baryons to photons

$$\eta = \frac{n_b}{n_\gamma}, \tag{15.61}$$

the number of relativistic degrees of freedom $g_*$, and the weak interaction rates for the neutron, which depend on the nuclear matrix element

$$|M|^2 \sim G_F^2(1+3g_A^2). \tag{15.62}$$

Hence, the main source of error in the prediction for $Y$ is the lifetime of the neutron $\tau_n$

$$\tau_n = 885.3 \pm 2.0\ s, \tag{15.63}$$

since the axial vector coupling constant $g_A$ is obtained from the neutron lifetime [44]

$$\tau_n^{-1} = \frac{1.636\ G_F^2}{2\pi^3}(1+3g_A^2)\ m_e^5. \tag{15.64}$$

At around 0.1 MeV, the equilibrium distribution of the neutron to proton ratio should have been $n_n/n_p = 1/74$, but since their interaction rate went out-of-equilibrium, their actual distribution remains much higher, almost $n_n/n_p = 1/7$. The formation of heavier elements is prevented strongly because of the nonexistence of stable atomic masses 5 and 8 nuclei and the Coulomb barriers. This explains why the abundance of $^7$Li is suppressed by 10 orders of magnitude compared with the helium abundance. Abundance of any heavier elements is also strongly suppressed for the same reason.

**Nucleosynthesis Results**

The abundance of a nucleus $A(Z)$ depends on $\eta^{A-1}$ as shown in the defining equation (15.52) for the relative mass fraction. The experimental values of the abundances of elements could be used to estimate the value of the baryon-to-photon ratio $\eta$, which comes out to be [44]

$$4.7 < \eta_{10} < 6.5 \qquad 98\%\ \text{c.l.}, \tag{15.65}$$

where $\eta_{10} = \eta \times 10^{10}$. A highly refined value of $\eta_{10}$ is possible from an analysis of the cosmic microwave background anisotropy data, which will be discussed in the next section.

To estimate the amount of baryonic matter in the universe, we use the relation $\eta = 2.73 \times 10^{-8}\ \Omega_b h^2$ for the present temperature, where

$$h = \frac{H_0}{100\ km\ s^{-1}\ Mpc^{-1}}. \tag{15.66}$$

The range of $\eta$, as obtained from nucleosynthesis, then gives [44]

$$0.017 \leq \Omega_b h^2 \leq 0.024. \tag{15.67}$$

For a value of the Hubble constant ($h \sim 0.7$) this implies $0.035 \leq \Omega_b \leq 0.049$, i.e., the amount of baryonic matter constitutes only about 3 to 5% of the critical density of the universe. On the other hand, the density of optically luminous matter is $\Omega_{lum} \sim 0.0024\, h^{-1}$, which is only 0.35% of the critical density of the universe.

The temperature fluctuations measured in cosmic microwave background radiation experiments (this will be discussed in the next section) on angular scales smaller than the horizon at last scattering, can also provide an independent value for the baryon-to-photon ratio $\eta$. This comes from the amplitudes of the acoustic peaks in the cosmic microwave background (CMB) angular power spectrum. If photons are created or destroyed between the nucleosynthesis era and the CMB decoupling, that would distort the CMB angular power spectrum. The CMB anisotropies independently test the nucleosynthesis predictions for $\Omega_b h^2$.

One of the most interesting predictions of the primordial nucleosynthesis is the number of neutrino species that were active at the time of nucleosynthesis, which is $N_\nu = 3$ [44]. This bound results from the constraint on the parameter involved in nucleosynthesis, the number of effective degrees of freedom $g_*$. However, this bound is not too strong and a fourth generation neutrino is not ruled out completely as yet. If this result is combined with the present LEP constraint that the number of neutrinos with mass less than $m_Z/2$ is 3, it allows one sterile neutrino (a singlet under the standard model), which can mix with the ordinary neutrinos and contribute to the nucleosynthesis bound. However, there is no constraint on the number of sterile neutrinos that do not mix with the ordinary neutrinos.

## 15.4 Cosmic Microwave Background Radiation

One of the most crucial predictions of the standard big-bang cosmological model is the CMB radiation [246]. Another otherwise successful cosmological model, the steady state theory [247], could be discriminated from the observation of the CMB radiation. Only the standard cosmological model predicts remnant isotropic and homogeneous background radiations, which have been observed.

The universe was radiation dominated in the early stage. When the universe cooled down and the average energy per particle came down below 1 eV, electrons began combining with nuclei to form neutral atoms. Until this recombination era, photons were interacting strongly with the charged electrons, protons and other nuclei, and the mean free path of radiation was small. Once neutral atoms were formed, radiation decoupled from matter and they could travel long distances almost unperturbed until the present day. The cosmic microwave background is the radiation coming from this surface of last scatter, which is the radiation from the time when radiation decoupled

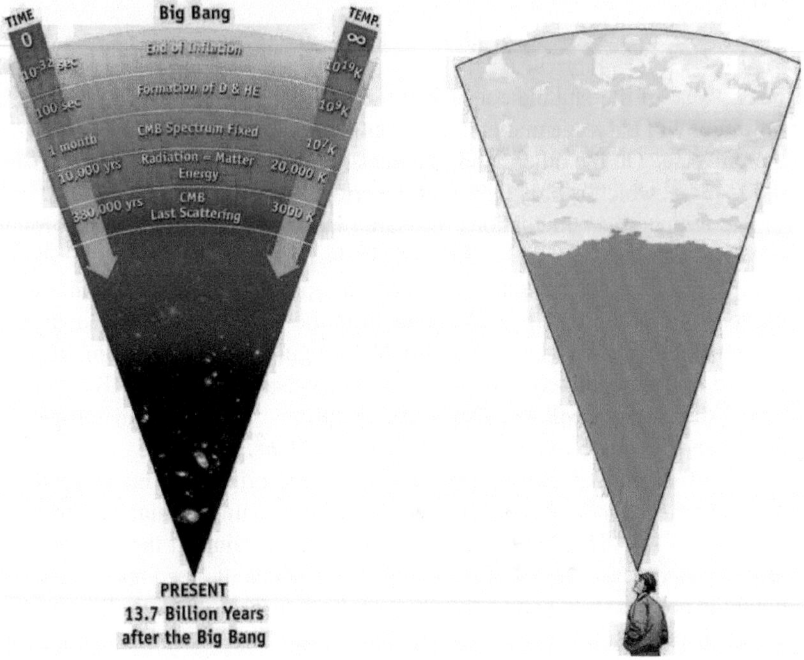


**FIGURE 15.2**
On a cloudy day we can see light from the surface of the cloud where light was last scattered by water molecules forming the cloud. Similarly there is a surface of last scatter in the cosmic microwave background radiation. (Reproduced from http://map.gsfc.nasa.gov/ContentMedia/990053sb.jpg courtesy NASA/WMAP Science Team)

from matter and the universe became matter dominated. To understand this, consider the light coming from the sky on a cloudy day. On a clear day, we can see a large distance if we look toward the sky. But since light scatters from water molecules, on a cloudy day the farthest point we can see toward the sky is the lower surface of the clouds (see figure 15.2). This lowest surface of cloud is similar to the surface of last scattering of the cosmic microwave background radiation.

When radiation decoupled from matter, the temperature of the universe was about 3000 $^{\circ}K$. So the background radiation at that time was gamma radiation with very short wavelength. Since the scale factor increases in the expanding universe, the wavelength of radiation also increases (see figure 15.3). Thus, the high energy gamma rays emitted from the surface of last scattering after the recombination era reach us as microwave background radiation at present.

Before atoms were formed, the interaction of radiation with charged particles was in equilibrium. During this time, radiation interacted with matter very strongly

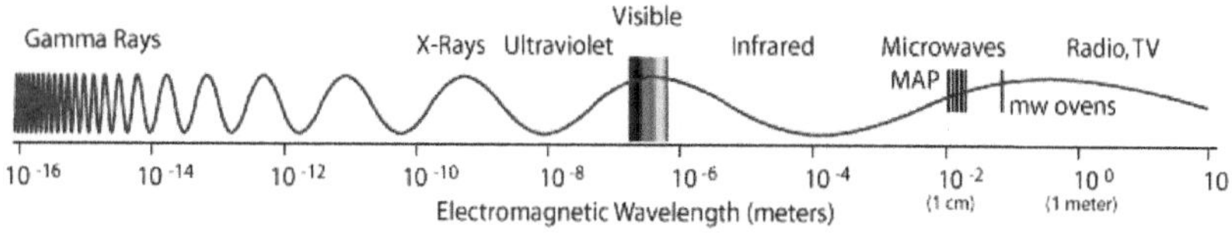


**FIGURE 15.3**
The microwave background radiation we see today, started off as high energy gamma radiation when radiation decoupled from matter after formation of atoms. (Reproduced from http://map.gsfc.nasa.gov/ContentMedia/MicrowavesGuide_b.jpg courtesy NASA/WMAP Science Team)

through three processes:

*Double Compton Scattering:* $e^- + \gamma \leftrightarrow e^- + \gamma + \gamma$, which dominates at temperatures above 1 keV.

*Compton scattering:* $\gamma + e^- \leftrightarrow \gamma + e^-$, photons transfer momentum and energy to the electron, which dominates during 1 keV to 90 eV.

*Thermal bremsstrahlung:* In the presence of ions electrons get accelerated and emit electromagnetic radiation, $e^- + X \leftrightarrow e^- + X + \gamma$, which is the primary process between 90 eV and 1 eV.

These processes are in thermal equilibrium since they occur at a faster rate than the expansion rate of the universe. This radiation will maintain the thermal spectrum and the intensity distribution will correspond to a blackbody spectrum,

$$\rho_0(\nu)d\nu = \frac{8\pi\nu^3\, d\nu}{\exp^{E/T_0} - 1}, \tag{15.68}$$

where $T_0 = T(t_R)a(t_R)/a_0$ and $t_R$ is the time of recombination. Due to the expansion of the universe, both the frequency and the temperature would decrease. However, both of them will decrease as $(1+z)$ and hence the form of the blackbody spectrum will remain unchanged.

Measurements of the cosmic background radiation give the distribution $\rho_0(\nu)$ for any particular frequency $\nu$ and the temperature $T_0(\nu)$ that corresponds to the blackbody spectrum for that frequency $\nu$ is reported. This temperature $T_0(\nu)$ is taken as the temperature of the cosmic background radiation independent of $\nu$. Theoretical estimate of the background temperature was about 5°K, and the first measurement of the background radiation [246] in the microwave region at the wavelength $\lambda = 7.35$ cm reported a value of $T_0 = (3.5 \pm 1)$°K. This cosmic microwave background radiation has been measured by many other experiments at different frequencies. The most accurate measurement by the FIRAS (Far Infrared Absolute Spectrometer) instrument of the COBE (cosmic background explorer) satellite reported a temperature of [248]

$$T_0 = 2.726 \pm 0.01\ ^\circ\text{K}. \tag{15.69}$$

The COBE data also confirmed the perfect blackbody spectrum of the cosmic microwave background radiation and gave a photon number density of

$$n_\gamma = 412 \pm 2\ \text{cm}^{-3}. \tag{15.70}$$

This background radiation came from the time of recombination.

To estimate the time of recombination or when the radiation decoupled from matter, we start with the scattering rate of $e^-$ with $\gamma$,

$$\Gamma_\gamma = n_e \sigma, \tag{15.71}$$

where the scattering cross-section $\sigma$ can be approximated by the Thomson scattering $\sigma = 6.65 \times 10^{-25}$ cm$^2$ and the equilibrium number density of electrons $n_e$ can be obtained from

$$n_i = g_i \left(\frac{m_i T}{2\pi}\right)^{3/2} \exp^{(\mu_i - m_i)/T}. \tag{15.72}$$

Here $n_i$ is the equilibrium number density of the particle $i = e, p, H$ with mass $m_i$, chemical potential $\mu_i$ and statistical weight $g_i$. We shall now consider the chemical equilibrium of $p + e \to H + \gamma$, which gives the relation $\mu_p + \mu_e = \mu_H$, to write the hydrogen equilibrium number density as

$$n_H = n_p n_e \left(\frac{m_e T}{2\pi}\right)^{-3/2} \exp^{B/T}, \tag{15.73}$$

where we used $g_H = 2g_p = 2g_e = 4$. The binding energy $B$ of hydrogen is defined as $B \approx m_p + m_e - m_H = 13.6$ eV. We can simplify it further by considering the charge neutrality condition, $n_e = n_p$.

We shall next define the ionization fraction $X_e$, in terms of the total baryon number density $n_B = n_p + n_H = \eta n_\gamma$, as $X_e = n_p/n_B$. Using the hydrogen equilibrium number density, the Saha ionization formula for the equilibrium ionization fraction then gives

$$\frac{1 - X_e^{eq}}{(X_e^{eq})^2} = \frac{4}{3}\eta n_e \left(\frac{m_e T}{2\pi}\right)^{-3/2} \exp^{B/T} = \frac{4}{3}\eta \frac{n_H}{n_p}. \tag{15.74}$$

Solving this equation it is possible to estimate the time of recombination, when 90% of the electrons combined with protons to form hydrogen atoms during a red-shift of $z_{rec} = 1200$ to 1400.

Around the time of recombination, the radiation-dominated universe had also changed over to a matter-dominated universe. The scale factor varies with time as $a \sim t^{1/2}$ in the radiation-dominated universe and as $a \sim t^{2/3}$ in the matter-dominated universe. The change over in the evolution of the scale factor took place when the matter density and the radiation density were equal. This took place at around a red-shift of $z_{eq} = 1500$. It is, therefore, usually assumed that the recombination had already taken place in a matter-dominated universe.

Soon after the formation of hydrogen atoms, since the number density of charged particles was reduced drastically, the mean free paths of photons increased and the universe became optically transparent. This decoupling of radiation from matter took place during the red-shift of $z_{dec} = 1100$ to $1200$. After decoupling, this radiation remained as background radiation, which reached us as the cosmic microwave background radiation.

## 15.5 Formation of Large Scale Structures

The cosmic microwave background radiation is almost isotropic and homogeneous and shows very little anisotropy, which means that the universe was very smooth at the time of decoupling. But if the universe was so smooth, it would have been difficult to have structures as we observe. We see stars around us, which are grouped in galaxies. These galaxies are again grouped in clusters of galaxies and clusters arrange themselves in superclusters. In between the galaxies, clusters, or even superclusters, there are vast voids, which seems to be inconsistent with the homogeneous and isotropic universe as indicated by the cosmic microwave background radiation. It is thus expected that as the universe expands, small quantum fluctuations that went out of the horizon size before the inflation enter our horizon and the gravitational amplification of these small fluctuations create the large scale structures.

We start with a discussion of Jeans or gravitational instability [249], which could amplify any small fluctuations that were present after the recombination. This would give an indication of how such small density perturbations at the time of decoupling could be amplified by gravitational instability to form the large scale structure we observe today (see figure 15.4).

Let us consider a perfect fluid with density $\rho$, velocity $\vec{v}$ and pressure $p$, which is described by the Euler equation and the equation of continuity as

$$\frac{\partial \vec{v}}{\partial t} + (\vec{v}\cdot\vec{\nabla})\vec{v} + \frac{1}{\rho}\vec{\nabla}p + \vec{\nabla}\phi = 0$$
$$\frac{\partial \rho}{\partial t} + \vec{\nabla}\cdot(\rho\vec{v}) = 0. \tag{15.75}$$

The Newtonian gravitational force $\vec{g}$ will satisfy the equations

$$\vec{\nabla}\cdot\vec{g} = -4\pi G\rho$$
$$\vec{\nabla}\times\vec{g} = 0. \tag{15.76}$$

The last equation gives $\vec{g} = -\vec{\nabla}\phi$, where $\phi$ is the Newtonian gravitational potential.

We shall now consider adiabtic perturbations and expand $X \equiv \rho, p, \vec{v}, \phi$ as

$$X = X_0 + X_1, \tag{15.77}$$

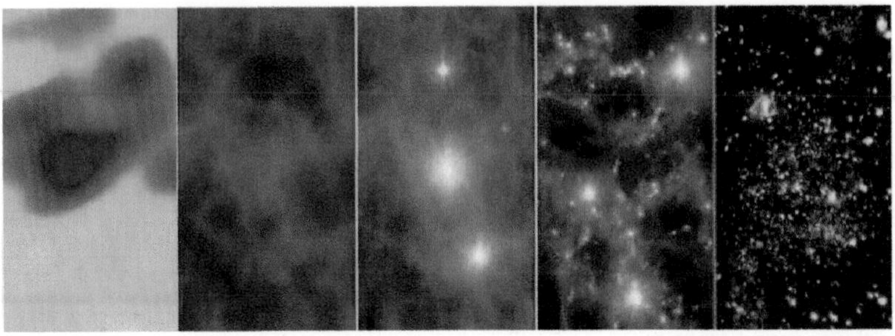

**FIGURE 15.4**
Simulation showing gravitational amplification of small fluctuations to large scale structures (left to right). Frame 1 shows temperature fluctuations, as observed in the cosmic microwave background radiation by WMAP, while the subsequent frames show how clouds clustered in the denser regions and finally stars, galaxies and large scale structures were formed. (Reproduced from http://map.gsfc.nasa.gov/m_or/m_or3.html courtesy NASA/WMAP Science Team).

where $X_0$ corresponds to the static solution and $X_1$ represents small perturbations. This would mean, $\vec{v}_0 = 0$ and $p_0, \rho_0 =$ const and we can define the adiabatic sound speed as

$$v_s^2 = \left(\frac{\partial p}{\partial \rho}\right) = \frac{p_1}{\rho_1}. \tag{15.78}$$

Since the perturbations are small, we can simplify these equations by writing, e.g., $\vec{\nabla} \cdot (\rho \vec{v}) = \rho_0 \vec{\nabla} \cdot \vec{v}_1$ and combine these equations in a second-order differential equation in $\rho_1$ as

$$\frac{\partial^2 \rho_1}{\partial t^2} - v_s^2 \vec{\nabla}^2 \rho_1 = 4\pi G_N \rho_0 \rho_1. \tag{15.79}$$

The solution to this equation is

$$\rho_1(\vec{r},t) = \delta(\vec{r},t)\rho_0 = A\exp^{(-i\vec{k}\cdot\vec{r}+i\omega t)}\rho_0 \tag{15.80}$$

satisfying the dispersion relation

$$\omega^2 = v_s^2 k^2 - 4\pi G_N \rho_0, \tag{15.81}$$

where $k = |\vec{k}|$. For $k$ less than some critical value,

$$k_J = \left(\frac{4\pi G_N \rho_0}{v_s^2}\right)^{1/2}, \tag{15.82}$$

defined as the Jeans wavenumber, $\omega$ will be imaginary and the density $\rho_1$ will grow (decay) exponentially. Otherwise for $k > k_J$, $\omega$ will be real and the perturbation will oscillate as sound waves. The mass contained within a sphere of radius $\lambda_J/2 = \pi/k_J$ is called the Jeans mass

$$M_J = \frac{4\pi}{3}\left(\frac{\pi}{k_J}\right)^3 \rho_0 = \frac{\pi^{5/2}}{6}\frac{v_s^3}{G_N^{3/2}\rho_0^{1/2}}. \tag{15.83}$$

Any structures with $M > M_J$ are unstable against gravitational collapse.

It is convenient to study the density fluctuations in terms of the Fourier transform of the density contrast $\delta(\vec{r},t) = \rho_1/\rho_0 = \delta\rho/\rho$, given by

$$\delta(\vec{r},t) = \frac{V}{(2\pi)^3}\int \delta_k(t)\exp^{-i\vec{k}\cdot\vec{r}} d^3k. \tag{15.84}$$

$|\delta_k|^2$ is known as the power spectrum and it enters into statistical quantities such as the correlation function, where

$$\delta_k(t) = V^{-1}\int \delta(\vec{r},t)\exp^{i\vec{k}\cdot\vec{r}} d^3x. \tag{15.85}$$

$V$ is the normalization factor.

We shall now include the effect of expansion of the universe by considering the scale factor in these equations. The first order perturbations now satisfy the equations

$$\begin{aligned}
&\frac{\partial \vec{v}_1}{\partial t} + \frac{\dot{a}}{a}\vec{v}_1 + \frac{\dot{a}}{a}(\vec{r}\cdot\vec{\nabla})\vec{v}_1 + \frac{v_s^2}{\rho_0}\vec{\nabla}\rho_1 + \vec{\nabla}\phi_1 = 0 \\
&\frac{\partial \rho_1}{\partial t} + 3\frac{\dot{a}}{a}\rho_1 + \frac{\dot{a}}{a}(\vec{r}\cdot\vec{\nabla})\rho_1 + \rho_0\vec{\nabla}\cdot\vec{v}_1 = 0 \\
&\vec{\nabla}^2\phi_1 = 4\pi G_N\rho_1.
\end{aligned} \tag{15.86}$$

We can express the density contrast $\delta(\vec{r},t) = \rho_1/\rho_0$ in Fourier transform including the scale factor as

$$\delta(\vec{r},t) = \frac{1}{(2\pi)^3}\int \delta_k(t)\exp^{-i(\vec{k}\cdot\vec{r})/a(t)} d^3k. \tag{15.87}$$

Similarly we expand the velocity $v_1$ and the potential $\phi_1$ to solve these equations. The first order equation for the density perturbation becomes

$$\frac{\partial^2\delta_k}{\partial t^2} + 2\frac{\dot{a}}{a}\frac{\partial\delta_k}{\partial t} = \delta_k\left(4\pi G_N\rho_0 - \frac{v_s^2k^2}{a^2}\right). \tag{15.88}$$

This result is similar to static universe ($\dot{a} = 0$) with physical wavenumber $k/a$. The Jeans wavenumber now becomes $k_J^2 = 4\pi G\rho_0 a^2/v_s^2$, so for $k \gg k_J$ the perturbation oscillates.

The solution is unstable for $k \ll k_J$. For a spatially flat ($k = 0$) matter-dominated universe, the universe is curvature dominated $a \sim t$, and we can write

$$4\pi G_N\rho = \frac{2}{3t^2} \quad \text{and} \quad \frac{\dot{a}}{a} = \frac{2}{3t}. \tag{15.89}$$

The density contrast then satisfies the equation

$$\ddot{\delta}_k + \frac{4}{3t}\dot{\delta}_k - \frac{2}{3t^2}\delta_k = 0, \tag{15.90}$$

which has two solutions, one growing and the other decaying. Since we are looking for solutions that can amplify the small density fluctuations, we ignore the decaying solution. The growing solution is given by

$$\delta_k \sim t^{2/3} \sim a = (1+z)^{-1}. \tag{15.91}$$

Thus, compared with the static case, with the inclusion of the scale factor the growth is no longer exponential. The expansion of the universe slows down the growth from an exponential behavior to a power law growth. In the relativistic case the power law becomes $\delta_k \sim t \sim a^2 = (1+z)^{-2}$.

We now consider the Jeans mass as a function of time. In the radiation-dominated phase, assuming that the universe consists of baryons and photons ($\rho = \rho_B + \rho_\gamma$), the pressure is provided by the photons since the baryons are interacting strongly with photons. Then the Jeans mass grows like $a^3$. The Jeans mass grows from about a solar mass to about $M = 10^{16} M_\odot$ by the time the universe grows from a red-shift of $10^{10}$ until the time of recombination.

After recombination, radiation decouples from matter and only the nonrelativistic hydrogen atoms maintain the pressure, for which $v_s^2 = (5/3)(T/m_H)$. This decrease in the velocity of sound causes a sudden reduction of the Jeans mass to

$$M_J = 1.3 \times 10^5 \left(\frac{z}{1100}\right)^{3/2} (\Omega_b h^2)^{-1/2}\, M_\odot. \tag{15.92}$$

During decoupling the Jeans mass drops from $10^{16} M_\odot$ to $10^6 M_\odot$, the typical mass of the globular clusters, the oldest objects in the universe. All heavier masses become gravitationally unstable; $\delta_k$ grows proportional to $a$. This tells us that density fluctuations have grown since the time of recombination by a factor $(t_0/t_a)^{2/3} = 1 + z \approx 10^3$.

In this simple discussion we have not included some factors. Before the Jeans instability becomes effective, the weakly interacting particles such as the neutrinos could escape from the dense regions without interaction, smearing out some of the perturbations. In actual calculations this factor of free streaming of neutrinos and other weakly interacting particles has to be included in the Boltzmann equations. This free streaming will not allow any structure formation at very small scales, but it will not affect the large scale structure formation. Another damping is the collisional damping or Silk damping [250]. At the time of recombination, the photon mean free path increases and there can be damping due to collisions. Here also small scales are washed out, i.e., inhomogeneities in the photon–baryon plasma are washed out.

## 15.6 Cosmic Microwave Background Anisotropy

The universe was initially radiation dominated, but they were interacting strongly with charged electrons and protons and heavier nuclei. At around the red-shift of $z = 1500$, the radiation and matter density became equal and then the formation of atoms began. During the period $z = 1200$ to $1400$ most of the charged particles combined to form atoms. Since the number of charged particles reduced drastically, radiation decoupled from matter after this recombination at around a red-shift of $z = 1100$. The light coming from this decoupling time, which is the surface of last scattering, reaches us now from all directions evenly as the cosmic microwave background radiation.

Although the cosmic microwave backgrond (CMB) radiation is almost isotropic and homogeneous, a small anisotropy has been observed, which can explain the formation of large scale structures in the universe. The gravitational amplification of the small fluctuations in the CMB radiation of the order of $\Delta T/T \sim 10^{-6}$ could explain the formation of even the largest objects. In addition to explaining the formation of large scale structures, the CMB anisotropy data could be probed to obtain precision values for several cosmological parameters.

The CMB anisotropy comes from a measurement of the temperature anisotropy with very high precision. A differential measurement of the full sky with a complex sky scan pattern allows the data to limit systematic measurement errors. The most precise data on the anisotropy of the temperature of the cosmic microwave background radiation have been provided by WMAP, as shown in figure 15.5. The CMB temperature at two different points separated by an angle $\theta$ is measured and then by taking the map average of $\Delta T/T$, the correlation function $C(\theta)$ is calculated, which contains most of the required information.

We can then define a statistical correlation function $C(\theta)$ in terms of the temperatures $T(n)$ and $T(n')$ measured in the two directions $n$ and $n'$ separated by an angle $\theta$ as

$$C(\theta) = \langle \Delta T(n) \Delta T(n') \rangle, \tag{15.93}$$

where $\cos\theta = n \cdot n'$. Since we are interested in measuring a deviation from spherical symmetry, it is convenient to expand the anisotropy parameter in terms of spherical harmonics,

$$T(n) = T_0 \sum_{l,m} a_l^m Y_l^m(\theta, \phi), \tag{15.94}$$

so the correlation function can be expressed in terms of the Legendre polynomials $P_l(\cos\theta)$ as

$$C(\theta) = \frac{1}{4\pi} \sum_l (2l+1) C_l P_l(\cos\theta). \tag{15.95}$$

The monopole term ($l = 0$) corresponds to the spherically symmetric case and gives the mean CMB radiation temperature

$$T_0 = 2.725 \pm 0.001 \text{ °K}. \tag{15.96}$$

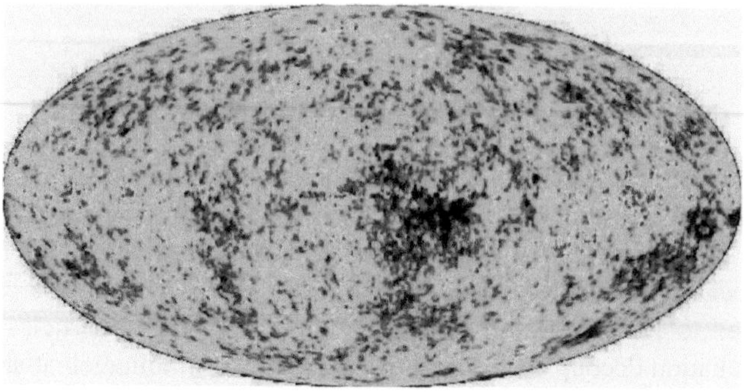

**FIGURE 15.5**
The anisotropy of the temperature of the CMB radiation, as measured by WMAP. Temperature variation is restricted between $-270.4252°\ C$ and $-270.4248°\ C$. (Reproduced from http://map.gsfc.nasa.gov/m_ig/020598/020598_ilc_640.jpg courtesy NASA/WMAP Science Team)

The dipole anisotropy ($l = 1$) is interpreted as the Doppler effect caused by our relative movement through the cosmic background. Measurement of the dipole anisotropy by COBE gives [251]

$$\frac{\Delta T}{T} = 3.346 \pm 0.017 \times 10^{-3}\ °\mathrm{K}, \tag{15.97}$$

which corresponds to a velocity of the sun compared with the cosmic background of $(368 \pm 2)$ km s$^{-1}$.

The anisotropies in CMB maps at higher multipoles ($l \geq 2$) correspond to the perturbations in the density at the time of decoupling. In the analysis of CMB anisotropy, $l(l+1)C_l/2\pi$ is plotted against multipole $l$. A 10-parameter fit to the data allows precise predictions for all parameters.

To understand the physics of anisotropies, we divert ourselves to explain what is meant by the horizon. The horizon at any time $t$ means the region of the universe around us from where light could have reached us within this time. It will then be a sphere of radius $dH(t)$, if light could travel a distance $dH(t)$ in time $t$. Since light can travel a distance $dt/a(t)$ in time $dt$, the distance travelled by light in time $t$ will be

$$d_H(t) = a(t) \int_t^{t_0} \frac{dt'}{a(t')} = \frac{t}{1-n}, \tag{15.98}$$

when the scale factor grows as $a(t) = t^n$ with $n < 1$. Thus, the size of the horizon expands faster than the scale factor. For a matter-dominated universe, $n = 2/3$, so $d_H/a \sim t^{1/3}$. Consider the microwave radiation we see today.

As the universe expands, the wave length for a mode $\delta_k$ increases as

$$\lambda = \frac{2\pi}{k} a(t) \propto \lambda_0 \left(\frac{a(t)}{a_0}\right) \sim t^n. \tag{15.99}$$

$\lambda_0$ is some characteristic scale for each wavelength $\lambda$. At a time $t$, some modes $\delta_k$ with wavelength $\lambda(t)$ may become larger than the Hubble radius $H^{-1}(t)$. The spectrum of density fluctuations is thus discussed at a time $t_H$ when it crossed the horizon, $\lambda \approx H^{-1}$.

The small-scale anisotropy corresponds to fluctuations within the horizon at the time of decoupling. When $l(l+1)C_l/2\pi$ is plotted against $l$, the first few multipoles corresponding to $2 \leq l \leq 100$ represent the fluctuations within the horizon. For $n = 1$, $l(l+1)C_l$ is almost constant and the anisotropy plot shows a flat plateau, called the Sachs–Wolfe plateau. This plateau becomes prominent only when $l$ is plotted in a logarithmic scale. For the lowest few $l$, the $C_l$s get an upward turn if the equation of state deviates from $\omega = 0$.

Anisotropy on the large scale correspond to the fluctuations outside the horizon at the time of decoupling. The COBE data corresponds to a temperature of $15.3^{+3.8}_{-2.8}\ \mu$K and an anisotropy of [252]

$$\frac{\Delta T}{T} \simeq 6 \times 10^{-6}. \tag{15.100}$$

This anisotropy directly reflects the primordial perturbation spectrum.

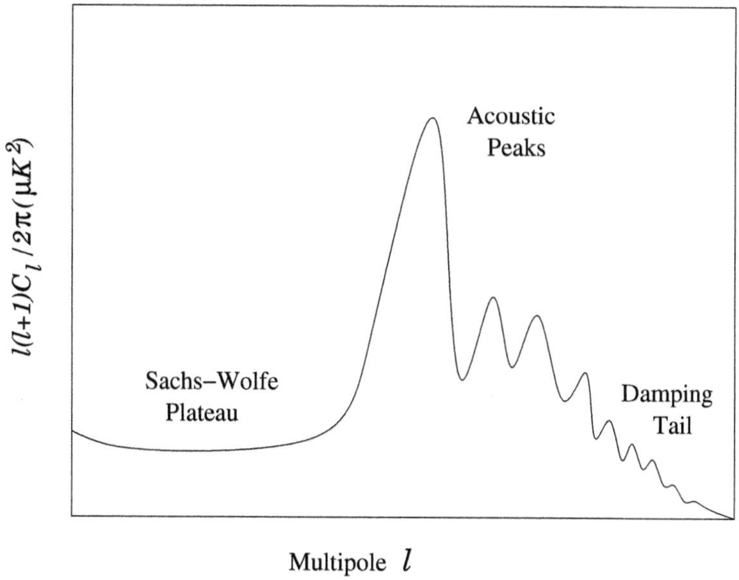


**FIGURE 15.6**
Sketch of theoretical CMB anisotropy power spectrum.

In the $l(l+1)C_l/2\pi$ versus $l$ power spectrum plot (see figure 15.6), the large scale anisotropy corresponds to $l \geq 100$ and contains imprints from the gravity-driven acoustic oscillations. Before the formation of atoms, the universe was filled with charged particles and the radiation was interacting strongly with matter. The gravitational potential dominated by the dark matter and dark energy components allowed the small perturbations $\sim (10^{-5})$ to evolve linearly. Each of the Fourier modes now oscillates with the characteristic frequency determined by the adiabatic sound speed in the fluid. The Fourier modes that were inside the horizon at the time of decoupling would smear out these perturbations. But the Fourier modes that were outside the horizon at the time of decoupling would retain these phases when they entered our horizon. This gives the acoustic peaks in the power spectrum plot for $z \geq 100$. However, since the surface of last scattering has a finite thickness, the oscillations of the anisotropies will be damped for large $l$. This acoustic peak has been observed in the power spectrum and the analysis of the anisotropy data could give us precise values for the various parameters in the standard model [73].

To understand the evolution of the different modes, we have to specify the spectrum. We have to specify the amplitude of a given mode when it enters the horizon. Since it is not possible to predict the spectrum without knowing how the fluctuations came into existence, some simple forms are considered. One popular spectrum is a power law spectrum

$$\left(\frac{\delta\rho}{\rho}\right)_{hor} \simeq \frac{k^{3/2}|\delta_k|}{\sqrt{2}\,\pi} = AM^{-\alpha}, \tag{15.101}$$

where $|\delta_k|^2 \propto k^{3(2\alpha-1)}$. The inflationary models predict the Harrison–Zeldovich spectrum [253] of constant curvature perturbation, $\alpha = 0$, which corresponds to the Harrison–Zeldovich parameter $n = 1$. The COBE data yield a value for the Harrison–Zeldovich parameter [254]

$$n = 1.2 \pm 0.3. \tag{15.102}$$

Since inflation predicts $n = 1$, this may be an indication that the universe went through an inflationary phase in the early universe. Formation of galaxies and the measured CMB anisotropy gives a bound $\alpha > -0.1$, while the overproduction of black holes, and hence excess radiation that has not been observed, gives the constraint $\alpha < 0.2$.

Specifying the amplitude of different modes at the time they cross the horizon amounts to specifying time-dependent amplitudes. To avoid that one uses a spectrum at a given time, which may then be related as

$$\left(\frac{\delta\rho}{\rho}\right)_t = \left(\frac{M}{M_{hor}(t)}\right)^{-2/3} \left(\frac{\delta\rho}{\rho}\right)_{hor}, \tag{15.103}$$

where $M_{hor}$ is the mass of the nonrelativistic particles within the horizon.

The time considered for specifying the spectrum is when the structure formation begins, $t = t_{eq}$. This depends on the type of dark matter of the universe. In the case of hot dark matter (like neutrinos) the washing out of the fluctuations takes a long time and structures form when the particles become nonrelativistic. On the other hand, the cold dark matter (weakly interacting massive particles such as the neutralino LSP) became nonrelativistic at a very early stage and introduced a gravitational potential for the baryonic matter since the time of matter dominance. In these models baryonic matter fell into the gravitational potential created by the cold dark matter soon after recombination and formed the large scale structures.

The anisotropy in the cosmic microwave background radiation temperature probes the density fluctuations in the early universe on comoving scales greater than $\sim 100$ Mpc. There are several models of structure formation, which represent the range of cosmological parameters. Each model predicts how a power spectrum of infinitesimal density perturbations in the early universe develops into cosmic microwave background radiation anisotropies and inhomogeneities in the galaxy [255]. Most of the models are consistent with the constraints on baryon density from the big-bang nucleosynthesis. Although earlier observations were consistent with both hot and cold dark matter, for the present observations the best fit corresponds to the cold dark matter with a cosmological constant.

## 15.7 Inflation

In spite of all the successes of the standard model of cosmology, to solve some of its problems it was proposed that the universe went through an inflationary phase in the early universe. The CMB anisotropy data now indicates that the universe had indeed gone through an inflationary phase. We shall now briefly mention how an inflationary model works, after mentioning the problems of the standard model of cosmology that required an inflation.

### Flatness Problem

The curvature of our universe is a topological property of the metric. So, if the universe started with a positive, negative, or vanishing curvature ($k = \pm 1, 0$), it should have been maintaining it at all times. As we discussed earlier, the mass density ($\Omega$), which is directly related to the curvature ($k$) is given by

$$\Omega - 1 = \frac{\rho + \rho_v - \rho_c}{\rho_c} = \frac{k}{H^2 R^2}. \tag{15.104}$$

A positive curvature ($k = +1$) implies $\Omega > 1$, a negative curvature ($k = -1$) implies $\Omega < 1$, and a vanishing curvature ($k = 0$) implies $\Omega = 1$.

During the radiation- or matter-dominated era the density of the universe grew as $\rho \sim R^{-4}$ or $\rho \sim R^{-3}$, respectively. If the universe started with a matter density of $\Omega < 1$, around the Planck time ($t \sim M_{Pl}^{-1}$), the curvature of the universe would have been

$$|\Omega - 1| \ll 10^{-50}. \tag{15.105}$$

However, if the universe would have started with $\Omega = 1$, then it would have been the same all the time, which is the most natural solution. Present observations also indicate that $\Omega_{tot} \sim 1$ (see equation (17.13)).

### Horizon Problem

The second question is concerned with the event horizon, the distance from which light could reach us today starting from the beginning of the universe. This distance is given by

$$d_H(t) = R(t) \int_0^t \frac{dt'}{R(t')}. \tag{15.106}$$

Since the scale factor grows as $R \sim t^n$ with $n < 1$, the event horizon expands as

$$d_H = \frac{t}{1-n}, \tag{15.107}$$

which is faster than the increase in the scale factor. For a matter-dominated universe, $n = 2/3$, so $d_H/R \sim t^{1/3}$. Consider the microwave radiation we see today. This radiation was emitted at the time of decoupling $t_{dec} \approx 0.37 \times 10^6$ y. At that time the universe was much smaller than its present size. The most distant microwave radiation we could observe today corresponds to the present age of the universe of $t_0 = 13.7 \times 10^9$ y. Then the radiation we observe today was not causally connected at the time of its emission. In fact, from the time of decoupling until now the universe has grown by a factor of $(t_0/t_{dec})^{1/3} \sim 35$, and hence, the radiation we see today was about 35 horizon lengths apart at the time of decoupling when it was emitted. Then the isotropy of the background radiation cannot be explained. If we assume it was isotropic at all lengths at all times, then the formation of large scale structures could not be explained.

### Monopole Problem

Another problem of the standard cosmological model is the monopole problem. During the phase transition of the grand unified theories, the symmetry breaking will form regions with $<\phi>= v$ and $<\phi>= 0$. Then after the phase transition is over, with each causally connected region of space, there will be at least one region with $<\phi>= 0$, which are the t'Hooft–Polyakov monopoles [256] with color and magnetic properties and mass $\sim O(M_X/\alpha)$. This gives a number density of the monopoles to be [257]

$$n_M \sim d_H^{-3} \sim \frac{M_X^6}{M_{Pl}^3}. \tag{15.108}$$

Taking the scale of grand unification to be around $10^{15}$ GeV, this gives a mass density of the monopoles to be around

$$\rho_{monopoles} \approx 10^9 \, \rho_c, \tag{15.109}$$

which is far beyond any permissible value. Similarly other topological objects such as domain walls or cosmic strings could also be present with unacceptably high contribution to the mass density of the universe.

**Models of Inflation**

The models of inflation [258] were proposed to solve these problems of horizon, flatness, and monopole, but these models can have several other interesting features. Present experiments also support the idea of inflation. During inflation the universe goes through a phase of extremely fast expansion when the universe is dominated by the vacuum energy density $\rho_v$. The scale factor evolves exponentially fast in this de Sitter phase [259] (equation (15.35))

$$a(t) \sim a(0)\exp^{Ht} \quad \text{and} \quad H^2 = \frac{8\pi G_N}{3}\rho_v. \tag{15.110}$$

During this phase a small smooth causally coherent region of space of size less than $H^{-1}$ can grow to an enormous comoving volume, which becomes all of our observable space at present. Then the horizon problem does not appear any more since at present we see only a small part of the much larger causally connected space. So, the smoothness of the small region before the inflation now appears as the isotropic part of our entire observable space. The flatness problem is solved since the fast expansion during the inflationary phase could flatten the universe to the extent that it would appear almost flat at all times. Moreover, most inflationary models give the density of the universe to be the same as that of the critical density. Since only one monopole is expected in one causally connected region in space and in the inflationary models much larger regions of space are causally connected, the number density of monopoles and other topological defects becomes much less and consistent with observations.

The main ingredient for the inflation is a scalar field $\phi$ with a particular potential, so after the field acquires a *vev* the inflationary dynamics follows. The singlet scalar field $\phi$ that drives inflation is called the inflaton. Consider a Higgs scalar $\phi$, whose *vev* $<\phi>= v \neq 0$ breaks a symmetry at the scale $v \sim M$. This will then produce a vacuum energy of $\rho_v = V(\phi = 0) \sim M^4$, and at a temperature below the critical temperature, the vacuum energy will dominate. In the early universe the phase transition took place in a sequence. At a temperature much higher than the critical temperature ($T_c$), the potential has only one minimum corresponding to $<\phi>= 0$. As the temperature approaches the critical temperature, the potential starts developing a second minimum at $<\phi>= v$. However, since the two minima are separated by a barrier, the universe still remains in the minimum corresponding to restored symmetry. At the critical temperature the free energy of both minima becomes equal and allows the universe to stay in any one of the minima. So, while most of the universe remains in the old minimum, some bubbles start developing where inside the bubble the minimum corresponds to the new minimum at $<\phi>= v$.

As the universe cools down below the critical temperature, the new minimum becomes the lowest energy minimum and, hence, the true vacuum, while the origin becomes the false vacuum. At this time more and more particles start penetrating the barrier and the bubbles start growing. Finally after a while the barrier height vanishes, the false vacuum decays completely to the true vacuum, and the bubbles grow at the speed of light. During this time the inflaton field rolls down slowly to the

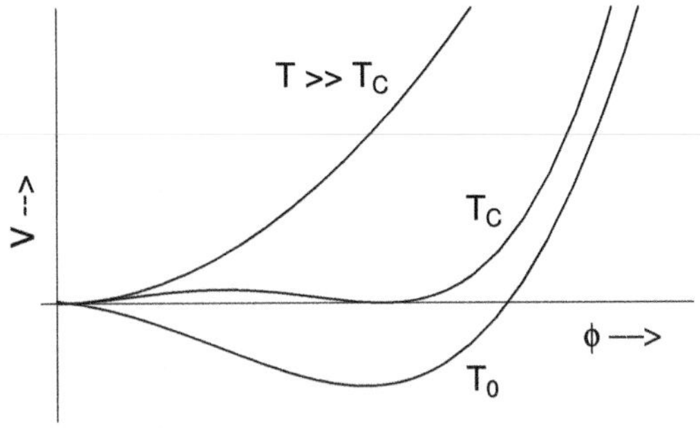


**FIGURE 15.7**
Temperature dependence of the inflaton potential.

now true vacuum following the classical equation

$$\ddot{\phi} + 3H\dot{\phi} + V'(\phi) = 0. \tag{15.111}$$

This is similar to the motion of a particle rolling down against the force of friction, where the expansion of the universe is giving the force of friction slowing down the motion. During the time the inflaton field rolls down, the universe is dominated by the vacuum energy, and it remains in the de Sitter phase and grows exponentially fast. If, for example, it takes $\Delta t = 100H^{-1}$ time to evolve to $< \phi >= v$, then during this time the universe would expand by a factor of $\exp^{100} \sim 10^{43}$. If the phase transition is the grand unification phase transition, so the energy scale is $M \sim 10^{14}$ GeV, the expansion time scale is $t \sim H^{-1} \sim 10^{-34}$ s, which means the roll time is $10^{-32}$ s. As the universe expands exponentially fast, it supercools and the temperature also falls exponentially fast, $T \propto \exp^{-Ht}$, with the entropy remaining the same.

Finally when the inflaton falls into the true vacuum, it oscillates coherently around the minimum of the potential until it cools down releasing energy. The oscillation period is decided by the curvature of the potential. The enormous vacuum energy of the inflaton field is then released in the form of lighter decay products, which when thermalized reheat the universe to the original temperature $T_{RH} \sim T_c \sim M$. During the time of reheating, the entropy increases exponentially. This entropy increase connects large distances causally, solving the problem of large scale smoothness and flatness, and reducing the number density of the topological defects such as monopoles exponentially.

The flat part of the potential is highly crucial for the slow roll down of the inflaton, since this is the time when the universe supercools and grows exponentially fast. This is achieved by the temperature-dependent Coleman–Weinberg potential [260]

$$V_{CW}(\Phi) = \frac{1}{2}B\sigma^4 + B\Phi^4\left(\ln\frac{\Phi^2}{\sigma^2} - \frac{1}{2}\right), \tag{15.112}$$

where $B$ is a parameter related to the gauge coupling constant corresponding to the unbroken symmetry. For nonzero temperature a barrier exists with a height of

$\Phi \sim T^4$. If the phase transition temperature is $M \sim 10^{14}$ GeV, the barrier becomes unstable only at a temperature of about $T \sim 10^9$ GeV. However, the original Coleman–Weinberg scenario has problems, because it assumes a massless Higgs scalar. There are other scenarios such as the chaotic inflationary models, where a potential of the form [261]

$$V(\Phi) = \lambda \Phi^4 \tag{15.113}$$

is considered. The initial values $\Phi_i$ of the field $\Phi$ are distributed chaotically around space and the minimum lies at $< \Phi >= 0$. There is no symmetry breaking involved and the field is introduced just to drive the inflation.

In the de Sitter phase, the inflatons roll over a fairly flat potential. During this time the scalar is essentially massless. A massless, minimally coupled scalar field in the de Sitter space can be decomposed into its Fourier components, which characterize the spectrum of its quantum mechanical fluctuations [262]

$$(\Delta\Phi)_k^2 \equiv \frac{k^3 \, |\delta \phi_k|^2}{2\pi^2 V} = \left(\frac{H}{2\pi}\right)^2, \tag{15.114}$$

where $\delta\phi_k = \int d^3x \exp^{i\vec{k}\cdot\vec{x}} \Phi(\vec{x})$. The size of quantum fluctuations in $\Phi$ is set by $H/2\pi$. As each mode $k$ crosses outside the horizon, it decouples from microphysics and freezes as classical fluctuations. Since the potential energy density depends upon the scalar field $\Phi$, fluctuations in $\Phi$ produce perturbations in the energy density. The power spectrum does not depend on the mode $k$ and corresponds to Harrison–Zeldovich spectrum with constant-curvature fluctuations, i.e., $\alpha = 0$. This is a generic feature of all inflationary models, which relies on the fact that the potential is almost flat.

All massless or very light particles are also excited similarly to the inflaton in the de Sitter space. When these excited modes re-enter the horizon, they propagate as particles. In other words, all de Sitter space excitations ultimately result in particle production after reheating. In supersymmetric models, in the de Sitter phase gravitinos are produced with an abundance proportional to the reheating temperature. These gravitinos can dissociate the light elements produced during nucleosynthesis and alter the light element abundances causing a problem, known as the gravitino problem. A solution to this requires the reheat temperature $T_{RH} \lesssim 10^9$ GeV [263]. In the case of stable gravitinos, a limit on $T_{RH}$ can be derived from the closure limit of the universe [264], which could be as large as $10^8$ to $10^{11}$ GeV depending on the mass of the gravitino. For unstable gravitinos, the upper bound on $T_{RH}$ from primordial nucleosynthesis becomes [265]

$$\begin{aligned} T_{RH} &\leq 10^9 \text{ GeV} & m_{3/2} &< 1 \text{ TeV} \\ T_{RH} &\leq 10^{12} \text{ GeV} & 1 \text{ TeV} < m_{3/2} &< 5 \text{ TeV}. \end{aligned} \tag{15.115}$$

Although the inflationary models solve several problems of the standard cosmological model, at present there is no standard model of inflation. However, with more and more data from different observations, we now understand the problem better. At present there is no explanation for the origin of the inflaton. It could be scalar giving rise to some symmetry breaking, or it could be some other fields introduced just to drive the inflation. It is also not clear what determines the initial values in the inflationary models. The dynamics of the inflationary model are yet to be understood. The gravitino problem prefers a lower reheating scale, although it depends on the model. All these drawbacks are mostly due to our lack of knowledge of the very early universe and also of the possible extensions of the standard model of particle physics. If a better understanding of the particle physics models beyond the standard model can come from the next generation accelerators, then with the improvements in our understanding of the early universe with all the new experiments that are coming up due to advancements of technology, we hope to resolve these issues in future.

# 16

# *Baryon Asymmetry of the Universe*

The interplay between particle physics and cosmology is opening up newer areas of research, which are some of the topics in astroparticle physics. The motivations behind these studies are to explain some problems of cosmology by particle physics models and constrain models of particle physics using the astrophysical observations. The baryon asymmetry of the universe is one of these issues in cosmology.

There are many models of baryogenesis, which can explain the present baryon asymmetry of the universe. However, some interactions are severely constrained by the requirement of the baryon asymmetry of the universe. The most popular model of baryogenesis is through lepton number violation, which is called leptogenesis. In leptogenesis, the lepton number violation required to explain the small neutrino masses generates a lepton asymmetry of the universe, which gets converted to a baryon asymmetry of the universe before the electroweak phase transition. If any fast baryon or lepton-number violating interactions had been present in the early universe, their existence could have erased the existing baryon asymmetry of the universe. This constrains the strengths of any baryon or lepton-number violating interactions. The constraints on the lepton number violation is directly related to the neutrino masses, which makes these models verifiable. In this chapter we shall highlight some of the activities in this field.

## 16.1 Baryogenesis

All astrophysical observations indicate that our visible universe is dominated by matter and there is very little antimatter. This matter-dominance in our universe is most crucial for our existence, but an understanding of why there is more matter than antimatter is difficult. A natural assumption is that the universe was neutral with respect to all conserved charges, including baryon number. Then a locally baryon-symmetric universe would imply that the number density of baryons and antibaryons in a comoving volume around the time of nucleosynthesis would be around $n_b/s \approx n_{\bar{b}}/s \approx 10^{-20}$.

The primordial abundances of light elements depend crucially on the amount of baryonic matter ($\eta$) present at the time of nucleosynthesis, which requires a value of the matter density higher than the antimatter density. The present value for the matter density per comoving volume is

$$\frac{n_b}{s}(t = \text{present}) = \eta\ \frac{n_\gamma}{s} = (6.1^{+0.3}_{-0.2}) \times 10^{-10}\ \frac{n_\gamma}{s}, \qquad (16.1)$$

where $s/n_\gamma = 1.8g_{*s}(\text{present}) = 7.04$, while the antibaryon density is around $n_{\bar{b}}/s \sim 10^{-20}$. At an early time before nucleosynthesis $t < 10^{-6}$ sec, the number densities of baryons and antibaryons would have been much higher $n_b/n_\gamma \approx n_{\bar{b}}/n_\gamma \approx 1$ due to quark–antiquark pair production and annihilations. However, the net baryon number density per comoving volume would have been the same as the present value

$$\frac{n_B}{s}(t < 10^{-6}s) = \frac{(n_b - n_{\bar{b}})}{s} = (6.1^{+0.3}_{-0.2}) \times 10^{-10}\frac{n_\gamma}{s}.$$

This implies that at that time a very small amount of baryon asymmetry in the universe was present

$$\frac{n_b - n_{\bar{b}}}{n_b}(t < 10^{-6}s) \sim \frac{n_B}{n_\gamma} \sim \frac{n_B}{s} \cdot \frac{s}{n_\gamma} \sim 10^{-8}, \qquad (16.2)$$

taking $g_{*s}(t < 10^{-6}s) \sim 100$. This tiny baryon asymmetry at an early time produced the large difference between the baryon density ($n_b/s \sim 10^{-10}$) and the antibaryon density ($n_{\bar{b}}/s \sim 10^{-20}$) at the time of nucleosynthesis at around 1 sec.

In other words, before nucleosynthesis both the baryon-to-photon ratio and anti-baryon-to-photon ratio were of the order of 1, but they should have cancelled each other to 8 decimal points so their difference was about $10^{-8}$. This small primordial baryon asymmetry is not natural. Since it is expected that the universe started with an equal amount of baryons and antibaryons, some interactions of particle physics should have generated this small baryon asymmetry of the universe before nucleosynthesis. Starting from a baryon-symmetric universe, the process of generating this small amount of baryon asymmetry is called *baryogenesis*. In 1967 Sakharov proposed a mechanism for generating such small baryon asymmetry starting from a baryon-symmetric universe [266]. This proposal requires some elementary interactions to satisfy three criteria to generate a baryon asymmetry of the universe starting from a baryon-symmetric universe, which are

(i) *baryon number violation,*

(ii) *C and CP violation* and

(iii) *departure from thermal equilibrium.*

We shall now discuss these criteria in some details.

**Baryon Number Violation**

If baryon number $B$ is conserved, then any interactions that generate a baryon from a $B = 0$ state will also generate an antibaryon so the net baryon number will always be zero. The starting assumption, that the universe is neutral with respect to any conserved charge, will imply that in equilibrium the number density of particles with nonzero baryon number $B$ would be the same as the antiparticle number density. This follows from the fact that the expectation value of the conserved charge $B$ vanishes. The expectation value of any conserved baryon number $B$ can be written as

$$<B>= \frac{\mathrm{Tr}\left[B \exp^{-\beta H}\right]}{\mathrm{Tr}\left[\exp^{-\beta H}\right]}. \tag{16.3}$$

Since particles and antiparticles have opposite baryon number, $B$ has odd eigenvalues under $C$ operation, but even eigenvalues under $\mathscr{P}$ and $\mathscr{T}$ operations. Thus, any conserved baryon number $B$ is odd while $H$ is even under $CPT$ transformation. Thus, $CPT$ conservation implies vanishing of this expectation value. This implies the first condition that baryon number should be violated for the generation of the baryon asymmetry of the universe.

### *C* and *CP* Violation

Let us consider an interaction in which the initial state $i$ goes to the final state $f$ where baryon number $B$ is violated. In general, $CP$ invariance implies

$$M(i \to j) = M(\bar{i} \to \bar{j}) = M(j \to i). \tag{16.4}$$

The condition that the amount of violation of baryon number in process $i \to j$ should not be compensated by the equal amount of violation in the interactions of the antiparticle $\bar{i} \to \bar{j}$ requires $CP$ violation. In other word, the number of left-handed particles generated in any process will be different from the number of right-handed antiparticles (which are the conjugates of the left-handed particles) only when $CP$ is violated. In addition, $C$ violation is also required, since the generation of the left-handed particles should not compensate the generation of the left-handed antiparticles (which are the conjugates of the right-handed particles).

### Departure from Thermal Equilibrium

Equation (16.3) and $CPT$ conservation tell us that in equilibrium the total baryon number of the universe vanishes since $B$ is odd and $H$ is even under $CPT$, unless there is a nonvanishing chemical potential. Assuming chemical equilibrium then means that the only way one can have a nonvanishing baryon number density is by going away from the equilibrium distribution when equation (16.3) is no longer valid. This is achieved when the interaction rate is very slow compared with the expansion rate of the universe.

The out-of-equilibrium condition can be justified by considering the equilibrium number densities of particles and antiparticles. $CPT$ theorem tells us that the masses of particles and their antiparticles are the same. If we assume that the chemical potential associated with $B$ is zero and $CPT$ is conserved, then in thermal equilibrium the phase space densities of baryons and antibaryons, given by $n_b \sim n_{\bar{b}} \sim [\exp(\sqrt{p^2+m^2}/T)+1]^{-1}$, are identical (we assume that there are no scalar fields with nonvanishing baryon number, although this result is true for scalars also). This will mean that there cannot be any baryon asymmetry of the universe in thermal equilibrium.

Although the departure from equilibrium manifests itself in the Boltzmann equations depending on the parameters in the problem, a crude way to put the out-of-equilibrium condition is to say that the universe expands faster than the interaction rate. For example, if some B-violating interaction is slower than the expansion rate of the universe, this interaction may not bring the distribution of baryons and antibaryons of the universe into equilibrium. As the heavy particle decays, the decay product will move apart before it could participate in the inverse decay, causing a departure from equilibrium. In other words, before the chemical potentials of the two states become equal, they move apart from each other. Thus, we may state the out-of-equilibrium condition as

$$\Gamma(T) < H(T) = 1.66\sqrt{g_*}\frac{T^2}{M_{Pl}}, \tag{16.5}$$

where $\Gamma$ is the baryon-number violating interaction rate under discussion, $g_*$ is the effective number of degrees of freedom available at that temperature $T$ and $M_{Pl}$ is the Planck scale.

One more factor to influence the generation of baryon asymmetry is that the total decay width of any particle is always the same as the total decay width of its antiparticle by $CPT$. So, for the generation of a baryon asymmetry of the universe, we need at least two particles whose decay violates baryon number. Then the partial decay widths of these particles can differ if there is $CP$ violation, and that can generate an asymmetry. We shall now give an example of generating a baryon asymmetry of the universe, which will illustrate these points.

**Example of Baryogenesis**

We illustrate the mechanism of baryogenesis with an example. Consider two scalar particles $X$ and $Y$, which interact with the standard model fermions $A, B, C$, and $D$ through the Yukawa couplings

$$\mathscr{L} = f_x^{ab}\bar{A}BX + f_x^{cd}\bar{C}DX + f_y^{ac}\bar{A}CY + f_y^{bd}\bar{B}DY + H.c.. \tag{16.6}$$

The decay modes of these particles, as shown in figure 16.1, are given by

$$X \to A + B^*, \quad \text{and} \quad X \to C + D^*,$$
$$Y \to A + C^*, \quad \text{and} \quad Y \to B + D^*,$$

which violates baryon number, if these fermions carry different baryon numbers. For example, one possibility could be $B_A = -B_C = 1$ and $B_B = B_D = 0$, then the two final states of $X$ have $B = \pm 1$, while for $Y$ it is $B = 2, 0$. If both final states in the decays of $X$ carry the same $B$ number, then we can always assign a $B$ number for $X$ and check that there is no $B$ violation. So, for baryon number violation we need at least two decay modes of $X$ and $Y$. The total decay rates of $X$ and $\bar{X}$ are same by

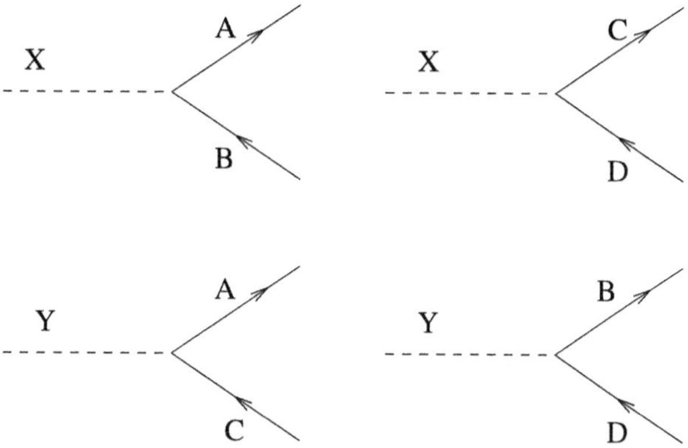


**FIGURE 16.1**
Baryon number violating decays of the heavy bosons into fermions.

$CPT$, but the partial decay rate of $X$ and $\bar{X}$ to one of these decay channels can be different, which gives the $CP$ asymmetry and, in turn, the baryon asymmetry.

We now define the asymmetry parameters for the decays of $X$ and $Y$ as

$$\varepsilon_X = \sum_f B_f \frac{\Gamma(X \to f) - \Gamma(\bar{X} \to \bar{f})}{\Gamma_X}$$
$$\varepsilon_Y = \sum_f B_f \frac{\Gamma(Y \to f) - \Gamma(\bar{Y} \to \bar{f})}{\Gamma_Y}, \qquad (16.7)$$

where $f$ corresponds to the final states with baryon number $B_f$, and $\Gamma_{X,Y}$ are the total decay widths of these particles. The nonzero contribution to the asymmetries comes from an interference of the tree-level diagrams (figure 16.1) with the one-loop diagrams (figure 16.2). As mentioned above, if $B_f$ is the same for both decay channels of $X$, the asymmetry vanishes since the total decay rates of $X$ and $\bar{X}$ are the same. For the decays $X \to A + B^*$ and $\bar{X} \to A^* + B$ these interference terms are given by

$$\Gamma(X \to A + B^*) = f_x^{ab} f_x^{cd*} f_y^{ac*} f_y^{bd}\, \mathrm{I}_{XY} + (f_x^{ab} f_x^{cd*} f_y^{ac*} f_y^{bd}\, \mathrm{I}_{XY})^*$$
$$\Gamma(\bar{X} \to A^* + B) = f_x^{ab*} f_x^{cd} f_y^{ac} f_y^{bd*}\, \mathrm{I}_{XY} + (f_x^{ab*} f_x^{cd} f_y^{ac} f_y^{bd*}\, \mathrm{I}_{XY})^*, \qquad (16.8)$$

where $\mathrm{I}_{XY}$ is the finite part of the one-loop integral. Including the other decay modes of $X$, we get the final asymmetry to be

$$\varepsilon_X = \frac{4}{\Gamma_X}\, \mathrm{Im}[\mathrm{I}_{XY}]\ \mathrm{Im}[f_x^{ab*} f_x^{cd} f_y^{ac} f_y^{bd*}]\, (B_A + B_D - B_B - B_C)\,. \qquad (16.9)$$

Similarly for $Y$ we get an asymmetry which is $\varepsilon_Y = -\varepsilon_X$.

In this expression for the asymmetry, $J_{abcd} = \mathrm{Im}[f_x^{ab*} f_x^{cd} f_y^{ac} f_y^{bd*}]$ is the measure of $CP$ violation, equivalent to the Jarlskog invariant for the $CP$ violation entering

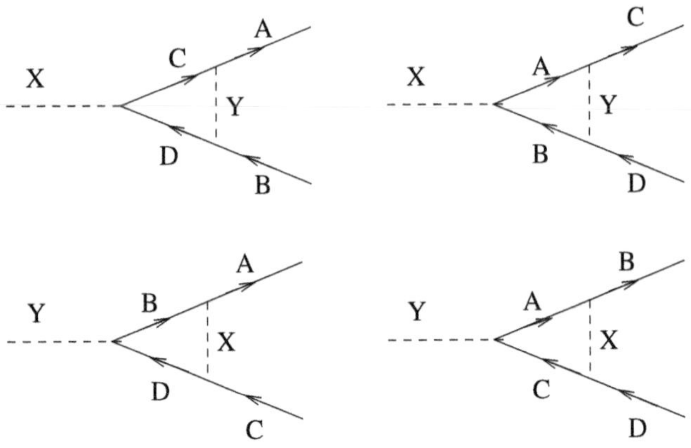


**FIGURE 16.2**
One-loop diagrams contributing to $CP$ violation in heavy boson decays.

in the quark mass matrices. If the theory has $CP$ violation, this quantity will be nonzero. Rephasing of the light fields will not affect this quantity. Under rephasing of the fields

$$\psi_i \to \exp^{i\phi_i}\psi_i,$$

where $i = A, B, C, D$ and $\psi_A$ represents the particle $A$, the Yukawa couplings transform as

$$f^{ij}_{x,y} \to \exp^{i(\phi_j - \phi_i)} f^{ij}_{x,y},$$

and some of the Yukawa couplings can be made real. However, $[f_x^{ab*} f_x^{cd} f_y^{ac} f_y^{bd*}]$ transforms to itself and $J_{abcd}$ remains invariant under this rephasing. So, if there is $CP$ violation in the theory, this quantity will be nonvanishing for any allowed phase transformations of the light fields.

Since the imaginary part $\mathrm{Im}[\mathrm{I}_{XY}]$, i.e., the absorptive part of the integral, enters the calculation of the asymmetry parameter, this integral is nonvanishing when the fermions in the loop are lighter than the scalars, In this case the fermions could be on mass shell and we could cut the diagram along the fermion lines. Both the scalars should have baryon-number violating interactions, otherwise one of the $\varepsilon$'s will vanish implying vanishing of the other. If $X$ and $Y$ are degenerate then there are two problems, both of which give a vanishing asymmetry. First the loop integral vanishes when these masses are degenerate. Second, the contribution to the asymmetry generated by one of them will be erased by that of the other, since $\varepsilon_X = -\varepsilon_Y$. If their masses are different, say, $M_Y > M_X$, then at a temperature $M_Y > T > M_X$, the number density of particles $Y$ will be less than the photon number density, while the number density of $X$ will be the same as the photon number density. Then the contribution to the asymmetry in the decay of $Y$ will have a Boltzmann suppression factor compared with the decay of $X$. This will make the sum of the two contributions nonvanishing.

**Boltzmann Equation**

The amount of asymmetry that can be generated will depend on the various interaction rates, the asymmetry parameter, and can be obtained by solving the Boltzmann equations [267]. Since we are writing the interaction at a temperature $T < M_Y$, the inverse decay rate of $Y$ is very small, $Y$ has already decayed and the number density of $Y$ has exponentially fallen off. Then we have to consider the interactions of only the particles $X$, which are the decay rate $\Gamma_D$, inverse decay rate $\Gamma_{ID}$ and the $2 \leftrightarrow 2$ scattering $A + \bar{B} \leftrightarrow C + \bar{D}$ mediated by $X$ and $\bar{X}$ scalar exchange $\Gamma_S$. There will also be scattering processes mediated by $Y$ and $\bar{Y}$, but they will be suppressed compared with the $X, \bar{X}$ scalar exchange. These rates are given by

$$\Gamma_D = \alpha m_X \begin{cases} \frac{m_X}{T} & T \geq m_X \\ 1 & T \leq m_X \end{cases}$$

$$\Gamma_{ID} = \Gamma_D \begin{cases} 1 & T \geq m_X \\ \left(\frac{m_X}{T}\right)^{3/2} \exp^{-m_X/T} & T \leq m_X \end{cases}$$

$$\Gamma_S = n\sigma = T^3 \alpha^2 \frac{T^2}{(T^2 + m_X^2)^2}, \tag{16.10}$$

where $\alpha = f^2/4\pi$ is the coupling strength of the $X$-bosons.

We now assume that $A$ and $B^*$ are quarks, so $X$ decays into two light quarks $qq$ with baryon number $B_{qq} = +2/3$ and $\bar{X}$ decays into two antibaryons $\bar{q}\bar{q}$ carrying baryon number $B_{\bar{q}\bar{q}} = -2/3$. The decays of $Y$ also violate $B$, but the details are not important for this discussion. In grand unified theories, $C$ could be a quark and $D$ could be a lepton, but this detail is not required for this analysis. We also assume that the Yukawa coupling constants are complex so there is $CP$ violation, which should show up in the interference of the tree-level and one-loop diagrams. Then the amplitudes for the $X$ and $\bar{X}$ decays can be written as

$$|M(X \to bb)|^2 = |M(\bar{b}\bar{b} \to \bar{X})|^2 = \frac{1}{2}(1 + \varepsilon_X)|M_0|^2$$

$$|M(\bar{X} \to \bar{b}\bar{b})|^2 = |M(bb \to X)|^2 = \frac{1}{2}(1 - \varepsilon_X)|M_0|^2.$$

We further assume that $Y$ is heavier than $X$ and at around $m_X$, the inverse decay of $Y$ is not allowed, and hence, the number density of $Y$ is negligible, but the number density of $X$ and $\bar{X}$ is controlled by the decays and the inverse decays of $X$ and $\bar{X}$ and is given by their equilibrium distribution. With this consideration we can now write the Boltzmann equation of these scalars as

$$\dot{n}_X + 3Hn_X = -\Gamma_D(n_X - n_X^{eq}), \tag{16.11}$$

where the second term comes due to the expansion of the universe and $n_X^{eq}$ is the equilibrium number density of $X$ and is given by

$$n_X^{eq} = \begin{cases} s\, g_*^{-1} & T \gg m_X \\ \frac{s}{g_*}\left(\frac{\pi}{2}\right)^{1/2}\left(\frac{m_X}{T}\right)^{3/2} \exp^{-m_X/T} & T \ll m_X\,. \end{cases} \tag{16.12}$$

$\bar{X}$ will satisfy the same equations. The out-of-equilibrium condition implies that the number density $n_X$ remains different from $n_X^{eq}$ at temperatures $T \leq m_X$. Then the decay of the excess $X$'s can generate an asymmetry. The second term on the left-hand side is due to the expansion of the universe and $H$ is the Hubble constant. The negative sign on the right implies that the decay decreases the number density of these scalars $X$ [267].

Similarly we can write the Boltzmann equation for the baryons and antibaryons. The difference gives the Boltzmann equation for the excess baryon number of the universe $n_B = n_b - n_{\bar{b}}$,

$$\dot{n}_B + 3Hn_B = \varepsilon_X \Gamma_D (n_X - n_X^{eq}) - n_B \left(\frac{n_X^{eq}}{n_\gamma}\right) \Gamma_D - 2n_B n_b \langle\sigma|v|\rangle. \qquad (16.13)$$

The first term is the crucial $CP$ violating contribution to the asymmetry, which comes from $|M(X \to bb)|^2 - |M(X \to \bar{b}\bar{b})|^2$. This is the only term that generates an asymmetry when the scalars $X$ decay out-of-equilibrium. All other contributions deplete the asymmetry generated by this term. The second term comes from the inverse decay of the $X$ bosons, while the last term comes from the matrix elements for $2 \leftrightarrow 2$ baryon-number violating scattering with the part due to the real intermediate state $X$ removed, which has already been included in the decay and inverse decay of the $X$ bosons. The thermally averaged scattering cross section now becomes

$$\langle\sigma|v|\rangle \sim \frac{A\alpha^2 T^2}{(T^2 + m_X^2)^2}, \qquad (16.14)$$

where $A$ is a numerical factor giving the number of scattering channels.

We simplify these equations by defining a dimensionless quantity $z = m_X/T$ and expressing these equations in terms of the number of particles per comoving volume, $Y_i = n_i/s$, and making use of the relation $t = z^2/2H(z=1)$. We can now define a measure of the out-of-equilibrium condition as

$$K = \frac{\Gamma_D(z=1)}{2H(z=1)} = \frac{\alpha M_{Pl}}{3.3\, g_*^{1/2} m_X}, \qquad (16.15)$$

and write the new form of the Boltzmann equations as [267]

$$\begin{aligned}\frac{dY_B}{dz} &= \varepsilon_X K z \gamma_D (Y_X - Y_X^{eq}) - K z \gamma_B Y_B \\ \frac{dY_X}{dz} &= -K z \gamma_D (Y_X - Y_X^{eq}),\end{aligned} \qquad (16.16)$$

where

$$\begin{aligned}\gamma_D &= \frac{\Gamma_D(z)}{\Gamma_D(z=1)} = \begin{cases} z/2 & z \ll 1 \\ 1 & z \gg 1 \end{cases} \\ \gamma_B &= g_* Y_X^{eq} \gamma_D + \frac{2n_\gamma < \sigma|v| >}{\Gamma_D(z=1)} \\ &= \begin{cases} z/2 + A\alpha/z & z \ll 1 \\ z^{3/2}\exp^{-z} + A\alpha z^{-5} & z \gg 1\,. \end{cases}\end{aligned} \qquad (16.17)$$

For $K \ll 1$, these equations may be solved exactly to get

$$Y_B = \frac{n_B}{s} = \frac{\varepsilon_X}{g_*}. \tag{16.18}$$

This represents a situation when the interaction rate is very slow compared with the expansion rate of the universe. In this case, there is an excess of $X$ bosons compared with their equilibrium distribution $X^{eq}$, since these scalars did not get the time to reach their equilibrium distribution as the universe was expanding faster than their inverse decay rate. So, they slowly decay and generate an asymmetry given by the amount of $CP$ violation, as shown in figure 16.3.

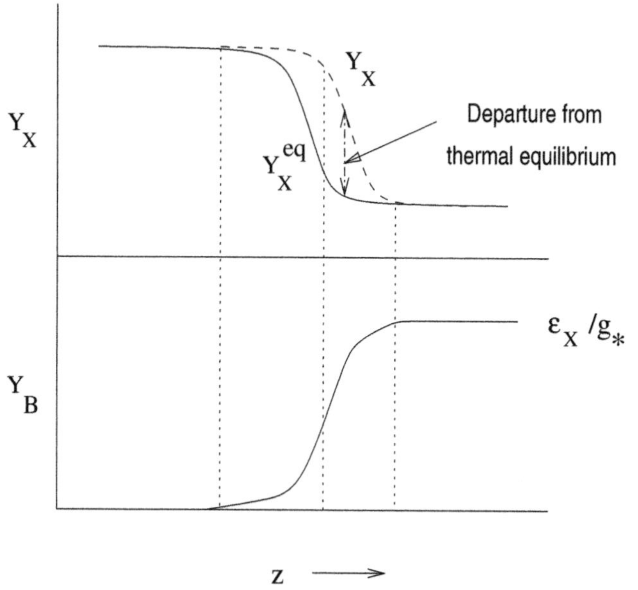


**FIGURE 16.3**
Generation of baryon asymmetry during departure from equilibrium for $K \ll 1$.

When $K$ approaches 1, the interaction rate becomes comparable with the expansion rate of the universe. In this case, as the scalars decay they generate an asymmetry, which approaches its asymptotic value faster than previously. However, now the inverse decay rate remains comparable even after the asymmetry has been generated. This inverse decay of the scalars starts depleting the asymmetry exponentially quickly. However, as the universe cools down the inverse decay rate becomes slow compared with the expansion rate of the universe, and the baryon asymmetry reaches an asymptotic value, which is lower than the maximum amount of $CP$ asymmetry (which is $Y_B = \varepsilon_X/g_*$, corresponding to $K < 1$). For large $K$, the asymptotic value of the asymmetry does not fall off exponentially and instead it falls off almost linearly. The asymptotic value of the baryon asymmetry for large $K$ comes out to be approximately

$$Y_B \approx \frac{0.3\varepsilon_X}{g_* K (\ln K)^{0.6}}. \tag{16.19}$$

The scattering rate remains small compared with the inverse decay, and does not affect the generation of the baryon asymmetry. However, scattering processes may become important and deplete any existing baryon asymmetry, when the number density of the scalar is small (usually at $T \ll m_X$), and the decays and inverse decays of the scalar become insignificant. In that case the scattering term depletes the primordial asymmetry exponentially quickly, if the scattering rate is fast compared with the expansion rate of the universe, i.e., the scattering rate does not satisfy the out-of-equilibrium condition. However, at the time of generating an asymmetry usually one may neglect the scattering processes. The numerical solutions of the Boltzmann equations are presented in figure 16.4 for different values of $K$ [267].

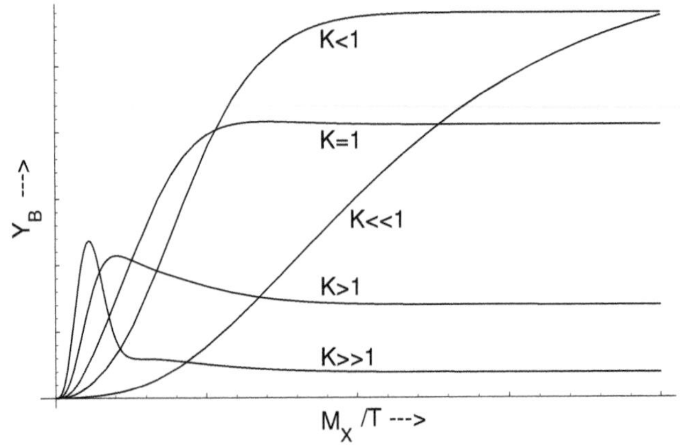


**FIGURE 16.4**
Solutions of the Boltzmann equations for different values of $K$.

### GUT Baryogenesis

A natural realization of baryogenesis became possible in the grand unified theories [268]. The quark–lepton unification provided the required baryon number violation, which manifests itself in the decays of some of the heavy bosons (for details about grand unified theories, see chapter 8). Since fermions belong to the chiral representations, $C$ is maximally violated. The couplings of the heavy bosons, whose decays violate baryon number, could be considered to be complex. There will then be interference between the tree-level and one-loop diagrams for the decays of the heavy bosons that could give the required $CP$ asymmetry and, hence, baryon asymmetry. Departure from thermal equilibrium is also naturally satisfied in these models since the scale of unification is sufficiently high and the universe was expanding very rapidly in that epoch. So, a coupling of $\alpha \sim .01$ would mean that for the unification scale $m_U \leq 10^{16}$ GeV, the heavy bosons could decay satisfying the out-of-equilibrium condition [268].

Although baryogenesis was considered to be a major success of the grand unified theories, later it was realized that there are problems in these scenarios. In $SU(5)$ grand unified theory, both the baryon number ($B$) and the lepton number ($L$) are violated, but a combination ($B-L$) remains conserved globally. In $SO(10)$ grand unified theories, $B-L$ number is a local symmetry of the theory, which is broken at some intermediate scale. So the baryon number violation at the scale of grand unification in both the $SU(5)$ and $SO(10)$ grand unified theories conserves ($B-L$). This can also be seen from the operator analysis for proton decay, which says that to the lowest order all baryon number violations in grand unified theories conserve $B-L$. It was later pointed out that any primordial baryon asymmetry, which conserves $B-L$, will be washed out by fast baryon number violation due to the chiral anomaly before the electroweak phase transition [269].

In the standard model the global baryon ($B$) and lepton ($L$) number symmetries have anomalies [100] (see section 4.4). A global $B$ anomaly comes from a triangle loops with the fermions in the loop, the two $SU(2)_L$ gauge bosons at the two external vertices, and a global $B$ current at the third vertex. The sum over all the fermions gives the total $SU(2)_L$ global $B$ anomaly. A nonvanishing global $B$ anomaly in the standard model means that although the baryon number $B$ is classically a global symmetry in the standard model, quantum effects violate this global symmetry. Thus, quantum–mechanical tunneling will allow transitions from one vacuum to another with differing baryon numbers, which will violate baryon number.

In the standard model, both baryon and lepton numbers are global symmetries, which are violated due to anomalies, but the anomaly corresponding to a combination $B-L$ vanishes. So, the anomaly induced vacuum transitions will conserve the combination $B-L$. In fact, the amount of baryon number violation ($\Delta B$) and lepton number violation ($\Delta L$) are the same, and hence, $\Delta(B-L)=0$ although $\Delta(B+L)=2\Delta B=2\Delta L\neq 0$. This $B+L$ violating process is highly suppressed by the quantum tunnelling probability at zero temperature [100]. But at high temperature this process was very fast due to an enhancement by the Boltzmann factor in the presence of an instanton-like solution, the sphaleron [269]. In a typical sphaleron-induced process, a vacuum with $B+L=6$ and $B-L=0$ can decay into another vacuum with $B+L=0$ and $B-L=0$ releasing $uude^-ccs\mu^-ttb\tau^-$. This sphaleron-induced $B+L$ violating process was in equilibrium for a long time before the electroweak phase transition, which would have washed out any baryon asymmetry that conserves $B-L$. Thus, the baryon asymmetry generated at the scale of grand unification is erased leaving a baryon symmetric universe in the presence of the sphalerons before the electroweak phase transition.

**Electroweak Baryogenesis**

Since the baryon asymmetry created at the time of grand unification phase transition was washed out by the sphaleron-induced transitions, attempts were made to generate a baryon asymmetry of the universe making use of the sphaleron transitions [270]. In this mechanism there was no need to introduce any new physics beyond the standard model. There is *CP* violation coming from the quark mixing matrix in the standard model. The only hurdle was the out-of-equilibrium condition, which is difficult to satisfy. A new mechanism was then suggested by considering the electroweak phase transition to be of first order. In a first order phase transition, when the temperature reaches the critical point, the free energies corresponding to the true vacuum and the false vacuum become equal. At this point, both the vacua coexist and bubbles start forming all over the place, where inside the bubble the electroweak symmetry is broken and sphaleron transitions are not allowed. Since there is no baryon number violation inside the bubbles, any baryon number violation along the bubble wall can generate a $B+L$ asymmetry inside the bubbles, which will not be affected by the fast sphaleron transitions outside the bubbles. This makes it possible to satisfy the out-of-equilibrium condition. *CP* violation takes place along the wall of the bubble only, which can generate the required baryon asymmetry of the universe. This model is highly attractive since it includes parameters entering in the standard model.

Although the models of electroweak baryogenesis were predictive and calculable, the simplest models did not work. The amount of *CP* violation coming from the quark mass mixing alone in the standard model is not enough and one requires extensions of the standard model with new sources of *CP* violation to explain the required baryon asymmetry of the universe. There is a more serious problem with these models, which comes from the fact that after the phase transition is over, the generated baryon asymmetry is not washed out by the remaining fast baryon number violation. After the phase transition is over, the false vacuum decays into the true vacuum and all the bubble walls expand at a speed of light and finally cover the entire space. If the baryon-number violating interactions have not ceased to exist by this time, they will make the universe baryon symmetric again. This depends on the dynamics of the model and gives a very strong constraint on the mass of the Higgs scalar [271]. With the present experimental limit on the Higgs mass most of the models of electroweak baryogenesis have been ruled out. In the supersymmetric models, the uncertainties in the parameters allow electroweak baryogenesis for a slightly higher Higgs mass. But the predictability of the electroweak baryogenesis is lost in these models.

There are many other mechanisms of baryogenesis [272], but the most popular of these mechanisms is the leptogenesis. In this mechanism the lepton number violation due to the Majorana masses of the neutrinos generate a lepton asymmetry of the universe, which then gets converted to a baryon asymmetry before the electroweak phase transition in the presence of the sphalerons [273].

## 16.2 Leptogenesis

Leptogenesis is the mechanism of generating a lepton asymmetry of the universe before the electroweak phase transition, which can then be converted to the required baryon asymmetry of the universe in the presence of the sphalerons. The fast sphaleron-induced $B+L$ violating transitions converts any $L$ asymmetry and, hence, $B-L$ asymmetry into a baryon asymmetry of the universe. We shall now discuss how the sphaleron-induced transitions due to global $B$ and $L$ anomaly relate the $L$ asymmetry to the $B$ asymmetry before the electroweak symmetry breaking.

Let us consider a global symmetry $U(1)_X$ in the standard model. The global $SU(2)_L$ anomaly of this $U(1)_X$ symmetry will then mean that in a triangle-loop diagram there are $SU(2)_L$ gauge bosons at the two vertices and at the third vertex there is a global current corresponding to the $U(1)_X$ symmetry. So, the group theoretical factor for a fermion in the loop transforming as doublet under $SU(2)_L$ and carrying $U(1)_X$ quantum number $x$ will be $\mathrm{Tr}\{T^aT^a\}\cdot x$. For a doublet of $SU(2)$, the quadratic invariant may be normalized to $\mathrm{Tr}\{T^aT^a\}=1/2$, thus the $X$ anomaly will be proportional to $x/2$. Fermions, which transform as singlets under $SU(2)_L$, do not contribute to the anomaly. The total $X$ anomaly will be sum over all the fermions in the theory.

The quarks carry baryon number $1/3$ and only the left-handed quarks transform as doublets under $SU(2)_L$. Thus, the $SU(2)_L$ $B$ anomaly gets contribution only from the left-handed doublet quarks and the corresponding group theoretical factor is $\mathscr{A}_B = 3\times 1/2\times 1/3 = 1/2$ (the factor 3 comes because there are three colored doublets). On the other hand the $L$ anomaly comes from the left-handed leptons carrying $L=1$, which is $\mathscr{A}_L = 1/2 \Rightarrow \mathscr{A}_B = \mathscr{A}_L \Rightarrow \mathscr{A}_{B-L}=0$ and $\mathscr{A}_{B+L}=1$.

In the standard model both $B$ and $L$ are global symmetries, but due to the $SU(2)_L$ global anomaly, the $B+L$ symmetry is broken. The chiral nature of the weak interaction makes the $B+L$ symmetry anomalous and the sum over all the fermions gives a nonvanishing axial current [100]

$$\delta_\mu j^{\mu 5}_{(B+L)} = 6[\frac{\alpha_2}{8\pi}W_a^{\mu\nu}\tilde{W}_{a\mu\nu} + \frac{\alpha_1}{8\pi}Y^{\mu\nu}\tilde{Y}_{\mu\nu}], \tag{16.20}$$

which breaks the $B+L$ symmetry. There is no anomaly corresponding to the $B-L$ charge. Because of this anomaly, baryon and lepton numbers are broken during the electroweak phase transition,

$$\Delta(B+L) = 2N_g\frac{\alpha_2}{8\pi}\int d^4xW_a^{\mu\nu}\tilde{W}_{a\mu\nu} = 2N_g\nu, \tag{16.21}$$

but their rate is very small at zero temperature, since they are suppressed by the quantum tunnelling probability, $\exp[-\frac{2\pi}{\alpha_2}\nu]$, where $\nu$ is a topological quantum number called the Chern–Simmons number. At finite temperature, this interaction rate becomes strong in the presence of an instanton-like solution, the sphalerons [269], and the quantum tunnelling factor is replaced by the Boltzmann factor $\exp[-\frac{V_0}{T}\nu]$, where the potential or the free energy $V_0$ is related to the mass of the sphaleron field, which is about TeV. As a result, at temperatures between

$$10^{12} GeV > T > 10^2 GeV, \tag{16.22}$$

the sphaleron mediated $B+L$ violating interactions will be very strong.

For the simplest scenario of $\nu = 1$, the sphaleron-induced processes are $\Delta B = \Delta L = 3$, given by

$$|vac> \longrightarrow [u_L u_L d_L e_L^- + c_L c_L s_L \mu_L^- + t_L t_L b_L \tau_L^-]. \tag{16.23}$$

These baryon and lepton-number violating fast processes will exponentially deplete any primordial $B+L$ asymmetry of the universe. As a result, the number density of both particles and antiparticles with $B+L$ number will become the same as their equilibrium distribution. On the other hand, if there is any $B-L$ asymmetry of the universe, that will be converted to a baryon asymmetry of the universe before the electroweak phase transition. This can be seen from an analysis of the chemical potential [274].

We shall now analyze the chemical potential assuming all particles are ultra-relativistic, which is the case above the electroweak symmetry breaking. The particle asymmetry, i.e., the difference between the number of particles ($n_+$) and the number of antiparticles ($n_-$), can be given in terms of the chemical potential of the particle species $\mu$ (for antiparticles the chemical potential is $-\mu$) as

$$n_+ - n_- = n_d \frac{gT^3}{6} \left(\frac{\mu}{T}\right), \tag{16.24}$$

where $n_d = 2$ for bosons and $n_d = 1$ for fermions.

Before the electroweak phase transition during the period (16.22), the sphaleron-induced $B+L$ violating interactions will be in equilibrium along with the other interactions. In table 16.1, we present all other interactions and the corresponding relations between the chemical potentials. Using these relations we eliminate some of the chemical potentials to relate the baryon and lepton numbers with the $B-L$ quantum number during this interaction. In the third column we give the chemical potential which we eliminate using the given relation. We start with chemical potentials of all the quarks ($\mu_{uL}, \mu_{dL}, \mu_{uR}, \mu_{dR}$); leptons ($\mu_{aL}, \mu_{\nu aL}, \mu_{aR}$, where $a = e, \mu, \tau$); gauge bosons ($\mu_W$ for $W^-$, and 0 for all others); and the Higgs scalars ($\mu_-^\phi, \mu_0^\phi$).

The chemical potentials of the neutrinos always enter as a sum and for that reason we can consider them as one parameter. We can then express all the chemical potentials in terms of the following independent chemical potentials,

$$\mu_0 = \mu_0^\phi; \;\; \mu_W; \;\; \mu_u = \mu_{uL}; \;\; \mu = \sum_i \mu_i = \sum_i \mu_{\nu iL}. \tag{16.25}$$

We can further eliminate one of these four potentials by making use of the relation given by the sphaleron processes (16.23). Since the sphaleron interactions are in

**TABLE 16.1**
Relations among the chemical potentials, arising from the interactions that are in chemical equilibrium.

| Interactions | $\mu$ relations | $\mu$ eliminated |
|---|---|---|
| $D_\mu\phi^\dagger D_\mu\phi$ | $\mu_W = \mu_-^\phi + \mu_0^\phi$ | $\mu_-^\phi$ |
| $\overline{q_L}\gamma_\mu q_L W^\mu$ | $\mu_{dL} = \mu_{uL} + \mu_W$ | $\mu_{dL}$ |
| $\overline{l_L}\gamma_\mu l_L W^\mu$ | $\mu_{iL} = \mu_{\nu iL} + \mu_W$ | $\mu_{iL}$ |
| $\overline{q_L}u_R\phi^\dagger$ | $\mu_{uR} = \mu_0 + \mu_{uL}$ | $\mu_{uR}$ |
| $\overline{q_L}d_R\phi$ | $\mu_{dR} = -\mu_0 + \mu_{dL}$ | $\mu_{dR}$ |
| $\overline{l_{iL}}e_{iR}\phi$ | $\mu_{iR} = -\mu_0 + \mu_{iL}$ | $\mu_{iR}$ |

equilibrium, we can write the following $B+L$ violating relation among the chemical potentials for three generations,

$$9\mu_u + 6\mu_W + \mu = 0. \tag{16.26}$$

We then express the baryon number, lepton numbers and the electric charge and hypercharge number densities in terms of these independent chemical potentials,

$$B = 12\mu_u + 6\mu_W \tag{16.27}$$
$$L_i = 3\mu + 6\mu_W - 3\mu_0 \tag{16.28}$$
$$Q = 24\mu_u + (12+2m)\mu_0 - (4+2m)\mu_W \tag{16.29}$$
$$Q_3 = -(10+m)\mu_W, \tag{16.30}$$

where $m$ is the number of Higgs doublets $\phi$.

At temperatures above the electroweak phase transition, $T > T_c$, both $Q$ and $Q_3$ must vanish. These conditions and the sphaleron-induced $B-L$ conserving, $B+L$ violating condition can be expressed as

$$<Q>=0 \Longrightarrow \mu_0 = \frac{-12}{6+m}\mu_u \tag{16.31}$$
$$<Q_3>=0 \Longrightarrow \mu_W = 0 \tag{16.32}$$
$$\text{Sphaleron transition} \Longrightarrow \mu = -9\mu_u \tag{16.33}$$

Using these relations we can now write the baryon number, lepton number and their combinations in terms of the $B-L$ number density as

$$B = \frac{24+4m}{66+13m}(B-L) \tag{16.34}$$

$$L = \frac{-42-9m}{66+13m}(B-L) \tag{16.35}$$

$$B+L = \frac{-18-5m}{66+13m}(B-L). \tag{16.36}$$

Below the critical temperature, the $SU(2)_L$ and the $U(1)_Y$ groups are broken leaving the $U(1)_Q$ symmetry. Thus, $Q$ should vanish since the universe is neutral with respect to all conserved charges. However, since $SU(2)_L$ is broken below the critical temperature, we can consider $\mu_0^\phi = 0$ and $Q_3 \neq 0$. This gives us

$$< Q >= 0 \Longrightarrow \mu_W = \frac{12}{2+m}\mu_u \tag{16.37}$$

$$\langle \phi \rangle \neq 0 \Longrightarrow \mu_0 = 0 \tag{16.38}$$

$$\text{sphaleron transition} \Longrightarrow \mu = \frac{-90-9m}{2+m}\mu_u, \tag{16.39}$$

which then allows us to write the baryon and lepton numbers as some combinations of $B-L$ as

$$B = \frac{32+4m}{98+13m}(B-L) \tag{16.40}$$

$$L = \frac{-66-9m}{98+13m}(B-L) \tag{16.41}$$

$$B+L = \frac{-34-5m}{98+13m}(B-L)\,. \tag{16.42}$$

Thus, before the electroweak phase transition, any $B-L$ asymmetry will be related to the baryon asymmetry of the universe, while any $B+L$ asymmetry will be wiped off.

Any lepton number violation at some intermediate scale within the period $10^{12} - 10^2$ GeV, can thus affect the baryon asymmetry of the universe in the two cases:

- If there is $L$ violation at a faster rate than the expansion rate of the universe, any $L$ asymmetry will be washed out. Since $B+L$ asymmetry is also washed out by the sphaleron transition, together they will wash out any baryon asymmetry $[B = (B+L) - L]$ of the universe before the electroweak phase transition [274, 275].

- If there is $L$ violation satisfying the out-of-equilibrium condition and if there is enough $CP$ violation, an $L$ asymmetry can be generated. In there is no primordial $B$ asymmetry, the $L$ asymmetry will give us an equal amount of $B-L$ asymmetry. This $B-L$ asymmetry will then give us a baryon asymmetry of the universe

$$B = \frac{24+4m}{66+13m}(B-L)\,.$$

The generation of the baryon asymmetry of the universe originating from a lepton number violation is known as leptogenesis [273, 274].

In the rest of this chapter we shall consider these two cases and study the interplay between the baryon asymmetry of the universe and the lepton-number violating Majorana masses of the neutrinos.

## 16.3 Models of Leptogenesis

Among the several generic mechanisms of neutrino mass, only the see-saw mechanism and the triplet Higgs mechanisms can generate a lepton asymmetry of the universe naturally in the simplest models. In the radiative models, there are some versions where it may be possible to generate a lepton asymmetry of the universe, but most of the simple versions of these models are inconsistent with leptogenesis. Some other mechanisms have to be invoked to explain the baryon asymmetry of the universe in these models. We shall now demonstrate how, in the minimal version of the see-saw and the triplet Higgs models, it is possible to generate a lepton asymmetry of the universe.

### See-Saw Leptogenesis

In the see-saw mechanism [55] for neutrino masses, the standard model is extended to include right-handed neutrinos ($N_{Ri}, i = e, \mu, \tau$). The interaction Lagrangian for the right-handed neutrinos is given by

$$\mathcal{L}_{int} = h_{\alpha i}\, \overline{\ell_{L\alpha}\phi}\, N_{Ri} + M_i\, \overline{(N_{Ri})^c}\, N_{Ri}, \tag{16.43}$$

where $\ell_{L\alpha}$ are the light leptons with $\alpha = 1, 2, 3$ as the generation index and $h_{\alpha i}$ are the Yukawa couplings, which could be complex. Without loss of generality we work in a basis in which the Majorana mass matrix of the right-handed neutrinos is real and diagonal with eigenvalues $M_i$ and assume a hierarchy $M_3 > M_2 > M_1$.

The Majorana nature of the right-handed neutrinos will imply lepton number violation and hence the right-handed neutrinos can decay into a lepton and an antilepton violating lepton number. The lepton-number violating decays of the right-handed neutrinos

$$\begin{aligned} N_{Ri} &\to \ell_{jL} + \bar{\phi}, \\ &\to {\ell_{jL}}^c + \phi \end{aligned} \tag{16.44}$$

can then generate a lepton asymmetry of the universe if there is enough *CP* violation and these interactions satisfy the out-of-equilibrium condition [273]. This $L$ asymmetry will be same as the $B - L$ asymmetry, which will then become the baryon asymmetry of the universe in the presence of the sphalerons.

There are now two independent sources of *CP* violation. In the original proposal for baryogenesis and all subsequent articles, only the decay type *CP* violation was considered, which comes from interference of tree-level and vertex diagrams [266].

In the language of $K$-physics, it is the *CP* violation coming from the penguin diagram in $K$-decays which is denoted by $\varepsilon'$. Leptogenesis was first proposed with only this contribution [273]. The tree-level decays of the right-handed neutrinos interfere with the one-loop vertex diagrams of figure 16.5. In the decay of the right-handed neutrinos, the one-loop diagram involves a right-handed neutrino of another generation. Since both right-handed neutrinos violate lepton number in their decays and they have different masses, they can give the required *CP* violation [273, 276]. This is similar to the example we discussed and the asymmetry parameter now comes out to be

$$\begin{aligned}\varepsilon' &= \frac{\Gamma(N \to \ell\phi^\dagger) - \Gamma(N \to \ell^c\phi)}{\Gamma(N \to \ell\phi^\dagger) + \Gamma(N \to \ell^c\phi)} \\ &= -\frac{1}{8\pi}\frac{M_1}{M_2}\frac{\mathrm{Im}[\sum_\alpha (h^*_{\alpha 1}h_{\alpha 2})\sum_\beta (h^*_{\beta 1}h_{\beta 2})]}{\sum_\alpha |h_{\alpha 1}|^2}, \end{aligned} \tag{16.45}$$

in the limit of large mass difference between $M_2$ and $M_1$. This result is for two generations. When they are degenerate, the asymmetry parameter vanishes.

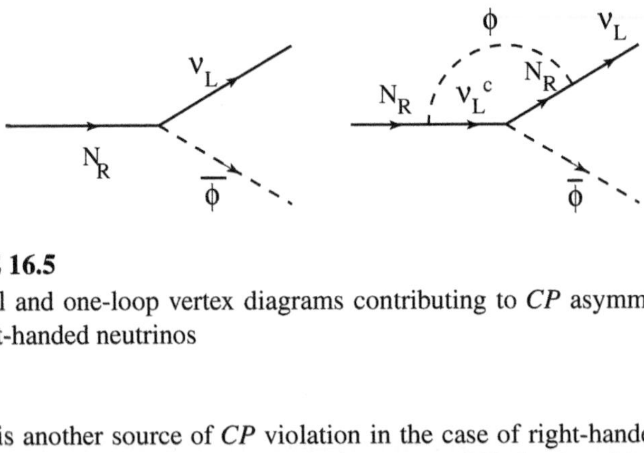


**FIGURE 16.5**
Tree-level and one-loop vertex diagrams contributing to *CP* asymmetry in models with right-handed neutrinos

There is another source of *CP* violation in the case of right-handed neutrino decays, which is more interesting. In the language of $K$-physics, this is the *CP* violation entering into the mass matrix denoted by $\varepsilon$ and comes from the box diagram giving the $K - \bar{K}$ mixing. This type of *CP* violation coming from the mixing matrix was not considered in earlier baryogenesis models. In the case of right-handed neutrino decay, this type of *CP* violation comes from the Majorana mass matrix of the heavy right-handed neutrinos [277] through a self-energy diagram [278, 279]. The interference of the tree-level diagrams for the right-handed neutrino decays with the self-energy diagram of figure 16.6, in which the two heavy neutrinos belong to two different generations, gives rise to this new oscillation type *CP* violation. In this case unitarity implies that the decay of one of the heavy neutrinos may cancel the asymmetry from the decay of the other neutrino entering into the self-energy diagram. But when the heavy neutrinos decay satisfies the out-of-equilibrium condition, the number densities of the two neutrinos differ during their decay, and hence, this cancellation does not take place. This is the first example of an oscillation type *CP* violation entering in the mass matrix of the heavy particles, which is used for generating a baryon asymmetry of the universe [277].

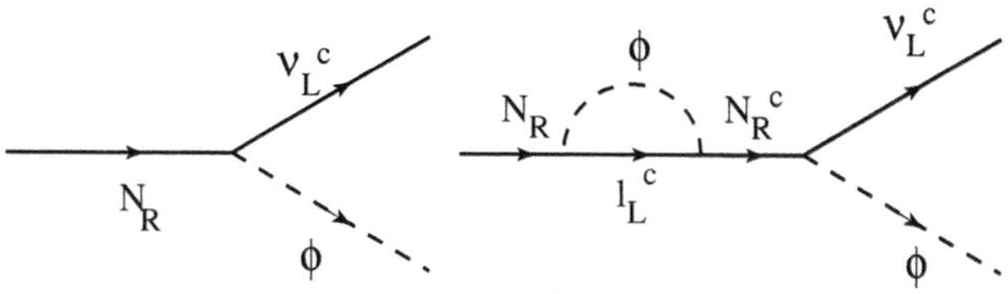

**FIGURE 16.6**
Tree-level and one-loop self-energy diagrams contributing to the generation of lepton asymmetry in models with right-handed neutrinos

The oscillation type *CP* violation, which comes from the self-energy diagram, makes the study of leptogenesis more fascinating. It may be interpreted as the right-handed neutrinos oscillating into antineutrinos of different generations and since the rates $\Gamma[\textit{particle} \to \textit{antiparticle}] \neq \Gamma[\textit{antiparticle} \to \textit{particle}]$, an asymmetry in the right-handed neutrinos is generated before they decay.

We shall now explain how the *CP* violation enters in the Majorana mass matrix, which can then generate a lepton asymmetry. A Majorana particle is its own antiparticle, but when *CP* is violated, it is convenient to consider the particle and the antiparticle independently. The Hermitian conjugate terms give the interactions of the *CP* conjugate states, which now have different couplings since there is *CP* violation. So, we shall denote the right-handed neutrinos as $N_i, i = 1,2$ (we shall restrict our present discussion to two generation only) and their *CP* conjugate antiparticles as $N_i^c, i = 1,2$. The Yukawa couplings of these fields can be written as

$$\begin{aligned}
\mathscr{L}_{int} &= \sum_i M_i[\overline{(N_{Ri})^c} N_{Ri} + \overline{N_{Ri}}(N_{Ri})^c] \\
&+ \sum_{\alpha,i} h^*_{\alpha i} \overline{N_{Ri}} \phi^\dagger \ell_{L\alpha} + \sum_{\alpha,i} h_{\alpha i} \overline{\ell_{L\alpha}} \phi N_{Ri} \\
&+ \sum_{\alpha,i} h^*_{\alpha i} \overline{(\ell_{L\alpha})^c} \phi^\dagger (N_{Ri})^c + \sum_{\alpha,i} h_{\alpha i} \overline{(N_{Ri})^c} \phi (\ell_{L\alpha})^c
\end{aligned} \tag{16.46}$$

where $M_i$ are the diagonal Majorana masses and real. *CP* violation comes from the complex phases in $h_{\alpha i}$. We have also written the Hermitian conjugate terms explicitly.

The states $|N_i\rangle$ and $|N_i^c\rangle$ are states with definite *CP*, but due to the complex phases in the Yukawa couplings $h_{\alpha i}$, the physical states will be different from these states. Since the physical states will evolve with time, their decays can violate *CP*. For example, $|N_i\rangle$ can decay into leptons, while $|N_i^c\rangle$ can decay into antileptons. But the physical states will be admixtures of $|N_i\rangle$ and $|N_i^c\rangle$, and hence, both the physical states could decay into both leptons and antileptons, violating $CP$.

We can write the effective mass matrix in the basis $(|N_1^c\rangle \quad |N_2^c\rangle \quad |N_1\rangle \quad |N_2\rangle)$ as

$$\mathscr{M}^{(0)} = \begin{pmatrix} 0 & 0 & M_1 & 0 \\ 0 & 0 & 0 & M_2 \\ M_1 & 0 & 0 & 0 \\ 0 & M_2 & 0 & 0 \end{pmatrix}. \tag{16.47}$$

The one-loop self-energy diagram will introduce small perturbation to this effective mass matrix, given by

$$\mathscr{M}^{(1)} = \begin{pmatrix} 0 & 0 & M_{11}^{(1)} & M_{12}^{(1)} \\ 0 & 0 & M_{12}^{(1)} & M_{22}^{(1)} \\ \widetilde{M}_{11}^{(1)} & \widetilde{M}_{12}^{(1)} & 0 & 0 \\ \widetilde{M}_{12}^{(1)} & \widetilde{M}_{22}^{(1)} & 0 & 0 \end{pmatrix}, \tag{16.48}$$

with

$$M_{ij}^{(1)} = M_{ji}^{(1)} = \left[M_i \sum_\alpha h_{\alpha i}^* h_{\alpha j} + M_j \sum_\alpha h_{\alpha i} h_{\alpha j}^*\right] (g_{\alpha ij}^{dis} - \frac{i}{2} g_{\alpha ij}^{abs}) \tag{16.49}$$

$$\widetilde{M}_{ij}^{(1)} = \widetilde{M}_{ji}^{(1)} = \left[M_i \sum_\alpha h_{\alpha i} h_{\alpha j}^* + M_j \sum_\alpha h_{\alpha i}^* h_{\alpha j}\right] (g_{\alpha ij}^{dis} - \frac{i}{2} g_{\alpha ij}^{abs}) \tag{16.50}$$

and

$$M_{ii}^{(1)} = \widetilde{M}_{ii}^{(1)} = \left[2M_i \sum_\alpha h_{\alpha i} h_{\alpha i}^*\right] (g_{\alpha ij}^{dis} - \frac{i}{2} g_{\alpha ij}^{abs}). \tag{16.51}$$

The dispersive part of the integral $g_{\alpha ij}^{dis}$ is absorbed in the renormalization of bare parameters, while the absorbtive part $g_{\alpha ij}^{ab}$ enters in the asymmetry parameter, which is given by

$$g_{\alpha ij}^{abs} = \frac{1}{16\pi}, \tag{16.52}$$

neglecting terms of order $O\left(m_\alpha^2/p^2\right)$, $O\left(m_\phi^2/p^2\right)$ with $p^2 \geq M_i^2$.

In general, all the physical states $\psi_i, i = 1,2,3,4$ are now combinations of all the four states $|N_1^c\rangle, |N_2^c\rangle, |N_1\rangle$ and $|N_2\rangle$, and decays of the physical states $|\Psi_i>$ generate the lepton asymmetry. The asymmetry parameter can, thus, be defined as

$$\Delta = \sum_{i=1}^{2} \frac{\Gamma_{\Psi_i \to l} - \Gamma_{\Psi_i \to l^c}}{\Gamma_{\Psi_i \to l} + \Gamma_{\Psi_i \to l^c}}. \tag{16.53}$$

In the limit $|M_2 - M_1| \gg |M_{ij}^{(1)}|$ or $|\widetilde{M}_{ij}^{(1)}|$, we get

$$\delta = \frac{1}{8\pi} \frac{M_1 M_2}{M_2^2 - M_1^2} \frac{\mathrm{Im}\left[\sum_\alpha (h_{\alpha 1}^* h_{\alpha 2}) \sum_\beta (h_{\beta 1}^* h_{\beta 2})\right]}{\sum_\alpha |h_{\alpha 1}|^2}. \tag{16.54}$$

For three generations there is a factor of $3/2$. When the mass difference is large compared with the width, the *CP* asymmetries generated through the mixing $\delta$ (through the self-energy diagram) of the heavy neutrinos and the decays $\varepsilon'$ (through the vertex diagram) are comparable. However, for small mass difference $\delta$ becomes very large compared with $\varepsilon'$ by orders of magnitude [277] (as shown in figure 16.7).

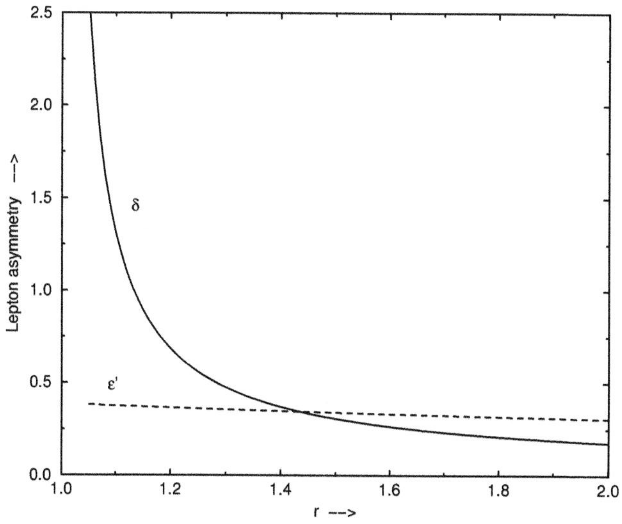


**FIGURE 16.7**
Comparison of the two *CP*-violating contributions to the lepton asymmetry, where $r = M_2^2/M_1^2$. The vertex contribution to *CP* violation is denoted by $\varepsilon'$ and the self-energy contribution is denoted by $\delta$ (from [277]).

The self-energy contribution to the *CP* violation $\delta$ becomes very large when the two right-handed neutrinos are almost degenerate, i.e., when the mass difference is comparable with their width and there is a resonance effect [277] (as shown in figure 16.8). As a result the amount of lepton asymmetry generated through this mechanism could be orders of magnitude larger than the asymmetry generated by the usual vertex-type *CP* violation. This phenomenon is called *resonant leptogenesis* [277, 279].

The lightest right-handed neutrino $N_1$ decays generate the asymmetry, so it should satisfy the out-of-equilibrium condition

$$\frac{|h_{\alpha 1}|^2}{16\pi} M_1 < 1.66\sqrt{g_*}\frac{T^2}{M_P} \qquad \text{at } T = M_1, \tag{16.55}$$

which gives a bound on the mass of the lightest right-handed neutrino to be $m_{N_1} >$

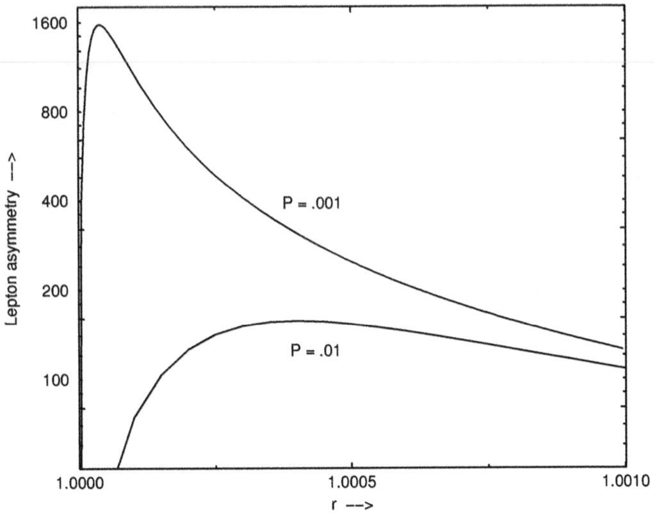


**FIGURE 16.8**
Lepton asymmetry generated for a small mass difference between two heavy neutrinos, where $\mathrm{P} = \sum_\alpha h^*_{\alpha 1} h_{\alpha 1}$ and $\mathrm{r} = M_2^2/M_1^2$. The resonant leptogenesis occurs when the masses $M_1$ and $M_2$ are almost degenerate (from [277]).

$10^8 GeV$ [280]. Although this out-of-equilibrium condition can give us some estimate, for actual calculation of the asymmetry one needs to solve the Boltzmann equation, taking into consideration both lepton-number conserving and lepton-number violating processes mediated by the heavy neutrinos [276, 277, 278, 279].

In the present case we define the lepton number asymmetry as $n_L = n_l - n_{l^c}$ and write its time evolution equation

$$\frac{dn_L}{dt} + 3Hn_L = (\varepsilon' + \delta)\Gamma_{\psi_1}[n_{\psi_1} - n^{eq}_{\psi_1}] - \left(\frac{n_L}{n_\gamma}\right) n^{eq}_{\psi_1}\Gamma_{\psi_1} - 2n_\gamma n_L \langle\sigma|v|\rangle. \quad (16.56)$$

Here, $\Gamma_{\psi_1}$ is the decay rate of the physical state $|\psi_1\rangle$, and $\langle\sigma|v|\rangle$ describes the thermally averaged cross-section of $l + \phi^\dagger \longleftrightarrow l^c + \phi$ scattering. The number density of $\psi_1$ satisfies

$$\frac{dn_{\psi_1}}{dt} + 3Hn_{\psi_1} = -\Gamma_{\psi_1}(n_{\psi_1} - n^{eq}_{\psi_1}). \quad (16.57)$$

We now define the parameter $K = \Gamma_{\psi_1}(T = M_{\psi_1})/H(T = M_{\psi_1})$ as a measure of the deviation from equilibrium. For $K \ll 1$ we can find an approximate solution

$$\frac{n_L}{s} = \frac{1}{g_*}(\varepsilon' + \delta), \quad (16.58)$$

where $g_*$ denotes the total number of massless degrees of freedom.

The bound on the lightest right-handed neutrino mass comes from the out-of-equilibrium condition and the constraint on the Yukawa couplings from the required amount of lepton asymmetry. In the case of resonant leptogenesis, since there is large enhancement of asymmetry, the constraints on the Yukawa couplings is relaxed. Thus, the Boltzmann suppression will be compensated by the resonant enhancement and the scale of leptogenesis could be lowered to as low as TeV [279].

The lepton asymmetry generated is the same as the $B-L$ asymmetry of the universe ($n_{B-L}$), since there is no primordial baryon asymmetry at this time

$$\frac{n_{B-L}}{s} = -(\varepsilon' + \delta)\,\eta\,\frac{n^{eq}_{\psi_1}(T \gg m_{N_1})}{s}, \tag{16.59}$$

where, $n^{eq}_{\psi_1}(T \gg m_{N_1})/s = 135\zeta(3)/(4\pi^4 g_*)$, with $g_* = 106.75$ for the standard model and $\eta$ is a measure of a departure from thermal equilibrium. When we lower the scale, $\eta$ becomes very small, but the increase in $\delta$ due to resonant oscillation could compensate and the required amount of $B-L$ asymmetry may be generated. The sphaleron interactions now convert this $B-L$ asymmetry to a baryon asymmetry of the universe

$$\frac{n_B}{s} = \frac{24+4n_H}{66+13n_H}\frac{n_{B-L}}{s}, \tag{16.60}$$

where $n_H$ is the number of Higgs doublets in the standard model. It is then possible to explain the baryon asymmetry of the universe, as required by the big-bang nucleosynthesis and measured by the WMAP, assuming the standard cold dark matter cosmological model including a nonvanishing cosmological constant,

$$\frac{n_B}{n_\gamma} = (6.15 \pm 0.25) \times 10^{-10} \quad \text{with} \quad s = 7.04 n_\gamma. \tag{16.61}$$

In some cases it is possible to relate the neutrino masses with the amount of generated baryon asymmetry. This would then allow us to predict the absolute neutrino mass from the observed value of the baryon asymmetry.

### Triplet Higgs Leptogenesis

We now discuss leptogenesis in the triplet Higgs mechanism of neutrino masses [57]. In this scenario one adds two complex $SU(2)_L$ triplet Higgs scalars, which carry a $U(1)_Y$ charge $-1$ ($\xi_a \equiv [1,3,-1]; a = 1,2$). The interactions of these Higgs scalars that break lepton number explicitly and are relevant for leptogenesis are given by

$$\mathscr{L}_{int} = f^a_{ij}\xi_a \ell_i \ell_j + \mu_a \xi^\dagger_a \phi\phi. \tag{16.62}$$

These give the lepton-number violating decays of the triplet Higgs $\xi_a$,

$$\xi_a^{++} \to \begin{cases} l_i^+ l_j^+ & (L=-2) \\ \phi^+\phi^+ & (L=0) \end{cases} \tag{16.63}$$

These interactions can generate a lepton asymmetry of the universe if they are slow enough and there is enough $CP$ violation. In this case there are no vertex corrections

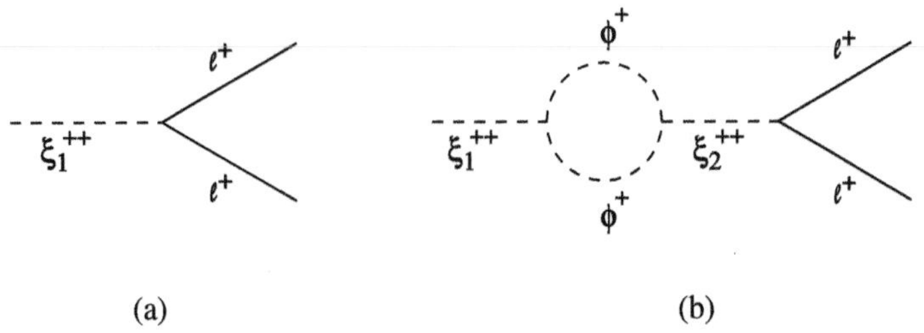


**FIGURE 16.9**
The decay of $\xi_1^{++} \to l^+ l^+$ at tree-level (a) and in one-loop order (b), whose interference gives *CP* violation.

that can introduce *CP* violation. The only source of *CP* violation is the self-energy diagrams of figure 16.9.

If there is only one Higgs triplet $\xi$, the relative phase between any $f_{ij}$ and $\mu$ can be made real. Hence, a lepton asymmetry cannot be generated. With two $\xi$'s, even if there is only one lepton family, one relative phase must remain nonvanishing. As for the possible relative phases among the $f_{ij}$'s, they cannot generate a lepton asymmetry because they all refer to final states of the same lepton number.

In the presence of the one-loop diagram, the mass matrices ${M_a}^2$ and $M_a^{*2}$ become different. This implies that the rate of $\xi_b \to \xi_a$ no longer remains the same as $\xi_b^* \to \xi_a^*$. Since by the *CPT* theorem $\Gamma[\xi_b^* \to \xi_a^*] \equiv \Gamma[\xi_a \to \xi_b]$, this means

$$\Gamma[\xi_a \to \xi_b] \neq \Gamma[\xi_b \to \xi_a]. \tag{16.64}$$

This is a different kind of *CP* violation compared with the *CP* violation in models with right-handed neutrinos or all other models of decays of scalars. This *CP* violation is analogous to the lepton-number conserving *CP* violation, which may enter in the neutrino oscillation experiments, where *CP* violation manifests itself as $\Gamma[\nu_a \to \nu_b] \neq \Gamma[\nu_b \to \nu_a]$. The oscillation type *CP* violation in the decays of right-handed neutrinos involve transition between particles and antiparticles, which is also different from the present case. However, the analyses are similar in both cases where *CP* violation comes from the self-energy diagrams, so we shall not present the details here.

If we consider that $\xi_2$ is heavier than $\xi_1$, then the decay of $\xi_1$ will generate a lepton asymmetry given by

$$\begin{aligned}\delta &= \frac{\Gamma(\xi \to \ell\ell) - \Gamma(\xi^\dagger \to \ell^c \ell^c)}{\Gamma(\xi \to \ell\ell) + \Gamma(\xi^\dagger \to \ell^c \ell^c)} \\ &= \frac{Im\left[\mu_1 \mu_2^* \sum_{k,l} f_{kl}^1 f_{kl}^{2\,*}\right]}{8\pi^2 (M_1^2 - M_2^2)} \left[\frac{M_1}{\Gamma_1}\right],\end{aligned} \tag{16.65}$$

where

$$\Gamma_{ab}M_b = \frac{1}{8\pi}\left(\mu_a\mu_b^* + M_aM_b\sum_{k,l} f_{kl}^{a\,*} f_{kl}^{b}\right)$$

and $\Gamma_a = \Gamma_{aa}$. In this model the out-of-equilibrium condition is satisfied when the masses of the triplet Higgs scalars are of the order of $10^{13}$ GeV.

The decay of the Higgs triplet will generate a lepton asymmetry, which becomes the $B-L$ asymmetry. This $B-L$ asymmetry would convert itself to the baryon asymmetry of the universe before the electroweak phase transition in the presence of sphalerons. A neutrino mass in the range of $0.001 - 1$ eV can allow a baryon asymmetry of the universe of $n_B/s \sim 10^{-10}$ as desired [57].

The see-saw mechanism and the triplet Higgs mechanism of neutrino masses can solve the problem of smallness of the Majorana neutrino masses and the baryon asymmetry of the universe simultaneously. This makes these two scenarios popular compared with the other models of neutrino masses. In the left–right symmetric models, both right-handed neutrinos and the triplet Higgs scalars are present. So, both these mechanisms can compete with each other for the generation of neutrino masses and the baryon asymmetry of the universe.

## 16.4 Leptogenesis and Neutrino Masses

We shall now consider the constraints on the neutrino masses coming from leptogenesis. The first assumption is that the neutrino masses are Majorana masses, so lepton number violation is associated with the neutrino masses. Any natural extensions of the standard model thus imply a large lepton-number violating scale required for the small neutrino masses. Depending on the strength of the lepton number violation, this can either generate a baryon asymmetry of the universe through leptogenesis or wash out any existing baryon asymmetry of the universe.

The sphaleron-mediated $B+L$ violating transitions are in equilibrium during the period $10^{12} - 10^{2}$ GeV. If the lepton number violation satisfies the out-of-equilibrium condition and there is *CP* violation, this can generate a lepton asymmetry of the universe. This lepton asymmetry will be the same as the $B-L$ asymmetry, which can then generate the required baryon asymmetry in the presence of the sphalerons. On the other hand, if there is fast lepton number violation at any time before the electroweak phase transition, that will erase any lepton asymmetry of the universe. Erasure of the lepton asymmetry, together with the sphaleron-induced erasure of the $B+L$ asymmetry will leave us with a baryon symmetric universe. The requirement of the baryon asymmetry at the time of nucleosynthesis thus constrains any lepton number violation. The lepton number violation should not be too fast so it washes out any baryon asymmetry. This can then constrain the mass of the neutrinos and put stronger constraints on some specific models of neutrino masses [275, 281, 282]. In fact, it will constrain any lepton-number violating processes taking place before the electroweak phase transition. In this section we shall discuss some general constraints on lepton-number violating interactions.

The lepton number violation associated with any neutrino mass can be described by the interaction [275]

$$L = \frac{2}{M} \ell_L \ell_L \phi \phi, \tag{16.66}$$

where $\ell_L$ are lepton doublets, $\phi$ is the Higgs doublet, and $M$ is the lepton-number violating scale in the theory. Without specifying the origin of this scale, we can estimate a neutrino mass from this operator to be $m_\nu \sim < \phi >^2 /M$. We are not including the couplings involved, which can change the details of the predictions. But to show how it works, let us write the lepton-number violating scattering process allowed by this interaction:

$$\sigma(\nu_L + \phi^\dagger \to \bar{\nu}_L + \phi) \sim \frac{1}{\pi} \frac{1}{M^2}. \tag{16.67}$$

This process should not be faster than the expansion rate of the universe before the electroweak phase transition, so as not to erase the baryon asymmetry of the universe, which means

$$\Gamma_{\Delta L \neq 0} \sim \frac{0.12}{\pi} \frac{T^3}{M^2} < H = 1.66\sqrt{g_*}\frac{T^2}{M_{Pl}} \quad \text{at } T \sim 100 \text{ GeV}, \tag{16.68}$$

where $H$ is the Hubble constant at that time, characterizing the expansion rate of the universe. This gives a constraint on the lepton-number violating mass scale,

$$M > 10^9 \left( \frac{T_c}{100 \text{ GeV}} \right)^{1/2} \text{GeV}, \tag{16.69}$$

where $T_c$ is the critical temperature of the electroweak phase transition. This in turn can give a bound on the mass of the neutrinos to be

$$m_\nu < 50 \text{ keV}. \tag{16.70}$$

Although this is not a stringent limit, it gives an indication of how survival of the baryon asymmetry of the universe can constrain the neutrino masses. This bound can be improved slightly considering the scattering processes

$$W^\pm + W^\pm \to e^\pm_\alpha + e^\pm_\alpha, \tag{16.71}$$

where $\alpha = e, \mu, \tau$. Soon after the electroweak phase transition starts, this process will become operational and should be slow enough, which gives a constraint of $m_\nu < 20$ keV [283].

There exist strong general constraints on any $B - L$ number violating interactions and, hence, lepton-number violating processes [281]. These constraints originate from the fact that any fast $B - L$ number violating interactions will make the universe $B - L$ symmetric which in turn will make the universe baryon symmetric in the presence of the sphalerons. However, these constraints can be evaded in several ways [282]. One common solution is that in the electroweak anomaly mediated processes, $B-3L_i$ is conserved [282]. To see this, consider the sphaleron-mediated process,

$$|vac> \longrightarrow [uude^- + ccs\mu^- + ttb\tau^-], \tag{16.72}$$

where the vacuum can decay into any one of these states. So, if any one of the three combinations $B-3L_i$ $(i=e,\mu,\tau)$ is conserved after the electroweak phase transition, then the baryon asymmetry will not be erased even if the other two interactions are fast. However, in practice it is difficult to make such models phenomenologically consistent and to get the proper neutrino mixing.

Let us now consider the constraints on the $R$-parity violating models of neutrino masses. There are lepton-number violating interactions

$$L_i + Q_j \to (\tilde{d}_k^c)^* \to H_1 + Q_l, \tag{16.73}$$

which should not be too fast and should satisfy

$$\frac{\lambda'^2 T}{8\pi} \lesssim 1.7\sqrt{g_*}\frac{T^2}{M_{Pl}} \quad \text{at } T = M_{SUSY}. \tag{16.74}$$

Assuming that the supersymmetry breaking scale $M_{SUSY}$ is $10^3$ GeV, we find [284]

$$\lambda' \lesssim 2\times 10^{-7}, \tag{16.75}$$

which is very much below the typical minimum value of $10^{-4}$ needed for radiative neutrino masses [166]. The bound of equation (16.75) cannot be evaded even if one uses the bilinear terms for neutrino masses. So, in all $R$-parity violating models one needs to find some way to generate a baryon asymmetry of the universe.

The radiative Zee-type models also suffer from a similar problem [284]. In these models one requires a charged singlet Higgs scalar, whose couplings break lepton number explicitly. Consider a Zee-type radiative model, where the charged scalar $\eta^-$ violates lepton number explicitly through its couplings

$$\mathscr{L} = \sum_{i<j} f_{ij}(\nu_i e_j - e_i \nu_j)\eta^+ + \mu(\phi_1^+\phi_2^0 - \phi_1^0\phi_2^+)\eta^-. \tag{16.76}$$

Lepton-number is violated in the above by two units, and a neutrino mass is generated radiatively. If these lepton number violating interactions are too fast, they will make the universe lepton-symmetric before the electroweak phase transition, which will then erase any baryon asymmetry of the universe in the presence of the sphalerons. This requires $\eta^-$ to be heavy [285],

$$\frac{M_\eta}{f_{ij}^2} \gtrsim 10^{15}\text{GeV} \quad \text{or} \quad \frac{M_\eta}{\mu^2} \gtrsim 10^{15}\text{GeV} \quad \text{and} \quad \frac{M_\eta}{\mu^2 f_{ij}^2} \gtrsim 10^{16}\text{GeV}. \tag{16.77}$$

In this model the required neutrino mass cannot satisfy these out-of-equilibrium conditions and generate a baryon asymmetry of the universe.

These radiative models through $R$-parity violation or by the introduction of new scalars can be made to work by introducing new physics, but the simplest models cannot generate a lepton asymmetry of the universe. Only the see-saw mechanism and the triplet Higgs models are mechanisms for neutrino masses which can generate a lepton asymmetry of the universe without any extra input.

The condition that the lepton number violation should not be too fast implies a strong bound on the lepton-number violating couplings in several other models. The doubly charged leptons $L^{--}$ will also have similar lepton-number violating interactions, $h_1 L^{++} l_R l_R + h_2 M_L L^{++} \chi^- \chi^-$, which will also be constrained by baryogenesis.

All the bounds coming from the survival of the baryon asymmetry of the universe may be avoided to some extent, if a baryon asymmetry of the universe is generated after the heavy scalars (whose interactions violate baryon number) have all decayed away. However, in this case, there will still be a bound on the masses of these heavy scalars, which is the scale of baryogenesis. If these scalars are lighter than the scale at which baryon asymmetry of the universe is generated, then their interactions will erase the asymmetry thus generated. Hence the bounds from the constraints of baryogenesis can at most be made milder by generating a baryon asymmetry of the universe at a lower energy scale.

In the see-saw mechanism of neutrino masses, leptogenesis takes place through decays of the right-handed neutrinos. Without loss of generality, it is possible to work in the basis where the right-handed neutrino Majorana mass matrix $M$ is real and diagonal with the hierarchy $M_3 > M_2 > M_1$. Then at a temperature $T \sim M_1$, the number densities of $N_{2,3}$ will be much less, and only the decays of $N_1$ (with mass $M_1$) will contribute to leptogenesis. It can be shown that the amount of lepton asymmetry generated in these models can be related to the low-energy neutrino masses [286, 287].

We parameterize the light neutrino masses in terms of the measured quantities, the atmospheric neutrino mass-squared difference $\Delta m^2_{atm}$, the solar neutrino mass-squared difference $\Delta m^2_{sol}$, and the total mass squared

$$\bar{m}^2 = m_1^2 + m_2^2 + m_3^2,$$

where $m_{1,2,3}$ are the mass-eigenvalues of the light neutrinos. All neutrino masses can then be expressed in terms of these three mass parameters. We also define an effective mass $\tilde{m}_1$, which enters the process of leptogenesis through decays of $N_1$,

$$\tilde{m}_1 = \frac{(m_D^\dagger m_D)_{11}}{M_1},$$

where $m_D = hv$ is the Dirac mass matrix of the neutrinos, $h \equiv h_{\alpha i}$ are the Yukawa couplings, and $v$ is the *vev* of the electroweak symmetry breaking Higgs doublet.

It has been shown that the amount of $CP$ violation in the decays of $N_1$ can be expressed in terms of these low-energy parameters and $M_1$ [286] as

$$\varepsilon_1 = -\frac{3}{16\pi} \frac{M_1}{(h^\dagger h)_{11}} \operatorname{Im}\left[(h^\dagger h) \frac{1}{M} (h^T h^*)\right]_{11}. \tag{16.78}$$

Corrections of order $\sim O(M_1/M_{2,3})$ have been neglected. This *CP* asymmetry satisfies an upper bound [280] $\varepsilon_1 \leq \varepsilon(M_1, \bar{m})$, where

$$\varepsilon(M_1, \bar{m}) = \frac{3}{16\pi} \frac{M_1}{v^2} \frac{\Delta m^2_{atm} + \Delta m^2_{sol}}{m_3} \tag{16.79}$$

and the maximum baryon asymmetry produced by $N_1$ decay is given by [287, 288]

$$\eta^{max}_{B0}(\tilde{m}_1, M_1, \bar{m}) = 1.38 \times 10^{-3} \varepsilon(M_1, \bar{m}) \kappa_0(\tilde{m}_1, M_1, \bar{m}), \tag{16.80}$$

where $\kappa_0$ is the efficiency factor that measures the number density of $N_1$ with respect to the equilibrium value, the out-of-equilibrium condition at decay and the thermal corrections to $\varepsilon_1$. $m_3$ may be expressed in terms of $\bar{m}$, $\Delta m^2_{atm}$ and $\Delta m^2_{sol}$ for the different scenarios of neutrino mass hierarchy. For the normal hierarchy, it is $m_3 = (\bar{m}^2 + 2\Delta m^2_{atm} + \Delta m^2_{sol})/3$, while for the inverted hierarchy, it is $m_3 = (\bar{m}^2 + \Delta m^2_{atm} + 2\Delta m^2_{sol})/3$. These relations can then be used to obtain bounds on the mass of $M_1$ and the light neutrinos considering thermal productions of right-handed neutrinos. Taking into account all the interactions, one solves the Boltzmann equations numerically to obtain bounds on the neutrino masses for which successful leptogenesis is possible.

The see-saw models of neutrino masses allow leptogenesis through decays of heavy right-handed neutrinos only for a small range of light neutrino masses [287]. The upper bound on the neutrino mass comes out to be $\bar{m}_\nu < 0.20$ eV, which translates into individual neutrino masses for the normal hierarchical neutrinos to be

$$m_{1,2} < 0.11 \text{ eV} \quad \text{and} \quad m_3 < 0.12 \text{ eV}.$$

In the case of inverted hierarchy the bounds remain similar with $\bar{m}_\nu < 0.21$ eV and for individual neutrinos

$$m_1 < 0.11 \text{ eV} \quad \text{and} \quad m_{2,3} < 0.12 \text{ eV}.$$

Above these values, leptogenesis is not possible. There is also a lower bound on the neutrino mass below which leptogenesis will not be possible in these models, but those bounds are model dependent and weak.

The lower bound on the neutrino mass comes from the fact that at the time when the right-handed neutrinos decay, their number density should not be too small to generate the required lepton asymmetry; however, it should not be too fast so after the asymmetry is generated it is washed out. In models of thermal leptogenesis, the right-handed neutrinos are produced through scattering in the thermal bath. Including all their interactions at temperatures above $T = M_1$, it can be shown that leptogenesis is possible for neutrino masses above $m_\nu > 10^{-3}$ eV [287]. The range of mass-squared difference suggested by the atmospheric neutrinos $\sqrt{\Delta m^2_{atm}} \sim 0.046$ eV and solar neutrinos $\sqrt{\Delta m^2_{sol}} \sim 9 \times 10^{-3}$ eV is well suited to explain thermal leptogenesis in the see-saw mechanism of neutrino masses. These considerations also give a lower bound on the lightest right-handed neutrino mass of $M_1 \geq 10^8$ GeV [280].

The upper bounds on the light neutrino masses are consistent with the bound coming from the WMAP result of $\sum m_\nu < 0.69$ eV, and this result implies that for the values of the neutrino Majorana mass, as obtained in the neutrinoless double beta decay, thermal leptogenesis is only marginally possible in the see-saw model. However, a more general analysis [289] including the precise computation of the thermal leptogenesis [288] relaxed these bounds substantially. Although the bound for the hierarchical heavy neutrinos remains similar, when the two heavy right-handed neutrinos become almost degenerate the enhancement due to resonant leptogenesis relaxes the upper bound on the neutrino masses to as high as 1 eV, thus making the neutrinoless double beta decay observations consistent with resonant thermal leptogenesis [277]. For the thermal leptogenesis with triplet Higgs scalars [57], or when the triplet Higgs in left–right symmetric models contribute mostly to the neutrino masses, there is no constraint on the neutrino mass [290].

## 16.5 Generic Constraints

If there is fast baryon or lepton number violation before the electroweak phase transition, that will wash out any baryon asymmetry of the universe in the presence of the sphaleron-induced fast $B+L$ violation. This can, in turn, constrain the rate of baryon and lepton-number violating interactions. Let us now consider the possible effective operators, which can contribute to the baryon and lepton number violation in the standard model [54]. The lowest-dimensional $B$ violating operators are all of the form $QQQL$. This operator $QQQL$ has $B-L=0$ and $B+L=2$, and hence, all processes allowed by this operator conserve $B-L$ but violate $B+L$. The baryon number violation arising from this operator could generate a baryon asymmetry of the universe in grand unified theories but will be washed out by the fast sphaleron-induced $B+L$ violating transitions.

The lowest-dimensional $B-L$ violating operator, which violates baryon number, is $QQQ\bar{L}\phi$, where $\phi$ is the usual Higgs doublet in the standard model. However, compared with the $B-L$ conserving processes, this process will be suppressed by $(<\phi>/M_U)^2 \sim 10^{-25}$. In some grand unified theories, the scale of $B-L$ violation ($M_{B-L}$) could be much lower than the scale of grand unification. In these theories it is possible to have $B-L$ violating baryon and lepton number violations. The lowest-dimensional $B-L$ violating operator is of the form $LL\phi\phi$, which contributes to the neutrino mass. There are also higher-dimensional $B-L$ violating effective operators, which are $QQQQQQ$ and $QQQLLL$. These operators give rise to neutron–antineutron oscillation and a three-lepton decay mode of the proton. Explicit realization of all these processes would require new Higgs scalars as well as new gauge bosons. Some of these new particles could be very light and may have other low-energy consequences.

**Leptoquarks, Diquarks and Other Exotics**

In grand unified theories proton decay is mediated by leptoquark and diquark gauge bosons, which mix with each other. In addition, there could be leptoquark and diquark scalars, which can mix with each other and mediate proton decay. So, the present proton lifetime requires all these particles to be very heavy. However, if there are only leptoquarks or only diquarks or if there is no mixing between the two, these particles could remain light. In general, all possible *scalar bilinears*, which are scalars that couple to two of the standard model fermions, could be observed at low energy [291, 292]. Direct searches for these particles by accelerators have given bounds on the mass to coupling ratio of these particles. These bounds are not very strong. However, there are indirect constraints on these particles, including the constraints from baryogenesis, which are much stronger [285, 291, 293].

The standard model has one such scalar bilinear, the usual Higgs doublet that couples to $\bar{q}_{iL}u_R$, $\bar{q}_{iL}d_R$ and $\bar{\ell}_{iL}e_R$ (generation indices are suppressed). Other scalar bilinears have been considered to understand the neutrino masses, which are the triplet Higgs scalar [57] or the charged dileptons [65], which transform under $SU(3)_c \times SU(2)_L \times U(1)_Y$ as $(1,3,-1)$ or $(1,1,-1)$, respectively. All possible scalar bilinears that could exist in theories beyond the standard model are listed in table 16.2.

**TABLE 16.2**
Exotic scalar bilinears that could couple to two fermions of the standard model.

| Representation | Notation | $qq$ | $\bar{q}\bar{l}$ | $q\bar{l}$ | $ll$ |
|---|---|---|---|---|---|
| $(1,1,-1)$ | $\chi^-$ | | | | $\times$ |
| $(1,1,-2)$ | $L^{--}$ | | | | $\times$ |
| $(1,3,-1)$ | $\xi$ | | | | $\times$ |
| $(3^*,1,1/3)$ | $Y_a$ | $\times$ | $\times$ | | |
| $(3^*,3,1/3)$ | $Y_b$ | $\times$ | $\times$ | | |
| $(3^*,1,4/3)$ | $Y_c$ | $\times$ | $\times$ | | |
| $(3^*,1,-2/3)$ | $Y_d$ | $\times$ | | | |
| $(3,2,1/6)$ | $X_a$ | | | $\times$ | |
| $(3,2,7/6)$ | $X_b$ | | | $\times$ | |
| $(6,1,-2/3)$ | $\Delta_a$ | $\times$ | | | |
| $(6,1,1/3)$ | $\Delta_b$ | $\times$ | | | |
| $(6,1,4/3)$ | $\Delta_c$ | $\times$ | | | |
| $(6,3,1/3)$ | $\Delta_L$ | $\times$ | | | |
| $(8,2,1/2)$ | $\Sigma$ | | | | |

There are three dileptons, which are singlets under the $SU(3)_c$ group. They can take part in baryon-number violating interactions, through their interactions with other scalar bilinears. The fields $Y$ take part in baryon number violation, since they can couple with two quarks or a quark and a lepton. The scalars $X$ are purely leptoquarks while the $\Delta$ are purely diquarks. They can allow baryon number violation only through their mixing.

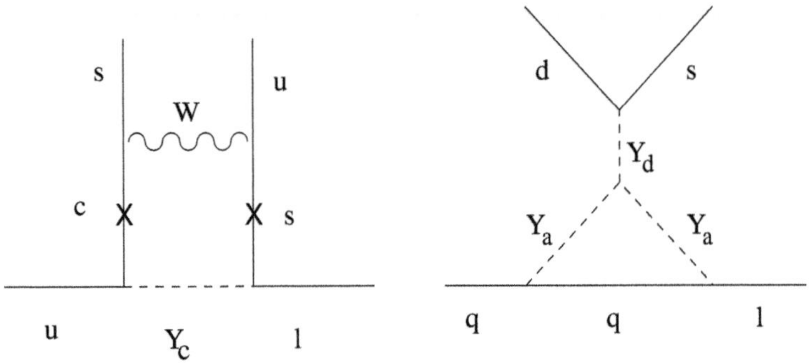


**FIGURE 16.10**
One-loop proton decay diagrams with scalar bilinears.

There are two types of scalar diquarks, one having antisymmetric coupling to two quarks ($3^* \subset 3 \times 3$, $\varepsilon_{abc} q^a q^b$) and the other symmetric ($6 \subset 3 \times 3$, $q^a q^b$, symmetric in $[ab]$). The $3^*$ may also couple to an antiquark and an antilepton as shown in table 16.2. This means that proton decay is always possible, either at the tree level [292], or through one-loop diagram as shown in figure 16.10.

There are also some general constraints, which are applicable to all the baryon and lepton-number violating interactions allowed by any of these scalars. These interactions should not be too fast to prevent erasure of the primordial baryon asymmetry of the universe. This gives very strong constraints for the couplings of these exotic scalars for all generations of fermions. All the baryon-number violating interactions we studied with the exotic scalar bilinears violate $B-L$, and hence, all these scalars and all their couplings should satisfy

$$\frac{M_X}{f^2} \gtrsim 10^{15}\text{GeV}. \tag{16.81}$$

The couplings in this expression apply to all generations, unlike the bounds from the proton decay, where the bound is only for the first generation.

# 17

# *Dark Matter and Dark Energy*

The evolution of the universe depends crucially on the total matter density of the universe. Combined fit to all cosmological measurements now implies that the total matter density equals the critical density of the universe, so the universe is flat. Out of this, a contribution of about 70% comes from the vacuum energy (also called the dark energy) or the cosmological constant

$$\Lambda \sim 0.75 \times 10^{-56} \text{ cm}^{-2}. \tag{17.1}$$

Baryonic and nonbaryonic matter in the universe amount to only about 30% of the critical density, out of which only about 5% is baryonic matter (see section 15.2). The amount of nonbaryonic dark matter is then about 1/4 of the critical density of the universe, i.e.,

$$\rho_{DM} = 1.7 \times 10^{-30} \text{ g cm}^{-3}. \tag{17.2}$$

Neutrinos could be a hot (relativistic) dark matter candidate, but the present limits on the neutrino masses imply that they cannot contribute to more than 1% of the dark matter. Moreover, the large scale structure formation and the cosmic microwave background anisotropy together require that most of the dark matter is nonrelativistic (cold dark matter). Thus, the ΛCDM (nonvanishing cosmological constant and cold dark matter) model is the most favored cosmological model at present, and the matter content implies that the universe is flat and accelerating.

Although it is established that the universe is dominated by dark matter and dark energy or cosmological constant, we have yet to find satisfactory solutions to these two problems. The dark matter problem requires new particles, which would depend on the model of particle physics beyond the standard model. So, detection and identification of the dark matter candidates may provide us new inputs to physics beyond the standard model. The dark matter events reported by DAMA and the indirect evidence for dark matter reported by EGRET could be the first indication. However, the cosmological constant problem poses a new challenge for any theory. One now expects to get a solution to this problem from a theory which includes gravity. Two very promising solutions come from the string theory and the brane world scenarios, although neither of these solutions was complete until now. Another interesting observation that the scale of the cosmological constant and the neutrino masses are of the same order of magnitude motivated a neutrino dark energy solution. After discussing the present status of dark matter candidates and searches, we shall mention some of these solutions of the cosmological constant problem and mention their drawbacks.

## 17.1 Dark Matter and Supersymmetry

The existence of dark matter in the universe has been established from many observations. The main criterion for any particle to be dark matter is that it should not interact too fast and it must give us the observed dark matter relic density of $\Omega_d\, h^2 \sim 0.12$. There are many possible candidates of dark matter, originating from different particle physics models. These dark matter candidates may be classified into two categories: hot and cold dark matter. If the dark matter candidate is a very light particle that can remain relativistic and freeze out at about 1 MeV, we call it a hot dark matter. The most probable candidates for hot dark matter are the neutrinos, but the present experimental upper bound on the neutrino mass implies that they can at most contribute to 0.3% of the critical density, whereas the latest observations in cosmology require a contribution of dark matter amounting to about 25% of the critical density.

Heavier particles with mass of GeV and higher can be considered as cold dark matter candidates, provided they interact weakly so they are present until now and they are highly nonrelativistic. Since the abundance of these particles is inversely proportional to their annihilation cross-section, their survival until today requires them to be very weakly interacting. That is why these particles are also called the weakly interacting massive particles (WIMP). A popular explanation is that WIMPs are supersymmetric particles [294].

Any interaction in supersymmetric models with conserved $R$-parity contains at least two superpartners of ordinary particles. As a result, whenever any superpartners decay, the decay product contains at least one supersymmetric particle. As a result, the lightest supersymmetric particle (LSP) cannot decay into any other particles. One of the neutralinos could be the LSP and its interaction rate is also very small and hence there could be relic neutralinos which contribute to the dark matter of the universe. At an early time when the temperature was much higher than the supersymmetry breaking scale, neutralinos were in thermal equilibrium with the primordial cosmic soup of particles and radiation. As the universe cooled down to a temperature below the mass of the neutralino, neutralinos could no longer be produced, but they could still annihilate away. Finally, the relic density of neutralinos is fixed when the expansion rate of the universe exceeds the annihilation rate. This relic density has been estimated for different supersymmetric models. Their abundance depends on their annihilation cross-section, which should be comparable to the weak interactions for them to contribute to the dark matter. This will also allow us to detect the neutralinos directly in scattering experiments.

The relic density

$$\Omega_\chi h^2 = \frac{\rho_\chi}{\rho_c} h^2$$

of the neutralino ($\chi$) is calculated [295, 296] by first determining the particle density

$n$ using the Boltzmann equation

$$\frac{dn}{dt} + 3Hn = -(\sigma v)_{ann}(n^2 - n_{eq}^2), \quad (17.3)$$

where $H$ is the Hubble parameter, $n_{eq}$ is the equilibrium abundance and $(\sigma v)_{ann}$ is the thermally averaged product of the annihilation cross-section and velocity, taking all channels of the $\chi - \chi$ annihilation. Since the neutralinos are mixtures of gauginos and Higgsinos, the annihilation can occur both via s-channel exchange of the $Z^0$ and Higgs bosons and via t-channel exchange of a scalar particle.

The present observations require a neutralino relic density $\Omega_\chi$ in the range of $0.1 < \Omega_\chi h_0^2 < 0.3$, taking into account that the universe is flat and accelerating and about $2/3$ of the total matter in the universe is in the form of dark energy. If there are other cold dark matter candidates contributing partly to the matter density in the universe, then the neutralino relic density could be $\Omega_\chi < 0.1$. This decrease in $\Omega_\chi$ will imply a larger elastic scattering cross-section and the detection rate will be reduced due to the reduced density of LSP in the galactic halo. However, for all neutralino cold dark matter calculations, it is assumed that there are no other sources of cold dark matter, so $\Omega_\chi \geq 0.1$ [297].

The neutralino dark matter candidates are highly model dependent and the details depend on the supersymmetry breaking mechanism. A typical neutralino mass is considered in the range of 50 GeV and a few TeV. The direct dark matter search experiments concentrate on the LSP. In the minimal supersymmetric standard model, the lightest neutralino is a probable candidate for the LSP with a mass in the range of 50 GeV to a TeV. The main uncertainty in the theoretical estimate of supersymmetric dark matter comes from the neutralino–nucleon cross-section and usually large $\tan\beta$ solutions are preferred. However, it is not possible to make any more definite comments at this stage. Only in the supergravity-inspired models and some other choice of a constrained parameter space, the dark matter search might promise the possibility of detection of supersymmetric particles even before the direct search experiments in colliders [297]. In all other supersymmetric models the uncertainties in the parameters and the supersymmetry breaking mechanism is too large to make any predictions.

In the absence of a complete theory one makes certain assumptions about the grand unification scale physics and extrapolates them to low-energy. The minimal supersymmetric standard model is assumed to be valid at all energies from the electroweak symmetry breaking scale to the grand unification scale. This is motivated by the original $N = 1$ supergravity model, where $N = 1$ supersymmetry is broken in the hidden sector using a Polonyi potential at an intermediate scale of $M_I \sim 10^{10}$ GeV, which is then communicated to the observable sector via gravitational coupling at the Planck scale $M_{Pl}$ [158]. This gives the low-energy soft supersymmetry breaking terms in addition to the minimal supersymmetric standard model. In the observable sector all supersymmetry breaking terms are soft terms and of the order of $M_{susy} \sim M_I^2/M_{Pl}$, which is in the range of a few TeV. This scale of supersymmetry breaking is also in the favored range for a cold dark matter particle of 50 GeV to a TeV. At the GUT scale this requires a common mass for all scalars $m_0$ and a common trilinear coupling $A_0$ [155, 157].

In addition to the gauge coupling unification, the gaugino coupling unification to a value $m_{1/2}$ at the GUT scale $M_U$ is assumed in these theories. The weak scale spectrum of the supersymmetric particles is derived from renormalization group evolution of the soft terms. Radiative breaking of the electroweak symmetry imposes further constraints. In this model a minimal set of parameters [155, 157],

$$m_0,\ m_{1/2},\ A_0,\ \tan\beta,\ \text{and sign}(\mu),$$

determines the sparticle masses and couplings.

In one mSUGRA analysis [298], the lightest neutralino $\chi \equiv \tilde{\chi}_1^0$ is the LSP in the model. It is mainly a Bino (the $U(1)_Y$ gaugino) for the choice of parameters considered, taking into account the present laboratory constraints on the minimal supersymmetric standard model parameter space. This follows from the large Yukawa coupling of the top quark, which drives the radiative electroweak symmetry breaking. In case of Bino LSP, the relic density of the neutralino LSP depends inversely on the cross-section of $\chi$ annihilation process $\chi\chi \to \ell^+\ell^-$ ($\ell = e, \mu, \tau$) through the exchange of sleptons in the $t$- or $u$- channel. Since the charged leptons carry the largest hypercharge among all the fermions and since they are the lightest in most models, this is the dominant process in the $\chi$ annihilation. The final relic density is given by [296]

$$\Omega_\chi h^2 = \frac{\left(m_\chi^2 + m_{\tilde{l}_R}^2\right)^4}{10^6\ \text{GeV}^2\ m_\chi^2 \left(m_\chi^4 + m_{\tilde{l}_R}^4\right)}.$$

The condition $\Omega_\chi h^2 < 0.3$ then requires $m_\chi, m_{\tilde{l}_R} \leq 200$ GeV. This is a welcome result since this few hundred GeV supersymmetry breaking scale is independently preferred to solve the gauge hierarchy problem.

The lightest neutralino in the minimal supersymmetric standard model, which is the LSP, is a mixture of four superpartners of gauge and Higgs bosons (Bino, Wino and two Higgsinos):

$$\chi = N_{11}\widetilde{B}^0 + N_{12}\widetilde{W}^0 + N_{13}\widetilde{H}_1^0 + N_{14}\widetilde{H}_2^0. \tag{17.4}$$

For different choice of the parameters, the LSP could be different combinations of the neutralinos. For $P \equiv N_{11}^2 + N_{12}^2 > 0.9$, the LSP $\chi$ is gaugino-like, which is mostly a combination of different gauginos. For $P < 0.1$, the LSP $\chi$ is Higgsino-like, otherwise the LSP $\chi$ is mixed-type. In general, the parameters $N_{ij}$ depend on the soft supersymmetry breaking terms $m_1$, $m_2$ and $\mu$ given in equation (10.7). However, in the mSUGRA model these parameters are not independent and the gaugino mass unification relates them. In some interesting limiting cases, the LSP becomes a pure state:

When $\mu \to 0$, the state $\widetilde{S}^0 = \widetilde{H}_1\ \sin\beta + \widetilde{H}_2\ \cos\beta$ is the LSP with mass $m_{\tilde{S}} = \mu \sin 2\beta$;

When $M_2 \to 0$, the photino is the LSP with mass $m_{\tilde{\gamma}} = \frac{8}{3}\frac{g_1^2}{g_1^2+g_2^2}M_2$;

When $M_2$ is large and $M_2 \ll \mu$, the Bino is the LSP with mass $m_{\tilde{B}} = M_1$;

When $\mu$ is large and $\mu \ll M_2$, the Higgsino is the LSP with mass $\mu$ (modulo the sign).

In the different cases, the neutralino–nucleon cross-sections could be different leading to differing predictions. For a Bino LSP to be the relic neutralino, the neutralino–nucleon cross-sections are very small in the mSUGRA allowed space so direct and indirect searches for dark matter may not probe this mSUGRA parameter space. When $|\mu|$ decreases, the Higgsino components $N_{13}$ and $N_{14}$ of $\chi$ increase, and hence, $P$ decreases. In this case the spin-independent cross-section increases, resulting in good prospects for the detection of such a neutralino as a dark matter (see section 10.1). So, the composition of the neutralino LSP becomes a deciding factor for its detection as a dark matter candidate. There are other variants of the CMSSM (see section 10.2), which can also lead to an enhanced neutralino–nucleon cross-section by making the soft terms nonuniversal [159]. The large $\tan\beta$ regime was also considered as a source of higher cross-sections and is usually preferred. There are other approaches, in which multiTeV scalars were considered satisfying the naturalness criterion.

The neutralino–nucleon cross-sections can improve in the mSUGRA models, if we assume that the scale of universality of the soft terms is $M_I$, which is different from the GUT scale $M_U$. Now the universality of the scalar masses $m_i(M_I) \equiv m_0$ and the universality of the trilinear scalar couplings $A^{l,u,d}(M_I) \equiv A_0$, are at some intermediate scale $M_I$, which may have its origin in the string theory and can have a wide range of choice [159]. The universality scale $M_I$ then changes the size of the neutralino–nucleon cross-section. This class of models has the main drawback that the new universality scale is not determined by any theory. One more phenomenological variant of this model was then considered, where the universality criterion is further relaxed and the parameters are determined at the electroweak scale [294, 299]. This effective low-energy phenomenological supersymmetric model is referred to as effMSSM. In this approach it has been found that most of the MSSM parameter space allows a gaugino-like neutralino with a small admixture of the Higgsino component, which could be detected with the present level of sensitivity of the ongoing and planned experiments. It is important that experimental observations are analyzed in a joint spin-independent–spin-dependent approach [299, 300].

Soft terms arising from M-theory models also accommodate an LSP dark matter candidate [301]. However, in gauge-mediated supersymmetry breaking models, there is no scope of dark matter candidate. The LSP is the gravitino and is very light and the next-to-lightest supersymmetric particle (NLSP) is no longer stable. NLSP is now heavier than the gravitino but lighter than any other sypersymmetric particles and cannot be stable. It now decays into the gravitino and ordinary particles and cannot remain as relics. Similarly, in the $R$-parity violating models, there is no stable LSP that can become the dark matter candidate. The $R$-parity breaking trilinear interactions would allow a superparticle to decay into two ordinary nonsupersymmetric particles. As a result the LSP can also decay into ordinary particles and is not stable. So there is no dark matter candidate in $R$-parity violating supersymmetric models.

## 17.2 Models of Dark Matter

There are many possible candidates for the cold dark matter. The most popular one is the $R$-parity conserving supersymmetric models, which seems to provide a candidate for cold dark matter naturally. However, we have not evidenced any signal of supersymmetry and this prompted us to consider other extensions of the standard model that can provide dark matter candidates naturally. Since the only evidence we have for physics beyond the standard model is the neutrino masses, there are now attempts to look for dark matter candidates originating from the models of neutrino masses. At one time neutrinos were considered to be the strongest dark matter candidate, but the present upper limit on the neutrino masses ruled out this possibility. In recent times it has been noticed that some radiative models of neutrino masses require Higgs scalars that do not acquire any *vev* and can be candidates for dark matter of the universe. There are also some exotic dark matter candidates, which appear in some models of particle physics, including the models with extra dimensions.

### Dark Matter and Neutrinos

The right-handed sterile neutrinos could be a dark matter candidate depending on their mass and mixing with active neutrinos [302]. However, the usual see-saw mechanism for neutrino masses requires the mass of the right-handed neutrinos to be very high with extremely small mixing with the active neutrinos, making their contribution to the matter density of the universe to be negligible. In general, it is difficult to satisfy all the low-energy constraints on the neutrino masses and mixing and consistently explain the cold dark matter by the right-handed neutrinos alone.

Recently an interesting possibility of radiative neutrino masses has been proposed, in which the generation of neutrino masses requires an additional Higgs doublet. There is an extra $Z_2$ symmetry, under which the right-handed neutrinos ($N_i$, i = 1,2,3) and the new Higgs doublet $\eta$ are odd, while all other particles are even. In this model the neutral component of the new Higgs scalar, $\eta$, could account for the required cold dark matter of the universe [303]. The new Higgs scalar with this exact $Z_2$ discrete symmetry does not interact with other fermions, and hence, it can modify the bound on the standard model Higgs doublet $\phi$ mass coming from the naturalness and the electroweak precision test observables.

The scalar potential for the fields $\phi$ and $\eta$ is given by

$$V = \mu_1^2|\phi|^2 + \mu_2^2|\eta|^2 + \lambda_1|\phi|^4 + \lambda_2|\eta|^4 + \lambda_3|\phi\eta|^2 + \lambda_4|\phi^\dagger\eta|^2 + \frac{\lambda_5}{2}[(\phi^\dagger\eta)^2 + H.c.]. \tag{17.5}$$

We now assume that only the standard model Higgs $\phi$ acquires a *vev*,

$$\langle\phi\rangle \equiv \begin{pmatrix} 0 \\ v \end{pmatrix} \quad \text{and} \quad \langle\eta\rangle \equiv \begin{pmatrix} 0 \\ 0 \end{pmatrix}, \tag{17.6}$$

so the components of the scalar fields can be expressed as

$$\phi = \begin{pmatrix} \phi^+ \\ v + (\phi_R^\circ + i\phi_I^\circ)/\sqrt{2} \end{pmatrix} \quad \text{and} \quad \eta = \begin{pmatrix} \eta^+ \\ (\eta_R^\circ + \eta_I^\circ)/\sqrt{2} \end{pmatrix}. \tag{17.7}$$

The fields $\eta^+$ and $\eta_I^\circ$ will become the longitudinal modes of the gauge bosons making them massive and the usual physical Higgs scalar $\phi_R^\circ$ can now become as heavy as 400 to 600 GeV. The scalars $\eta^+$, $\eta_R^\circ$ and $\eta_I^\circ$ do not interact with the light fermions and remain inert. It is now possible to choose the coupling constants of the scalar potential, which will make one of the neutral components of $\eta$ the lightest inert particle (LIP). The $Z_2$ symmetry will make the LIP a stable particle and, hence, a candidate for the dark matter of the universe.

The parameters of the scalar potential can be constrained by the present limits on the dark matter. Direct detection of halo dark matter requires $\lambda_5 \neq 0$, so $\eta_R^\circ$ and $\eta_I^\circ$ do not become degenerate. The electroweak precision test parameters are modified by the LIP, which allows a heavier physical Higgs scalar $\phi_R^\circ$. For the required dark matter density, the LIP should have a mass of around 50 to 70 GeV with a splitting between the mass of $\eta_R^\circ$ ($m_R$) and the mass of $\eta_I^\circ$ ($m_I$) around 8 to 9 GeV, which is consistent with $\lambda_5 \neq 0$.

The interaction $(\phi^\dagger\eta)^2$ and, hence, $\lambda_5 \neq 0$ are also required for the neutrino masses. The $Z_2$ symmetry allows the coupling

$$\mathscr{L}_Y = h_{\alpha i}\,(\nu_\alpha\eta^\circ - \ell_\alpha\eta^+)N_i + H.c. + M_i N_i N_i\,, \tag{17.8}$$

but the Higgs $\phi$ does not couple to the right-handed neutrinos $N_i$ and hence there is no Dirac mass term for the neutrinos as in the canonical see-saw mechanism. Without loss of generality we assume that the right-handed neutrino mass matrix is real and diagonal. The neutrino mass matrix is then given by

$$\mathscr{M}_{\nu\alpha\beta} = \sum_i \frac{h_{\alpha i}h_{\beta i}M_i}{16\pi^2}\left[\frac{m_R^2}{m_R^2 - M_i^2}\ln\frac{m_R^2}{M_i^2} - \frac{m_I^2}{m_I^2 - M_i^2}\ln\frac{m_I^2}{M_i^2}\right]. \tag{17.9}$$

In principle, the lightest of the right-handed neutrinos could also be a dark matter candidate. But from phenomenological constraints this possibility is not favored. The main constraint comes from the branching fraction of $\mu \to e\gamma$. This leaves the LIP to be a natural candidate for the dark matter.

### Exotic Dark Matter

Several other extensions of the standard model predict new dark matter candidates. The main constraint on any weakly interacting massive particles (WIMP) is that the WIMPs should not overclose the universe, that is, their contribution to the matter density of the universe should not be more than the critical density of the universe. This gives an upper limit on the mass of any stable particle to be 340 TeV [304]. The same consideration also gives bounds on the mass of the stable neutrinos to be greater than 2 GeV [305] or the sum of masses of all the neutrinos to be less than 94 eV [306]. When neutrinos were considered as dark matter candidates, these constraints were important. However, the present upper bound on the absolute mass of the neutrinos rules out any possibility of neutrino dark matter. Another dark matter candidate, the axion, was also constrained by the consideration of overclosure of the universe. These considerations work in an inverted manner for the axions. While for neutrinos the intermediate range is forbidden, for axions only an intermediate window is allowed. From cosmological and astrophysical considerations axions can have a mass in the range of $10^{-2}$ to $10^{-5}$ eV with a very weak coupling (see section 7.3). If the axion could interact strongly, it would have come into thermal equilibrium before the QCD phase transition leading to a background sea of invisible axions in analogy to the neutrinos. Their interaction is required to be so weak that they could never come into thermal equilibrium. But still coherent oscillations of the axions could be excited and this could contribute to the mass density. For the axion mass to be in the range of a few $\mu$eV to about an meV, the axion could contribute substantially to the mass density of the universe [307].

Another class of dark matter candidates is very weakly interacting superheavy particles. These particles were so weakly interacting that they were never in thermal equilibrium in the early universe. Their abundance is also not determined by their annihilation cross-section. They were produced during reheating after the inflation with very little abundance. This would allow a superheavy dark matter candidate with mass in the range of $10^{12}$ to $10^{16}$ GeV, which has to be stable or have a lifetime greater than the age of the universe, known as the WIMPZILLAs [308]. Their annihilation cross-section should be weaker than the expansion rate of the universe at the end of the inflation. Another stable dark matter candidate is the baryonic Q-balls [309]. However, they cannot account for the entire dark matter since the total baryon density is constrained by the limit from nucleosynthesis to be around 5% of the critical density. Electrically charged Q-balls will lose their energy in atomic collisions and could be detected in gravitational lensing experiments. Nonobservation of such events gives a lower limit on the baryon number of the Q-balls to be $10^{21}$. For electrically neutral Q-balls the lower limit on the baryonic charge is about $10^{23}$. There is one more interesting dark matter candidate, the primordial black holes. The string theory and M theory allow superheavy particles, the cryptons, which are stable or metastable bound states of matter in the hidden sector and do not have any direct couplings to ordinary matter; hence they are weakly interacting. Kaluza–Klein excited states associated with extra dimensions could also be dark matter candidates. States with mass of around $10^{12}$ GeV were considered as possible superheavy dark matter candidates [310].

The new developments with large extra dimensions and low scale strong gravity promises new dark matter candidates. The higher-dimensional gravitons will have several Kaluza–Klein modes in four dimensions, when the extra dimensions are compactified. Although these Kaluza–Klein states can decay in our brane, it is possible that they are created in our brane but then are captured on a different brane where gravity propagates. It can then become a natural dark matter candidate [311]. Although this dark matter is dark in our brane, it interacts very strongly in its own brane. In some models the Kaluza–Klein excited states of the axions can also be dark matter candidates, particularly if they are trapped in extra dimensions. Another possible candidate of dark matter in models of extra dimensions is a gas of fundamental strings of submillimeter length, which could have been produced at a temperature around MeV but could be as heavy as $10^{10}$ GeV or larger. These are the strings, which connect two branes that sit on each other at higher temperature. There is no analogous object in ordinary theories. Only at low temperature do the branes move apart leaving the heavy strings as dark matter.

## 17.3 Dark Matter Searches

Although several experiments are going on to detect dark matter, so far only a very small range of the parameter space could be covered by all the experiments. These experiments can be classified as direct or indirect searches. In the direct detection experiments, the elastic scattering of a WIMP off nuclei would be detected by detecting the energy loss of the recoil nuclei through ionization and thermal processes. The indirect detections plan to find events that could be due to the presence of WIMPs trapped in the sun or Earth or contributing to the galactic density. If WIMPs populate the halo of our galaxy, they could be detected directly by low-background experiments or indirectly through their annihilation products in the halo, the center of the sun or Earth.

The direct dark matter search experiments are very difficult. They look for a very few events with a fairly large background. For neutralinos ($\chi$) with masses between 10 GeV and 1 TeV, the deposited energy is below 100 keV. The event rate for such experiments is approximately

$$R_{direct} = \frac{1}{m_N}\phi_\chi \sigma_{\chi N} = \frac{1.4}{m_\chi/\mathrm{GeV}}\ \mathrm{kg}^{-1}\ \mathrm{yr}^{-1}, \tag{17.10}$$

where $\sigma_{\chi N}$ is the neutralino–nucleon interaction and $\phi_\chi$ is the neutralino flux with mass $m_\chi$. The requirement of a large detector mass, a low-energy threshold, a low background, and an effective background discrimination technique make these experiments extremely difficult.

The direct detection experiments are based mostly on the spin-independent interactions of the WIMP because the cross-section of the spin-independent interaction with heavier nuclei is proportional to the atomic number of the target nucleus and, hence, large. However, experiments based on the spin-dependent interactions of the WIMP have higher sensitivity. As a result, both the spin-dependent and spin-independent detectors span the same parameter space of supersymmetric models[299], even though the spin-dependent interaction cross-section of the WIMP is small.

When the scattering does not depend on other factors such as isospin of the nucleons in the spin-independent case, the interaction rate for all the $A$ numbers of nucleons will add up coherently. In this case the total cross-section will be proportional to $I_{SI} = A^2$. Then one may compare different target elements by dividing the total cross-section in any nuclei by $I_{SI}$ to get the rate ($R_0$) corresponding to a single nucleon. In a more realistic situation, this factor $I_{SI}$ will have a form

$$I_{SI} = [(A-Z) + \varepsilon Z]^2,$$

where $\varepsilon = (1 - 4\sin^2\theta_W) \sim 0.08$ with the weak mixing angle $\sin^2\theta_W$. However, a reasonable method of comparing results is to renormalize by a factor $A^2$ or $(A-Z)^2$. In the spin-dependent case this interaction factor becomes

$$I_{SD} = C^2\lambda^2 J(J+1),$$

where $C$ depends on the quark spin content of the nucleon.

The DAMA experiment in Gran Sasso detected a variation of the event rate due to the movement of the sun in the galactic halo and the Earth rotation around the sun using NaI detectors [300]. From the analysis of about 295 kg y in terms of a WIMP annual modulation signature, a positive signal has been reported. This result should be confirmed by an independent experiment looking for annual modulation. Ongoing experiments [312, 313] do not have the required sensitivity to look for the modulation effect (see [300]).

The indirect search experiments for dark matter considers two important factors: the gravitational capture rate of the WIMPs and their annihilation rate to neutrinos, which can be detected on Earth. The solar gravitational capture cross-section is given by [314]

$$\sigma_{sun} = \mathrm{f}(1.2 \times 10^{57})\ \sigma_{XN},$$

where an order of magnitude estimate of the WIMP-nucleon interaction cross-section $\sigma_{XN}$ is given by

$$\sigma_{XN} \approx G_F^2 \frac{m_N^4}{m_W^2} \approx 6 \times 10^{-42}\ \mathrm{cm}^2$$

and the focusing factor f ($\sim 10$) is the ratio of kinetic to potential energy of the WIMP near the sun. Then the flux of WIMP annihilation products (generally neutrinos) from the sun is given by

$$\phi_{sun} = \frac{1}{4\pi d^2}\ \phi_X\ \sigma_{sun},$$

where $d$ is the Earth–sun distance and $\phi_X$ is the WIMP flux. Assuming that the WIMPs constitute the major part of the measured dark matter halo density, the resulting neutrino flux at Earth is then [314]

$$\phi_\nu = 0.1\ \phi_{sun} = \frac{3 \times 10^{-5}}{m_X/\mathrm{GeV}}\ cm^{-2}s^{-1}. \qquad (17.11)$$

The event rate for the indirect WIMP search is then given by

$$R_{indirect} = \phi_\nu P = 1.8\ (m_X/\text{GeV}) yr^{-1}\ km^{-2}, \tag{17.12}$$

where $P \sim 2 \times 10^{-13}\ (m_X/\text{GeV})^2$ is the probability to detect a neutrino with a neutrino telescope. For this purpose one looks for muons produced by neutrinos, e.g., in the Antarctic muon and neutrino detector array (AMANDA) experiment, which is located in the ice of the Antarctics [315].

Recently an indirect evidence of dark matter signal from galactic high-energy gamma ray has been reported. Using the Energetic Gamma Ray Experiment Telescope (EGRET) aboard the Compton gamma ray observatory scientists have carried out all-sky survey in high-energy gamma rays ($E > 30$ MeV) [316]. The observed diffuse emission has a galactic component and a uniformly distributed extragalactic component. A recent analysis of the excess in the diffuse galactic gamma ray data above 1 GeV claims that this excess is consistent with a signal from dark matter annihilation [317]. It is claimed that the excess shows all features expected from dark matter WIMP annihilation and the WIMP mass expected is in the range 50 to 100 GeV compatible with supersymmetry.

## 17.4 Dark Energy and Quintessence

At present all astrophysical observations indicate that the total density of the universe is close to its critical density [44] $\Omega_{tot} = 1.003^{+0.013}_{0.017}$ and about 70% of the total density is in the form of cosmological constant or dark energy. This implies that the universe is flat and accelerating.

The present value of the cosmological constant corresponds to a very small mass scale, comparable with the neutrino mass scale. In a different form it can be written as

$$|\Lambda| \sim \left[10^{-2}\ eV\right]^4.$$

While the contribution of the cosmological constant to the total density of matter is almost one order of magnitude larger than that of the visible matter of the universe, compared with the mass scales in particle physics this is too small. Compared with the electroweak symmetry breaking scale, this is 13 orders of magnitude smaller:

$$\frac{|\Lambda|^{1/4}}{\langle\phi\rangle} \sim 10^{-13}.$$

Thus, the electroweak phase transition would induce a cosmological constant, which is 52 orders of magnitude larger than the present observations.

An explanation of this cosmological constant problem can assume that the universe started with a small cosmological constant and the effects of the different phase transitions are tuned to zero. But this requires most severe fine tuning in particle physics to all orders of perturbation. Even after the contributions from the electroweak phase transition is cancelled by fine tuning of parameters, radiative corrections will again generate similar contributions. Now the fine tuning has to be more severe. In every order of perturbation theory this fine tuning will get worse and is highly unnatural. A more popular solution to the cosmological constant problem is the quintessence solution [318]. Although a fully consistent quintessence solution has yet to be found, this appears to be the most likely solution to the cosmological constant problem.

Varieties of models have been considered as a solution to this problem and differ by their predictions for the equation of state of the dark energy, $\omega = p/\rho$, where $p$ and $\rho$ are the pressure and the density of the dark energy. When $\omega$ is included in the $\Lambda$CDM model to interpret some observations, degeneracy in some parameters appears and it becomes difficult to interpret the data. However, a combined fit of all the data sets gives [73]

$$\omega = -0.98 \pm 0.12, \tag{17.13}$$

which supports the simplest model that the equation of state for the dark energy is $\omega = -1$. It could then be either a cosmological constant or dark energy. The cosmological constant is a constant term in the Einstein equation, which remains the same during the entire evolution of the universe. On the other hand a dark energy comes from the nonvanishing of the trace of the energy–momentum tensor in vacuum, which could vary with time from its large value in the early universe to its present value.

When the cosmological constant or the dark energy is constant in time, the fine tuning can be neglected if some symmetry can protect the vanishing of the cosmological constant to all orders in perturbation. The symmetry breaking can then induce a small cosmological constant as required. Supersymmetry became the obvious choice for this purpose, since supersymmetry implies vanishing of the vacuum energy. Let us consider a global $N = 1$ supersymmetry generated by the Majorana fermion $Q_M$. The anticommutation relation with $\bar{Q}_M$ is given by

$$\{Q_M, \bar{Q}_M\} = 2\gamma^\mu P_\mu. \tag{17.14}$$

The condition that the vacuum is supersymmetric $Q_M\,|0\rangle = 0$ then implies that the vacuum energy should vanish

$$\langle 0|H|0\rangle = \langle 0|P_0|0\rangle = 0. \tag{17.15}$$

This ensures that as long as the global supersymmetry is exact, the vacuum energy and the cosmological constant vanish. However, this solution could protect the cosmological constant only up to the scale of supersymmetry breaking. When supersymmetry is broken, a cosmological constant proportional to the fourth power of the supersymmetry breaking scale will again be generated.

A time varying dark energy starts with a very high value in the early universe depending on the natural scale in the theory. The scale for the dark energy in the early universe could be the QCD phase transition scale, electroweak symmetry breaking scale, or scale of inflation. In supersymmetric theories the largest scale would be the supersymmetry breaking scale. This dark energy would vary with time and at present it is given by the small observed value depending on the dynamics.

The quintessence solutions [318] of the cosmological constant problem assumes that the mass density of a scalar field $\phi$ gives the dark energy. If this field has certain typical potential that allows slow variation of the field, the dark energy decreases with time. If the dark energy density were dominating over the matter density in the nucleosynthesis era, it could have spoiled the nucleosynthesis predictions. So, the decrease of the dark energy is slower than the mass densities of matter and radiation, so the universe was dominated by matter and radiation for most of the time. Only recently has the cosmological constant become more than the mass density in the universe. Moreover, although the initial mass density of the scalar field was much higher, given by the vacuum expectation value or a condensate of the scalar field, and could be as high as the scale of inflation or the electroweak symmetry breaking scale, in recent times the value of the dark energy has become comparable to the matter density. The present value of the dark energy is twice as much as the matter density of the universe.

There are varieties of quintessence models, each of which has some interesting features and differing predictions for the evolution of the universe. Similar to the different models of inflation, different models of quintessence can imply a substantially different cosmological model. In all of these models a scalar singlet field $\phi$ is introduced. It should not transform nontrivially under any gauge groups or the Lorentz group, so it does not have any strong coupling with any matter. The scalar field could be the inflaton field that drives the inflation or the cosmon field associated with the dilatation symmetry [318], or pseudo-Nambu–Goldstone bosons, that is presently relaxing in its vacuum state [319], or some other scalar field depending on the model. The potential of the scalar field determines its evolution and, hence, the dynamics of the quintessence field. The direct coupling of this scalar quintessence field to ordinary matter should be further suppressed to make it phenomenologically consistent, since no such particles have been observed. This can lead to a new long range force, weaker than gravity, which may have some observable effect.

A typical Lagrangian for the quintessence field can be of the form

$$\mathscr{L}_q = V(\phi) + \frac{1}{2}k(\phi)\partial_\mu\phi\partial^\mu\phi \tag{17.16}$$

in units of reduced Planck mass $M_{Pl} = 1$. The potential for the quintessence field $\phi$ should be slowly varying. Usually an exponential potential $V(\phi) = \exp^{-\phi}$ or a power-law potential $V(\phi) = \phi^{-\alpha}$ for large $\phi$ can allow the decrease of the mass density $\rho_\phi$ to be slower than the mass densities of matter and radiation. The kinetic function $k(\phi)$ parameterizes a class of models such as the cosmon models, while it is taken to be unity in other models. In the cosmon models, for small almost constant $k$ the dark energy density is small and varies very slowly, while for large $k$ the universe is dominated by the scalar field. In a realistic model $k$ changes from small to large values after structure formation. A few generic features of all the quintessence models include a tiny time varying mass of the scalar quintessence field, which interacts with ordinary matter very weakly giving rise to a new long-range interaction. Some exotic features of some specific models include negative kinetic energy terms ($\omega < -1$, phantom cosmology) or higher derivative kinetic terms (K-essence), but none of these models explains why the cosmological constant becomes important in the present epoch.

Since the cosmological constant originates in the Einstein equation as a constant term or in the energy-momentum tensor, any theory of gravity attempts to solve the cosmological constant problem by identifying some existing field with the scalar quintessence field. String theory also attempts to solve the cosmological constant problem with the fields already present in the theory. In string theory dimensional quantities can be expressed by two parameters, the string tension and the *vevs* of scalar fields. Thus, to explain the cosmological constant any mechanism involving dynamics of some scalar fields is most natural. One considers a scalar field $\phi$ slowly evolving in a runaway potential which decreases monotonically to zero as $\phi \to \infty$.

Runaway potential is also very common in dynamical supersymmetry breaking scenarios. In any higher-dimensional theories the dilaton field is always present, which comes as a scaling factor in the higher-dimensional metric, when expressed in terms of the four-dimensional metric. This field can have the correct property of the quintessence field in some models. A small value of the dilaton field $s = \langle\phi\rangle/M_{Pl}$ provides a strong coupling, which in turn breaks supersymmetry dynamically. The coupling effectively goes to zero together with the supersymmetry breaking effects for $\phi \to \infty$ and the potential decreases monotonically. Such dynamical symmetry breaking with runaway potential is considered to explain the smallness of the electroweak symmetry breaking coming from superstring theory.

When supersymmetry is broken by gaugino condensation in an effective superstring theory, the gauge coupling constant $g_0^2 \sim 1/s$ at the superstring scale $M_s$ becomes strong at a scale

$$\Lambda \sim M_s \exp^{-1/2bg_0^2} = M_s \exp^{-s/2b}, \tag{17.17}$$

where $b$ is the one-loop beta function. Gauginos condensate at this scale with a value $\langle\bar{\lambda}\lambda\rangle \sim \Lambda^3$, which in turn, produces an effective potential for the dilaton field

$$V \sim \left|< \bar{\lambda}\lambda >\right|^2 \propto \exp^{-3s/b}. \tag{17.18}$$

This potential has the required behavior for a quintessence solution and could explain the smallness of the cosmological constant [318]. Moduli fields in superstring theories, whose *vevs* describe the radius of compactification, could also provide similar solutions.

We shall now discuss this problem in a little more detail and point out the problems with these solutions. Let us consider an example of a real scalar field $\phi$, minimally coupled to gravity and described by a potential

$$V(\phi) = V_0 \exp^{-\lambda\phi/m_P}, \tag{17.19}$$

where $V_0$ is a positive constant. The energy density and pressure due to the scalar field are now

$$\rho_\phi = \frac{1}{2}\dot{\phi}^2 + V(\phi) \quad \text{and} \quad p_\phi = \frac{1}{2}\dot{\phi}^2 - V(\phi). \tag{17.20}$$

Neglecting the spatial curvature ($k \sim 0$), the equation of motion becomes

$$\ddot{\phi} + 3H\dot{\phi} = -\frac{dV}{d\phi}, \tag{17.21}$$

with

$$H^2 = \frac{1}{3m_P^2}(\rho_B + \rho_\phi). \tag{17.22}$$

A solution with the behavior $\rho_\phi \propto a^{-n_\phi}$ or $\dot{\rho}_\phi/\rho_\phi = -n_\phi H$ is of particular importance, which gives $\phi$ as a logarithmic function of time

$$\phi = \phi_0 + \frac{2}{\lambda}\ln(t/t_0). \tag{17.23}$$

The equation of state now reads

$$p_\phi = \omega_\phi \rho_\phi, \tag{17.24}$$

with $\omega_\phi = (\lambda^2/3) - 1$. For sufficiently small $\lambda$, the field $\phi$ can play the role of quintessence. However, in a realistic situation we have to include the effect of the background matter and radiation.

Let us now include the background matter and radiation energy density $\rho_B$ with pressure $p_B$ satisfying a standard equation of state

$$p_B = \omega_B \rho_B, \tag{17.25}$$

where $\omega_B$ is positive and lies between 0 and 1/3. We also have to consider $\lambda^2 > 3(1+\omega_B)$ and then the solution satisfies the condition

$$\begin{aligned} \Omega_\phi \equiv \frac{\rho_\phi}{\rho_\phi + \rho_B} &= \frac{3}{\lambda^2}(1+\omega_B) \\ \omega_\phi &= \omega_B. \end{aligned} \tag{17.26}$$

This is certainly not the required quintessence solution, which requires $\omega_\phi \sim -1$. Thus, in realistic scenarios, consistent with observations that $\rho_\phi$ should be subdominant during nucleosynthesis, a quintessence solution solving the cosmological constant problem is difficult in these theories [320, 321].

A few solutions to this problem have been suggested [322], but none of them is completely satisfactory. In one case a tracker field is introduced with the potential

$$V(\phi) = \lambda\,\frac{\Lambda^{4+\alpha}}{\phi^\alpha}, \tag{17.27}$$

with $\alpha > 0$. This provides a solution to a good approximation. But in this case, the fact that the present value of $\phi$ is of order $M_{Pl}$ becomes a source of problem. In another solution the quintessence component comes out of the background energy density, where the potential has a local minimum (a bump). When the field approaches the bump, it slows down and then allows for the background energy density. There are also models, known as quintessential inflation or deflation, where one tries to use the same field for quintessence as well as to drive inflation. There exists another class of models where the scalar field has not yet reached its stable ground state and is still evolving in its potential [323]. The idea is very close to quintessence.

Although the quintessence solutions appear to be a solution of the cosmological constant problem, there are problems in implementing them into realistic theories [321]. The main problem comes from the weak coupling of the quintessence with ordinary matter. The coupling of matter with the quintessence field would generate higher-order corrections such as $\lambda_d \phi^d$, which can destabilize the slow roll over, unless we impose very stringent conditions on the coupling $\lambda_d$. The quintessence fields are also supposed to be very light. Supersymmetric corrections will make them heavy, unless the coupling is extremely weak. Another problem is the *vev* of $\phi$ of the order of $M_{Pl}$. This would jeopardize the positive definiteness of the scalar potential and hence the quintessence solution. Supersymmetric corrections may also cause problems in this case. Thus, the quintessence problems have shifted the problem of small cosmological constant to why the quintessence fields couple to matter so weakly. Another problem is why the dark energy dominates over the dark matter only in the recent times.

## 17.5 Neutrino Dark Energy

It is interesting to note that the scale of dark energy is almost the same as the neutrino masses. This observation leads us to consider whether there is any connection between the two phenomena. In an attempt to utilize this coincidence a neutrino mass varying solution to the dark energy problem has been suggested [324]. In this scenario the neutrino mass becomes a function of a scalar field, the acceleron, which drives the universe to a late time accelerating phase. Although the proposal is not complete, it has several interesting features.

The basic idea behind the neutrino dark energy ($\nu DE$) models is that the neutrino mass varies as a function of a light scalar field $\mathscr{A}$ (the acceleron) [324]. In the nonrelativistic limit, the thermal background of the neutrinos and antineutrinos $n_\nu$ gives a contribution to the potential amounting to $m_\nu n_\nu$. This neutrino source term for $\mathscr{A}$ should not significantly vary spatially, at least over a distance of the inter-neutrino spacing. For thermal relic neutrinos at the present time this implies that the mass of the acceleron should be less than $\sim (10^{-4}\ \text{eV})$.

In addition there is a vacuum energy $V_0(m_\nu)$ attributed to the acceleron, which also depends on the varying neutrino mass. Thus, the effective potential

$$V(m_\mu) = m_\nu n_\nu + V_0(m_\nu) \tag{17.28}$$

tell us that the first term will drive $m_\nu$ to a lower value, while the second term will force $m_\nu$ to large value. The minima of this potential

$$V'(m_\mu) = n_\nu + V_0'(m_\nu) = 0 \tag{17.29}$$

would then correspond to a point with intermediate $m_\nu$ and nonvanishing vacuum energy $V_0(m_\nu)$. The relationship of $\omega(t)$ with the scalar potential is given by

$$\omega + 1 = -\frac{m_\nu\, V_0'(m_\nu)}{V} = \frac{\Omega_\nu}{\Omega_\nu + \Omega_{DE}} \tag{17.30}$$

for a simple equation of state $p(t) = \omega(t)\ \rho(t)$. The observed value of $\omega \approx -1$ implies that the energy density in neutrinos is much smaller than the total energy density, which corresponds to $m_\nu\ V_0' \ll V(m_\nu)$ and, hence, a flat potential. This could be satisfied with a flat potential for $\mathscr{A}$ or if the dependence of $m_\nu$ on $\mathscr{A}$ is rather steep. The flat potential could be taken to be

$$V_0(m_\nu) = \Lambda^4 \log\left(1 + \left|\frac{M_1(\mathscr{A})}{\mu}\right|\right). \tag{17.31}$$

The form of $M_1(\mathscr{A})$ determines the dynamics of the model. Two extreme possible forms have been discussed, which are a linear form $M_1(\mathscr{A}) \sim \lambda \mathscr{A}$ and an exponential form $M_1(\mathscr{A}) \sim \exp^{\mathscr{A}^2/f^2}$. In the former case the potential is flat, while for the second case the potential is quadratic.

The origin of the acceleron field was not specified in the original model. Subsequently it was pointed out that a pseudo-Nambu–Goldstone boson (pNGB), originating from lepton flavor violation, could be the acceleron field [325]. We demonstrate with a two-generation example.

We start with the Yukawa couplings of the right-handed neutrinos $N_i, i = 1,2$ with some scalar singlet fields $\Phi_i, i = 1,2$, given by

$$\mathscr{L}_N = \frac{1}{2}\alpha_1 \bar{N}_1 N_1^c \Phi_1 + \frac{1}{2}\alpha_2 \bar{N}_2 N_2^c \Phi_2. \tag{17.32}$$

This model has a $U(1)_1 \times U(1)_2$ symmetry, under which these fields transform as $N_1 \equiv (1,0)$, $N_2 \equiv (0,1)$, $\Phi_1 \equiv (2,0)$ and $\Phi_2 \equiv (0,2)$, respectively. When the fields $\Phi_i$ acquire *vev*, both these symmetries are broken and there will be two Nambu–Goldstone bosons.

We also assume that the left-handed doublets $\ell_{\alpha L}^T \equiv (\nu_\alpha \quad \alpha), \alpha = e, \mu$ transform under $U(1)_1 \times U(1)_2$ as (1,0) and (0,1), respectively. To maintain the $U(1)_1 \times U(1)_2$ symmetry, by all the renormalizable dimension-4 terms, we introduce three Higgs doublet fields $H_a, a = 0,1,2$, all of which transforming as (1,2,1/2) under the standard model. Under the $U(1)_1 \times U(1)_2$ symmetry, these fields transform as $(0,0), (1,-1)$

and $(-1,1)$, respectively. Then the $U(1)_1 \times U(1)_2$ invariant Yukawa interactions are given by

$$\mathscr{L}_{mass} = f_{11}\bar{N}_1\ell_1 H_0 + f_{12}\bar{N}_1\ell_2 H_1 + f_{21}\bar{N}_2\ell_1 H_2 + f_{22}\bar{N}_2\ell_2 H_0. \qquad (17.33)$$

After the electroweak symmetry breaking, these interactions will allow all the Dirac mass terms for the neutrinos,

$$\mathscr{L}_{mass} = m_{ij}\bar{N}_i \nu_j.$$

These mass terms are soft terms, which break the $U(1)_1 \times U(1)_2$ symmetry explicitly. This soft symmetry breaking will give a small mass to one of the Nambu–Goldstone bosons (which now becomes the acceleron field $\mathscr{A}$), while the other Nambu–Goldstone boson corresponding to the total lepton number will still remain massless leading to a singlet-Majoron.

The mass of the acceleron $\mathscr{A}$ can be computed by evaluating the one-loop Coleman–Weinberg potential, which gives a small finite mass of the acceleron of the order of $m_{\mathscr{A}} \sim m^2/M$, which is of the order of the neutrino mass. After replacing the fields $\Phi_i$ by their *vevs* and absorbing all the phases, it can be shown that the Majorana masses of the right-handed neutrinos become dependent on the acceleron $M_i \equiv M_i(\mathscr{A})$. We can then write the complete mass term as

$$\mathscr{L}_\mu = \frac{1}{2}M_i(\mathscr{A})\bar{N}_i^c N_i + m_{ij}\bar{N}_i \nu_j + H.c. \qquad (17.34)$$

It is interesting to note that $M_i(\mathscr{A})$ is specified and the effective neutrino mass also varies $m_\nu(\mathscr{A}) = m^T M^{-1}(\mathscr{A})m$. We, thus, arrive at an effective potential

$$-\mathscr{L}_{eff} = m_{ij}^T[M_i(\mathscr{A})]^{-1} m_{ij}\ \nu_i \nu_j + H.c. + \Lambda^4 \log(1 + |M_1(\mathscr{A})/\mu|), \qquad (17.35)$$

whose minima implies $\omega = -1$.

This model has some problems [326], but some of them have been taken care of in certain variants of this model [327, 328]. This scenario can be embedded in models with triplet Higgs scalars [327], in which the naturalness problem of the present model can be removed. For example, naturalness requires $M_i$ to be of the order of eV, but in the triplet Higgs models the heavy scale could be of the order of TeV, which provides interesting phenomenology. Although the models of mass varying neutrinos are not yet fully consistent in their present form, they provide interesting phenomenology that could be tested in the near future.

# 18

# *Cosmology and Extra Dimensions*

The extra dimensions were introduced to unify gravity with other gauge interactions. Since we do not experience any new dimensions beyond our four space–time dimensions, these extra dimensions were assumed to be compact. This would make these extra dimensions invisible at our present energies. Any probe having energy of the order of the compactification scale, which is the inverse of the radius of compactification, can only see these extra dimensions. Usually the compactification scale is considered to be about the Planck scale, so all new excited states become heavy and decouple from low-energy physics.

In a new class of theories with extra dimensions, gravitons are assumed to propagate along the extra dimensions and all other interactions become blind to these extra dimensions. This allows the construction of these theories with extra dimensions with a very low compactification scale, as low as a few TeV. Since gravitational interaction can see all the dimensions, in our four space–time dimensions the effect of gravity is suppressed by the volume of the extra dimensions. Although gravity is weak in our world, it is very strong in extra dimensions even at our present energies. Around the scale of compactification, the effects of quantum gravity may not be neglected. These studies have opened up new interesting phenomenological possibilities as well as cosmological consequences.

In these theories with TeV scale extra dimensions there is no large scale beyond the scale of compactification. Moreover, since quantum gravity effects cannot be ignored around the compactification scale, we cannot extend our theory of early universe much beyond the scale of compactification. There can be new physics appearing around this scale. For example, one can use the holography principle to construct a universe [329] which is a dense gas of black holes. Only at a later time did they decay through Hawking radiation to a radiation dominated era. In this scenario inflation becomes redundant. There is also the standard brane cosmology, where it is expected that the universe starts as a higher-dimensional theory and then all matter gets concentrated to the branes and the bulk becomes almost empty [330]. Since the subject is still developing, many of these concepts may change drastically. For example, the fine tuning problem for the cosmological constant was first solved [331, 332], but then another problem was noticed [333]. We shall, thus, restrict ourselves to the basic concepts, which should not change. For example, in any brane cosmologies the predictions of the standard big-bang cosmology should remain valid. We shall first discuss the constraints on the models of extra dimensions arising from cosmology and astrophysics. Keeping these constraints in mind we then proceed to discuss some generic features of the brane cosmology.

## 18.1 Bounds on Extra Dimensions

The models with large extra dimensions have one common feature that there are infinite towers of Kaluza–Klein excitations of the graviton, all moving in the bulk. At present energies only the zero mode of the graviton can be present, and hence, the interaction of gravity with ordinary matter is weak in our four space–time dimension. In addition, if there are other bulk particles such as the axion, then there will be another tower of axion excitations, all interacting weakly with matter. We shall restrict ourselves to gravitons only, which are inevitable.

Consider an $n$-dimensional space–time, in which $d = n - 4$ dimensions are compact. The gravitons propagate in all the dimensions, so after compactification of the extra dimensions, in our four dimensional space–time the graviton will have all the Kaluza–Klein excited states. All these modes of graviton will interact with matter with the same strength

$$\mathscr{L} = -\frac{\kappa}{2} \sum_n \int d^4x h^n_{\mu\nu} T^{\mu\nu}, \tag{18.1}$$

where $\kappa = \sqrt{16\pi G_N}$, $T^{\mu\nu}$ is the energy-momentum tensor of the standard model and $h^n_{\mu\nu}$ represents the n-th excited mode of the Kaluza–Klein graviton state $G^n$, with mass $m_n = n/R$. Although at our present energy most of these states cannot be produced, in the early universe at high temperature all these states could have been produced. For an order-of-magnitude estimate, we assume that the interaction strength is approximated by $1/M_{Pl}^2$.

We shall now try to check the consistency of these models of large extra dimensions in terms of our present knowledge of the universe without restricting ourselves to any particular model or going into the details of the evolution of the universe [311]. At present we can predict the light element abundances from the big-bang nucleosynthesis. Our first concern will be to prevent any new physics that destroys the predictions of nucleosynthesis. At high temperatures if the Kaluza–Klein excited graviton states $G^n$ are produced in large number in the early universe, that would destroy the prediction for the big-bang nucleosynthesis. As we shall see next, such destructions can be prevented if the maximum temperature of the universe in these models of large extra dimensions can be constrained.

Due to the lack of any fully consistent theory of quantum gravity, we start our discussions on the early universe from the time of the Planck scale, when gravity starts becoming weak. Similarly, in models of large extra dimensions we start from temperatures below the Planck scale in our brane, when quantum gravity effects start smearing out. This is the fundamental scale in the theory of extra dimensions $T < M_*$. For preserving the nucleosynthesis predictions of the standard cosmological model, we assume that the extra dimensions do not influence the expansion of the 3-brane at the time of nucleosynthesis. Below a certain temperature $T_*$, our universe expands in the standard way with the Hubble parameter given by

$$H = 1.66\,\sqrt{g_*}\,\frac{T^2}{M_{Pl}}, \tag{18.2}$$

where $g_*$ is the effective number of degrees of freedom. The evolution of the extra dimensions becomes independent of the evolution of our 4-dimensional universe. This dictates the concept of a new maximum temperature $T_*$ in these theories [311]. At this maximum temperature $T_*$ the universe starts evolving in the standard way, the bulk becomes virtually empty and the radii of the extra dimensions get stabilized to their present value.

In these models of extra dimensions, the history of the early universe before the maximum temperature $T_*$ becomes drastically different from the standard model of cosmology. The maximum temperature $T_*$ is usually taken to be the reheating temperature after inflation $T_{RH}$ and this is constrained mainly by the effects of the Kaluza–Klein gravitons $G^n$. Production of these Kaluza–Klein gravitons $G^n$ cools the universe at very high temperature. Given the production rate for a species $G^n$ with mass $m_n < T$ to be

$$\Gamma \sim \frac{T^7}{M_{Pl}^2}, \tag{18.3}$$

the rate of decrease of radiation energy density is given by the rate of creation of $G^n$

$$\frac{dn}{dt} \sim -\frac{d\rho}{dt} \sim \frac{T^7}{M_{Pl}^2}(TR)^d \sim \frac{T^{d+7}}{M_*^{d+2}}. \tag{18.4}$$

We use the relation $M_{Pl} = M_*\ (M_*R)^{d/2}$. This cooling rate should be less than the cooling of the universe due to expansion, which dictates the condition,

$$T_* < M\left(\frac{M_*}{M_{Pl}}\right)^{\frac{1}{d+1}}. \tag{18.5}$$

For $d = 2$ and $M_* = 1$ TeV, the bound is $T_* < 10$ TeV, which is a weak bound since $T_* > 0.7$ MeV is acceptable within the nucleosynthesis scenario.

Another bound comes from the overclosure of the universe by the Kaluza–Klein gravitons. Even with very low production rate, the total number of Kaluza–Klein gravitons created in the Hubble time $H^{-1}$ is fairly large,

$$n(T) \sim T^2\,M_{Pl}\left(\frac{T}{M}\right)^{d+2}. \tag{18.6}$$

If all of these states survive, they can modify the evolution of the universe and give a very strong bound on these models. The lifetime of the n-th mode is

$$\tau_n \sim \frac{1}{\Gamma_n} \sim \frac{M_{Pl}^2}{m_n^3} \sim 10^{31}\ \text{yr}\left(\frac{1\ \text{eV}}{m_n}\right)^3, \tag{18.7}$$

which is greater than the age of the universe, and hence, the Kaluza–Klein gravitons can overclose the universe.

At the maximum temperature $T_*$, Kaluza–Klein gravitons with mass of order $T_*$ are created. They become nonrelativistic ($n \sim T^3$) below the maximum temperature $T_*$. Consistency with the big-bang nucleosynthesis then requires that the mass density at the nucleosynthesis epoch

$$\rho_{grav}(T_{NS}) \sim \left(\frac{T_{NS}}{T_*}\right)^3 T_* \, n(T) \sim T_{NS}^3 M_{Pl} \left(\frac{T_*}{M}\right)^{d+2} \tag{18.8}$$

should be less than the energy density of one massless species. This constrains the maximum temperature $T_*$ to be

$$T_* < M \left(\frac{T_{NS}}{M_{Pl}}\right)^{\frac{1}{d+2}}. \tag{18.9}$$

For $d = 2$ and $M = 1$ TeV, the bound is $T_* < 10$ MeV. If one requires that $\rho_{grav}$ should be less than the critical density of the universe, a stronger bound of $T_* < 0.2$ MeV is obtained for $d = 2$ and $M = 1$ TeV. To avoid conflict with nucleosynthesis, we get a bound on the fundamental scale of $M_* > 10$ TeV.

The maximum temperature $T_*$ can be in the range of 1 MeV to 1 GeV without conflicting with the nucleosynthesis, but in this case generating a baryon asymmetry of the universe and models for inflation becomes extremely difficult. One suggestion to avoid this is to raise the maximum temperature $T_*$ without affecting nucleosynthesis predictions. This can be done by allowing faster decay of the gravitons in the bulk to something massless in the bulk or in another brane. This would allow the energy density to scale as $T^4$, although the maximum temperature $T_*$ is still required to be fairly low.

We shall now discuss the astrophysical bounds coming from stellar evolution, particularly from the sun and supernovae [334]. Light gravitons $G^n$ produced in the stars through

(a) $\gamma\gamma \rightarrow G^n$, photon–photon annihilation,

(b) $e^+ e^- \rightarrow G^n$, electron–positron annihilation,

(c) $e^- \gamma \rightarrow e^- G^n$, gravi-Compton Primakoff scattering,

(d) $e^- (Ze) \rightarrow e^- (Ze) G^n$, gravi-bremsstrahlung in a static electric field, and

(e) $N N \rightarrow N N G^n$, nucleon–nucleon bremsstrahlung

can carry away energy from the interior of stars. If they are the main source of cooling, then they can change the stellar evolution making it inconsistent with observation.

The nonobservation of shortening of the neutrino emission from the supernovae SN1987a gives the strongest bound on the fundamental scale $M_*$ [311, 334, 335]. In a supernovae explosion the neutrinos have a very short mean free path and cannot escape. Slowly they thermalize and the mean free path increases and eventually they may escape. But in models of extra dimensions gravitons can allow very fast cooling, since they would efficiently transport energy out of the explosion. This would reduce the spread in the arrival time of the neutrinos, so from the observation of the supernovae SN1987a neutrinos, we get a bound of $M_* > 30$ TeV.

## 18.2 Brane Cosmology

The standard model of cosmology has to be modified in theories with large or small extra dimensions and TeV scale quantum gravity. In these classes of theories with extra dimensions, we live in a 3-brane or in a world of four space–time dimensions attached to a wall of the larger dimensions. Only gravity propagates in the bulk and hence all the dimensions. In these theories the universe should evolve in a different way than in ordinary models of particle physics where the Planck scale is at $M_{Pl}$. Due to the dynamics in these theories of extra dimensions, inflation should also take place at very low temperature. So new mechanisms for inflation have to be considered. There are many new open questions in this scenario, which are yet to be answered. So, developments in this field of research may change our present knowledge very rapidly. We shall thus try to mention only a few generic issues of the present state of brane cosmology.

The earliest epoch we can explore in the early universe is at the end of the quantum gravity era, which is the Planck scale in the standard model of cosmology. In models of extra dimensions, this becomes the fundamental scale of about a TeV

$$t_0 \sim M_*^{-1}. \tag{18.10}$$

Another major difference between the standard model of cosmology and the brane cosmology is the introduction of the maximum temperature $T_*$. We understand the abundance of elements starting from nucleosynthesis fairly well in the standard model of cosmology. But the Kaluza–Klein excited graviton $G_n$ production can destroy this prediction in any theories with extra dimensions. To preserve the predictions of nucleosynthesis the universe should be allowed to evolve as in the ordinary theory after a maximum temperature $T_*$, which depends on the model [311]. Depending on the number of extra dimensions, this maximum temperature $T_*$ could be between MeV and a few GeV. For any number of dimensions, the highest value for the maximum temperature $T_*$ could be the fundamental scale $T = M_*$. Below the maximum temperature $T_*$ the extra dimensions should not affect our 4-dimensional universe and hence the brane cosmology becomes identical to the ordinary standard model of cosmology.

In any brane cosmology scenario the Kaluza–Klein gravitons should not be produced in too many numbers, and the contributions to the energy density due to the

Kaluza–Klein gravitons should satisfy

$$\rho_{KK} \ll M_*^4. \tag{18.11}$$

This allows an estimate of the Hubble parameter at the earliest epoch $t_0$ (equation (18.10))

$$H \sim \left(\frac{8\pi V(\phi)}{3M_{Pl}^2}\right)^{1/2} < \frac{M_*^2}{M_{Pl}} \ll M_*. \tag{18.12}$$

The inflation should now occur within the period of $T = M_*$ and the maximum temperature $T_*$ at a time scale $t_i \sim H^{-1} \gg M_*^{-1}$.

We should impose one more severe constraint, that the bulk matter should not influence the expansion of our universe [330]. The total energy in the bulk should be much less than the total energy in our brane

$$\rho_{bulk} \ll M_*^{d+4}\left(\frac{M_*}{M_{Pl}}\right)^2. \tag{18.13}$$

In the ideal situation one assumes that the bulk is empty after the inflation and all matter is concentrated in our brane. Before the temperature corresponding to the fundamental scale, all dimensions were equal and there was equal distribution of matter in all directions in the higher-dimensional world. So, some specific mechamism must have decoupled all matter from all extra dimensions leaving a multidimensional empty space at some time in the past. After this time all matter in the universe concentrated in our brane.

It is difficult to make the bulk essentially empty. In the early evolution of the universe we have to take into consideration some new higher-dimensional phenomena. One suggestion is based on the spontaneous creation of our brane in a higher-dimensional space, where there were no branes or excitation at the beginning [336]. Let us consider a 5-dimensional example with a four-form $B$ (which is a four index antisymmetric tensor field) and branes with nonzero and zero charges with respect to $B$. The coupling of the branes with the 4-form is

$$S_{int} = e\int B_{ABCD}d\Omega^{ABCD}. \tag{18.14}$$

The $B$ field with nonvanishing constant field strength at the initial state of the 5-dimensional universe will decay spontaneously into pairs of charged branes. A neutral brane could also be associated with the creation of the charged branes, which will then become our universe. After the spontaneous creation, the three volumes of the branes increase and the junction between the charged and neutral ones moves away leaving our universe as a homogeneous 3-dimensional space in an empty bulk.

Another important question we need to understand in these theories is the brane inflation. We shall discuss one interesting possibility where the inflation originates from the interaction of two parallel branes [337]. Consider the inflationary models in ordinary theories, where the scalar field, whose *vev* drives the inflation, requires an extremely flat potential. In general, the quantum gravity corrections may destabilize the solution. If the Hubble parameter at that epoch is $H$, the quantum gravity correction could be $\sim H^2$, and this can destabilize the scalar potential and prevent slow roll over of the scalar field to its minimum, required for inflation. If $V(\phi)$ is the scalar potential of the inflaton field $\phi$, the curvature of the potential should be less than the Hubble parameter

$$H^2 \sim V(\phi)/3M_{Pl}^2$$

at some region. Then quantum gravity corrections would allow low-energy terms such as $\bar{\phi}\phi V(\phi)/3M_{Pl}^2$, which will spoil the condition for inflation.

This problem of conventional theories of inflation has a natural solution in the brane inflation scenario with two interacting parallel branes. In this scenario the inflaton $\phi$ parameterizes the distance $r$ between two parallel brane worlds, one of which is our brane. This scale of about $r \sim \mu m$ is too large compared with the quantum gravity scale of $\text{TeV}^{-1} \sim 10^{-16}$ mm, and all the quantum effects are negligible at this distance scale $r$. The potential is now given in terms of the distance between the branes $r$ as

$$V(r) = M^d \left( a + b_i r^l \exp^{-m_i r} - \frac{1}{(M_* r)^{d-2}} \right), \tag{18.15}$$

where $a$ and $b_i$ are constants and $m_i$ are masses of heavy modes of $\phi$. At large distances this potential has a power-law behavior and this translates to a flat potential for the inflaton field $\phi = M_*^2 r$ naturally.

In the brane world scenarios, the question of the baryon asymmetry of the universe could be addressed in different ways [338]. While it is possible to generate a baryon asymmetry of the universe with higher-dimensional bulk fields or using the Affleck–Dine mechanism, a higher-dimensional mechanism originating from the brane world scenario may also be invoked. In another possibility the collision of two parallel branes in the early universe could also produce a matter–antimatter asymmetry of the universe. If there was *CP* violation during the collision of the parallel branes, some of the antimatter from our world could have moved to the other parallel brane. If the collision took place after the inflation, then there would be excess matter in our brane and equal amount of excess of antimatter in the other brane. It is also possible that after the inflation when our brane is freely expanding, baby branes containing all the antimatter would have formed due to deformations of the world volume, which then moves apart from our brane making our brane baryon asymmetric.

There are also new sources of dark matter in the universe coming from the models of extra dimensions [339]. In some cases the Kaluza–Klein gravitons and Kaluza–Klein axions could be the dark matter candidates. One interesting possible candidate is a gas of fundamental strings of submillimeter length. They could have been produced after the inflation at a temperature around MeV, but could be as heavy as $10^{10}$ GeV or larger. In ordinary cosmology production of such objects at low temperature is not allowed, but in brane cosmology they can come from new brane physics. The strings could connect two branes, which sit on each other at higher temperature. Only at low temperature do the branes move apart leaving the heavy strings as dark matter.

If such strings break down at some point of time, they will release their energy in the form of very high energy particles.

Brane cosmologies have several other features, which are not present in the ordinary standard model of cosmology. For example, inflation could take place in the bulk and in our brane at different times leading to new consequences. The maximum temperature $T_*$ is fairly low which may lead to inflation after the electroweak symmetry breaking. In that case it may not be possible to extract much information about the physics beyond the standard model from cosmology, since all such information would have been wiped off during inflation. The main disadvantage of the brane cosmologies is that the predictions may not be tested for a long time, although these models may provide us with many interesting possibilities.

## 18.3 Dark Energy in a Brane World

Astrophysical observations confirmed that about 2/3 of the total matter density of the universe is in the form of dark energy or cosmological constant. Explanation of this dark energy is one of the most serious problems for any theories of particle physics. It is the most severe naturalness problem in particle physics that is yet to be solved. While the present observations require the amount of dark energy to be nonvanishing and extremely small, the *vevs* of the scalar fields breaking gauge symmetries always act as vacuum energy contributing a large amount to the dark energy. Supersymmetry appeared to solve this problem, but then supersymmetry breaking was found to generate a larger cosmological constant. Superstring theory provided a consistent theory of the dark energy as a quintessence solution, which, however, led to the fine tuning problem of the coupling constant of the quintessence field. In models with extra dimensions there are new problems. The first solution appeared to be highly successful, but later it was found that the solution had only shifted the problem from one place to another. There are now new solutions to the dark energy problem in models with extra dimensions, but they are yet to be established.

The problem of dark energy can be stated in a different way. The vacuum expectation value corresponding to the electroweak symmetry breaking scale implies a large energy density of the vacuum, which would pose a high curvature to all the dimensions, which has not been observed. The observed value of the dark energy or the cosmological constant corresponds to an order of magnitude smaller curvature. In models of extra dimensions one solution suggests that it is possible that the strong curvature would be a feature of only the extra dimensions, keeping our brane flat. The cosmological constant problem in the brane cosmology then gets reformulated to a problem of why the vacuum energy density has no effect on the curvature induced on our brane.

In the brane world scenario it is possible to have a scalar living in the bulk with coupling to our brane, which can then imply that the vacuum energy density affecting all the dimensions occurs in a way that the metric induced in our brane is almost flat. It will then be possible to make the cosmological constant vanishing by adjusting the zero modes of the scalar [332]. In the case of nonfactorizable geometry it may be possible to maintain the 4-dimensional Poincaré invariance even when the vacuum energy density induces a nontrivial warp factor.

Let us consider a bulk scalar field $\phi$ in 5 dimensions, which has couplings in our brane. The total action has two parts

$$S_{tot} = S_{bulk} + S_{brane},$$

which are given by

$$S_{bulk} = \int \mathrm{d}y\mathrm{d}^4x\sqrt{-G}\left(-\frac{1}{2\kappa_5}R^{(5)} + \frac{b}{2}G^{MN}\partial_M\phi\partial_N\phi\right)$$
$$S_{brane} = \int \mathrm{d}^4x\sqrt{-g}\exp^{4\kappa_5\phi(0)}\mathscr{L}(\psi,\, g_{\mu\nu}\exp^{\kappa_5\phi(0)}), \tag{18.16}$$

where $\mathscr{L}$ is the Lagrangian in the brane, $\psi$ represents all matter, $g_{\mu\nu}$ is the induced 4-dimensional metric, and $\phi(0) = \phi(x, y = 0)$ is the induced scalar field. Representing all the vacuum energy by $\varepsilon_{vac}$, including quantum effects, we can write the interaction of $\varepsilon_{vac}$ with gravity and the scalar field $\phi$ as

$$S_{brane} = \varepsilon_{vac}\int \mathrm{d}^4x\sqrt{-g}\exp^{4\kappa_5\phi(0)}. \tag{18.17}$$

In this case there exist solutions preserving Poincaré invariance in 4 dimensions. This solution would be valid for any value of $\varepsilon_{vac}$ with the same sign.

Let us now consider an explicit example of a metric, which is restricted by the 4-dimensional Poincaré invariance

$$\mathrm{d}s^2 = a^2(y)\eta_{\mu\nu}\mathrm{d}x^\mu\mathrm{d}x^\nu - \mathrm{d}y^2. \tag{18.18}$$

The warp factor $a(y)$ then comes out to be

$$a(y) = \left(1 - \frac{2}{3\kappa_5^{2/3}}\exp^{2\kappa_5\phi_0}|y|\right)^{1/4}, \tag{18.19}$$

where $\phi_0$ comes out as a constant of integration in the solution with the 4-dimensional Poincaré invariance and is related to $\phi(x, y = 0)$ by

$$\phi_0 = \phi(0) + \frac{1}{2\kappa_5}\ln(\varepsilon_{vac}\,\kappa_5^{8/3}). \tag{18.20}$$

Thus, we get a solution to the problem, which is highly promising, and the cosmological constant problem seems to have been solved [332].

When we look into the problem more carefully, we find that near the branes these models have a singularity. In the above example, the singularity lies at

$$|y_{sing}| = \frac{3}{2}\kappa_5^{2/3}\exp^{-2\kappa_5}. \tag{18.21}$$

This singularity can be given a physical interpretation in the presence of an additional source term. The curvature singularity shows no finite extensions of the branes into the transverse directions, like the black holes for point-like source. With these sources the vanishing of the cosmological constant requires a consistency condition for the brane world setup. This imposes an additional fine tuning of parameters of the model, which is of the same order as the fine tuning of the cosmological constant.

Let us consider a parallel brane scenario as an example, where the second brane is situated at $|y| < |y_{sing}|$ and we impose a orbifold symmetry. Between the two branes there is no singularity in this case, which is the entire orbifold. However, in this case the consistency condition requires that the metric should be a solution to the Einstein equations. This will then imply that the tension of the second brane should be fine tuned, which is not better than the fine tuning of the cosmological constant.

The brane cosmology appeared to provide a solution to the dark energy problem, but it led to a new problem which requires similar amount of fine tuning as the cosmological constant problem. There is another suggestion to the cosmological constant problem in the context of brane world making use of the anthropic principle. Compared with the usual scenario with anthropic principle, now the smallness of the parameters is protected by some discrete symmetry. Another interesting idea being studied now is the possibility of a brane world with infinite volume extra dimension [340]. In these theories gravity is modified at large distances and the universe is accelerating due to gravity leaking into the extra dimensions. The dark energy problem has yet to be solved completely, but the models of extra dimensions have at least provided us with some new possibilities with interesting consequences.

# *Epilogue*

The standard model of particle physics has met with all success so far. Although there are indirect indications, except for the neutrino masses, we have not evidenced any signals for physics beyond the standard model. On the other hand, from theoretical considerations there are many reasons to believe that there is new physics beyond the standard model.

The gauge coupling constants for the strong, the weak and the electromagnetic interactions are free parameters and are widely different. So, an extremely attractive proposition would be to consider a theory in which all three forces get unified into a grand unified theory, so a single coupling constant can explain all three forces at some very high energy. One major problem associated with any grand unified theory is the gauge hierarchy problem. The electroweak phase transition takes place at a much lower scale compared with the scale of grand unification. So the quadratic divergences originating from the heavy particles in the theory tend to make the light particles as heavy as the grand unification scale. This problem could be solved by making the theory supersymmetric.

Supersymmetry is a fascinating theory, a symmetry between a fermion and a boson, which promises unification of space–time symmetry with internal symmetries and, in addition, provides a natural solution to the gauge hierarchy problem. Since every supermultiplet contains both fermions and bosons, every supermultiplet corresponding to the known particles (quarks, leptons, gauge bosons and Higgs scalars) also contain their corresponding superpartners (squarks, sleptons, gauginos and Higgsinos). Moreover, every interaction will be associated with new interactions, in which the particles are replaced by their corresponding superpartners. Supersymmetric theories promise very rich phenomenology. Since supersymmetry eliminates the quadratic divergences, to solve the gauge hierarchy problem supersymmetry must survive until low energy, making it verifiable in the next generation accelerators, such as LHC or ILC.

On the theoretical side the superstring theory is emerging as the most consistent theory of quantum gravity. Several new ideas have evolved in this direction following the second string revolution with the advent of duality conjectures. A new class of phenomenological models is emerging, in which the Planck scale could be lowered and gravity could become strong at very low scale. The most interesting features of this class of models is that the next generation accelerators may find signals for the extra dimensions, grand unification, or even quantum gravity. In these models there are new extra dimensions, in which only gravity could propagate and we are confined at one end point in these extra dimensions. Then, at energies accessible at

LHC or ILC, the Kaluza–Klein excitations of the usual particles will show up. There are other signals of these models, such as the missing energy. Since gravitons can take away energy from ordinary particles and vanish in the extra dimensions, there will be events with missing energy. All this rich phenomenology makes these models with extra dimensions highly attractive.

Another approach to test these attractive models of particle physics at high energy is to look for cosmological consequences. We try to find if some information of these high energy theories is imprinted in the astrophysical observations coming from the early universe, when the universe was hot and average energy per particle was much higher than our next generation accelerators can reach. We now have a satisfactory understanding of all four fundamental interactions and the underlying fundamental principles. With this knowledge we can explain the evolution of the universe in the framework of the standard big-bang cosmological model. The predictions of the nucleosynthesis and the cosmic microwave background radiation have established the big-bang model as the standard model of cosmology. The parameters of the standard model of cosmology have been measured with fairly good accuracy. Different types of measurements are leading to the same values of the cosmological parameters, establishing the reliability of the results. Already we have evidenced a major breakthrough result that the cosmological constant is required to be nonvanishing by most of the precision observations, implying that the universe is flat and accelerating. The total matter density of the universe is equal to the critical density, 2/3 of which is cosmological constant in the form of dark energy and almost 1/3 is in the form of dark matter. About 5% of the total matter is baryonic matter, and only a fraction of it is visible. Although a large fraction of total matter is in the form of dark energy, in particle physics scale this amount is too small to be explained naturally from any models of particle physics.

There are also nonaccelerator particle physics experiments, such as the proton decay, neutron–antineutron oscillation, dark matter searches, axion searches, and neutrino physics experiments, which could also provide some information about the nature of new physics we should expect at higher energies. Another clue for physics beyond the standard model may come from the Higgs sector. If the Higgs scalar is lighter than 138 GeV, then it would mean that either there are more Higgs or the standard model breaks down at around 1 TeV. On the other hand, if the Higgs scalar is found to be heavier than 130 GeV, then it will mean that probably there is no supersymmetry in nature. If we cannot find any Higgs scalars at LHC or ILC, then we have to study the Higgsless models more seriously.

Thus, we have now reached a very exciting juncture in physics. A new era is about to begin. Although supersymmetry and extra dimensions are the two most interesting possibilities, the reality could be something different. We have to wait and watch; nature may surprise us with some drastically new observations. We are about to start our new journey. Although we have no idea about this new epoch, one thing we have realized from our experience is that *particle* and *astroparticle physics* have to work hand in hand to reveal the beauty of nature in this new era.

# *Bibliography*

While preparing this manuscript I extensively followed some of the books and reviews, which are listed in this bibliography. I am sure there are many other books, which were not accessible to me, but which are very good books. The first few books were used for several chapters, so they have been listed separately under the heading "Particle Physics and Cosmology." The remaining books and reviews are listed subject-wise.

**Particle Physics and Cosmology**


T.-P. Cheng and L.-F. Li, *Gauge Theory of Elementary Particle Physics*, Oxford University Press, Oxford, UK, 1984.

P.D.B. Collins, A.D. Martin and E.J. Squires, *Particle Physics and Cosmology*, John Wiley and Sons, New York, 1989.

H.V. Klapdor-Kleingrothaus and U. Sarkar, *Contemporary Particle Physics and Cosmology*, incomplete and unpublished.

H.V. Klapdor-Kleingrothaus and K. Zuber, *Particle Astrophysics*, Institute of Physics, Bristol, 1997.

R.N. Mohapatra, *Unification and Supersymmetry*, Springer Verlag, Heidelberg, 1992.


**Field Theory**


J.D. Bjorken and S.D. Drell, *Relativistic Quantum Mechanics*, McGraw-Hill, New York, 1964.

J.D. Bjorken and S.D. Drell, *Relativistic Quantum Fields*, McGraw-Hill, New York, 1965.

M. Chaichian, *Introduction to Gauge Field Theories*, Springer, Heidelberg, 1984.

M. Kaku, *Quantum Field Theory, A Modern Introduction*, Oxford University Press, New York, 1993.

M.E. Peskin and D.V. Schroeder, *An Introduction to Quantum Field Theory*, Perseus Books, Reading, MA, 1995.

**Group Theory**

H. Georgi, *Lie Algebras in Particle Physics*, Benjamin/Cummings Publishing Co., Reading, MA, 1982.

R. Slansky, Group Theory for Unified Model Building, *Phys. Rep.* **79**, 1 (1981).

B.G. Wybourne, *Classical Groups for Physicists*, John Wiley & Sons, New York, 1974.

**Standard Model**

E.S. Abers and B.W. Lee, Gauge theories, *Phys. Rep.* **9**, 1 (1973).

G. Altarelli, Partons in quantum chromodynamics, *Phys. Rep.* **81**, 1 (1982).

E.D. Commins and P.H. Bucksham, *Weak Interaction of Leptons and Quarks*, Cambridge University Press, Cambridge, 1983.

H. Georgi, *Weak Interactions and Modern Particle Physics*, Benjamin-Cummings, Reading, MA, 1984.

E.A. Paschos, *Electroweak Theory*, Cambridge University Press, Cambridge, 2006.

S. Weinberg, *The Quantum Theory of Fields, Modern Applications* vols I and II, Cambridge Univrsity Press, Cambridge, 1996.

**Neutrino Physics**

J.N. Bahcall, *Neutrino Astrophysics*, Cambridge University Press, Cambridge, 1989.

S.M. Bilenky and S.T. Petcov, Massive neutrinos and neutrino oscillations, *Rev. Mod. Phys.* **59**, 671 (1987).

F. Boehm and P. Vogel, *Physics of Massive Neutrinos*, Cambridge University Press, Cambridge, 1987.

M. Fukugita and A. Suzuki, (eds.), *Physics and Astrophysics of Neutrinos*, Springer Verlag, Heidelberg, 1994.

B. Kayser, F. Gibrat-Debu and F. Perrier, *The Physics of Massive Neutrinos*, World Scientific, Singapore, 1989.

R.N. Mohapatra and P. Pal, *Massive Neutrinos in Physics and Astrophysics*, World Scientific, Singapore, 1991.

**CP Violation**

A. Ali, CKM phenomenology and B meson physics: present status and current issues, *hep-ph/0312303* (2003).

I.I. Bigi and A.I. Sanda, *CP Violation*, Cambridge University Press, Cambridge, 1999.

C. Jarlskog, (ed.), *CP Violation*, World Scientific, Singapore, 1989.

Y. Nir and H.R. Quinn, CP violation in B physics, *Nucl. Sci.* **42**, 211 (1992).

E.A. Paschos and U. Türke, Quark mixing and CP violation, *Phys. Rep.* **178**, 145 (1989).

G.G. Raffelt, Astrophysical methods to constrain axions and other novel particle phenomenon, *Phys. Rep.* **198**, 1 (1990).

G.G. Raffelt, *Stars as Laboratories for Fundamental Physics*, Univ. of Chicago Press, Chicago, 1996.

L.J. Rosenberg and K.A. van Bibber, Searches for invisible axions, *Phys. Rep.* **325**, 1 (2000).

M.S. Turner, Windows on the axion, *Phys. Rep.* **197**, 67 (1990).

**Grand Unified Theory**

C. Kounnas, A. Masiero, D.V. Nanopoulos and K.A. Olive, *Grand Unification with and without Supersymmetry and Cosmological Implications*, World Scientific, Singapore, 1993.

P. Langacker, Grand unified theories and proton decay, *Phys. Rep.* **72**, 185 (1981).

G.G. Ross, *Grand Unified Theories*, Addison-Wesley, Reading, MA, 1984.

## Supersymmetry and Supergravity

R. Arnowitt, A. Chamseddine and P. Nath, *Applied N=1 Supergravity*, World Scientific, Singapore, 1984.

S.P. Misra, *Introduction to Supersymmetry and Supergravity*, Wiley Eastern, New Delhi, 1992.

A. Salam and J. Strathdee, Supersymmetry and superfields, *Fortschr. Phys.* **26**, 57 (1978).

P. Van Nieuwenhuizen, Supergravity, *Phys. Rep.* **68**, 189 (1981).

J. Wess and J. Bagger, *Supersymmetry and Supergravity*, Princeton University Press, Princeton, NJ, 1983.

P. West, 1986 *Introduction to Supersymmetry and Supergravity*, World Scientific, Singapore, 1986.

J. Gunion, H. Haber, G. Kane, and S. Dawson, *The Higgs Hunter's Guide*, Addison Wesley, Menlo Park, CA, 1990.

S. Dawson, *Supersymmetry, Supergravity and Supercolliders, TASI97*, (ed. J. Bagger) World Scientific, Singapore, 1998.

K.A. Olive, Introduction to supersymmetry: astrophysical and phenomenological constraints, *hep-ph/9911307* (1999).

## String Theory

D. Bailin and A. Love, *Supersymmetric Gauge Field Theory and String Theory*, Institute of Physics, Bristol, 1994.

D. Bailin and A. Love, Orbifold compactification of string theory, *Phys. Rep.* **315**, 285 (1999).

M. Green, Connection between M theory and superstrings, *Nucl. Phys. Proc. Suppl.* **68**, 242 (1998).

M. Green, J.H. Schwarz and E. Witten, *Superstring Theories* vols I and II, Cambridge University Press, Cambridge, 1986.

M. Kaku, *Introduction to Superstrings*, Springer-Verlag, New York, 1998.

J.H. Schwarz, From superstring to M theory, *Phys. Rep.* **315**, 107 (1999).

J.H. Schwarz, Introduction to M(atrix) theory and noncommutative geometry, *Phys. Rep.* **360**, 353 (2002).

A. Sen, An introduction to nonperturbative string theory, *hep-th/9802051* (1998).

**Extra Dimensions**

T. Appelquist, A. Chodos and P. Freund, *Modern Kaluza-Klein Theories*, Benjamin-Cummings, New York, 1988.

M. Besancon, Experimental introduction to extra dimensions, *hep-ph/0106165* (2001).

S. Förste, Strings, branes and extra dimensions, *Fortsch. Phys.* **50**, 221 (2001).

Y.A. Kubyshin, Models with extra dimensions and their phenomenology, *hep-ph/0111027* (2001).

D. Langlois, Cosmology with an extra-dimension, *astro-ph/0301021* (2003).

A. Perez-Lorenzana, Theories in more than four-dimensions, *AIP Conf. Proc.* **562**, 53 (2001).

A. Perez-Lorenzana, An introduction to the brane world, *hep-ph/0406279* (2004).

V.A. Rubakov, Large and infinite extra dimensions: an introduction, *Phys. Usp.* **44**, 871 (2001); *Usp. Fiz. Nauk*, **171**, 913 (2001).

**Cosmology**

P. Coles and F. Lucchin, *Cosmology: The Origin and Evolution of Cosmic Structure*, John Wiley and Sons, New York, 1998.

J.R. Ellis, Particle physics and cosmology, *astro-ph/0305038* (2003).

A. Guth, *The Inflationary Universe, The Quest for a New Theory of Cosmic Origin*, Helix Books, Addison-Wesley, Reading, MA, 1997.

E.W. Kolb and M.S. Turner, *The Early Universe*, Addison-Wesley, Reading, MA, 1989.

A. Lin, *Particle Physics and Inflationary Cosmology*, Harvard University Press, Boston, MA, (1990).

K. Olive, Inflation, *Phys. Rep.* **190**, 307 (1990).

P.J.E. Peebles, *The Large Scale Structure of the Universe*, Princeton University Press, Princeton, NJ, 1980.

D.N. Schramm and M.S. Turner, Big bang nucleosynthesis enters the precision era, *Rev. Mod. Phys.* **70**, 303 (1998).

S. Weinberg, *Gravitation and Cosmology, Principles and Applications of the General Theory of Relativity*, John Wiley and Sons, New York, 1972.

**Astroparticle Physics**

V. Barger and C. Kao, Neutralino relic density in the minimal supergravity model, *Phys. Rep.* **307**, 207 (1998).

P. Binetruy, Cosmological constant versus quintessence, *Int. J. Theor. Phys.* **39**, 1859 (2000).

W. Buchmuller and M. Plumacher, Matter antimatter asymmetry and neutrino properties, *Phys. Rep.* **320**, 329 (1999).

G. Dvali, Submillimeter extra dimensions and TeV-scale quantum gravity, *Int. J. Theor. Phys.* **39**, 1717 (2000).

H.B. Kim, Cosmology of Randall-Sundrum models, *hep-ph/0102182* (2001).

H.V. Klapdor-Kleingrothaus, ed., *Dark Matter in Astrophysics and Particle Physics*, Institute of Physics, Bristol, 1998.

D. Langlois, Brane cosmology: An Introduction, *Prog. Theor. Phys. Suppl.* **148**, 181 (2003).

# References


[1] P.A.M. Dirac, *Proc. Roy. Soc. Loc.* **A 114**, 243 (1927).

[2] S. Tomonaga, *Prog. Theor. Phys.* **1**, 27 (1946); *Phys. Rev.* **74**, 224 (1948); J. Schwinger, *Phys. Rev.* **75**, 651 (1949); *Phys. Rev.* **76**, 790 (1949); R. Feynman, *Phys. Rev.* **76**, 749 (1949); *Phys. Rev.* **76**, 769 (1949).

[3] F.J. Dyson, *Phys. Rev.* **75**, 486 (1949); *Phys. Rev.* **75**, 1736 (1949).

[4] W. Gordon, *Z. Phys.* **40**, 117 (1926); O. Klein, *Z. Phys.* **41**, 407 (1927).

[5] P.A.M. Dirac, *Proc. Roy. Soc. Loc.* **A 117**, 610 (1928); *Proc. Roy. Soc. Loc.* **A 126**, 360 (1930).

[6] E. Noether, *Kgl. Ges. Wiss. Nachrichten. Math. Phys. Klasse*, Göttingen p 235 (1918).

[7] W. Pauli and V. Weisskopf, *Helv. Phys. Act.* **7**, 709 (1934).

[8] W. Pauli, *Phys. Rev.* **58**, 716 (1940).

[9] S.N. Gupta, *Proc. Roy. Soc. Loc.* **A 63**, 681 (1950) K. Bleuler, *Halv. Phys. Acta.* **23**, 567 (1950).

[10] G.C. Wick, *Phys. Rev.* **80**, 268 (1950).

[11] H. Lehmann, K. Symanzik and W. Zimmerman, *Nuovo Cim.* **1**, 1425 (1955).

[12] R.E. Cutkowsky, *J. Math. Phys.* **1**, 429 (1960).

[13] C. Møller, *Ann. Phys.* **14**, 531 (1932).

[14] H.J. Bhabha, *Proc. Roy. Soc. Loc.* **A 154**, 195 (1935).

[15] S. Mandelstam, *Phys. Rev.* **B 94**, 266 (1968).

[16] F. Bloch and A. Nordseick, *Phys. Rev.* **52**, 54 (1937).

[17] W. Pauli and F. Villars, *Rev. Mod. Phys.* **21**, 433 (1949).

[18] J.C. Ward, *Phys. Rev.* **78**, 182 (1950); Y. Takahashi, Nuovo Cim. **6**, 371 (1957).

[19] J. Schwinger, *Phys. Rev.* **73**, 416L (1948).

[20] N.N. Bogoliubov, and O. Parasiuk, *Acta. Math.* **97**, 227 (1957);
K. Hepp, *Comm. Math.* **2**, 301 (1966);
W. Zimmerman, *Comm. Math.* **11**, 1 (1968); *Comm. Math.* **15**, 208 (1969).

[21] S. Weinberg, *Phys. Rev.* **118**, 838 (1960).

[22] J.H. Christenson et al., *Phys. Rev. Lett.* **13**, 138 (1964).

[23] G. Lüders, *Kgl. Dansk. Vidensk. Selsk. Mat. Fiz. Medd.* **28**, 5 (1954); *Annals of Physics* **2**, 1 (1957);
W. Pauli, *Niels Bohr and the Development of Physics*, McGraw-Hill, New York, 1955.

[24] J. Goldstone, *Nuovo Cim.* **9**, 154 (1961);
Y. Nambu, *Phys. Rev. Lett.* **4**, 380 (1960);
J. Goldstone, A. Salam and S. Weinberg, *Phys. Rev.* **127**, 965 (1962).

[25] Y. Chikashige, R.N. Mohapatra and R.D. Peccei, *Phys. Rev. Lett.* **45**, 1926 (1980); *Phys. Lett.* **B 98**, 265 (1981);
H. Georgi, S. Glashow and S. Nussinov, *Nucl. Phys.* **B193**, 297 (1981);
G. Gelmini, S. Nussinov and T. Yanagida, *Nucl. Phys.* **B219**, 31 (1983);
J. Moody and F. Wilczek, *Phys. Rev.* **D 30**, 130 (1984).

[26] G. 't Hooft, *Nucl. Phys.* **B 33**, 173 (1971); *Nucl. Phys.* **B 35**, 167 (1971);
B.W. Lee, *Phys. Rev.* **D 5**, 823 (1972).

[27] P.W. Higgs, *Phys. Lett.* **12**, 132 (1964); *Phys. Rev. Lett.* **13**, 508 (1964); *Phys. Rev.* **145**, 1156 (1966);
F. Englert and R. Brout, *Phys. Rev. Lett.* **13**, 321 (1964);
G.S. Guralnik, C.R. Hagen and T.W.B. Kibble, *Phys. Rev. Lett.* **13**, 585 (1964);
T.W.B. Kibble, *Phys. Rev.* **155**, 1554 (1967);
S. Weinberg, *Phys. Rev. Lett.* **18**, 507 (1967).

[28] S. Adler, *Phys. Rev.* **177**, 2426 (1969);
J. Bell and R. Jackiw, *Nuovo Cim.* **51A**, 47 (1969);
W. Bardeen, *Phys. Rev.* **184**, 1848 (1969).

[29] U. Sarkar, *Int. J. Mod. Phys.* **A 22**, 931 (2007).

[30] C.N. Yang and R.L. Mills, *Phys. Rev.* **96**, 191 (1954);
R. Shaw, *Problem of particle types and other contributions to the theory of elementary particles*, Ph.D. Thesis, Cambridge University Press, Cambridge, 1955.

[31] M. Gell-Mann, *Phys. Lett.* **8**, 214 (1964).
G. Zweig, CERN preprint 8182/Th.401 (1964), reprinted in S.P. Rosen and D. Lichtenberg, (ed.) *Developments in Quark Theory of Hadrons*, vol I, Hadronic Press, MA, 1980.

[32] C.G. Callan, *Phys. Rev.* **D 2**, 1541 (1970);
K. Symanzik, *Comm. Math. Phys.* **18**, 227 (1970);
D. Gross and F. Wilczek, *Phys. Rev. Lett.* **30**, 1343 (1973); *Phys. Rev.* **D 8**, 3497 (1973);
H.D. Politzer, *Phys. Rev. Lett.* **30**, 1346 (1973).

[33] K. Wilson, *Phys. Rev.* **D3**, 1818 (1971).

[34] R.P. Feynman, *Photon-Hadron Interactions*, Benjamin, Reading, MA, 1972;
J.D. Bjorken and E.A. Paschos, *Phys. Rev.* **185**, 1975 (1969).

[35] E. Fermi, *Z. Phys.* **88**, 161 (1934).

[36] T.D. Lee and C.N. Yang, *Phys. Rev.* **104**, 254 (1956).

[37] C.S. Wu, E. Ambler, R.W. Hayward, D.D. Hoppes and R.P. Hudson, *Phys. Rev.* **105**, 1413 (1957).

[38] E.C.G. Sudarshan and R.E. Marshak, in *Proc. Padua-Venice Conference on Mesons and Newly Discovered Particles* 1957;
R.P. Feynman and M. Gell-Mann, *Phys. Rev.* **109**, 193 (1958);
E.C.G. Sudarshan and R.E. Marshak, *Phys. Rev.* **109**, 1860(L) (1958).

[39] S. Glashow, *Nucl. Phys.* **22**, 579 (1961);
A. Salam, in *Elementary Particle Theory*, ed. N. Svartholm, Almquist and Forlag, Stockholm, 1968;
S. Weinberg, *Phys. Rev. Lett.* **19**, 1264 (1967).

[40] *UA1 Collaboration :* G. Arinson et al., *Phys. Lett.* **B 122**, 103 (1983);
*UA1 Collaboration :* G. Banner et al., *Phys. Lett.* **B 122**, 476 (1983).

[41] N. Cabibbo, *Phys. Rev. Lett.* **10**, 531 (1963);
M. Kobayashi and T. Maskawa, *Prog. Theor. Phys.* **49**, 652 (1973).

[42] S. Glashow, J. Illiopoulos and L. Maiani, *Phys. Rev.* **D 2**, 1285 (1970).

[43] L. Wolfenstein, *Phys. Rev. Lett.* **51**, 1945 (1983).

[44] W.-M. Yao et al. (Particle Data Group), *J. Phys.* **G 33**, 1 (2006); Latest update available at: http://pdg.lbl.gov/.

[45] F. Parodi, in *Proc. of XXIXth Int. Conf. on High Energy Physics*, Vancouver, BC 1998.

[46] A. Ali, in *Proc. 13th Topical Conf. on Hadron Collider Physics*, TIFR, Bombay, India 1999; Report no. *hep-ph/9904427* (1999).

[47] C. Jarlskog, *Phys. Rev. Lett.* **55**, 1039 (1985); *Z. Phys.* **C 29**, 491 (1985);
I. Dunietz, O.W. Greenberg and D. Wu, *Phys. Rev. Lett.* **55**, 2935 (1985);
O.W. Greenberg, *Phys. Rev.* **D 32**, 1841 (1985);
D. Wu, *Phys. Rev.* **D 33**, 860 (1986).

[48] W. Pauli, *(open letter to the participants of the conference in Tübingen)*, 1930; Recorded in W. Pauli, *Wissenschaftlicher Briefwechsel*, Band II, Springer, Berlin, 1985; the first publication was given in W. Pauli, Septieme Conseil de Physique Solvay: *Noyaux Atomiques*, Paris 1934, p.324f (1933).

[49] E. Majorana, *Nuovo Cim.* **14**, 171 (1937).

[50] H. Weyl, *Z. Phys.* **56**, 330 (1929).

[51] S.M. Bilenky and B. Pontecorvo, *Lett. Nuovo Cim.* **17**, 569 (1976);
V. Barger, P. Langacker, J.P. Leveille and S. Pakvasa, *Phys. Rev. Lett.* **45**, 692 (1980);
J. Schechter and J.W.F. Valle, *Phys. Rev.* **D 22**, 2227 (1980).

[52] J. Schechter and J.W.F. Valle, *Phys. Rev.* **D 23**, 1666 (1981); *Phys. Rev.* **D 25**, 774 (1982);
L. Wolfenstein, *Nucl. Phys.* **B 186**, 147 (1981);
D. Wyler and L. Wolfenstein, *Nucl. Phys.* **B 218**, 205 (1983);
C.N. Leung and S.T. Petcov, *Phys. Lett.* **B 125**, 461 (1983); *Phys. Lett.* **B 145**, 416 (1984).

[53] L. Wolfenstein, *Phys. Lett.* **B 107**, 77 (1981);
J. Schechter and J.W.F. Valle, *Phys. Rev.* **D 24**, 1883 (1981); *Phys. Rev.* **D 25**, 2951 (1982);
J. Nieves, *Phys. Rev.* **D 26**, 3152 (1982); *Phys. Lett.* **B 147**, 375 (1984);
B. Kayser, *Phys. Rev.* **D 30**, 1023 (1984).

[54] S. Weinberg, *Phys. Rev. Lett.* **43**, 1566 (1979); *Phys. Rev.* **D 22**, 1694 (1980);
F. Wilczek and A. Zee, *Phys. Rev. Lett.* **43**, 1571 (1979);
F. Weldon and A. Zee, *Nucl. Phys.* **B 173**, 269 (1980).

[55] P. Minkowski, *Phys. Lett.* **B 67**, 421 (1977);
M. Gell-Mann, P. Ramond and R. Slansky, in *Supergravity, Proc of the Workshop*, Stony Brook, New York, ed. P. van Nieuwenhuizen and D. Freedman, North-Holland, Amsterdam, 1979;
T. Yanagida, in *Proc of the Workshop on Unified Theories and Baryon Number in the Universe*, Tsukuba, Japan, ed. A. Sawada and A. Sugamoto, (KEK Report No. 79-18, Tsukuba) 1979.

[56] J. Schechter and J.W.F. Valle, *Phys. Rev.* **D 22**, 2227 (1980);
R. N. Mohapatra and R. E. Marshak, in *Proc of the Orbis Scientiae Conf.*, Coral Gables, Fl., ed. A. Perlmutter and L.F. Scott, Plenum Press, NY, 1980;
T. P. Cheng and L. F. Li, *Phys. Rev.* **D 22**, 2860 (1980);
G.B. Gelmini and M. Roncadelli, *Phys. Lett.* **B 99**, 411 (1981);
C. Wetterich, *Nucl. Phys.* **B 187**, 343 (1981);
G. Lazarides, Q. Shafi and C. Wetterich, *Nucl. Phys.* **B 181**, 287 (1981);

[57] E. Ma and U. Sarkar, *Phys. Rev. Lett.* **80**, 5716 (1998).

[58] G. Lazarides and Q. Shafi, *Phys. Rev.* **D 58**, 071702 (1998);
W. Grimus, R. Pfeiffer and T. Schwetz, *Eur. Phys. J.* **C 13**, 125 (2000);
T. Hambye, E. Ma and U. Sarkar, *Nucl. Phys.* **B 602**, 23 (2001).

[59] J.C. Pati and A. Salam, *Phys. Rev. Lett.* **31**, 661 (1973); *Phys. Rev.* **D 8**, 1240 (1973); ibid. **D 10**, 275 (1974).

[60] R.N. Mohapatra and J.C. Pati, *Phys. Rev.* **D 11**, 566, 2558 (1975);
R.N. Mohapatra and G. Senjanovic, *Phys. Rev.* **D 12**, 1502 (1975).

[61] R.E. Marshak and R.N. Mohapatra, *Phys. Rev. Lett.* **44**, 1316 (1980).

[62] R.N. Mohapatra and G. Senjanovic, *Phys. Rev. Lett.* **44**, 912 (1980); *Phys. Rev.* **D 23**, 165 (1981).

[63] G. Senjanovic, *Nucl. Phys.* **B 153**, 334 (1979);
G. Dvali, Q. Shafi and Z. Lazarides, *Phys. Lett.* **B 424**, 259 (1998);
K.S. Babu, J.C. Pati and F. Wilczek, *Nucl. Phys.* **B 566**, 33 (2000);
B. Brahmachari, E. Ma and U. Sarkar, *Phys. Rev. Lett.* **91**, 011801 (2004);
U. Sarkar, *Phys. Lett.* **B 594**, 308 (2004).

[64] S.M. Barr, *Phys. Rev. Lett.* **92**, 101601 (2004);
S.M. Barr and C.H. Albright, *Phys. Rev.* **D 69**, 073010 (2004); *Phys. Rev.* **D 70**, 033013 (2004);
B.R. Desai, G. Rajasekaran and U. Sarkar, *Phys. Lett.* **B 626**, 167 (2005);
U. Sarkar, *Phys. Lett.* **B 622**, 118 (2005).

[65] A. Zee, *Phys. Lett.* **B 93**, 389 (1980);
L. Wolfenstein, *Nucl. Phys.* **B 175**, 93 (1980).

[66] B. Pontecorvo, *Zh. Eksp. Teor. Fiz.* **33**, 549 (1957); [*Sov. Phys. JETP* **6**, 429 (1958)]; *Zh. Eksp. Teor. Fiz.* **34**, 247 (1957); [*Sov. Phys. JETP* **7**, 172 (1958)].

[67] Z. Maki, M. Nakagawa and S. Sakata, *Prog. Theor. Phys.* **28**, 870 (1962);
S. Eliezer and D.A. Ross, *Phys. Rev.* **D 10**, 3088 (1974);
S.M. Bilenky and B. Pontecorvo, *Phys. Lett.* **B 61**, 248 (1976);
H. Fritzsch and P. Minkowski, *Phys. Lett.* **B 62**, 72 (1976).

[68] *CHOOZ Collaboration:* M. Apollonio et al., *Phys. Lett.* **B 466**, 415 (1999).

[69] L. Wolfenstein, *Phys. Rev.* **D 17**, 2369 (1978);
S.P. Mikheyev and A. Smirnov, *Yad. Fiz.* **42**, 1441 (1985) [*Sov. J. Nucl. Phys.* **42**, 913 (1986)].

[70] V. Barger, S. Pakvasa, R.J.N. Phillips and K. Whisnant, *Phys. Rev.* **D 22**, 2718 (1980);
P. Langacker, *Nucl. Phys.* **B 282**, 589 (1987).

[71] *Super-Kamiokande Collaboration:* Y. Fukuda et al., *Phys. Lett.* **B 433**, 9 (1998); *Phys. Rev. Lett.* **81**, 1562 (1998); ibid. **93**, 101801 (2004).

[72] *KamLAND Collaboration:* K. Eguchi et al., *Phys. Rev. Lett.* **90**, 021802 (2003);
T. Araki et al., *Phys. Rev. Lett.* **94**, 081801 (2005).

[73] D.N. Spergel et al., *Astrophys. J. Suppl.* **148**, 175 (2003).

[74] H.V. Klapdor-Kleingrothaus, A. Dietz, H. Harney and I. Krivosheina, *Mod. Phys. Lett.* **A 16**, 2409 (2001).

[75] *LSND Collaboration:* A. Athanassopoulos et al., *Phys. Rev. Lett.* **77**, 3082 (1996); *Phys. Rev.* **C 58**, 2489 (1998); *Phys. Rev. Lett.* **81**, 1774 (1998).

[76] *MiniBOONE Collaboration:* A.A. Aguilar-Arevalo et al., *Phys. Rev. Lett.* **98**, 231801 (2007).

[77] *K2K Collaboration:* S.H. Ahn et al., *Phys. Lett.* **B 511**, 178 (2001); *Phys. Rev. Lett.* **90**, 041801 (2003).

[78] J.N. Bahcall and M.H. Pinsonneault, *Rev. Mod. Phys.* **67**, 781 (1995); *Phys. Rev. Lett.* **92**, 121301 (2004);
J.N. Bahcall, S. Basu and M.H. Pinsonneault, *Phys. Lett.* **B 433**, 1 (1998).

[79] R. Davis, Jr., *Phys. Rev. Lett.* **12**, 303 (1964).

[80] B.T. Cleveland et al., *Nucl. Phys. Proc. Suppl.***B 38**, 47 (1995).

[81] *Kamiokande Collaboration:* Y. Fukuda et al. *Phys. Rev. Lett.* **77**, 1683 (1996); *Super-Kamiokande Collaboration:* Y. Fukuda et al., *Phys. Rev. Lett.* **81**, 1158 (1998); ibid. **82**, 1810 (1999); ibid. **86**, 5656 (2001); *Phys. Rev.* **D 69**, 011104 (2004).

[82] *SNO Collaboration:* Q.R. Ahmad et al. *Phys. Rev. Lett.* **87**, 071301 (2001); ibid. **89**, 011301 (2002); ibid. **89**, 011302 (2002); ibid. **92**, 181301 (2004).

[83] A. Halprin, S.T. Petcov and S.P. Rosen, *Phys. Lett.* **B 125**, 335 (1983);
C.W. Kim and H. Nishiura, *Phys. Rev.* **D 30**, 1123 (1984);
W.C. Haxton, *Prog. Part. Nucl. Phys.* **12**, 409 (1984).

[84] H.V. Klapdor-Kleingrothaus and U. Sarkar, *Proc. Ind. Nat. Sc. Acad.*, **70**, 251 (2004).

[85] A. Pierce and H. Murayama, *Phys. Lett.* **B 581**, 218 (2004).

[86] W. Hu, D.J. Eisenstein and M. Tegmark, *Phys. Rev. Lett.* **80**, 5255 (1998).

[87] O. Elgaroy et al., *Phys. Rev. Lett.* **89**, 061301 (2002).

[88] S. Hannestad, *JCAP* **0305**, 004 (2003).

[89] A. Burrows and J. Lattimer, *Astrophys. J.* **307**, 178 (1986); *Astrophys. J.* **318**, L63 (1986);
R. Mayle, J.R. Wilson and D. Schramm, *Astrophys. J.* **318**, 288 (1987).

[90] M. Aglietta et al., *Europhys. Lett.* **3**, 1321 (1987);
E.N. Alekseev et al., *Phys. Lett.* **B 205**, 209 (1988); *JETP Lett.* **45**, 589 (1987);
K. Hirata et al., *Phys. Rev. Lett.* **58**, 1490 (1987);
R. Bionta et al., *Phys. Rev. Lett.* **58**, 1494 (1987).

[91] R. Barbieri, L.J. Hall and A. Strumia, *Phys. Lett.* **B 445**, 407 (1999);
E. Ma, U. Sarkar and D.P. Roy, *Phys. Lett.* **B 444**, 391 (1998);
E. Ma and D.P. Roy, *Phys. Rev.* **D 59**, 097702 (1999);
E. Akhmedov, *Phys. Lett.* **B 467**, 95 (1999); *Nucl. Phys. Proc. Suppl.***87**, 321 (2000);
S.F. King, *Rep. Prog. Phys.* **67**, 107 (2004).

[92] H.V. Klapdor-Kleingrothaus, H. Päs and A.Yu. Smirnov, *Phys. Rev.* **D 63**, 073005 (2001);
H.V. Klapdor-Kleingrothaus and U. Sarkar, *Mod. Phys. Lett.* **A 16**, 2469 (2001).

[93] A. Alavi-Harati. et al., *Phys. Rev. Lett.* **83**, 22 (1999);
V. Fanti et al., *Phys. Lett.* **B 465**, 335 (1999).

[94] *BELLE Collaboration:* K. Abe et al., *Phys. Rev. Lett.* **87**, 091802 (2001).

[95] T.T. Wu and C.N. Yang, *Phys. Rev. Lett.* **13**, 380 (1964).

[96] L. Wolfenstein, *Phys. Rev. Lett.* **13**, 562 (1964);
T.D. Lee, *Phys. Rep.* **9**, 145 (1974);
S. Weinberg, *Phys. Rev. Lett.* **34**, 657 (1976).

[97] *ARGUS Collaboration:* H. Albrecht et al. 1987 *Phys. Lett.* **B 192**, 245 (1987).

[98] A.B. Carter and A.I. Sanda, *Phys. Rev. Lett.* **45**, 952 (1980); *Phys. Rev.* **D 23**, 1567 (1981);
I.I. Bigi and A.I. Sanda, *Nucl. Phys.* **B 193**, 85 (1981).

[99] S. Weinberg, *Phys. Rev.* **D 11**, 3583 (1975).

[100] G. 't Hooft, *Phys. Rev. Lett.* **37**, 8 (1976); *Phys. Rev.* **D 14**, 3432 (1976).

[101] C.G. Callan, R. Dashen and D. Gross, *Phys. Lett.* **B 63**, 334 (1976);
R. Jackiw and C. Rebbi, *Phys. Rev. Lett.* **37**, 172 (1976).

[102] R.D. Peccei and H.R. Quinn, *Phys. Rev. Lett.* **38**, 1440 (1977); *Phys. Rev.* **D 16**, 1791 (1977).

[103] V. Baluni, *Phys. Rev.* **D 19**, 2227 (1979);
R. Crewther, P. DiVecchia, G. Veneziano and E. Witten, *Phys. Lett.* **B 88**, 123 (1979).

[104] K. Choi, C.W. Kim and W.K. Sze, *John Hopkins report*, JHU-TIPAC 8804 (1988).

[105] S. Weinberg, *Phys. Rev. Lett.* **40**, 223 (1978);
F. Wilczek, *Phys. Rev. Lett.* **40**, 271 (1978).

[106] Y. Asano et al. *Phys. Lett.* **B 107**, 159 (1981).

[107] F. Wilczek, *Phys. Rev. Lett.* **39**, 1304 (1977).

[108] S. Yamada, in *Proc. Int. Symp. on Lepton and Photon Interactions at High Energy*, (ed. Cassel D G and Kreinick), Cornell University, 1983.

[109] M. Dine, W. Fishler and M. Sredniki, *Phys. Lett.* **B 104**, 199 (1981);
A.P. Zhitnitskii, *Sov. J. Nucl. Phys.* **31**, 260 (1980).

[110] J.E. Kim, *Phys. Rev. Lett.* **43**, 103 (1979);
M.A. Shifman, A.I. Vainshtein and V.I. Yakharov, *Nucl. Phys.* **B 166**, 493 (1980).

[111] D. Demir and E. Ma, *Phys. Rev.* **D 62**, 111901 (R) (2000); *J. Phys. G: Nucl. Part. Phys.* **27**, L87 (2001);
D. Demir, E. Ma and U. Sarkar, *J.Phys. G: Nucl. Part. Phys.* **26**, L117 (2000).

[112] A. Akhoury, I. Bigi and H. Haber, *Phys. Lett.* **B 135**, 113 (1984).

[113] K. Choi and J.E. Kim, *Phys. Lett.* **B 154**, 393 (1985).

[114] J. Frieman, S. Dimopoulos and M. Turner, *Phys. Rev.* **D 36**, 2201 (1987).

[115] G.G. Raffelt and D.S.P. Dearborn, *Phys. Rev.* **D 36**, 2211 (1987);
D.S.P. Dearborn, D.N. Schramm and G. Steigman, *Phys. Rev. Lett.* **56**, 26 (1986).

[116] R. Brinkmann and M. Turner, *Phys. Rev.* **D 38**, 2338 (1988);
G. Raffelt and D. Seckel, *Phys. Rev. Lett.* **60**, 1793 (1988);
M.S. Turner, *Phys. Rev. Lett.* **60**, 1797 (1988);
K. Choi, K. Kang and J.E. Kim, *Phys. Rev. Lett.* **62**, 849 (1989).

[117] J. Preskill, M. Wise and F. Wilczek, *Phys. Lett.* **B 120**, 127 (1983);
L.F. Abbott and P. Sikivie, *Phys. Lett.* **B 120**, 133 (1983);
M. Dine and W. Fishler, *Phys. Lett.* **B 120**, 137 (1983);
J.E. Kim, *Phys. Rep.* **150**, 1 (1987);
M.S. Turner, *Phys. Rev.* **D 33**, 819 (1986).

[118] P. Sikivie, *Phys. Rev. Lett.* **51**, 1415 (1983); *Phys. Rev.* **D 32**, 2988 (1985).

[119] G.G. Raffelt and L. Stodolsky, *Phys. Rev.* **D 37**, 1237 (1988).

[120] J. Nieves and P. Pal, *Phys. Rev.* **D 36**, 315 (1987).

[121] A. Kusenko and R. Shrock, *Phys. Lett.* **B 323**, 18 (1994);
G.C. Branco, T. Morozumi, B.M. Nobre and M.N. Rebelo, *Nucl. Phys.* **B 617**, 475 (2001);
G.C. Branco, R. GonzalezFelipe, F.R. Joaquim, I. Masina, M.N. Rebelo and C.A. Savoy, *Phys. Rev.* **D 67**, 073025 (2003);
S. Singh and U. Sarkar, *Nucl. Phys.* **B 771**, 28 (2007).

[122] P. Pal and J. Nieves, *Phys. Rev.* **D 64**, 076005 (2001); *Phys. Rev.* **D 67**, 036005 (2003).

[123] V. Barger, R.J.N. Phillips and K. Whisnant, *Phys. Rev. Lett.* **45**, 2084 (1980).

[124] H. Georgi and S. Glashow, *Phys. Rev. Lett.* **32**, 438 (1974).

[125] H. Georgi, H. Quinn and S. Weinberg, *Phys. Rev. Lett.* **33**, 451 (1974).

[126] H. Georgi and C. Jarlskog, *Phys. Lett.* **B 86**, 297 (1979).

[127] Super-Kamiokande Collaboration: Y. Hayato et al., *Phys. Rev. Lett.* **83**, 1529 (1999);
K.S. Hirata et al., *Phys. Lett.* **B 220**, 308 (1989);
*IMB Collaboration*: W. Gajewski et al., *Nucl. Phys. Proc. Suppl.***A 28**, 1610164 (1992);
*FREJUS Collaboration*: C. Berger et al., *Phys. Lett.* **B 245**, 305 (1991).

[128] J.C. Pati, A. Salam and U. Sarkar, *Phys. Lett.* **B 133**, 330 (1983).

[129] R.N. Mohapatra and R.E. Marshak, *Phys. Lett.* **B 91**, 222 (1980);
A. Davidson, *Phys. Rev.* **D 20**, 776 (1979).

[130] U. Amaldi, W. de Boer and H. Furstenau, *Phys. Lett.* **B 260**, 447 (1991).

[131] C.T. Hill, *Phys. Lett.* **B 135**, 47 (1984);
C. Wetterich and Q. Shafi, *Phys. Rev. Lett.* **52**, 875 (1984);
C. Panagiotakopoulos and Q. Shafi, *Phys. Rev. Lett.* **52**, 2336 (1984);
M.K. Parida, P.K. Patra and A.K. Mohanty, *Phys. Rev.* **D 39**, 316 (1989).

[132] B. Brahmachari, P.K. Patra, U. Sarkar and K. Sridhar, *Mod. Phys. Lett.* **A 8**, 1487 (1993).

[133] J. Ellis and M.K. Gaillard, *Phys. Lett.* **B 88**, 315 (1979).

[134] H. Georgi, in *Particles and Fields*, ed. C.E. Carlson, AIP, NY, (1975);
H. Fritzsch and P. Minkowski, *Ann. Phys.* **93**, 193 (1975).

[135] D. Chang, R.N. Mohapatra and M.K. Parida, *Phys. Rev. Lett.* **52**, 1072 (1984); *Phys. Rev.* **D 30**, 1052 (1984);
D. Chang, R.N. Mohapatra, J. Gipson, R.E. Marshak and M.K. Parida, *Phys. Rev.* **D 31**, 1718 (1985).

[136] Riazuddin, R.E. Marshak and R.N. Mohapatra, *Phys. Rev.* **D 24**, 1310 (1981).

[137] W. Caswell, J. Milutinovic and G. Senjanovic, *Phys. Rev.* **D 26**, 161 (1982).

[138] S. Rao and R. Schrock, *Phys. Lett.* **B 116**, 238 (1982).

[139] S.P. Misra and U. Sarkar, *Phys. Rev.* **D 28**, 249 (1983);
J. Pashupathy, *Phys. Lett.* **B 114**, 172 (1982).

[140] Chetyrkin et al., *Phys. Lett.* **B99**, 358 (1981);
P.G. Sandars, *J. Phys.* **G 6**, L161 (1980);

Riazuddin, *Phys. Rev.* **D 25**, 885 (1982);
C. Dover, M. Gal and J. Richards, *Phys. Rev.* **D 27**, 1090 (1983).

[141] M. Baldo-Ceolin et al., *Z. Phys.* **C 63**, 409 (1994).

[142] D. Volkov and V.P. Akulov, *JETP Lett.* **16**, 438 (1972); *Phys. Lett.* **B 46**, 109 (1973);
J. Wess and B. Zumino, *Nucl. Phys.* **B 70**, 39 (1974); *Phys. Lett.* **B 49**, 52;
A. Salam and J. Strathdee, *Nucl. Phys.* **B 76**, 477 (1974); *Phys. Lett.* **B 51**, 353 (1974); *Nucl. Phys.* **80**, 317 (1974).

[143] E. Witten, *Nucl. Phys.* **B185**, 513 (1981);
M. Dine, W. Fischler, and M. Srednicki, *Nucl. Phys.* **B 189**, 575 (1981);
S. Dimopoulos and S. Raby, *Nucl. Phys.* **B 192**, 353 (1981);
L.E. Ibanez and G. Ross, *Phys. Lett.* **B 105**, 439 (1981);
R. Kaul, *Phys. Lett.* **B 109**, 19 (1982);
R. Kaul and Majumdar P, *Nucl. Phys.* **B 199**, 36 (1982);
J. Polchinski and L. Susskind, *Phys. Rev.* **D26**, 3661 (1982).

[144] S. Coleman and J. Mandula, *Phys. Rev.* **159**, 1251 (1967).

[145] R. Haag, J.T. Lopuszanski and M.F. Sohnius, *Nucl. Phys.* **B 88**, 257 (1975).

[146] L. O'Raifeartaigh, *Nucl. Phys.* **B 96**, 331 (1975).

[147] P. Fayet and J. Illiopoulos, *Phys. Lett.* **B 51**, 461 (1974).

[148] S. Dimopoulos and H. Georgi, *Nucl. Phys.* **B193**, 150 (1981);
S. Dimopoulos, S. Raby and F. Wilczek, *Phys. Rev.* **D 24**, 1681 (1981);
N. Sakai, *Z. Phys.* **C 11**, 153 (1981);
L.E. Ibanez and G. Ross, *Phys. Lett.* **B 105**, 439 (1981);
P. Fayet, *Phys. Lett.* **B 69**, 489 (1977); *Phys. Lett.* **B 84**, 416 (1979);
Kuroda M, Report no. *hep-ph/9902340* (1999).

[149] S. Weinberg, *Phys. Rev.* **D 26**, 287 (1982);
N. Sakai and T. Yanagida, *Nucl. Phys.* **B 197**, 533 (1982);
C.S. Aulakh and R.N. Mohapatra, *Phys. Lett.* **B 119**, 136 (1982);
S. Dimopoulos, S. Raby and F. Wilczek, *Phys. Lett.* **B 112**, 133 (1982);
J. Ellis, D. Nanopoulos and S. Rudaz, *Nucl. Phys.* **B 202**, 43 (1982).

[150] L. Hall and M. Suzuki, *Nucl. Phys.* **B231**, 419 (1984);
T. Banks, Y. Grossman, E. Nardi and Y. Nir, *Phys. Rev.* **D 52**, 5319 (1995);
B. de Carlos and P. White, *Phys. Rev.* **D54**, 3427 (1996);
E. Nardi, *Phys. Rev.* **D 55**, 5772 (1997).

[151] G. Farrar and P. Fayet, *Phys. Lett.* **B76**, 575 (1978);
F. Zwirner, *Phys. Lett.* **132B**, 103 (1983);
J. Ellis, G. Gelmini, C. Jarlskog, G. Ross and J.W.F. Valle, *Phys. Lett.* **B150**, 142 (1985);
G. Ross and J.W.F. Valle, *Phys. Lett.* **B 151**, 375 (1985);

S. Dawson, *Nucl. Phys.* **B261**, 297 (1985);
R. Barbieri and A. Masiero, *Nucl. Phys.* **B 267**, 679 (1986);
S. Dimopoulos and L. Hall, *Phys. Lett.* **B 207**, 210 (1988).

[152] J.A. Casas et al., *Nucl. Phys.* **B 436**, 3 (1995); (E) **B 439**, 466 (1995);
M. Carena and C.E.M. Wagner, *Nucl. Phys.* **B 452**, 45 (1995).
M. Carena, J.R. Espinosa, M. Quiros and C.E.M. Wagner, *Phys. Lett.* **B 335**, 209 (1995);
M. Carena, M. Quiros and C.E.M. Wagner, *Nucl. Phys.* **B 461**, 407 (1996);
H.E. Haber, R. Hempfling and A.H. Hoang, *Z. Phys.* **C 75**, 539 (1997).

[153] A. Sopczak, in *Proc. of the First International Workshop on Non-Accelerator Physics*, Dubna, *Phys. Atom. Nucl.* **61**, 938 (1998).

[154] H. Baer, C. Kao and X. Tata, *Phys. Rev.* **D48**, 2978 (1993);
H. Baer, J. Ellis, G. Gelmini, D. Nanopoulos and X. Tata, *Phys. Lett.* **B 155**, 175 (1985);
H. Baer, V. Barger, D. Karatas and X. Tata, *Phys. Rev.* **D36**, 96 (1987);
R. Barnett, J. Gunion and H. Haber, *Phys. Rev.* **D37**, 1892 (1988); *Phys. Lett.* **B315**, 349 (1993).

[155] H. Baer, C. Chen, F. Paige and X. Tata, *Phys. Rev.* **D52**, 1565; 2746 (1995);
*Phys. Rev.* **D54**, 5866 (1996); *Phys. Rev.* **D53**, 6241 (1996).

[156] S. Dimopoulos and D. Sutter, *Nucl. Phys.* **B 452**, 496 (1995);
D.W. Sutter, *The Supersymmetric flavor problem and* $\mu \to e^+\gamma$ Ph.D. thesis, Stanford Univ., Report no. *hep-ph/9704390* (1997).

[157] L. Hall, J. Lykken and S. Weinberg, *Phys. Rev.* **D 27**, 2359 (1983).

[158] A. Chamseddine, R. Arnowitt and P. Nath, *Phys. Rev. Lett.* **49**, 970 (1982);
S. Weinberg, *Phys. Rev. Lett.* **50**, 387 (1983).

[159] L.E. Ibanez and D. Lüst, *Nucl. Phys.* **B 382**, 305 (1992);
B. de Carlos, J.A. Casas and C. Munoz, *Phys. Lett.* **B 299**, 234 (1993);
V. Kaplunovsky and J. Louis, *Phys. Lett.* **B 306**, 269 (1993).

[160] M. Dine and A. Nelson, *Phys. Rev.* **D 48**, 1277 (1993);
M. Dine, A. Nelson and Y. Shirman, *Phys. Rev.* **D51**, 1362 (1995);
M. Dine, A. Nelson, Y. Nir and Y. Shirman, *Phys. Rev.* **D53**, 2658 (1996);
C. Kolda, *Nucl. Phys. Proc. Suppl.***62**, 266 (1997).

[161] S. Ambrosanio, S. Heinemeyer, B. Mele, S. Petrarca, G. Polesello, A. Rimoldi and G. Weiglein, CERN yellow report CERN-TH-2000-054 Report no. *hep-ph/0002191* (2000).

[162] M. Dine, N. Seiberg and E. Witten, *Nucl. Phys.* **B289**, 585 (1987);
J. Attick, L. Dixon and A. Sen, *Nucl. Phys.* **B 292**, 109 (1987).

[163] L. Randall and R. Sundrum, *Nucl. Phys.* **B 557**, 79 (1998);
G.F. Guidice, M. Luty, H. Murayama and R. Rattazi, *JHEP* **9812**, 27 (1998).

[164] I. Jack and D.R.T. Jones, *Phys. Lett.* **B 482**, 167 (2000).

[165] E. Ma and P. Roy, *Phys. Rev.* **D 41**, 988 (1990);
A. Joshipura and M. Nowakowski, *Phys. Rev.* **D 51**, 2421 (1995);
R. Hempfling, *Nucl. Phys.* **B 478**, 3 (1996);
F.M. Borzumati, Y. Grossman, E. Nardi and Y. Nir, *Phys. Lett.* **B 384**, 123 (1996);
B. Mukhopadhyay and S. Roy, *Phys. Rev.* **D 55**, 7020 (1997);
A. Diaz, J.C. Romao and J.W.F. Valle, *Nucl. Phys.* **B 524**, 23 (1998);
B. Mukhopadhyaya, S. Roy and F. Vissani, *Phys. Lett.* **B 443**, 191 (1998).

[166] M. Drees, S. Pakvasa, X. Tata and T. ter Veldhuis, *Phys. Rev.* **D 57**, R5335 (1998).

[167] R.N. Mohapatra, *Phys. Rev.* **D 34**, 3457 (1986);
V. Barger, G.F. Guidice and T. Han, *Phys. Rev.* **D 40**, 2987 (1989);
I. Hinchliffe and T. Kaeding, *Phys. Rev.* **D 47**, 279 (1993);
R.M. Godbole, P. Roy and X. Tata, *Nucl. Phys.* **B 401**, 67 (1993);
M. Hirsch, H.V. Klapdor-Kleingrothaus and S.G. Kovalenko, *Phys. Rev. Lett.* **75**, 17 (1995); *Phys. Rev.* **D 53**, 1329 (1996);
K.S. Babu and R.N. Mohapatra, *Phys. Rev. Lett.* **75**, 2276 (1995);
G. Bhattacharyya, J. Ellis and K. Sridhar, *Mod. Phys. Lett.* **A 10**, 1583 (1995);
G. Bhattacharyya, D. Choudhury and K. Sridhar, *Phys. Lett.* **B 355**, 193 (1995);
J.L. Goity and M. Sher, *Phys. Lett.* **B 346**, 69 (1995).

[168] Y. Grossman and H.E. Haber, *Phys. Rev. Lett.* **78**, 3438 (1997);
M. Hirsch, H.V. Klapdor-Kleingrothaus and S.G. Kovalenko, *Phys. Lett.* **B 398**, 311 (1997); *Phys. Rev.* **D 57**, 1947 (1998);
L.J. Hall, T. Moroi and H. Murayama, *Phys. Lett.* **B 424**, 305 (1998).

[169] R.N. Mohapatra, *Phys. Rev.* **D 34**, 3457 (1986);
G. Bhattacharyya, H.V. Klapdor-Kleingrothaus and H. Päs, *Phys. Lett.* **B463**, 77 (1999).

[170] G. Bhattacharyya, H. Päs, L. Song and T.J. Weiler, *Phys. Lett.* **B 564**, 175 (2003);
H.V. Klapdor-Kleingrothaus and U. Sarkar, *Mod. Phys. Lett.* **A 18**, 2243 (2003).

[171] S. Dimopoulos, S. Raby and F. Wilczek, *Phys. Rev.* **D24**, 1681 (1981);
U. Amaldi et al., *Phys. Rev.* **D36**, 1385 (1987);
J. Ellis, S. Kelley and D. Nanopoulos, *Phys. Lett.* **B260**, 447 (1991);
M. Carena, S. Pokorski and C. Wagner, *Nucl. Phys.* **B406**, 59 (1993);
P. Langacker and N. Polonsky, *Phys. Rev.* **D47**, 4028 (1993).

[172] L.E. Ibanez, *Phys. Lett.* **B118**, 73 (1982); *Nucl. Phys.* **B218**, 514 (1983);
L.E. Ibanez and G. Ross, *Phys. Lett.* **B110**, 215 (1982);
J. Ellis, D. Nanopoulos and K. Tamvakis, *Phys. Lett.* **B121**, 123 (1983);

L. Alvarez-Gaume, J. Polchinski and M. Wise, *Nucl. Phys.* **B221**, 495 (1983);
B. Ananthanarayan, G. Lazarides and Q. Shafi, *Phys. Rev.* **D 44**, 1613 (1991);
V. Barger, M. Berger and P. Ohmann, *Phys. Rev.* **D49**, 4908 (1994).

[173] A. Einstein, Zur allgemeinen Relativitätstheorie, *Sitzungsber, Preuss. Akad. Wiss.*, Berlin, 778 (1915);
A. Einstein, Zur allgemeinen Relativitätstheorie (Nachtrag), *Sitzungsber, Preuss. Akad. Wiss.*, Berlin, 799 (1915);
A. Einstein, Die Feldgleichungen der Gravitation, *Sitzungsber, Preuss. Akad. Wiss.*, Berlin, 844 (1915).

[174] D. Hilbert, Die Grundlagen der Physik, *Königl. Gesell d. Wiss. Götingen, Natr. Math. Phys. Kl.*, 395 (1915).

[175] K. Schwarzschild, Über das Gravitationsfeld eines Masspunktes nach der Einsteinschen Theorie, *Sitzber. Deutsch. Akad. Wiss. Berlin, Kl. Math. Phys. Tech.*, 189 (1916).

[176] F.W. Dyson, A.S. Eddington and C. Davidson, *Phil. Trans. Roy. Soc.* **A 220**, 291 (1919).

[177] C.C. Counselman III et al., *Phys. Rev. Lett.* **33**, 1621 (1974);
E.B. Fomalont and R.A. Sramek, *Ap. J.* **199**, 749 (1975).

[178] A. Einstein, *Sitzungsber, Preuss. Akad. Wiss.*, Berlin, 831 (1915);
G.M. Clemance, *Rev. Mod. Phys.* **19**, 361 (1947);
C.W. Will, *Ap. J.* **196**, L3 (1975);
J.H. Taylor, L.A. Fowler and R.M. McCullach, *Nature* **277**, 437 (1979).

[179] D.Z. Freedman, P. van Nieuwenhuizen and S. Ferrara, *Phys. Rev.* **D 13**, 335 (1976);
S. Desser and B. Zumino, *Phys. Lett.* **B 62**, 335 (1976).

[180] E. Mach, *The Science of Mechanics*, Chicago, Open Court (1893).

[181] Th. Kaluza, Sitzungsber. Akad. Wiss. Berlin *Math. Phys. Kl.* 966 (1921);
O. Klein, *Z. Phys.* **37**, 895 (1926).

[182] J. Scherk and J.H. Schwarz, *Phys. Lett.* **B 57**, 463 (1975);
E. Cremmer and J. Scherk, *Nucl. Phys.* **B 103**, 393 (1976); *Nucl. Phys.* **B 108**, 409 (1976);
Z. Horvath, L. Palla, E. Cremmer and J. Scherk, *Nucl. Phys.* **B 127**, 57 (1977);
Z. Horvath and L. Palla, *Nucl. Phys.* **B 142**, 327 (1978);
A. Salam and A. Strathdee, *Phys. Rev.* **184**, 1750 (1969); *Phys. Rev.* **184**, 1760 (1969).

[183] E. Witten, *Nucl. Phys.* **B 186**, 412 (1981).

[184] W. Rarita and J. Schwinger, *Phys. Rev.* **60**, 61 (1941).

[185] K.S. Stelle and P. West, *Nucl. Phys.* **B 145**, 175 (1978);
P. van Nieuwenhuizen and S. Ferrara, *Phys. Lett.* **B 76**, 404 (1978);

E. Cremmer, B. Julia, J. Scherk, S. Ferrara, L. Girardello and A. van Proeyen, *Nucl. Phys.* **B 147**, 105 (1979);
K. Inoue et al., *Prog. Theor. Phys.* **68**, 927 (1982);
L.E. Ibanez and G. Ross, *Phys. Lett.* **B110**, 227 (1982);
E. Cremmer, S. Ferrara, L. Girardello and A. van Proeyen, *Phys. Lett.* **B 116**, 231 (1982); *Nucl. Phys.* **B 212**, 413 (1983);
J. Bagger, *Nucl. Phys.* **B211**, 302 (1983);
L. Alvarez-gaume, M. Polchinski and M. Wise, *Nucl. Phys.* **B250**, 495 (1983).

[186] J.Polonyi, Budapest preprint no. *KFKI-1977-93* (1977).

[187] P.G.O. Freund and M. Rubin, *Phys. Lett.* **B 97**, 233 (1980);
M. Duff and C.N. Pope, *Supersymmetry and Supergravity '82* ed. S. Ferrara et al. (World Scientific, Singapore), 1983.

[188] M. Green and J.H. Schwarz, *Phys. Lett.* **B 149**, 117 (1984).

[189] P. Candelas, G.T. Horowitz, A. Strominger and E. Witten, *Nucl. Phys.* **B 258**, 46 (1985);
E. Witten, *Nucl. Phys.* **B 258**, 75 (1985).

[190] M. Dine, V. Kaplunovsky, M. Mangano, C. Nappi and N. Seiberg, *Nucl. Phys.* **B 259**, 519 (1985);
del Aguila F, G. Blair, M. Daniel and G.G. Ross, *Nucl. Phys.* **B 272**, 413 (1986);
J.P. Deredings, L.E. Ibanez and H.P. Niles, *Phys. Lett.* **B 155**, 65 (1985);
V. Kaplunovsky, *Phys. Rev. Lett.* **55**, 1036 (1985);
S. Nandi and U. Sarkar, *Phys. Rev. Lett.* **56**, 564 (1986);
R.N. Mohapatra, *Phys. Rev. Lett.* **56**, 561 (1986);
A. Joshipura and U. Sarkar, *Phys. Rev. Lett.* **57**, 33 (1986);
E. Ma, *Phys. Rev. Lett.* **58**, 969 (1987).

[191] D. Gepner, *Phys. Lett.* **B 199**, 380 (1987); *Phys. Lett.* **B 311**, 191 (1988); *Nucl. Phys.* **B 296**, 757 (1988);
B. Greene, C. Vafa and N.P. Warner, *Nucl. Phys.* **B 324**, 371 (1989);
B. Greene, C. Lutken and G. Ross, *Nucl. Phys.* **B 325**, 101 (1989);
A.N. Shellekens and Yankeilowicz, *Phys. Lett.* **B 242**, 45 (1990);
A.B. Zamolodchikov and V.A. Fateev, *Sov. J. Nucl. Phys.* **43**, 657 (1986).

[192] P. Ramond, *Phys. Rev.* **D 3**, 2415 (1971).

[193] A. Neveu and J.H. Schwarz, *Nucl. Phys.* **B 31**, 86 (1971).

[194] F. Gliozzi, J.H. Scherk and D. Olive, *Phys. Lett.* **B 65**, 282 (1976); *Nucl. Phys.* **B 122**, 253 (1977).

[195] D.J. Gross, J.A. Harvey, E. Martinec and R. Rohm, *Phys. Rev. Lett.* **54**, 502 (1985); *Nucl. Phys.* **B 256**, 253 (1985).

[196] G. Tian and S.T. Yau, in *Proc. Argonne Symp. on Anomalies, Geometry and Topology*, (World Scientific, Singapore), 1985.

[197] B.R. Greene, K.H. Kirklin, P.J. Miron and G.G. Ross, *Nucl. Phys.* **B 278**, 667 (1986).

[198] F. Gursey, P. Ramond and P. Sikivie, *Phys. Lett.* **B 60**, 117 (1976);
Y. Achiman and B. Stech, *Phys. Lett.* **B 77**, 389 (1987);
F. Gursey and M. Serdaroglue, *Nuovo Cim.* **A 65**, 337 (1981).

[199] Y. Hosotani, *Phys. Lett.* **B 126**, 309 (1983); *Phys. Lett.* **B 129**, 193 (1983);
E. Witten, *Phys. Lett.* **B 126**, 351 (1984).

[200] K.R. Dienes, *Phys. Rep.* **287**, 447 (1997);
L.E. Ibanez, *Class. Quan. Grav.* **17**, 1117 (2000);
J. Ellis, Report no. *hep-ph/9804440* (1998).

[201] A. Sen, *Nucl. Phys.* **B 475**, 562 (1996);
C. Vafa, *Nucl. Phys.* **B 469**, 403 (1996);
K. Dasgupta and S. Mukhi, *Phys. Lett.* **B 385**, 125 (1996).

[202] A. Sen, Report no. *hep-ph/9810356* (1998).

[203] L. Susskind and J. Uglam, *Phys. Rev.* **D 50**, 2700 (1994);
J. Russo and L. Susskind, *Nucl. Phys.* **B 437**, 611 (1995);
A. Sen, *Mod. Phys. Lett.* **A 10**, 2081 (1995);
A. Strominger and C. Vafa, *Phys. Lett.* **B 379**, 99 (1996);
C. Callan and J. Maldacena, *Nucl. Phys.* **B 472**, 591 (1996);
A. Dhar, G. Mandal and S. Wadia, *Phys. Lett.* **B 375**, 51 (1996);
S. Das and S. Mathur, *Phys. Lett.* **B 375**, 103 (1996);
J. Maldacena and A. Strominger, *Phys. Rev.* **D 55**, 861 (1997).

[204] S. Hawking, *Nature* **248**, 30 (1974); *Comm. Math. Phys.* **43**, 199 (1975);
J. Bekenstein, *Nuovo Cim.* **4**, 737 (1972); *Phys. Rev.* **D 7**, 2333 (1973); *Phys. Rev.* **D 9**, 3192 (1974).

[205] E. Witten, *Nucl. Phys.* **B 443**, 85 (1995);
J. Polchinski and E. Witten, *Nucl. Phys.* **B 460**, 335 (1996);
C. Hull and P. Townsend, *Nucl. Phys.* **B 451**, 525 (1995);
A. Font, L.E. Ibanez, D. Lust and F. Quevedo, *Phys. Lett.* **B 249**, 35 (1990);
S. Rey, *Phys. Rev.* **D 43**, 35 (1991);
A. Sen, *Int. J. Mod. Phys.* **A 9**, 3707 (1994).

[206] M. Rocek and E. Verlinde, *Phys. Lett.* **B 373**, 630 (1992).

[207] J. Maldacena, *Adv. Theor. Math. Phys.* **2**, 231 (1998).

[208] S.S. Gubser, I.R. Klebanov and A.M. Polyakov, *Phys. Lett.* **B 428**, 105 (1998);
E. Witten, *Adv. Theor. Math. Phys.* **2**, 505 (1998).

[209] G. 't Hooft, *gr-qc/9310026.*
L. Susskind, *J. Math. Phys.* **36**, 6377 (1995).

[210] A. Sen, *Nucl. Phys.* **B 475**, 562 (1996);
T. Banks, M. Douglas and N. Seiberg, *Phys. Lett.* **B 387**, 278 (1996);
A. Hanany and E. Witten, *Nucl. Phys.* **B 492**, 152 (1997).

[211] N. Arkani-Hamed, S. Dimopoulos and G. Dvali, *Phys. Lett.* **B429**, 263 (1998);
I. Antoniadis, S. Dimopoulos and G. Dvali, *Nucl. Phys.* **B 516**, 70 (1998).

[212] I. Antoniadis, *Phys. Lett.* **B 246**, 377 (1990);
J.D. Lykken, *Phys. Rev.* **D 54**, 3693 (1996);
P. Horava and E. Witten, *Nucl. Phys.* **B 460**, 506 (1996).

[213] J. Polchinski, *Phys. Rev. Lett.* **75**, 4724 (1995);
E. Witten, *Nucl. Phys.* **B 460**, 335 (1996);
A. Strominger, *Phys. Lett.* **B 383**, 44 (1996);
A.A. Tseytlin, *Nucl. Phys.* **B 469**, 51 (1996).

[214] K.R. Dienes, E. Dudas and T. Gherghetta, *Phys. Lett.* **B436**, 55 (1998); *Nucl. Phys.* **B537**, 47 (1999).

[215] C.D. Hoyle et al., *Phys. Rev. Lett.* **86**, 1418 (2001).

[216] V.A. Rubakov and M. Shaposhnikov, *Phys. Lett.* **B 125**, 136 (1983).

[217] G. Dvali and M. Shifman, *Phys. Lett.* **B 396**, 64 (1997).

[218] W.D. Goldberger and M.B. Wise, *Phys. Rev. Lett.* **83**, 4922 (1999).

[219] G.F. Giudice, R. Rattazzi and J.D. Wells, *Nucl. Phys.* **B 544**, 3 (1999); *Nucl. Phys.* **B 595**, 250 (2001);
T. Han, J.D. Lykken and R.-J. Zhang, *Phys. Rev.* **D 59**, 105006 (1998).
E. Mirabelli, M. Perelstein and M. Peskin, *Phys. Rev. Lett.* **82**, 2236 (1999);
P. Mathews, S. Raychaudhuri and K. Sridhar, *Phys. Lett.* **450**, 343 (1999);
J.L. Hewett, *Phys. Rev. Lett.* **82**, 4765 (1999);
T. Rizzo, *Phys. Rev.* **D 59**, 115010 (1999).

[220] N. Arkani-Hamed and S. Dimopoulos, *Phys. Rev.* **D 65**, 052003 (2002);
N. Arkani-Hamed, H.-C. Cheng, B.A. Dobrescu and S. Dimopoulos, *Phys. Rev.* **D 62**, 096006 (2000);
N. Arkani-Hamed, L.J. Hall, D. Smith and N. Weiner, *Phys. Rev.* **D 61**, 116003 (2000).

[221] N. Arkani-Hamed and M. Schmaltz, *Phys. Rev.* **D 61** (2000), 033005;
E.A. Mirabeli and M. Schmaltz, *Phys. Rev.* **D 61**, 113011 (2000);
G. Dvali and M. Shifman, *Phys. Lett.* **B 475**, 295 (2000).

[222] N. Arkani-Hamed, S. Dimopoulos, G. Dvali and J. March-Russell, *Phys. Rev.* **D 65**, 024032 (2002);
K.R. Dienes, E. Dudas and T. Gherghetta, *Nucl. Phys.* **B537**, 25 (1999).

[223] G. Dvali and A.Yu. Smirnov, *Nucl. Phys.* **B 563**, 63 (1999);
R. Barbieri, P. Creminelli and A. Strumia, *Nucl. Phys.* **B 585**, 28 (2000);

A. Ionnisian and A. Pilaftsis, *Phys. Rev.* **D 62**, 066001 (2000);
A. de Gouvea, G.F. Giudice, A. Strumia and K. Tobe, *Nucl. Phys.* **B 623**, 395 (2002).

[224] E. Ma, M. Raidal and U. Sarkar, *Phys. Rev. Lett.* **85**, 3769 (2000); *Nucl. Phys.* **B 615**, 313 (2001).

[225] R.N. Mohapatra, A. Perez-Lorenzana and C.A. de S Pires, *Phys. Lett.* **B 491**, 143 (2000);
R.N. Mohapatra, S. Nandi and A. Perez-Lorenzana, *Phys. Lett.* **B 466**, 115 (1999);
E. Ma, *Phys. Rev. Lett.* **86**, 2502 (2000).

[226] Y. Grossman and M. Neubert, *Phys. Lett.* **B 474**, 361 (2000).

[227] S. Chang, S. Tazawa and M. Yamaguchi, *Phys. Rev.* **D 61**, 084005 (2000);
K.R. Dienes, E. Dudas and T. Gherghetta, *Phys. Rev.* **D 62**, 105023 (2000).

[228] L. Di Lella, A. Pilaftsis, G. Raffelt and K. Zioutas, *Phys. Rev.* **D 62**, 125011 (2000).

[229] E. Ma, M. Raidal and U. Sarkar, *Phys. Lett.* **B 504**, 296 (2001).

[230] M. Chaichian and A.B. Kobakhidze, *Phys. Rev. Lett.* **87**, 171601 (2001).

[231] L. Randall and R. Sundrum, *Phys. Rev. Lett.* **83**, 3370 (1999).

[232] L. Randall and R. Sundrum, *Phys. Rev. Lett.* **83**, 4690 (1999).

[233] Y. Kawamura, *Prog. Theor. Phys.* **103**, 613 (2000); 2001 ibid. **105**, 999 (2001);
L.J. Hall and Y. Nomura, *Phys. Rev.* **D 64**, 055003 (2001);
A.B. Kobakhidze, *Phys. Lett.* **B 514**, 131 (2001);
G. Altarelli and F. Feruglio, *Phys. Lett.* **B 511**, 257 (2001);
A. Hebecker and J. March-Russell, *Nucl. Phys.* **B 613**, 3 (2001);
Y. Nomura, D.R. Smith and N. Weiner, *Nucl. Phys.* **B 613**, 147 (2001);

[234] N. Arkani-Hamed and S. Dimopoulos, *JHEP* **0506**, 073 (2005);
N. Arkani-Hamed, S. Dimopoulos, G.F. Giudice and A. Romanino, *Nucl. Phys.* **B 709**, 3 (2005);
G.F. Giudice and A. Romanino, *Nucl. Phys.* **B 699**, 65 (2004); (Err) ibid. **B 706**, 65 (2005).

[235] L. Anchordoqui, H. Goldberg and C. Nunez, *Phys. Rev.* **D 71**, 065014 (2005);
B. Mukhopadhyay and S. Sengupta, *Phys. Rev.* **D 71**, 035004 (2005);
J.L. Hewitt, B. Lillie, M. Masip and T.G. Rizzo, *JHEP* **0409**, 070 (2004);
A. Arvanitaki, C. Davis, P.W. Graham and J. G. Wacker, *Phys. Rev.* **D 70**, 117703 (2004);
A. Pierce, *Phys. Rev.* **D 70**, 075006 (2004);
W. Kilan, T. Plehn, P. Richardson and E. Schmidt, *Eur. Phys. J.* **C 39**, 229 (2005).

[236] U. Sarkar, *Phys. Rev.* **D 72**, 035002 (2005).

[237] J. Scherk and J. Schwarz, *Nucl. Phys.* **B 153**, 61 (1979).

[238] D.E. Kaplan and N. Weiner, Report no. *hep-ph/0108001* (2001).

[239] C. Csaki, C. Grojean, H. Murayama, L. Pilo and J. Terning, *Phys. Rev.* **D 69**, 055006 (2004).

[240] C. Csaki, C. Grojean, L. Pilo and J. Terning, *Phys. Rev. Lett.* **92**, 101802 (2004).

[241] Y. Nomura, *JHEP* **0311**, 050 (2003);
C. Csaki, C. Grojean, J. Hubisz, Y. Shirman and J. Terning, *Phys. Rev.* **D 70**, 015012 (2004);
R. Barbieri, A. Pomarol and Rattazzi R, *Phys. Lett.* **B 591**, 141 (2004);
S. Gabriel, S. Nandi and G. Seidl, *Phys. Lett.* **B 603**, 74 (2004);
T. Nagasawa and M. Sakamoto, *Prog. Theor. Phys.* **112**, 629 (2004).

[242] H. Georgi, *Phys. Rev.* **D 71**, 015016 (2005).

[243] H.P. Robertson, *Astrophys. J.***82**, 284 (1935); *Astrophys. J.***83**, 257 (1936);
A.G. Walker, *Proc. Lond. Math. Soc.* (2) **42**, 90 (1936).

[244] A. Friedmann, *Z. Phys.* **10**, 377 (1922); *Z. Phys.* **21**, 326 (1924);
A.G. Lemaitre, *Mon. Not. Roy. Astron. Soc.* **91**, 483 (1931). (Translated from the original paper *Annales de la Societe Scientifique de Bruxelles*, **XLVII A**, 49 (1927)).

[245] G. Gamow, *Phys. Rev.* **70**, 527 (1946);
R.A. Alpher, H. Bethe and G. Gamow, *Phys. Rev.* **73**, 803 (1948);
F. Hoyle and R.J. Taylor, *Nature* **203**, 1108 (1964);
P.J.E. Peebles, *Astrophys. J.* **146**, 542 (1966);
L. Kawano, D.N. Schramm and G. Steigman, *Astrophys. J.* **327**, 750 (1988).

[246] A.A. Penzias and R.W. Wilson, *Astrophys. J.* **142**, 419 (1965).

[247] H. Bondi and T. Gold, *Mon. Not. R. Astron. Soc.* **108**, 252 (1948);
F. Hoyle, *Mon. Not. R. Astron. Soc.* **108**, 371 (1948);
F. Hoyle and J.V. Narlikar, *Proc. Roy. Soc.* **A282**, 191 (1964); **A294**, 138 (1966).

[248] J.C. Mather et al., *Astrophys. J.* **420**, 439 (1994).

[249] J.H. Jeans, *Phil. Trans. Roy. Soc.* **A 199**, 49 (1902).

[250] G. Efstathiou and J. Silk, *Fund. Cosmic Phys.* **9**, 1 (1983).

[251] A. Kogut et al., *Astrophys. J.* **464**, L5 (1996);
D.J. Fixsen et al., *Astrophys. J.* **508**, 123 (1998).

[252] G.F. Smoot et al., *Astrophys. J.* **396**, L1 (1992);
E.L. Wright et al., *Astrophys. J.* **396**, L13 (1992);
C.L. Bennett, *Nucl. Phys. Proc. Suppl.***B 38**, 415 (1995).

[253] E.R. Harrison, *Phys. Rev.* **D 1**, 2726 (1970);
Y. Zeldovich, *Mon. Not. Roy. Astron. Soc.* **160**, 1 (1972).

[254] K.M. Gorski et al., *Astrophys. J.* **464**, L11 (1996).

[255] E. Gawiser and J. Silk, *Science* **280**, 1405 (1998);
J. Primack, *astro-ph/9707285*.

[256] G. t'Hooft, *Nucl. Phys.* **B 79**, 276 (1974);
A. Polyakov, *Pis'ma Zh. Eksp. Teor. Fiz.* **20**, 430 (1974); *Sov. Phys. JETP Lett.* **20**, 194 (1974).

[257] T.W.B. Kibble, *J. Phys. A : Math. Gen.* **9**, 1387 (1976).

[258] A.H. Guth, *Phys. Rev.* **D 23**, 347 (1981);
A.D. Linde, *Phys. Lett.* **B 108**, 389 (1982); *Rep. Prog. Phys.* **47**, 925 (1984); *Prog. Theor. Phys.* **85**, 279 (1985);
A. Albrecht and P.J. Steinhardt, *Phys. Rev. Lett.* **48**, 1220 (1982);
S.-Y. Pi, *Phys. Rev. Lett.* **52**, 1725 (1984);
Q. Shafi and A. Vilenkin, *Phys. Rev. Lett.* **52**, 691 (1984);
R. Holman, P. Ramond and G.G. Ross, *Phys. Lett.* **B 137**, 343 (1984).

[259] W. de Sitter, *Proc. Akad. Weteusch. Amsterdam* **19**, 1217 (1917).

[260] S. Coleman and E. Weinberg, *Phys. Rev.* **D 7**, 1988 (1973);
S. Weinberg, *Phys. Rev.* **D 9**, 3357 (1974).

[261] S. Dimopoulos and L. Hall, *Phys. Lett.* **B 196**, 135 (1987).

[262] T. Bunch and P.C.W. Davies, *Proc. Roy. Soc. Lond.* **A 360**, 117 (1978).

[263] J. Ellis, J.E. Kim and D. Nanopoulos, *Phys. Lett.* **B 145**, 181 (1984);
M.Yu. Khlopov and A.D. Linde, *Phys. Lett.* **B 138**, 265 (1984);
J. Ellis, G.B. Gelmini, J.L. Lopez, D.V. Nanopoulos and S. Sarkar, *Nucl. Phys.* **B 373**, 399 (1992);
M. Kawasaki and T. Moroi, *Progr. Theor. Phys.* **93**, 879 (1995);
T. Moroi, Ph.D. Thesis, *Effects of the gravitino on the inflationary universe*, Report no. *hep-ph/9503210* (1995);

[264] M. Bolz, W. Buchmüller and M. Plumacher, *Phys. Lett.* **B 443**, 209 (1998).

[265] E. Holtmann, M. Kawasaki, K. Kohri and T. Moroi, *Phys. Rev.* **D 60**, 023506 (1999).

[266] A.D. Sakharov, *Pis'ma Zh. Eksp. Teor. Fiz.* **5**, 32 (1967).

[267] J.N. Fry, K.A. Olive and M.S. Turner, *Phys. Rev. Lett.* **45**, 2074 (1980); *Phys. Rev.* **D 22**, 2953; 2977;
E.W. Kolb and S. Wolfram, *Nucl. Phys.* **B 172**, 224 (1980).

[268] M. Yoshimura, *Phys. Rev. Lett.* **41**, 281 (1978); (E) *Phys. Rev. Lett.* **42**, 7461 (1979);

[269] V. Kuzmin, V.A. Rubakov and M.E. Shaposhnikov, *Phys. Lett.* **B 155**, 36 (1985).

[270] M.E. Shaposhnikov, *Nucl. Phys.* **B 287**, 757 (1987);
M.E. Shaposhnikov, *Nucl. Phys.* **B 299**, 797 (1988);
N. Turok and J. Zadrozny, *Phys. Rev. Lett.* **65**, 2331 (1990);
L. McLerran, M.E. Shaposhnikov, N. Turok and M. Voloshin, *Phys. Lett.* **B 256**, 451 (1991);
A.D. Dolgov, *Phys. Rep.* **222**, 309 (1992);
V.A. Rubakov and M.E. Shaposhnikov, *Usp. Fiz. Nauk* **166**, 493 (1996); *Phys. Usp.* **39**, 461 (1996);
A. Kazarian, S. Kuzmin and M.E. Shaposhnikov, *Phys. Lett.* **B 276**, 131 (1992).

[271] A.I. Bochkarev, S.V. Kuzmin and M.E. Shaposhnikov, *Phys. Lett.* **B 244**, 275 (1990);
M. Dine, P. Huet and R. Singleton, Jr., *Nucl. Phys.* **B 375**, 625 (1992);
M.E. Shaposhnikov, *Phys. Lett.* **B 316**, 112 (1993).

[272] M.S. Turner, *Phys. Lett.* **B 89**, 155 (1979);
D. Lindley, *Mon. Not. Roy. Astron. Soc.* **196**, 317 (1981);
L.L. Krauss, *Phys. Rev. Lett.* **49**, 1459 (1982);
I. Affleck and M. Dine, *Nucl. Phys.* **B 249**, 361 (1985).

[273] M. Fukugita and T. Yanagida, *Phys. Lett.* **B 174**, 45 (1986).

[274] S.Yu. Khlebnikov and M.E. Shaposhnikov, *Nucl. Phys.* **B 308**, 885 (1988);
J.A. Harvey and M.S. Turner, *Phys. Rev.* **D 42**, 3344 (1990).

[275] M. Fukugita and T. Yanagida, *Phys. Rev.* **D 42**, 1285 (1990);
S.M. Barr and A. Nelson, *Phys. Lett.* **B 246**, 141 (1991).

[276] P. Langacker, R.D. Peccei and T. Yanagida, *Mod. Phys. Lett.* **A 1**, 541 (1986);
M. Luty, *Phys. Rev.* **D 45**, 445 (1992);
H. Murayama, H. Suzuki, T. Yanagida and J. Yokoyama, *Phys. Rev. Lett.* **70**, 1912 (1993);
A. Acker, H. Kikuchi, E. Ma and U. Sarkar, *Phys. Rev.* **D 48**, 5006 (1993);
W. Buchmüller and M. Plümacher, *Phys. Lett.* **B 389**, 73 (1996).

[277] M. Flanz, E.A. Paschos, and U. Sarkar, *Phys. Lett.* **B 345**, 248 (1995);
M. Flanz, E.A. Paschos, U. Sarkar and J. Weiss, *Phys. Lett.* **B 389**, 693 (1996).

[278] A. Ignatev, V. Kuzmin and M.E. Shaposhnikov, *JETP Lett.* **30**, 688 (1979);
F.J. Botella and J. Roldan, *Phys. Rev.* **D 44**, 966 (1991).
J. Liu and G. Segre, *Phys. Rev.* **D 48**, 4609 (1993);
L. Covi, E. Roulet and F. Vissani, *Phys. Lett.* **B 384**, 169 (1996).

[279] A. Pilaftsis, *Phys. Rev.* **D 56**, 5431 (1997);
L. Covi, E. Roulet and F. Vissani, *Phys. Lett.* **B 424**, 101 (1998);
W. Buchmüller and M. Plümacher, *Phys. Lett.* **B 431**, 354 (1998);

M. Flanz and E.A. Paschos, *Phys. Rev.* **D 58**, 113009 (1998);
T. Hambye, J. March-Russell and S.M. West, *JHEP* **0407**, 070 (2004);
A. Pilaftsis and T.E.J. Underwood, *Nucl. Phys.* **B 692**, 303 (2004).

[280] S. Davidson and A. Ibarra, *Phys. Lett.* **B 535**, 25 (2002).

[281] W. Fischler, G.F. Giudice, R. Leigh and S. Paban, *Phys. Lett.* **B 258**, 45 (1991);
W. Buchmüller and T. Yanagida, *Phys. Lett.* **B 302**, 240 (1993);
B. Campbell, S. Davidson, J. Ellis and K. Olive, *Phys. Lett.* **B 256**, 457 (1991).

[282] H. Dreiner and G.G. Ross, *Nucl. Phys.* **B 410**, 188 (1993);
J.M. Cline, K. Kainulainen and K.A. Olive, *Phys. Rev.* **D 49**, 6394 (1994);
A. Ilakovac and A. Pilaftsis, *Nucl. Phys.* **B 437**, 491 (1995).

[283] U. Sarkar, *Phys. Lett.* **B 390**, 97 (1997).

[284] E. Ma, M. Raidal and U. Sarkar, *Phys. Lett.* **B 460**, 359 (1999).

[285] E. Ma, M. Raidal and U. Sarkar, *Eur. Phys. J.* **C 8**, 301 (1999).

[286] W. Buchmüller, P. Di Bari and M. Plumacher *Nucl. Phys.* **B 643**, 367 (2002).

[287] W. Buchmüller, P. Di Bari and M. Plumacher, *Phys. Lett.* **B 547**, 128 (2002);
W. Buchmüller, P. Di Bari and M. Plumacher, *Nucl. Phys.* **B 665**, 445 (2003).

[288] G.F. Giudice, A. Notari, M. Raidal, A. Riotto and A. Strumia, *Nucl. Phys.* **B 685**, 89 (2003).

[289] T. Hambye, Y. Lin, A. Notari, M. Papucci and A. Strumia, *Nucl. Phys.* **B 695**, 169 (2003).

[290] T. Hambye and G. Senjanovic, *Phys. Lett.* **B582**, 73 (2004).

[291] F. Cuypers and S. Davidson, *Eur. Phys. J.* **C 2**, 503 (1998).

[292] J.P. Bowes, R. Foot and R.R. Volkas, *Phys. Rev.* **D 54**, 6936 (1996).

[293] M. Leurer, *Phys. Rev.* **D 49**, 333 (1994);
S. Davidson, D. Bailey and B.A. Campbell, *Zeit. Phys.* **C 61**, 613 (1994).

[294] J. Ellis, J. Hagelin, D. Nanopoulos, K. Olive and M. Srednicki, *Nucl. Phys.* **B 238**, 453 (1984);
K. Griest, *Phys. Rev.* **D 38**, 2357 (1988);
J. Ellis and R. Flores, *Phys. Lett.* **B 300**, 175 (1993);
A. Bottino et al., *Astroparticle Phys.* **2**, 67; (1994)
T. Falk, K. Olive and M. Srednicki, *Phys. Lett.* **B 339**, 248 (1994);
V. Bednyakov, H.V. Klapdor-Kleingrothaus and S.G. Kovalenko, *Phys. Lett.* **329**, 5 (1994); *Phys. Rev.* **D 55**, 503 (1997).

[295] H. Goldberg, *Phys. Rev. Lett.* **50**, 1419 (1983);
J. Ellis, J. Hagelin, D.V. Nanopoulos and M. Srednicki, *Phys. Lett.* **B 127**, 233 (1983);

K. Griest, M. Kamionkowski and M. Turner, *Phys. Rev.* **D 41**, 3565 (1990);
J. Ellis, and L. Roszkowski, *Phys. Lett.* **B 283**, 252 (1992);
P. Nath and R. Arnowitt, *Phys. Rev. Lett.* **70**, 3696 (1993);
V. Barger, C. Kao, P. Langacker and H.-S. Lee, *Phys. Lett.* **B 600**, 104 (2004);
H. Baer and M. Brihlik, *Phys. Rev.* **D 53**, 597 (1996).

[296] M. Drees and M.M. Nojiri, *Phys. Rev.* **D 47**,376 (1993);
M. Drees, Report no. *hep-ph/9703260* (1997); *Pramana* **51**, 87 (1998).

[297] J. Ellis, A. Ferstl and K.A. Olive, *Phys. Lett.* **B 481**, 304 (2000).

[298] J. Ellis et al., *Phys. Lett.* **B 388**, 97 (1996); ibid. **B 413**, 355 (1997); *Phys. Rev.* **D 58**, 095002 (1998).

[299] V.A. Bednyakov and H.V. Klapdor-Kleingrothaus, *Phys. Rev.* **D 62**, 043524 (2000); ibid. **D 66**, 115005 (2002); ibid. **D 70**, 096006 (2004).

[300] R. Bernabei et al., *Phys. Lett.* **B 389**, 757 (1996); **408**, 439 (1996); **424**, 195 (1996); *AIP Conf. Proc.* **698**, 328 (2003); *astro-ph/0405282*; *Neutrinoless double beta decay*, ed. V.K.B. Kota and U. Sarkar, Narosa, Delhi, 121 (2007);
A. Incicchitti et al., *Proc. TAUP 2005*, Zaragoza, Spain (2005).

[301] D. Bailin, G.V. Kraniotis and A. Love, *Phys. Lett.* **432**, 90 (1998); *Nucl. Phys.* **B 556**, 23 (1999).

[302] K. Abazajian, G.M. Fuller and M. Patel, *astro-ph/0101524.*

[303] R. Barbieri, L.J. Hall and V.S. Rychkov, *Phys. Rev.* **D 74**, 015007 (2006);
E. Ma, *Mod. Phys. Lett.* **A 21**, 1777 (2006).

[304] K. Griest and M. Kaminowski, *Phys. Rev. Lett.* **64**, 615 (1990).

[305] B.W. Lee and S. Weinberg, *Phys. Rev. Lett.* **39**, 165 (1977);
P. Hut, *Phys. Lett.* **B 69**, 85 (1977);
K. Sato and H. Kobayashi, *Prog. Theor. Phys.* **58**, 1775 (1977);
M.I. Vysotski, A.I. Dolgov and Ya.B. Zel'dovich, *JETP Lett.* **26**, 188 (1977).

[306] R. Cowsik and J. McClelland, *Phys. Rev. Lett.* **29**, 669 (1972);
G. Gerstein and Ya.B. Zel'dovich, *Zh. Eksp. Teor. Fiz. Pis'ma Red.* **4**, 174 (1966).

[307] G.G. Raffelt, *Dark Matter in Astro- and Particle Physics*, ed. H.V. Klapdor-Kleingrothaus, Springer-Verlag, Heidelberg, 60 (2001).

[308] D. J. H. Chung, E. W. Kolb and A. Riotto, *Phys. Rev.* **D 59**, 023501 (1999).

[309] A. Kusenko, *Dark Matter in Astro- and Particle Physics*, ed. H.V. Klapdor-Kleingrothaus, Springer-Verlag, Heidelberg, 306 (2001).

[310] K. Benakli, J. Ellis and D.V. Nanopoulos, *Phys. Rev.* **D 59**, 047301 (1999).

[311] N. Arkani-Hamed, S. Dimopoulos, G. Dvali, *Phys. Rev.* **D 59**, 086004 (1998).

[312] H.V. Klapdor-Kleingrothaus, *Nucl. Phys. Proc. Suppl.***110**, 58 (2002); *Eur. Phys. J.* **C 33**, S962 (2004).

[313] *CDMS Collaboration:* R. Abusaidi et al., *Phys. Rev. Lett.* **84**, 5699 (2000); D.S. Akerib et al., *Nucl. Instrum. Meth.* **A 20**, 105 (2004).
*Edelweiss Collaboration*: L. Chabert et al., *Eur. Phys. J.* **C 33**, S965 (2004).

[314] D. Hooper, *Nucl. Phys. Proc. Suppl.***B 101**, 347 (2001);
V. Barger, F. Halzen, D. Hooper and C. Kao, *Phys. Rev.* **D 65**, 075022 (2002).

[315] *AMANDA Collaboration*: J. Ahrens et al., *Eur. Phys. J.* **C 33**, S953 (2004); *Nucl. Phys.* **A 721**, 545 (2003); *Phys. Rev.* **D 66**, 032006 (2002).

[316] *EGRET Collaboration*: P. Sreekumar et al., *Astrophys. J.* **494**, 523 (1998); R.C. Hartman et al., *Astrophys. J.* **123**, 79 (1999).

[317] W. de Boer, *astro-ph/0506447.*

[318] C. Wetterich, *Nucl. Phys.* **B 302**, 668 (1988);
B. Ratra and P.J.E. Peebles, *Astrophys. J.* **325**, L17 (1988).

[319] C.T. Hill, D.N. Schramm and J.N. Fry, *Nucl. Part. Phys.* **19**, 25 (1989);
J.A. Freeman, C.T. Hill and R. Watkins, *Phys. Rev.* **D 46**, 1226 (1992);
A.K. Gupta, C.T. Hill, R. Holman and E.W. Kolb, *Phys. Rev.* **D 45**, 441 (1992);
J.A. Freeman, C.T. Hill, A. Stebbins and I. Waga, *Phys. Rev. Lett.* **75**, 2077 (1995).

[320] R.D. Peccei, J. Sola and C. Wetterich, *Phys. Lett.* **B 195**, 183 (1987);
P. Ferreira and M. Joyce, *Phys. Rev. Lett.* **79**, 4740 (1997); *Phys. Rev.* **D 58**, 023503 (1997).

[321] S.M. Caroll, *Phys. Rev. Lett.* **81**, 3067 (1998);
C. Kolda and D. Lyth, *Phys. Lett.* **B 458**, 197 (1999).

[322] I. Zlatev, L. Wang and P.J. Steinhardt, *Phys. Rev. Lett.* **82**, 896 (1999);
A.R. Liddle and R.J. Scherrer, *Phys. Rev.* **D 59**, 023509 (1998).

[323] J. Frieman, C. Hill, A. Sebbins and I. Waga, *Phys. Rev. Lett.* **75**, 2077 (1995).

[324] R. Fardon, A.E. Nelson and N. Weiner, *JCAP* **10**, 005 (2004).

[325] R. Barbieri, L.J. Hall, S.J. Oliver and A. Strumia, *Phys. Lett.* **B 625**, 189 (2005);
C. Hill, I. Mocioiu, E.A. Paschos and U. Sarkar, *Phys. Lett.* **B 651**, 188 (2007).

[326] R. D. Peccei, *Phys. Rev.* **D71**, 023527 (2005);
N. Afshordi, M. Zaldarriaga, K. Kohri, *Phys. Rev.* **D72**, 065024 (2005);
X.-J. Bi, B. Feng, H. Li, and X. Zhang, *Phys. Rev.* **D72**, 123523 (2005);
A. W. Brookfield, C. van de Bruck, D. F. Mota, and D. Tocchini-Valentini, *Phys. Rev. Lett.* **96**, 061301 (2006), *Phys. Rev.* **D 73**, 083515 (2006);
H. Li, B. Feng, J.-Q. Xia, and X. Zhang, *Phys. Rev.* **D 73**, 103503 (2006).

[327] E. Ma and U. Sarkar, *Phys. Lett.* **B 638**, 358 (2006);
P. Gu, H. He and U. Sarkar, *arXiv:0704.2020 [hep-ph]* (2007).

[328] R. Takahashi and M. Tanimoto, *Phys. Lett.* **B633**, 675 (2006);
R. Fardon, A. E. Nelson, and N. Weiner, *JHEP* **0603**, 042 (2006).

[329] T. Banks and W. Fischler, *An Holographic cosmology*, Report no. *hep-th/0111142* (2001).

[330] N. Kaloper and A. Linde, *Phys. Rev.* **D 59**, 101303 (1999).

[331] V.A. Rubakov and M.E. Shaposhnikov, *Phys. Lett.* **B 125**, 139 (1983).

[332] N. Arkani-Hamed, S. Dimopoulos, N. Kaloper and R. Sundrum, *Phys. Lett.* **B 480**, 193 (2000);
S. Kachru, M. Schulz and E. Silverstein, *Phys. Rev.* **D 62**, 085003 (2000).

[333] S. Förste, Z. Lalak, S. Lavignac and H.P. Niles, *Phys. Lett.* **B 461**, 360 (2000).

[334] V. Barger, T. Han, C. Kao and R.-J. Zhang, *Phys. Lett.* **B 461**, 34 (1999).

[335] S. Cullen and M. Perelstein, *Phys. Rev. Lett.* **83**, 268 (1999).

[336] A. Gorsky and K. Selivanov, *Phys. Lett.* **B 485**, 271 (2000).

[337] G. Dvali and Tye S-H, *Phys. Lett.* **B 450**, 72 (1999).

[338] A. Pilaftsis, *Phys. Rev.* **D 60**, 105023 (1999);
G. Dvali and G. Gabadadze, *Phys. Lett.* **B 460**, 47 (1999);
A. Mazumdar and A. Perez-Lorenzana, *Phys. Rev.* **D 65**, 107301 (2002);
Z. Berezhiani, A. Mazumdar and A. Perez-Lorenzana, *Phys. Lett.* **B 518**, 282 (2001);
R. Allahverdi, Enqvist K, A. Mazumdar and A. Perez-Lorenzana, *Nucl. Phys.* **B 618**, 277 (2001).

[339] G. Dvali, *Phys. Lett.* **B 459**, 489 (1999).

[340] C. Deffayet, G. Dvali and G. Gabadadze, *Phys. Rev.* **D 65**, 044023 (2002);
G. Dvali, G. Gabadadze and M. Porrati, *Phys. Lett.* **B 485**, 208 (2000);
G. Dvali and G. Gabadadze, *Phys. Rev.* **D 63**, 065007 (2001).

# Index